Günter Fandel · Günther Zäpfel (Eds.)

Modern Production Concepts

Theory and Applications

Proceedings of an International Conference,
Fernuniversität, Hagen, FRG, August 20-24, 1990

With 206 Figures

Springer-Verlag
Berlin Heidelberg New York
London Paris Tokyo
Hong Kong Barcelona

Professor Dr. Günter Fandel
Fernuniversität
Faculty of Economics
Department of Production Management
Feithstraße 140 / AVZ II
D-5800 Hagen, FRG

Professor Dr. Günther Zäpfel
University of Linz
Institute of Production Management
Altenbergerstraße 69
A-4040 Linz, Austria

ISBN-13:978-3-642-76403-5 e-ISBN-13:978-3-642-76401-1
DOI: 10.1007/978-3-642-76401-1

Softcover reprint of the hardcvoer 1st edition 1991

2142/7130-543210

Preface and Guide through the Book

The International Conference on Modern Production Concepts - Theory and Applications - was held at the Fernuniversität Hagen, Federal Republic of Germany, August 20 - 24, 1990. The aim of this Conference consisted in reviewing the current problems and concepts of Production Management.

The topics of the conference, treated in this volume, can be classified in the following way:

I. Production Planning and Control

1. Constitutive aspects

2. Aspects and elements of production planning and control (PPC)-systems

3. Just-in-time (JIT)

4. Inventory problems

5. Scheduling problems

II. Production as a Competitive Weapon

6. General concepts

7. Designing new production systems

8. Computer integrated manufacturing (CIM)

1. Constitutive Aspects

By designing production planning models and concepts in a scientific manner, it is necessary to consider the underlying microeconomic theory, the logistical concept, and the used managerial accounting system as well as the flexibility concept. Hierarchical production planning can also be regarded as an essential basis for designing PPC-systems.

In the paper by FANDEL it is shown, how activity analysis may serve as a suitable platform for joining up the theory of production with production planning. This suggests the idea to found production planning including its many subproblems and the information interdependences existing between them upon activity analysis, whereby, along with the production relationships between input and output, the planning structures in the form of the information flows connected with this and the availability of data are also described as input-output relations. In this way information in the form of relevant data concerning the production goods would - at least partly - be explicitly brought in as the fourth production factor in the production function, extending them to so-called "Informational Production Functions".

The development of PPC-systems in practice is at the present time challenged by two logistics concepts: material stock optimization versus material flow optimization. The concept of material stock optimization attempts to guarantee quick deliveries, circumvent the unpleasant consequence of equipment breakdowns, ensure constant utilization of production units etc., through optimization of inventory. In the concept of material flow optimization, stock-keeping is seen as the "root of all evil", because inventories often prevent eliminating waste in the operating system, e.g. long set up times etc. In the article by ZÄPFEL the two fundamental logistics concepts are compared by having recourse to a model of production logistics. It is pointed out in detail, that there is a trade-off between stock-keeping and overtime, which must be considered by designing the PPC-model.

Because profits and costs are important objectives in production planning, many production models involve cost parameters, calculated within the system of cost accounting. The fundamental problem is how to get the exact cost parameters. That's no easy task. The reason for these problems originates in the separation of decisions. Therefore we can argue, that simultaneous models should be used. The problem with general simultaneous models is that they only give a theoretical but no practical solution. The article by KÜPPER illustrates the relations between production planning and managerial accounting by means of an example of multi-period programme planning. The critical cost parameter of this problem is depreciation. The author demonstrates a way to determine variable depreciation by managerial accounting in order to coordinate one-period programme planning with the long-term objective and planning.

In order to comply with the need of flexibility in production, a lot of steps must be realized. One necessary feature is the flexibility of production control systems, which is not yet defined clearly. TAKAHASHI/HIRAKI/SOSHIRODA clarify the factors affecting the flexibility of production control systems and discuss the performance measures for flexibility in their paper. The authors especially analyze the amplifications of production quantity and inventory level of MRP and JIT, and compare them with each other in order to evaluate the flexibility of both concepts.

Flexibility is one special feature of production control systems. In general, flexibility plays a much more important role in production, but often the term is not correctly understood in a broader context. SCHNEEWEISS and KÜHN derive a measure of flexibility starting with notions like variability, reactivity and elasticity. The obtained measure of flexibility is higly operational. An illustrative example is presented.

In practice, production planning and control is based on decomposition of the planning process, aggregation of data and decision variables and a successive structure of the planning system. Therefore the hierarchical model building process can be seen as a theoretical basis for PPC-systems. In the article of KISTNER and STEVEN the elements, the fields of application, and the process of hierarchical production planning are explained and theoretical foundations of decomposition principles and aggregation theory as well as the mechanisms of coordination are presented in a concise manner.

2. Aspects and Elements of Production Planning and Control (PPC)-Systems

An important scientific task in our context consists in developing hierarchical planning concepts respectively PPC-concepts, like MRP, JIT, OPT etc., for practical problems and to compare existing concepts or elements of those.

In their paper TABUCANON and KONGRIT develop a hierarchical concept for a mosaic tile manufacturing company. The production process is characterized by very high setup and holding costs due to a wide range of product variety. The formulated models are presented on the basis of integer programming to find a set of orders to be produced, and linear programming is also employed to find the optimal allocation of pressed materials to kiln zones. Individual products on each machine are determined by a capacitated lot sizing model. Finally, the actual sequences are found by applying an assignment problem algorithm.

In his paper ZIJM elaborates a hierarchical production control system for a printed circuit board (PCB) assembly line for an electronic company. A series of production preparation and control problems are addressed in an almost fully automated assembly line for PCB's. These boards are widely applied both in the consumer electronics (e.g. video) and in the professional industry (e.g. medical equipment). A hierarchical system of combinatorial models is worked out, which ranges from global workload balancing to component feeder assignment and placement head routing problems.

WIJNGAARD and MILTENBURG describe the design of a production control concept for a coffee roaster and packager in their paper. One of the bottlenecks are the buffer facilities for roasted coffee. The control concept implies a distinction between fast movers and slow movers at roasted coffee level. Certain buffer facilities are allocated completely to each of the fast movers. The other buffer facilities are roasted and stored on a day-schedule. The roasting of fast movers is only based on the buffer content. For fast movers a JIT-like way of control is used. The control concept leads to a higher throughput, a higher flexibility and a reduced complexity.

MRP is still the most common system for production planning in use today. To implement such a system in practice, the user should know, how forecast errors of end items, the number of specials in the product structure, the buffering strategy (safety stock or safety lead time, worker flexibility, capacity cushions), the magnitude of lot sizes impact inventory and past-due demand. In their paper RITZMAN and KING analyze these relationships by developing and using a large scale computer simulation model, which reflects the complex structure and dynamics found in real multistage production systems.

Management control of the production planning process can be achieved by a formal development of the production plan, which fixes the annual commodity plan for capacity, inventory and workforce units, and the master production schedule (MPS). The capacity-constrained master production schedule determines the lot sizing and timing of family or product production. MPS "drives" the material requirements explosion process. In his paper, VITT shows the formal structure of hierarchical manufacturing planning, taking into consideration the above mentioned elements and presents an integrated example.

The MRP-system has been advocated as a tool for material coordination, that means the activity which coordinates the order release decisions for successive stages in a multi-echelon production/inventory system. In practice, MRP has a number of drawbacks (e.g. "nervousness" of the plans based on MRP, large inventories etc.). New concepts of material coordination are therefore of special interest. In the paper of VAN DONSELAAR two alternative methods for material coordination have been developed: line requirements planning (LRP), which is similar to base stock control, and flexible integral requirements planning strategy, which combines the strong features of MRP and LRP.

In some MRP II-systems a heuristic method of load-oriented job release, commonly referred to as "Belastungsorientierte Auftragsfreigabe" (BORA), has been added. Especially for the potential users of BORA it is important to know what the main idea behind BORA is, how this procedure works, what the connections to other job release concepts are, which benefits can be derived by applying this method, which deficiencies of BORA must be stated? The answers to these questions can be found in the paper of KNOLMAYER.

3. Just-in-Time (JIT)

Today, JIT-production is one of the most discussed concepts. For the user it is ultimately important to know the exact behaviour of such a system. In the corresponding papers fundamental results are given in this context.

The contribution of CARLSON/OSGOOD/KONOVALOV handles the implementation of JIT in a low volume, hi-tech production environment. The intent of a firm, implementing JIT, can be seen in improving quality, reducing work and space, reducing lead time for product enhancements etc. and finally regain market share. The lessons learnt in the implementation process and tools used for planning a JIT-project (simulation experiments) can be found in this paper.

JALALI/MORAS/DUDEK also describe a study of JIT application in their paper. The problem considered is characterized by a study of a production line, operating in a near JIT mode, demand and supply arrivals are being stochastic. The specific objectives consist of developing a predictive system model, to investigate the effects of four system parameters on the production line idle time and the carrying time of inventory. A simulation study combined with regression analyses is utilized to determine the operating policy of a near JIT-production system. Interaction effects of various parameters on the production line idle time and the carrying time for inventory are shown in detail.

The continuous improvement of quality in the organization is a major step in the implementation of JIT-systems. In his paper EBRAHIMPOUR clarifies the quality management practices in JIT and Non-JIT firms. A number of characteristics, which are necessary to implant and implement a continuous quality improvement system in an enterprise, are handled in detail.

JIT purchasing is also becoming more and more important. In this context new problems arise and must be solved. For example, if the supplier also wishes to produce his goods as late as possible, a deviation from his optimal production rate can be necessary and will increase his unit costs of production because of overtime, extra shifts etc. Starting the production earlier with optimum production rate leads to higher inventory costs, but lower unit production costs. As a result there is a trade-off between the production costs and the inventory costs. In his paper REESE discusses this trade-off and gives some deeper hints by means of marginal analysis and discrete optimization.

4. Inventory Problems

Risks from uncertain demands are generally covered by interstage safety stock piles. In his paper INDERFURTH studies the problem of cost-minimizing determination of safety stocks in a multi-stage production system with variable lead times. The author presents an approach for simultaneous optimization of safety stock piles and processing times for divergent production structures, which can be found in chemical production processes like those of active ingredients in the pharmaceutical industry.

In an assembly-to-order environment there are a number of end items, each of which requires an assembly of several components. Some of the components are common to two or more final products. Components must, due to their long leadtimes, be ordered prior to knowing the precise end-product-demands. The final assemblies can be quickly completed after realizing the demand. The problem now consists of maximizing the expected profit subject to a budget constraint on the component inventory investment. In their paper JÖNSSON and SILVER develop the optimal allocation for the special case of a commonality structure with normally distributed end item demand levels, and give some heuristics and upper bounds for more general common component inventory problems.

Standardization of hierarchy parts structure of product (SHPSP) has been proposed to improve product quality and productivity, that is, to reduce the level of safety stock and costs. But SHPSP is problematic in that it can decrease the possibility of fulfilling the various users' needs, with the result that the market share of the product group decreases. In their paper ISHII and SUMINOKURA develop an evaluation method of the SHPSP which clarifies the effects of "amplifications" on production ordering and term-end inventory levels of products under the same conditions of product demands, production facilities and production ordering systems. Numerical examples show that the higher loading ratio and autocorrelation of a product demands and the larger variation of a down time in production facilities, the more effective is the SHPSP.

In-process-inventory influences the utilization of expensive machinery in production processes. The paper of RAMANI discusses a reliability-theoretical approach to solve an in-process-inventory problem in a concrete brick manufacturing plant. The central problem consists in evaluating and optimizing the bunker capacity for holding lime.

In practice it is common that an enterprise manages several parallel regional warehouses when distributing products over a large geographical area. Each warehouse is then providing service to a certain local market. The warehouses normally replenish from a common outside supplier. In the event of a stockout it may be advantageous to replenish from a nearby warehouse to improve the service-performance of the total logistics system through transshipments. AXSÄTER gives a detailed problem formulation for modelling such lateral transshipments and describes a technique for numerical solution in his paper.

5. Scheduling Problems

Current applications for computer-assisted production planning and control often lack adequate support for shop-floor control, especially in the case of decentral production control. To close this gap decision support systems (DSS) for shop-floor control have been created recently. WOERLEE presents a framework for a decision support system, which should be powerful enough to handle a broad variety of scheduling environments, partly determined by fuzzy restrictions and human guess-work. The author defines how scheduling problems have to be defined, how the communication between the DSS and the user should be specified and explains the components within the DSS.

The development of new concepts for production organization (automatic (flexible) manufacturing systems, autonomous production units etc.) leads to new decentral production control systems. In this context the "Leitstand" is becoming more and more important. The term "Leitstand" is an electronic data processing system which is characterized by its user friendly man machine interface showing a Gantt chart and is based on an interaction style which is oriented to support, not to automate the thoughts of the decision maker required for scheduling and shop floor control applications. In their paper, SCHEER and HARS describe the new requirements for decentral production control, show the characteristics of a "Leitstand" and give a view on future trends.

Real-time scheduling for automatic production systems requires new algorithms. In their paper AHN and KUSIAK present two algorithms for scheduling problems with and without alternative process plans. The effect of alternative process plans on the performance of the schedules generated is studied in detail. The computational studies show that the quality of schedules (e.g. makespan) improves if alternative process plans are used.

In many manufacturing processes work processing times may vary over the ranges of durations bounded by maximal and minimal value of duration. In such situations a scheduler may have certain continuously-devisible limited resources (e.g. energy, raw material) at his disposal, which may be applied to effectively reduce the processing times of individual work. The scheduler does not only have to determine the order in which tasks should be processed, but also the distribution of limited resources to work stations or to machines on which they are processed. JANIAK deals with three manufacturing processes (in steel and copper industry) facing these problems and outlines suitable solution algorithms.

A succesful implementation of a flexible manufacturing system (FMS) requires an efficient shop floor control system. In their paper JOSHI and WYSK demonstrate a generic, flexible, cost effective architecture to plan, schedule, and control shop floor actions. The paper addresses in detail: 1) an architecture for a multipass expert based control system, 2) an on-line procedure to detect and avoid system deadlocks, and 3) an automatic software generator for FMS control.

Studying the dynamics of discrete manufacturing systems, very few mathematical tools are available. One important tool for modelling of job-shop and assembly systems is based on event graphs. HARHALAKIS/LAFTIT/PROTH explain in detail how to model job-shops and assembly systems using event graphs respectively Petri nets. The authors demonstrate in particular, that in case the demand is expressed as a product mix, models are always strongly connected event graphs.

In today's production environment the job-shop scheduling problem with deadlines is characterized by the following policy: all jobs must be delivered on time, one must follow a published due date policy, the objective is to minimize the cost of delivering jobs on time. Random variations in workload are accommodated through undertime and overtime, not through late deliveries. In his paper HALL investigates this policy with respect to scheduling rule, capacity optimization and due date policy for a work-conserving single-stage queue.

In order to gain deep insights into multi-product machine shop control, the production interaction of sales with operating variables such as inventory, work in progress, production order and completion rates, lead time, safety stock etc. should be identified. In their paper, CHEEMA and TOWILL illustrate these interactions by using some of the concepts of industrial dynamics. A manufacturing system of a producer of orthopedic joints and supporting devices is modelled.

6. General Concepts

Production is an important element for gaining a competitive advantage. Therefore it is of great importance for an enterprise nowadays to form a cohesive manufacturing strategy which should include the following basic cornerstones: just-in-time (JIT), total quality control (TQC), factory automation (FA), and total productive maintanance (TPM). A cohesive strategy requires to look at the interrelationships of these cornerstones. HUANG provides a brief introduction to each of the cornerstones and discusses the importance of these concepts, and clarifies the interrelationships with each other in his paper.

The ability of an enterprise to compress time in all phases of the product delivery cycle is a key factor for business success today. Speed in manufacturing must be coupled with short-time development of new products to stay abreast of competition. In the paper of BLACKBURN, methods of time compressing for the new product introduction process are investigated. There is a comparison of Western and Japanese practices in manufacturing and new product development.

Strategic thinking tells us that production is not an end in itself, it is a means of delivering superior product attributes to meet customer expectations. Appropriate product attributes must be the focus for improvement design, manufacturing and product support. In his paper FABRYCKY shows how this concept can be implanted and implemented in an enterprise by life cycle engineering for the optimization of products, systems and structures. Examples for visualizing this concept can also be found in this paper.

7. Designing New Production Systems

Enterprises seek competitive advantages by designing certain characteristics for their products or production systems such as low cost, high quality, fast delivery speed, flexibility etc. Firms which compete in multiple markets often find themselves in a position of producing products with differing subsets of these characteristics on the same manufacturing process. This can lead to conflicting manufacturing tasks, e.g. achieving low costs might require large batch runs on dedicated equipment, while providing product flexibility might require low-volume runs on general purpose machinery. In this case a strategic formation of manufacturing cells is especially important. In their article SHEU and KRAJEWSKI handle this problem by splitting up a manufacturing plant into a number of manufacturing units and organizing them, so that (1) each unit focuses on a limited number of manufacturing tasks, (2) the assigned manufacturing tasks link with marketing strategy. Models and heuristics for solving this problem are presented by the authors.

When one looks at current developments in manufacturing, there is a general trend away from the organization of manufacture around simple series flow with considerable division of labor to new forms of organization, which give each worker substantially more work content and place the responsibility for quality on the worker. Another trend in manufacturing consists in the increasing degree of product variety which a manufacturing system is required to handle, the so-called scope of production. Coping with volume, variety and variability is a central task. In the paper of BUZACOTT the influence of variability within the manufacturing system itself is intensively discussed and conclusions for organization of manufacture are drawn by looking at the results of some stochastic models.

In the design phase of a flexible manufacturing system (FMS), quick and approximately exact solutions should be determined. The CQN (classical closed queueing network)-model is a well-known solution methodology in this context. But this method neglects some essential features, e.g. in the process of a FMS considerable utilization losses arise, which are due to limited local and/or central buffer space and the limited velocity of the material handling system. In their paper TEMPELMEIER/KUHN/TETZLAFF look at the phenomenon of starving (a downstream machine is ready to process a piece and this unit is still located elsewhere), occuring in FMS with limited local buffers. For this case the authors develop an approximate performance model which is based on the iterative application of the CQN-model.

In his paper DEKLEVA also proposes a procedure for designing FMS. A two-phase procedure is developed to obtain the best possible performance. Job shop simulation is used for FMS modelling or - to be more precise - a modified non-delay schedule is offered. The usefulness of this heuristic schedule generation for FMS simulation was tested by comparing its results with those of mean value analyses of queues and discrete simulation.

Process planning in FMS is a key problem in its design and planning. An efficient process planning is a prerequisite for an FMS to attain a good machine load balance and maintaining a high utilization rate. In the paper of CHENG the FMS process planning problem is formulated by a non-linear 0-1 integer programme, considering as an objective the total cost of part-type production and part-type transportation among the machines. The constraint sets include part-type assignment, machine capacity as well as machine load balancing restrictions. Several heuristics are developed by the author to obtain approximate solutions to the problem.

Automated guided vehicle systems (AGVS) play an important role in the modern manufacturing process. Primary vehicle management functions are dispatching, routing and scheduling. Dispatching policies determine which vehicle should be assigned to perform a task to satisfy objectives set by the system planner. In the design phase the level of control automation, pickup and delivery points of loads served by a vehicle, carrying capacity of vehicles etc. have to be determined. Routing is another task and consists of selecting specific paths taken by vehicles to reach the destinations. CO and TANCHOCO prepare a comprehensive review of research on these AGV management systems in their paper.

The modelling and design of flexible manufacturing systems is another challenging problem nowadays. The paper of ADAN/WESSELS/ZIJM proposes new "extended shortest queue models" in designing flexible assembly systems. This approach uses mixtures of the product form as approximative solutions for the equilibrium distribution of the Markov chain.

The design of the logistics system of an enterprise also makes it necessary to fix the distribution network system. The questions which have to be answered are: where should the firm produce the products or components, should the firm stock products at production centres, supply centres or sales organizations, how should the firm replenish stock piles at the stock points, what mode of transport should the firm choose etc.? VAN DOREMALEN and FLEUREN design and report on an implementation of a decision support tool for the quantitative analysis of a number of different scenarios for the physical distribution of an enterprise. An important aspect of the analysis is the determination of the stock levels at the decoupling point in the logistics chain.

8. Computer Integrated Manufacturing (CIM)

CIM is one of the most challenging problems of production management nowadays. The implementation of a CIM system is much more complex than all the other previous application systems because all functional areas of an enterprise are concerned. The personal capacity, the necessary knowledge to realize an enterprises-wide CIM system, are often not available in a single firm. In his paper GRODITZKI shows how CIM can be systematically developed, implemented and continuously adapted to the fast changing environments of the marketplace. The concept described by the author deals with engineering methods and tools that can be used to build and maintain complex application systems (CASE - computer aided software engineering), and expert systems, in order to adapt generic systems intelligently to particular user requirements, and data base technology.

Inadequate specification of a CIM system is a cause that the achievements of investment are less than anticipated. To avoid these shortcomings CARRIE proposes computer aided software engineering tools for specifying both computer software systems and material flow systems. In the paper the author specifically points out the relationship between CASE tools and simulation modelling. It should be possible to develop the specification of a manufacturing system, to create simulation models of the manufacturing system at any stage during the specification process and to develop the control software to manage the system.

The integration aspect in the framework of CIM is essential and involves a much higher degree of complexity than the design of other application systems. Integration calls for new approaches to modelling problems. In his paper PUN presents an integrated conceptual reference model (INCOREMO). The author suggests forming this model with two parts: one organic part representing the functional structures of the organs and one behavioural part representing the operational knowledge of the organs.

A tool for realizing a CIM system can be seen in implementing local area computer networks (LAN). In their paper PAVLIDOU and PAPATHANASSIU discuss design factors for interconnected

LANs and study in particular hierarchical routing techniques aiming at performance evaluation of the internet. Special emphasis has been given to keeping the routing tables at a reasonable size.

The conference could not have been realized without the generous financial support we received from our sponsors. For this we want to express our deep gratitude to all of them. We especially want to mention the Chancellor of the Fernuniversität, Mr. RALF BARTZ, who continously encouraged us throughout the uncertain phases of our preparations even when there were no substantial financial contributions.

Finally, we also would like to thank our colleagues Mr. KLAUS GUBITZ and Mr. JÖRG SCHNEIDER for all their organizational support in the two years of the preparation of the conference as well as for helping us to put together these proceedings.

Hagen and Linz,
November 1990

GÜNTER FANDEL
GÜNTHER ZÄPFEL

Table of Contents

2. Aspects and Elements of Production Planning and Control (PPC)-Systems

3. Just-in-Time (JIT)

4. Inventory Problems

5. Scheduling Problems

Part I: Production Planning and Control

A Theoretical Basis for the Rational Formation of Production Planning and Control (PPC) Systems

GÜNTER FANDEL

1. Preliminary remarks

The analytic importance of the economic theory for the area of production management can be shown by many references. From the methodical-theoretical point of view, however, production management and production theory have experienced the most fruitful prospects for expansion through the activity analysis developed by KOOPMANS (1951) which has gradually been introduced into the considerations of business management (WITTMANN 1968, FANDEL 1987, DYCKHOFF 1988). The efficiency of the activity analysis as an examination instrument and modelling concept is founded on many circumstances and advantages.

- Firstly, the activity analysis is a very general approach for the study of production relationships and for examining their principles with respect to the transformation of inputs into outputs. In doing this the analysis uses only a few notions, definitions and assumptions and its analytic power lies in the formal stringency of its way of proceeding. This is what makes it attractive for use as a closed examination concept for practical problems (FANDEL/REESE 1988).

- Secondly, some authors see in the activity analysis, or in the consideration of production correspondences closely connected with it, a suitable platform for joining up the theory of production with production planning (SHEPHARD et al. 1977, WITTMANN 1979). Independent of this direct frame of reference, from the management science point of view the argument of the necessity of this type of connection has been put forward many times. SCHNEIDER (1964) wanted to have the theory of production understood in the sense of a theory of production planning. KÜPPER (1979) regarded the drawing in of problems of operations organization in production process planning as important and attempted to integrate them accordingly in his dynamic production function. MATTHES (1979) has a similar approach when he refers to the necessity of taking process and structurally conditioned characteristics of operational production more explicitly into account when formulating models and bases his proposal for a dynamic production function on this.

- Finally, this type of consideration suggests the idea of preparing production planning itself, with its branches of program planning, choice of procedure, procurement planning and process planning, as well as the multitude of information interdepences existing between these subproblems, in an activity analysis. Along with the production relationships between input and output, the planning structures in the form of the information flows connected with this and the availability of data are also included as input-output relationships. If production planning is structured in the sense of production planning and control systems (PPC, SCHEER 1988) the planning structures, with their sub-modules and the information and data processing which has to be carried out therein, show amazing similarities with the structures of multistage multi-product manufacture. The attempt could thus be made on the basis of the activity analysis - by analogy with the integration of the technical characteristics of the goods taking part in production in Engineering Production Functions - to integrate the information characteristics of these goods with regard to the requirements of production planning in the input-output relationships and to represent them in the form of so-called "Informational Production Functions". In this way information in the form of relevant data concerning the production goods would - at least partly - be brought in explicitly as the fourth production factor in the production function.

The further considerations are devoted to the question of how far production planning and control (PPC) systems can be presented from an activity analysis point of view and thus systematically formed analogously to production theoretical models.

2. Production planning and control (PPC) systems from the point of view of activity analysis

2.1. Necessity and forms of including information activities in production planning

The necessity to include information activities - i.e. gathering and processing information - explicitly into the solution of business administration tasks and thus especially into the theory of production and production planning has always been stressed by many authors from differing perspectives and has led to very different proposals without showing a clearly developing trend.

- SCHEER (1985) points to the danger that in future business economics will not be able to arrive at any sufficient competence in the formation of actual business processes if care is not taken to include information technologies and data processing in the theory of business economics. He sees a possible way out in the EDP-oriented theory of business economics which he himself introduced into the discussion (SCHEER 1984).

- MÜLLER-MERBACH (1985) regards the development of an information-oriented theory of business economics starting from the information structures and sequences of information processing in a company as a necessary alternative and complement to the concepts of certain function-oriented business economics theories which have been formed up to now. He is of the opinion that this is a most important sensible connection between the theory of business economics and computer science.

- MÜLLER (1987) regrets that the central areas of a functionally oriented theory of business economics only take aspects of information and its effects insufficiently or in a simplified form into account, and counters this unacceptable neglect of information problems with his own suggestion for a model of information production into which methodical elements of an input-output analysis certainly flow, whereby, however, the procedural concept of information production still remains unclear in parts.

In this context it is necessary to clear up the question of which conceptual definition of the word "information", which is by no means treated in a uniform manner, is to be used in the following pages.

JAEGER (1980) points out that in communications research a message serving a purpose can be understood as information. SHANNON/WEAVER (1964) regard information as the degree of freedom when selecting a message. HIRSHLEIFER (1973) defines information as a message which leads to the person making a decision changing his a priori probabilities of the relevant environmental circumstances. The International Federation for Information Processing (IFIP) defines information as the meaning which a person assigns to data by the agreed conventions used in its representation (POENSGEN 1978). On the other hand data are conceptually differentiated in that they are the formalized representation of facts or ideas with which they can be passed on or used in some sort of process. WITTMANN (1959) described information as special-purpose knowledge; this definition is still used in German-language business economics research. Information will be understood here in this sense as well, whereby a mingling with the definition of data may be permissible in so far as information in computer-supported production planning and manufacturing control consists anyway only of data relevant to decision-making.

The proposals for the integration of information or data into the decision-making in business economics are also extremely varied. If in the following short examples a few forms of integrating possibilities are pointed out, the choice has been made above all because these proposals let a way of putting into operation be seen.

- MARSCHAK (1964, 1971) sees a logical extension of traditional microeconomic theory in the incorporation of information problems in decision-making models and, using problems taken from the area of financial investments, how an optimal allocation of resources of this kind can be brought about by incorporating the problems of gathering and processing information. Transferring this to planning problems in production suggests itself. The essential connections are characterized in such a way that, because of the additional information function brought in, data from randomly distributed environmental states result, which in their turn lead to actions by way of the decision function. A result function then assigns a particular result to each combination of action and environmental constellation. The information function and the decision function must be chosen in such a way that the expected value of the results is maximized (BÖSSMANN 1978).

- In their production-theory approach to the representation and execution of data-processing processes MÜLLER/ECKERT (1978) and MÜLLER (1987) suggest, as in the activity analysis, differentiating according to production factors, outputs, processes of information production and carriers of information processing, combinatorial rules, and information processing rules. Input information, information processing appliances and other physical resources can be regarded as production factors. Outputs are also informations which come into existence from the inputs by reason of the combination rules and the information processing rules. A formal description of this type of information process of the production of information through the corresponding definition of an activity for this type of case does not, however, take place, so that the generalization of the material theory of production regarding the absorption of information in a model of production planning is not yet recognizable from the point of view of an activity analysis. In addition the operation of information processing has not yet been developed procedurally.

- Looking at a satisfactory settling of this type of open question WITTMANN (1979) sees the greatest chances in an extension of the activity analysis to production planning. The flexibility of this method makes it seem especially suitable to serve as a representational basis for planning, whereby the production of knowledge or information can also be modelled within the framework of this concept. According to WITTMANN's considerations, this concerns "a statement of stocks and a production calculation (information activity), whereby the problem of conceiving of knowledge as a good is faced with difficulties which are in part unsolvable. Indirectly conceivable in part is an assignment to technical progress or to sales or production programmes (whereby the construction of a production plan is represented as the result of an information activity). Information stocks appear in accounting, in customer files, drawings, technical regulations etc." To represent this type of planning procedure he favours forms of the network planning technique which have been used by SHEPHARD et al. (1977) and SHEPHARD (1983) in the concrete case of shipbuilding. Here, information exists in the production structure and in the chronological sequence of subactivities for carrying out manufacturing. From the methodical point of view, however, this procedure can be opposed by saying that it combines a general planning situation too much with a special concept for solutions.

- One possibility of making the advantages of the activity analysis approach referred to above useful, without being bound from the start too closely to one method of solving problems, lies in the modelling of production planning and control systems (PPC) (ADAM 1988, SCHEER 1988, ZÄPFEL 1989a, ZÄPFEL 1989b). These all attempt to strip down the complex task of production planning into module-type subtasks, which are then subjected with the help of computers to an iterative problem-solving procedure for optimal formation and mutual coordination, whereby very different models, concepts or methods may be brought into operation to solve individual subtasks. The attempt of a systematic activity analytical formalizing of the procedures of PPC systems under the aspect of the explicit integration of the utilization and transformation of information for optimal production planning is to be the subject of the next two sections. The formal structure of PPS systems will first be observed; there then follows a detailed discussion of their basic elements from the point of view of an activity analysis.

2.2. Task and structure of PPC systems

PPC systems are aimed at the optimal, step-by-step solution of the complex problem of production planning and control with its multiplicity of sub-problems and the existing functional and time interdependences which lie between them. They can be regarded as practical solution procedures for mastering programme planning, choice of procedure, procurement planning and process planning, whereby the abundance of data which have to be dealt with makes the use of computers necessary. Production planning and control, which really have to be planned simultaneously to achieve an overall optimum, are broken down into successive planning by means of the formulation of visible subtasks which each have to be treated separately. The coordination of these subtasks and their optimal solution in the interactive mode with the computer and the possibility of being able at the same time to make running improvements make a permanent exchange of information between the submodules of a PPC system necessary. The representation of the subtasks of a PPC system, their informatory input-output relationships to one another and the interactive solution process is illustrated by the flow chart in Fig. 1 (FANDEL/FRANÇOIS 1988).

The first four subtasks A_n, $n=1,...,4$, of this type of PPC system are usually considered to be planning, the last three subtasks A_n,

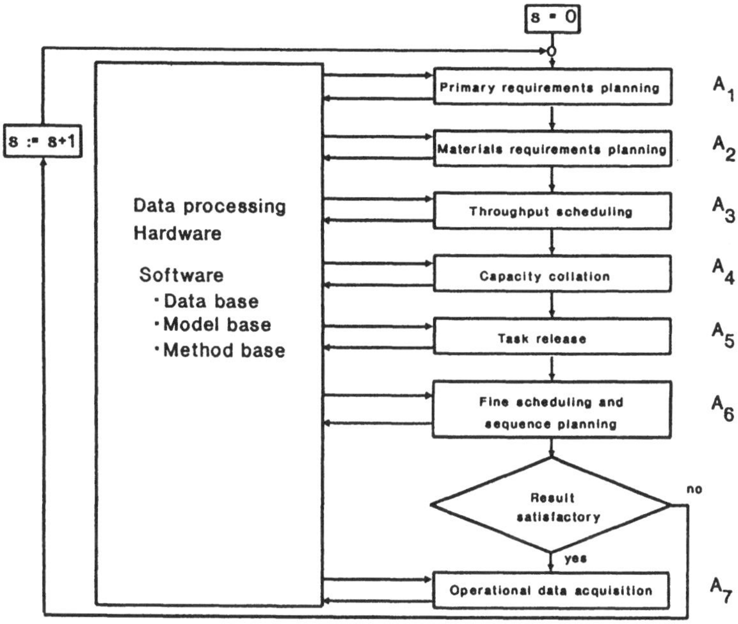

Figure 1

n=5,...,7, on the other hand are regarded as control. The degree of particularization of the subtasks increases from top to bottom, the time horizon for their mastering becomes shorter.

Normally, production planning and control which is shaped procedurally and with regard to data processing through a PPC system begins with planning primary requirements. The data outputs gathered go into the subsequent tasks of materials requirements planning as inputs, or are temporarily stored until further use. Here the data processing facility with its hardware and software functions as an instrument of the data processing routine. All the subtasks are worked out one after another analogously. The linear flow of information or data between the subtasks on the one hand and the data processing on the other does not represent a practical restriction as the task circulation can be run through several times interactively until the decision maker is satisfied with the results of the planning and control achieved up to then and realizes them. In this way the interdependences between the subtasks can be taken into account, whereby it is not absolutely necessary that all substeps must be carried out again when a new run-through of the tasks takes place.

The PPC system joins up the theory of production with production planning on the basis of data processing, as the production technology is implicitly contained in the data. Structurally it shows several similarities with multistage manufacturing with loops, only here goods are not brought into action, produced and processed further, but data on the goods taking part in production. A study and formalisation of data processing in a PPC system in a step-by-step, activity analytical manner suggests itself and thus to see information by the course of production planning as the fourth factor in production. A basis for the rational formation of PPC systems could be developed from this. Here it is necessary to consider how in this context the basic elements of an activity analysis, broadly-based with regard to PPC systems, should be specified. The input-output character of the data transformation in the subtasks of a PPC system is illustrated by way of example in Fig. 2 for the module of sequence planning.

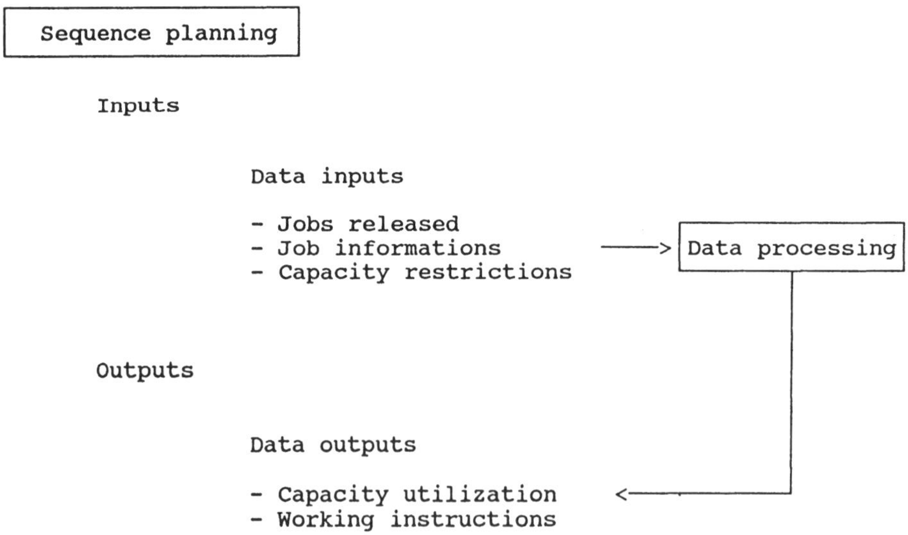

Figure 2

2.3. Basic elements of an activity analysis of PPC systems

A PPC system may be understood in the following as a linkage of information activities to be carried out one after the other to master subtasks A_n, n=1,...,N, in production planning (cf. Fig. 1 with N=7), whereby the circular flow of tasks usually has to be run several times before a satisfactory solution is reached. For every run s, s=1,2,..., let an information total activity v^s be defined through the composition of the information subactivities $v^{s,n}$ for tasks n, so that

$$v^{s,n}=(v_1^{s,n},\dots,v_{k_n}^{s,n})'=(i_1^{s,n},\dots,i_{\tilde{k}_n}^{s,n},\ o_1^{s,n},\dots,o_{\hat{k}_n}^{s,n})',$$

$$v^s=(v^{s,1},v^{s,2},\dots,v^{s,N})'=(v_1^s,\dots,v_K^s)',$$

with $\tilde{k}_n+\hat{k}_n=k_n$ and $\sum_{n=1}^{N}k_n=K$, $v^s\epsilon D$, $D=\mathop{\text{X}}_{k=1}^{K} D_k$,

whereby D_k is the range of the k-th data element.

The information total and subactivities are therefore vectors of information or data which in the sense of knowledge for a purpose are relevant to decision-making for production planning or for solving their subtasks. The amount of data which has in each case to be processed and to be generated may be finite and denoted by K. An information activity consists of data inputs $i_\chi^{s,n}$ and data outputs $o_\mu^{s,n}$, $\chi=1,\dots,\tilde{k}_n$, $\mu=1,\dots,\hat{k}_n$, n=1,...,N, s=1,2,...; the data variables should be quantifiable, i.e. measurable on cardinal or ordinal scales. The following qualify as data variables:

- quantity variables; these include for example the quantities of goods to be used and manufactured in production;

- time quantities; among others the delivery times of the resources ordered as parts bought-in, the processing times of the jobs or the calculating time which has passed up to the generating of a suggestion for a solution;

- money quantities; for example, profit rates, fixed lot costs and inventory costs.

Among data inputs and outputs as well as among the goods of the material production a difference can be made between long-lived data (analogous with the utilization factors and consumer durables) and short-lived data (analagous with consumption factors and consumer disposables). Long-lived data are the parent data in production planning such as for example structured bills of materials and machine sequences for the jobs conditioned by technology. Shortlived data, for example, would be capacity utilization and batch sizes.

On this basis an information technology of production planning for a manufacturing company could be formulated. This should consist of the set of all possible sequences of information total activities which result from the repeated run of the submodules for solving production planning and control in the company by way of the S iteration steps up to the satisfactory discontinuance of the planning process. The information technology contains therefore the information activities which the company can generate within its own PPC system on the basis of its technical knowledge of the production and the accompanying technique of data processing.

$$T^I=\{[v^s]_{s=1}^S \epsilon D^S \mid [v^s]_{s=1}^S \text{ can be generated and realized by the company in production planning, } S \epsilon N \wedge S<\infty\}.$$

The question of how the concept of efficiency is to be understood in this context is more difficult as the components of the data vectors do not have to be so naturally ordered in their quantities as the quantities of goods of an activity component in the case of the static-deterministic theory of production. In the following suggestions for the definition of efficiency will be put forward for discussion.

Let

$$v=(v_1,\ldots,v_K)' \in D$$

be a data vector. Its components k, k=1,...,K, can be divided as regards their range D_k into

- those on which an order is defined by the decision-maker, such as will be the case for operating quantities, manufacturing quantities and processing time. Let the index set of these vector components be given by

$\tilde{O}=\{k\epsilon\{1,...,K\} \mid \text{an order is defined on } D_k\}$;

- and those on which no order is defined by the decision-maker, as may be the case for the starting and ending times of the processing of jobs. Let this index set be characterized by

$\bar{O}=\{k\epsilon\{1,...,K\} \mid \text{no order is defined on } D_k\}$.

Let

$\tilde{O}\cap\bar{O}=\phi$ and $\tilde{O}\cup\bar{O}=\{1,...,K\}$

still be valid.

The following definitions of efficiency will be introduced here:

- Definition of efficiency 1 (primary efficiency)

A data vector $v\epsilon D$ is called primary efficient iff there is no other data vector $w\epsilon D$, $w\neq v$, with

1) $w_k \circledge v_k$ for all $k\epsilon\tilde{O}$ and
2) $w_k \circledgt v_k$ for one $k\epsilon\tilde{O}$.

$\circledge$ or $\circledgt$ is the operator of the order ranking on D_k, whereby $\circledgt$ means "better than" and depending on the analogous interpretation can correspond to the mathematical "<" or ">". Data vectors $w\epsilon D$ which satisfy the above conditions are called primary dominating.

Through the primary efficiency the conventional concept of efficiency is maintained which manages without any additional evaluation. Starting from this general form of the definition of efficiency the following two concepts of efficiency can be derived through reduction or specification respectively in the components of the information activity.

- Definition of efficiency 2 (quantity efficiency)

Let $M\subset\{1,...,K\}$ be the subset of the component indices of the information activity $v \in D$, whose data elements represent the input and output quantities of the goods taking part in the planned production. The data vector is then quantity efficient iff there is no data vector $w \in D$, $w \neq v$, with

1) $w_k \geq v_k$ for all $k \in M$ and

2) $w_k > v_k$ for one $k \in M$.

Data vectors $w \in D$ which satisfy the above conditions are called quantity dominant.

- Definition of efficiency 3 (information efficiency with respect to criterium Z)

  Let g^Z: $D \to E$, $\dim E = L$, be a function which assigns the vector $g^Z(v) \in E$ to each information activity $v \in D$ according to criterium Z. The data vector $v \in D$ is information efficient regarding criterium Z iff there is no data vector $w \in D$, $w \neq v$, with

  1) $g^Z(w) \circledgeq g^Z(v)$ and

  2) $g^Z(w)_l \circledgt g^Z(v)_l$ for one $l \in \{1, \ldots, L\}$.

  Data vectors $w \in D$ which satisfy the above conditions are named information dominant.

Quantity efficiency corresponds to the efficiency concept of the theory of production. Information efficiency permits an evaluation of the information activities under various slants. For example, the financial "profit efficiency" could be derived as a special case with information on which profit is linked with the input-output sets of the production for $g^Z: D \to R$ with $L=1$ and the monetary evaluation of the quantity data.

The definitions of efficiency shown here or the definitions of dominance implicit in them respectively can now be applied optionally to $v^{s,n}$, v^s or $[v^s]_{s=1}^{S}$. This results in the following considerations:

(1) A PPC procedure is certainly superior to another when it is primarily dominant with regard to $[v^s]_{s=1}^{S}$. The same applies when it achieves the same or a better result in fewer iterations S.

(2) A PPC module is better than another when it is primarily dominant with regard to $v^{s,n}$ for all s=1,...,S.

(3) By limiting to the quantity efficiency for $v^{S,N}$ the information problem of production planning is neglected.

(4) Information efficiency offers the possibility of evaluating PPC systems according to different criteria; further examinations will certainly be necessary regarding the suitable choice of criteria for evaluation.

For practical considerations only the knowledge of the information activity $v^{S,N}$ is of interest which indicates which production was finally realized on completion of production planning. The sequence $[v^s]_{s=1}^{S}$ allows here for the first time the explicit knowledge of how the dispositive factor effects the goods in the production by way of the information. If all sequences of primarily efficient information activities which in the end lead to a quantity efficient information activity $v^{S,N}$ are understood as "Informational-Production-Functions", this links the information processing in production planning with the theory of production.

Formally this means: let

$$f^I: \mathcal{D}^S \to R,$$

then f^I should be described as an "Informational-Production-Function" when

$$f^I([v^s]_{s=1}^{S})=0 \text{ applies iff}$$

$v^{S,N}$ is information efficient according to criterium Z.

The explicit formulation of these Informational-Production-Functions is an interesting field for further research in the area of operational production. Being able to form them empirically for use in concrete cases would bring information-oriented production economics a decisive step further.

References:

Adam, D. (ed.): Fertigungssteuerung I, Grundlagen der Produktionsplanung und -steuerung, Wiesbaden 1988.

Bössmann, E.: Information, in: Handwörterbuch der Wirtschaftswissenschaft, Vol. 4, 1978, pp. 184-200.

Dyckhoff, H.: Production Theoretic Foundation of Cutting and Related Processes, in: Fandel, G., Dyckhoff, H., and Reese, J. (eds.): Essays on Production Theory and Planning, Berlin-Heidelberg-New York-London-Paris-Tokyo 1988, pp. 151-180.

Fandel, G.: Produktion I, Produktions- und Kostentheorie, Berlin-Heidelberg-New York-London-Paris-Tokyo 1987.

Fandel, G., and François, P.: Rational Material Flow Planning with MRP and Kanban, in: Fandel, G., Dyckhoff, H., and Reese, J. (eds.): Essays on Production Theory and Planning, Berlin-Heidelberg-New York-London-Paris-Tokyo 1988, pp.43-65.

Fandel, G., and Reese, J.: Kostenminimale Kraftwerksteuerung in einem Industriebetrieb, in: Zeitschrift für Betriebswirtschaft, 1988, pp. 137-152.

Hirshleifer, J.: Where are we in the theory of information?, in: The American Economic Review, 1973, Papers and Proceedings, pp. 31-39.

Jaeger, H.: Die quantitative Analyse von Informationsflüssen in Organisationen, Bochum 1980.

Koopmans, T.C.: Analysis of Production as an Efficient Combination of Activities, in: Koopmans, T.C. (ed.): Activity Analysis of Production and Allocation, New York-London 1951, pp. 33-97.

Küpper, H.-U.: Dynamische Produktionsfunktion der Unternehmung auf der Basis des Input-Output-Ansatzes, in: Zeitschrift für Betriebswirtschaft, 1979, pp. 93-106.

Marschak, J.: Problems in Information Economics, in: Bonini, C.P., et al. (eds.): Management Controls: New Directions in Basic Research, New York 1964, pp. 38-74.

Marschak, J.: Economics of Information Systems, in: Frontiers of Quantitative Economics, Amsterdam 1971, pp. 32-107.

Matthes, W.: Dynamische Einzelproduktionsfunktion der Unternehmung, Betriebswirtschaftliches Arbeitspapier Nr. 2/1979, Seminar für Fertigungswirtschaft der Universität zu Köln, Köln 1979.

Müller, W.: Zur informationstheoretischen Erweiterung der Betriebswirtschaftslehre - Ein Modell der Informationsproduktion, in: Adam, D. (ed.): Neuere Entwicklungen in der Produktions- und Investitionspolitik, Wiesbaden 1987, pp. 119-136.

Müller, W., and Eckert, J.: Informationsproduktion und Entscheidungsprozeß, in: Helmstädter, E. (ed.): Neuere Entwicklungen in den Wirtschaftswissenschaften, Berlin 1978, pp. 455-478.

Müller-Merbach, H.: Ansätze zu einer informationsorientierten Betriebswirtschaftslehre, in: Ballwieser, W., and Berger, K.-H. (eds.): Information und Wirtschaftlichkeit, Wiesbaden 1985, pp. 117-144.

Poensgen, O.H.: Kommunikation, in: Handwörterbuch der Wirtschaftswissenschaft, Vol. 4, 1978, pp. 466-477.

Scheer, A.-W.: EDV-orientierte Betriebswirtschaftslehre, Berlin-Heidelberg-New York-Tokyo 1984.

Scheer, A.-W.: Wirtschaftlichkeitsfaktoren EDV-orientierter betriebswirtschaftlicher Problemlösungen, in: Ballwieser, W., and Berger, K.-H. (eds.): Information und Wirtschaftlichkeit, Wiesbaden 1985, pp. 89-114.

Scheer, A.-W.: CIM - Der computergesteuerte Industriebetrieb, Berlin-Heidelberg-New York-London-Paris-Tokyo 1988.

Schneider, D.: Produktionstheorie als Theorie der Produktionsplanung, in: Liiketaloudellinen Aikakauskiirja, 1964, pp. 199-229.

Shannon, C., and Weaver, W.: The Mathematical Theory of Communication, Urbana 1964.

Shephard, R.W.: Dynamic Production Networks, in: Eichhorn, W., Henn, R., Neumann, K., and Shephard, R.W. (eds.): Quantitative Studies on Production and Prices, Würzburg-Wien 1983, pp. 113-128.

Shephard, R.W., Al-Ayat, R.A., and Leachman, R.C.: Shipbuilding Production Function, in: Albach, H., Helmstädter, E., and Henn, R. (eds.): Quantitative Wirtschaftslenkung, Festschrift für W. Krelle, Tübingen 1977, pp. 627-654.

Wittmann, W.: Unternehmung und unvollkommene Information, Köln-Opladen 1959.

Wittmann, W.: Produktionstheorie, Berlin-Heidelberg-New York 1968.

Wittmann, W.: Aktivitätsanalytische Ansätze dynamischer Produktionstheorie und ihre Beziehungen zur Planung, in: Mellwig, W. (ed.): Unternehmenstheorie und Produktionsplanung, Festschrift für H. Koch, Wiesbaden 1979, pp. 273-304.

Zäpfel, G.: Strategisches Produktions-Management, Berlin-New York 1989 (1989a).

Zäpfel, G. (ed.): Neuere Konzepte der Produktionsplanung und -steuerung, Linz 1989 (1989b).

Comparison of Two Production Logistics Concepts

GÜNTHER ZÄPFEL

1. Introduction

Logistical problems arise when departments confine themselves to considering the material-flows only in their own functional areas (purchasing, production and marketing) and ignore the effects of their decisions upon the enterprise as a whole (figure 1).

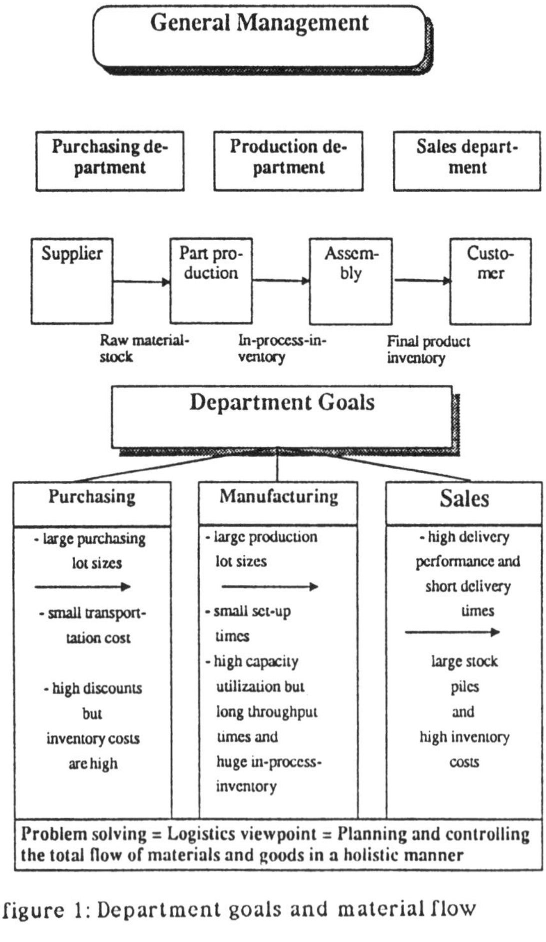

figure 1: Department goals and material flow

Based on the traditional functions of purchasing, production and marketing, the logistical system is generally divided into three areas of competence:

- the material flow in purchasing and the therewith associated information flow
- the flow of material and goods through production and the therewith associated flow of information
- the flow of goods from the firm to its customers and the therewith associated information flow.

The isolated pursuit of functional objectives enhances the probability that solutions arrived at will be merely suboptimal ones. That can mean high inventories and long throughput times. Considering this weakness of traditional organization structure, we see the necessity for viewing the chain of material- and goods-flow in a holistic manner. This is the fundamental concern of logistics; the role of business logistics is to plan and control the total flow of materials and goods from the acquisition of the raw materials up to the delivery of the finished products to the ultimate users, and the related counter-flows of information that both record and control the movements of materials. Logistics can therefore be seen as the art of managing the flow of material and goods - getting the products where they are needed, at the time they are wanted, and at reasonable costs (Magee/Copacino/Rosenfield 1985).

Logistics Management can be divided in strategic, tactical and operational logistics.
The strategic and tactical logistics management has to design the flow of materials and goods and determine a course of action that enables the enterprise to gain competitive advantages. In this context the following problems have to be dealt with:

- Determination of the logistics strategy. Logistics is an important part of the competetive strategy. If the enterprise chooses a strategy of cost leadership, the logistics must insure that the required logistical performance will be attained with minimal costs. In pursuing a differentiation strategy, logistics has to support the differentiation process by improving the logistical performance and at the same time keeping costs in line.

- Design of products from a logistical point of view (this means considering the number of variants and the extent of parts variety).

- Selection of the number of production processes (this is especially associated with the kind and number of production and inventory units within the enterprise).

- Selection of the supplier structure (this requires, that the number and the locations of the suppliers and the mode of the partnerships be determined).

- Design of the location and distribution system of the enterprise.

- Selection of the concept for the (operational) planning of the material and goods flows.

Operational logistics is responsible for the current planning and realization of the real flows of materials and goods. In this context complex tasks arise concerning the coordination of the material flows within enterprises which produce a variety of products in many production stages. Examples of this kind of enterprise are producers of parts which assemble these parts to form final products (for instance in the machinery or electronic engineering industry), etc.

Based on planned or realized orders, (operational) production logistics has the task of determining the materials and goods flow by fixing the product types in quantity and time through all production stages, so that certain objectives will be attained, and advancing, controlling and securing these plans when deviations occur. Production logistics fixes the throughput times of the products, the in-process-inventories and the capacity utilization of the production units. The content of (operational) production logistics therefore agrees with the task of production planning and control (PPC).
The objective of PPC is to plan and control the operating system, so that customer wants are satisfied and at the same time an efficient utilization of resources is guaranteed. An operating system is a configuration of resources combined for the function of manufacture, transport supply or service (Wild 1980).

In the following section we will discuss the fundamental concepts of production logistics.

2. Two concepts of production logistics

As we have seen, production logistics spans all activities which are concerned with planning and controlling the material and goods flows from the raw material inventory to production and between all the production units (including the in-process-inventories) up until the final-product-inventories.

We can distinguish two different logistical philosophies (or concepts) for the realization of these tasks:

* the concept of material stock optimization
* the concept of material flow optimization.

The traditional concept of material stock optimization tries to attain a certain customer service level by investing in inventories. The benefit of high inventories is that goods can be quickly delivered to the customer, in the case the right products are at hand. For a given service level the stock should be as minimal as possible. That is the reason for the name: material stock optimization.

The drawbacks of the concept are:

- High material- and product-stock piles demand a great deal of turnover capital and negatively affect the liquidity of the enterprise.

- Inventories increase the risks of having unsellable products, since forecasts sometimes prove incorrect.

- High inventories prevent the weaknesses of production processes (e.g. high setup times, unsynchronized production processes, imperfect flexibility in production) from becoming apparent. Therefore waste is not eliminated.

  Inventory can therefore be judged from two viewpoints:

- In general the defenders of the concept of material stock optimization regard inventory in a positive manner. Inventory makes it possible to guarantee a smooth production and quick deliveries and to avoid the negative consequences of break-downs, they enable a constant utilization of production units, etc.

- According to the opponents of this concept, stock piles are the "root of all evil", because they prevent the elimination of long setup times, unsynchronized production processes, high reject rates, etc. These opponents prefer the concept of material flow optimization to determine the modes of action. These involve a continous process of improvements. Examples are 1) minimizing of setup times, which enables smaller lot sizes to become economical, 2) new kinds and layouts of machines (flexible manufacturing cells or flexible manufacturing systems) which are able to produce parts or even products completely, 3) the synchronization of capacities, 4) the optimization of variants, (5) quality assurance, etc. A production-synchronous-procurement is also an example of this kind of rationalization. This reduces the throughput time dramatically, eliminates break-downs and lowers stock piles.

In summary this mode of action should make a Just-in-Time-Production possible; at any given time only the materials and products which are immediately needed at certain production stages or by customers are produced.

Now we have to pose the questions:

Under which conditions is a JIT-Production economical? Can inventory be justified under certain circumstances?

In the following chapter we will try to answer these questions by taking a theoretical-logistical model as a basis.

3. Theoretical foundation of production logistics

In parts and assembly industries, planning and controlling the flow of materials and goods is an especially complex task. The products consist of many parts and processing takes place in several production stages. The jobs must share capacities. An important function of management is therefore the coordination and control of complex activities, including the machine sequencing problem. The well-known objectives - minimal throughput times, low inventories, high capacity utilization - are ultimately associated with the goal of maximizing the sum of the differences of revenues minus expenses for all periods.
In this sense Goldratt/Fox (1986) defines the goal of production control in a very direct manner: to make money. The objectives expressed in time or quantity dimensions (e.g. throughput time, in-process-inventory etc.) are only subsidiary ones. The problem of production logistics can therefore be defined more exactly as in figure 2.

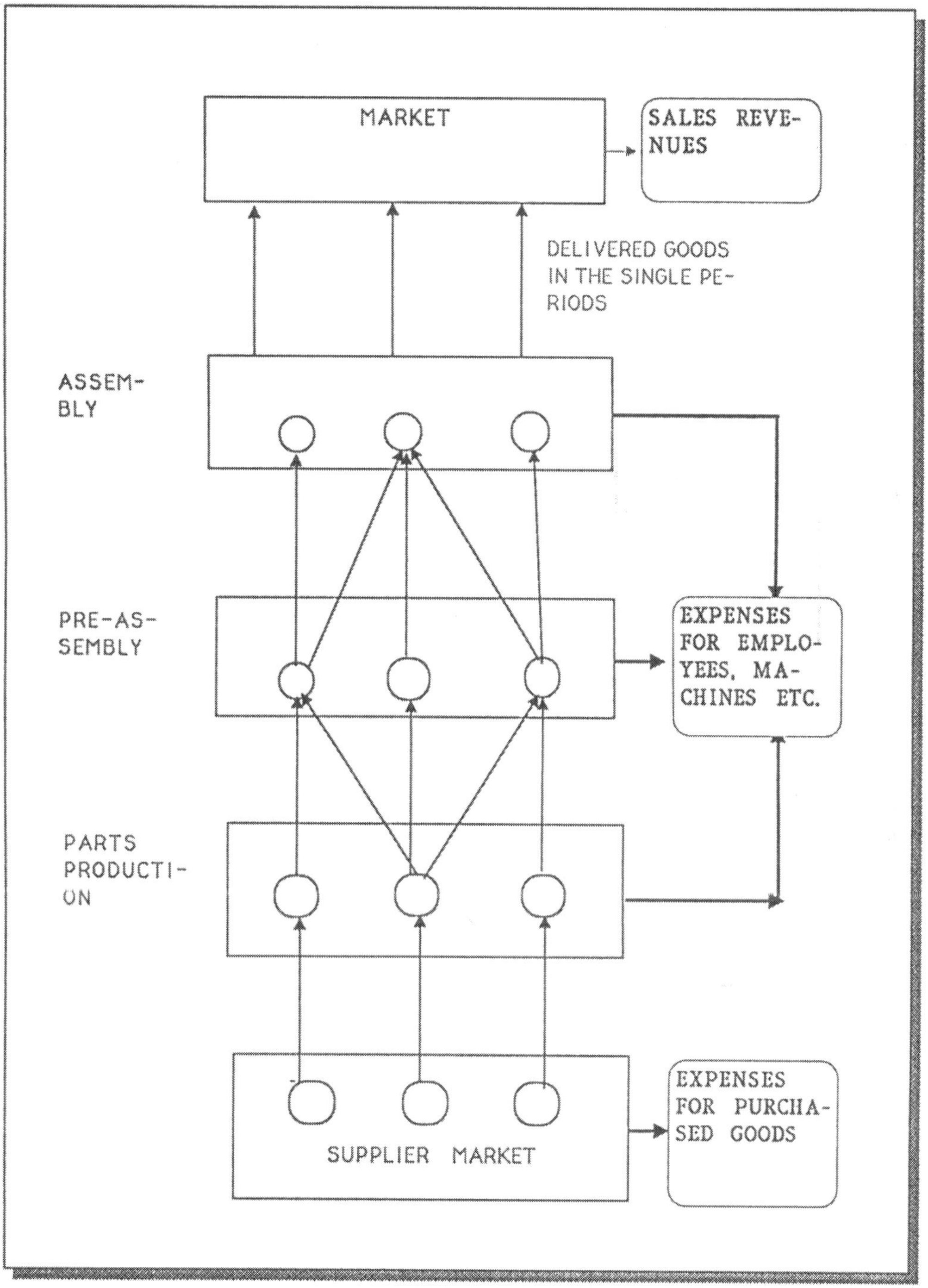

figure 2: Material flow chain and related money flow

By determining the material and goods flow, the cash inflows and cash outflows during the periods of the planning horizon become fixed.

Cash inflows result from the delivered outputs, which come in the form of revenues from the customers. Expenses arise through the purchasing of raw materials and the maintenance of the equipment. The timing and extent of the expenses for raw materials depend upon when the orders come through and also upon their sizes. The expenses for the equipment of the enterprise (e.g. salaries) are rather fix in time.

By choosing alternative material and goods flows, the difference between revenues and expenses in the single periods will be changed. If a positive difference (cash flow) arises in a given period, funds can be invested and an income (interest) results. A negative cash flow during a period makes it necessary to borrow money and expenditures will result (payment of interest). This view is advantageous because it enables us to avoid determining inventory costs. These can be considered as creating expenses at an earlier stage and therefore altering the cash flow as a result of interest payments.

Taking these assumptions as a basis, we can formulate a model of production logistics.

The following notation is used to describe the model in mathematical terms:

Decision Variables:

x_{it}: quantity of production units of product i in period t
- $i \in P_3$: final products
- $i \in P_2$: assembly products
- $i \in P_1$: parts

x_{lt}: quantity in period t of purchased goods l, $l \in P_0$

y_{it}: inventory level of product i at the end of period t

O_{mt}: necessary additional capacity (overtime) of production stage m in period t

$$\delta(x_{it}) = \begin{cases} 1, & \text{if the product i in period t is selected} \\ 0, & \text{otherwise} \end{cases}$$

DF_t^+, DF_t^- positive or negative payment difference in period t

CF_t^+, CF_t^- (positive or negative) cash balance at the end of period t including interest payments

Parameters:

b_{it}: expected demand for final product i in period t

q_{ji}: the number of units of product j which are required for one unit of the immediate parent product i

a_i: the number of hours required per unit of product i

r_i: total setup time required by product i

u_{mt}: payments for overtime capacity per time unit (e.g. hours) of production stage m in period t

N_{mt}: regular capacity of production stage m in period t, expressed in time units (hours)

G_i: a large number greater than or equal to the value of $\max_t \{x_{it}\}$

e_i: selling price per unit for final product i

F_t: fixed payments in period t (e.g. salaries)

s_l: payments per unit for purchased goods l

$\bar{O}_{mt}$: maximum available overtime capacity of production stage m in period t, expressed in time units

R_{mt}: maximum available setup capacity at production stage m in period t, expressed in time units

z^-: debt interest

z^+: credit interest ($z^- > z^+$)

L_l: delivery time for purchased goods l, expressed in periods

y_{io}: inventory level of product i at the beginning of the planning horizon

CF_o: cash balance at the beginning of the planning horizon

A simple version of the production logistics model involves maximizing the cash balance at the end of the planning horizon, that is

Maximize CF_T

subject to

(1) <u>material flow restrictions</u>

$$\left.\begin{array}{lll} \text{(1 a)} & y_{it} = y_{it-1} + x_{it} - b_{it} & i \in P_3 \\ \text{(1 b)} & y_{jt} = y_{jt-1} + x_{jt} - \sum_{i \in P_3} q_{ji} \cdot x_{it} & j \in P_2 \\ \text{(1 c)} & y_{kt} = y_{kt-1} + x_{kt} - \sum_{j \in P_2} q_{kj} \cdot x_{jt} & k \in P_1 \\ \text{(1 d)} & y_{lt} = y_{lt-1} + x_{l,t-L_l} - \sum_{k \in P_1} q_{lk} x_{kt} & l \in P_0 \end{array}\right\} t = 1,\dots\dots,T$$

The coordination of the flow of material with the flow of goods has to be guaranteed. It is assumed that the throughput time through all production stages encompasses only one period, because the load can't exceed the capacity. The delivery times span L_l periods.

(2) Capacity restrictions

(2 a) $$\sum_{i \varepsilon P_m} (a_i . x_{it} + r_i . \delta(x_{it})) \leqslant N_{mt} + O_{mt},$$

whereby

(2 b) $$O_{mt} \leqslant \bar{O}_{mt}, \qquad m = 1,...,3 \quad t = 1,...,T$$

and

(2 c) $$x_{it} \leqslant G_i . \delta(x_{it}), \qquad i \varepsilon P_m, \; m = 1,...,3 \quad t = 1,...,T$$

(2 d) $$\sum_{i \varepsilon P_m} r_i . \delta(x_{it}) \leqslant R_{mt} \qquad m = 1,...,3 \quad t = 1,...,T$$

In every single production stage (m =1,...,3) the load caused by the quantity of production, which consists of production and set up times, may not exceed the regular and added capacity (restrictions (2 a) and (2 b)). If production takes place, setups occur (restrictions (2 c)). Only a certain number of setups can be executed in every period, because the crew available for accomplishing the setup operations is limited (restrictions (2 d)).

(3) <u>Financial restrictions</u>

$$\text{(3 a)}\quad DF_t^+ - DF_t^- = CF_{t-1} + \sum_{i \varepsilon P_3} b_{it} \cdot e_i - F_t - \sum_{l \varepsilon P_0} x_{l,t-L_l} s_l - \sum_{m=1}^{3} u_{mt} \cdot O_{mt}$$

$$\text{(3 b)}\quad CF_t = CF_t^+ - CF_t^- = (1 + z^+)DF_t^+ - (1+z^-)DF_t^-$$

The payment difference of a period is determined by the cash balance at the beginning of the period plus the sales per period minus the fixed and variable expenses including overtime costs (restrictions (3 a)). If the payment difference is positive, credit interest arises; a negative difference causes debt interest (restrictions (3 b)).

4) Non-negativity-restrictions

$x_{it} \geqslant 0 \quad i \; \varepsilon \; P_3,$

$y_{it} \geqslant 0 \quad i \; \varepsilon \; P_2, \quad t = 1,\ldots,T$

$\quad i \; \varepsilon \; P_1$

$\quad i \; \varepsilon \; P_0$

$DF_t^+, DF_t^- \geqslant 0 \quad t = 1,\ldots,T$

$CF_t^+, CF_t^- \geqslant 0 \quad t = 1,\ldots,T$

$\delta(x_{it}) \; \varepsilon \; \{0,1\}$ for all i and t

$O_{mt} \geqslant 0 \quad m = 1,\ldots,3$

$\quad t = 1,\ldots,T$

This logistical model defines a mixed-integer programming problem by including set up times according to restriction 2. In parts and assembly industries there are tens of thousands of final products, assembly products as well as parts, and this can make it difficult to apply the above model. If we neglect the set up times, we end up with a huge linear programming model.

In this paper we are not primarily interested in a practical solution for the model. Our aim is rather to arrive at conclusions concerning the applicability of the "production or logistics philosophies".

It is possible to prove the following theorem:

If a feasible solution $\hat{y}_{it} = 0,\ i \in P_0 \cup P_1 \cup P_2 \cup P_3$ *and* $t = 1,\dots,T$ *exists, that is the restrictions (1) - (4) and especially (2 a) - (2 b) are satisfied as well as* $\hat{O}_{mt} = 0$ *for all m and t, then this is the optimal solution; it is designated* $\widehat{CF}_t$.

The proof of this theorem can be found in the Appendix.

What are the implications of this theorem? A Just-in-Time-Production is optimal if the defined conditions are valid. A production-synchronous-procurement and a sales-synchronous-production is optimal. Stock piles should be avoided as much as possible.

The theorem is valid as long as the expense per unit for all procurement goods is constant in the planning horizon and doesn´t depend on the amount for single deliveries. In other words, discounts for single deliveries are excluded and price rises are not considered.

A typical attribute of a production-synchronous-procurement is a skeleton contract with suppliers about an anticipated quantity in a pre-determined time horizon and a pre-specified price per unit. It depends on the size of the planned deliveries and not on the current quantity released. In this case, the underlying assumptions of the theorem are met.

In a production-synchronous-procurement (PSP), the number of transports can increase, and this leads to higher expenses for these activities. Within the framework of PSP, the enterprise attempts to keep the expenses for transportation in line. This can be done in several ways: the number of suppliers for each material can, for example, be dramatically reduced, or suppliers can be chosen which are located near the enterprise. The materials from different suppliers in a region can also be collected by a forwarding agent and delivered together to the enterprise.

To avoid an increase in quality-control activities, which would lead to added quality control costs, the enterprise shifts these activities to the suppliers. In the case of PSP, it is important that the suppliers only deliver parts with no defects, because safety stocks are small and the delivered quantities are immediately used in production. If an increase in the supplier's costs thereby arises, a new price can be negotiated between the buyer and the supplier.

The fundamental difference between the traditional concept of inventory optimization and PSP can be seen in the following: the traditional concept strives for an optimization of the purchase quantities by devoting considerable attention to quantity discounts and price speculation. If necessary, the enterprise changes suppliers. In PSP, close connections (partnership) are established with one or a few suppliers for each PSP-material. The buyer has to pay a defined price per unit for a pre-determined amount in the planning horizon (whereas a limit to the variability of the amount is also agreed upon) and the buyer is authorized to recall material in the actual required quantity within the planning horizon. The benefit for the suppliers consists in warranting a pre-determined amount for the length of the contract, which assures the supplier a certain capacity utilization.

A further condition expressed in the theorem implies that the time for the planned production quantities may not exceed the regular capacity. This is fulfilled, if the sale of the products is uniform in the planning horizon and the resulting capacity demand doesn't exceed the regular capacity. It is also possible to obtain a uniform capacity utilization if the sales of some products increase in time, but other products decrease synchronously.

Especially when seasonal variations in demand occur, the capacities must be at the level of peak demands; otherwise frequent adaptations of capacities (e.g. overtime) will arise. High capacities imply large fixed expenses, that is huge capital assets. The alternative of frequent adaptations leads to high variable expenses, since overtime is required. Therefore we have to consider that a tradeoff between fixed and variable expenses can occur. Furthermore there is a tradeoff between different variable expenses. One must either opt for stock-keeping and therefore produce goods before they are needed (which leads to stock-keeping expenses) or for producing the goods exactly when they are needed by adaptation of capacities, which leads to expenses for overtime. To clarify the tradeoff let's look at a simplified practical problem.

4. Considerations of production logistics in JIT: tradeoff-curves

The problem is described in figure 3.

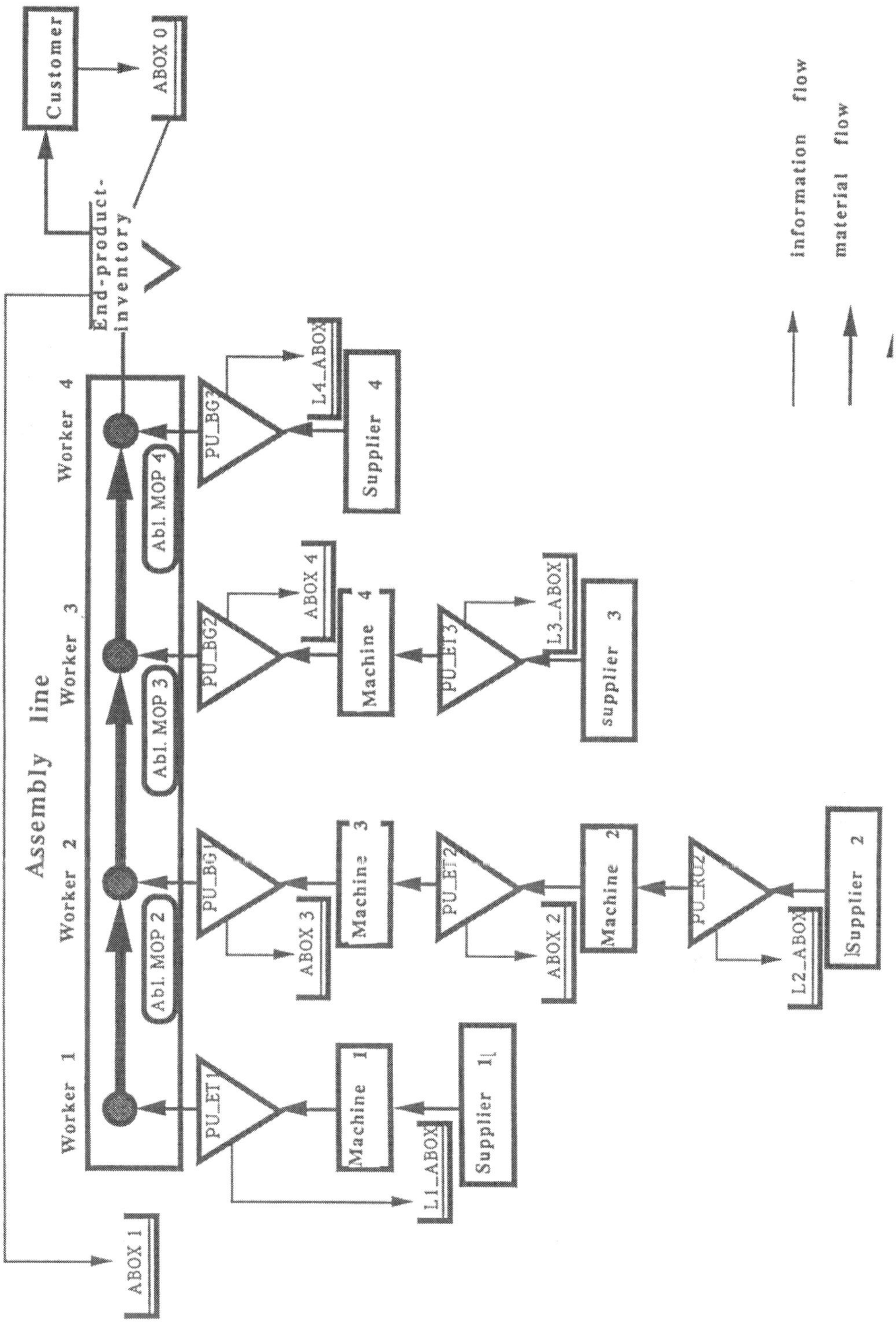

figure 3: An example for a material and an information flow

Several variants of an electronic product are processed by an assembly line, which is fed by four supplementary production units producing different parts and sub-assemblies for the line. There are several buffer stores for the production units, designated as PU- ET1 - PU BG3. An end product inventory is also installed.
Five variants are processed. In the figure 4 two examples of the product structure of the variants can be seen. The variants differ by the quantity of parts, which goes into the final product. R1 and R2 stand for raw materials, ET1 to ET3 are parts, BG1 to BG3 designate sub-assemblies, and P_{ij} represents variant j of product i.

Product 1 - Variant 1:

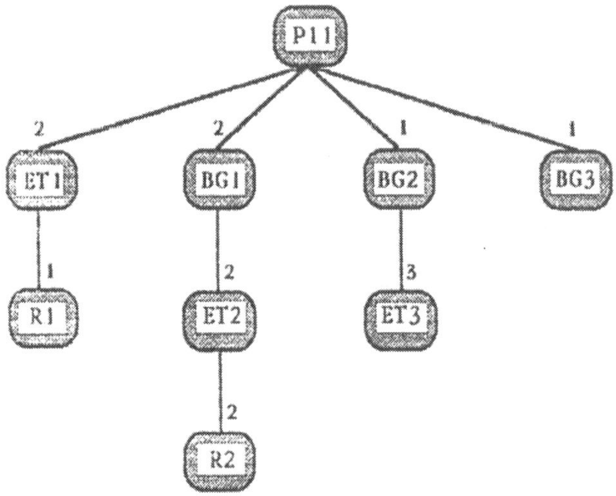

Product 2 - Variant 2:

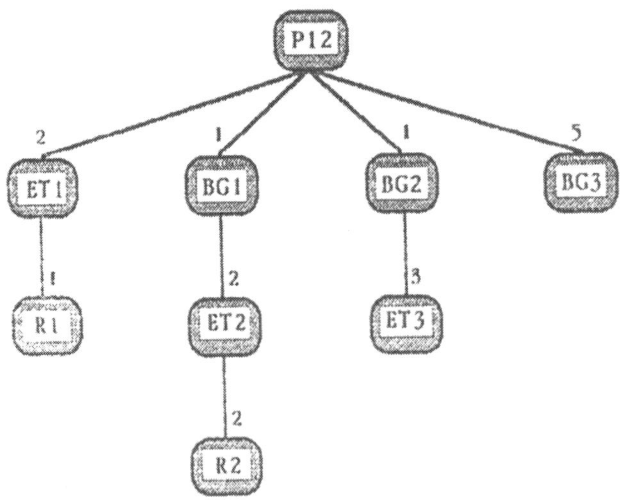

figure 4: Product structure of two variants

The business policy is to maintain a capability for delivering the variants to the customer within a fixed delivery lead time, e.g. a day. For this a planning concept like JIT has to be implemented. This can be accomplished, because the suppliers´ delivery times and the throughput times of parts production and pre-assembly is usually short enough to enable the realization of the products within this horizon. The operation and delivery times are shown in figure 5.

Delivery- and operation times

Assembly Line

- Worker 1	25 minutes/unit
- Worker 2	28 minutes/unit
- Worker 3	30 minutes/unit
- Worker 4	15 minutes/unit

Parts Production

- Machine 1	25 minutes/unit
- Machine 2	15 minutes/unit
- Machine 3	30 minutes/unit
- Machine 4	25 minutes/unit

Suppliers

- Supplier 1	15 minutes/delivery
- Supplier 2	5 hours/delivery
- Supplier 3	5 hours/delivery
- Supplier 4	2 hours/delivery

figure 5: Delivery- and operation times

If the regular daily capacities aren't sufficient to satisfy the actual demand, overtime is necessary. The problem consists of determining the in-process-inventories and the end-product-inventory. JIT strives to attain minimal inventories. But if the stock piles are too small, every variation in demand causes problems, and capacity adjustments become necessary.
The effects of different demand patterns (constant demand versus demand fluctuations of the variants in the time process) and different stock piles on the production system can be determined by simulation experiments. For this special case, SIMAN was used as a simulation language. Detailed information can be found in the work of Hödelmoser (1990).

In figure 6 we can see the relation between overtime and stock-keeping, when the utilization rate of the production system differs and the demand is **constant**. That means that every day the same variants (two pieces of every variant) are produced with a constant work content.

figure 6:

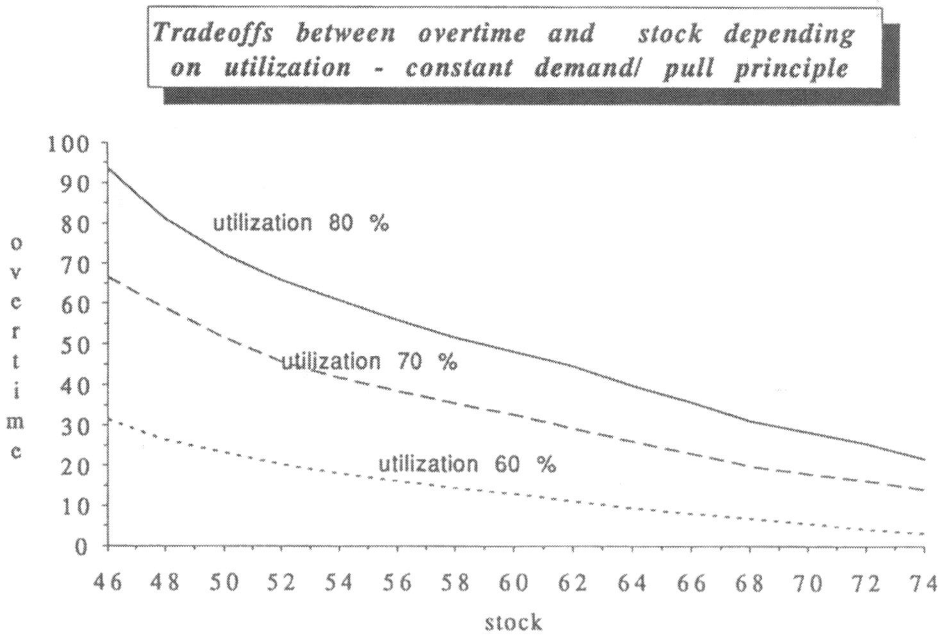

What are the basic findings of this simulation? A tradeoff-curve results. If the stock becomes greater, overtime decreases and vice versa. This is valid for different utilization rates of the production system. To obtain the curves we make several simulation runs (29 times per curve). For a certain number of customer orders (and therefore a fixed load on the production system) we define buffer capacities and, by a simulation run, the overtime of all the production units and the bottlenecks are determined. In the next step the capacity of a bottleneck is increased by one unit. Now the overtime and the new bottlenecks are determined by a new simulation run and the buffer stock for the bottleneck is again increased, etc. (figure 7).

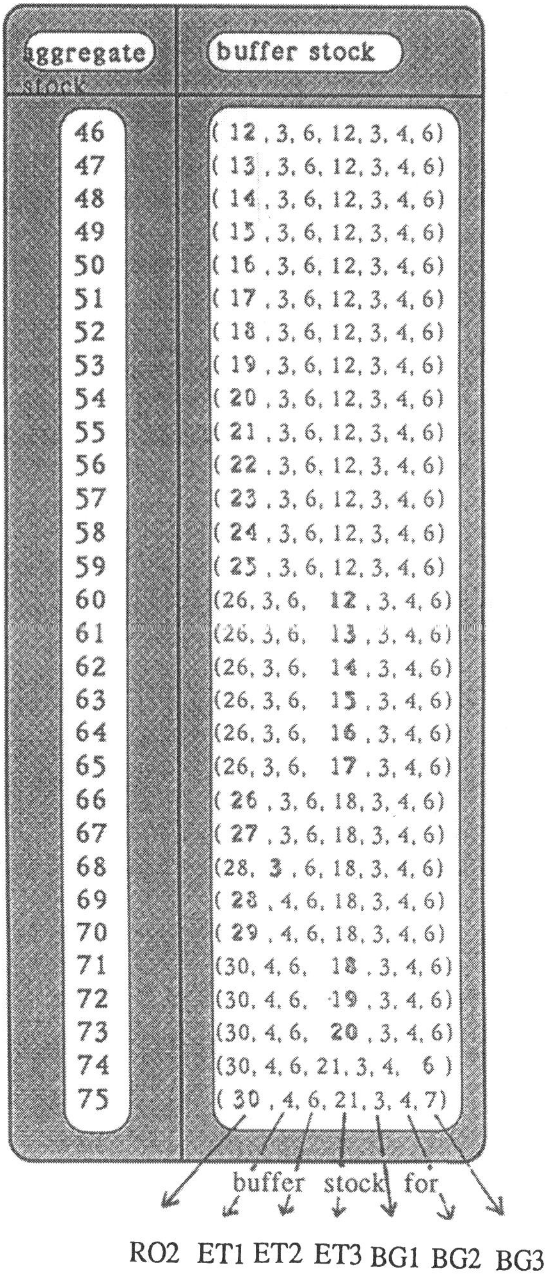

figure 7: The combination of the stock piles

One assumption is that we tried to increase only buffer stock piles for actual bottlenecks. If we arbitrarily increase the stock, the tradeoff-curve is less smooth (figure 8). Buffering is more important for the bottleneck stations. Buffer stock is not so important for non-bottleneck stations, because faster machines act somewhat like buffers (Conway/Maxwell/McClain/Thomas 1988, p. 237).

If **demand variability** exists, and therefore the load can change every day, we get the following figure 9. In our special case we supposed that each day a certain number of customer orders arrives at the enterprise. On the first day a particular lot of variant P11, the second day a lot of variant P12, and so on. The simulation experiments show that there is again a tradeoff between overtime and stock, depending on the utilization rates of the whole production system. But the effects are not as impressive. It should be noticed, that in reality the trade-off curve is sometimes not applicable over the total range, because the employee contract limits the amount of overtime.

The above results are in accordance with those of Huang/Rees/Taylor (1983). Similar conclusions are arrived at in a somewhat different context by Gupta/Gupta(1989), Deleersynder/Hodgson/Muller(-Malek)/O`Grady (1989), Lee (1987) and others. Lee shows that an increase in the production Kanban size and minimum output level improves the mean job tardiness (which could be avoided by overtime) but causes an escalation in the output Kanban inventory level.

The conclusions from these tradeoff-curves are:
The extent of overtime depends on 1) the size of the stock piles and 2) the total utilization rate of the production system, in turn, is determined by the size and the mix of the incoming customer orders. By increasing the buffer stock of the actual bottleneck capacities, overtime can be decreased. It is not surprising, that the higher the utilization of the whole production system is, the higher the overtime becomes. As also can be seen, the higher the utilization rate of the production system is, the more sharply overtime falls, when buffer stock piles increase. The effect is not so strongly marked in the case of variable demand rates.

Up until now we have taken the *pull principle* as a basis: The succeeding stage demands and takes in-process units from the preceding stage only at the rate in which the succeeding production stage consumes the items. The removal of inventory at the preceding stage authorizes the manufacture of an additional unit. Since manufacturing cannot occur without such authorization, each stage produces just in time. Thus, we have a serial ordering system of sucessive production orders, extended to a number of suppliers.

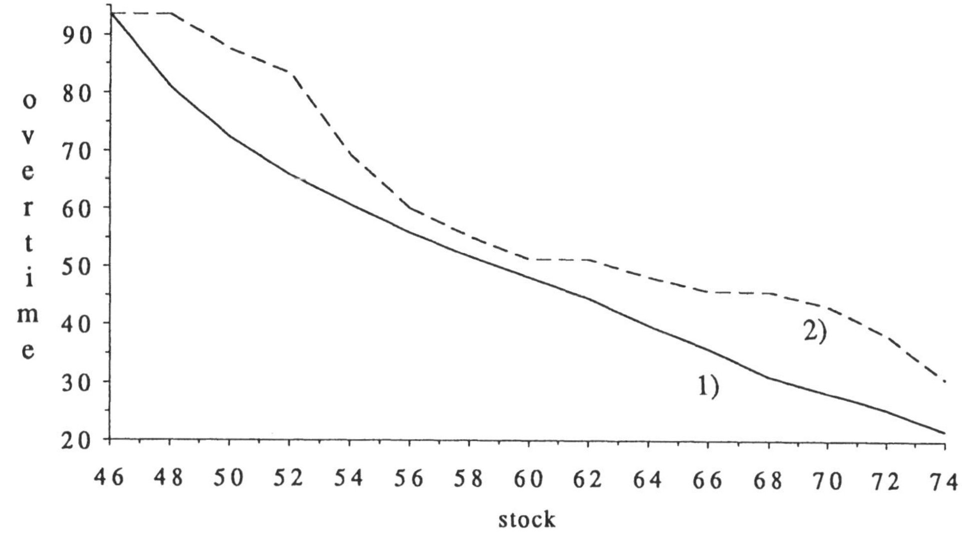

figure 8: Comparison of two tradeoff-curves
1) increasing only buffer stock piles for actual bottlenecks
2) arbitrarily increasing the stock

figure 9:

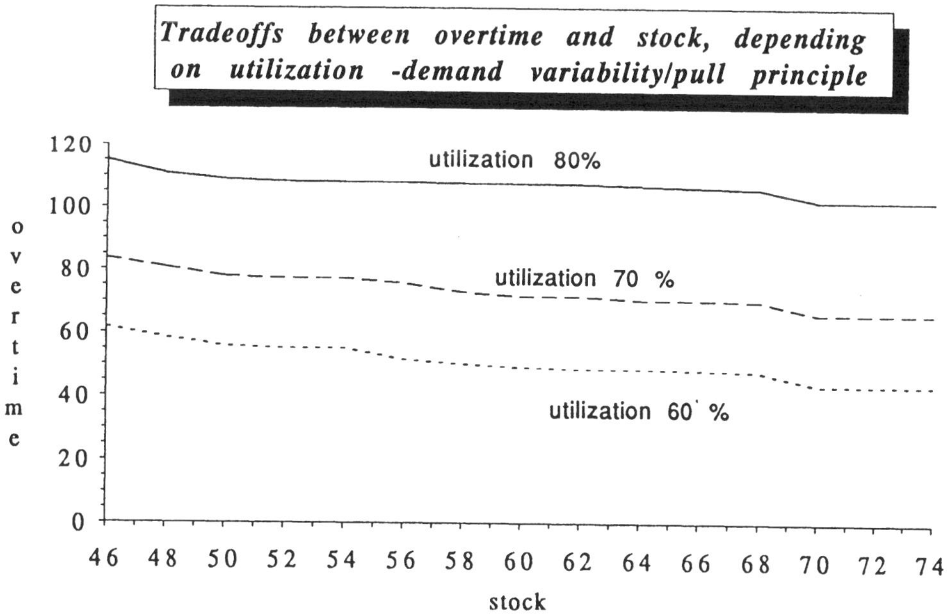

The production system considered here can also be managed with the traditional *push principle*: Information regarding demand forecasts or customer orders is processed to all production stages. Usually, production orders are released at the first stage and then the order is pushed through the production system. In our special case a customer order immediatly leads to an order release at all production stages. When *the push-principle for constant demand* is applied, the tradeoff curve is similar to the curve for the pull-principle (figure 10). But, in comparison with the pull principle, the necessary overtime is much smaller for a given utilization rate of the total production system.
Demand variability leads to a greater requirement for overtime in comparision with constant demand, but less overtime is needed in the case of the push-principle (figure 11). It should be noticed that there exists no maximal limits for the buffer stock piles like those associated with the pull principle.

figure 10:

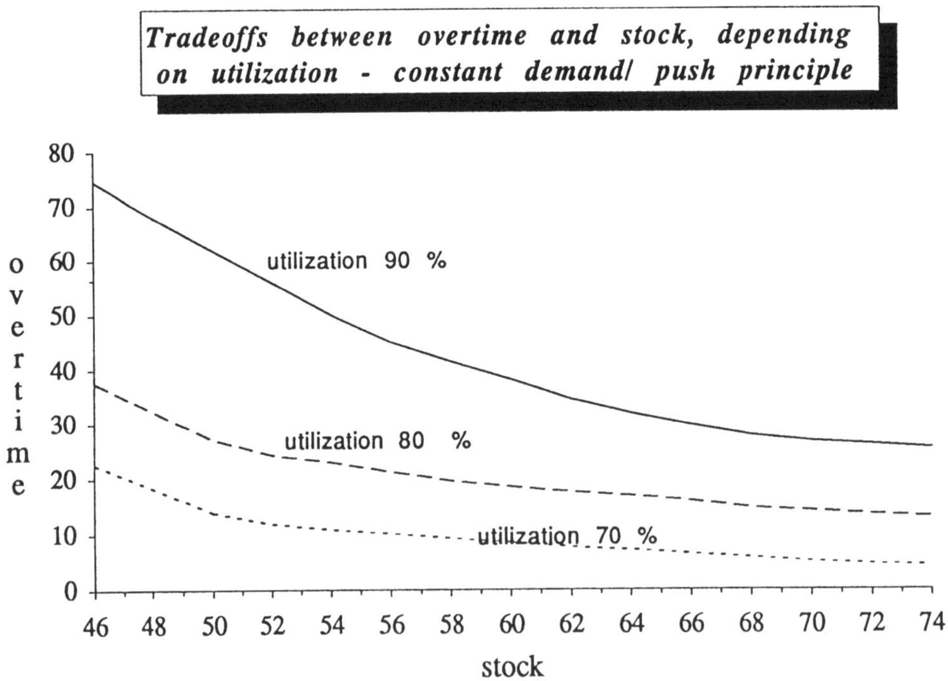

figure 11:

Tradeoffs between overtime and stock, depending on utilization - demand variability/ push principle

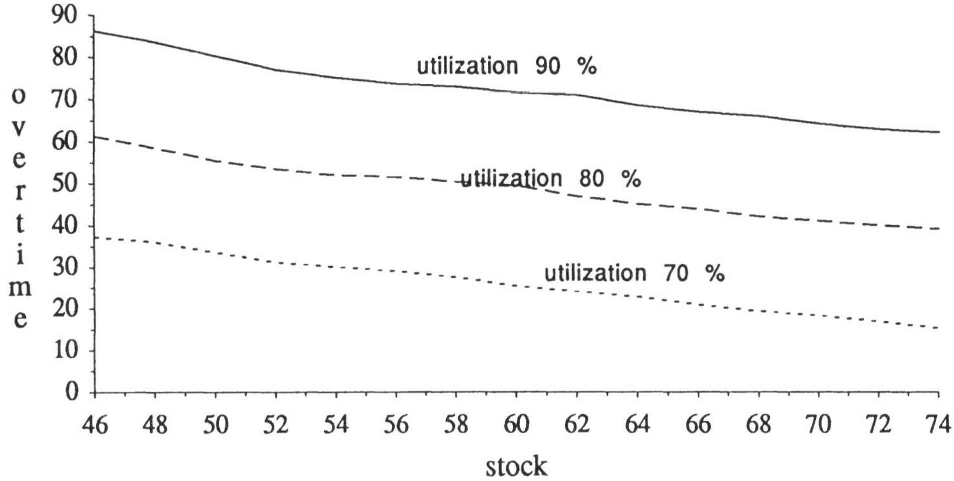

5. Conclusions

For a Just-in-Time-Production, it is proposed to invest in over-capacities and not in stock-keeping. The ideal is a stockless production. By closing inventories fixed expenses can also be reduced. In the case of seasonal variations of sales, the trade-offs must be taken into consideration in the choice of a particular logistical philosophy.

These trade-offs must already be considered in the strategic planning, especially in arriving at a decision as to how capacities are to be structured. The enterprise can opt for either:

1) Choosing a dimension of capacities which is orientied at the peak level of demand and enables the synchronization of the amounts of production with the sales quantities during all periods. The advantage of high-dimensioned capacities is that overtime expenses can be avoided. Short delivery times can be attained without stock keeping. If various products are manufactured, short set up times are also required. A reduction of the set up times is specifically important for a JIT-Production. Set up times narrow the availability of capacities (equations (2 a)) and the set-up-capacity constrains the number of set ups (equations (2 d)). Longer set up times make a JIT-Production impractical. Shortening of set up times results in a reduction of expenses for stock piles and for overtime: there is more time for productive utilization. One disadvantage of high-dimensioned capacities is that high fixed expenses occur during all periods. This means a high risk when demand declines.

2) Choosing a "moderate" dimension of capacities. This necessitates overtime capacity in periods with peak levels of demand or alternatively requires the manufacturer to produce in advance of demand realization. In the latter case, stock-keeping becomes unavoidable. Both alternatives involve added expenses. The advantage of moderate dimensions of capacities is that the fixed expenses are low. The risk upon demand decline is therefore lower than with high-dimensioned capacities.

The statements show that the efficiency of a JIT-Production depends on the selection of the dimension of capacities. In cases where significant seasonal variations, long lead times or huge set-up times exist, stock-keeping can be more advantageous than a JIT-Production.

6. Summary

An important feature of the traditional art of managing the flow of materials in enterprises is the division of the responsibility into several functional subsystems (e.g. purchasing, different production areas). A major disadvantage of this method is that sometimes optimization efforts remain confined to the individual subsystem; effects upon the enterprise as a whole are not taken into consideration. The consequences are frequently high inventories and long throughput times.
Logistics can help us to avoid these effects: it involves the consideration of the total flow of materials from the receipt of raw materials through manufacturing and processing stages up to the delivery of the finished products.

There are two different concepts for managing the flow of material and goods in production:

* the concept of material stock optimization
* the concept of material flow optimization.

With the concept of material stock optimization one can, through optimization of inventory, attempt to guarantee quick deliveries, circumvent the unpleasant consequence of equipment break downs, insure constant utilization of production units even when seasonal variations of demand exist, etc.
The traditional systems of production planning and control are based on this concept. Complex operating systems require highly sophisticated production planning and control systems.

Recently the concept of material flow optimization has attained importance. Stock-keeping is now seen as the "root of all evil"; when inventories are available, the management has little incentive to improve the operating system, e.g. to reduce set up times, to increase process and product quality reliability, etc.
Therefore the aim is first to eliminate waste (e.g. high set up times) in the operating system and afterwards to install new planning concepts, e.g. JIT. Although a stockless production is the ultimate goal, trade offs (expenses for stock-keeping versus expenses for capacity adaptions) must be taken into consideration. Finally, production logistics has to carefully evaluate the consequences of these trade-offs, thereby increasing the competitive advantage of an enterprise.

Selected Bibliography

Adam, D. (1988): Fertigungssteuerung I. Grundlagen der Produktionsplanung und -steuerung, Wiesbaden

Bowersox, D.J.(1974): Logistical Management. A Systems Integration of Physical Distribution Management, Materials Management and Logistical Coordination, New York-London

Conway,R./W.Maxwell/J.O.McClain/L.J.Thomas(1988): The Role of Work-In-Process Inventory In Serial Production Lines, in: Operations Research, Vol. 36, No. 2, pp. 229-241

Deleersynder, J.L./T.J. Hodgson/H. Muller(-Malek)/ P.J. O`Grady (1989): Kanban Controlled Pull Systems: An Analytic Approach, in: Management Science, Vo. 35, No. 9, pp. 1079-1091

Fandel, G./P. Francois (1989): Just-in-Time-Produktion und -Beschaffung. In: Zeitschrift für Betriebswirtschaft, Nr. 5, S. 531-544

Goldratt, F.M./ J. Fox (1984): The Goal, Boston

Gupta, Y.P./ M.C. Gupta (1989): A System Dynamics Model for a Multi-stage Multi-line Dual-card JIT-Kanban System, in: International Journal of Production Research, Vol 27, No. 2, pp. 309-352

Harhen, J. (1988): MRP/MRP II. In: Computer-Aided Production Management, ed. by A. Rolstadas, Berlin, Heidelberg, New York

Heinrich, C.E. (1989): Das MRP II-Planungskonzept (Manufacturing Resource Planning) und dessen Realisierung mit Standard-Software, dargestellt am System R/2 der Sap AG. In: Neuere Konzepte der Produktionsplanung und -steuerung, hrsg. von G. Zäpfel, S. 95 - 112

Helberg, P. (1987): PPS als CIM-Baustein: Gestaltung der Produktionsplanung und -steuerung für die computerintegrierte Produktion, Berlin

Hödelmoser, R. (1990): Simulation von JIT-Systemen mit SIMAN, Master Thesis at the Institute of Industrie und Fertigungswirtschaft (Prof. Dr. Zäpfel), University of Linz

Huang, Ph.Y./Rees, L.P./Taylor, B.W. (1983): A Simulation Analysis of The Japanese JUST-IN-TIME Technique (with Kanbans) For a Multiline, Multistage Production System, in: Decision Sciences, Vol. 14, pp. 326-344

Ingersoll Engineers (1985): Flexible Fertigungssysteme, Berlin, Heidelberg, New York

Krajewski, L.J./L.P. Ritzman (1990): Operations Management. Strategy and Analysis, 2.Ed., Reading, Massachusetts

Lee, L.C. (1987): Parametric Appraisal of the JIT System, in: International Journal of Production Research, Vol 25, No. 10, pp. 1415-1429

Magee, J.F./W.C. Copacino/D.B. Rosenfield (1985): Modern Logistics Management. Integrating Marketing, Manufacturing, and Physical Distribution, New York -Sydney

Pfohl, H. Chr. (1985): Logistiksysteme. Betriebswirtschaftliche Grundlagen. Berlin-Heidelberg-New York-Tokyo

Silver, E.A./ Peterson, R.(1985): Decision Systemes For Inventory Management and Production Planning, New York, Chichester, Brisbane, Toronto, Singapore

Tersine R.J. (1980): Production/Operations Management: Concepts, Structure and Analysis, New York-Oxford

Wiendahl, H.P. (1987): Belastungsorientierte Fertigunggssteuerung. Grundlagen - Verfahrensaufbau - Realisierung. München - Wien

Wild, R.(1980): Operations Management. A Policy Framework, Oxford-New York-Toronto

Wildemann, H. (1984): Flexible Werkstattsteuerung durch Integration von KANBAN-Prinzipien, München

Wildemann, H. (1989): Reorganisation von Produktion und Zulieferung zur Einführung von Just-In-Time: In: Neuere Konzepte der Produktionsplanung und -steuerung, hrsg. von G. Zäpfel, Linz, S. 139 -179

Zäpfel, G. (1989 a): Strategisches Produktions-Management, Berlin- New York

Zäpfel, G. (1989 b): Taktisches Produktions-Management, Berlin- New York

Zäpfel, G./H. Missbauer (1987): PPS-Systeme mit Bestandsregelung zur Senkung von Durchlaufzeiten und Beständen. In: Information Management 1, S. 65-73

Appendix

Proof of the theorem:

For a given sequence of interests z_t, $t = 1,\ldots,T$, ($z_t = z+$ or $z_t = z-$) and a given period s, employing the equations (3a) and (3b) of the model, we obtain:

$$(1.1) \quad CF_s = \{[CF_0 \prod_{t=1}^{s-1} (1 + z_t) + \sum_{t=1}^{s-1} RF_t \prod_{t'=t}^{s-1} (1 + z_{t'})] + RF_s \} \; (1 + z_s)$$

for all $s = 1,\ldots,T$

The term in the square bracket corresponds to CF_{s-1} and $z_s = z+$ ($z_s = z-$) is fulfilled, if $CF_{s-1} + RF_s \geqslant 0$ ($CF_{s-1} + RF_s < 0$).
The cash flow RF_t in every period t is defined by:

$$RF_t = \sum_{i \varepsilon P_3} b_{it}\, e_i \; - F_t - \sum_{l \varepsilon P_0} x_{l,t-L_l} \cdot s_l \; - \sum_{m=1}^{3} u_{mt}\, O_{mt}$$

For distinct feasible solutions the $RF_1,\ldots,RF_T$ differ only by the following expenses:

$$\sum_{l \varepsilon P_0} x_{l,t-L_l}\, s_l \quad \text{and} \quad \sum_{m=1}^{3} u_{mt}\, O_{mt}$$

We label the purchase quantities $\hat{x}_{l,t-L_l}$ for the optimal feasible solution $\hat{y}_{it} = 0$ and $\hat{O}_{mt} = 0$. Then we get by taking into consideration restrictions (1a) - (1c) of the model:

$$(1.2) \quad \hat{x}_{l,t-L_l} = \sum_{k \varepsilon P_1} q_{lk} \sum_{j \varepsilon P_2} q_{kj} \sum_{i \varepsilon P_1} q_{ji} \cdot b_{it}, \quad l \varepsilon P_0$$

It is obviously valid that for any other feasible solution $\bar{x}_{l,t-L_l}$ the following equation applies:

$$(1.3) \quad \sum_{t=1}^{s} \bar{x}_{l,t-L_l} \geqslant \sum_{t=1}^{s} \hat{x}_{l,t-L_l}, \quad s = 1,.....T; \quad l \varepsilon P_0,$$

because in every period the cumulated quantities of purchased goods must at least be sufficient to meet the cumulative demand up to a period s. Notice that $\hat{x}_{l,t-L_l}$ is the smallest quantity which can satisfy the demand. This can be seen by looking at restrictions 1.

From (1.3) and taking into consideration $\hat{O}_{mt} = 0$ we get for the cash flows

$$(1.4) \quad \sum_{t=1}^{s} \overline{RF}_t \leqslant \sum_{t=1}^{s} \widehat{RF}_t, \quad s = 1,.....,T.$$

The payment difference Δ_s up to the period s (s = 1,....,T) for the solution $\hat{x}_{l,s-L_l}$ ($\bar{x}_{l,s-L_l}$) can be written according to (1.1):

$$\Delta_s = \overline{CF}_{s-1} + \overline{RF}_s - \widehat{CF}_{s-1} - \widehat{RF}_s$$

For s = 1 we obtain:

$$\Delta_1 = CF_0 + \overline{RF}_1 - CF_0 - \widehat{RF}_1 .$$

With (1.4) we get:

$$(1.5) \quad \Delta_1 \leqslant 0.$$

If $CF_0 + \widehat{RF}_1 \geqslant 0$, then $\hat{z}_1 = z+$. Is also $CF_0 + \overline{RF}_1 \geqslant 0$ valid so $\bar{z}_1 = z+$ and by (1.5) the inequality (1.6.) is fulfilled:

$$(1.6) \quad (CF_0 + \overline{RF}_1)(1 + \bar{z}_1) \leqslant (CF_0 + \widehat{RF}_1)(1 + \hat{z}_1) \text{ or } \overline{CF}_1 \leqslant \widehat{CF}_1.$$

But if $CF_0 + \overline{RF}_1 < 0$, so (1.6) is valid in any case.

On the contrary, if $CF_0 + \widehat{RF}_1 < 0$, then $\hat{z}_1 = z-$. But because of (1.5) we also obtain $\bar{z}_1 = z-$ and again the inequality (1.6) is fulfilled.

For the case s = 2 the difference results:

$$\Delta_2 = \overline{CF}_1 + \overline{RF}_2 - \widehat{CF}_1 - \widehat{RF}_2 .$$

From (1.4) and (1.6) we get again $\Delta_2 \leqslant 0$.

In analogy to the above it can be seen that the following inequality is true:

$(\overline{CF}_1 + \overline{RF}_2)(1+\bar{z}_2) \leqslant (\widehat{CF}_1 + \widehat{RF}_2)(1+\hat{z}_2)$, that is $\overline{CF}_2 \leqslant \widehat{CF}_2$.

Altogether by complete induction we obtain the result:

(1.7) $\overline{CF}_s \leqslant \widehat{CF}_s$, $s = 1,\ldots,T-1$.

From (1.4) and (1.7) we finally get

$\overline{CF}_T \leqslant \widehat{CF}_T$.

q.e.d.

Multi-Period Production Planning and Managerial Accounting

HANS-ULRICH KÜPPER

1. Relations Between Production Planning and Managerial Accounting

Profits and costs are important objectives in production planning. Therefore many production models involve *cost parameters*. Their values must be calculated within the system of cost accounting.

Production models are frequently constructed without paying attention to the problems in calculating these cost parameters. Sometimes it is very difficult to find their correct values. For example lot size models include parameters for set up costs, storage costs and - rarely - shortage costs. The values of these parameters cannot be identified directly. Set up costs consist of personnel costs for preparing the machine and sometimes of energy costs. The calculation of storage and especially of shortage costs is still more difficult. Storage costs include personnel costs for the workers in the storage and interest costs on capital. Do they also include parts of the costs for the storage equipment? Shortage costs are opportunity costs. They cannot be derived from expenses as they reflect the loss of revenues.

The fundamental reason for these problems originates in the *separation of decisions*. The size and the equipment of a storage is determined by the stocks of several products and the expectations of the further needs of inventory space. A single lot only has minimal influence on these factors.

In general there are two ways how to solve these problems. The first way is to *estimate* those cost parameters by allocating costs. Often this can only be done arbitrarily. The other way is shown by *simultaneous* models. In some of these models, for example models of production and investment planning, the expenses for machines and workers can be related directly to variables of their utilization times [5, p. 259]. The relations between product quantities, lot sizes, sequencing and the utilization of machines as well as personnel are worked out within the model [10]. The calculation of the cost parameters becomes easier. There are fewer problems of allocating full costs. On the other hand the planning models get more complicated.

In consequence we perceive some sort of *symmetric* relation between planning and managerial accounting. If independent partial models are used for different problems of production planning as product quantities, lot sizes, staff assignment, sequencing etc., it is very difficult to determine the required cost parameters correctly. These parameters express the mutual dependencies of isolated decisions. In managerial accounting their determination becomes very important for the coordination between the decisions. Hence, the problems of coordination are partially shifted to managerial accounting.

Simultaneous models facilitate the calculation of those cost parameters. Such complex models however can frequently not be used for *practical* problems. Lot sizes and sequencing often lead to mixed integer models which cannot be solved and require extensive data. Furthermore central planning is not convenient for many organizations.

The symmetric relation between planning and managerial accounting displays a *dilemma*: simultaneous models only give a theoretical, but not a practical solution. The problem of separation and coordination between several decision problems has to be solved by other means of planning and managerial accounting.

One approach will be demonstrated in the following chapters. It illustrates the relations between production planning and managerial accounting by means of an example of multi-period programme planning. The critical cost parameter of this problem is depreciation. We will try to find a way to determine variable depreciation by managerial accounting in order to coordinate one-period programme planning with the long-term objective and planning.

2. Integration of One-Period and Multi-Period Production Planning

2.1. Central Elements of the Investment Approach

The calculation of *depreciation* is an important and difficult problem of cost accounting [3, 15]. In recent years a new approach for managerial accounting has been developed [2, 4, 6]. It is based on *investment theory*. The central idea of this approach is to derive the theory of managerial accounting from the models of investment theory. The investment approach proceeds on the assumption of a long-term plan. In the simplest case infinite identical investment repetitions are assumed, according to the principle of "going concern". The objective of managerial accounting is to determine the implications of short-term decisions on long-term profit objectives. The common objective of all profit-oriented calculations is the net present value. It is calculated in terms of *cash-flows*.

In order to get the relevant information it is necessary to know how the net present value varies with the decision variable. Given that, it is possible to identify the influence of short-term changes of the decision variable on profit. These changes are the relevant costs.

For example: The net present value G_t of a good at time t is a function of the cash-flows caused by input of material, labor, and equipment. If its value depends on lifetime and cumulative utilization Y_t, we get the *net present value* of equation 1:

$$G_t = G(t, Y_t) \tag{1}$$

The information which is relevant for a decision on utilization is determined by infinitely small *changes of utilization*. This is given within the total differential:

$$\frac{dG_t}{dt} = \frac{\partial G_t}{\partial t} + \frac{\partial G_t}{\partial Y_t} \cdot \frac{dY_t}{dt} \tag{2}$$

The degree of utilization is embodied in this approach by means of cumulative utilization. This might be the miles done by a truck, the time of utilization for an engine or the working time for labor. The differential ratio of the function of the net present value represents the costs, which are relevant to short-term decisions.

2.2. Multi-Period Planning of the Production Programme

As a consequence of this approach the relations between one-period and multi-period planning of the production programme can be demonstrated. Let us consider an example [7]. Three types of products have to be produced by three machines. In long-term planning the periodical product quantities will be fixed simultaneously with the life and replacement time of the machines. For the purpose of simplification we presume constant data and infinite identical investments. *Table 1* shows the data of the example. There are the acquisition and the residual values of the three machines. These machines cause *current expenses* C for energy, maintenance etc. Their values may be determined by the variables time t, periodical utilization y_t and cumulative utilization Y_t. The *capacity* of each machine in each period t is limited. The *contribution margins* CM of the products are calculated without machinery costs.

The *objective* of long-term planning is to maximize the present value of the cash-flows G at t=0. They are the sum of the present values of the contributions for the continuously sold products and the present values of the expenses for the machines. To achieve an optimal solution we should maximize this non-linear objective function with linear capacity constraints.

In order to find an easier way to get a solution we only consider the *corners* of the decision space in a first step (see *Figure 1*). The present values of the two best alternatives are shown in *Table 2*. For both alternatives the periodical operating times of the machines A and B are equal. There is only one difference concerning machine C. Consequently both alternatives include the same present value of the cash-flows and the same replacement times for the machines A and B.

A better solution might be achieved if a *switching* between the alternatives is allowed. To examine this point we only regard the utilization of machine C and a switching between the two best alternatives. The switching between the alternatives and therefore the change of the periodical utilization of machine C may occur at τ. If y_I denotes the *periodical utilization* of the first selected alternative and y_{II} that of the second alternative the *cumulative utilization* Y_t before and after the switching time τ can be calculated as follows:

$$t < \tau \rightarrow \quad Y_t = y_I \cdot t \tag{3.1}$$

$$t > \tau \rightarrow \quad Y_t = y_{II} \cdot t + (y_I - y_{II}) \cdot \tau = y_{II} \cdot t + \Delta y \cdot \tau \tag{3.2}$$

In consequence the *current costs* of this machine before (C_I) and after (C_{II}) the switching in τ are equal to

Machine	A	B	C
Acquisition Value	$A_A = 600$	$A_B = 400$	$A_C = 500$
Residual Value	$L_A = 600 \cdot e^{-0,1T}$	$L_B = \frac{400}{T + 2} + \frac{1200}{Y_T + 6}$	$L_C = \frac{20000}{Y_T + 40}$
Current Expenses For Machine Per Time Unit $t<\tau$ / $t\geq\tau$	$C_A=3t + y_t + 0,1Y_t$ $C_A=3t + y_t + 0,1Y_t + 0,1\Delta Y$	$C_B=2t + 3y_t + 0,2Y_t$ $C_B=2t + 3y_t + 0,2Y_t + 0,2\Delta Y$	$C_C=4t + 1,8y_t + 0,3Y_t$ $C_C=4t + 1,8y_t + 0,3Y_t + 0,3\Delta Y$
Machine Operating Time	$2 \cdot x_1 + 2 \cdot x_2 + 1,6 \cdot x_3 \leq 32$	$1,5 \cdot x_1 + 1 \cdot x_2 + 1,5 \cdot x_3 \leq 21$	$1,5 \cdot x_1 + 2 \cdot x_2 + 1,5 \cdot x_3 \leq 30$
Contribution Before Variable Machinery Costs	$CM_1 = 39$; $CM_2 = 35$; $CM_3 = 35$		

Table 1: Data of an example of production programme planning

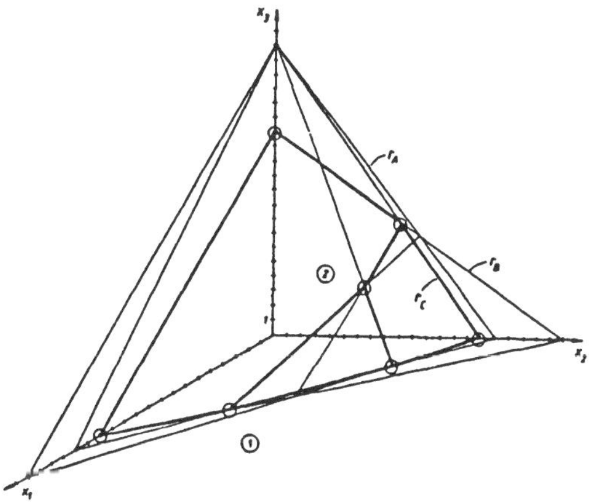

Figure 1: Solution space of the example of production programme planning

Alternative	Product Quantities		Machines A	Machines B	Machines C	Net Present Value of Current Costs	Net Present Value of Contribution Margin	Net Present Value of Profits
1	$x_1 = 10$	y	32	21	27	4629,46	6000	1370,54
	$x_2 = 6$	T	10,908	12,970	9,300			
	$x_3 = 0$	K	1399,43	1486,30	1743,73			
2	$x_1 = 3$	y	32	21	30	4722,49	6070	1347,51
	$x_2 = 9$	T	10,908	12,970	8,998			
	$x_3 = 5$	K	1399,43	1486,30	1836,77			

Table 2: Best long-term programme alternatives without change of capacity utilization (i = 0,10)

$$C_I = \alpha \cdot t + \beta \cdot y_I + \varepsilon \cdot Y_t \tag{4.1}$$
$$C_{II} = \alpha \cdot t + \beta \cdot y_{II} + \varepsilon \cdot Y_t = \alpha \cdot t + \beta \cdot (y_I - \Delta y) + \varepsilon \cdot Y_t \tag{4.2}$$

Until τ we get the contribution CM_I of the first selected alternative, after τ the contribution CM_{II}. At the end of the service time of machine C we return to CM_I and so on. The *objective function* consists of the present values of the following terms:

- the *contribution* CM_I and *current costs* C_I of the first alternative from zero until the switching in τ,
- the *contribution* CM_{II} and *current costs* C_{II} of the second alternative from the switching in τ until the replacement in T,
- the *acquisition value* A for machine C and
- the *residual value* of machine C at replacement in T.

The objective function for the net present value of the profit G at the beginning of the whole planning period t=0

$$G = \frac{1}{1-e^{-iT}} \cdot \left[\int_0^{\tau} (CM_I - C_I) \cdot e^{-it} dt + \int_{\tau}^{T} (CM_{II} - C_{II}) \cdot e^{-it} dt - A + L(Y_T) \cdot e^{-iT} \right] \tag{5}$$

depends on two variables, the switching time τ and the replacement time T of machine C. To get the optimum of this function we must differentiate by these variables. With respect to *replacement time* T we get

$$\frac{\partial G}{\partial T} = \frac{1}{1-e^{-iT}} \cdot \left[(CM_{II} - C_{II}) \cdot e^{-iT} + \frac{dL}{dY_T} \cdot \frac{dY_T}{dT} \cdot e^{-iT} - i \cdot e^{-iT} \cdot L(Y_T) - i \cdot e^{-iT} \cdot G \right] = 0 , \quad \frac{\partial^2 G}{\partial T^2} < 0 , \tag{6.1}$$

or

$$CM_{II} - C_{II}(T) + \frac{dL}{dY_T} \cdot \frac{dY_T}{dT} - i \cdot L(Y_T) - i \cdot G = 0 \tag{6.2}$$

The equation 6.2 corresponds to the well-known condition of optimal replacement in investment theory.

The differentiation by τ gives the optimal *switching time*:

$$\frac{\partial G}{\partial \tau} = \frac{1}{1-e^{-iT}} \cdot \left[(CM_I - C_I(\tau)) \cdot e^{-i\tau} - (CM_{II} - C_{II}(\tau)) \cdot e^{-i\tau} - \int_{\tau}^{T} \frac{\partial C_{II}}{\partial \tau} \cdot e^{-it} dt + \frac{dL}{dY_T} \cdot \frac{dY_T}{d\tau} \cdot e^{-iT} \right] = 0 , \quad \frac{\partial^2 G}{\partial \tau^2} < 0 , \text{ or} \tag{7.1}$$

$$(CM_I - C_I(\tau)) - (CM_{II} - C_{II}(\tau)) = e^{i\tau} \cdot [\int_{\tau}^{T} \frac{\partial C_{II}}{\partial \tau} \cdot e^{-it} dt - \frac{dL}{dY_T} \cdot \frac{dY_T}{d\tau} \cdot e^{-iT}] \quad (7.2)$$

The equations 4.1 and 4.2 lead to

$$C_I - C_{II} = \beta \cdot (y_I - y_{II}) = \beta \cdot \Delta y, \quad (8)$$

$$\frac{\partial C_{II}}{\partial \tau} = \varepsilon \cdot \Delta y = \frac{\partial C}{\partial Y_t} \cdot \Delta y \quad (9) \quad \text{and} \quad \frac{dY_t}{d\tau} = \Delta y \quad (10)$$

Now we can transform the equation 7

$$CM_I - CM_{II} - \Delta y \cdot \beta = e^{i\tau} \cdot \Delta y \cdot [\int_{\tau}^{T} \frac{\partial C}{\partial Y_t} \cdot e^{-it} dt - \frac{dL}{dY_T} \cdot e^{-iT}] \quad (7.3)$$

Setting

$$e^{i\tau} \cdot [\int_{\tau}^{T} \frac{\partial C}{\partial Y_t} \cdot e^{-it} dt - \frac{dL}{dY_T} \cdot e^{-iT}] = d_N \quad (11)$$

we finally get:

$$CM_I - y_I \cdot (\beta + d_N) = CM_{II} - y_{II} \cdot (\beta + d_N) \quad (12)$$

Each side of this equation stands for the difference between the periodical contributions and costs of machine C. The optimal switching time is reached as soon as the *periodical profits* become equal. Both sides represent a one-period objective function. As β denotes the costs proportional to utilization time y_t the parameter d_N may be interpreted as variable depreciation.

If we are able to determine this cost parameter in managerial accounting, we can solve the problem in one-period planning. It has to be shown that this way is given by the investment approach.

2.3. One-Period Planning with Variable Depreciation

In the investment approach we calculate the depreciation corresponding to equation 2. In order to get the variable depreciation we have to differentiate the objective function $G_t(t, y_t, Y_t)$ for the net present value *at each time* t by the *cumulative utilization* variable Y_t:

$$\frac{\partial G_t}{\partial Y_t} = e^{it} \cdot [e^{-i\tau} \cdot \frac{d\tau}{dY_t} \cdot \{CM_I - CM_{II} - C_I(\tau, Y_\tau) + C_{II}(\tau, Y_\tau)\}$$

$$+ e^{-iT} \cdot \frac{dT}{dY_t} \cdot \{CM_{II} - C_{II}(T, Y_T) + \frac{dL}{dY_T} \cdot \frac{dY_t}{dT} - i \cdot L(Y_T) - i \cdot G\} \tag{13.1}$$

$$- \int_t^\tau \frac{\partial C_I}{\partial Y_s} \cdot e^{-is} ds - \int_\tau^T \frac{\partial C_{II}}{\partial Y_s} \cdot e^{-is} ds - \int_t^\tau \frac{\partial C_I}{\partial \tau} \cdot \frac{\partial \tau}{\partial Y_s} \cdot e^{-is} ds - \int_\tau^T \frac{\partial C_{II}}{\partial \tau} \cdot \frac{\partial \tau}{\partial Y_s} \cdot e^{-is} ds]$$

Because of

$$Y_T = Y_t + (T-t) \cdot y \quad (14) \qquad \text{and} \qquad \frac{dY_T}{dY_t} = 1 \tag{15}$$

we get:

$$\frac{dL}{dY_T} \cdot \frac{dY_t}{dT} = \frac{dL}{dY_T} \cdot \frac{dY_T}{dT} \tag{16}$$

In the optimum we can add the equations 7.2 and 6.2 and come to:

$$\frac{\partial G_t}{\partial Y_t} = e^{it} \cdot [\frac{d\tau}{dY_t} \cdot \{ \int_\tau^T \frac{\partial C_{II}}{\partial \tau} \cdot e^{-it} dt - \frac{dL}{dY_T} \cdot \frac{dY_T}{d\tau} \cdot e^{-iT}\}$$

$$- \int_t^T \frac{\partial C}{\partial Y_s} \cdot e^{-is} ds - \int_t^\tau \frac{\partial C_I}{\partial \tau} \cdot \frac{\partial \tau}{\partial Y_s} \cdot e^{-is} ds - \int_\tau^T \frac{\partial C_{II}}{\partial \tau} \cdot \frac{\partial \tau}{\partial Y_t} \cdot e^{-it} dt] \tag{13.2}$$

Because of equations 4 and 3 it follows:

$$\frac{\partial C_I}{\partial \tau} = 0 \quad (17) \qquad \text{and for } t < \tau \qquad \frac{d\tau}{dY_t} \cdot \frac{dY_T}{d\tau} = -1 \tag{18}$$

and we finally get:

$$\frac{\partial G_t}{\partial Y_t} = e^{it} \cdot [\int_\tau^T \frac{\partial C_{II}}{\partial \tau} \cdot \frac{d\tau}{dY_t} \cdot e^{-it} dt + \frac{dL}{dY_T} \cdot e^{-iT}$$

$$- \int_t^T \frac{\partial C}{\partial Y_s} \cdot e^{-is} ds - \int_\tau^T \frac{\partial C_{II}}{\partial \tau} \cdot \frac{\partial \tau}{\partial Y_t} \cdot e^{-it} dt]$$

$$= - e^{it} \cdot [\int_t^T \frac{\partial C}{\partial Y_s} \cdot e^{-is} ds - \frac{dL}{dY_T} \cdot e^{-iT}] = -d_N \tag{13.3}$$

This equation gives us the negative *variable depreciation.* In consideration of equation 12 this depreciation leads to the optimal switching time τ. If we calculate the variable depreciation in the investment approach, the maximization of the objective function of the *periodical profits*

$$G_t^* = CM - y \cdot (\beta + d_N) \tag{19}$$

for one-period planning leads to the optimum of the multi-period planning. *Figure 2* shows that the periodical profits of the alternatives change over time. Their point of intersection delivers the optimal switching time τ. *Table 3* demonstrates this result for the established example.

But there is one *problem.* To calculate the depreciation by means of the investment approach of the equation 13.3 we must know the optimal replacement time T of machine C. Its value must be fixed corresponding to equation 6.2. The solution of this optimization condition depends on the switching time τ. Therefore the exact solution of the one-period optimization problem presumes the solution of the multi-period problem. This dilemma is well-known from dual variables in linear programming.

The example of *Table 3* shows that approximate values of the replacement time T also lead to a good solution. Therefore this dilemma can reasonably be evaded by working with realistic *approximate* values of T. The investment approach seems to be a suitable concept to determine cost parameters which may coordinate short-term and long-term planning.

3. The Investment Approach as a General Concept for Managerial Accounting

3.1. Fundamental Characteristics of the Investment Approach

The investment approach forms a theoretical foundation of managerial accounting. It points out the way to generate relevant information for decision problems. There are four important characteristics [6]:

(1) The investment approach integrates *short- and long-term* profit calculations into one accounting system. Its basis is the long-term investment theory.

(2) The investment approach takes into account the aspect of time. Therefore it has to be developed as a *dynamic theory* [9].

(3) All calculations are based on payments or expenses and revenues, i.e. on *observable* figures. These are cash-flows with respect to profit. The other variables, such as periodical costs, should be derived from those by clear rules.

(4) All profit-oriented calculations are directed towards a *common objective.* The same profit objective should exist for all decisions. Hence, the point of departure is the long-term profit objective.

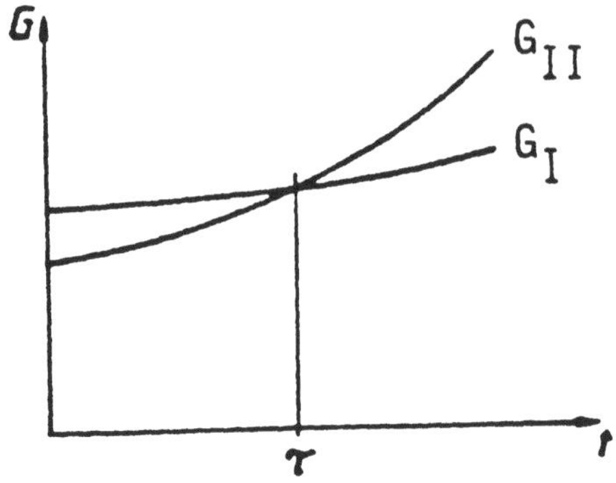

Figure 2: Development of short-term periodical profits in consideration of depreciation

Case	T_C	t	Short-term Objective Functions	Periodical Profits Alt. 1	Alt. 2	Chosen Alternative	Net Present Value of Profit
Without Depre- ciation	8,998	∀ t	$29,8x_1+26,4x_2+26,2x_3$	Alt. 2: 458		2	1.347,51
With	9,306 (Opti- mal)	0,5 8,146 9,0	$23,6x_1+20,0x_2+20,3x_3$ $27,4x_1+24,1x_2+23,9x_3$ $28,0x_1+24,8x_2+24,5x_3$	356,041 418,108 428,415	352,126 418,107 429,062	1 switching 2	1.370,88
Depre-	9,300 (Alt. 1)	0,5 8,146 9,0	$23,7x_1+20,1x_2+20,3x_3$ $27,4x_1+24,1x_2+23,9x_3$ $28,0x_1+24,8x_2+24,5x_3$	357,158 418,141 428,445	353,245 418,144 429,097	1 switching 2	1.370,88
ciation	8,998 (Alt. 2)	0,5 8,146 8,998	$23,7x_1+20,1x_2+20,4x_3$ $27,5x_1+24,2x_2+24,0x_3$ $28,2x_1+25,0x_2+24,6x_3$	357,984 419,816 431,358	354,162 420,005 432,334	1 2 2	1.370,55

Table 3: Short-term objective functions with and without regard to variable depreciation (i = 0,10)

3.2. The Determination of Several Cost Types in the Investment Approach

One-period and multi-period production planning are linked by depreciation. As a general concept for managerial accounting the investment approach has to prove its performance in determining the different types of costs and by solving decision problems of managerial accounting.

It can be shown for several types of costs how relevant costs are derived. The basic approach is the same for all types of costs: Firstly, the *cash-flows* have to be determined. Secondly, it has to be found out how the net present value depends on the *decision variable* and other variables. Thirdly, the *function* of the present value of cost has to be *differentiated* by the decision variable.

In order to derive *material costs*, for example, an infinite repetition of purchases can be assumed in the simplest case. The present value K_t of material input is determined by taking material quantities β, price q and manufacturing time π per product unit as well as interest rate i into account. The planned purchasing interval may be T^* and the production quantity x. Then we get the net present value of material input at time t as follows:

$$K_t = \beta \cdot q \cdot \frac{T^*}{\pi} \cdot \frac{e^{-i[T(x)-t]}}{1 - e^{-iT^*}} \tag{20}$$

If a further product unit, which has not been included in long-term planning, shall be produced, the next purchasing point T(x) will be reached earlier and the residual chain of cash-flows will be realized sooner. That causes a variation of net present value, which can be interpreted as material costs:

$$\frac{dK_t}{dx} = \frac{\partial K_t}{\partial T} \cdot \frac{dT}{dx} = \frac{\beta \cdot q \cdot T^* \cdot i \cdot e^{-i(T-t)}}{1 - e^{-iT^*}} \tag{21}$$

These costs seem to differ significantly from the *traditional* material costs. This is only caused by interest. Converging interest rate i or planned purchasing interval T^* towards zero the traditional material costs show up as marginal value of the present value:

$$\lim_{i \to 0} \frac{dK_t}{dx} = \lim_{i \to 0} \beta \cdot q \cdot T^* \cdot \frac{e^{-i(T-t)} - i \cdot (T-t) \cdot e^{-i(T-t)}}{T^* \cdot e^{-iT^*}} = \beta \cdot q \tag{22}$$

$$\lim_{T^* \to 0} \frac{dK_t}{dx} = \lim_{T^* \to 0} \beta \cdot q \cdot i \cdot \frac{e^{-i(T-t)}}{i \cdot e^{-iT^*}} = \beta \cdot q \qquad t \leq T \leq T^* \tag{23}$$

Hence, direct material costs are *marginal values* of the investment approach when interest is treated as a separate type of cost.

The determination of *depreciation* has been illustrated above. It is possible to show that total depreciation under certain conditions converges to a linear depreciation. Once again a well known application proves to be the marginal value of the presented theory.

Other types of costs like *personnel costs* [6], *replacement costs* [16], *interest costs* etc. can be determined in the same manner. It enables us to test how far traditional approaches of cost accounting provide the relevant costs for decision problems.

3.3. Application of the Theory of Managerial Accounting to Typical Decision Problems

The theory of managerial accounting has to show which costs are relevant to a decision problem in order to arrive at an optimal solution. In this respect the performance of the theory of managerial accounting can only be demonstrated by solving typical decision problems. In order to illustrate the performance of the investment approach *two* other typical decision problems for managerial accounting may be sketched, the determination of optimal order quantities and the determination of minimum prices.

In applying the investment approach to the determination of the optimal *order quantity* one has to focus on those cash-flows that are caused by the purchasing decision [12, 14]. The present value of purchases can be calculated under the usual assumptions. In order to derive the optimal order quantity, one has to determine the minimum of the present value function. We assume that the quantity x will be ordered for a cycle of w periods. In each period r units are required, each purchase causes fixed ordering expenses F and variable expenses $q \cdot r \cdot w$ for the delivered units, with q representing the price per unit. The storage costs without interest are c per unit of quantity and per unit of time. These parameters lead to the net value of the expenses for *one order cycle* K_C:

$$K_C = F + q \cdot r \cdot w + \int_0^w c \cdot r \cdot (w-t) \cdot e^{-it} dt \tag{24}$$

In the simplest case we assume an infinite identical repetition of order cycles. Therefore we get the net present value K for *all cycles*:

$$K = \frac{K_C}{1-e^{-iw}} = \frac{1}{1-e^{-iw}} \cdot (F+qrw+crw) \cdot \int_0^w e^{-it} dt - c \cdot r \cdot \int_0^w t \cdot e^{-it} dt \tag{25}$$

The order quantity x depends on the requirements r per period and the duration w of one order cycle:

$$x = r \cdot w \tag{26}$$

To derive the optimal order quantity x we have to minimize the function of the net present value depending on the order cycle w. We get the optimum condition:

$$\frac{dK}{dw} = 0 \tag{27a}$$

$$r \cdot (q \cdot i + c) \cdot \frac{e^{iw} - 1}{i} = i \cdot F + q \cdot r \cdot i \cdot w + c \cdot r \cdot w \tag{27b}$$

Because of the compound interest this equation is difficult to solve analytically. Therefore we need a close-up solution. It is given by the first two parts of the *Taylor approximation*:

$$\frac{e^{iw} - 1}{i} = \int_0^w e^{it}\, dt \approx w + \frac{i \cdot w^2}{2} \tag{28}$$

As equation 29 demonstrates, this approximation neglects the compound interest:

$$w + i \cdot [(0{,}5 + w - 1) + (0{,}5 + w - 2) + \ldots + (0{,}5 + w - w)] = w + \frac{i \cdot w^2}{2} \tag{29}$$

A revision shows that this approximation is satisfying for relatively low costs of interest and/or small order cycles (see *Table 4*).

Interest Rate i	Purchasing Interval w	Accurate Value $\frac{e^{iw}-1}{i}$	Approximate Value $w + \frac{i \cdot w^2}{2}$
0,002	3	3,009	3,009
0,002	5	5,025	5,025
0,002	10	10,101	10,100
0,1	1	1,052	1,050
0,1	10	17,183	15,000
0,5	10	294,826	35,000

Table 4: Revision of approximation

When applying the approximation to the optimizing equation we get:

$$r \cdot (q \cdot i + c) \cdot \left(w + \frac{i \cdot w^2}{2}\right) = i \cdot F + q \cdot r \cdot i \cdot w + c \cdot r \cdot w \tag{30}$$

Now the *optimal order cycle* w* can be determined by:

$$w^* = \sqrt{\frac{2 \cdot F}{r \cdot (q \cdot i + c)}} \tag{31}$$

One arrives at the well-known traditional formula for *the optimal order quantity* x*:

$$x^* = r \cdot w^* = \sqrt{\frac{2 \cdot F \cdot r}{q \cdot i + c}} \tag{32}$$

The way of approximation and the neglection of compound interest determine the deviation from the long-term profit objective.

The determination of *minimum prices* is another link to integrate the investment approach and other accounting approaches [6]. In this determination the whole lifecycle of a product including research, design, equipment purchase and manufacturing has to be considered. The minimum price is the premium on top of the profit contribution, which sets the present value to zero.

For example, one can assume that the expenses for *research* E_R have to be paid at the beginning of the first period, the expenses for *design* E_D in t=1 and the expenses for *equipment* E_M in t=2. Production and sale will start after two periods and will amount to x product units per period. Considering only *one product cycle*, production will last until t=T=8. During this time fixed costs F have to be paid at the beginning of each production period, whereas variable costs k per unit have to be incurred continuously.

The minimum prices will be calculated as a proportional charge α % to variable costs k. The net present value G of one product cycle is given by equation 33:

$$G = - E_R - E_D \cdot e^{-i} - E_M \cdot e^{-2i} - F \cdot \sum_{t=2}^{T-1} e^{-it} + k \cdot \frac{\alpha}{100} \cdot x \cdot \int_{t=2}^{T} e^{-it} dt \qquad (33)$$

In order to get the minimum price α_0^* at t=0 we have to solve this equation for G=0:

$$\alpha_0^* = \frac{E_R + E_D \cdot e^{-i} + E_M \cdot e^{-2i} + F \cdot \sum_{t=2}^{T-1} e^{-it}}{k \cdot x \cdot \int_{t=2}^{T} e^{-it} \, dt} \cdot 100 \qquad (34)$$

The minimum prices α_t^* vary over time. They are calculated for several periods in *Table 5a*. Apparently they go down to the variable costs. In this case, the minimum price of the last period obtained by the investment approach is equal to the well known absolute minimum price of variable accounting.

If, in contrast, one assumes *succeeding products* with identical cash-flows, the lower price limit varies around an average value (see *Table 5b*). As *Figure 3* illustrates this value is equal to the value of full costing. When converging the interest costs towards zero or making the payments continuous during the whole product lifecycle this value results as a satisfying approximation. The solutions of full and direct costing prove to be special cases of the investment approach.

3.4. The Investment Approach as a Special Case of the Control Theory

If the investment approach is applied to the model of control theory under identical assumptions, the same costs per good and period are obtained. This is illustrated by an example in *Table 6* [9]. The investment approach is therefore a special case of the control theory [8].

Thus it is linked with an useful instrument to analyze dynamic problems [11, 13]. However, there are already difficulties with simple models. The application of control theory requires the full understanding of

Table 5: Determination of minimum prices (i = 0,10)

t	0	0	1	1	2	2	3	3	4	4
α^*_t	65,94	55,35	55,35	43,31	43,31	8,30	10,48	7,97	10,48	7,48
t	5	5	6	6	6,5	7	7	8		
α^*_t	10,48	6,65	10,48	4,99	6,82	10,48	0	0		

Table 5a: Minimum prices of one product cycle

t	0	0	1	1	2	2	3	3	3,5			
α^*_t	66,78	62,17	62,17	56,93	56,93	41,68	45,84	44,89	47,08			
t	4	4	5	5	6	6	7	7	8	8	9	9
α^*_t	49,38	48,43	53,27	52,32	57,55	51,83	57,47	42,22	56,93	41,68	45,84	44,89
t	10	10	11	11	12	12	13	13	14	14	15	15
α^*_t	49,38	48,43	53,27	52,32	57,55	51,83	57,47	42,22	56,93	41,68	45,84	44,89

Table 5b: Minimum prices of infinitely repeated, identical product cycles

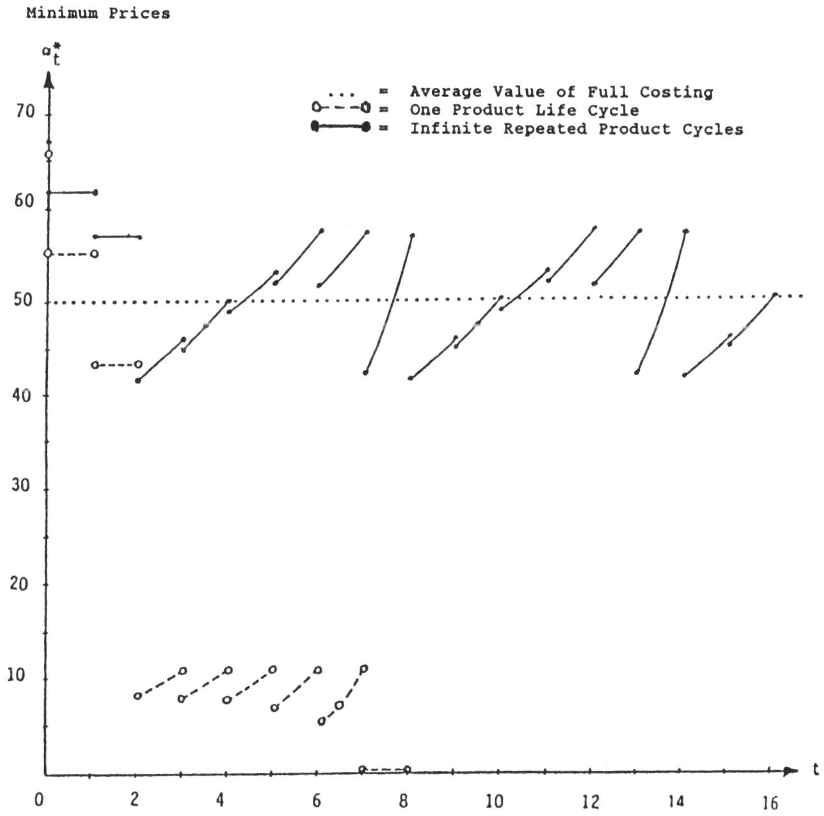

Figure 3: Example of the determination of minimum prices

	Control Theory					Investment Approach	
(1)	(2)	(3)	(4)	(5)	(6)	(7)	(8)
	Condition of Equipment	Change of Condition	Dual Price		Costs per Period	Future Value	Change of Future Value
t	Z(t)	$\frac{dZ}{dt}$	p(t)	$p(t) \cdot \frac{dZ}{dt}$	$\int_a^b p \cdot \frac{dZ}{dt}\, dt$	G_t	$\frac{dG_t}{dt}$
	$=60e^{-0,2t}+140$	$=-12e^{-0,2t}$	$=50 - 5,09 \cdot e^{0,3t}$	$=61,10e^{0,1t} -600e^{-0,2t}$	$=610,95 \cdot (e^{0,1a}-e^{0,1b}) + 3000 \cdot (e^{-0,2a}-e^{-0,2b})$	$=610,95e^{0,1t} +3000e^{-0,2t} +20900$	$=61,10e^{0,1t} -600e^{-0,2t}$
0	200,00					24510,95	
0,5		-10,86	44,08	-478,68	479,55		-478,68
1	189,12					24031,40	
1,5		- 8,89	42,01	-373,51	374,22		-373,51
2	180,22					23657,18	
2,5		- 7,28	39,22	-285,47	286,05		-285,47
3	172,93					23371,13	
3,5		- 5,96	35,45	-211,25	211,71		-211,25
4	166,96					23159,42	
4,5		- 4,88	30,36	-148,13	148,49		-148,13
5	162,07					23010,93	
5,5		- 4,00	23,49	- 93,83	94,12		- 93,83
6	158,07					22916,81	
6,5		- 3,27	14,21	- 46,49	37,34		- 46,49
6,70	155,71	- 3,14	12,00	- 37,70		22879,47	- 37,70
			L(T) = 1868,52		= 1631,48	24510,95 - 22879,74 = 1631,48	A - L(T) = 1631,48

Table 6: Example of comparing the control theory to the investment approach

rather sophisticated instruments. "Analytical solutions to practical dynamic decision problems are rare ... " [1, p. 3]. A direct application of control theory to practical problems seems not to be possible.

Therefore the investment approach is important as a *simpler* concept to coordinate long-term and short-term planning. It avoids difficulties in solving differential equations by identifying the past utilization of the equipment with the cumulative utilization. The marginal-principle behind it is easier to understand than the Hamilton-Function and the maximum-principle.

4. Implications of the Dynamic Theory for Managerial Accounting

4.1. Performance of the Presented Theory

The presented theory accomplishes the combination of managerial accounting and investment theory. It provides the *theoretical basis* for integrated managerial planning. That seems to be very important for the relations between managerial accounting and production planning. Long-, medium- and short-term planning calculations can be derived from the investment approach. For this purpose separation theorems have to be formulated, which indicate the conditions under which simplified approaches can be used. At the same time the investment approach accomplishes the link to control theory. By this a connection to an overall dynamic theory is found.

The investment approach enables to determine the relevant information for short-term decision situations. This becomes particularly obvious with depreciations. The most important contrast to the traditional approaches is the different *way of thinking*. Within the investment approach you need not ask how to split up costs. The only important thing is the impact of decisions on *future* cash-flows. Therefore we search for the functions of the relationship between decision variables and the future cash-flows. Even if these relationships are not exactly known, the information obtained is more relevant to the decision than the information obtained by traditional systems.

Under certain assumptions the traditional approaches can be derived from the presented theory. It therefore enables the assessment of the *traditional systems* of managerial accounting. One can show under which conditions the systems of marginal costing and full costing have to be used. It becomes obvious that the systems of full costing can be useful for medium- and long-term decisions. Under special conditions they approximate the precise investment theory to a satisfying extent.

4.2. Perspectives for Further Development of Managerial Accounting

Besides the continuing work on existing elements, two directions seem to be important for further development:

(1) From a long-term perspective, *uncertainty* becomes more and more important. Therefore one has to leave the premise of perfect information. Under this assumption it might be possible to gain new knowledge about the derivation of relevant information and systems for decision making.

(2) Other important aspects of managerial accounting, besides the planning aspect, are *behavioral influence and control*. These have to be included in order to arrive at an overall theory of managerial accounting.

REFERENCES

1. *Feichtinger, G./R. F. Hartl*: Optimale Kontrolle ökonomischer Prozesse, Berlin, New York 1986.

2. *Hotelling, H.*: A General Mathematical Theory of Depreciation, in: The Journal of the American Statistical Association (20) 1925, p. 340 ff.

3. *Kilger, W.*: Flexible Plankostenrechnung und Deckungsbeitragsrechnung, 9. ed., Wiesbaden 1988.

4. *Kistner, K.-P./ A. Luhmer*: Zur Ermittlung der Kosten für Betriebsmittel in der statischen Produktionstheorie, in: Zeitschrift für Betriebswirtschaft (51) 1981, p. 165 - 179.

5. *Küpper, H.-U.*: Interdependenzen zwischen Produktionstheorie und der Organisation des Produktionsprozesses, Berlin 1980.

6. *Küpper, H.-U.*: Investitionstheoretische Fundierung der Kostenrechnung, in: Zeitschrift für betriebswirtschaftliche Forschung (37) 1985, p. 26 - 46.

7. *Küpper, H.-U.*: Investitionstheoretischer Ansatz einer integrierten betrieblichen Planungsrechnung, in: Information und Wirtschaftlichkeit, ed. by W. Ballwieser and K.-H. Berger, Wiesbaden 1985, p. 405-432.

8. *Küpper, H.-U.*: Investitionstheoretische versus kontrolltheoretische Abschreibung: Alternative oder gleichartige Konzepte einer entscheidungsorientierten Kostenrechnung, in: Zeitschrift für Betriebswirtschaft (58) 1988, p. 397-415.

9. *Küpper, H.-U.*: Gegenstand und Ansätze einer dynamischen Theorie der Kostenrechnung, in: Zeitaspekte in betriebswirtschaftlicher Theorie und Praxis, ed. by H. Hax, W. Kern and H.-H. Schröder, Stuttgart 1989, p. 43-59.

10. *Küpper, H.-U.*: Process Organization and Scheduling, in: Handbook of German Business Management, ed. by E. Grochla et ai., Stuttgart et al. 1990, col. 1850-1862.

11. *Luhmer, A.*: Maschinelle Produktionsprozesse. Ein Ansatz dynamischer Produktions- und Kostentheorie, Opladen 1975.

12. *Rieper, B.*: Die Bestellmengenrechnung als Investitions- und Finanzierungsproblem, in: Zeitschrift für Betriebswirtschaft (56) 1986, p. 1230-1255.

13. *Roski, R.*: Einsatz von Aggregaten - Modellierung und Planung, Berlin 1986.

14. *Schramm, K.*: Über die Kapitalwertfunktion des klassischen Losgrößenmodells, in: Zeitschrift für Betriebswirtschaft (57) 1987, p. 465-482.

15. *Schweitzer, M./ Küpper, H.-U.*: Systeme der Kostenrechnung, 4. ed., Landsberg 1986.

16. *Zhang, S.*: Instandhaltung, Betriebskosten und Einsatzplanung von Produktionsanlagen, Wiesbaden 1989.

Flexibility of Production Control Systems

KATSUHIKO TAKAHASHI, SHUSAKU HIRAKI and MICHIO SOSHIRODA

1 INTRODUCTION

Recently, manufacturing flexibility becomes a topic to be researched all over the world. The production systems with flexibility, such as FMS or CIM, have been developed, and a lot of studies with respect to the flexible systems have been researched. In addition, the flexibility of production systems itself has been also researched.

Flexibility is a measure to evaluate responses to changes, so the flexibility of production systems must evaluate the response of production control systems against the changes surrounding production system because the response of production systems is planned and controlled by production control systems. In considering the flexibility of production systems, therefore, it is necessary to clarify the changes and responses of production control systems, and to consider the flexibility of those responses. However, the main objectives of the literature on manufacturing flexibility are not to clarify the factors of flexibility but to classify the resulting types of flexibility.

This paper first clarifies the factors affecting the flexibility of production control systems after reviewing the literature on flexibility, and discusses the performance measures for flexibility. Next, this paper constructs a model of production system considering with demand fluctuation and machine breakdown as changes, also constructs models of production control systems, MRP and JIT, as responses for changes. And this paper analyzes the flexibility of the systems. Finally the concluding remarks summarize the findings of our research.

2 REVIEW OF LITERATURE

Being stimulated by active developments of FMS, manufacturing flexibility has been addressed, and many researchers point out the necessity of flexibility. In particular, DeMeyer et al.[7] point out that flexibility will become an advantage in the competitive environment of companies from the results of survey for companies in

North America, Japan, and Europe.

And a lot of researchers have studied on flexibility. Recently, Gupta and Goyal[11] review the literature on flexibility. They classify the literature based on the ways researchers have defined flexibility and the approaches used in measuring it.

In this paper, we classify the literature on flexibility into the following four categories based on the objective of each literature.

The literature on

FMS	: compare FMS with previous production systems, or develop control methods for FMS
The typology of flexibility	: develop the framework for classifying flexibility
The measures of flexibility	: propose measures to evaluate flexibility
The factors affecting flexibility	: clarify the structure of flexibility with specifying the factors affecting flexibility

Classifying the literature on flexibility into these categories and discussing its problems, we clarify the directions to construct the framework of flexibility of production control systems.

2.1 The literature on FMS

There are a great number of literature on this category. So we outline only some of the literature on reviewing the researches of this category.

Buzacott and Shantikumar[4], Buzacott and Yao[5], Kalkunte et al.[14], Van Looveren et al.[26], and Wilhelm and Sarin[27] review the researches on FMS. In particular, Van Looveren et al.[26] classify the planning method of FMS into three levels, strategic, tactical, and operational level. Also they classify the problems of this three levels into screening/selection, batching/loading, and releasing/dispatching, respectively. In addition, based on the techniques of analysis for these problems, they classify the literature into queuing network, mathematical programming, simulation analysis, and other models.

From this literature, it becomes clear that the problem of FMS can be or must be classified based on its objective.

2.2 The literature on the typology of flexibility

As mentioned in the preceding subsection, the problems of FMS planning were classified into some categories in reviewing the literature on FMS. There are also a lot of literature on the typology of flexibility itself.

Browne et al.[1] point out that all of the systems called FMS do not necessarily have same characteristics although there are a lot of systems identically labeled FMS. And they classify flexibility into eight types, that is, machine flexibility, process flexibility, product flexibility, routing flexibility, volume flexibility, expansion flexibility, operation flexibility, and production flexibility.

Gerwin[9] also classifies flexibility into similar types proposed by Browne et al.[1]. And Falkner[8] and Carter[6] classify flexibility with reference to the classifications proposed by Browne et al.[1] and Gerwin[9].

They propose various types of flexibility, however, they do not address the relationships among the types of flexibility. Thus, the following researchers discuss the relationships among the types of flexibility, and rearrange the types.

Slack[23] classifies flexibility into long-term and short-term flexibility, also into range and response flexibility. Next he clarifies the relationships among the types of flexibility proposed by Gerwin[9] and Slack[22] according to the degree of variability and uncertainty. And Taymaz[25] points out that different types of flexibility can be unified in a hierarchical structure, and clarifies the structure with three levels of components level, operations level, and system level. Also he constructs a model of the relationships between machine flexibility as component level and volume flexibility as operations level.

From the literature on the typology of flexibility, it can be pointed out the necessity to clarify not only the relationships among the types of flexibility but also the factors affecting flexibility. Also it can be pointed out the necessity to discuss on the performance measures evaluating flexibility.

2.3 The literature on the measures of flexibility

In the literature on the measures of flexibility, the followings propose some flexibility measures based on some measures relating to flexibility.

Pye[19] and Kumar[15] propose entropy measures of flexibility as functions of the degree of uncertainty or freedom of selection. And Mandelbaum and Buzacott[16] formulate a performance measure of flexibility with introducing utility function for decision making of the selection of systems.

On the other hand, Son and Park[24], Miltenburg and Krinski[17], and Huchinson and Sinha[13] propose economic measures to evaluate flexibility of FMS. Of course, economic measures cannot be neglected in evaluating FMS finally, however, these are influenced not only by the factors directly relating to flexibility but also by cost parameters. Therefore, it is necessary to investigate the degree of influence of each factor. Or, if some physical measures relating to flexibility can be evaluated, these can be translated into economic measures afterward, and the physical measure has an advantage that it can evaluate flexibility without suffering the influence of cost parameters. From this reason, the following researchers propose physical measures to evaluate flexibility.

Muramatsu et al.[18] propose the number of items produced by production system, production lead time, inventory level, and the amplifications of production quantity and inventory level, as the measures to evaluate the flexibility of production systems. And, Young and Murrey[28] point out that the criteria evaluating FMS are grouped into productivity, quality, reliability, diagnostics, flexibility, material management, and economics, and they propose the changeability from a certain component to others and its lead time as flexibility measures.

2.4 The literature on the factors affecting flexibility

The literature on flexibility typology or on flexibility measures do not address the problem itself. On the other hand, the following researchers discuss the factors affecting flexibility in order to construct the framework of problems of flexibility.

First, Buzacott and Mandelbaum[3] clarify the changes influencing flexibility and classify these changes into internal and external changes.

Gupta and Buzacott[10] classify the changes into short, medium, and long range changes according to these time frame. And for evaluating the overall flexibility, they propose a schematic model that the flexibility of a low level is evaluated with setting the changes of its upper levels at certain values and the flexibility of an upper level is evaluated with aggregating the variables of its lower level.

Because flexibility is a measure to evaluate responses to changes, changes and responses are two factors affecting flexibility. And it becomes clear that the changes can be classified into internal and external changes, also grouped according to time frame. And it can be pointed out the necessity to classify not only the changes but also the responses of production control systems. In addition, it can be pointed out the necessity to consider the relationships among the changes and the responses, that is, which category of the changes must be considered in investigating the flexibility of response of a certain category.

3 FLEXIBILITY OF PRODUCTION CONTROL SYSTEMS

In this section, we clarify the flexibility of production control systems on the basis of the findings obtained at the preceding section. For this purpose, we define production control systems at first, and clarify the flexibility of production control systems with considering its factors, that is, changes, responses, and performance measures.

3.1 Production Control System in Production Management System

Here we clarify the flexibility of production control systems through clarifying the framework of production management system that is a total system including production control system, the customers' needs for production management system, and the objectives of production management systems.

(1) Framework of production management system

Production management system can be considered to manage product, process, and production by the following subsystems in order to satisfy customers' needs.

Product Control System	: Research and develop the products that satisfy customers' needs, plan and control of model change.
Process Control System	: Plan and control the processes where produce the

products developed by product control system, that is Man(worker), Material, and Machine(equipment). And, if necessary, propose the improvement of product to product control system.

Production Control System : Plan and control the production of products with specification and quality developed by product control system, at the production processes planned by process control system, in order to satisfy customers' needs about volume and delivery time.

In these subsystems, we focus on production control system. And we consider products and processes as conditions for production control.

(2) Customers' needs from market

Customers' needs from market to production systems are classified into three categories as follows.

Specification and Quality : shape, function, performance, color, etc.
Volume and Delivery Time : quantity of entry order, due date request, etc.
Price : price acquiring product, usage cost, etc.

The changes that must be considered in production control systems are mainly relevant to volume/delivery time. And specification/quality or price is much more relevant to other control systems, such as product control system, process control system, or price control system(this is not included in production management system), rather than production control system.

(3) Objectives of production management systems

The performance measures of production management systems, that evaluate the management in responding customers' needs, can be classified as follows. And the objectives of production management systems are defined as raising some or all of the performances of production systems.

Service level : the level of customers' needs to be satisfied
Reliability : the rate to assure the service level
Productivity : the degree of effectiveness in utilizing production system to satisfy customers' needs
Flexibility : the ability to respond the changes of customers' needs and production process
Cost : the cost to maintain the necessary performances of production systems

Previously, service level for customers' needs, reliability, productivity, and cost are considered as performance measures of production management systems. Recently, however, it becomes necessary to evaluate production management systems not only by these measures but also by flexibility because of the diversity and the uncertainty of customers' needs(see Slack[22]). And, as DeMeyer et al.[7] pointed out, flexibility will become an advantage in the competitive environment. Flexibility is a measure to evaluate responses to changes, so the flexibility of production management systems must be a measure to evaluate the response planned and controlled by production management systems against the changes occurred in and/or out of production systems.

3.2 Changes Affecting Flexibility

The flexibility of production systems and production control systems can be defined as a measure to evaluate the responses of production systems or production control systems against the changes of customers' needs or production processes. So the changes are main factors affecting flexibility. It can be considered that a lot of changes affect production control systems, however, these can be classified into two categories based on their sources, also classified into three according to the time frame controlled by production control systems. Various classifications based on the time frame can be considered, however, it is very much convenient if the time frame corresponds to the control range of production control systems

For this reason, we classify the changes based on their sources and according to control ranges as follows.

(1) Changes of product
- Long Range Changes : product life, model change, demand trend
- Mid Range Changes : seasonal demand, change of mix, random fluctuation
- Short Range Changes: due date change, emergent order

(2) Changes of process
- Long Range Changes : hire/fire, new equipment, new material
- Mid Range Changes : learning, maintenance time, breakdown, default
- Short Range Changes: variation of processing time, machine interfere, default

Where, there are three categories in customers' needs mentioned above, however, we limit our attention to volume/delivery time because production control systems mainly related to that category of needs. When we do not limit attention to production control and consider not only it but also product control and process control, some changes considered here as conditions may be considered as control parameters. For example, product mix and maintenance time can be considered as control parameters to product control and process control respectively. However, we regard their changes only as conditions of production control systems because we limit our attention to production control systems in production management systems. Also we limit our attention of the changes of process to the changes only affecting volume/delivery time.

Where, the range of changes, long, mid, and short range, corresponds to the control range of production control systems.

3.3 Responses of Production Control System

Here we clarify the responses of production control system against the changes. Production control systems are considered not as single systems but as the systems that are composed by some subsystems, and that those subsystems organize a hierarchical structure. So we define production control systems as systems composed by three levels and show the techniques developed for each control level.

Long Range Control : Capacity Planning, Aggregate Planning
Mid Range Control : MRP(Material Requirements Planning), JIT(Just-in-time), Classical Ordering Techniques(e.g. Reorder Point Systems, such as S-s System, S-R System, etc.)
Short Range Control : Scheduling, JIT, OPT

Long range controls plan and control the production for long term, that is about one to five years, based on demand forecast from market. Also, long range controls plan and control production capacity to perform its production plan. The techniques for long range control are Capacity Planning and Aggregate Planning. Under the condition of production plan and capacity of long range control, mid range controls plan and control the production for each production period based on more accurate demand forecast and the information from production process and inventory point. The techniques for mid range control are MRP(Material Requirements Planning), JIT(Just-in-time), and classical ordering techniques such as S-s System or S-R System. And short range controls plan and control production sequence and its schedule to accomplish the production plan of mid range control. The techniques for short range controls are JIT, OPT(Optimal Production Technology) and other sequencing and scheduling techniques. Where JIT is considered not only as a technique for mid range control but also as that for short range control. For example, the production for each process in a plant is planned and controlled by JIT as a short range control, and the production for each plant is planned and controlled by JIT as a mid range control.

The difference of control level or control range affects the factors to be considered by each production control subsystem. The range of changes mentioned in the preceding subsection corresponds to this control range.

By the way, the responses to changes are also classified into two types, that is, performing new actions and performing only prepared actions. This type of response causes the conditions of flexibility that must be considered as follows(see Buzacott[2] or Slack[22]).

Action Flexibility : How flexible are the new actions that revise production systems or production control systems in responding to changes? For example, how short or how economic is the action?
State Flexibility : How flexible are the prepared responses that do not alter production systems and production control systems?

In production control systems, new actions are considered as comparing some techniques in a certain control level, selecting and implementing one of those. And after the decision of implementation, various comparisons of the values of parameters of selected technique are considered as control of state. The difference between this action and state influences upon the changes to be considered. New actions can be carried out at all levels of production control systems. However, when the new action of a certain level is considered, it is necessary to consider the relationships between its level and its upper level, and it is necessary to consider the changes of those levels, because the performance of new action may depend on the values of the changes of its upper level. On the other hand, when the state of a certain level is considered, it is necessary to consider the relationships between its level and its lower level, and it is necessary to consider the changes of those levels, because the state of a certain level has suffered the influence of the changes of its lower level.

3.4 Performance Measures for Flexibility

In measuring flexibility, the degree of the influence of changes must be evaluated because flexibility is a measure to evaluate responses to changes. So, the variables that are related to changes can be considered as performance measures. These variables can be classified into two categories. That is, state variables that suffer the influence of changes, and control variables that can be controlled and have some effects on state variables. The variables of production control systems can be classified into these two categories as follows.

Control Variables: No. of items, production capacity, inventory point, production order, No. of set-up, safety stock level, order sequence, start time, etc.

State Variables : allowance of capacity, No. of stockout, production quantity, inventory level, backlog quantity, completion time, delay, etc.

Some of those variables or some statistics of those variables may be considered as the performance measures of flexibility. And a total measure integrating some of those measures, for example, economic measure, can be also considered, however, only evaluating physical measures is sufficient because those can be translated into economic measures and those do not include cost parameters. For example, the rate of production capacity allowance (it can be controlled by the item to be produced, and production capacity) and stockout rate are considered as the performance measures of flexibility at long range control. At mid range control, the amplification of production quantity or inventory level is considered as the performance measure of flexibility. Also throughput time, setup time, waiting time, and delaying time are considered as the performance measures of flexibility at short range control.

4 ANALYSIS OF FLEXIBILITY FOR PRODUCTION CONTROL SYSTEMS

In this section, we deal with the problem of selecting new action at mid range control, that is, designing a mid range production control system. For this purpose, we construct models of two representative techniques, that is, MRP and JIT, and analyze the amplifications of production quantity and inventory level of those techniques as the performance measures of flexibility.

Where, in order to evaluate the action flexibility of mid range control, we consider not only demand fluctuation and machine breakdown as mid range changes but also production capacity as long range variables because the performance of new action may depend on the values of the changes of its upper level as mentioned above.

4.1 Modeling of Production Control Systems

Here we construct the models of a production system and production control systems, MRP and JIT.

(1) Assumptions of Model

We assume the following conditions about product, process and production considered in our models.

#1 <u>Product</u> : The production system is defined to deal with a single item for simplicity. The demand of product at each planning period is a steady-state time series with exponentially autocorrelation, that is, the autocorrelation of demand can be expressed by an exponential function such as $R(i)=\lambda^i$. Where λ is the parameter of autocorrelation function. This time series of demand includes not only random fluctuation but also short term trend or seasonal fluctuation.

#2 <u>Demand forecast</u> : The demands of product at future periods are forecasted by the following method, that is called Autocorrelation Method, because the forecasting error of that method is minimum for the steady-state time series with exponentially autocorrelation.

$$\hat{D}_{t,t+i}=\lambda^i\cdot D_t+(1-\lambda^i)\cdot\mu_D \tag{1}$$

Where $\hat{D}_{t,t+i}$: the demand forecast of (t+i)th period forecasted at the end of t-th period

D_t : the actual demand during the t-th period

μ_D : the mean value of product's demand at each period

#3 <u>Production process</u> : Production processes are defined to connect serially. That is, the production system in this paper is a multi-stage production system. And each process is called 1st, 2nd, ..., or n-th stage according to go upstream the stage from the final production process.

#4 <u>Production Capacity</u> : The production capacity of the final production stage, C, restricts the production at the stage, and other stages have sufficient capacity. So some part of production order may be backlogged only at the final stage.

#5 Machine Breakdown : At the final stage, machine breakdown occurs and the actual production capacity is decreased and fluctuated because of the fluctuation of the down time at each period, X_t.

This paper analyzes the influences of changes upon the fluctuations of production quantity and inventory level. The changes considered in this paper are not only demand fluctuation as a change of product but also down time fluctuation as a change of process. Machine breakdown may occur at every production stage, however, we assume, for simplicity, that machine breakdown occurs only at the final stage, also that the production capacity restricts the production only at the final stage.

#6 Production control : The production at each stage is controlled periodically by an equal length and synchronized period.

#7 Production lead time : Each production process invariably takes one period to accomplish the production order of each period regardless of ordered volume. That is, the production order at the end of t-th period for the n-th stage, $O_t^{(n)}$ is performed during (t+1)st period, and is supplied to the inventory point of the stage at the beginning of (t+2)nd period. At the final stage, however, some part of production order may be backlogged and carried forward because of a production capacity constraint and machine breakdown.

From the conditions #4, #5, #6, and #7, the production quantities at each stage, $P_t^{(n)}$, are formulated as follows.

$$P_t^{(1)}=\min\{O_{t-1}^{(1)},\ C-X_t\} \tag{2}$$

$$P_t^{(n)}=O_{t-1}^{(n)} \qquad (n=2,3,\dots) \tag{3}$$

And from the conditions #7, the inventories of each stage at the end of t-th period, $I_t^{(n)}$, are formulated as follows.

$$I_t^{(1)}=I_{t-1}^{(1)}+P_{t-1}^{(1)}-D_t \tag{4}$$

$$I_t^{(n)}=I_{t-1}^{(n)}+P_{t-1}^{(n)}-P_t^{(n-1)} \qquad (n=2,3,\dots) \tag{5}$$

#8 Safety stock : Each inventory point has sufficient safety stocks, $S^{(n)}$, in order to neglect the constraint of inventory.

#9 Performance measures : The changes and the conditions of production capacity influence upon and make fluctuate the state variables of mid range production control systems, that is, production quantity and inventory level. The magnitudes of fluctuations of these state variables can be considered as the flexibility of production control systems, so we considered the amplifications of those variables as the performance measures of flexibility. Where, the amplification of production quantity or inventory level is defined as the variance of production quantity or inventory level standardized by the variance of product's demand as follows(see Muramatsu et al.[18]).

$$Amp(P^{(n)})=\lim_{t\to\infty}\{V(P_t^{(n)})/V(D_t)\} \tag{6}$$

$$Amp(I^{(n)})=\lim_{t\to\infty}\{V(I_t^{(n)})/V(D_t)\} \tag{7}$$

Where, V(*) : the variance of *.

The amplification of production quantity is a measure to evaluate the magnitude of the fluctuation of production quantity under the influence of the fluctuations of demand and down time, and to evaluate the flexibility that the production process are possessed by a production control system. This amplification can be used to calculate the production capacity of the production process, and dampening it leads to increase economy and productivity.

On the other hand, the amplification of inventory level is a measure to evaluate the magnitude of the fluctuation of inventory level under the influence of the fluctuations of demand and down time, and to evaluate the flexibility that inventory point is possessed by a production control system. This amplification can be used to calculate the buffer stock level of the inventory point, and dampening it leads to increase economy.

(2) MRP (Material Requirements Planning)

In order to compare MRP with JIT, we assume that Lot Sizing Rule is Lot-for-Lot and that Master Production Schedule for the final product is planned directly from demand forecasts. Under these conditions, MRP can be considered to calculate production orders for every stage as follows.

$$O_t^{(n)}=\hat{D}_{t,t+n+1}+\sum_{m=1}^{n}\{(\hat{D}_{t,t+m}-O_{t-1}^{(m)})+(S^{(m)}-I_t^{(m)})\} \tag{8}$$

Considering the constraint of production capacity and substituting a released production order to an actual production quantity, it can be transformed as follows.

$$O_t^{(n)}=\hat{D}_{t,t+n+1}+\sum_{m=1}^{n}\{(\hat{D}_{t,t+m}-P_t^{(m)})+(S^{(m)}-I_t^{(m)})\} \tag{9}$$

(3) JIT(Just-in-Time)

JIT, that is developed by Japanese automobile manufacturer TOYOTA, is considered as a company-wide slogan, a production ordering technique with order card called Kanban, and a dispatching technique. And it has been studied all over the world at various viewpoints. However in this paper, we construct its model as a technique of mid range production control system and compare its performance with that of MRP.

Where, in order to compare JIT with MRP, we assume that one sheet of Kanban means one piece of product or part, and that there are a lot of sheets of Kanban. Also we assume that Kanban is withdrawn and used to production order at the intervals equal to a production control period, and that the equation for calculating the production order for the final stage is same as that of MRP.

Under these conditions, JIT can be considered to calculate the production orders for the preceding stages as follows.

$$O_t^{(n)}=P_t^{(n-1)} \qquad (n=2,3,...) \qquad (10$$

4.2 Analysis of Flexibility

Some researchers consider MRP and JIT as certain techniques of Push type and Pull type ordering systems respectively, and analyze these performances(see Muramatsu et al.[18], Rees et al.[20], Sarker and Fitzsimmons[21]). However, they could not consider the changes of process, consider MRP and JIT not as the techniques of mid range control but as the techniques of short range control, or analyze not physical performances but an economic measure as a total performance.

Here we analyze the amplifications of production quantity and inventory level of MRP and JIT, and compare with each other in order to evaluate the flexibility of both techniques.

(1) Method for Analysis

If all of the equations of a production control system can be expressed by linear differential equations, the amplifications of production quantity and inventory level can be analyzed theoretically through solving those equations, expressing the production quantity and the inventory level by the function of demand, and calculating the variances of those. In our model, however, the production quantity at the final stage is expressed by a non-linear equation under the influence of the production capacity constraint, and theoretical analysis can not be applied. Therefore we estimate the amplifications with simulation experiments.

In performing simulation experiments and estimating performances from their outputs, the autocorrelation of time series and the initialization bias can be considered as problems. So after warming up during a certain duration, we experiment during a predetermined duration, and replicate its experimentation with warming up. Next, we exclude warm-up duration from replicated time series data obtained from these experiments, and estimate the amplifications at every period. Finally, we estimate the steady state amplifications by averaging the estimated amplifications at every period.

The numerical conditions of our experiments are as follows.

Demand time series : $D_t \sim N(100, 10^2)$, λ=0.0, 0.4, 0.8
Time series of down time : $X_t \sim N(100, \sigma_X^2)$, σ_X=0, 10, 20
Production capacity level : C=205, 210, 215 ($\Delta C = C - \mu_D - \mu_X$=5, 10, 15)
Duration of an experiment : T=1000 (excluding warm-up duration, 200 periods)
Number of replications : R=100

(2) Results

The results of estimating the amplifications are shown in Figure 1 through Figure 3. In Figure 1, the amplifications at each level of the autocorrelation parameter of demand are shown. Also, the amplifications at each level of the standard deviation of down time, and the amplifications at each level of the production capacity are shown in Figure 2, and Figure 3, respectively.

We obtain the following findings from these figures.

#1 From Figure 1, both amplifications of MRP become more amplified in going upstream the stage as the autocorrelation parameter of demand increases. On the other hand, both amplifications of JIT also increase, however, these do not amplify in going upstream the stage unlike MRP.

#2 From Figure 2, in MRP, the amplification of production quantity at the final stage and the amplifications of the inventory levels at the final stage and its immediately preceding stage increase as the standard deviation of down time at the final stage increases. And all of the amplifications except for these do not increase in spite of the condition of down time. On the other hand, all of the amplifications of JIT increase as the standard deviation of down time increases.

#3 From Figure 3, the behavior of the amplifications of JIT in decreasing the production capacity is significantly dependent on the conditions of down time. That is, when the standard deviation of down time is low, both amplifications decrease as the production capacity decreases (see Figure 3(1)), however, when that is high, both amplifications increases(see Figure 3(2)). On the other hand in MRP, the production capacity has an influence only upon the amplification of production quantity at the final stage and the amplifications of the inventory levels at the final stage and its immediately preceding stage.

#4 Comparing the amplifications between MRP and JIT, the amplifications of inventory levels of MRP are smaller than those of JIT. However, as the autocorrelation parameter of demand increases, those of MRP become much amplified and larger than those of JIT. On the other hand, the amplifications of production quantities of JIT are smaller than those of MRP. However, as the standard deviation of down time increases, those of JIT increase and become larger than those of MRP(see Figure 1 - 3)

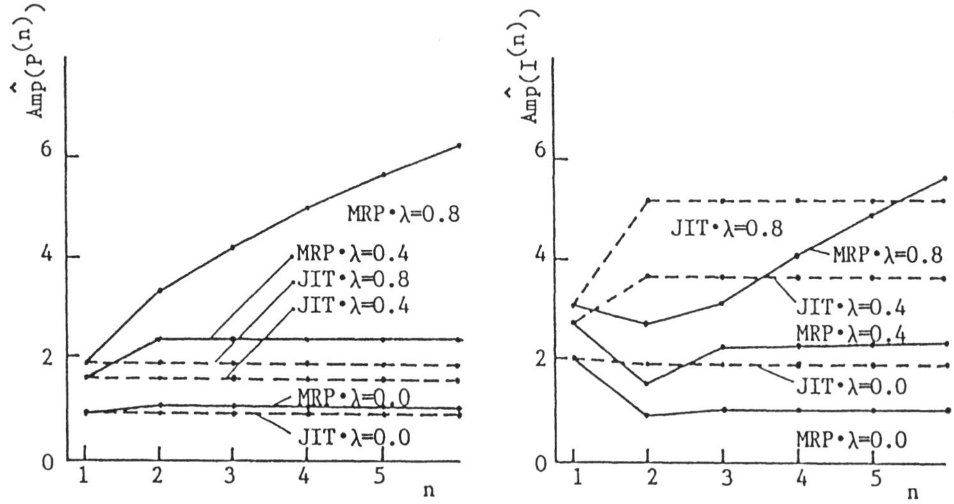

Figure 1 The amplifications of production quantity and inventory level at each level of the autocorrelation parameter of demand (λ)
(The production capacity C=215, The standard deviation of down time σ_x=0)

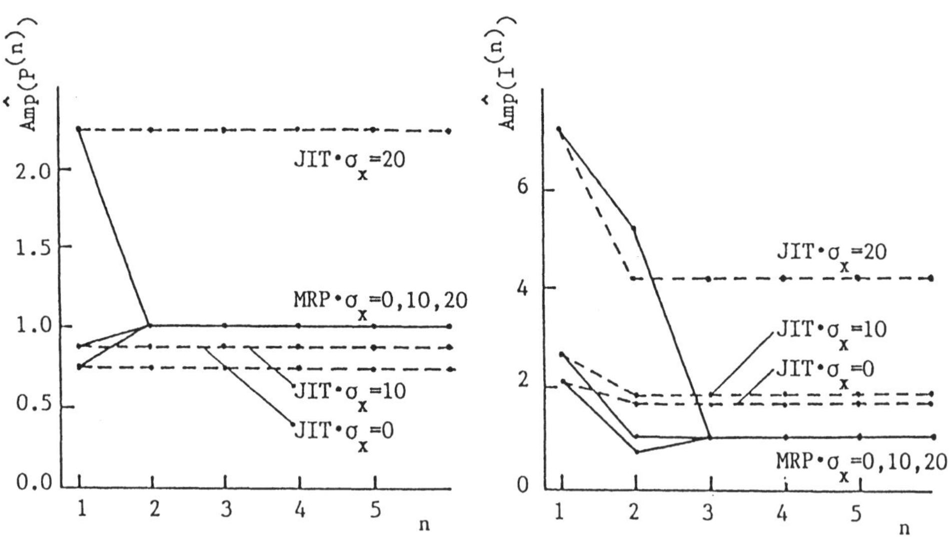

Figure 2 The amplifications of production quantity and inventory level at each level of the standard deviation of down time (σ_x)
(The autocorrelation parameter of demand λ=0.0, The production capacity C=210)

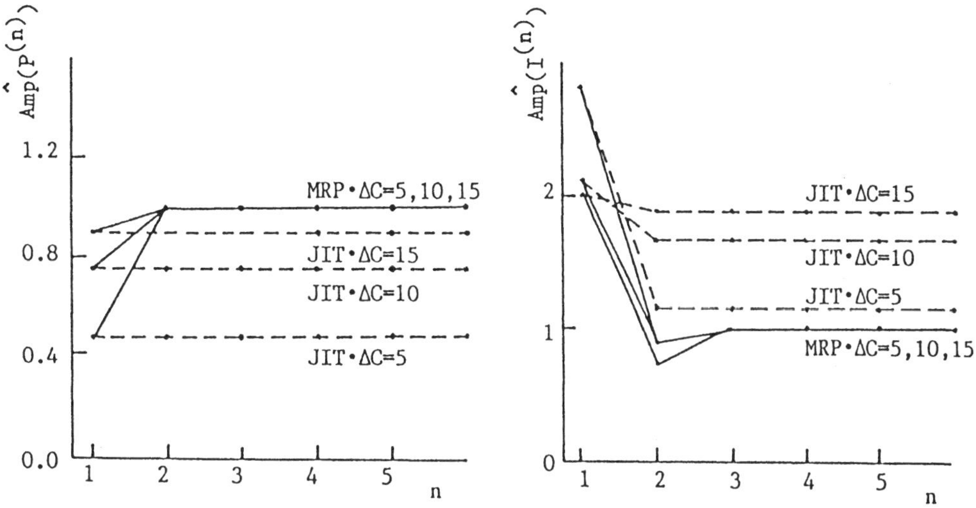

(a) The standard deviation of down time $\sigma_x=0$

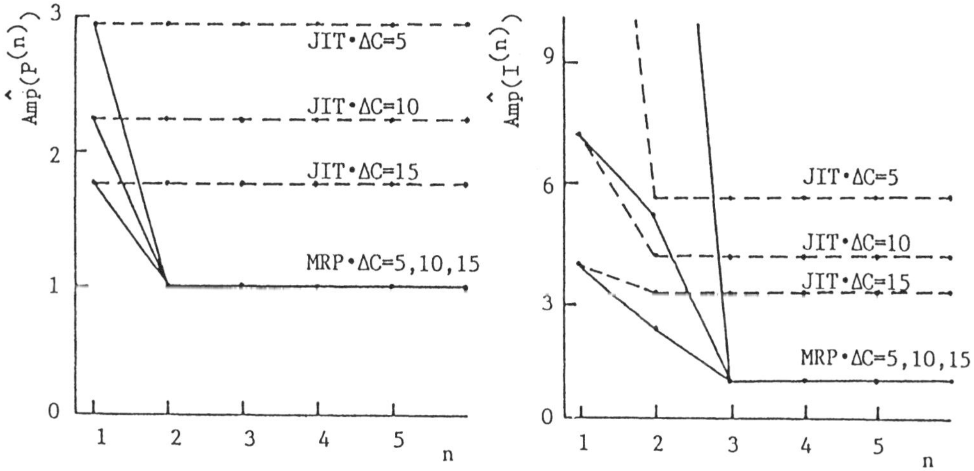

(b) The standard deviation of down time $\sigma_x=20$

Figure 3 The amplifications of production quantity and inventory level at each level of the production capacity (C)
(The autocorrelation parameter of demand $\lambda=0.0$, $\Delta C=C-\mu_D-\mu_X$)

5 CONCLUSIONS

This paper dealt with the flexibility of production control systems, that is, reviewed the literature on flexibility, clarified the framework of production control systems and its flexibility. Also this paper modeled and analyzed the flexibility of MRP and JIT as techniques of mid range production control systems.

As results, we conclude as follows.

(1) From reviewing the literature on flexibility, we point out the problems in clarifying flexibility, that is, changes and responses as the components of flexibility and these relationships must be clarified.

(2) With consideration of the findings obtained by the review of literature, we clarified the changes to be considered and the responses to be decided against the changes, also discussed the performance measures of the flexibility of production control systems.

(3) From modeling and analyzing the techniques of production control systems, we clarified that the flexibility of MRP and JIT, as the techniques of mid range production control systems, changes with respect to the conditions of the changes in and out of production systems and the conditions of long range control. In particular, MRP has suffered the influences of the changes out of production systems, and MRP becomes less flexible than JIT when these influences are significant. On the other hand, JIT has suffered the influences of the changes in production systems, and JIT becomes less flexible than MRP when these influences are significant.

(4) As results, it can be pointed out that the modeling and analysis of the flexibility of production control systems are necessary in designing flexible production control systems. Also it can be pointed out that changes and responses must be clarified and necessary factors must be considered as shown in our research.

REFERENCES

[1] Browne,J., Dubois,D., Rathmill,K., Sethi,S.P., Stecke,K.E., "Classification of Flexible Manufacturing Systems", The FMS Magazine, Vol.2,No.2,pp.114-117 (1984)

[2] Buzacott,J.A., "The Fundamental Principles of Flexibility in Manufacturing Systems", Proc. the 1st Intl Conf. FMS, pp.13-22 (1982)

[3] Buzacott,J.A. and Mandelbaum,M., "Flexibility and Productivity in Manufacturing Systems", Proc. 1985 Annual IIE Conf., pp.404-413 (1985)

[4] Buzacott,J.A. and Shantikumar,J.G., "Models for Understanding Flexible Manufacturing Systems", AIIE Trans., Vol.12,No.4, pp.339-350 (1980)

[5] Buzacott,J.A. and Yao,D.D., "Flexible Manufacturing Systems: A Review of Analytical Models", Management Science, Vol.32,No.7, pp.890-905 (1986)

[6] Carter,M.F., "Designing Flexibility into Automated Manufacturing Systems", Proc. 2nd ORSA/TIMS Conf. FMS: Oprs Res. Models and Appl.(eds. Stecke,K.E. and Suri,R.), Elsevier, pp.107-118 (1986)

[7] DeMeyer,A., Nakane,J., Miller,J.G., Ferdows,K., "Flexibility: The Next Competitive Battle the Manufacturing Futures Survey", Strat. Mgt., Vol.10, No.2, pp.135-144 (1989)

[8] Falkner,C.H., "Flexibility in Manufacturing Plants", Proc. 2nd ORSA/TIMS Conf. FMS: Oprs Res. Models and Appl. (eds. Stecke,K.E. and Suri,R.), Elsevier, pp.95-106 (1986)

[9] Gerwin,D., "Do's and Don'ts of Computerized Manufacturing", Harvard Business Review, Vol.60,No.2,pp.107-116 (1982)

[10] Gupta,D. and Buzacott,J.A., "A Framework for Understanding Flexibility of Manufacturing Systems", J. Mfg Sys., Vol.8,No.2, pp.89-87 (1989)

[11] Gupta,Y.P. and Goyal,S., "Flexibility of Manufacturing Systems: Concepts and Measurements", Euro. J. Oprl Res., Vol.43,No.2, pp.119-135 (1989)

[12] Gustavsson,S., "Flexibility and Productivity in Complex Production Processes", Intl J. Prod.Res., Vol.22,No.5,pp.801-808 (1984)

[13] Huchinson,G.K. and Sinha,D., "A Quantification of the Value of Flexibility", J. Mfg Sys., Vol.8,No.1, pp.47-57 (1989)

[14] Kalkunte,M.V., Sarin,S.C., Wilhelm,W.E., "Flexible Manufacturing Systems: A Review of Modeling Approaches for Design, Justification and Operation", In: Flexible Manufacturing Systems: Methods and Studies (ed. Kusiak,A.), Elsevier, pp.3-25 (1986)

[15] Kumar,V., "Entropic Measures of Manufacturing Flexibility", Intl J. Prod. Res., Vol.25,No.7, pp.957-966 (1987)

[16] Mandelbaum,M. and Buzacott,J.A., "Flexibility and its Use: A Formal Decision Process and Manufacturing View", Proc. 2nd ORSA/TIMS Conf. FMS: Oprs Res. Models Appl. (eds. Stecke,K.E. and Suri,R.), Elsevier, pp.119-130 (1986)

[17] Miltenburg,G.J. and Krinski,I., "Evaluating Flexible Manufacturing Systems", IIE Trans., Vol.19,No.2,pp.222-233 (1987)

[18] Muramatsu,R., Ishii,K., Takahashi,K., "Some Ways to Increase Flexibility in Manufacturing Systems", Intl J. Prod. Res., Vol.23,No.4,pp.691-703 (1985)

[19] Pye,R., "A Formal Decision-Theoretic Approach to Flexibility and Robustness", J. Oprl Res. Soc.,Vol.29,No.3, pp.215-227 (1978)

[20] Rees,L.P., Huang,P.Y., Taylor III,B.W., "A Comparative Analysis of an MRP System for a Multi-stage Production Operation", Intl J. Prod. Res., Vol.27,No.4,pp.1427-1443 (1989)

[21] Sarker,B.R. and Fitzsimmons,J.A., "The Performance of Push and Pull Systems: A Simulation and Comparative Study", Intl J. Prod. Sys., Vol.27,No.10,pp.1715-1731 (1989)

[22] Slack,N., "Flexibility as a Manufacturing Objective", Intl J. Oprs Prod. Mgt., Vol.3,No.3,pp.4-13 (1983)

[23] Slack,N., "The Flexibility of Manufacturing Systems", Intl J. Oprns Prod. Mgt., Vol.7,No.4, pp.35-45 (1987)

[24] Son,Y. and Park,C.S., "Economic Measure of Productivity, Quality and Flexibility in Advanced Manufacturing Systems", J. Mfg Sys., Vol.6,No.3, pp.193-207 (1987)

[25] Taymaz,E., "Types of Flexibility in a Single-Machine Production System", Intl J. Prod. Res.,Vol.27,No.11, pp.1891-1899 (1989)

[26] Van Looveren,A.J., Gelders,L.F., Van Wassenhove,L.N., "A Review of FMS Planning Models", In: Modelling and Design of Flexible Manufacturing Systems (ed. Kusiak,A.), Elsevier, pp.3-25 (1986)

[27] Wilhelm,W.E. and Sarin,S.C., "Models for the Design of Flexible Manufacturing Systems", Proc. 1983 Annual IIE Conf. pp.564-574 (1983)

[28] Young,A.R. and Murray,J., "Performance and Evaluation of FMS", Intl J. Oprns Prod. Mgt., Vol.6,No.5, pp.57-62 (1986)

A Quantitative Measure for Flexibility

CHRISTOPH SCHNEEWEISS and MARTIN KÜHN

1. Introduction

During the last decade flexibility has become a key notion of modern life and sciences. Especially for management science dealing with the dynamics of systems, flexibility turns out to be of particular importance. Expressions like flexible manufacturing, flexible workforce, and flexible market behaviour indicate its popularity.

Notions like robustness, elasticity and reactivity, or action variability are often used synonymously with flexibility. Moreover, the term flexibility often signifies rather different facts. As shall be pointed out later, for the above expressions "flexible manufacturing and flexible workforce", e.g., the term "flexible" has a quite different meaning.

There is a host of literature discussing different concepts of flexibility. Most of this literature, however, is concerned with manufacturing systems (see e.g. BUZACOTT/MANDELBAUM (1985), KUMAR (1987) and the recent survey of GUPTA/GOYAL (1989)). There are only a few papers deriving a quantitative measure of flexibility from a somewhat broader point of view, like HANSSMANN (1981), JACOB (1974) PARK/THUESEN (1979), or the recent paper of MANDELBAUM and BUZACOTT (1990).

Flexibility is a notion of general systems theory. Intuitively it characterizes the ability of a dynamic system to cope with alterations of its environment. Hence, ultimately flexibility is an important property of a system describing its ability to survive.

Of course, for such a property it may not be easy to obtain a simple quantitative measure as with costs, profits, rentabilities or productivities. Nevertheless, constructing such a measure and gaining a clear understanding of what is meant by flexibility seems to be urgently important. To illustrate some of the difficulties which have to be resolved consider two systems: one, reacting rather slowly to a wide variety of disturbances, and the other, reacting extremely fast but being able only to cope with a small variety of disturbances. Which system, we might ask, is more flexible?

In trying to answer this seemingly simple question we will finally arrive at an acceptable measure for flexibility. In doing so, we first expound in Sec. 2 the properties such a measure should possess. These properties will then in Sec. 3 be made more concise in terms of dynamic systems theory, leading to the definition of notions like reactivity and elasticity. Relying on these properties we will then be in a position to give a pre-measure (Sec. 4) which, in Sec. 5, after a proper normalization will result in the final measure of flexibility. An illustrating example will be added in Sec. 6 giving a clear understanding of the notions used so far. The next section (Sec.7) will further discuss the introduced flexibility measure relating it particularly to notions like robustness and service level. Then, in Sec. 8, flexibility will be further enriched by an important aspect: Flexibility should not only measure the reactivity of a system to external disturbances but also to alterations of its preference structure. Thus the measure is fully expounded and it seems to be appropriate in Sec. 9 to add a short survey of the relevant literature. Finally, Sec.10 gives a summary and some general remarks on the measurability properties required for the proposed measure.

2. Preliminary Considerations

As already mentioned flexibility describes the ability of a system to react to (stochastic) disturbances. Hence properties like
(1) variability and
(2) reactivity of actions
will play an important role. These two properties, however, are not sufficient to describe what is meant by flexibility. Variability and reactivity of actions could result

in a "blind actionism". Therefore, reacting to a disturbance, should be "rational" with respect to the system's performance structure, hence
(3) rationality
is a further requirement.

Finally, flexibility also depends on the
(4) stochastics
of the environment the system is responding to.

With these four properties it will be possible to define a suitable measure for flexibility. There is, however, an important aspect one still has to account for. The preference structure, by which the system's actions are evaluated may change. This kind of "uncertainty" may be called
(5) preference variability.
Hence, the system should be able to react suitably to such changes. Finally, the measure will depend on the
(6) planning ability
by which the resources of a system are utilized.

With these six properties we are now in a position to define a measure of flexibility and to deliniate it from notions like robustness and elasticity.

3. Elasticity as a Partial Aspect of Flexibility

Le us first give exact definitions for the two terms, variability and reactivity of actions.

Generally, a system may be described by its elements, attributes and relations. The state $z_t \epsilon Z_t$ of a system describes the information one may (at time t) possess. Therefore a dynamic system is characterized by the general state transition equation

$$z_{t+1} = G_t (z_t, a_t, r_t)$$

actions by enlarging the action spaces A_t. Flexible manufacturing, on the other hand, only has an effect on the system's reactivity. If, however, also production capacities can be changed variability will be involved as well, and in total one would have a higher elasticity.

Elasticity as a measure for flexibility has two disadvantages. First it does not take into account point (3) ("rationality") of Sec. 2. Secondly, it is only to a limited degree operational. Only if (for the same stochastic environment) the state space of a technology $\Theta=\Theta_1$ completely contains the state space of Θ_2:

$$Z(\Theta_1) \supset Z(\Theta_2)$$

one would have (at least) an ordinal measure for technologies.

4. Defining a Pre-Measure of Flexibility

Taking into account property (3) of Sec. 2 we may now characterize the flexibility of a system not by its elasticity, i.e. by all its strategies (implying $Z(\Theta)$), but only by those strategies $a=a^*$ which optimize the expectation $E\{V(a)\}$ of a performance measure $V(a)$

$$(1) \qquad V^* := V^*(\Theta) := \operatorname*{opt}_{a \in \Theta} E\{V(a)\}.$$

Assuming $V(a_1,...,a_T)$ to be separable (in the sense of stochastic Dynamic Programming) one has

$$(2) \qquad V^*(\Theta) := \operatorname*{opt}_{(a_1,\ldots,a_T) \in \Theta} E\{\sum_{t=1}^{T} V_t(a_t(z_t))\}.$$

In fact, if there has only one performance measure to be considered, $V^*(\Theta)$ is a much better measure for flexibility than $Z(\Theta)$. V^* comprises properties (1) to (4) of Sec. 2 and may therefore be called a **pre-measure** for flexibility. However, thinking of the intuitive meaning of flexibility, which has much to do with the ability of a system to cope with stochastics, in merely taking the expectation of a performance measure, property (4) is not taken into account in its full significance. This will be made up for in the next section by properly normalizing the measure $V^*(\Theta)$. (Sometimes, however, already $V^*(\Theta)$ is taken as a measure for flexibility; see e.g. WEBSTER/ TYBERGHEIN (1980)).

Before doing so let us close this section with a general remark on the nature of strategies $\{a_1,a_2,...,a_T\}$. Usually for managerial systems a strategy may be split up in a strategic decision a_1 building up a so-called **flexibility potential** and a remaining "truncated" strategy $\{a_2,...,a_T\}$ which operationally uses the flexibility potential. It is this general feature which will be most important for a deeper understanding of the flexibility measure to be defined in the next section.

5. The Final Derivation of a Flexibility Measure

Formally, in normalizing V^* one would have to introduce a suitable origin and a unit of scale. Semantically, in order to model the intuitive meaning of flexibility, one should relate V^* to a

- completely inflexible and an
- extremely flexible behaviour.

Complete inflexibility could be described by a behaviour a^0 showing no reaction to external disturbances. The corresponding performance measure would be given by $V^-:=V(a^0)$, so that flexibility should be measured by its deviation from complete inflexibility: V^*-V^-. Notice that defining V^-, is not always unambiguous. We shall come back to this point in Sec. 7.

Ideal flexible behaviour on the other hand, can be described by the so-called "prophetic" value V^+. This value is defined as follows. Imagine that all (future) disturbances $\{r_t'\}:=\{r_1',...,r_T'\}$ are known. In that case one could calculate an optimal value $V^P\{r_t'\}$. In relating this value to V^*, however, one would not capture the meaning of flexibility. In fact, flexibility has strategically not to be build up only for one special time series of disturbances but for a large variety of possible stochastic environments. Therefore $V^P\{r_t'\}$ has to be averaged, giving the desired prophetic value

$$(3) \qquad V^+ := E_{\{r_t\}} \operatorname*{opt}_{a \epsilon \Theta} V^P_{\{r_t\}}(a).$$

We are now in a position to define as our **final measure of flexibility**

$$(4) \qquad F_\Theta := \frac{V^* - V^-}{V^+ - V^-}.$$

F_Θ is normalized between 0 and 1, giving 0 in case of complete inflexibility: $V^* = V^-$, and 1 for the ideal prophetic situation $V^* = V^+$. Obviously, the better a system is informed about its environment, the smaller will be the difference between V^* and V^+ resulting in a high flexibility (near $F=1$).

6. An illustrative Example

A simple example will be presented to illustrate the terms introduced thus far. Consider the following investment-production problem given by the decision tree of Fig. 1. Suppose, in period 1, there is the option to purchase machine M1 or M2 in order to produce and satisfy demand in Period 2. Machine M_1 can produce lot sizes $q=0;1;2;3$ or 4 whereas M_2 is restricted to lots of size 0 or 4. Thus the three main branches of the decision tree represent the three flexibility potentials which can be built up by the technology.

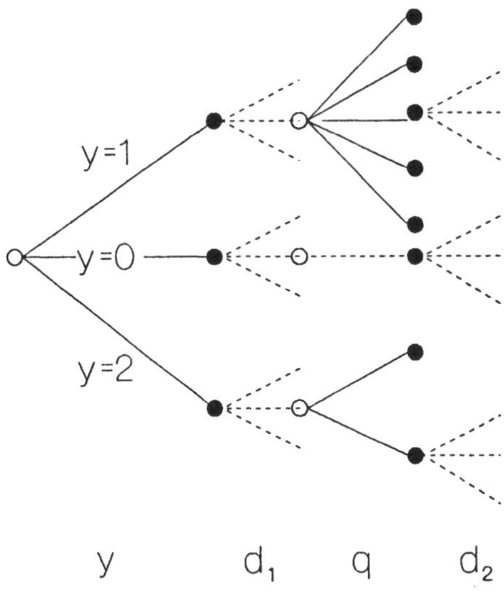

Fig. 1: Decision tree of the example

In periods 1 and 2 demands d_1 and d_2 occur. They are identically distributed and stochastically independent. The demands have to be satisfied at the end of period 2. Their probability distribution is given by

	0	1	2
P(d)	0,25	0,50	0,25 .

Fig. 2 illustrates the situation from a somewhat different point of view. The variable y indicates the investment decision of period 1 ($y=a_1$) and q denotes the (operative) decision to produce in period 2 ($q=a_2$). D indicates the total demand $D:=d_1+d_2$; D has to be delivered at the end of period 2, i.e. there is no delivery in period 1.

period		1		2	
	1		2		3
point of time	⌊	———	⊥	———	⌋
state	(0,0)		$(y,-d_1)$		$(y,q-D)$
disturbance		d_1		d_2	
decision	y		q		

Fig. 2: Definition of some of the terms used in the example

Procurement cost (AC), production cost (PC) and shortage cost (SC) are given in Table 1.

cost	M_1		M_2
AC	5	3,5	
PC	0	if	$q = 0$
	$3 + 2q$	"	$q > 0$
SC	$10(d_1+d_2-q)$	"	$d_1+d_2 \geq q$
	0	"	$d_1+d_2 < q$

Table 1: The cost functions of the example

The variability A (of admissible actions) is given by

$$A := \{A_1(z_1), A_2(z_2{}^1), A_2(z_2{}^2)\}$$

with

$A_1(z_1) := \{y : y = 0;1;2\}$,

$A_2(z_2{}^1) := \{q : q = 0;1;2;3;4\}$,

$A_2(z_2{}^2) := \{q : q = 0;4\}$.

The reactivity G is defined by the state transition equations (see Fig. 2) G:{G1,G2} with

G_1: (0,0) —> ($y,-d_1$);
G_2: ($y,-d_1$) —> ($y,q-D$).

The total cost C serves as a performance measure

$C = E\{C_1 + C_2\}$ => min with
C_1 = AC(y) and
C_2 = PC(q) + SC(q,D).

For the prophetic solution Equ. (3) one obtains $C^+ = 11{,}4$ and for the optimal (stochastic DP) solution $C^* = 14$. Since $a^0 = \{y=0,q=0\}$ (s. Sec. 5) Table 1 gives $C^- = 10E(D) = 20$.

Hence, the flexibility of cost is given by

$$F_C = \frac{C^* - C^-}{C^+ - C^-} = \frac{14 - 20}{11{,}4 - 20} = 0{,}70 .$$

Using as an approxmation instead of C^* a deterministic DP solution having strategy s:={y=1,q=2} one obtains $F^S = 0{,}49$ showing a substantial loss of flexibility. (All details of calculations may be found in KÜHN (1989)).

7. Further Discussion of the Flexibility Measure F(ϴ)

After the illustrative example of the last section let us now investigate somewhat more closely the general significance of our flexibility measure Equ. (4). In particular we are interested in the behaviour of F(ϴ) with respect to

(1) the state of information
(2) the exploitation of a given flexibility potential
(3) the dependance on different flexibility potentials.

Ad (1)
The first aspect has to do with the procurement of information. If one is well informed about future developments, i.e. if stocastics is low, the flexibility measure will have high values. By its refence value V^+ our measure gives an important hint as to the amount of information the system should reasonably be provided with.

Ad (2)
The operative usage of flexibility is another important aspect being measured by $F(\theta)$. The above example clearly illustrates this fact by the different values of $F_C = 0{,}70$ and $F_C^S = 0{,}49$. If this loss of flexibility turns out to be too large one will try to improve the planning procedure which may even result in the construction of a new flexibility potential (a problem being addressed in (3)).

Ad (3)
As we know, for a technology θ, different flexibility potentials can be constructed. Considering only two potentials, i.e. the truncated technologies θ_1 and θ_2, one obtains flexibility measures

$$F(\theta_1|\theta) = \frac{V^*(\theta_1)-V^-}{V^+(\theta)-V^-} \quad \text{and} \quad F(\theta_2|\theta) = \frac{V^*(\theta_2)-V^-}{V^+(\theta)-V^-}$$

with $V^*(\theta_i)$ being the optimal performance measure with fixed strategic decision $a_1 = a_1(\theta_i)$ $(i=1,2)$.

The two measures $F(\theta_i|\theta)$ $(i=1,2)$ stress again the following fact:
1. $V^+(\theta)$ is a reference value for the whole technology θ. It only depends on the technology and not on the special modelling of the system.

2. The ambiguity in the choice of the origin V^- is only of minor importance. Since one is often merely interested in a comparison of flexibilities, only $V^*(\Theta_1)$ and $V^*(\Theta_2)$ have to be known.

In contrary to the well defind value V^+, the origin V^-, i.e. the value of the performance measure for complete inflexibility requires a proper consideration for each particular class of problems. Sometimes it turns out to be reasonable to take aspiration levels or to set $V^- = O$. For a performance measure to be maximized: $V^- < V^* < V^+ > 0$ this turns out not to be an unrealistic choice resulting in the flexibility ratio

$$F = F^Q := \frac{V^*}{V^+} \quad .$$

F^Q has a simple intuitive meaning: It shows how much an optimal plan with complete information (V^+) is better than that with incomplete information (V^*). Even more intuitive would be the difference measure $F = F^D := V^+ - V^*$ giving the absolute loss of optimality caused by incomplete information. However, F^D is not normalized and depends on the dimension of V.

F^Q shows an interesting simularity with the β-service level (see e.g. SCHNEEWEISS /KÜHN(1990) or SCHNEIDER (1981)) of inventory theory. This service level is defind by the following ratio of expectation values

$$\beta := \frac{E\ \{\text{satisfied demand per period}\}}{E\ \{\text{total demand per period}\}}$$

Taking as a performance measure: "satisfied demand" (for the optimal plan) β can be written

$$\beta := \frac{V^*}{\overline{V}}$$

with $\bar{V}$ denoting expected total demand per period. Generally $\bar{V}$ will not be identical with V^+. Although V^+ has full information a complete satisfaction of demand (as is implicitely assumed by $\bar{V}$) is not necessarily guaranteed. This is due to the fact that in calculating V^+ constraints have possibly to be observed. Yet, in comparing β with F^Q the analogy is obvious.

This analogy will even be more convincing if, parallel to our flexibility measure, we define a measure of **robustness.** As is well known, many different definitions of robustness exist which, confusingly enough, are also used to measure flexibility (ROSENHEAD et al. (1972)).

Often in case of uncertainty (i.e. no probability distributions can be specified) robustness is defined by a regret measure. For subjective probabilites which we reasonably assume to be given, a simple robustness measure R(a) can be defined by the quantile

(5) $R_\Theta(a) := \text{Prob}(V \geq V^{AL} \mid \Theta).$

That is, the robustness of a strategy $a \epsilon \Theta$ is given by the probability by which a performance measure V still reaches acceptable values, i.e. values being above an aspiration level V^{AL}. Note that with this definition also the most flexible strategy a^* has a certain robustness $R(a^*)$, stressing again that both measures have a different meaning.

Taking again V to be satisfied demand, $R_\Theta(a)$ turns out to be the well known α-service-level (SCHNEIDER (1981)), showing the intimate relationship between flexibility and robustness.

8. Taking into account Preference Elasticity

Up to now our measure F_Θ is defined for only one performance measure. Usually, however, the performance of a system is described by more than one measure. Hence, it seems only natural to extend F_Θ to a multi-criterion situation. There is, however, still another reason for such an extension. Preferences may change and this is best described by switching from one performance measure to another. For a flexible system such a change should be "easily possible". We shall model this property by calculating not just one (most) flexible plan but a whole set of efficient plans.

As was pointed out earlier, for the pure technical part the system activated its elasticity to absorb external disturbances. A similar property holds for "disturbances" of the preference structure. Thus the set of all (stochastic) values of all performance measures will be called **performance elasticity.**

Deriving a flexibility measure for performance measures $V_k (k=1,...,K)$ one now has to determine the vector of efficient values $V^e := \{V_1{}^e,...,V_k{}^e\}$ which again can be obtained using stochastic Dynamic Programming. Consequently let us define a partial flexibility by

$$(6) \qquad F_k(V_k^e | V^e) := \frac{V_k^e - V_k^-}{V_k^+ - V_k^-}, \quad k=1,...,K.$$

(Note that for each efficient strategy one obtains a different F_k). F_k not only is a function of $V^e{}_k$ but via V^e of all performance measures. Total flexibility will be defined by the set of all partial flexibilities

$$(7) \qquad F := \{F_k(V_k{}^e | V^e);\, V^e,\, k=1,...,K\}.$$

As in the case of one performance measure approximations may again be necessary replacing $V_k{}^e$ and V^e by some $V_k{}^s$ and V^s respectively.

9. A short Survey of Literature

We are now in a position to compare our measure of flexibility with some definitions given in literature. In most cases only part of the six postulates stated in Sec. 2 are taken into account. Furthermore, most definitions are concerned with manufacturing systems and they all do not take into account a possible preference variability. Particularly, in discussing flexibility in some broader sense, the reactivity of a system with respect to this last type of "disturbances" should be taken into consideration.

A paper which only takes into account **variability** (property (1)) is that of LASSERRE/ROUBELLAT (1985). Lasserre and Roubellat measure the flexibility of production programmes by the volume of the polyeder spanned by linear constraints. For polyeders containing one another this measure is at least ordinal. Measures which take into accont elasticity (i.e.variability and reactivity) are given by MARSCHAK/NELSON (1962) and JONES/OSTROY (1984). For Marschak and Nelson a decision a_1 is more flexible than a decision a_2 if (in our terms) the flexibility potential $\Theta(a_1)$ contains $\Theta(a_1)$. Measures which also take into account the stochastics are defind by PYE (1978) and KUMAR (1987). They take an entropy measure for flexibility. Definitions which also take into account certain performance aspects are those of ROSENHEAD et al. (1972). However, robustness and flexibility are measured synonymously. Finally WEBSTER/TYBERGHEIN (1980) define flexibility simply by the optimal value of V* of a performance measure, i.e. by our pre-measure of flexibility (see Sec.4). A measure which is not too far from that we proposed, is given by HANSSMANN (1987). However, property (5) is not taken into account and formally his measure is not correct (s. the discussion in SCHNEEWEISS/KÜHN (1990).

An interesting approach is presented by a recent paper of MANDELBAUM and BUZACOTT (1990) which could be best explained by referring to the notions of modelling theory (see SCHNEEWEISS (1987)): Consider a certain abstraction level of a real world problem containing all possible uncertainties. Call this model a mastermodel M (true model) having a performance measure L. Relax all stochastic variables in taking unbiased estimates to obtain a mastermodel $\bar{M}$ with performance measure $\bar{L}$. Let L and $\bar{L}$ be optimized by a* and a^0 respectively. Then the "modelling gap" (see SCHNEEWEISS (1987))

$$G_{M,\bar{M}} := L(a^*)-L(a^0)$$

my be used to measure flexibility. Since a^0 can be considered as a laisez-faire strategy, one of the main results of MANDELBAUM and BUZACOTT is a flexibility measure being based on laissez-faire strategies, whereas our measure relies more on the prophetic value.

10. Summary and Concluding Remarks

Using a stepwise procedure we derived a measure of flexibility starting with notions like variability, reactivity and elasticity. The obtained measure is highly operational. It should in its non-approximated versions (4) or (6) allow for the application of Dynamic Programming; i.e. it should be possible to formulate Bellman's functional equations. A sufficient condition for these equations to hold is given by the additivity of the performance measure with respect to time:

$$V = \sum_{t=1}^{T} V_t \ .$$

This in turn is guaranteed for a decision maker being utility independent with respect to time periods (FRENCH (1988), KEENEY/RAIFFA (1976)).

Of course, at a first glance, this high level of measurability seems to restrict the flexibility measure to systems being comparatively well described. However, the following aspects should be taken into consideration:

1) Often a quantitative measure of flexibility will only be needed in well described situations. For more "soft" problems one will be content with variabilities, reactivities or elasticities.
2) The application of stochastic Dynamic Programming is not indispensible. In most cases, as has been pointed out, approximations will be necessary. Therefore if preferential independence is not given, one could completely avoid the application of Dynamic Programming.
3) On the other hand, even if the decision maker is not preferentially independent, measures like (4) and (6) might still be of some relevance. (Note that using

mathematical analysis does not imply performance measures to be monetary. Also "qualitative" quantities measured by utility scores may be used.)

Of course, flexibilities F_k will not be the only measures to characterize a system's behaviour. Robustness measures and the performance measures V_k themselves will be of importance, all taken into account within the framework of a multi-criterion decision problem: $\{F_k, V_k, R_k\}$ => opt.

Typically one would consider non-cost flexibility measures together with their adjoint costs. I.e. costs of flexibility have to be determined within a multi-criterion decision problem. This is typically the situation of determining simultanously minimal costs and a maximal value of a service level for an inventory problem. Describing costs by V_1 and flexibility of a non-cost performance measure V_2 by V_2/V_2^+ one would have the multi-criterion decision problem

$$\{ V_1, \frac{V_2}{V_2^+} \} \Rightarrow \text{opt.}$$

Hence, in summarizing, flexibility and robustness turn out to be generalized service-levels. Besides the "profit-criteria" they are of particular importance since they measure the (short and long term) ability of a system to "survive" in an uncertain invironment.

References

BUZACOTT, J.A., MANDELBAUM, M.: Flexibility and Productivity in Manufacturing Systems, in: Proceedings of the IIE Conference (1985), 403-413, Chicago

FRENCH, S.: Decision Theory - An Introduction to the Mathematics of Rationality, New York 1988

GUPTA, Y.P., GOYAL, S.: Flexibility of Manufacturing Systems: Concepts and Measurements, in: European Journal of Operational Research 43(1989), 119-135

HANSSMANN, F.: Einführung in die Systemforschung, München-Wien 1987

JACOB, H.: Unsicherheit und Flexibilität, in: ZfB 44(1974), S. 299-326, S. 403-448, S. 505-526

JONES, R.A., OSTROY, J.M.: Flexibility and Uncertainty, in: Review of Economic Studies (1984) LI, 13-32

KENNEY, R., RAIFFA, H.: Decisions with Multiple Objectives, New York 1976

KUMAR, V.: Entropic Measures of Manufacturing Flexibility, in: Int. J. Prod. Res. 25(1987), 957-966

KÜHN, M.: Flexibilität in logistischen Systemen, Heidelberg (1989)

LASSERRE, J.B., ROUBELLAT, F.: Measuring Decision Flexibility in Production Planning, in: IEEE Transactions on Automatic Control, 30/5(1985), 447-452

MANDELBAUM, M., BUZACOTT, J.: Flexibility and Decision Making, in: EJOR 44(1990), 17-27

MARSCHAK, T., NELSON, R.R.: Flexibility, Uncertainty and Economic Theory, in: Metroeconomica 14(1962), 45-59

PARK, C.S., THUESEN, G.J.: The Effectiveness of Using Uncertainty Resolution to Measure Investment Flexibility in Sequential Capital Rationing, in: AIIE Transactions, 11/3(1979), 207-220

PYE, R.: A Formal Decision Theoretic Approach to Flexibility and Robustness, in: Journal of the Operations Society 29(1978), 215-227

ROSENHEAD, J., Elton, M., Gupta, S.K.: Robustness and Optimality as Criteria for Strategic Decisions, in: Operational Research Quarterly 23(1972), 413-431.

SCHNEEWEISS, Ch.: On a Formalisation of the Process of Quantitative Model Building, EJOR, Vol. 29, No. 1 (1987), 24-41

SCHNEEWEISS,Ch.: Einführung in die Produktionswirtschaft, 3. Auflage, Berlin, Heidelberg, New York (1989)

SCHNEEWEISS, CH., KÜHN, M.: Zur Definition und gegenseitigen Abgrenzung der Begriffe Flexibilität, Elastizität und Robustheit, in: ZfbF 42(5/1990), 377-395

SCHNEIDER, H.: Effect of Service-Levels on Order-Points or Order-Levels in Inventory Models, in: Int.J.Prod. Res. 19(1981), 615 ff.

WEBSTER, D.B., TYBERGHEIN, M.B.: Measuring Flexibility of Job-Shop Layouts, in: Int. J. Prod. Res. 18/1(1980), S. 21-29

Applications of Operations Research in Hierarchical Production Planning

KLAUS-PETER KISTNER and MARION STEVEN

A. Hierarchical Production Planning and the Crisis of Operations Research in Production Planning

Production is one of the traditional fields of application of operations research. The following examples may indicate that decision models describing problems of production planning and control and deriving optimal solutions or good approximations have been developed quite early.

The first quantitative model of *inventory control* has been formulated by HARRIS in 1913; the structure of modern inventory models is based on the model of ARROW/HARRIS/MARSCHAK [1956] for the stochastic case, and on the model of WAGNER/WHITIN [1958] for the deterministic case.

Optimizing models of *scheduling* and *sequencing* have been introduced by MANNE [1959] for the *job shop scheduling*, by IGNALL/SCHRAGE [1965] for the *flow shop scheduling* problem; the *assembly line* problem has been considered by BOWMAN [1960]. A comprehensive survey of scheduling models has been given by CONWAY/MAXWELL/MILLER [1967]. Due to the importance of small batch sizes in the german industry, integrated models of lot sizing and *lot sequencing* met comparatively strong interest in the German literature (cf. ADAM [1963], DINKELBACH [1964], STÄHLY [1964]).

Applications of *input-output analysis* to material flows within the enterprise (cf. VASZONYI [1955], PICHLER [1966]) may be considered as one of the major roots of *material requirements planning*.

Last but not least, production planning is one of the traditional applications of *linear programming*. The importance of this technique for production planning has been pointed out by KOOPMANS already in 1951. Decision models for the planning of the production program on a detailed level, e.g. the choice of products to be produced and of processes to be used, the matching of production and demand, as well as capacity allocation have been considered by DORFMAN/SAMUELSON/SOLOW [1958, p. 130 ff.], KILGER [1973], JACOB [1971] and others. *Aggregate production planning* uses linear programming techniques (cf. HANSSMANN/HESS [1960]) as well as the approach of *linear decision rules* (cf. HOLT/MODIGLIANI/SIMON [1955], HOLT/MODIGLIANI/MUTH/SIMON [1960]).

If this amount of theoretical work stipulating applications of operations research in production planning, and the plethora of quantitative models describing various aspects of its problems are considered, its acceptance is quite disappointing: Despite occasional reports on real world applications of operations research, the impartial observer has to notice that these concepts are almost completely ignored in the practice of production planning. Traditional procedures were not replaced by OR-based, computer supported decision methods; instead, PPS-systems were introduced which emphasize the aspects of data handling, but neglect optimizing features and the modelling point of view. Usually, these systems do not search for optimal recommendations, but supply only feasible solutions, at best. It is quite obvious that traditional PPS-systems perform unsatisfactory; the completion-time syndrome is common to most of these systems (cf. ZÄPFEL/MISSBAUER [1988, S. 77], KISTNER/STEVEN [1990, S. 268]). Attempts to eliminate their failings do not incorporate integrated models, guaranteeing feasibility, and striving for optimality, but introduce local modifications, giving rise to new insufficiencies. Consequently, simple heuristics like KANBAN (cf. MONDEN [1983]) replace PPS-systems and any other attempt to base production planning and control on scientific methods and computer support.

In order to master this crisis of operations research and other quantitative methods in production planning, and to improve acceptance of these approaches, reasons for their failure have to be examined. In the past, lack of acceptance has been due to two major reasons:

(1) Insufficient communication between designers and users of a system and inadequate implementations of operations research models.

(2) Intrinsic difficulties of quantitative models in production planning.

Frequently, operations research is considered to be a special branch of mathematics; hence, standards of mathematical argumentation, esp. generality and brilliance of argumentation, dominate requirements of applied sciences, namely adequacy and applicability. They influence model building as well as presentation: presentation emphasizes technical assumptions, which might be important for the designer of quantitative models, but confuse the potential user, and intensify his doubts in the applicability of quantitative approaches.

Furthermore, the claim of operations research to provide "optimal solutions" for decision problems gives rise to misinterpretations of its significance: It suggests that the solution of an optimizing model is an imperative, which has to be followed by every rational decision maker. However, a manager, responsible for the results of his decisions, will not accept blindly this command, but will definitely reject a tool urging him to do so. In fact, it has to be stressed that quantitative models do not replace, but support decision making.

Even if these flaws in implementations of quantitative models are not longer relevant and will be avoided in future, they deeply influenced the image of operations research and its acceptance.

Intrinsic difficulties appropriate to the operations research approach are even more important: Except for linear programming, most of the models quoted above consider more or less single aspects of the production planning process. For problems of this type, efficient solution methods or at least good heuristics are available; they neglect, however, relations and interdependencies between various other aspects of production planning.

The importance of interrelations between different areas of production planning has been stressed by modern decision theory; insufficiencies of traditional PPS-systems support this notion. Linear programming models of production planning constitute an appropriate theoretical framework to consider these interdependencies. However, the integration of operative management - e.g. inventory planning and scheduling - into production planning on the tactical level - viz. the planning of the production program and capacity allocation - implies the introduction of binary variables. Hence, the complexity of monolithic models, incorporating both aspects of production planning, obstructs the applicability of this type of models.

(a) Usually, the effort needed to record and update all data required by monolithic models is considered to be prohibitive. But as PPS-systems, which are used quite frequently, already provide and update most of these operative data, this argument becomes more or less obsolete.

(b) It is well known that large mixed-integer programs are hard to solve. As operative control implies detailed modelling with a large number of variables and restrictions, monolithic models of production planning can usually not be solved numerically with reasonable efforts of time and computer capacity.

(c) Monolithic models assume centralized decision making and neglect established organizational structures. Decentralized departments tend, however, to impede the implementation of a decision system questioning their competence and authority.

Hierarchical production planning provides an encouraging approach to overcome these difficulties of implementation, and to bridge the gap between the potential of operations research methods and the requirements of their application. The basic idea of this procedure proposed by HAX/MEAL [1975] is to reduce problem complexity by decomposition of the planning process into isolated subproblems, which may be solved using appropriate models. Subproblems are coupled by few, controlled interfaces. Hence, the approach considers interdependencies appropriately and at the same time reduces problem complexity. For defining subproblems, existing organizational structures may be considered adequately.

Furthermore, PPS-systems may be integrated in order to use their data basis within the framework of hierarchical production planning: The coordination of data, which is achieved by PPS-systems, may be completed by the coordination of decision making attained by the hierarchical planning approach.

Methods of operations research are important tools of this procedure: As complexity of models is reduced and the amount of data needed is diminished, special models may be designed to solve subproblems of production planning and control adequately. They are coordinated using procedures of operations research. Furthermore, these methods can be applied to give a satisfactory theoretical foundation of the hierarchical planning process.

Since the seminal paper of HAX/MEAL [1975], a lot of research in hierarchical production planning has been achieved. Most papers report about case studies based more or less on the HAX/MEAL model; only few consideration is given, however, to general aspects of this approach: DEMPSTER et al. [1981] give a first survey of fields of application, and show that this type of problems may be modelled as a multistage stochastic program. STADTLER [1987] presents a survey of applications of operations research to solve different subproblems, SWITALSKI [1988,1989] resp. STEVEN [1989] and KISTNER/STEVEN [1989] analyze basic elements of this approach.

The aim of this paper is twofold: Some general aspects of hierarchical production planning will be examined and the importance of operations research for this approach will be stressed. The paper is organized as follows: First, the basic ideas will be sketched; in the second part, applications of operations research to solve planning problems which arise in subsystems of production control will be reviewed. Finally, some theoretical aspects of hierarchical production planning will be considered.

B. Hierarchical Production Planning and Operations Research

I. The Basic Ideas of Hierarchical Production Planning

1) Elements of Hierarchical Production Planning

The concept of hierarchical production planning is based on four elements:

(1) Decomposition of the planning process.
(2) Hierarchical structure of the decision system.
(3) Aggregation of data and decision variables.
(4) Coordination of decisions.

a) Decomposition

Decomposition of a planning process may be defined as the division of a complex decision problem into simple subproblems. It requires the identification of distinct planning areas with little interactions which may be coordinated by few, well-defined interfaces. In production planning, the following aspects may be used to separate subproblems:

(1) Length of the decision horizon and level of the decisions:

ANTHONY [1965] distinguishes between strategic, tactical and operative decisions; even if he stresses differences between this classification of decision levels and the distinction between long range planning, medium term decisions and short run control, there is a high degree of correspondence between both criteria.

(2) Organizational principles:

Workshops, assembly lines, autonomous manufacturing units, and flexible manufacturing systems may be considered as examples of subsystems defined according to the organization of the production process; product lines exemplify a separation according to marketing aspects.

(3) Properties of products:

The classical HAX/MEAL approach which combines articles to product families, sharing common set-up costs, and to product groups with similar unit costs, and which plans production of these aggregates separately, may be considered as an example for a decomposition with respect to properties of the products.

These aspects of decomposition are not exclusive, usually different principles have to be combined.

b) Hierarchical Structure of the Decision Process

Hierarchical planning is characterized by the fact that separated subproblems are linked by a hierarchical structure. In principle, coordination is achieved by top-down control: Decisions on a higher level set data for subordinate planning; tactical production planning defines targets and production requirements as well capacity allocations etc. which have to be met by operative control.

c) Aggregation

Aggregation is a procedure to simplify decision problems by condensing data and decision variables: It achieves not only a reduction of model complexity, and of the amount of data needed, but also a simplification of forecasting. Aggregation may be based on the following criteria:

(1) Aggregation with respect to *time*: The time grid of the decision process is condensed.

(2) Aggregation according to characteristics of the *production process*:

  (a) Aggregation may be performed with respect to properties of the *product*; an example of this strategy is the classical aggregation of articles to product families, and of product families to product groups, as proposed by HAX/MEAL [1975].

  (b) Instead, machines and *equipment* may be aggregated, too. In the context of hierarchical production planning, STEVEN [1989] considers the aggregation of machines to a flexible manufacturing system; an other example is the aggregation of machines to a workshop.

These strategies of aggregation may be used separately or in parallel. Choosing the criteria and the degree of aggregation, hierarchical production planning considers usually the following principles:

(1) The higher the hierarchical level, the higher is the degree of aggregation of data and decision variables; the length of the planning horizon increases, the time grid becomes coarser; on the superior levels, the range of competence is broader, whereas decisions are less detailed.

(2) In order to improve acceptance, aggregation should consider existing organizational structures; planning and decision making units should coincide.

(3) Furthermore, aggregation has to consider aspects of solubility: Models of decision problems and algorithms to solve them should be available, means of coordination should exist; and errors due to aggregation should be minimized.

d) Coordination

In principle, hierarchical planning is based on top-down control. However, infeasibilities on the lower levels, or errors due to disaggregation may give rise to modifications. According to the degree of interactions between higher and subordinated levels, SWITALSKI [1989] distinguishes the following mechanisms of coordination:

(1) Top-down control without any further coordination.
(2) Feedback within a rolling horizon setting.
(3) Coordination by costs and prices.
(4) Introduction of slack.
(5) Decision models of subordinate levels used to forecast their reactions on given targets.
(6) Tuning of decisions by coordination committees.
(7) Interactive planning and control.

The rolling horizons approach is an important tool to coordinate decisions in hierarchical planning systems: Infeasibilities on the subordinate level will be reported to the higher level, and trigger replanning at this level. In order to cope with different time grids at various levels of the decision process, the following modifications of the rolling horizon concept can be introduced:

(1) Simple two-stage rolling horizons:

At the superior level a rolling decision process with a coarse time grid is introduced. According to the usual procedure, decisions on the first period are considered to be fixed, whereas planning for the following periods is tentative. Only fixed decisions set data for the planning process on the subordinate level. According to the aggregation principles stated above, on this level a refined time grid is used; hence, targets and restrictions have to be broken down adequately. Based on disaggregated data for the refined time grid, a rolling decision process is introduced also on the subordinate level. Infeasibilities and divergences are reported to the higher level, and trigger replanning of tentative decisions on this level.

As the length of the fixed planning period at the higher level is given, the planning horizon at the subordinate level cannot be shifted according to the rolling horizon concept. So the length of its planning period is reduced in the course of time, and effects of artificial terminal conditions increase (STADTLER [1987]).

(2) Rolling horizon with anticipation:

In order to meet these problems, GÜNTHER [1985, 1986] suggests that three types of decisions on the superior level have to be considered:

- Fixed plans
- Frozen plans
- Tentative plans

The subordinate level can use frozen plans to anticipate future targets; hence, the planning horizon and the tentative decisions at this level can exceed the period limit of the fixed plan on the higher level.

2) Fields of Application of Hierarchical Production Planning

The basic model of HAX/MEAL [1975] and most case studies in the field of hierarchical production planning consider problems arising for mass production or for production of large batches. Usually they are solved by two stage models:

- Aggregate planning of the production program.
- Disaggregate planning of lot sizes.

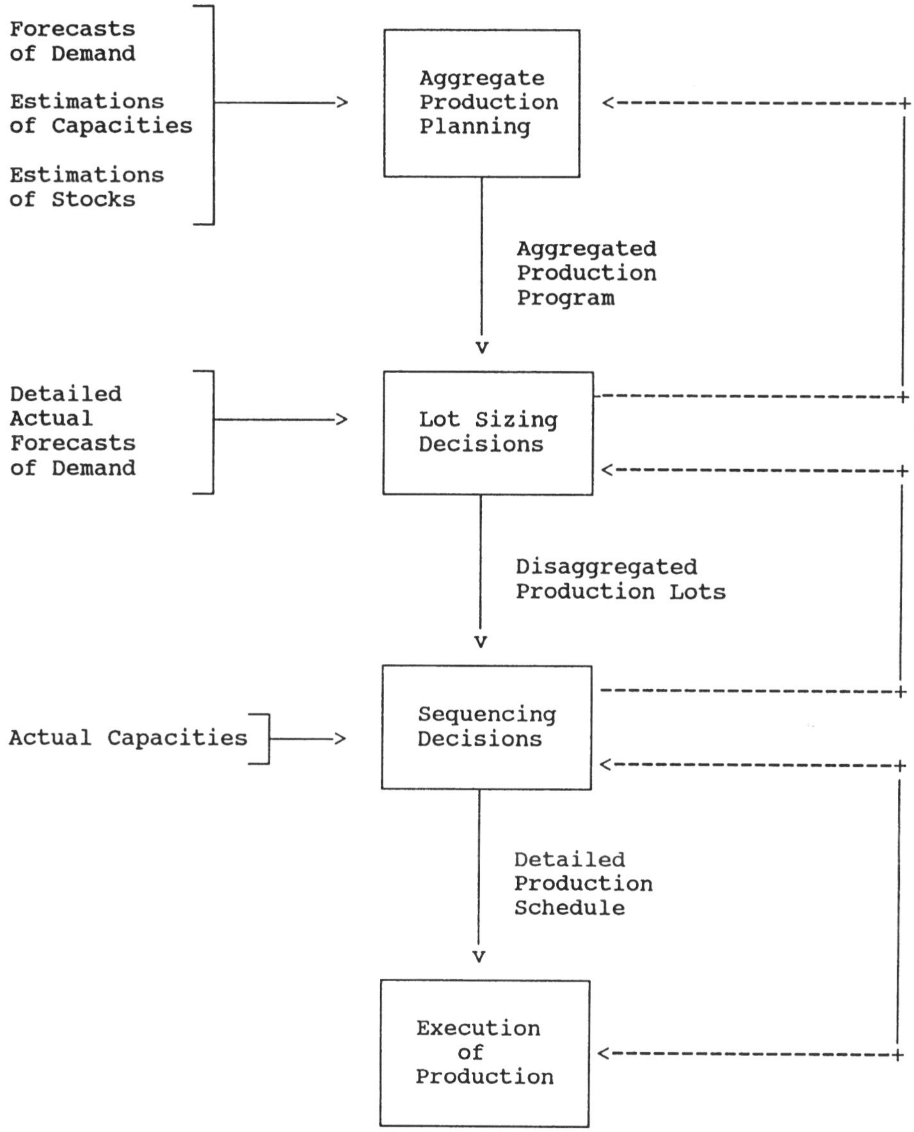

Figure 1: Three Stage Hierarchical Production Planning System

Hierarchical planning of small batches (cf. STADTLER [1987], SWITALSKI [1989]) usually requires three stage models:

- Aggregate planning of the production program.

- Disaggregate planning of lot sizes.
- Sequencing and scheduling of batches.

Both approaches rely on product-based aggregation; as an example of equipment-based models, STEVEN [1990] investigates a hierarchical programming model of flexible manufacturing systems which may be considered as aggregates of CNC- or DNC-machines. This model uses a two-stage approach:

- Assignment of orders or lots to a flexible manufacturing system.
- Scheduling of batches within the system..

3) The Process of Hierarchical Production Planning

Modules of a three stage hierarchical production planning model and flows of information in this system are shown in figure 1 above.

II. Applications of Operations Research in Hierarchical Production Planning

As a comprehensive survey of case studies and applications of operations research models in hierarchical production planning is given by STADTLER [1987], it will be sufficient in the following to sketch types of OR-models, which may be applied to solve subproblems on different levels of this planning system. The presentation follows the logic of Figure 1.

1) Aggregate Production Planning on the Tactical Level

Most hierarchical systems apply models of aggregate production planning to set targets for operative control, to allocate capacities, and to plan employment. Data and decision variables are aggregated as well as the time grid of the decisions. Usually, aggregate production planning covers a planning horizon of one year with decision periods of one month.

Aggregation simplifies the planning process in three aspects:

(1) Reduction of complexity by decreasing the amount of data and the number of decision variables.

(2) Simplification of forecasting by aggregation of data.

(3) Decoupling of lot sizing and scheduling makes it possible to avoid binary decision variables and integer programming formulations.

Due to the last aspect, linear programming models may be used to support aggregate production planning.

Furthermore, models of production smoothing, e.g. linear decision rules introduced by HOLT/MODIGLIANI/MUTH/SIMON [1955,1960] and by DEZIEL/EILON [1967], or the management coefficient rule suggested by BOWMAN [1963], may be applied in aggregate production planning.

2) Lot Sizing

Lot sizing requires that targets set by aggregate planning are disaggregated, and that demand or planned quantities of final products and parts are given for each subperiod. The hierarchical planning process achieves the following simplifications of the problem:

(1) Decoupling of scheduling problems.

(2) Separation of lot sizing from aggregate planning of production quantities.

(3) The problem of capital allocation is already solved by aggregate production planning.

Due to the first two aspects, special models of inventory theory may be applied. If capacity restrictions are disregarded, the dynamic programming algorithm proposed by WAGNER/WHITIN [1958] or simple lot sizing heuristics such as the part-period or the least unit cost method, the Silver-Meal heuristic (cf. SILVER/MEAL [1973]) or Groff's heuristic (cf. GROFF [1979]) may be applied. If capacity restrictions have to be considered, optimizing algorithms (cf. FLORIAN/KLEIN [1971])) usually will be too complex, but efficient heuristics like the Dixon-Silver heuristic (cf. DIXON/SILVER [1981]) are available. Using results of ZANGWILL [1966] and LOVE [1971], heuristics to solve multistage inventory models have been developed (cf. CROWSTON/WAGNER [1973], GRAVES [1981], HEINRICH [1987]). Furthermore, traditional approaches of material requirements planning may be applied (cf. ORLICKY [1975]).

3) Scheduling Problems

Predetermination of lot sizes reduces the problem of simultaneous lot sizing and sequencing to the classical problem of job shop or flow shop scheduling. Despite these simplifications, the problem remains hard in the sense of complexity theory.

Hence, optimizing models, usually formulated as mixed integer programs, can be solved only for small problems. Sometimes hierarchical planning techniques may be used to reduce scheduling problems to a manageable size. For example, ESCUDERO [1989] schedules a flexible manufacturing system in real-time, using MPSX.

In general, however, heuristics have to be applied to solve scheduling problems. Aside from priority rules and simple greedy heuristics, intelligent procedures and branch-and-bound heuristics are available.

III. Theoretical Foundation of Hierarchical Production Planning

Beside the importance of operations research for the solution of subsystems of hierarchical production planning, its significance for the theoretical foundation of this planning method has to be stressed. At least three aspects have to be considered:

(1) The decomposition principle of linear programming and its heuristic modifications to incorporate integer variables can be used to justify the iterative approach of hierarchical production planning.

(2) Results of aggregation theory show possibilities and limitations of complexity reduction by condensation of data.

(3) Models of operations research may be used as means of coordination.

1) Hierarchical Production Planning and Heuristic Decomposition

a) The Decomposition Principle of Linear Programming

It is well-known that the decomposition principle of linear programming introduced by DANTZIG/WOLFE [1960] may be interpreted as a coordination procedure between centralized planning units and decentralized departments (cf. BAUMOL/FABIAN [1964], H. HAX [1965]) which applies a pricing mechanism for the use of common capacities based on opportunity costs:

Starting with tentative opportunity costs for the use of scarce capacities, departments solve subproblems, which ignore central capacity restrictions, and report tentative department plans to the head office. These propositions are reduced according to capacity restrictions, and are combined to a central program; using dual variables of centralized planning, opportunity costs which are charging the use of capacities, are modified. Based on these modified costs, departmental plans are revised until no further improvement can be found. The procedure terminates after a finite number of steps with an optimal solution, if an optimal solution of the problem exists. In contrast to classical approaches of decentralized planning, the final decision has to be calculated by the head office as a convex combination of all departmental plans, derived in the course of the iterative process.

This procedure requires that all variables are real. To model set-ups and sequencing decisions, integer variables have to be introduced; hence, the approach cannot be applied to production planning immediately. Even if procedures like BENDERS' decomposition (cf. BENDERS 1962]) and LAGRANGIAN relaxation techniques (cf. GEOFFRION [1974]), may be used to partition the problem, the perfect coordination mechanism of the decomposition principle has to be replaced by heuristic procedures: The main ideas of decomposition - introduction of an hierarchy and coordination by opportunity costs as well as feed-back mechanisms - may, however, be applied to design appropriate decomposition heuristics.

b) Lagrangian Relaxation and Iterative Estimation of Opportunity Costs

These heuristic features of coordination in hierarchical production planning are stressed by the model of GRAVES [1982]. The basic structure of the model is equal to the HAX/MEAL-model: Targets for aggregated stocks of product families are set by tactical planning; detailed lot sizes for articles are determined by operative planning.

The coordination mechanism suggested by GRAVES starts with the determination of aggregated stocks, using a linear programming model. Using the WAGNER/WHITIN algorithm for the uncapacitated lot sizing problem or an appropriate heuristic, disaggregated lot sizes for each article are determined. In order to reduce differences between targets set by aggregate production planning and lot sizes for articles, these inconsistencies are charged with shadow prices derived by LAGRANGIAN relaxation techniques; essentially, these opportunity costs split holding costs between aggregate and disaggregate planning. Objective functions are modified accordingly, and aggregate and disaggregate decisions are adapted. This process is continued until remaining inconsistencies are tolerable.

A similar approach using BENDERS' cuts has recently been proposed by AARDAL/LARSSON [1990].

In contrast to the decomposition of linear programs, these approaches are heuristic and do not guarantee optimality. This is due to the following reasons:

(1) Integer variables cause a duality gap and prevent convergence of multipliers to exact opportunity costs; hence, matching of targets for stocks, set by tactical planning, and lot sizes, determined by operative control, is not guaranteed, and the optimal solution may be missed.

(2) Furthermore, the duality gap results in quite fuzzy termination criteria.

(3) Errors due to aggregation and disaggregation cannot be avoided.

2) Aggregation Theory

This observation leads to the second aspect of a theoretical foundation of hierarchical planning systems: In order to reduce complexity, data and decision variables on the upper levels have to be aggregated. Hence, to get insight into possibilities and limitations of this approach, results of aggregation theory have to be considered.

Initially, aggregation theory has been developed within the framework of macroeconomics (cf. FISHER [1969], EICHHORN [1978]). AXSÄTER et al. [1979, 1981, 1983, 1984, 1986] recognized its importance for managerial planning on an aggregate level and applied its results to hierarchical production planning.

The notion of *perfect aggregation* is one of the basic concepts of aggregation theory; in the context of decision making, it demands that planning based on aggregate data should yield the same results as the aggregation of decisions based on detailed data. Unfortunately, perfect aggregation is guaranteed only under very strict conditions: If products are aggregated, each product in a product group has to be assigned to the same group of machines; hence, results on perfect aggregation have occasionally been interpreted as "impossibility theorems" (WENIG [1973]). However, in the case of aggregation with respect to equipment, these conditions may easily be met: They are equivalent to the established principle that one product should be completely processed using just one flexible manufacturing system, or one manufacturing unit.

The concept of perfect aggregation may be of interest in macroeconomics; the claim that decision making and aggregation should be commutative operations seems to be of minor importance in management. In order to establish optimality of decisions, *exact aggregation* has to be guaranteed, that means, disaggregation of plans based on aggregate data should be identical with decisions based on disaggregated data. As aggregation always involves loss of information, exact aggregation will be possible only in trivial cases, when data lost in the process are irrelevant.

Therefore, aggregation theory has to concentrate on infeasibilities caused by inevitable aggregation errors and their consequences on decisions and objectives: FONTAN/IMBERT/MERCE [1985] and ERSCHLER/FONTAN/MERCE [1986] consider the problem of *consistency of aggregation* in a rolling horizon surrounding: A disaggregated production plan for a given period is consistent, if demand for each product can be settled in this period, and if feasibility of aggregated plans for the following periods is not effected. They show that the knapsack algorithm of BITRAN/HAAS/HAX [1982] in connection with the look-ahead rule guarantees consistency.

These results of aggregation theory show clearly the limitations of the procedure; they give, however, no general insight into the question how aggregation should be performed. This can be done only in a heuristic way depending on the problem at hand. Due to the great number of case studies, examples for various aggregation techniques are available.

3) Mechanisms of Coordination

The last aspect of applications of operations research in hierarchical production planning is coordination of decisions on various levels. One of the major problems to be solved is capacity allocation: Aggregate production planning needs informations on capacities available. Because of set-up times and idle times, net capacities are dependent on lot sizing and sequencing decisions, which are results of decisions on the operative level. Planning on this level has to rely, however, on targets to be set by aggregate production planning on the tactical level. To break this circle inherent in sequential planning, two types of approaches may be applied:

(1) Iterative approaches to reduce inconsistencies in a finite number of steps.

(2) Application of models of the subordinate level to estimate data needed for decisions on the higher level.

a) Coordination by Iterative Approaches

Coordination of decisions by heuristic decomposition as introduced by GRAVES [1982] and AARDAL/LARSSON [1990] have been analyzed above. The knapsack algorithm of BITRAN/HAX [1981] may be considered as another example of iterative approaches.

On the other hand, GELDERS/VAN WASSENHOVE [1982] proposed steering committees and computer based interactive coordination procedures.

b) Coordination Based on Decision Models

As mentioned above, DEMPSTER et al. [1981] showed that hierarchical production planning may be considered as a two-stage stochastic programming problem. This approach is primarily interesting from a theoretical point of view, as the compensation model of stochastic programming is not operable for large problems.

However, chance constrained programming may be applied to anticipate the probable behaviour of subordinate planning and to derive estimations of net capacities (cf. SWITALSKI [1989]): As idle times and set-up times are still unknown at the time of aggregate production planning, net capacities available on the operative level may be considered as random variables. Using data from past periods, distributions of net capacities and their parameters may be estimated. Hence, a linear programming model of aggregate production planning may be modified into a chance constrained model, requiring that capacity allocated must not exceed net capacities available with a given probability α. It is well-known that the deterministic equivalent of a chance constrained model with a stochastic right hand side is a linear program; its restriction coefficients equal their mean values reduced by a security deduction depending on the probability α (cf. CHARNES/COOPER [1959]). Hence, the chance constrained model gives a theoretical justification of coordination by slack.

Queueing theory supplies another type of stochastic models which may be used to estimate net capacities on the operative level. A similar problem arises in the ex ante evaluation of flexible manufacturing systems; a basic approach modelling flexible manufacturing systems as queueing networks (cf. SOLBERG [1977]), may be modified and applied in hierarchical production planning (cf. KISTNER/STEVEN [1990]): From the point of view of aggregate planning, the exact product mix is unknown. Hence, it may be assumed that processing times of articles are independent exponentially distributed random variables, that there are constant proportions between articles over time, and that each type of article has to be routed differently. This type of a job shop may be modelled by Markovian queueing networks. Open systems with Poisson-input may be applied to derive measures of performance of a production system as a function of the input rate, that is the desired employment of capacities; closed queueing networks may be applied to determine the maximum throughput. A different approach has recently be proposed by DE KOSTER [1989].

In order to derive general solutions for queueing networks, very strict assumptions have to be introduced; more general models of the behaviour on the operative level may be solved, however, using simulation models.

C. Results

Various methods of operations research may be applied to solve problems arising in hierarchical production planning:

Operations research models are applied for detailed planning on the operative level. Complexity is reduced by separation of decisions on the operative and tactical levels, and by splitting lot-sizing and sequencing decisions on the operative level. This facilitates application of traditional models such as linear programming for aggregate production planning and capacity allocation, or inventory theory and scheduling.

Furthermore, methods of operations research may be applied to coordinate different levels of an hierarchical decision process; for example, approaches of heuristic decomposition as well as the knapsack algorithm may be applied for tuning decisions on the tactical and the operative levels of the planning process. For tactical planning, stochastic models such as chance constrained programming, and queueing networks may be applied to estimate net capacities, available on the operative level.

Last not least, results of operations research concerning decomposition and aggregation supply theoretical foundations of hierarchical production planning.

References

Aardal, K., Larsson, T., A Benders Decomposition Based Heuristic for the Hierarchical Production Planning Problem. EJOR 45 (1990) 4-14

Adam, D., Simultane Ablauf- und Programmplanung bei Sortenfertigung mit ganzzahliger linearer Programmierung. ZfB 33 (1963) 233-245

Andler, K., Rationalisierung der Fabrikation und optimale Losgröße. München 1929

Anthony, R.N., Planning and Control Systems: A Framework for Analysis. Cambridge/Mass. (Harvard University Press) 1965

Arrow, K.T., Harris, T., Marschak, J., Optimal Inventory Policy. Econometrica 19 (1951), 250-272

Axsäter, S., On the Design of the Aggregate Model in a Hierarchical Production Planning System. Engineering and Process Economics 4 (1979), 89-97

Axsäter, S., Aggregation of Product Data for Hierarchical Production Planning. OR 29 (1981) 744-756

Axsäter, S., On the Feasibility of Aggregate Production Plans. OR 34 (1986) 796-800

Axsäter, S., Jönsson, H., Thorstenson, A., Approximate Aggregation of Product Data. Engineering Costs and Production Economics 7 (1983) 119-126

Axsäter, S., Jönsson, H., Aggregation and Disaggregation in Hierarchical Production Planning. EJOR 17 (1984) 338-350

Baumol, W.J., Fabian, T., Decomposition Principle for Decentralisation and External Economies. MS 11 (1964) 1-32

Benders, J.F., Partitioning Procedures for Solving Mixed-Variables Programming Problems. Numerische Mathematik 4 (1962) 238-252

Bitran, G.R., Hax, A.C., Disaggregation and Resource Allocation Using Convex Knapsack Problems with Bounded Variables. MS 27 (1981) 431-441

Bowman, E.H., The Schedule-Sequencing Problem. OR 7 (1959) 621-624

Bowman, E.H., Assembly-Line Balancing by Linear Programming. OR 8 (1960) 385-389

Bowman, E.H., Consistency and Optimality in Managerial Decision Making. MS 9 (1963) 310-321

Charnes, A., Cooper, W.W., Chance-Constrained Programming, MS 6 (1959) 74-79

Conway, R.W., Maxwell, W.L., Miller, L.W., Theory of Scheduling. Reading/Mass. (Addison-Wesley) 1967

Crowston, W.B., Wagner, M., Dynamic Lot Size Models for Multi-Stage Assembly Systems, MS 20 (1073) 14-21

Dantzig, G.B., Wolfe, P., Decomposition Principle for Linear Programs. OR 8 (1960) 101-110

de Koster, M.B.M., Capacity Orientated Analysis and Design of Production Systems. Berlin-Heidelberg-New York (Springer) 1989

Dempster, M.A.H., Fisher, M.L., Jansen, L., Lageweg, B.J., Lenstra, J.K., Rinnoy Kan, A.H.G., Analytical Evaluation of Hierarchical Planning Systems. OR 29 (1981), 707-716

Deziel, D.P., Eilon, S., A Linear Production-Inventory Control Rule. The Production Engineer 46 (1967) 93-104

Dinkelbach, W., Ein Problem der Produktionsplanung in Ein- und Mehrproduktunternehmen. Würzburg-Wien (Physica) 1964

Dixon, P.S., Silver, E.A., A Heuristic Solution Procedure for the Multi-Item, Single-Level, Limited Capacity, Lot-Sizing Problem. Journal of Operations Management 2 (1981) 23-39

Dorfman, R., Samuelson, P.A., Solow, R.M., Linear Programming and Economic Analysis. New York (McGraw Hill) 1958

Eichhorn, W., Functional Equations in Economics. Reading/Mass. (Addison-Wesley) 1978

Eilon, S., Five Approaches to Aggregate Production Planning. AIIE Transactions 7 (1975) 118-131

Eisenhut, P.S., A Dynamic Lot Sizing Algorithm with Capacity Constraints. AIIE Transactions 7 (1975) 170-176

Erschler J., Fontan, G., Merce, C., Consistency of the Disaggregation Process in Hierarchical Planning. OR 34 (1986) 464-469

Escudero, L.F., An Inexact Algorithm for Part Input Sequencing and Scheduling with Side Constraints in FMS. The International Journal of Flexible Manufacturing Systems 1 (1989) 143-174

Fisher, F.M., The Existence of Aggregate Production Functions. Econometrica 37 (1969), S. 553-577

Fontan, G., Imbert, S., Merce, C., Consistency Analysis in a Hierarchical Production Planning System. Engineering Costs and Production Economics 9 (1985) 193-199

Gelders, L.F., van Wassenhove, L.N., Hierarchical Integration in Production Planning: Theory and Practice. Journal of Operations Management 3 (1982) 27-35

Geoffrion, A.M., Lagrangian Relaxation for Integer Programming. Mathematical Programming Study 2 (1974) 82-114

Giffler B., Thompson, G.L., Algorithms for Solving Production-Scheduling Problems. OR 8 (1960) 487-503

Graves, S.C., Multi-Stage Lot Sizing: An Iterative Procedure. in: Schwarz, L.B., (ed.), Multi-Level Production and Inventory Control Systems. Amsterdam (North Holland) 1981, 95-110

Graves, S.C., Using Lagrangean Techniques to Solve Hierarchical Production Planning Problems. MS 28 (1982) 260-275

Groff, G.K., A Lot-Sizing Rule for Time-Phased Components Demand, Prod. and Inv. Management 20 (1979), S. 47-53

Günther, H.-O., Revidierende Produktionsplanung bei Sortenfertigung. Operations Research Proceedings Proceedings 1984, Berlin-Heidelberg-New York (Springer) 1985, S. 140-147

Günther, H.-O., The Design of an Hierarchical Model for Production Planning and Scheduling. in: Axsäter, S. et al. (Ed.), Multi-Stage Production Planning and Inventory Control. Berlin-Heidelberg-New York (Springer) 1986, S. 227-260

Hanssmann, F., Hess, S.W., A Linear Programming Approach to Production and Employment Scheduling. Management Technology 1 (1960) 46-51

Harris, F.W., How Many Parts to Make at Once. Factory, The Magazine of Management 10 (1913) 135-136, 152

Hax, A.C., Meal, H.C. Hierarchical Integration of Production Planning and Scheduling. in: Geisler M.A. (Ed.), Logistics, TIMS Studies in the Management Sciences, Amsterdam (North Holland) 1975, 53-69

Hax, H., Die Koordination von Entscheidungen. Köln (Heymanns) 1965

Heinrich, C.E., Mehrstufige Losgrößenplanung in hierarchisch strukturierten Produktionsplanungssystemen. Berlin-Heidelberg-New York (Springer) 1987

Heller, J., Logemann, G., An Algorithm for the Construction and Evaluation of Feasible Schedules. MS 8 (1961) 168-183

Holt, C., Modigliani, F., Simon, H.A., A Linear Decision Rule for Production and Employment Scheduling. MS 2 (1955) 1-30

Holt, C., Modigliani, F., Muth, J.F., Simon, H.A., Planning Production, Inventories, and Work Force. Englewood Cliffs (Prentice Hall) 1960

Ignall, E.J., Schrage, L., Application of the Branch and Bound Technique to Some Flow-Shop Scheduling Problems. OR 13 (1965) 400-412

Jacob, H., Investitionsplanung und Investitionsentscheidung mit Hilfe der Linearprogrammierung. Wiesbaden (Gabler), 2. Aufl. 1971

Kilger, W., Optimale Produktions- und Absatzplanung. Opladen (Westdeutscher Verlag) 1973

Kistner, K.-P., Switalski, M., Hierarchische Produktionsplanung. ZfB 59 (1989) 477-503

Kistner, K.-P., Steven, M., Produktionsplanung. Heidelberg (Physica) 1990

Kistner, K.-P., Steven/Switalski, M., Warteschlangen-Netzwerke in der hierarchischen Produktionsplanung. OR-Spektrum 12 (1990) 89-102

Florian, M., Klein, M., Determinstic Production Planning with Concave Costs and Capacity Constraints. MS 18 (1971) 12-20

Koopmans, T.C., Analysis of an Efficient Combination of Activities. in: Koopmans, T.C., Activity Analysis of Production and Allocation. New York (J. Wiley) 1951, S. 33-97

Lambrecht, M.R., Vanderveken, H., Heuristic Procedures for the Single Operation, Multi-Item Loading Problem. AIIE Transactions 11 (1979) 319-326

Love, S.F., A Facilities in Series Inventory Model with Nested Schedules. MS 18 (1971) 327-338

Manne, A.S., On the Job-Shop Scheduling Problem. OR 8 (1960) 219-223

Monden, Y., Toyota Production System: Practical Approach to Production Management. Atlanta (Inst. Industrial Engineering) 1983

Orlicky, J., Material Requirements Planning, New York (McGraw Hill) 1975

Pichler, O., Anwendungen der Matrizenrechnung bei der Betriebskostenüberwachung. in: Adam, A. u.a. (Ed.), Anwendungen der Matrizenrechnung auf wirtschaftliche und statistische Probleme. Würzburg (Physica) 1966, S. 74-111

Silver, E.A., Meal, H.C., A Heuristic for Selecting Lot Size Quantities for the Case of Deterministic Time Varying Demand Rate and Discrete Opportunities for Replenishment. Prod. and Inv. Management 14 (1973) No. 2, 64-74

Solberg, J.J., A Mathematical Model of Computing Manufacturing Systems. Proceedings of the Fourth International Conference on Production Research, Tokyo 1977, 1265-1275

Stadtler, H., Hierarchische Produktionsplanung bei losweiser Fertigung. Heidelberg (Physica), 1988

Stähly, P., Kurzfristige Fabrikationsplanung in der industiellen Werkstattfertigung, Würzburg (Physica) 1964

Steven, M., Hierarchische Produktionsplanung für flexible Fertigungssysteme. ZfbF 41 (1989) 1029-1047

Switalski, M., Hierarchische Produktionsplanung und Aggregation. ZfB 58 (1988) 381-396

Switalski, M., Hierarchische Produktionsplanung. Heidelberg (Physica) 1989

Vaszonyi, A., The Use of Mathematics in Production and Inventory Control, MS 1 (1955) 70-85

Wagner, H.M., Whitin, T.M., Dynamic Version of the Economic Lot Size Model. MS 5 (1958) 89-96

Wenig, A., Über das Aggregationsproblem in der ökonomischen Theorie. in: Grimm et al. (Ed.), Prismata, Pullach (Verlag Dokumentation) 1973, 361-367

Zangwill, W.I., A Deterministic Multiproduct, Multifacility Production and Inventory Model. OR 14 (1966) 486-507

Zäpfel, G., Missbauer, H., Traditionelle Systeme der Produktionsplanung und -steuerung in der Fertigungsindustrie. WiSt 17 (1988) 73-77

Hierarchical Approach to Production Planning and Scheduling: The Case of a Mosaic Tile Plant

MARIO TABUCANON and ANUKUL KONGRIT

I. INTRODUCTION

This paper describes a production planning and scheduling model developed for a large mosaic tile plant. The company has over 20 overseas customers. In the ordering process, the company gives a monthly production quota for each customer. The order in which each customer uses his quota is called "fixed date order" because the customer can determine the delivery date by himself. The order that does not use the quota system is called "tentative order" because the customer can only determine the delivery month. The delivery date of this order type will still have to be confirmed by the company.

The manufacturing process (Fig. 1) can be generally considered as composed of 5 major stages namely, preparation of dry powdered body; preparation of glaze; dry pressing and glazing; firing in tunnel kiln; and sorting, sticking and packing. The main value-adding stages considered in this paper are pressing, glazing and firing (see Fig. 2). Two tunnel kilns with capacities of 55 and 53 kiln cars/day are served by 4 pressing machines, 2 of 14,400 saggers/16 hrs. and 2 of 18,000 saggers/16 hrs. capacities. Sagger is the mosaic container for firing made from refractory. Each pressing machine is served by 3 stations of glazing machine for base coat, color coat and top coat. A kiln car is divided into zones (see Fig. 3). The total capacity of each kiln car is 1,140 saggers.

There are many types of products, i.e. 28 sizes, each having 7 colors, 9 shades and 7 glazing types. On the average, 80 product types are ordered each month. The management of the company has realized that the existing manual intuitive planning procedure based on experience is inadequate. There are several major factors attributing complicacy of the production

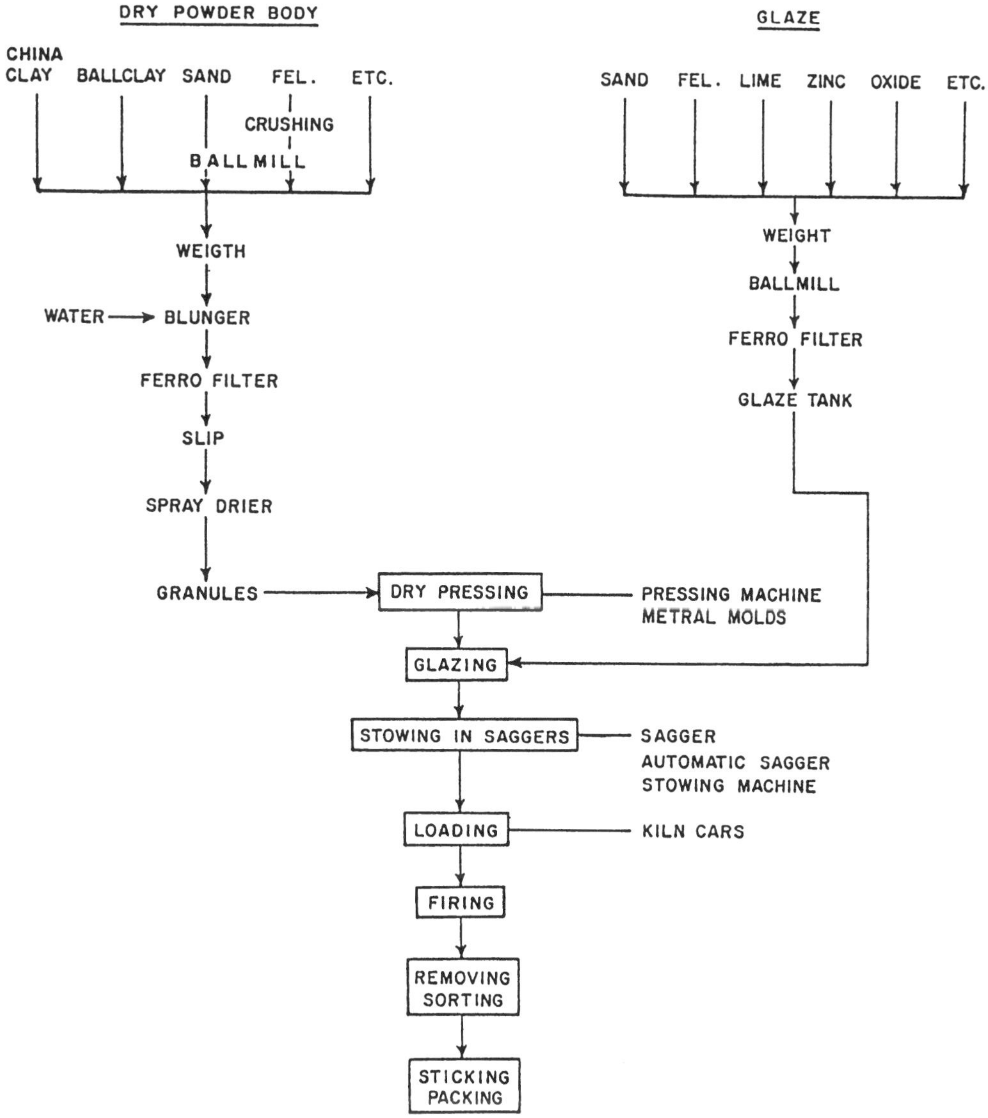

Fig. 1 Production Flow Process Chart

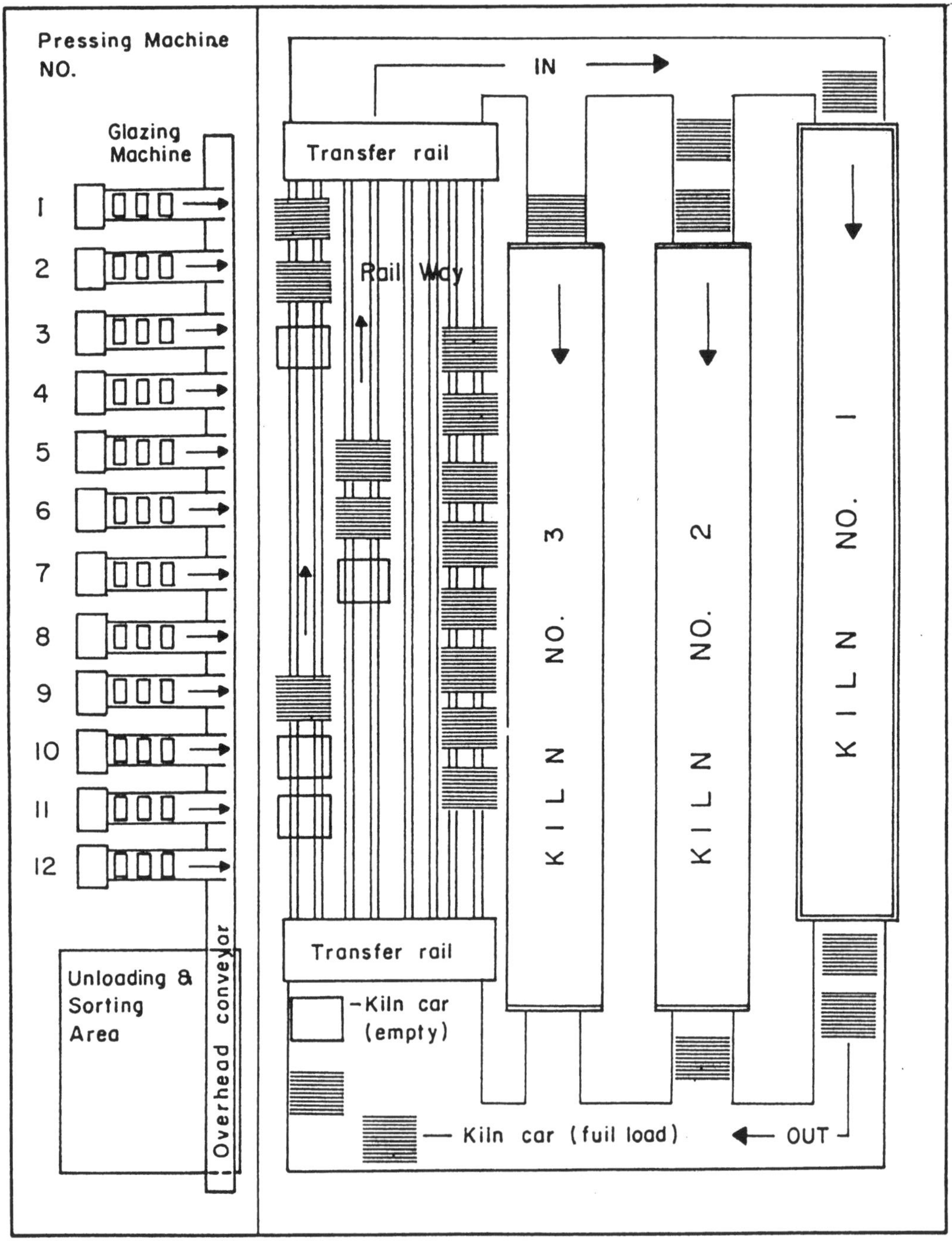

Fig. 2 Facility Layout

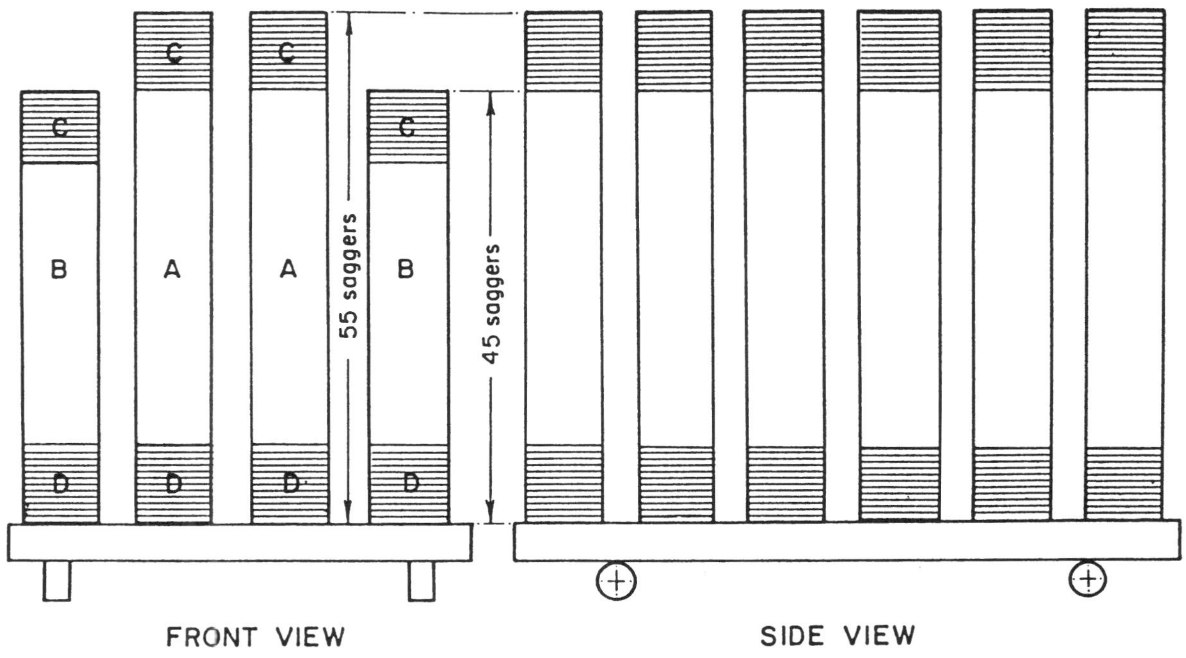

Inner	zone (a)	contains	480	saggers
Outer	zone (b)	contains	420	saggers
Top	zone (c)	contains	120	saggers
Bottom	zone (d)	contains	120	saggers
		Total	1,140	saggers / kiln car

Fig. 3 Kiln Car Zoning

planning procedure namely:

1) Pressing and glazing machines setup due to wide product range.
2) Color control; mosaic tile color is sensitive to heat transfer in kiln. Coloring is influenced by zoning in the kiln car.
3) Facility capacity.
4) Customer constraint (priority, demand and due date).

II. PROBLEM DESCRIPTION

A major problem in the existing system is in production planning. The management would like to maximize sales contribution, minimize production cost, and satisfy customers by delivering finished products on time. For these objectives, the following salient aspects had to be taken into account:

a) The factory is continuously working for 24 hours per day and 6 days per week. Kiln operation is stopped at the end of the week for preventive maintenance. In this case, the kiln temperature is reduced and held at a certain level for purposes of energy savings and easy start up.

b) For each product the following information are known:

- Order on hand for the next 3 months;
- Every product can be pressed on any machine;
- Allocation of products to kiln zones depends on their property of glazing chemical; and
- In all production facilities, sagger is the equivalent unit.

c) For each pressing machine the setup cost is known.

d) For each glazing machine the sequence-dependent setup cost is also known.

e) Pressing and glazing machines are independent in terms of setup.

f) By management policy, pressing and glazing machines are working in 2 shifts.

g) Kiln is the bottleneck in the production process.

This paper addresses planning and scheduling in a normal situation, i.e. not considering time of shortage of raw materials and machine breakdowns. The kiln constraints and the costs are considered to be known and constant. The firing loss for each type of product, as observed from experience, is assumed constant throughout the planning horizon.

III. PROBLEM ANALYSIS

Integer Linear Programming (ILP) can be employed in the first step, i.e. the 3-month planning problem. It selects a subset from the set of available orders on hand to be produced in each month for the next 3 months. In this step, the objective is that of maximizing sales contribution subject to demand requirement and plant capacity constraints.

The next step is determination of the monthly plan which can also be formulated as ILP. More specific constraints are considered such as machine setup, inventory holding, machine capacity, etc. The selected demand, output from the 3-month plan, is used as input data in this step. The objective here is to minimize total production cost. The mathematical model formulation is shown in problem P1:

$$\text{Min} \quad \sum_m s_m \left(\sum_i \sum_t \delta_{i,m,t}\right) + \sum_m c_{j,k} \sum_t \left(\sum_{j=1}^{c} \sum_{k=1}^{c} Y_{j,k,t}\right) \sigma_{c,m,t}$$

$$+ F\left[\sum_{\substack{z=1 \\ z \neq max}}^{z} \left((A_z)_{max} - A_z\right)\right] + h\, I_{i,c,t} \quad \cdots\cdots\cdots \qquad \text{(P1)}$$

Subject to

- Zone capacity constraint

$$\sum_{c \varepsilon S_z} \left(\sum_i \left(\sum_m x_{i,c,z,m,t}\right)\right) \leq L_{z,t} \qquad (1)$$

for all z and t

- Inventory constraint

$$I_{i,c,t-1} + \sum_m \sum_z x_{i,c,m,z,t} - I_{i,c,t} = d_{i,c,t} \qquad (2)$$

for all i,t and $c \varepsilon S_z$

- Pressing machine capacity constraint

$$\sum_i \sum_c \sum_z x_{i,c,m,z,t} \leq l_{m,t} \qquad (3)$$
$$c \varepsilon S_z$$

for all m and t

- Kiln firing constraint

$$A_z - r_z \left[\sum_{c \varepsilon S_z} \left(\sum_i x_{i,c,z,m,t}\right)\right] = 0 \qquad (4)$$

for all z

- Glazing machine setup constraint

$$\sum_{j=1}^{c} Y_{j,k,t} = 1 \qquad (5)$$

for all t and k, k=1,2,...,c

$$\sum_{k=1}^{c} Y_{k,j,t} = 1 \qquad (6)$$

for all t and j, j=1,2,...,c

where

$x_{i,c,z,m,t}$ = production of product i color c in zone z on machine m at time t

$d_{i,c,t}$ = demand of product i color c at time t

$l_{m,t}$ = capacity of machine m at time t

$I_{i,c,t}$ = inventory of product i color c at time t

s_m = set up cost of machine m

$c_{j,k}$ = set up cost of glazing machine from changing color j to color k

F = firing loss per kiln car

r_z = 1/(number of saggers in zone z in a kiln car)

S_z = set of products that can be produced in zone z

h = holding cost per unit

$\delta_{i,m,t} = 1$ if $\sum_z \sum_c (x_{i,c,z,m,t}) > 0$

$= 0$ otherwise

$\sigma_{c,m,t} = 1$ if $\sum_z \sum_i (x_{i,c,z,m,t}) > 0$

$= 0$ otherwise

$A_z = r_z \left[\sum_{\substack{c \\ c \in Sz}} \left(\sum_i \left(\sum_m x_{i,c,z,m,t} \right)\right)\right]$

$Y_{j,k} = [0,1]$, $i=1,2,\ldots,n$, $m=1,2,\ldots,8$, $z=1,2,\ldots,6$, $c=1,2,\ldots,C$, $t=1,2,\ldots,T$.

The unit production cost is assumed constant for each item. Therefore, the total production cost (excluding setup cost) is constant and hence not included in the model.

This type of problem is known to be NP-hard since the single item capacitated dynamic lot sizing problem is already NP-hard problem (Manne 1958). Optimal solution methods have failed to solve all but very small problems with reasonable computation times.

From the problem formulation two objectives are apparent: Minimize firing loss at the kiln; and Minimize total cost, i.e. holding cost plus setup cost. It is obvious that the problem can be broken down by considering the bottleneck operation first, that is the kiln firing operation. The purpose is to find an appropriate plan for allocating the range of products to kiln zones to minimize firing loss subject to kiln firing constraints. In order to minimize firing loss, every kiln car should be loaded at full capacity to avoid using dummy saggers as well as to balance degree of utilization among kiln zones. The objective can be addressed by minimizing the utilization variation among kiln zones.

Further analysis of this problem involves finding the optimal lot size for each product. There are 2 sets of 4 machines for 3 zones in each kiln; so this case is a multi-item, multi-machine capacitated lot sizing problem. Another complication is machine setup. Product size can be changed through setup at the pressing machine, and change in color is done through setup at the glazing machine. It is known that setup cost at the pressing machine is higher than at the glazing machine by more than 20 times so the setup cost at the glazing machine can be relaxed in this stage but will be considered later in determining the sequence of color setup in each lot after the suitable lot size is known.

The multi-item, multi-machine capacitated lot sizing problem (P1) can now be stated as follows:

$$\text{Min. } \sum_m S_m\left(\sum_i \sum_t Y_{i,m,t}\right) + h\, I_{i,z,t} \quad \ldots\ldots\ldots \quad \text{(P2)}$$

Subject to

- Demand constraint

$$I_{i,z,t-1} + \sum_m X_{i,m,z,t} - I_{i,z,t} = d_{i,z,t} \qquad (7)$$

for all i,z,t

- Capacity constraint

$$\sum_i \sum_z X_{i,m,z,t} \leq L_{m,t} \qquad (8)$$

for all m,t

- Setup constraint

$$M_{i,m,t} \quad Y_{i,m,t} \geq \sum_z X_{i,m,z,t}$$

$$M_{i,m,t} \geq 1 \qquad (9)$$

for all i,m,t

Linearize (9) by the equation as follows:

$$\sum_z x_{i,m,z,t} \leq \left(\sum_{t=1}^{T} \sum_z d_{i,z,t}\right) Y_{i,m,t} \qquad (9.1)$$

for all i,m,t

(i=1,2,...,n), (m=1,2,...,4), (z=1,2,3), (t=1,2,...,T)

$X_{i,m,z,t} \geq 0$, $I_{i,z,t} \geq 0$, $Y_{i,m,t} = (0,1)$

where

$X_{i,m,z,t}$ = production of product i in zone z on machine m at time t

$d_{i,z,t}$ = demand of product i in zone z at time t

$L_{m,t}$ = capacity of machine m at time t

$I_{i,z,t}$ = inventory of product i in zone z at time t

S_m = setup cost of machine m

h = inventory holding cost

$Y_{i,m,t}$ = 1 if there is a setup for product i on machine m at time t

$(\sum_z X_{i,m,z,t} > 0)$

= 0 otherwise

This model is also NP-hard. Moreover, the result from this model may not have the possibility of implementation without incurring extra redesign cost. The output from this model will necessitate the redesigning of facility layout and the kiln car loading sequence. The firm has already allocated fixed machines to zone and has designed the flow of materials to support this policy. The existing machine-to-zone allocation is shown in Table 1 and Figure 4.

Table 1 Pressing Machines and Kiln Zones Capacity Balancing

kiln no. / zone	requirement* / day	machine* capacity	slack*	utilization
1 top/bot	13200	14400	1200	92 %
in	26400	18000+9000	600	98 %
out	23100	14400+9000	300	99 %
2 top/bot	12720	14400	1680	89 %
in	25440	18000+9000	1560	94 %
out	22260	14400+9000	1140	95 %

average utilization = 95 %

*unit in sagger

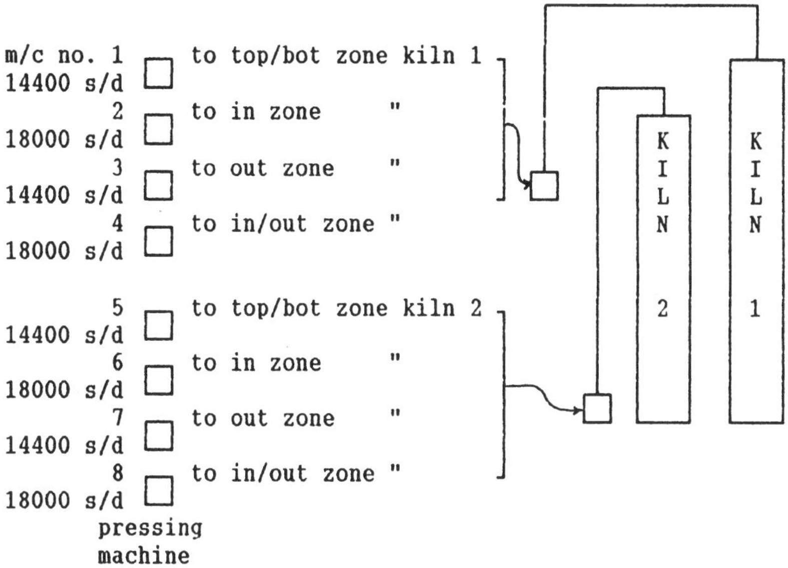

*** s/d = saggers/day

Figure 4 Material Flow Diagram

Eighty small examples using the existing demand patterns were tested by using ILP problem (P2) and the results were compared against single machine multiple item problem solved by Lambrecht and Vanderveken heuristic (see more details in next section). The results are shown in Figures 5 to 8.

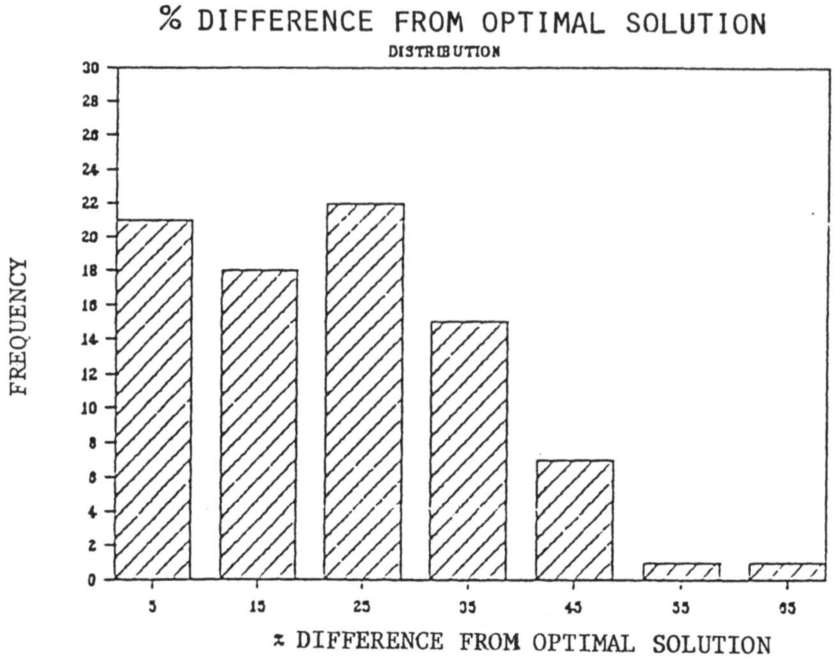

Figure 5 Distribution of Percent Difference from Optimal Solution of Heuristic

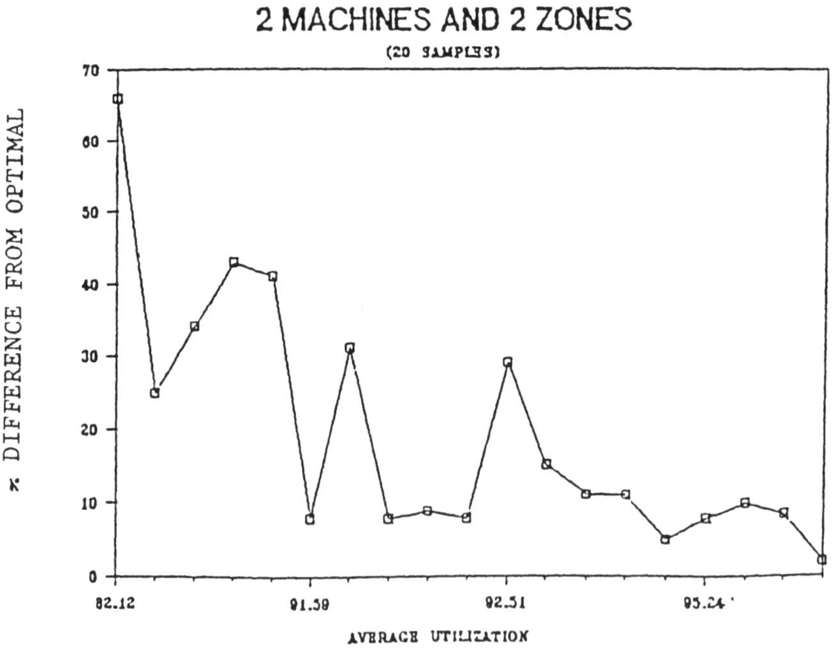

Figure 6 Percent Difference from Optimal Solution of Heuristic Versus Average Utilization of Machines (2M/2Z)

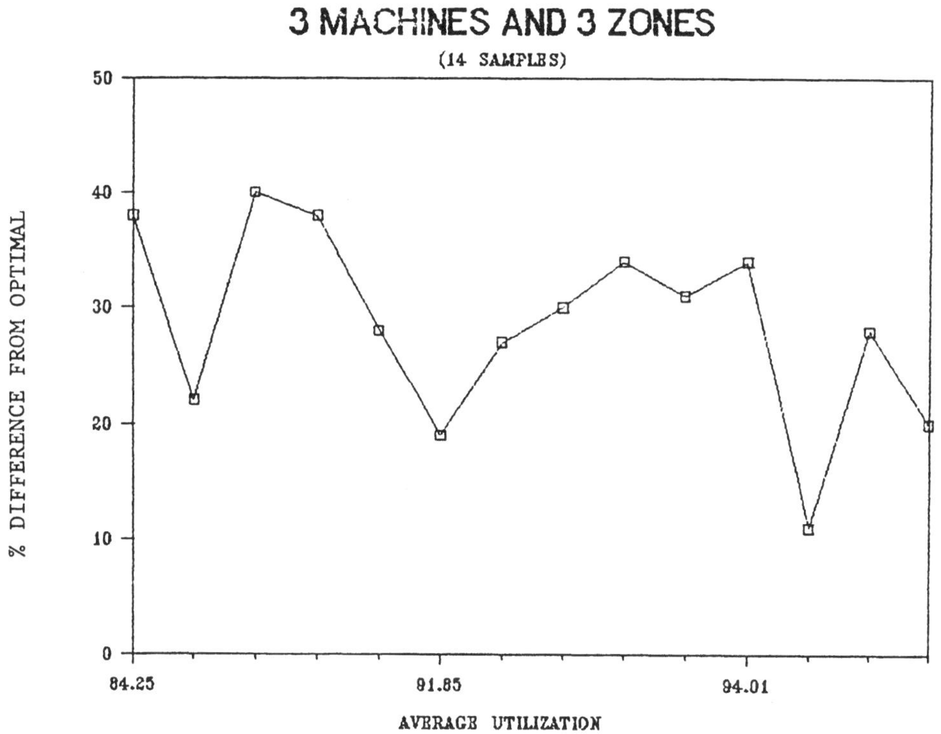

Figure 7 Percent Difference from Optimal Solution of Heuristic versus Average Utilization of Machines (3M/3Z)

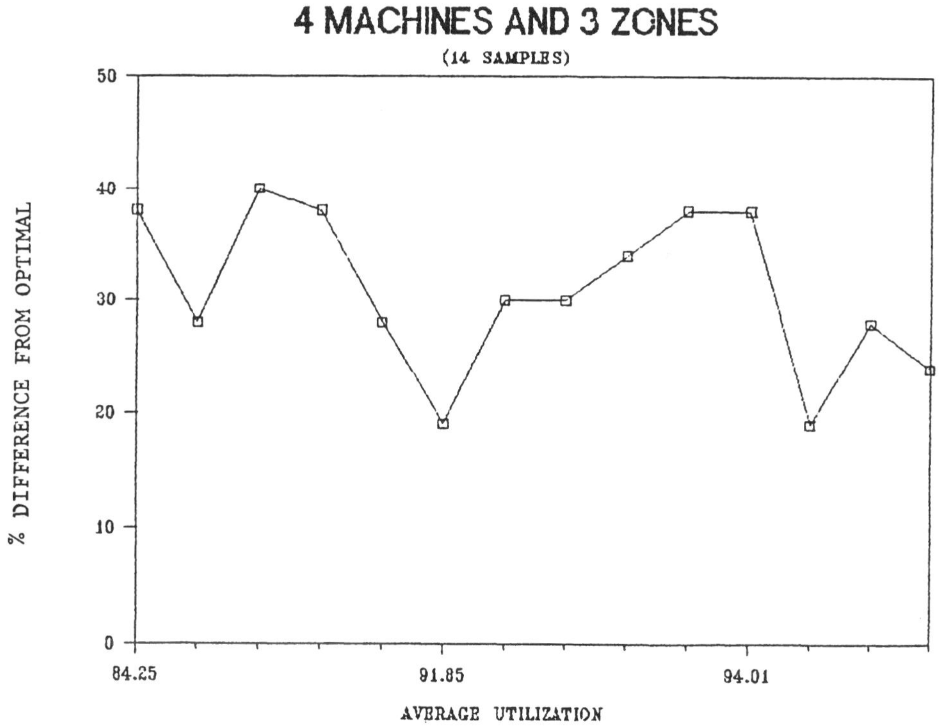

Figure 8 Percent Difference from Optimal Solution of Heuristic versus Average Utilization of Machines (4M/3Z)

From the results, it can be concluded that the percentage of the difference of output from P2 (optimal solution) and the existing solution depends on the utilization of the machine. In this case, the average utilization of the machine is 95% which means a low percentage difference. The result obviously shows that there is no reason to modify the facility layout and change the kiln car loading sequence.

The previous analysis logically leads to a ,hierarchical problem approach [Bitran et al (1981), Hax and Candea (1984), and Wassenhove and Derhenst (1983)]. The first step consists of allocating the pressed inprocess materials to kiln zones (planning at the bottleneck operation). The next step involves lot sizing on each machine by applying single-machine, multi-item approach trading off between setup cost and holding cost. Finally, given the production lot, a decision could be taken on the best color sequence to minimize glazing machine setup cost.

VI. PRODUCTION PLANNING AND SCHEDULING MODEL

The hierarchical scheme for solving the problem is given in Figure 9. It consists of four basic steps:

1. Select the order for production in the next 3 months
2. Allocate in products to kiln zone (at firing)
3. Determine the production lot size (at pressing)
4. Plan the sequence in each lot (at glazing)

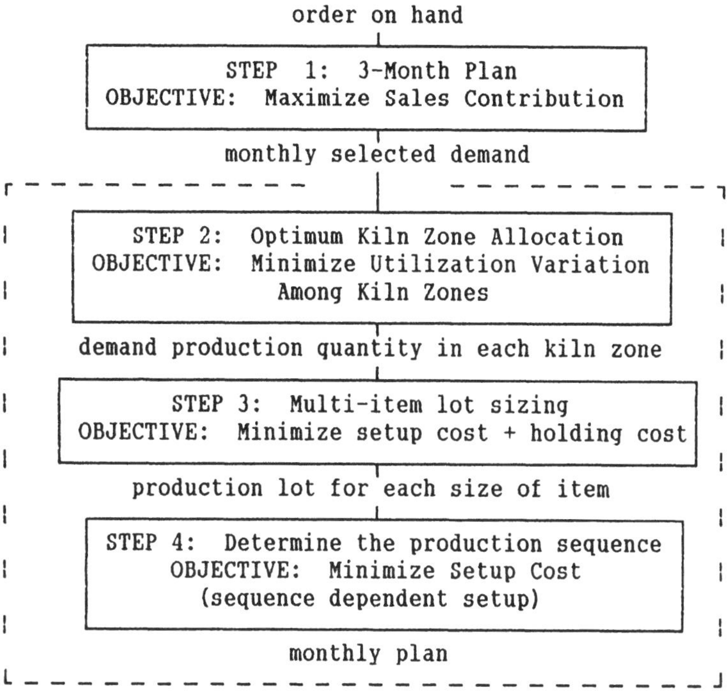

Figure 9 Hierarchical Modeling

Step 1: Select the order for production in the next 3 months.

The objective here is to develop a 3-month plan by maximizing sales contribution of the selected orders from all orders on hand. Mathematically, the problem can be stated as follows.

$$\text{Max}: \ \alpha_{a0}\Big(\sum_{i\varepsilon A_{t-1}} D_i Z_i\Big) + \alpha_{a1}\Big(\sum_{i\varepsilon A_t} D_i X_i\Big) + \alpha_{a2}\Big(\sum_{i\varepsilon A_{t+1}} D_i Y_i\Big) +$$

$$\beta_{b0}\Big(\sum_{i\varepsilon B_{t-1}} D_i Z_i\Big) + \beta_{b1}\Big(\sum_{i\varepsilon B_t} D_i X_i\Big) + \beta_{b2}\Big(\sum_{i\varepsilon B_{t+1}} D_i Y_i\Big) \quad .....(P3)$$

Subject to

- Capacity Constraint

Plant capacity is the capacity of the bottleneck facility, that is the capacity of the tunnel kilns, 1088 pallets/month.

$$\sum_{\substack{i\varepsilon A_{t-1}\\ B_{t-1}}} D_i Z_i + \sum_{\substack{i\varepsilon A_t\\ B_t}} D_i X_i + \sum_{\substack{i\varepsilon A_{t+1}\\ B_{t+1}}} D_i Y_i \leq C_t \quad \text{for all } t \; (t=1,2,3) \; (i=1,2,\ldots,n) \tag{10}$$

$$Z_i + X_i + Y_i \leq = 1 \quad \text{for all } i \tag{11}$$

$$Z_i = 0 \quad \text{for all } i\varepsilon\ A_0, B_0 \tag{12}$$

where

A = set of the order of first class (regular) customers

A_t = subset of the order of first class customers due in period t.

B = set of the order of second class (irregular) customers

B_t = subset of the order of second class customer due in period t.

α_{a0} = weight of contribution of production in class A if A_t is produced in period t-1.

α_{a1} = weight of contribution of production in class A if A_t is produced in period t.

α_{a2} = weight of contribution of production in class A if A_t is produced in period t+1.

β_{b0} = weight of contribution of production in class B if B_t is produced in period t-1.

β_{b1} = weight of contribution of production in class B if B_t is produced in period t.

β_{b2} = weight of contribution of produciton in class B if B_t is produced in period t+1.

D_i = demand in pallet of production order i.

C_i = production capacity of plant in period t.

X_i = 1 if product i is produced in period t

= 0 otherwise

Y_i = 1 if product i produce in period t+1

= 0 otherwise

Z_i = 1 if product i produce in period t-1

= 0 otherwise

The essential input data for this model is the weight of contribution. These values are based on customer priority and management policy. This is confidential information; the management approximated and normalized these values to be $\alpha_{a0} = 0.85$, $\alpha_{a1} = 1.00$, $\alpha_{a2} = 0.65$, $\beta_{b0} = 0.75$, $\beta_{b1} = 0.80$, $\beta_{b2} = 0.70$, as used in this paper.

This step is formulated as ILP and the optimal solution can be found within a few seconds using a mainframe computer. The output of this step is input to the monthly planning step, steps 2-4 in the hierarchical procedure.

Step 2: Allocate the production to kiln zone.

This is the first step in monthly planning by considering the bottleneck operation first. The purpose here is to allocate the production to the appropriate kiln zones by minimizing firing cost or, alternatively, minimizing utilization variation among zones. The planning model can be formulated as:

$$\text{Min}: \quad |A_1 - r_{12}A_2| + |A_1 - r_{13}A_3| + \ldots\ldots\ldots + |A_5 - r_{56}A_6|$$

$$(\ {}_2C_6 \text{ combinations }) \qquad\qquad (P4)$$

Subject to

- Capacity Constraint

$$\sum_{i \varepsilon S_z} X_{iz} \leq L_z \qquad \text{for all } z \quad (z=1,2,\dots,6) \tag{13}$$

- Demand Constraint

$$\sum_{z} X_{iz} = d_i \qquad \text{for all } i \varepsilon S_z \ (i=1,2,\dots,n) \tag{14}$$

$$\sum_{i \varepsilon S_z} X_{iz} = A_z \qquad \text{for all } z \tag{15}$$

$$X_{ij} \geq 0$$

The problem cannot be solved in this form so transformation from absolute value objective function to standard form of LP is essential as shown in problem P4.1 below.

$$\text{Min : } D_1 + D_2 + D_3 + \dots\dots\dots + D_{30} \tag{P4.1}$$

Subject to

- Capacity Constraint

$$\sum_{i \varepsilon S_z} X_{iz} \leq L_z \qquad \text{for all } z \quad (z=1,2,\dots,6) \tag{13a}$$

- Demand Constraint

$$\sum_{z} X_{ix} = d_i \qquad \text{for all } i \varepsilon S_z \ (i=1,2,\dots,n) \tag{14a}$$

$$\sum_{i \varepsilon S_z} X_{iz} = A_z \qquad \text{for all } z \tag{15a}$$

- Transformation Constraints

$$\begin{array}{l} A_1 - r_{12}A_2 + D_1 - D_2 = 0 \\ A_1 - r_{13}A_3 + D_3 - D_4 = 0 \\ \quad \vdots \\ A_5 - r_{56}A_6 + D_{29} - D_{30} = 0 \end{array} \tag{16}$$

$$X_{iz} \geq 0, \quad D_j \geq 0 \quad (j=1,2,\dots,30)$$

where

X_{iz} = amount of product i produced in zone z (sagger)

L_z = production capacity of zone z (saggers/week)

d_i = demand of product i (sagger)

A_z = allocation (in saggers) to zone z

S_z = set of product that can be produced in zone z

r_{ij} = ratio of zone capacity in kiln car zone i to zone j.

The output of this step is in fact the demand profile for each machine. After transformation the model becomes a simple LP. In this case, the MPSX package on mainframe computer IBM 3083 and Super Lindo package on NEC-PC were used. The demand profile of each zone serves as input to the lot sizing model used in step 3 of the hierarchical procedure.

Step 3: Multi-item Single Machine Lot Sizing

The purpose here is to plan production lots on each machine by trading off setup costs against inventory holding costs. Since there is fixed machine to fixed zone situation, the problem is solved by the multi-item single machine approach. The capacity of the kiln zones is used in the capacity constraint because the policy of management in this operation is not to allow in-process inventory to serve the following week's demand. The moisture in each pressed mosaic piece is lost by then. Such lead time could also create defects in production kiln firing. If the machine capacity is used in this way, it will create such a problem because the machine capacity is greater than kiln zone capacity. Management does not want to deteriorate product quality. Linear relationship between setup cost and machine capacity is assumed. The planning model for this problem can be formulated as follows.

$$\text{Min} \quad \sum_{i=1}^{N} \sum_{t=1}^{T} [S_i \delta(x_{it}) + h_i I_{it}] \qquad \ldots\ldots\ldots \qquad (P5)$$

Subject to

- Demand Constraint

$$I_{i,t-1} + x_{it} + I_{it} = d_{it}; \qquad i=1,\ldots,N \qquad t=1,\ldots,T \qquad (17)$$

- Capacity Constraint

$$\sum_{i=1}^{N} x_{it} \leq L_t \qquad t=1,\ldots,T \qquad (18)$$

$$x_{it} \geq 0, \; I_{it} \geq 0$$

- Setup Constraint

$$\delta(x_{it}) = \begin{cases} 0 & \text{if } x_{it} = 0 \\ 1 & \text{if } x_{it} > 0 \end{cases} \qquad (19)$$

where

d_{it} = Production requirements for product i in period t

x_{it} = Number of production units of product i in period t

I_{it} = Inventory of product i in period t

L_t = Capacity limitation of kiln zone in period t

S_t = Setup cost when production is started for product i

h_t = Holding cost per unit, per period.

The resulting nonlinear programming problem is, from a computational point of view, extremely difficult to solve. In fact, this problem is completely NP-hard. Hence, Lambrecht and Vanderveken's heuristic is suggested. It is a period by period procedure which can provide a very good solution (3% difference from the optimal solution on the average) and requires little computer time [Lambrecht and Vanderveken 1979]. It is based on a trade off between setup cost, cumulative holding cost and capacity absorption. Production lots are then assigned to the current period on the basis of part period balancing algorithm until the capacity constraint is hit. The procedure then moves to the next period. If an infeasibility occurs in period t, then feedback procedure can be used to transfer demand of that period to earlier periods at minimal additional cost. The algorithm is applied to each zone. Some salient details of the algorithm can be summarized as follows.

a) Order up to period. Before explaining the feedback procedure, note that the problem (P5) has a feasible solution if and only if

$$\sum_{k=1}^{t} L_k \geq \sum_{k=1}^{t} \left[\sum_{i=1}^{N} d_{ik} \right] \qquad t=1,2,\ldots,T. \tag{20}$$

In order to guarantee that the feedback mechanism will always find a feasible production plan the concept of "the maximum order up to period T* is introduced".

Consider the current planning period (through relabeling this is always period 1) and compute

$$B_t = \sum_{k=2}^{t} \left[\sum_{t=1}^{N} d_k - L_k \right] \qquad t=2,\ldots,T. \tag{21}$$

The first period for which the above term is positive is called the maximum order up to period. Call this period T* the cumulative requirements exceeding the cumulative capacity by T_{T*} units. B_{T*} is, in other words, a minimum ending inventory of period 1. If in T* an infeasibility occurs (which is not necessarily the case since requirements may be grouped because of a positive coefficient of (22)) feedback will always result in a feasible solution because

$$L_i - \sum_{i=1}^{N} d_{i1} \geq B_{T^*} = \sum_{k=2}^{T^*} [\sum_{i=1}^{N} d_{ik} - L_k] .$$

Otherwise (20) would not be satisfied.

$$U_i(t) = \{S_i - I_i(t) \}/(t^2 d_{it}) \tag{22}$$

$$I_i(t) = h_t \sum_{k=1}^{t} (k-1) d_{ik} \tag{23}$$

The importance of T* lies in the fact that it always guarantees a feasible solution if backtracking must be done. Therefore, a negative coefficient is attached to requirements after period T*.

b) <u>Feedback Procedure</u>. If infeasibility occurs in period t, i.e. if requirements exceed capacity in period t, then demands of that period is transfered to period t-1. If the infeasibility still remains after this phase is completed (that means the period t-1 is loaded up to capacity) further transfers must be made to the earlier period.

For the sake of simplicity the feedback mechanism is explained in terms of an arbitrary period t ignoring the relabel.

b1) Let INF_t be the amount by which the demand requirements exceed capacity in period t.

$$INF_t = \sum_{i=1}^{N} d_{it} - L_t \quad \text{if} \quad \sum_{i=1}^{N} d_{it} > L_t, \quad \text{otherwise} \quad INF_t = 0 \tag{24}$$

Determine

$$C_t = \min \{INF_t, d_{it}\}, \; C_t > 0 \tag{25}$$

$$\text{Select product } q \text{ for which } \min_{i \varepsilon m^*} \{a \, C_t \, h_t\} \tag{26}$$

M* = set of products for which there is a setup in period t-a.

In the first run set a=1.

Next determine

$E = \min(e_{t-a}, C_q)$ and transfer E units of the selected product q.

If period t becomes feasible, continue with the next planning period. If period t is still infeasible go to step b2.

b2) If for step b1 it is found that $E = e_{t-a}$, which means that it is impossible to transfer more units to period t-a without exceeding the capacity constraint.

Set $a = a + 1$ and to to step b1.

If in step b1 it is found that $E \neq e_{t-a}$, go to step b3.

b3) $E = C_q = d_{q,t}$, which means that in addition to the demand for product q, units of other products must be transferred to period t-a. Go to step b1 if $M^* \neq 0$. If $M^* = 0$ go to step b4.

b4) Since $M^* = 0$ create an additional setup in (t-a) in order to restore feasibility. The remaining products may be classified in 2 groups shown in Table 2.

Table 2

Group	Determine	If	Transfer	Additional cost
1 $d_{it} \geq INF_t$	$E_i = \min(d_{it}, e_{t-a})$	$E_i = d_{it}$ $E_i = e_{t-a}$	d_{it} e_{t-a}	$ah_t d_{it}$ $ah_t e_{t-a} + S_i$
2 $d_{it} > INF_t$	$B_i = \min(INF_t, e_{t-a})$	$B_i = e_{t-a}$ $B_i = INF_t$ and if $d_{it} < e_{t-a}$	e_{t-a} INF_t d_{it}	$ah_t e_{t-a} + S_i$ $ah_t INF_t + S_i$ $ah_t d_{it}$

Select the product with the minimal additional costs and transfer the corresponding units to period (t-a).

The output of this step specifies for each time period and for each zone the volume of production. This heuristic procedure finds the solution within a few seconds on IBM 3083. In the pressing machines 1 and 5 which are allocated to zones top/bottom of each kiln, the production load can be assigned from this step without any problem because actually the pressing machines (1 and 5) have greater capacity than zone (top/bottom of each kiln) requirement. The problem is in pressing machines 2,3,4 and 6,7,8, each set serving 2 zones, in and out, of each kiln. The next step is the extension of this phase to disaggregate the production of 2 zones to 3 machines.

As mentioned earlier, the sequence of the kiln car loading pattern cannot be changed. Hence, machines 2 and 3 serve zones in and out, respectively, and machine 4 serves both zones in kiln 1. The same method applies in kiln 2.

The objective here is to minimize setup cost subject to disaggregate load from the two main machines to the other (third one) under the capacity constraint. The assumption is that the disaggregate load from each machine can be produced in both zones. This is so because usually 90% of production that can be produced in the inner zone can be produced in the outer zone also. The assumption is stated for the prevention of shortage of demand in the inner or outer zones. Mathematically, the problem can be stated as follows.

$$\text{Min} \quad \sum_m S_m \left(\sum_i Y_{i,m}\right) \qquad \text{.........} \qquad \text{(P5-1)}$$

Subject to

- Demand Constraint

$$\sum_m x_{i,m,z} = D_{i,z}, \qquad \text{for all } i,z \qquad (27)$$

m=1,3 for zone 1

m=2,3 for zone 2

- Capacity Constraint

$$\sum_z \sum_i x_{i,m,z} = L_m \qquad \text{for all } m,\ m=1,2,3 \qquad (28)$$

z=1 for m=1

z=2 for m=2

z=1,2 for m=3

- Setup Constraint

$$Y_{i,m} = 1 \ \text{ if } \ \sum x_{i,m,z} > 0 \qquad z=1 \ \text{ for } \ m=1 \qquad (29)$$

z=2 for m=2

z=1,2 for m=3

$$= 0 \ \text{ otherwise}$$

$$x_{i,m,z} \geq 0,\ i=1,2,\ldots,n, \quad m=1,2,3, \quad z=1,2$$

where

$x_{i,m,z}$ = product i produced in machine m for zone z

$D_{i,z}$ = demand of production i in zone z

L_m = capacity of machine m

m = 1 is the machine that supports zone 1 only

= 2 is the machine that supports zone 2 only

= 3 is the machine that supports both zones 1 and 2

After trying to apply this extended step (P5-1) in each lot of production in the inner and outer zones by using MPSX package, it was found out that the process is very time consuming. Thus, a heuristic procedure was developed. The heuristic is summarized in Figures 10 and 11. There are two major phases. Firstly, try to reduce the cost by transfering the maximum number of common items from the two main machines to the third machine until the excess demand is satiated. If it cannot be completed by the first phase, then, continue disaggregation of the excess demand by increasing the extra cost at the minimum level.

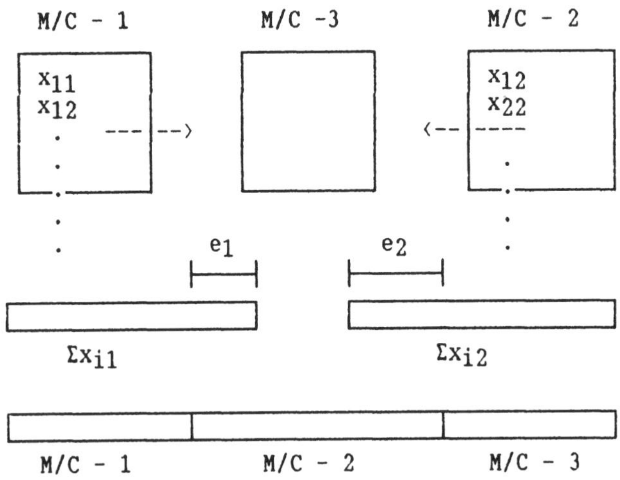

Figure 10 Demand Allocation from the Two Machines to the Third Machine

where

x_{im} = amount of product i on machine m

e_m = excess demand on machine m; m=1,2

or minimum amount of demands to transfer to machine 3

c_m = capacity of machine m

S_m = setup cost on machine m

S_1 & $S_2 \leq S_3$

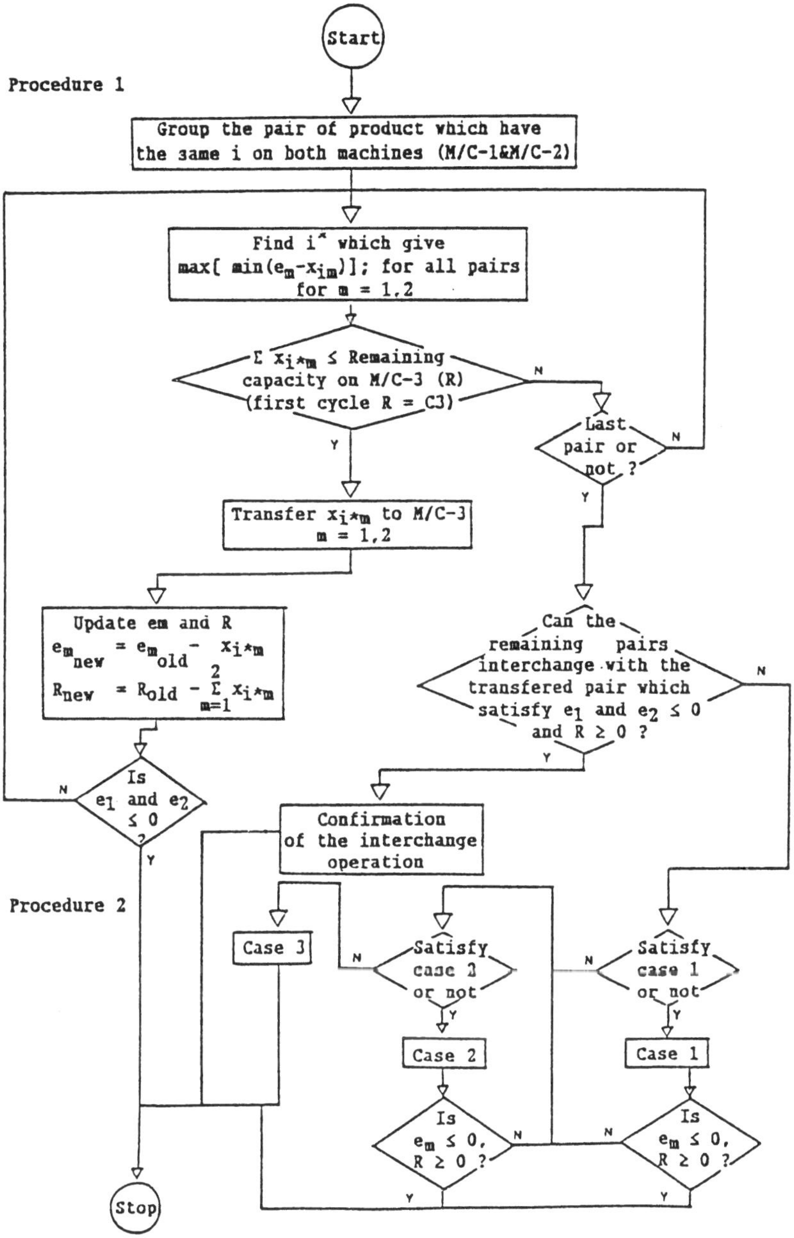

Figure 11 Demand Allocation from the two Machines to the Third Machine Heuristic Flow chart

Allocation of excess demand from two machines to the third machine is done using the heuristic procedure discussed and shown in Figure 8 as follows.

Phase 1: Cost Reduction

Transfer maximum number of pairs of products subject to the capacity of the third machine.

The saving of cost is $n\{(S_1 + S_2) - S_3\}$ where n is the number of pair of products that can be transfered.

Phase 2: Increase at Minimum Cost

Case 1: The pair of products which only one x_{im} can be transfered but the other one cannot. Transfer x_{im} and some amount of other x_{in}'s (for $n \neq m$)

The additional cost = $S_3 - S_m$

$m \varepsilon$ The one that can transfer the whole of x_{im}

Case 2: Transfer min [max $(x_{im} - e_m)$] on each machine

The additional cost = $S_3 - S_m$

Case 3: Transfer some of x_{im} on each machine

The additional cost = nS_3

n = number of transfered times

The lot size for each item to be produced on each machine is finally obtained. The production sequence for each lot on each machine is done in the next step.

Step 4: Determine a Production Sequence

Given the production lots in each machine (Step 2), the optimal production sequence can be found by solving an assignment problem for each time period on each machine. The application of the assignment problem to a sequence dependent setup machine problem can be mathematically formulated as follows.

$$\text{Min} \quad \sum_{i}\sum_{j,\ i \neq j} S_{ij} Y_{ij} \qquad \text{(P6)}$$

Subject to

$$\sum_{i,\ i \neq j} Y_{ij} = 1 \qquad \text{for all } j \qquad (30)$$

$$\sum_{j,\ j \neq i} Y_{ji} = 1 \qquad \text{for all } i \qquad (31)$$

$$Y_{ij} = [0,1]$$

where

S_{ij} = setup cost due to changing color i to color j

Solving the assignment problem requires very little computer time since often, contrary to a large range of product sizes, product color varieties are relatively few. The output of this step completes the hierarchical planning procedure.

V. CONCLUSION REMARKS

The existing procedure was applied and the results compared with those obtained using the existing procedure. The improvement is remarkable as shown in Table 3. The paper also demonstrates that hierarchical procedure is an effective way of handling large scale production planning and scheduling problem which would otherwise be extremely difficult to solve in the traditional way.

Table 3 Monthly Average Value Comparison

Significant Figure	Existing Procedure	Solution Procedure
Inventory level	5,943,700	1,658,190
Holding cost	297,185	82,902
No. of setup on pressing machines	71	61
No. of setup on glazing machines	254	303
Setup cost	89,475	76,840
Total cost	350,090	159,749
Sales contribution per pallet	1,302	1,475
Shipment	1,703,000	3,110,220

REFERENCES

[1] Bitran, G.R., Haas, E.A. and Hax, A.C.; "Hierarchical Production Planning: Single State System", Operations Research, Vol.29, No.4, 1981, pp. 717-743.

[2] Hax, A.C. and Candea, D.; Production and Inventory Management, Prentice-Hall., New Jersey, 1984.

[3] Lambrecht, M.R. and Vanderveken, H.; "A Heuristic Procedure for the Single Operation Multi-Item Loading Problem", AIIE Transactions, Vol.11, No.4, 1979, pp. 319-326.

[4] Manne, A.S.; "Programming of Economic Lot Size", Management Science, Vol.4, No.9, 1958 (Jan.), pp.115-135.

[5] Wassenhove, L.N.V. and Derhenst, P.V.; "Planning Production in a Bottleneck Department", European Journal of Operations Research, Vol.12, 1983, pp.127-147.

ACKNOWLEDGEMENT

Grateful acknowledgement is sincerely extended to those participants of the International Conference on Modern Production Concepts - Theory and Applications, August 19-24, 1990, Hagen, Federal Republic of Germany, who gave constructive comments that led to the improvement of this write-up.

Operational Control of Automated PCB Assembly Lines

HENK ZIJM

1. Introduction

The rapidly changing market demand characteristics, the increased pace of technological developments and fiercer competition has forced almost any industrial company to fundamentally rethink its manufacturing strategy (cf. Hayes and Wheelwright[1984], Skinner[1985]). In gaining competitive advantage quality, flexibility and speed of innovation have become important performance measures, next to being efficient (Bolwijn et al.[1986], Buzacott[1982], Solberg et al.[1985], Bullinger et al.[1986]). These changes have affected industrial organisations at almost every level: in logistics (MRP, DRP), marketing and product development (DFA, DFM) as well as in operations on the shopfloor itself, by the introduction of CNC machining centres and, finally, of Flexible Manufacturing Systems (FMS) and Flexible Assembly Systems (FAS).

FMS's and FAS's are characterized by a high degree of automation, not only in machining or assembly operations but also in transport and materials handling (cf. Ranky[1983]). At the same time it is realized that automation by itself does not solve manufacturing problems. The inherent versatility of flexible equipment leads to complex new problems related to the justification, to the design and layout, as well as to the planning, scheduling and control of such systems (cf. Stecke[1985], Zijm[1988]). As far as the Operations Research literature on FMS's is concerned, most papers deal with design, planning and scheduling issues, while applications, if mentioned at all, mainly concern the metal cutting area (cf. Stecke and Suri[1985, 1986], Kusiak[1986-a, 1986-b]). The amount of reported research on control problems in electronics manufacturing is much more limited, despite the very high degree of automation in this area. Indeed, the problems arising in electronics manufacturing often differ substantially (and are perhaps less familiar) than those experienced in metal cutting applications. For instance, before planning and scheduling a flexible flowline (cf. Wittrock[1985]), a carefull production preparation and process planning phase is needed to determine individual machine production rates, routings and the like.

This paper addresses a series of production preparation and control problems in an almost fully automated assembly line for Printed Circuit Boards (PCB's). These boards are widely applied in both consumer electronics (audio and video, personal computers) and in the professional industry (telecommunication systems, aircraft navigation, medical equipment). The report reviews and summarizes research motivated by consultancy projects carried out in a number of plants of Philips Electronics, The Netherlands. A hierarchical system of (combinatorial) models is developed to highlight several production preparation and numerical control problems in PCB-assembly lines. Part of the study reported here has resulted in an equipment oriented *production preparation system* for SMD's (Surface Mounted Devices, i.e. small integrated circuits which are first glued and then soldered on the board), which is now in use at several Philips' sites, as well as in other companies. For details concerning this part of the study, we refer to Van Laarhoven and Zijm[1989].

Some papers on optimization of *individual* placement machines have appeared. With respect to production preparation, sequencer mix problems (preparing the supply tape of axial lead component inserters, see section 2) are considered by Randhawa et al.[1985] and Fathi and Taheri[1989]. The optimal sequence of component insertions can be found as a solution to (a variant of) the well-known travelling salesman problem (cf. Ball & Magazine[1988], Chan & Mercier[1989], Leipälä & Nevalainen[1989]). Ahmadi, Grotzinger and Johnson[1988] explicitly address SMD insertion in their study on component allocation and partitioning for a single numerically controlled placement machine.

This paper differs from almost all previous ones in that it describes a *hierarchical, integrated control system*, not for an isolated machine but for a *system* of insertion machines. In section 2 we describe the basic technical processes in a PCB assembly line. Section 3 lists the hierarchy of control problems associated with the bottleneck part of such a line, a system of specialized SMD placement machines. In section 4, we describe models and algorithms developed to solve these problems. In section 5 we briefly discuss similar problems for other parts of a PCB assembly line, in particular for axial and radial lead component inserters and for a second SMD placement system. We end with some conclusions and suggestions for further research.

2. A Printed Circuit Board assembly line

The basic steps in the assembly of PCB's are the insertion processes for Axial Lead Components (ALC's), Radial Lead Components (RLC's) and Surface Mounted Devices (SMD's), the soldering process, a (manual) touch up, In Circuit Component Testing (ICCT) and finally the Functional Test (FT) of the complete board. Due to the rapid advance of new technologies (the replacement of ALC's and RLC's by SMD's), the ICCT phase is not always present anymore. Transport is usually carried out by carrierless, flexible conveyor systems or by AGV's (Automatic Guided Vehicles, transporting cassettes with e.g. 20 boards) or by both (for instance, AGV's between SLC and RLC insertion, a carrierless conveyor system to and between several SMD placement machines and through the wave soldering bath).

In the sequel, we concentrate on the first three operations, being the most relevant for the purpose of this paper, i.e. ALC, RLC and SMD insertion. An *ALC placement machine* basically consists of a placement head which mounts (axial lead) components on the board. The placement head is supplied with (all types of) components in exactly the right (previously determined) placement order by an externally prepared tape. Instead of directing the placement head to the different coordinate positions on the board (which would constitute mechanical difficulties due to the presence of the supply tape), the PCB itself is switching between every two insertions such that always the board coordinates of the component to be placed next matches with the (fixed) position of the placement head. It is possible that two heads are installed, each supplied by a different tape and each serving a different board, not necessarily of the same type.

Boards of several types may visit the ALC placement machine subsequently; in that case the placement head supply tape(s) must contain the relevant sets of components in the order in which different PCB's visit the machine.

A *sequencer* is a programmable machine producing the tapes which next are loaded to the ALC placement machines. Such a sequencer consists of a set of dispensing heads that cut the components from input reels (each containing components of only one type) according to the bill of materials and the insertion pattern. The dispensed components are fed on to a conveyor chain where they are spaced and remounted on an automatically fed tape. Due to still substantial changeover times, sequenced tapes are produced in lots which are generally significant larger than daily production plans for PCB types require. Hence, there is some build-up of inventory of sequenced tapes and therefore, the sequencing process is decoupled from the insertion process.

An *RLC placement machine* (fig. 1) basically performs the same functions as its ALC counterpart, of course on radial lead components. The main difference

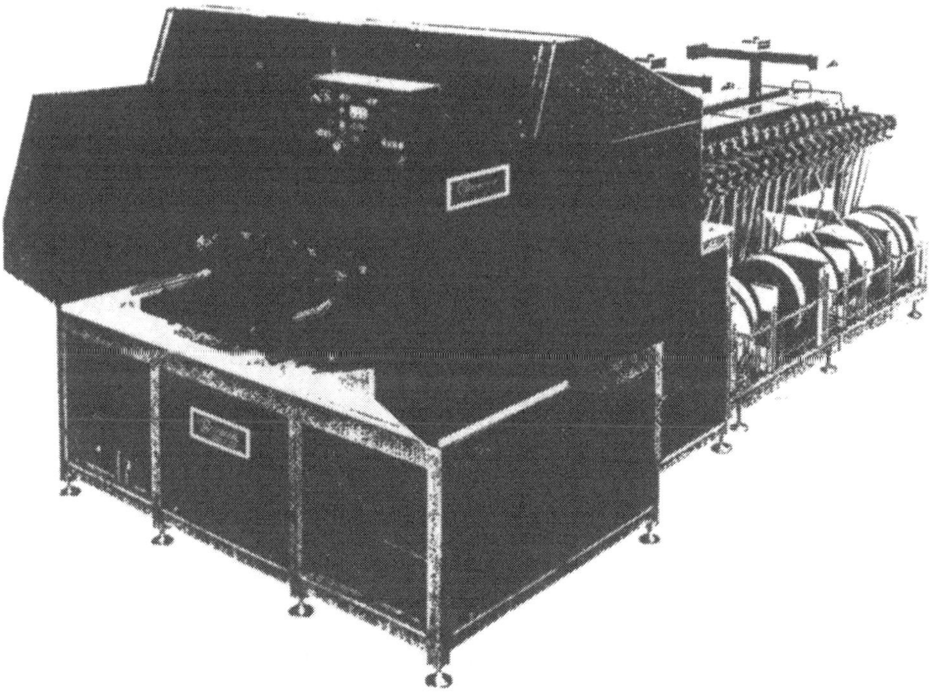

Fig. 1. A Radial Lead Component Placement Machine (with Sequencer)

is that the sequencer is integrated with the machine, i.e. components are dispensed to the placement head supply tape typically one or two minutes before they are actually mounted on the board. Again, these components are taken from (component-specific) feedertape reels loaded in the (machine-connected) sequencer. RLC sequencing therefore is an on-line operation, just preceding the insertion process.

SMD placement machines exist in a large number of varieties. We describe two typical representatives dedicated to high end (large volumes and a low diversity) and to low end (small to medium sized batches) market segments, respectively. An MCM is a typical high capacity (in terms of possible SMD placements per hour) machine which basically consists of a series of placement modules under *centralized* computer control. At each module (fig. 2), a software controlled placement unit (a beam) holds 32 pick and place pipettes; these pipettes simultaneously draw components from (type specific) tape reels placed in containers in front of the module. Next, the picked components are placed sequentially by moving the complete placement unit such that for each component insertion the relevant pipette is located exactly above the associated component coordinates on the PCB. If the components are to be glued, glue is applied simultaneously either to the components or their placement positions on the board.

Fig. 2. An MCM placement module with 32 pick-and-place pipettes

A CSM (fig. 3) is a small to medium volume SMD placement machine, equipped with two feeder banks (containing up to 60 type-specific tape reels or stick feeders), both at the front and rear. The actual placement unit is a robot arm containing three heads. During one pick and place move (called a *charge* in the sequel) the heads sequentially pick components after which the arm visits the board positions such that each head may place its component (sequentially again). Possible glue dots are placed in advance by a special machine.

Typically, a board mounted with SMD's runs through a sequence of MCM placement modules or a sequence of CSM placement machines. At each module (machine), a number of charges have to be executed before the board may travel to the next module (machine). Typical cycle times vary from 10 (MCM) to 90 seconds (CSM). The pipettes of an MCM are universal with respect to the component types, but not every pipette can reach every position on a PCB. The heads of a CSM placement unit are equipped with special nozzles and chucks which generally accomodate only a subset of all component types. These heads however can reach any position on the board.

Almost always, a carrierless (automatically controlled) rail system guides the PCB's through the line of SMD placement modules (machines) and subsequently through a wave soldering bath which finally fixes both ALC's and RLC's and the (already glued) SMD's definitely to their positions. We will not describe

Fig. 3. A CSM placement machine for SMD's

has to be performed on each board before it may travel to the next CSM. Typically also, a PCB needs so many components of so many different types that, due to the limited feeder bank capacity, indeed a number of CSM's have to be visited before all required SMD's are mounted.

Now, consider a line of m CSM's and suppose that we wish to minimize the *line cycle time* (or, to maximize the *line throughput*) for a production batch of *one* particular PCB-type (a *production batch* usually corresponds to a production order for one PCB-type, note however that the *transport lot size* within a CSM line equals *one board*). Clearly, workload between machines should be balanced as much as possible. More specifically, decisions have to be made with respect to the equipment (in terms of heads, feeders and components) for each machine and the operations (contents of the charges) at each machine. Below, we specify the control hierarchy designed by Van Laarhoven and Zijm[1989], given a line of m machines and a batch of *one type of PCB*.

(I) Choose 3m head equipments.

(II) For each component type i (i=1,...,n) and for each head equipment j (j=1,...,3m), determine the *number* of components of type i to be mounted by equipment j.

(III) Match the head equipments with the heads.

(IVa) For each component type i (i=1,...,n) and for each head (equipment) j (j=1,...,3m), determine *which* components of type i are to be mounted by head j (i.e. which coordinate positions have to be visited by head j).

(IVb) Determine which components are picked and placed together in one charge (such a group of three components is called a *cluster*)

(V) Assign the feeders to feeder positions on each machine.

(VIa) Determine the placement sequence *within* each component cluster.

(VIb) Determine a sequence for the clusters (charges) to be performed by each machine.

Note that problems (I)-(IVa) relate to the *line* of machines, whereas problems (IVb)-(VI) can be solved for each machine separately. Problems (IVa) and (IVb) are solved simultaneously, as well as problems (VIa) and (VIb).

All problems in the above hierarchy are of a combinatorial nature and most of them are generalizations of well known NP-hard problems, such as *multi-processor scheduling* and *travelling salesman*. In the next section we formulate approximation algorithms to solve the various subproblems (I)-(VI).

4. Models and algorithms

Let m be the number of machines and, for a given PCB, let n be the number of component types and N the number of components. n_i denotes the number of components of type i ($i=1,\ldots,n$); thus, $\Sigma n_i = N$. Q denotes the set of all head equipments. For each component type i, we are given a subset Q_i of Q, denoting the head equipments that can pick and place components of type i, as well as the coordinates of the board positions on which the n_i components of type i have to be mounted.

We assume to have available a real number w_i, estimating the average time necessary to mount components of type i. Obviously, the precise time differs per component and depends on the board location of the component and the location of the feeder the component is to be picked from; locations which are not known *a priori*. However, since the w_i's are composed of three terms, viz. the time to pick a component, the time to place a component and the time to travel between feeder and board positions, of which the first two are given for each component type (dependent on weight and size) and, in practice, the third one can be accurately estimated from previously mounted PCBs with similar component types, it is usually not hard to determine these numbers.

4.1. Choice of head equipments

For each element head equipment q ($q \in Q$), we define a *weight* h_q as the sum of the weights of the component types that can be mounted by q, i.e.

$$h_q = \sum_{(i \mid q \in Q_i)} n_i w_i . \qquad (1)$$

An equipment is called *redundant* iff it can be deleted from each Q_i in which it occurs without emptying Q_i. The following two steps now describe in which way the equipments are chosen:

i: Delete equipments that are redundant in ascending order of weight.

ii: For each equipment q remaining after step I, choose a number of equipments r_q proportional to $h_q / \Sigma h_q$, and such that $\Sigma r_q = 3m$.

4.2. Assignment of components to head equipments

Since we do not yet know which equipment is matched with each head on each machine (this constitutes problem III), we will assign components to equipments such that equipment workloads are balanced. This results in the following integer programming formulation

$$\text{minimize} \max_j \left(\sum_{i=1}^{n} s_{ij} w_i \right) + \lambda \left(\sum_{i=1}^{n} \sum_{j=1}^{m} \min(1, s_{ij}) \right) \quad (2)$$

$$\text{subject to} \quad \sum_{j=1}^{3m} s_{ij} = n_i, \quad i=1,\ldots,n, \quad (3)$$

$$s_{ij} \le M\, a_{ij}, \quad i=1,\ldots,n,\ j=1,\ldots,3m, \quad (4)$$

$$s_{ij} \text{ integer}, \quad i=1,\ldots,n,\ j=1,\ldots,3m, \quad (5)$$

where M is a suitably large number, λ a weighting factor and $a_{ij} = 1$ (0) if component type i ($i=1,\ldots,n$) can (cannot) be mounted by equipment j ($j=1,\ldots,3m$). Note that variable s_{ij} represents the number of components of type i assigned to equipment j; thus, the workload for equipment j is estimated by $\Sigma\, s_{ij} w_i$, and the solution of problem (2)-(5) is such that the workload for the bottleneck equipment is minimal if $\lambda=0$. The second term in the objective function (without the *control parameter* λ) is an upperbound on the total number of feeders used. Note that this problem boils down to a *multiprocessor scheduling problem* if we set $\lambda=0$ and furthermore drop constraint (4) (which would mean that all head equipments are universal). Consequently, the problem is NP-hard.

More general objective functions, related to a distinction between several feeder *types*, are presented in Van Laarhoven and Zijm[1989].

4.3. Matching of equipments with heads (and machines)

Recall that solutions to the previous problems specify a choice of head equipments as well as the *number* of components of each type, to be mounted by each head equipment. Next, we wish to match the 3m head equipments with the 3m heads on the m machines such that the workloads on the machines are more or less balanced. Let π denote a permutation of the set $\{1,\ldots,3m\}$, where $\pi(v)$, $v=1,\ldots,3m$, denotes the equipment matched with head v. Then we seek for a permutation π such that

$$C(\pi) = \max_{k=1,\dots,m} \left(\sum_{j=3k-2}^{3k} \left(\sum_i s_{i\pi(j)} w_i\right)\right) + \\ + \lambda \sum_{j=1}^{m} \sum_{i=1}^{n} \min(1, s_{i\pi(3j-2)}, s_{i\pi(3j-1)}, s_{i\pi(3j)}), \quad (6)$$

is minimized (again, λ is a weighting factor). Note that, for each permutation π, the numbers $s_{i\pi(j)}$ are determined in the previous section. In fact, we attempt to divide the 3m head equipments into m groups such that the largest "group weight" is minimized. The second term is related to the number of feeders assigned to each machine.

4.4. Construction of the component clusters

We now know *how many* components of each type are mounted by each head (since head equipments are assigned to heads now). However, it is still possible that the total number of components of type i (say), to be mounted on the PCB, is divided among more head equipments (possibly of the same type), assigned to different heads on, possibly, different machines. Hence, we still have to decide *which* components (identified now not only by type but also by their coordinate positions) are to be processed by each head.

By applying some simple sorting procedures, one can make sure that each head is assigned components with a wide range of x- and y-coordinates. This in turn will facilitate the subsequent clustering of components in charges since it is natural to attempt to place in one charge components with neighbouring coordinates. Let $t_{ij} = s_{i\pi(j)}$ $(i=1,\dots,n,\ j=1,\dots,3m)$, thus t_{ij} denotes the number of components of type i mounted by head j. The following 4 steps describe how the component clusters are chosen:

i: Rank all component types arbitrarily. For component type i with an even (odd) rank number, first sort its components in ascending (descending) order of x-coordinate. Sort those components of type i with identical x-coordinates in ascending (descending) order of y-coordinates.

ii: For each component type i, assign the first t_{i1} components from the sorted sequence of components of type i to head 1, the next t_{i2} components to head 2 etc. This yields an assignment of components to heads, where each component is uniquely identified by its board position.

iii: For each head j, sort the components assigned to this head in ascending order of x-coordinates (indifferent to type). Next, sort the components with identical x-coordinate in ascending order of y-coordinate. As a result of this step, the set of components assigned to each head by step ii is now a (sorted) list of components.

iv: On each machine, in the k-th charge each head now processes the k-th component from the list of components for that head (if the list in emptied, the corresponding head remains empty in that charge).

Note that after steps i and ii, which solve problem (IVa), we further may study each machine separately. Steps iii and iv finally solve problem (IVb).

4.5. Assignment of component feeders to feeder positions

At this stage, the heads as well as the component clusters for each machine have been chosen. As mentioned above we can therefore solve the problem of optimally assigning feeders to positions for each machine separately. Define for each feeder i (supplying components of type i) Γ_i as the set of component clusters containing components of type i. Note that if a cluster γ is in Γ_i, the arm has to travel from the position of feeder i to the board position of the component of type i in γ during the mounting process. We can thus minimize the total time for mounting by assigning feeder i 'close' to the board positions of the components in the clusters contained in Γ_i.

Let p_j denote the position of feeder j. For each cluster γ, let q_γ denote the coordinates of the centre of the board positions of the components in γ. Finally, let d(p,q) denote the Euclidean distance between positions p and q. Then we define the cost c_{ij} of assigning feeder i to position j by

$$c_{ij} = \sum_{\gamma \in \Gamma_i} d(p_j, q_\gamma) \tag{7}$$

The allocation of feeders to feeder positions is now easily determined by solving a *linear assignment problem* with cost coefficients c_{ij}.

4.6. Placement order within a cluster and sequencing of clusters

Knowing the position of each feeder in the feeder bank of each machine, we may solve problems (VIa) and (VIb). Consider two component clusters γ and δ.

Let $f_1(\gamma)$, $f_2(\gamma)$ and $f_3(\gamma)$ ($f_1(\delta)$, $f_2(\delta)$ and $f_3(\delta)$) denote the feeders from which the components in γ (δ) are picked and let $q1_\gamma$, $q2_\gamma$, and $q3_\gamma$ denote the board positions of the components in cluster γ. Suppose that δ is processed immediately after γ. After picking the last component from feeder $f_3(\gamma)$, the arm passes the positions $q1_\gamma$, $q2_\gamma$, $q3_\gamma$ (in an order to be chosen) and next moves to feeder $f_1(\delta)$ to pick the first component in cluster δ. For each of the six possible placement orders we can compute the distance to travel from the position of $f_3(\gamma)$ via the board positions to the position of $f_1(\delta)$. Thus, by enumeration we can also determine the best order. Let $z(\gamma,\delta)$ denote this order and $d(\gamma,\delta)$ the corresponding distance. Note that once we have determined the order in which to process the clusters, we have consequently determined the best order in which to place the components in each cluster as well.

Problem (VIb) is now solved by using the fact that it is an (asymmetric) travelling salesman problem (ATSP), in which the cities correspond to the component clusters and the distance between clusters γ and δ is given by $d(\gamma,\delta)$. Our TSP is asymmetric, because in general $d(\gamma,\delta) \neq d(\delta,\gamma)$. Once having solved this ATSP, the distances $d(\gamma,\delta)$ corresponding with the arcs on the minimum distance route determine the placement sequence in each cluster.

4.7. Solution procedures

The non-standard combinatorial problems have been solved by applying *simulated annealing* (Van Laarhoven and Aarts[1987]), combined with special heuristics. The sorting procedures present no difficulties. A linear assignment problem can be easily solved by e.g. the *Hungarian method* (cf. Papadimitriou and Steiglitz[1982]). To solve the ATSP, we have used the algorithm of Kanellakis and Papadimitriou[1980]. A typical medium sized real life problem (3 or 4 machines, 300 to 400 SMD's) takes just a few minutes of computing time on a PC/AT with an 80386 processor. For details, the reader is referred to Van Laarhoven and Zijm[1989].

The solution to problem (III) (section 4.3) may reveal some unbalance in certain special cases which make it necessary to carry out step (II) a second time (or, better, to force some components to be placed on another CSM). Similar loops may occur between steps (V) and (VI).

Computational experience is reported extensively in Van Laarhoven and Zijm[1989] on which section 4 is largely based. Here we only remark that an almost perfect balance on a number of real life applications (the production preparation for a CSM line in a factory in the UK) was obtained. More important, several operators have tried for almost two weeks to beat the algorithms. They did not succeed.

As mentioned earlier, the results in this section are not just a theoretical contribution. The hierarchical control structure, models and algorithms have been implemented as part of an overall control system (including product databases, interfaces to order entry systems and the like) which is now in use at several Philips plants, as well as in other companies.

5. Hierarchical control of classical component inserters and MCM's

As mentioned earlier, the control problems for ALC and RLC inserters are easily formulated, once having mastered the much more complex control structure for a line of SMD placement machines such as the CSM's. The relative simplicity is entirely due to the lower complexity of these machines.

Consider for instance a line of RLC inserters. Recall from the description in section 2 that each machine is equipped with exactly one, universal, placement head. Hence there is no need to consider the assignment of head equipments to heads. Also, since the placement head is supplied with components by a sequenced tape, filled in the *integrated* sequencer, it does not really matter how the component-specific tape feeders at one machine are assigned to feeder positions (the loading of the sequenced tape occurs software controlled; some places may be occupied while preceding places are still empty). Therefore, the main problem is to balance again the workload among several RLC inserters and next to place the components on each machine as efficiently as possible.

Consider a line of m RLC inserters. The *number* of components of each type to be placed on machine j is now determined by solving the following program:

$$\text{minimize} \max_j \left(\sum_{i=1}^{n} s_{ij} w_i \right) \tag{8}$$

$$\text{subject to} \quad \sum_{j=1}^{m} s_{ij} = n_i, \qquad i=1,\dots,n, \tag{9}$$

$$\sum_{i=1}^{n} \min(1, s_{ij}) \le b_j, \qquad j=1,\dots,m, \tag{10}$$

$$s_{ij} \text{ integer}, \quad i=1,\dots,n,\ j=1,\dots,m, \tag{11}$$

where, as before, n_i equals the total number of components of type i, to be mounted on the PCB, while b_j denotes the capacity of the sequencer of machine j. The variable s_{ij} represents the number of type i components to be mounted by machine j (hence a corresponding feeder tape should be loaded to the sequencer).

If a particular component type is processed by more than one machine we should, as before, determine *which* components (identified by their coordinate positions) are placed by each machine. Finally, having made that choice, the optimum placement order of all components by one placement unit (one machine) is easily determined by solving a *symmetric* travelling salesman problem (TSP), e.g. by the algorithm of Lin and Kernighan[1973]. However, the solutions to these TSP's (one for each machine) naturally depends on the above indicated ultimate choice of components. Let, for each component type i, A_{ij} be a set of *identified* (by their coordinates) type i components, to be mounted on machine j (hence the cardinality of A_{ij} equals s_{ij}). Let $T_j(A_{1j},\ldots,A_{nj})$ denote the optimum solution of the associated TSP, for each machine j. Then we seek for a choice of sets A_{ij}, i=1,...,n, j=1,...m, such that

$$\max_{j=1,..,m} [T_j(A_{1j},\ldots,A_{nj})]$$

is minimized. This is accomplished by interchanging components of the same type between machine, where we may base our selection on a most profitable interchange (e.g. selecting a pair of components whose interchange will decrease two calculated routes with a maximum distance).

We finally note that another way to solve the complete RLC control problem in one step (both the balancing and routing problems) can be accomplished by formulating it as a *multiple travelling salesmen problem* (cf. Lawler et al.[1985]) with additional capacity constraints on the total number of customer *types*, to be visited in one route (where customer types represent component types and the capacity constraint refer to feeder bank capacities).

The above remarks also apply to the ALC placement machine. However, due to the fact that the placement unit is now fed with an *externally* sequenced tape, it is generally possible to complete a PCB on a single ALC inserter. Off-line sequencers can be very large and therefore may load all ALC's required by one type of PCB. If not, the tape should pass two sequencers, resulting in a balancing problem similar to the formulation (8)-(11) (cf. Fathi and Taheri[1989]).

Production preparation for SMD insertion by an MCM is more difficult. Recall from section 2 that an MCM consists of a series of placement modules, where each pick-and-place unit holds 32 pipettes, picking components *simultaneously* from (SMD-specific) feeders in a feeder bank, after which these components are placed *sequentially*. Generally, a number of charges have to be performed at each module before a PCB moves to the next one. Since in each charge a large number of components are placed it is important to limit the number of charges at each module as much as possible. In the following we roughly sketch a hierarchical control system.

(I) For each *required* component type i, choose the number of feeders r_i such that $r_i \geq 1$ and furthermore proportional to the ratio $n_i / \Sigma n_i$, while finally $\Sigma r_i = 32m$ (with m the number of subsequent modules). This is done to decrease the total number of charges needed. Define w_i as the workload of a type i feeder, i.e. $w_i = n_i/r_i$. Naturally, this procedure may be modified to accomodate certain constraints on the number of feeders of a particular type as well as an overall fill rate constraint.

(II) To balance the workload among the m subsequent placement modules, solve the program defined by formulas (8)-(11) with n_i replaced by r_i for all i and $b_j=32$ for all j. The variables s_{ij} now represent the number of *feeders* of type i allocated to module j.

(III) For each component type i, divide the total number of n_i components into r_i equally sized groups (if n_i/r_i is not an integer, the group sizes may differ at most one) such that within each group the components have coordinate positions close to each other. Each group now represents a cluster to be served by one feeder.

(IV) Now, renumber the feeders by identifying each feeder with a cluster to be served. Note that each cluster contains components of only one type, however, feeders supplying the same component type in different clusters now have different numbers. Next, assign feeders to feeder positions at each module by solving a linear assignment problem. Similar to the procedure in section 4.5, we take the cost coefficients c_{ik} equal to the sum of the distances of feeder position k to the coordinate positions of the cluster served by feeder i.

(V) The number of charges needed on each module clearly equals the maximum cardinality, taken over the clusters of components to be placed at that module. The time needed to perform each charge is again a (symmetric) TSP, where the lengths of the arcs are obtained from certain translations (one should compensate the distances between coordinate positions of components on the board by the (fixed) distances between the pipettes placing these components). However, before formulating each TSP one should know which component in each cluster will be placed in the corresponding charge. As in the RLC case, this is accomplished simultaneously by interchanging components, now between charges, to minimize the *sum* of the charge-times needed on each module.

Note that the first three steps consider the complete MCM machine of m subsequent modules; the third step determines exactly which components (identified by their coordinate positions) will be processed by each placement unit (module). The last two steps concern isolated modules.

Summarizing, we again have defined a *hierarchical* control structure, now for the MCM. In general, SMD placement machines require a more complex control system than the relatively simple classical component inserters. In the final section we conclude this paper and suggest further research.

6. Conclusions and suggestions for further research

In this paper we have developed hierarchically structured models for the production preparation of several parts of a PCB assembly system, in particular for systems of ALC and RLC inserters, as well as for, more complex, SMD mounting lines. Part of the study has resulted in an operating Production Preparation system for a CSM line.

Several extensions are possible. First of all, new technologies, especially for SMD placement, demand for new control strategies. More important is the generalization of control systems as described in this paper to families of PCB's (instead of just one type). This is the subject of current research. The determination of sensible family structures, based on similarities of component requirements, can be accomplished by exploiting clustering techniques (see e.g. Kumar et al.[1986], Kusiak[1987]).

The effective operation of a PCB assembly line also highly depends on the efficiency of individual machining centres. Although this topic has been investigated on a macro level (see e.g. Buzacott and Hanifin[1978], Yao and Buzacott[1986], Altiok and Perros[1986]), the relationships on a micro level between the efficiency of a machine and the reliability of its isolated operations is seldomly well understood. In particular in high capacity mechanizations, such as the MCM's, one may find that an unreliability of 200 ppm (i.e. only 200 out of a million components are misplaced) may result in an overall efficiency of less than 60 %. Results of a study on these aspects are reported in Schuller and Zijm[1989].

References

Ahmadi, J., S. Grotzinger and D. Johnson[1988], *Component allocation and partitioning for a dual delivery placement machine*, Operations Research 36, pp. 176-191,

Altiok, T. and H.G. Perros[1986], *Open Networks of queues with blocking: Split and Merge Configurations*, IIE Transactions 18, pp. 251-261,

Ball, M.O. and M.J. Magazine[1988], *Sequencing of insertions in printed circuit boards*, Operations Research 36, pp. 192-201,

Bolwijn, P.T. et al.[1986], *Flexible Manufacturing: integrating technological and social innovation*, Elsevier, Amsterdam,

Bullinger, H.-J., H.J. Warnecke and H.-P. Lentes[1986], *Towards the Factory of the Future*, International Journal of Production Research 24, pp. 697-741,

Buzacott, J.A.[1982], *The fundamental principles of flexibility in manufacturing systems*, Proc. 1st Int. Conf. on FMS's, Brighton, U.K., pp. 13,

Buzacott, J.A. and L.E. Hanifin[1978], *Models of Automatic Transfer Lines with Inventory Banks, a review and comparison*, AIIE Transactions 10, pp. 197-207,

Chan, D. and D. Mercier[1989], *IC insertion: an application of the travelling salesman problem*, International Journal of Production Research 27, pp. 1837-1841,

Fathi, Y. and J. Taheri[1989], *A mathematical model for loading the sequencers in a printed circuit pack manufacturing environment*, International Journal of Production Research 27, pp. 1305-1316,

Hayes, R.H. and S.C. Wheelwright[1984], *Restoring our Competitive Edge: Competing Through Manufacturing*, Wiley, New York,

Kanellakis, P.C. and C.H. Papadimitriou[1980], *Local search for the asymmetric traveling salesman problem*, Operations Research 28, pp. 1086-1099,

Kumar, K.R., A. Kusiak and A. Vannelli[1986], *Grouping of parts and components in Flexible Manufacturing Systems*, European Journal of Operational Research 24, pp. 387-397,

Kusiak, A. (ed.)[1986-a], *Flexible Manufacturing Systems, Methods and Studies*, North Holland, Amsterdam,

Kusiak, A. (ed.)[1986-b], *Modelling and Design of Flexible Manufacturing Systems*, Elsevier, Amsterdam,

Kusiak, A.[1987], *The generalized group technology concept*, International Journal of Production Research 25, pp. 561-569,

Lawler, E.L. et al. (eds.)[1985], *The traveling salesman problem: a Guided Tour of Combinatorial Optimization*, Wiley, Chichester,

Leipälä, T. and O. Nevalainen[1989], *Optimization of the movements of a component placement machine*, European Journal of Operational Research 38, pp.167-177,

Lin, S. and B.W. Kernighan[1973], *An effective heuristic algorithm for the traveling salesman problem*, Operations Research 21, pp. 498-516,

Papadimitriou, C.H. and K. Steiglitz[1982], *Combinatorial Optimization: Algorithms and Complexity*, Prentice Hall, Englewood Cliffs,

Randhawa, S.U., E.D. McDowell and S.-D. Faruqui[1985], *An integer programming application to solve sequencer mix problems in printed circuit board production*, International Journal of Production Research 23, pp. 543-552,

Ranky, P.G.[1983], *The design and operation of Flexible Manufacturing Systems*, IFS Publications, Kempston, Bedford, U.K.,

Schuller, G. and W.H.M. Zijm[1989], *The design of mechanizations: reliability, efficiency and flexibility*, CQM-report 239, Centre for Quantitative Methods, Philips, Eindhoven (submitted for publication in International Journal of Production Research),

Skinner, W.[1985], *Manufacturing: the Formidable Competitive Weapon*, Wiley, New York,

Solberg. J.J. et al.[1985], *Factories of the Future: defining the target*, Research Report of CIDMAC (Computer Integrated Design Manufacturing and Automation Center), Purdue University, West Lafayette, Indiana 47907,

Stecke, K.E.[1985], *Design, Planning, Scheduling and Control problems of Flexible Manufacturing Systems*, Annals of Operations Research 3, pp. 3-12,

Stecke, K.E. and R. Suri (eds.)[1985], *Flexible Manufacturing Systems: Operations Research Models and Applications*, Annals of Operations Research 3, J.C. Baltzer AG Scientific Publishing Company, Basel,

Stecke, K.E. and R. Suri (eds.)[1986], *Flexible Manufacturing Systems: Operations Research Models and Applications*, Annals of Operations Research 15, J.C. Baltzer AG Scientific Publishing Company, Basel,

Van Laarhoven, P.J.M. and E.H.L. Aarts[1987], *Simulated Annealing: Theory and Applications*, Kluwer Academic Publishers, Dordrecht,

Van Laarhoven, P.J.M. and W.H.M. Zijm[1989], *Production Preparation and Numerical Control in PCB assembly*, CQM-note 093, Centre for Quantitative Methods, Philips, Eindhoven (submitted for publication to International Journal of Flexible Manufacturing Systems),

Wittrock, R.J.[1985], *Scheduling algorithms for flexible flow lines*, IBM Journal of Research and Development 29, pp. 401-412,

Yao, D.D. and J.A. Buzacott[1986], *Models of flexible manufacturing systems with limited local buffers*, International Journal of Production Research 24, pp. 107-118,

Zijm, W.H.M.[1988], *Flexible Manufacturing Systems: background, examples and models*, in: H. Schellhaas et al. (eds.), Operations Research Proceedings 1988, Springer Verlag, Heidelberg, pp. 142-161.

A Framework for Developing Production Control Systems: A Case in Coffee Roasting/Distributing

JACOB WIJNGAARD and JOHN MILTENBURG

1 Introduction

Production situations are unique. The way to control production has to be matched to its characteristics. The term production control concept stands for the heart of such a way of control. It is something like a proper view of the production situation; proper in the sense that on one hand the characteristics of the situation, as far as relevant for production control, are sufficiently taken into account, and on the other hand that it is suitable as a basis for developing rules, procedures, performance measures and information systems.
A production control concept is the way production control is stratified. To develop a production control concept requires care. Meal [1984] in his plea for a hierarchical approach of production planning and control stressed already the difficulty of the idiosyncracy of production situations. A too standard approach has to be avoided indeed. But by stressing the uniqueness of the situation too much it may get difficult to apply available general production control principles.

The aimed relevance of the paper is twofold.
In the first place, the paper stresses the process of developing a production control concept. The underlying conjecture is that in the theory of production planning and control more attention should be

paid on applying existing general control principles in (more or less) unique situations and maybe less attention on developing new control principles. The paper is meant to provide some evidence for that conjecture. In the second place, a production control concept is described for an actual coffee roaster and distributor (See Figure 1). In addition to illustrating the process of developing a production control concept, the particular production control concept developed may be applicable for other actual problems.
Other examples of the same kind of approach in developing a production control concept can be found in Bertrand et al. [1990].

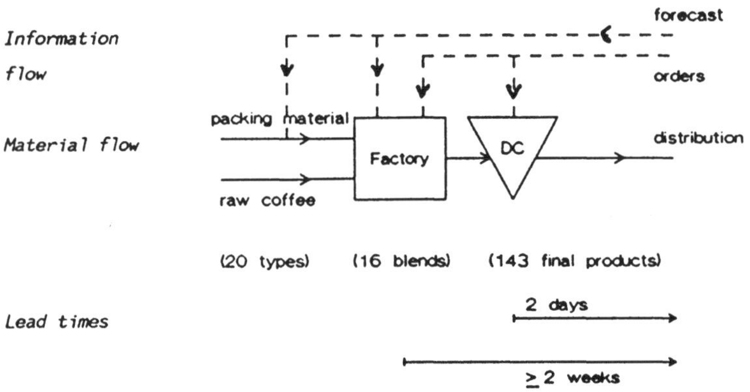

Stategy *The strategy of the firm is to be flexible, to be better able to follow short term sales actions. And to react to market opportunities, especially with respect to private labels and export. One wants a production control system to realize this and is prepared to change rules, procedures and information systems.*

Figure 1 Simplified Process Diagram for a Coffee Processor and Distributor

The process of developing a production control concept generally involves the following four phases.

Phase 1 Exploration (partly qualitative).
Phase 2 Process improvement.
Phase 3 Developing the production control concept itself.
Phase 4 Evaluation of the controlled system.

Exploration (phase 1) is meant to get a first impression of which characteristics of the production situation (if any) make it difficult to control. Such characteristics are important in stratifying the production control (phase 3) but are also important in scanning the possibilities to change the situation (phase 2) so that it is easier to control, with a better control performance. The fourth phase is to consider the performance of the controlled system. Developing a production control concept has to make the situation well structured. That means that thereafter it must also be possible to evaluate, at least partly, the performance of the controlled system. The four phases are discussed in the subsequent sections. Section 6 gives a summary and conclusions.

2 Exploration

With exploration is meant a general analysis of the production situation, to detect the elements that may make it difficult to control and so to get a first impression about the way to stratify production control. General characteristics that have to be considered here are:

- complexity of material structure
- complexity of routings
- restrictions with respect to volume
- restrictions with respect to variety
- market characteristics
- strategy with respect to customer order acceptance and delivery performance.

The importance of the characteristics varies from situation to situation. In the situation we use as an example here the material structure is rather simple for instance, while in an assembly factory the material structure is generally one of the main determinants of the production control concept.

In the coffee factory an important determinant turns out to be the restrictions with respect to variety and on the other hand the strategy to realize a high degree of flexibility in the market place. How the exploration was done by the coffee processor is described in Figure 2.

Strategy

Flexibility in the sense of high production variety and short reaction times is mainly determined here by the availability of packing material and by the factory itself, the process of blending, roasting, grinding and packing. The other two determining factors are the distribution process, and the availability of raw coffee. But the distribution process is completely as it should be in this market, and the procurement of raw coffee is heavily influenced by fluctuations in the market price and quality. It is decoupled therefore from the process of reacting on short term customer requirements and has no influence on the flexibility in that respect.
So, these two factors need not be considered further.

Packing material

The variety of final products is mainly determined by the way of packing. The availability of packing material is important therefore in realizing flexibility. Actual lead times of packing material vary from 5-10 weeks however. It is maybe possible to realize a certain reduction here. But the potential flexibility is nevertheless of an other order than the day to day flexibility aimed at in the factory. So there are two control loops.

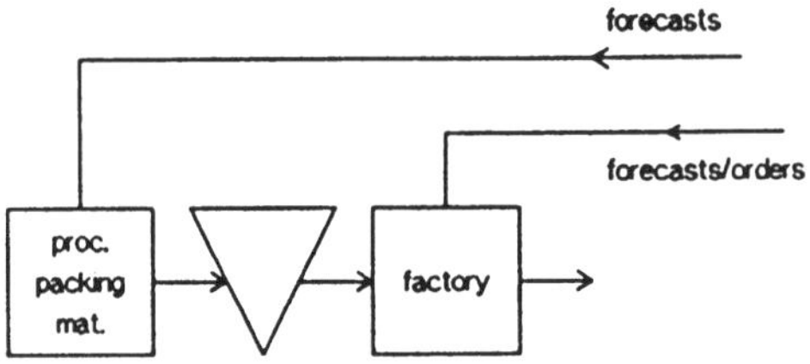

The control loop for the procurement of packing material is partly decoupled from the control loop for factory control. Attention is focussed on the factory control loop. For the procurement of packing material an MRP-type approach could be used. In developing a control concept for thefactory, it is assumed that packing material is available. Of course a check on that availability needs to be included.

Factory

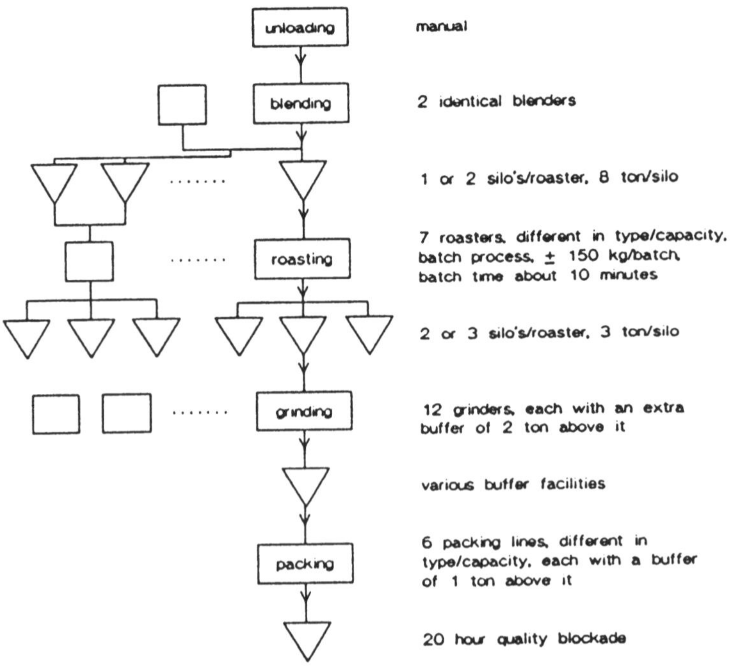

Capacity analyses reveal that the operator capacity for packing lines is a bottleneck. This is in fact not very surprising. The factory works with a two-shift system. Not all packing lines are always occupied. Using overtime is possible. People can be hired at short notice and mixed with experienced operators to stuff extra shifts. The medium term (months) flexibility of operator capacity is so high that one always aims at a utilization rate of about 100%. This means that in the short term (weeks) operator capacity is the bottleneck generally. See also chapter 10 in Bertrand et al. (1990).

The capacity of the grinders is significantly higher than the capacity of the packing lines. It is possible therefore in developing a production control concept to assume a hard coupling of packing lines and grinders: each packing line has two grinders coupled with it. That means that in the production control concept to develop, grinding and packing is considered as one combined production step, the capacity of which is determined by the packing lines. It has to be mentioned that this does not exclude the possibility to solve occasional operational problems by coupling grinders and packing lines in another way. The change over times of grinders and packing lines are about 15 minutes.

The packing lines are slightly different in type. There is substitutability however. Of course there is a pattern of always processing the same kind of products on the same packing line. That is useful in reducing the number of change overs. But the substitutability is sufficient to guarantee that a lack of capacity for one of the packing lines will never occur.

The roasters are not identical. Certain roasters are more suitable for certain blends. However there is a high substitutability. Furthermore, the changeover times are small (about 10 minutes). That means that the roasting capacity is very flexible. This flexibility is heavily blocked by the buffers above and below the roasters.

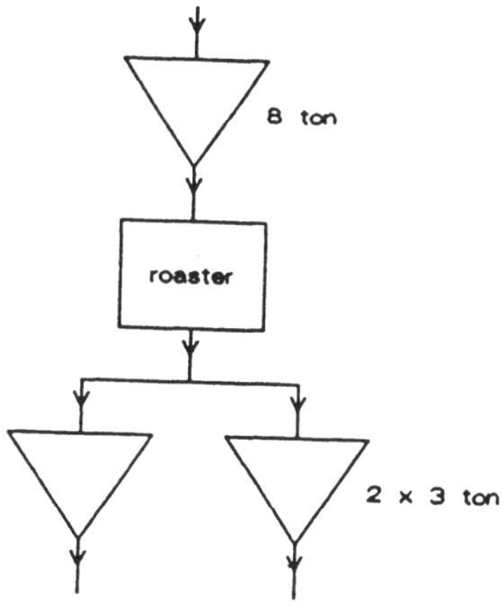

The roaster gets its raw material from the buffer above it. So it can only switch over to another blend if that buffer gets empty. That means that it is only possible to change over frequently if the buffer above the roaster is kept ratherempty. This can be realized by having frequent change overs in blending. This is technically possible. However, unloading is done manually. Another blend requires another pattern of unloading. That means that it is tempting to stick to the same blend until the buffer is completely filled.

In the next section process improvements will be discussed that lead to a reduction of the flexibility blocking of the upper buffer.The lower buffers also block the flexibility of the roasters. If a roaster is changed over to another blend, one of the lower silo's has to be empty. But this effect is less severe because there are more lower silo's and they are smaller. It is clear, however, that a roasting run of a "small" blend (small turnover) of more than 3 tons complicates the use of the roasting flexibility. Because then more than one silo has to be used, which blocks the flexibility severely. Or grinding and packing has to be executed partly simultaneous with roasting. And that requires a lot of planning effortand is anyhow difficult in case of disturbances in packing.

Figure 2 The Exploration Step at a Coffee Processor and Distributor

3 Process Improvement

Developing a production control concept is designing a way of living with, or making the best of, bottlenecks and inflexibilities. But first it should be tried to remove such points, to reduce inflexibilities and to increase capacity.

Thus, the second phase in the process of developing a production control concept is to consider possible process improvements, which should result in a simpler and more effective control concept (see Figure 3).

In this case straightforward potential improvements are in the upper part of the process, directed to circumventing that the upper buffers block the flexibility of the roasters. There are three possibilities:

- *Realize more connections between (upper) buffers and roasters,*
- *Change the manual patterns of unloading, or*
- *Automate unloading.*

The choice here was to automate unloading: to have a silo for each kind of raw coffee and automate to unload in the right mixture for each of the blends. In this way it is possible to unload and blend almost arbitrarily small quantities. A first effect of such an automation project of course is that the 8 ton buffers above the roasters may be replaced by much smaller ones.
In the rest of this example it is assumed that this process improvement is going to be implemented.

Realizing more connections between the roasters and the lower buffers is in principal also a useful improvement.

Figure 3 The Process Improvement Step at a Coffee Processor and Distributor

4 Developing the Production Control Concept

The production control concept for a production situation determines an appropriate way to stratify production control. Such a production control concept has to be taylor made. But of course there are certain general elements in it. Some of these general elements will be discussed here shortly. See Bertrand et al. [1990] for a more extensive discussion.

In the first place there is the concept of production unit. In the case of complex routings and many different capacities it may be appropriate to combine a number of capacity types and a subset of operations into such a production unit. Work orders have to be defined and the performance of the production units with respect to throughput and work order throughput time have to be formalized to so called operational characteristics. Such concepts are generally suitable for job-shop type environments and ideally the relationship between production units and (organizational) production departments is easy. The concept production unit is not very appropriate for the coffee factory considered here or rather the whole factory is one production unit.

In the second place in each production situation where resources are used for more products one needs to find a way to structure the interference between these products; all products can be made, but not simultaneously. For the products it is important to get a good coordination between the demand and the production in the various stages. The capacity use for each individual product is generally so small that if some product were given absolute priority it would be very easy to realize a high degree of coordination, so a short throughput time. The problem is the interference of the products. A good approach may be to have two levels of control, an aggregate level and a detailed level (see also Meal [1984]). At the aggregate level the product flows are aggregated to capacity content. But then it is crucial to design a good relationship between the two levels. Severe restrictions with respect to variety make that relationship difficult.

This point is going to be important in the coffee factory (see Figure 4).

In the third place there is the question of how to coordinate production and sales. For each organisation with limited resources it is important to guarantee in some way that commitments that are made to customers can be realized. It is important here which part of production is to-order. If all products are made to-stock, customer order acceptance can be restricted to checking the final inventory. If the whole production is to-order, it may be necessary to integrate customer order acceptance with production planning.
This point is given also attention in the example of the coffee factory (see Figure 4).

In the control concept the day to day planning of the factory is decoupled from the procurement of packing material and the procurement of raw coffee (see Figure 2). Of course the daily factory plans have to be checked for availability of raw coffee and packing material, but that availability is realized via other control loops, not considered in detail here. The procurement of packing material can be treated isolatedly, because it does not imply a struggle for scarce (internal) resources. So, it is not a very difficult part of the planning concept.
The quantities to produce daily may be based partly on final inventories and partly on customer orders. It is a mixture of make-to-stock and make-to-order. The make-to-order products have to be given priority. The final inventory of the make-to-stock products has to function aš a buffer for the fluctuations in the capacity required for make-to-order products.
That implies a high flexibility with respect to the make-to-order products and that means that customer order acceptance may be rather simple, also for these make-to-order products. A high integration of customer order acceptance and factory planning is not necessary.
Variety restrictions may make the relationship between the capacity level of control and the item level of control difficult. The main variety restrictors left here (after process improvement) are the silo's below the roasters and the packing operator capacity. The silo's below the roasters do that in the most complex way. So that will be given attention first.

On the use of silo's

It has been explained already that the flexibility of the roasters may also be blocked by the lower silo's. The problem would be solved of course if there were sufficient (dedicated) silo's for each blend and if the roasters were coupled with all or at least most silo's. There are 16 blends however and only 17 silo's. That means that for almost all blends there would be available only one silo. That requires a very precise adjustment of packing and roasting. And considering the unreliability of packing, it must be concluded that this is impossible. Extending the buffer capacity and making more connections is expensive. It would also lead to very inefficiently used silo's.

A better option is to only dedicate silo's to blends with a high turnover. The rest of the silo's can be used then for the other, smaller, blends. For these latter blends it is important to match the timing of packing and roasting, to keep the silo's as free as possible.

This is to be able to process the different blends as often as possible. And it is also important to take care not to process runs of more than 3 tons, because then it is sufficient for each run to use only one silo (see Figure 2). The length of time that a silo is occupied for the production of one run for a small blend and the number of silo's available for small blends determines the number of orders of small blends that can be processed per week or per day.

The occupation period of a silo for a production run of one blend is sketched below.

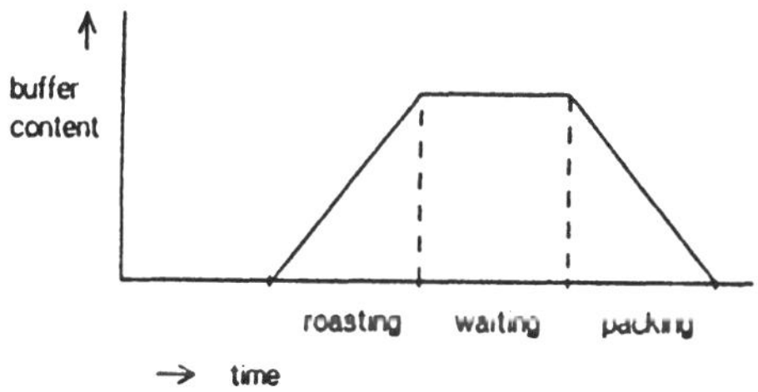

Buffer occupation period for one production run

Now, the way to realize a precise match of the timing of roasting and packing of small blends will be considered.

The concept for daily production planning

There are only two critical points left for daily production planning:

- effective use of packing operator capacity
- getting a precise match of the timing of roasting and packing of small blends.

This leads to a control concept providing daily packing plans for all final products and matched roasting and silo utilization plans for the smaller blends only. The matching of roasting and packing of the other (larger) blends is less critical. The silo's are dedicated and do not need to get empty after each run. Production is also more regular. Therefore the roasting of the larger blends is chosen to depend only on the content of the (dedicated) silo's, and not on the daily packing plans. The packing plan prescribes which products to make on which packing lines, how much of it and when. The quantities are based on final inventories and customer orders. This plan is translated to a roasting plan for the smaller blends. Each packing run for a smaller blend leads in principal to a separate roasting run. However in the case of very small packing runs of the same blend these may be clustered in the packing plan and combined to one roasting run. It has to be taken care that the roasting run does not exceed 3 tons (see above).

The smaller blends get priority in roasting. Since the roasting capacity needed for smaller blends is small compared to the total available roasting capacity, no problems may be expected in making that roasting plan. The packing plan and roasting plan together lead to a certain utilization pattern of the non-dedicated silo's. See below for a sketch.

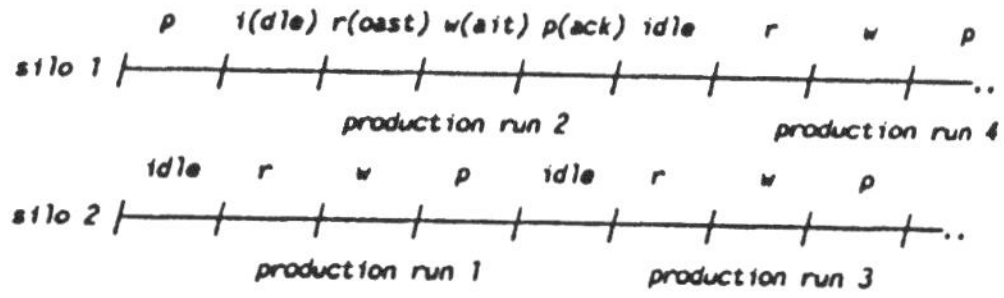

Silo utilization plan

Since the number of non-dedicated silo's is very restricted it may occur that a first packing-roasting plan leads to an infeasible silo utilization plan (overlapping production runs). In that case it may be necessary to choose other roasters to be

able to use silo's in another way. Or it may even be necessary to change the packing plan.
It may be useful to anticipate possible problems of this type by spreading the packing runs for different blends over the day. In the next section we will come back to how many non-dedicated silo's are necessary so that silo utilization plans can be realized.

The roasting capacity that is left can be used for the larger blends. The roasting of small blends may be seen as a disturbance of the roasting capacity available for larger blends. The capacity of the dedicated silo's has to be used to buffer such and other disturbances. In the next section it will be considered whether the available buffer capacity is sufficient for this.

Figure 4 Developing the Production Control Concept at a Coffee Processor and Distributor

5 Evaluation of the Controlled System

Once the production control concept has been developed and hence the production control is well structured it should be possible to evaluate the controlled system.
The general performance elements in production control are
- flexiblity to the market
- inventories and work in process
- production efficiency.

However, the flexibility to the market is too vague an aspect to evaluate directly. A production control concept should also make the performance aspects more operational. In the example of the coffee factory (and in many other process type production situations) a more operational performance variable is the production frequency that can be realized. The production frequency is the main determinant for the flexibility to the market, at least insofar as mix flexibility is concerned. It determines which products can be made to order and which have to be made to stock. It is also an important factor in the production efficiency in that it determines batch sizes.
The evaluation of the controlled system for the example of the coffee factory is described in Figure 5.

The main determinant for the performance of the controlled system is the production frequency. Here we consider the consequences of making all products every week. Of course another frequency can be considered in the same way. But this target was suitably concrete and attractive in the case situation under consideration.
Work-in-process and inventories of packing material and raw coffee will not be taken

into account. The influence of an increased production frequency on these inventories is positive. But work-in-process is anyway small compared to the final product inventories and a possible reduction of the inventories of packing material and raw coffee is prevented for the moment by the relative inflexibility of the procurement. Crucial consequences of a certain production frequency are:

- *capacity loss in packing*
- *utilization rate of non-dedicated silo's*
- *required buffer capacity in dedicated silo's*
- *inventory and flexibility.*

Capacity loss in packing lines

The number of change-overs per week is equal to the number of final products with a positive demand. Suppose each product has each week a positive demand (in reality this is less), then there are 143 change-overs per week. With a change over time of 15 minutes this implies a capacity loss of 33,25 hours. This is absolutely no problem for the machine capacity of the packing lines. But it implies also loss of operator capacity. There are more operators necessary than in case of a production frequency of 2 weeks. This loss has to be weighed against the extra flexibility realized in this way.

Utilization rate of non-dedicated silo's

The non-dedicated silo's are for the small blends. First the boundary between small blends and large blends must be determined. In this case the boundary is chosen at 1% of the total turnover. Of all 16 blends 9 appear to have a turnover of less than 1% of the turnover. These are by definition now the small blends. The 7 largest blends determine 95% of the turnover.
For each run for a small blend a lower silo is occupied during a certain time, consisting of roasting time, waiting time (until a packing line is available) and packing time. It is even possible to remove the packing time, because once a packing line is available the buffers above the corresponding grinders can be used to empty the silo. On the other hand the packing time may be longer if there are more final products per blend and if not all these products are processed on the same packing line. That may make it difficult to get the packing runs close together.
The roasting time is proportional to the quantity produced.
The waiting time is partly planned. The roasting has to be finished before packing starts, to avoid complications. And since the processes are not completely deterministic and a roaster has to be connected with one of the non-dedicated silo's one has to plan a positive delay between roasting and packing. This delay can get longer due to disturbances on the packing line.
Considering all these aspects a reasonable estimate for the silo occupation time for one run turns out to be 6 hours.
The number of runs is not much higher than the number of small blends. The largest of these small blends has an average week turnover of about 2 tons. Fluctuations in the weekly demand per blend may lead to weeks with a demand higher than 3 tons, requiring two runs. The average number of orders is estimated as 15.
This leads to a silo utilization of 15x6 = 90 hours per week. Each week there are 80 hours available per silo (two shift system). That means that 3 non-dedicated silo's are certainly sufficient.
That leaves 14 silo's for the larger blends.

Buffer capacity in dedicated silo's

For each of the larger blends there are one or more silo's dedicated to it. Roasting of these blends is not included in the daily plan. That is only the case for the roasting

of the smaller blends.
The roasting of the larger blends is based on the content of the silo's dedicated to it. That means that in the packing plan the availability of coffee of the right blend is not guaranteed. It is important then to check whether the capacity and flexibility of the roasters is sufficient to keep the silo's filled, independent of the actual use of the blends as following from the packing plan. Fluctuations in blend use are indeed determined by packing plan and execution. Fluctuations in available roasting capacity are determined by disturbances in the roasting capacity and by the use of roasting capacity for the smaller blends. These fluctuations have to be buffered by the silo content (see diagram).

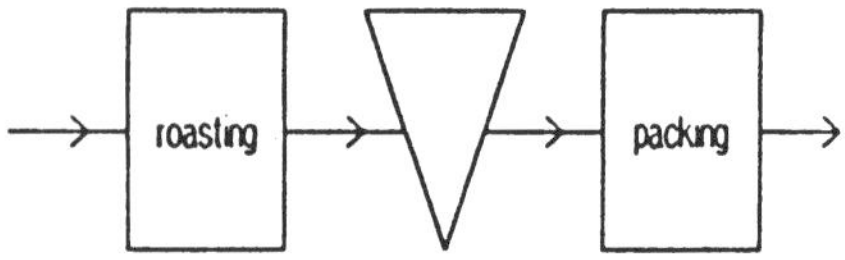

Buffering of roasting and packing

This suggests that there is only one type of material (one blend) while in fact there are more blends. However the diagram models the situation in the right way because the flexibility of the roasters is so high that it will never occur that there is a shortage of only one blend. A shortage of only one blend could easily be prevented by deploying more roasters for that blend. So a shortage of only one blend will never occur and that means that the only function of the buffer is to store roasting capacity. This implies also that the performance of the system is not very sensitive for the precise allocation of silo's to blends.
The effect of the buffer on the output of the combined roasting-packing system can be analyzed by models as developed by Buzacott (1967), De Koster (1989), Wijngaard (1979). Another possibility is to use the fact that there are no roasting capacity problems now while the packing runs are longer then they would be in case of a production frequency of 1 week and the use of blends less stable therefore. Here this last reasoning has been used indeed and it is concluded that buffer capacity for dedicated blends is no problem in case of a production frequency of once per week.

Inventory and flexibility

The foregoing discussion shows that a production frequency of once per week is feasible. That does not mean that that is the right frequency. A lower frequency is also possible. The decision has to be based on an evaluation of loss of (packing) operator capacity, inventory reduction and factory flexibility. Inventory and factory flexibility are partly substitutable. Increased flexibility can be evaluated by calculating the batch stock and safety stock reduction that is made possible so.
The safety stock is proportional to the uncertainty of the demand during the lead time (reaction time) and possible safety stock reduction can be determined easily if the proper data is available.
Inventory and factory inflexibility are not completely substitutable. They are only completely substitutable in case of a stable situation with given products and stationary demand. Factory flexibility is more valuable than the corresponding inventory in case of unexpected sales actions and other discontinuities in the demand.
This has to be taken into account in comparing the extra costs of extra change overs and the extra factory flexibility realized by a higher production frequency.

Figure 5 Evaluation of the Control System at a Coffee Processor and Distributor

6 Conclusion

Many papers have been written on production control principles (MRP, JIT, OPT), on information systems for production control and all kinds of decision support models for aspects of production control. The practitioner however who has to decide about the way to design an integrated production control for his own situation is left empty handed. He can adopt some general principle, but he knows his situation is different and he finds it difficult to project his experience in such general principles. He can try to apply certain decision support models but that is only possible if the whole situation is well structured already.
This paper is meant to give some support to the activity of structuring a production control situation. A four phase design process is presented and illustrated with an actual case study.
The process fits the unique features of a particular production situation into the general principles of production control, allowing an appropriate production control concept to be developed. The four phase design process, developed here, is a simple model that requires more research to better understand its strengths and limitations.

References

Bertrand, J.W.M., Wortmann, J.C., Wijngaard, J. (1990) "Production Control - A Structural and Design Oriented Approach", Elsevier Science Publishers.

De Koster, M.B.M. (1989) "Capacity Oriented Analysis and Design of Production Systems", Springer Verlag.

Buzacott, J.A. (1967) "Automatic Transfer Lines with Buffer Stocks", Int. Journal of Production Research 5, 183-200.

Wijngaard, J. (1979) "The Effect of Interstage Buffer Storage on the Output of Two Unreliable Production Units in Series, with Different Production Rates", AIIE-Transaction, 11, 42-47.

Meal, H.C. (1984) "Putting Production Decisions where they belong", Harvard Business Review, 84, 102-111.

The Impact of Forecast Errors in Multistage Production Systems

LARRY RITZMAN and BARRY KING

1. INTRODUCTION

In their pioneering work on aggregate production planning, Holt, Modigliani, and Simon (1955) and Bowman (1963) demonstrated that forecast errors can significantly increase manufacturing costs. Without having a perfect forecast, total costs increased by 10 percent. More recent work on the cost of forecast errors goes beyond the simplifying assumptions of a single product and single homogeneous work force. Forecast errors are being studied in the more realistic multistage manufacturing setting where parent-component relationships exist and there are capacitated work stations. However, the results to date imply that the impact of forecast errors depends on the manufacturing environment. The purpose of this paper is to find out more about these contingencies in realistic manufacturing settings, thereby adding to the growing literature base on how forecast errors affect manufacturing performance.

With multistage production systems, forecast errors occur only for end items, which are the finished products actually shipped to the customers. Intermediate items and purchased items are lower in the bills of material. These dependent-demand items do not experience forecast errors (assuming no service part demand), since their demand is derived from the production plans of their parents.

Forecast errors for end items have been introduced in prior studies with an error generating function. The error has typically been normally distributed, with a certain mean and standard deviation. Some studies have focused on the mean of this distribution (assuming the standard deviation to be zero), others have focused on the standard deviation (assuming the mean to be zero), and still others have allowed both parameters to be nonzero. For ease in presentation, we define two components of forecast error. Forecast variance occurs when the standard deviation of the distribution is nonzero. Forecast bias occurs when the mean is nonzero. A negative mean implies underforecasting and a positive mean implies overforecasting.

This paper begins with a review of the literature, with emphasis on unanswered questions and possible inconsistencies. Section 3 identifies the ways in which this study builds on, and in some cases departs from, prior studies. Section 4 identifies the six hypotheses which give this paper its focus. It also describes the simulator used, the manufacturing environments simulated, and the specific factor settings. The simulation results are discussed in Section 5, with the findings provided for each hypothesis. Finally, Section 6 summarizes what has been learned and suggests some implications for management.

2. LITERATURE REVIEW

Biggs and Campion (1982) were the first to investigate the real cost of forecast errors in a multistage MRP-based manufacturing system. They considered only forecast bias, assuming no forecast variance. The system simulated was one of the most complex ones studied to date, with 32 produced items and 3 capacitated work stations. They found that the cost function for underforecasting can be quite different than the one for overforecasting. Indeed, for the single manufacturing environment studied, overforecasting actually provided better overall performance than when there is no forecast bias. No interaction effects between forecast bias and the various elements of the manufacturing environment were considered.

Lee and Adam (1986), and later Lee, Adam, and Ebert (1987), went one step further by considering both components of forecast error. Four simple product structures were tested. Each had a single end item and from one to nine intermediate items. There was no shop structure, with one-week lead times assumed. Their findings are that both forecast variance and forecast bias, but particularly bias, impact system inventory costs and shortages. Generally, the greater the errors, the greater the shortages. Their findings are fairly much inconsistent with those of Biggs and Campion (1982), except that they found a slight overforecasting bias can improve overall performance. Of particular interest is the finding that there are significant interaction effects between forecast errors, lot sizing rules, and product structure.

The work by De Bodt and Van Wassenhove (1983), which was subsequently clarified by Wemmerlov (1985), showed how forecast errors increase lot-sizing costs. In these studies, there is neither a product structure with parent-component relationships nor a shop with capacitated work stations. Nevertheless, there are three relevant findings for our purposes. First, forecast variance was found to dramatically increase lot-sizing costs. Second, a buffer (in this case, safety stock) was introduced as a separate experimental factor rather than being confounded with forecast bias. The value of safety stock in buffering against forecast error was found to depend on the magnitude of lot sizes and the amount of demand variability. Third, as concluded by Wemmerlov (1989) in a followup study, research results in deterministic settings unfortunately are not automatically transferable to more realistic environments.

The study by Ritzman et al. (1984) and Krajewski et al. (1987), while not having forecast errors as its primary focus, also helps to orient our study. Guided by a panel of production and inventory managers from diverse plant environments, a comprehensive list of factors thought most important to manufacturing effectiveness was constructed. Low and high values of each factor were identified, to represent the range experienced in real plant environments. The manufacturing environments simulated had complex product and shop structures, with many elements of realism not present in prior studies. Forecast errors turned out to be less important to overall performance than such factors as the magnitude of lot sizes, scrap rates, worker flexibility, capacity cushion, and safety stock. However, only forecast variance (not bias) was considered as a factor, which might explain the only moderate impact of forecast errors. The single greatest determinant of inventory levels was the magnitude of lot sizes. Customer service was influenced by a combination of factors which work together to create uncertainties in demand, supply, and lead time. Of particular interest here is that each item in the product structure was classified as either a special or a standard. In all prior studies, items were always assumed to be standards, which means that they are produced or purchased to stock. A special, on the other hand, must be produced (or purchased) to order. Excess inventory is never purposely created for specials, and so it always is assigned a lot-for-lot (L4L) lot-sizing rule.

3. KEY FEATURES OF THIS STUDY

In order to address some unanswered questions, and yet build on prior research, this study has several key features. Each one is described and justified below.

MRP System

Sharma (1987) found that MRP is still the most common system in use today. It is used either "to a large extent" or "completely" by 58 percent of the plants in his survey. In contrast, the comparable number for JIT is only 11 percent. For this reason, and the desire to be consistent with prior studies, we use a weekly regenerative MRP system to plan and schedule production in the simulated plant. The master production schedule uses either the L4L rule or periodic order quantity (POQ) rule to calculate planned order releases for end items. The MRP explosion process calculates the planned order releases for intermediate items and purchased items, using either the L4L rule or fixed order quantity (FOQ) rule.

Forecast Variance and Bias

Both components of forecast errors are considered. Prior research implies that they are not equally important. Furthermore, a more realistic representation of bias is introduced here. In the earlier studies, the same level of overforecasting (underforecasting) bias was maintained over the full simulation run, which was 48 weeks in one case (Biggs and Campion 1982) and 120 periods in the other case (Lee and Adam 1986). It is more plausible that the manager will learn in a much shorter time span that a bias exists, and then make a new forecast. Furthermore, in one study (Lee and Adam 1986) the master schedule is made at the outset of the run for the full horizon to be simulated and is never changed. This approach assumes that managers are oblivious to the actual demands and inventory positions experienced as time moves forward. In contrast, the simulations in our study allow the mean bias to be reset several times over the simulated run, and allow the master production schedule to take advantage of information on actual demands experienced, as reflected in current inventory positions.

Specials

The product structure consists of both standards and specials. Specials are quite common in practice. For example, Sharma (1987) found that 21 percent of the plants surveyed make a customized product manufactured to customer specification. Another 30 percent make a standard product with components modified to customer specification. End items that are specials experience a different type of demand uncertainty than if they were standards. An order for a special often enters the production planning system with less than the normal "cumulative" lead time. Late orders, and the demand uncertainty that they create, are caused by optimistic marketing promises or engineering delays. More needs to be understood about the demand uncertainty associated with specials, and how it affects system performance.

Supply and Lead-Time Uncertainty

Taking heed of Wemmerlov's cautionary note, this study allows for the full range of uncertainties to more accurately gauge the impact of forecast errors. Uncertainty comes not just in demand, but also in supply and lead time. Supply uncertainty means that the number of units received in an order may not match the order quantity. Scrap losses to shop orders and short shipments from vendors are possible causes. Lead-time uncertainty means that actual lead times deviate from planned lead times. A primary cause is queuing delays in capacitated shops. They are a complex function of, among other things, the product structure and lot-sizing decisions of items processed at common work stations. Lead-time uncertainty also results from equipment downtime and variability in vendor lead times.

Such uncertainties, in turn, can create demand uncertainty for dependent-demand items. The timing and sizing of actual withdrawals of an item can differ from the gross requirements that were projected for it several weeks ago. For example, a co-component needed for a parent planned order release may not be available yet, which delays the items's own withdrawal.

Realistic Setting

The realism of the simulated environment can affect the findings. For example, studies on the best type of lot size always find the POQ rule superior to the FOQ rule (see Lee and Adam 1986) when there is no shop structure. However, when queuing delays are incorporated by having a shop structure, the FOQ rule becomes superior (see Biggs 1979). Every attempt is made here, therefore, to create a realistic manufacturing setting, beginning with having both a product structure and shop structure. The simulated plant has several bill-of-material levels, many items, the various types of uncertainties already discussed, inventory record inaccuracies, many work stations, limited capacities, and a representative amount of worker and machine flexibility. There is also some part commonality, with an item having more than one parent. Most prior work assumes no part commonality, even though Sharma (1987) found 67 percent of the plants surveyed have "some" to "very many" common parts and subassemblies.

Control for Buffers

In order to better understand the situational cost impact of forecast errors, we treat buffers as an experimental factor. There are really three types of buffers. The first type is the amount of safety stock for standards, and safety lead time for specials. This buffer is the one examined first by De Bodt and Van Wassenhove (1983). Ritzman et al. (1984) demonstrate that capacity cushion, defined here as the average percent of labor hours that are idle, is another essential one. Finally, worker flexibility and machine flexibility have recently been shown (Malhotra and Ritzman 1990) to be particularly helpful in environments characterized by high uncertainties, small capacity cushions, and large lot sizes. These three variables, if not carefully controlled, can easily be confounded with other variables. For example, it is not clear in Biggs and Campion (1982) and Lee and Adam (1986) how much of the results are really attributed to the benefits of buffers instead of an overforecasting bias. When there is overforecasting, the plant overbuilds and essentially creates safety stock as a buffer. Similarly, it appears in Biggs and Campion (1982) that the capacity cushion varied with the amount of bias and amount of allowable backordering.

Control for Lot Sizes

Lot sizes are also treated as an experimental factor, given the contingencies found by Biggs (1979) and Lee and Adam (1986). However, the focus here is on the magnitude of lot sizes, rather than the type of lot-size rule. De Bodt and Van Wassenhove (1983) found that, with the exception of the L4L rule, the cost differences between rules becomes rather insignificant when forecast errors are present. The L4L exception is noteworthy, since this rule produced considerably smaller lot sizes than the other rules. Its small lot sizes may very well explain the significant lot-size interaction effects found by Biggs (1979) and Lee and Adam (1986). Focusing on lot-size magnitudes also is more in keeping with the current interest in just-in-time production.

Performance Measures

The simulator used here provides many statistics on individual item cycle times, due-date performance, and inventories. It also reports shop information, such as equipment failures and the way individual work stations are utilized. In this paper, we focus on two aggregate performance measures. They are not combined into some overall measure of total cost, particularly since the impact from forecast errors on each one is quite different.

The first measure is the total average inventory investment, expressed as weeks of supply. This measure is equal to the average number of dollars tied up in inventories at all levels in the product structure, divided by the average dollar usage per week of all end items. The second measure is average past-due demand for all end items. Past-due demand occurs when the due date for a customer, as established by the master production schedule, is missed. As with the inventory measure, past-due demand is also expressed as weeks of supply. It is calculated by dividing the average weekly demand into the time-weighted average of late orders (in number of units).

There is a third aggregate performance measure, the average weekly labor hours, which measures work-force requirements. This measure can be calculated in advance for each cell in the experimental design. More resources obviously are needed for smaller lot sizes (more time lost to setups), more specials (also a lot-size effect), larger capacity cushions, higher scrap rates, and more equipment failures. We do not concentrate on this third measure because it is not a random variable and it is well understood.

4. RESEARCH HYPOTHESES AND RESEARCH DESIGN

Our research is conducted using the Manufacturing System Simulator (MASS), which has been fully tested and validated in prior studies (Krajewski, et al. 1987). MASS is a large-scale computer simulation model which reflects the complex relationships and dynamics found in multistage production systems.

Formal Hypotheses

Table 1 shows the hypotheses to be tested, and their source from the literature. They are tested here for the first time in an environment with the full range of demand, supply, and lead-time uncertainties. The first three hypotheses deal with the relative importance of different factors on performance. The fourth one singles out specials as a source of demand uncertainty, whereas

Hypothesis	Sources
1. Forecast bias has a bigger impact than forecast variance on inventory and past-due demand.	Biggs and Campion (1982) Lee and Adam (1986) Lee, Adam, and Ebert (1987)
2. Lot-size magnitude is the single biggest determinant of inventory.	Ritzman, et al. (1984) Krajewski, et al. (1987)
3. Forecast bias is the single biggest determinant of past-due demand.	Biggs and Campion (1982) Lee and Adam (1986) Lee, Adam, and Ebert (1987)
4. The number of specials in the product structure has a significant impact on inventory and past-due demand.	Ritzman, et al. (1984) Krajewski, et al. (1987)
5. The harmful impact of forecast bias on inventory and past-due demand is contingent on the buffering strategy.	De Bodt and Van Wassenhove (1983)
6. The harmful impact of forecast bias on inventory and past-due demand is contingent on the magnitude of lot sizes.	Biggs (1979) Lee and Adam (1986) Krajewski, et al. (1987)

TABLE 1. Hypotheses of Study.

the last two hypotheses look at the contingencies between forecast bias, buffers, and lot sizes.

Fixed Factors

The product structure to be simulated consists of 13 end items, 20 intermediate items, and 17 purchased items. Figure 1, which gives the indented bill of material for Item 5,

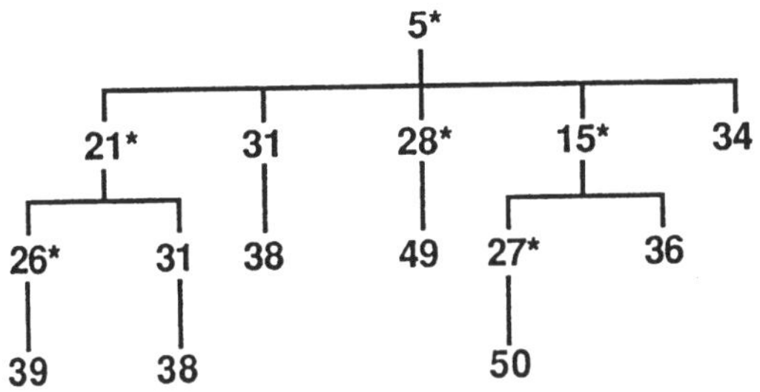

FIGURE 1. Product Structure for One of Thirteen End Items.

illustrates the amount of complexity involved. The asterisks designate which items are specials. All other items are standards. For this particular end item, there are four bill-of-material levels.

Material requirements planning (MRP) explosions and replanning are done weekly to launch new orders to the shop and suppliers. Figure 2 depicts one of two shop structures simulated. It has 50 individual work stations (machines) belonging to 18 machine groups, which in turn are assigned to 10 departments. Group 1, for example, has three identical work stations. It belongs, along with Group 2, to Department 1. The four workers at Department 1 can work at a machine in either group, depending on where the need is the greatest. However, they cannot work anywhere outside the departmental boundaries. The work stations within a given group are considered identical, as are the workers within a given department.

Each end item and intermediate item has a unique, pre-assigned routing through a specific sequence of groups. For example, the routing for end item 5 is 16-17-18; for intermediate item 20 it is 4-7-5-14-1. When a shop order for item 5 is released, it first gets into the job queue in front of Group 16. It waits there until one of the four work stations (WS 40 through WS 43) becomes free, one of the three workers in Department 9 gets assigned to this work station, and the order has the highest priority (earliest due date) of any job in the queue for Group 16. Once these conditions are met, the order is processed according to prespecified setup and per-unit processing times. The order then moves on to Group 17, and so forth.

Settings for many factors must be specified for the MASS simulator. Some factors deal with the MRP planning system, others deal with the product structure, and still others deal with the shop structure. Factor settings were established by a panel of managers for an earlier study. The managers identified

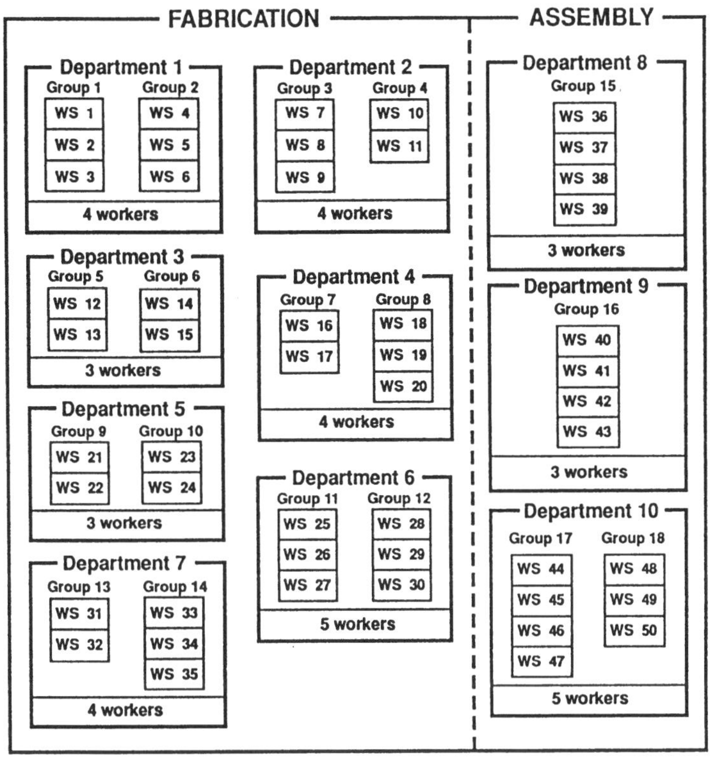

FIGURE 2. Shop Structure with Low Worker Flexibility.

low and high values for each factor that they considered to encompass the diversity of their individual plants across the country. For all factors, except the five experimental factors described in the next section, we essentially averaged their low and high settings to fix their settings for our experiments. See Krajewski, et al. (1987) for details.

Experimental Factors

Five clusters of experimental factors are included in our study. The first two, forecast bias (FB) and forecast variability (FV), are the two components of forecast error. To assess the situational impact of forecast errors, we also include three additional factor clusters: percent specials (PS), buffers (BF), and lot-sizes (LS). The low, medium, and high settings for these five clusters are summarized in Table 2. In general, a low setting was thought a priori to provide better on-time delivery than a high setting. Of course, some low settings can be worse in terms of inventory and weekly labor hours.

Forecast bias, which applies only to the end-item standards, is the mean of a normally distributed error generating function. The mean is randomly selected

A. **Forecast Bias (FB)**

1. Forecast Bias
   - Low: Mean error is uniformly distributed between -1% and +1% of true demand average. Bias is updated each week.
   - Medium: Mean error is between -35% and +35%, with update every 4 weeks.
   - High: Mean error is between -70% and +70%, with update every 12 weeks.

B. **Forecast Variability (FV)**

2. Forecast Variance (standards only)
   - Low: Standard deviation of normally distributed error function is 0.
   - Medium: Standard deviation is 5.5% of weekly demand average.
   - High: Standard deviation is 11% of weekly demand average.

3. Percent Late Orders (specials only)
   - Low: All orders received within normal cumulative lead time.
   - Medium: 15% of orders received late.
   - High: 30% of orders received late.

4. Demand Lumpiness (specials only)
   - Low: Average time between nonzero demand is 1 week.
   - Medium: Average time between nonzero demand is 2 weeks.
   - High: Average time between nonzero demand is 3 weeks.

C. **Percent Specials (PS)**

5. Percent Specials
   - Low: 10% of all items are specials.
   - High: 50% of all items are specials.

D. **Buffers (BF)**

6. Worker Flexibility
   - Low: Average of 17.5 work stations per worker.
   - High: Average of 5.2 work stations per worker.

7. Capacity Cushion
   - Low: 12% of worker time, on average.
   - High: 6% of worker time, on average.

8. Safety Factor
   - Low: For standards, safety stock is 50% of average weekly demand. For specials, safety lead time is 50% of the normal lead time.
   - High: For standards, safety stock is 25% of average weekly demand. For specials, safety lead time is 25% of the normal lead time.

E. **Lot Sizes (LS)**

9. Lot Sizes
   - Low: L4L for end items. FOQ equal to average of 1.6 weeks of supply for intermediate items and 3.1 weeks of supply for purchased items.
   - High: POQ (P = 4) for end items. FOQ equal to average of 6.6 weeks of supply for intermediate items and 12.1 weeks of supply for purchased items.

TABLE 2. Experimental Factors.

from a uniform distribution, with zero as its midpoint. Therefore, underforecasting and overforecasting are equally likely. Once the mean bias is selected, it is used to generate weekly demand forecasts for a particular end item over the whole planning horizon for each MRP explosion. Periodically the forecast is updated, so a new mean bias is selected from the same uniform distribution. In the low setting, the bias is negligible. The uniform distribution is from -1% to +1% of the true demand average, and the forecast bias is updated weekly. In the high setting, the bias is quite large. The uniform distribution is from -70% to +70% of the true demand average, and the forecast is updated quarterly.

Forecast variability is a cluster of three different factors: forecast variance, percent late orders, and demand lumpiness. The first one applies to end-item standards, while the last two apply to end-item specials. We defined forecast variance earlier as the standard deviation of the error generating function. It is zero at the low setting and 11 percent of the weekly demand average at the high setting. The percent late orders factor creates demand uncertainty for end-item specials. It is the percent of orders received in less than the "cumulative" lead time. If an order is late, the time remaining to its due date when it is first introduced as a gross requirement is randomly selected to be anywhere between 0 and 50% of its normal cumulative lead time. At the low setting there are no late orders; at the high setting, 30 percent of the orders are late. Demand lumpiness is the average number of weeks between nonzero demands for end-item specials. The longer the time interval, the more that demand has an on-again, off-again pattern so characteristic of specials. At the low setting there is some demand each week; at the high setting the time interval between demands averages three weeks.

Percent specials controls for the number of items in the product structure that are specials, rather than standards. At the low setting, 10 percent of the items are specials; it is 50 percent at the high setting. In either case, specials are concentrated at the higher levels of the bills of material. Furthermore, standard parents are more likely to have specials as components. Figure 1 represents the high setting for the part of the product structure involving Item 5.

Buffers is a cluster of three factors: worker flexibility, capacity cushion, and safety factors. Worker flexibility is measured as the number of different work stations that a worker can operate. Average worker flexibility ranges from 17.5 work stations in the low setting (representing considerable cross training) to 5.2 work stations at the high setting. The high setting is shown by Figure 2. We created the low setting simply by partitioning the 18 machine groups into only three (instead of 10) departments. For example, Department 1 would consist of all 17 work stations in machine groups 1 through 7, and have 12 workers. Capacity cushion reflects the percent of worker hours not used productively for processing, setups, and equipment downtime maintenance. It therefore is the complement of shop utilization. The safety factor assigns either safety stock or safety lead time to an item. At the low setting, safety stock is pre-assigned to standards in an amount equal to 50 percent of the average demand during the lead time. For a special, the normal value for its planned lead time is inflated by 50 percent, creating a safety lead time. At the high setting, the safety provision is only 25 percent.

The lot sizes factor controls the magnitude of lot sizes. At the low setting, end items have a L4L rule. Lower level items have a FOQ rule, with an average of 1.6 weeks of supply for intermediate items and 3.1 weeks of supply for purchased items. At the high setting, end items have a POQ rule (with P = 4). Lower level items have a FOQ rule, with an average of 6.6 weeks of supply for intermediate items and 12.1 weeks of supply for purchased items.

Four experimental factors affect the weekly labor requirement: capacity cushion, lot sizes, percent specials, and demand lumpiness. Increasing the cushion requires that labor hours be added to achieve the desired shop utilization. The last three factors have a lot-size effect; more hours are needed with the extra setups created by small lot sizes. For example, having more specials tends to reduce the average lot size, since they use the L4L rule. More labor capacity must be provided, so as to maintain the desired capacity cushion. Similarly, less demand lumpiness for specials creates smaller lot sizes, since the L4L rule is being used. In order to achieve the desired capacity cushion across all departments, we proportionately change the items' processing and setup times. This adjustment process is equivalent to adding or subtracting labor hours to a department's daily capacity, such as through the use of overtime. These adjustments are kept track of, and briefly reported in Section 5. However, our main concern is with past-due demand and inventory.

5. FINDINGS

We conducted a full factorial experiment which had 72 cells, since there are three levels for the first two factors and two levels for the other three factors. There were six replications per cell, resulting in a total of 432 observations. Each observation covered a 25-week simulation period, taken after a 100-week initialization period.

The SAS statistical package was used to analyze the data. ANOVA, step-wise regression, and cross tabulations were applied to the data. Table 3 gives an initial view of the results, showing the average performance at each setting for each experimental factor. Table 4 summarizes the ANOVA results by showing the statistical significance of each main effect and two-way interaction. The R^2 is 0.991 when inventory is the dependent variable; it is 0.787 when past-due demand is the dependent variable.

Hypothesis #1

The first hypothesis deals with the relative importance of forecast bias and forecast variability. As seen in Table 3, average performance is worse with high bias than with high variability. The impact of bias on past-due demand is particularly striking, jumping from 0.31 to 2.46 weeks of supply. When there is underforecasting for an item, not enough of it is produced. It takes some time to catch up even after a new forecast is made, and customer service suffers. As for inventory, bias is also more of a culprit. This is true even though high variability (through demand lumpiness) creates larger lot sizes. Having a bias simply means that the wrong items are being built. When an item's demand is

Experimental Factor	Setting	Inventory (Weeks)	Past-Due Demand (Weeks)	Labor Hours (Weekly)
1. **Forecast Bias (FB)**	Low	19.5	0.31	2579
	Medium	20.0	0.93	2579
	High	21.1	2.46	2579
2. **Forecast Variability (FV)**	Low	19.4	1.44	2619
	Medium	20.5	1.03	2583
	High	20.7	1.23	2554
3. **Percent Specials (PS)**	Low	24.8	1.14	2554
	High	15.6	1.33	2616
4. **Buffers (BF)**	Low	21.2	0.57	2693
	High	19.2	1.89	2477
5. **Lot Sizes (LS)**	Low	11.2	0.52	2683
	High	29.2	1.94	2487
	AVERAGES	20.2	1.23	2585

TABLE 3. Average Performance at Each Setting.

Independent Factors	Inventory			Past-Due Demand		
	Mean Square	R^2	F-test p-value	Mean Square	R^2	F-test p-value
Main Effects						
FB	107.0	0.0039	0.001	177.0	0.2291	0.0
FV	66.7	0.0024	0.001	6.2	0.0080	0.0005
PS	9177.9	0.1668	0.0	3.7	0.0024	0.0328
BF	457.5	0.0083	0.0	188.2	0.1218	0.0
LS	35095.9	0.6378	0.0	217.5	0.1408	0.0
Two-Way Interactions						
FB*FV	4.6	0.0003	0.0039	4.0	0.0105	0.0006
FB*PS	5.5	0.0002	0.0102	3.3	0.0043	0.0171
FB*BF	22.0	0.0008	0.0001	57.3	0.0741	0.0001
FB*LS	63.6	0.0023	0.0001	83.5	0.1081	0.0001
FV*PS	39.9	0.0014	0.0001	10.4	0.0134	0.0001
FV*BF	3.3	0.0001	0.0607	2.7	0.0035	0.0370
FV*LS	8.8	0.0003	0.0007	2.0	0.0026	0.0873
PS*BF	7.9	0.0001	0.0100	6.4	0.0042	0.0052
PS*LS	9145.6	0.1662	0.0	4.9	0.0032	0.0144
BF*LS	8.9	0.0002	0.0061	94.4	0.0611	0.0001
Model R^2		0.9913			0.7869	

TABLE 4. Summary of ANOVA Results.

overforecast, the excess produced must sit in inventory until demand catches up with supply. The excess production also robs capacity from shop orders that really do need to be made to avoid late shipments. Table 4 supports this first hypothesis by showing larger mean squares for FB with respect to inventory and past-due demand (107.0 and 177.0, respectively) than for FV (66.7 and 6.2, respectively). The high R^2 of FB for past-due demand (0.229) is particularly noteworthy, especially compared to that of FV (only 0.008).

Hypothesis #2

The second hypothesis receives support in Table 3, since the average jump in inventory between the low and high settings (11.2 to 29.2 weeks of supply) is greater for lot sizes than for any other factor. Stepwise regression was conducted, with inventory as the dependent variable. The LS factor entered first, with an R-square of 0.638. The sequence of factors to enter next were PS, BF, FB, and FV. The strength of these entries can be seen in the R^2 values of Table 4 for inventory. The LS, PS, and LS*PS effects alone account for 0.971 of the R^2 for the final model shown. As explained earlier, much of the impact of the PS factor, and even the FV factor, is really a lot-size effect.

Hypothesis #3

Table 3 implies that forecast bias is the biggest determinant of past-due demand, given the very large 2.46 weeks of supply at the high setting. To confirm this impression, stepwise regression was conducted using past-due demand now as the dependent variable. The FB factor entered first, with an R^2 of 0.216. It was followed next by LS and BF. PS and FV did not contribute enough to justify inclusion. When these three factors were entered, R^2 increased to 0.479. While forecast bias turned out to be most significant, it clearly is only one of several factors that create the demand, supply, and lead-time uncertainty so harmful to customer service. The interaction effects of bias with other factors are substantial, as demonstrated by the R^2 column of Table 4 for past-due demand. The impact of forecast variance, where an overforecast in one week is often cancelled out by an underforecast in the following week, is quite modest (FV $R^2 = 0.008$).

Hypothesis #4

Table 3 suggests that increasing the amount of product customization, as represented by the number of specials, helps reduce inventories. Notice the large mean square and R^2 values for PS and PS*LS in the inventory section of Table 4. These results are expected, since the L4L rule means smaller lot sizes.

The impact of PS on past-due demand is less clear-cut. As hypothesized, past-due demand gets worse at the high setting (see Table 3) and the change is statistically significant (see main effect and two-way interactions in Table 4). However, the impact of PS is dwarfed by other effects such as FB, LS, BF, and FB*LS. The PS factor has its highest interaction with FV, but even this effect explains only 1 percent of the variance ($R^2 = 0.013$). What can be concluded, however, is that demand uncertainty is every bit as troublesome with product structures having more specials (and fewer standards). Demand uncertainty remains, even when customized products are made to order rather than to forecast.

Hypothesis #5

Moving to more of a contingency theory on forecast errors, we find that the relative importance of forecast bias does indeed depend on the buffering strategy. The two-way interaction between forecast bias and buffers (see FB*BF in Table 4) is significant at the 0.0001 level for both inventory and past-due demand. Table 5 clarifies the exact nature of the contingencies. Inventory increases perceptibly with more bias and with larger buffers (due to the safety factor). However, the relative advantage of larger buffers diminishes with more forecast bias. For example, with a low bias the smaller buffers increase inventory by 15 percent (from 18.1 to 20.8 weeks of supply). With a large bias the impact is only 5 percent (from 20.6 to 21.7 weeks of supply). One reason for this dampening effect is that the three factors in the BF cluster work against each other relative to their inventory impact. Larger safety stocks and safety lead times increase inventory, but Malhotra and Ritzman (1990) show that larger capacity cushions and more worker flexibility reduce inventory (by reducing queuing delays and the frequency of component shortages).

Forecast Bias	Buffers*	Inventory (Weeks of Supply)	Past-Due Demand (Weeks of Supply)
Low	Small	18.1	0.44
	Large	20.8	0.18
	Percent Change	(+15%)	(-59%)
Medium	Small	18.9	1.42
	Large	21.2	0.44
	Percent Change	(+12%)	(-69%)
Large	Small	20.6	3.82
	Large	21.7	1.10
	Percent Change	(+5%)	(-71%)

*Small buffers correspond to a high setting of the BF factor.

TABLE 5. Interactions Between Forecast Bias and Buffers.

The contingencies are much more pronounced with past-due demand. With low bias, the larger buffers reduce past-due demand by 59 percent (from 0.44 to 0.18 weeks of supply). The impact grows to 71 percent when forecast bias is at its high setting (from 3.82 to 1.10 weeks of supply). Buffers become more and more important to customer service as forecast bias increases.

Hypothesis #6

The two-way interaction between forecast bias and lot sizes (see FB*LS in Table 4) also is significant at the 0.0001 level for inventory and past-due demand. Table 6 offers some insight on these contingencies. Large lot sizes inflate inventory somewhat more when bias is at the high setting (by 171 percent) than when there is a low bias (by 154 percent). Larger lot sizes commit the plant to inventory farther into the future; more time is needed to work off the excesses caused by overforecasts.

Forecast Bias	Lot Sizes*	Inventory (Weeks of Supply)	Past-Due Demand (Weeks of Supply)
Low	Small	11.0	0.25
	Large	27.9	0.36
	Percent Change	(+154%)	(+44%)
Medium	Small	11.1	0.40
	Large	28.9	1.45
	Percent Change	(+160%)	(+263%)
Large	Small	11.4	0.92
	Large	30.9	4.01
	Percent Change	(+171%)	(+336%)

*Small lot sizes correspond to a low setting of the LS factor.

TABLE 6. Interactions Between Forecast Bias and Lot Sizes.

The interaction between forecast bias and lot sizes is much more dramatic with past-due demand. Larger lot sizes increase past-due demand by only 44 percent when forecast bias is negligible. The impact grows to 336 percent when forecast bias is high. It takes considerable time in capacitated shops to recover from underforecasted demand.

6. CONCLUSIONS

This study analyzes the impact of forecast errors on total inventory and past-due demand in a multistage manufacturing environment. This environment has all three types of uncertainties: demand, supply and lead time. The simulation results support all six hypotheses. As with most research studies, the findings reported here are true only for the combinations of factors tested and for the assumptions made (such as the product and shop structure). We tried to use realistic settings, as established by a panel of managers for a prior study, and explicitly state all assumptions.

Several conclusions can be drawn from this research. They are listed below, along with their implications for management.

Concentrate on Forecast Bias

Forecast bias is much more critical to success than forecast variance. When monitoring forecast accuracy, managers should be concerned less with usual measures of forecast error stressed in the literature, such as average error (AE), mean absolute deviation (MAD), mean squared error (MSE), and mean absolute percent error (MAPE). A much more revealing statistic is the cumulative sum of forecast errors (CFE), because it measures the bias in the forecast. Tracking signals, which use CFE in the numerator, are a valuable tool, particularly when forecasts are being made automatically and the manager needs to know when to intervene. Another approach is simply to ask for more frequent forecast updates, with an eye to correcting a bias before it becomes too disruptive.

Use of Buffers

In addition to finding ways to reduce forecast bias, managers can also mask some of the problem with higher buffers. Safety stock and safety lead time are not the only buffering mechanisms. More worker flexibility and larger capacity cushions can also help. However, they come at a cost and do not fully solve the problem. Past-due demand can still be excessive, even with high buffers (see Table 5).

Reducing Inventory

If management's primary concern is system inventories, it must do something about lot sizes. This truth is closely allied with the current interest in just-in-time systems. The magnitude of lot sizes is the major determinant of overall inventory levels. Some secondary help can also come from reducing safety stocks and safety lead times, particularly if compensated for with the other buffers of work-force flexibility and capacity cushions. Otherwise, past-due demand becomes excessive.

Improving Customer Service

If management's primary concern is customer service, it should begin by bringing forecast bias down to acceptable levels. This step is particularly important in environments where large lot sizes and small capacity cushions are justified or necessary. All three factors work together to create the demand, supply, and lead-time uncertainties so harmful to customer service.

Dealing with Specials

Managers with plants making more customized products can expect lower inventories, since specials require the L4L rule. Furthermore, the problems of poor customer service and missed due dates can be as bad or worse when there are more specials, even though forecast bias is associated only with standards. Specials introduce a different type of demand uncertainty -- whether or not the production control analysts get the orders from engineering in enough time. A closer interface between manufacturing, engineering, and marketing can help. Specials are not well addressed in the literature, although they are quite common in practice.

In summary, the impact of forecast errors, and forecast bias in particular, is multifaceted. It is clear that the impact is situation dependent. No one step is the only avenue to improved performance. More research on the contingencies is needed, particularly relative to specials. In the meantime, managers must dig deeply for the real problems causing high inventories and a poor record of on-time delivery.

References

BIGGS, JOSEPH R., "Heuristic Lot-Sizing and Sequencing Rules in a Multistage Production-Inventory System," Decision Sci., 10, 1 (January 1979), 96-115.

BIGGS, JOSEPH R. AND WILLIAM M. CAMPION, "The Effect and Cost of Forecast Error Bias for Multiple-Stage Production-Inventory Systems," Decision Sci., 13, 4 (October 1982), 570-584.

BOWMAN, E. H., "Consistency and Optimality in Managerial Decision Making," Management Sci., 9, 2 (January 1963), 310-321.

DE BODT, MARC A. AND LUK N. VAN WASSENHOVE, " Cost Increases Due to Demand Uncertainty in MRP Lot Sizing," Decision Sci., 14, 3 (July 1983), 345-361.

HOLT, CHARLES C., FRANCO MODIGLIANI, AND HERBERT A. SIMON, "A Linear Decision Rule for Production and Employment Scheduling, "Management Sci., 2, 1 (October 1955), 10-30.

KRAJEWSKI, LEE J., BARRY E. KING, LARRY P. RITZMAN, AND DANNY S. WONG, "Kanban, MRP, and Shaping the Manufacturing Environment," Management Sci., 33, 1 (January 1987), 57-75.

LEE, T.S. AND EVERETT E. ADAM, JR., " Forecasting Error Evaluation in Material Requirements Planning (MRP) Production Systems," Management Sci., 32, 9 (September 1986), 1186 - 1205.

LEE, T.S., EVERETT E. ADAM, JR., AND RONALD J. EBERT, "An Evaluation of Forecast Error in Master Production Scheduling for Material Requirements Planning Systems," Decision Sci., 18, 2 (Spring 1987), 292-307.

MALHOTRA, MANOJ K. AND LARRY P. RITZMAN, "Resource Flexibility Issues in Multistage Manufacturing," Decision Sci., 21, 4 (Fall 1990).

SHARMA, DEVEN, "Manufacturing Strategy : An Empirical Analysis," Unpublished Ph.D. dissertation, The Ohio State University, 1987.

RITZMAN, LARRY P., BARRY E. KING, AND LEE J. KRAJEWSKI, "Manufacturing Performance -- Pulling the Right Levers," Harvard Business Rev., 84, 2 (March-April 1984), 143-152.

WEMMERLOV, URBAN, "Comments on 'Cost Increases Due to Demand Uncertainty in MRP Lot Sizing': A Verification of Ordering Probabilities," Decision Sci., 16, 4 (Fall 1985), 410-419.

WEMMERLOV, URBAN, "The Behavior of Lot-Sizing Procedures in the Presence of Forecast Errors," Journal of Operations Management, 8, 1 (January 1989), 37-47.

Integrated Manufacturing Planning

LAWRENCE VITT

INTRODUCTION

This paper will develop the formal structure of hierarchical manufacturing planning and present an integrated example. Management control of the planning process can be achieved by a formal development of the production plan and the master production schedule (MPS). A production plan sets the annual commodity plan for capacity, inventory and workforce units. The capacity and inventory data are passed to the capacity-constrained master production schedule which determines the lot sizing and timing of family/product production. The MPS data will "drive" the detailed material plan.

The sample production plan and sample master production schedule were calculated using commercial microcomputer-based linear programming software. A calculated production plan enhances the "what if" testing and evaluation of alternative manufacturing resource strategies. The output file for the production plan can be used as the input file for an optional financial planning program. A linear programming model for the master production schedule is generated using the parameters for each application.

One important aspect of a structured framework is that it can be used by most manufacturing firms. The same planning programs and planning software can be used if a "commodity" structure is used. A commodity structure is essential for the hierarchical planning process. A production plan and master production schedule are prepared for each commodity.

HIERARCHICAL PLANNING

Hierarchical planning is a term used to describe the development of an aggregate production plan and then passing the same plan on to the more detailed master production schedule.

PRODUCTION PLAN -- The one-year plan for the commodity which sets the capacity (production rate) for each period.

MASTER PRODUCTION SCHEDULE (MPS) -- Lot sizing and timing of production for the major products/families.

The production plan is passed directly and indirectly to the MPS. A production plan is prepared using the inventory, capacity and demand data for the commodity. Capacity data for each period are passed directly to the MPS. The same total demand, each period, appears in both plans; but the production plan uses commodity demand and the MPS uses product demand. The detailed inventory plan is "left open" in the MPS. However, with both planning levels having the same capacity and demand data--the MPS inventory plan will "follow" the production plan.

The components of hierarchical planning addressed in this paper are presented in Figure 1.

Figure 1
THE HIERARCHICAL PLANNING FRAMEWORK

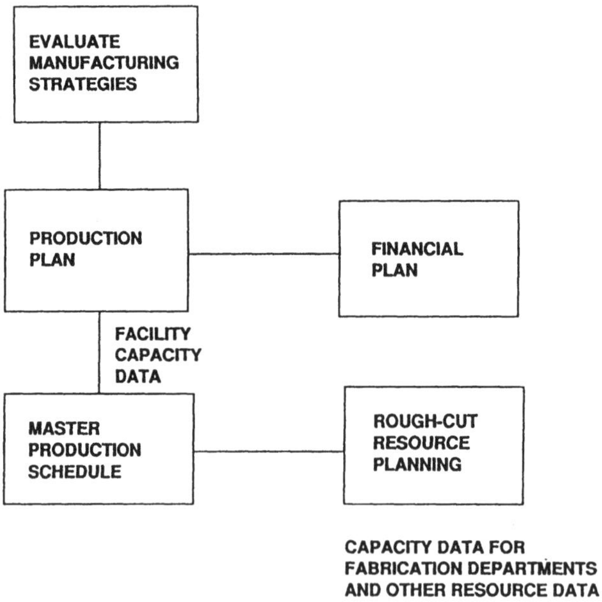

The hierarchical planning process actually continues down another level. The master production schedule will "drive" the detailed material plan. This is most evident in the formal planning of material requirements planning (MRP). Here the MPS data will appear in the LEVEL ZERO records and drive the material requirements explosion process. (This planning level is not addressed in this paper.) One very important benefit of good front-end planning is a more reliable and stable MPS.

Production plans can be prepared using twelve months or thirteen four-week periods. The sample production plan will use thirteen four-week periods. These time buckets remove the problem of more/fewer working days per month and provide a better calculation of actual changes in capacity rates and workforce levels. Four-week periods are also easier to disaggregate into weekly MPS data.

The sample production plan was prepared using commercial micro-computer-based linear programming software BLP87/88. (The BLP87/88 software is a proprietary program product of Eastern Software Products, Inc.) Calculated production plans provide an enhanced capability for "what if" testing and the evaluation of alternative manufacturing strategies. Capital investment proposals, shift structures, workforce and inventory strategies can be evaluated using the model. An optional financial plan is based on the one-year production plan. The linear programming output file is used as the input data file for the financial plan. Thus, the financial impact of alternative manufacturing strategies can be easily and quickly assessed.

The sample master production schedule was also calculated using the BLP87/88 software. A linear programming model was generated for this set of parameters. Thirteen four-week periods are used in the sample MPS to illustrate the hierarchical data flow. A more common scenario would be thirteen or twenty-six weekly time buckets in an MPS with a rolling planning horizon. (No changes required in the disaggregation procedures.) An optional rough-cut resource planning (RCRP) module is used to ensure that the MPS does not overload fabrication departments, vendor capacities, or other resources. Having the MPS and RCRP modules in one integrated linear programming model will ensure that the sample master production schedule is calculated within the RCRP constraints.

A SAMPLE PRODUCTION PLAN

The planning horizon for the production plan will start at the beginning of the slack demand season. Major components of the production plan are: the commodity inventory, the schedule for units produced each period, and the workforce plan. The linear programming model minimized the sum of the following aggregate programming costs:

- anticipation inventory carrying costs
- production costs for shift premiums and overtime premiums
- cost of changing the size of the workforce.

The optional workforce data have not been presented for the sample plan. The workforce size, each period, is related to the units produced (overtime production excluded). Shift efficiency factors will have an impact on the number of workers required.

Figure 2

A SAMPLE PRODUCTION PLAN FOR COMMODITY A
WITH 13 FOUR-WEEK PERIODS

	ANTICIPATION INVENTORY						PRODUCTION VARIABLES RELATED TO CAPACITY CONSTRAINTS								To The Master Production Schedule
PERIOD	Regular Inventory	Units/In Regular Inventory	Units/Out Regular Inventory	Extra Inventory	Units/In Extra Inventory	Units/Out Extra Inventory	Holiday first shift	Holiday second shift	Vacation first shift	Vacation second shift	Production first shift	Production first shift overtime	Production second shift	Production second shift overtime	Total production
0	150														
(Sept) 1	689	539					52	49			988		745		1733
2	1058	369					52	49			988		746		1734
3	1183	125		153	153						1040		794		1834
4	1183			557	404		52	49			988		746		1734
5	950		233	557			104	98	260	247	676		449		1125
6	950			599	42						1040		794		1834
7	950			599							1040		890		1930
8	950			584		15					1040		935		1975
9	950			271		313					1040		934		1974
10	950			131		140					1040	81	934	54	2109
11	231		719			131			260	247	780	81	688	54	1603
12			231				52	49			988	130	885	87	2090
(Aug) 13											1040	130	934	31	2135

NOTES: Beginning inventory of 150 units was specified.
Company inventory limit of 950 units was used for periods 5 to 10.
Upper limits used for all production variables.
Two Weeks of vacation time was required ("placed" in periods 5 and 11).
Holiday schedule was specified.
The efficiency factor for the second shift is 95 percent.
Zero units were backordered.

Only data for the anticipation commodity inventory appear in the production plan. The sample plan uses two types of anticipation inventory--regular and extra. Cost data can be applied to the variables for units in/out each period. The regular inventory may represent the factory warehouse inventory. The extra inventory could represent inventory in regional or public warehouses. With this scenario, a very powerful integrated manufacturing and distribution model can be prepared.

The major link to the master production schedule is the total number of units produced each period (refer to the last column presented in Figure 2). The sample plan uses two shifts and schedules some overtime production for each shift. Shift vacation and holiday allowances are stated in units of "product" (commodity). These allowances reduce the amount of available capacity in selected periods.

The planning model is based on period material flow equations and period capacity upper limits.

MATERIAL FLOW EQUATION

Beginning Inventory	+	Sum of "Production" Variables	-	Ending Inventory	=	Adjusted Demand for the Production Plan

CAPACITY UPPER LIMIT

Sum of "Production" Variables	+	Vacation Allowance	+	Holiday Allowance	≤	Capacity Data

Unit cost data for the linear programming model are fairly easy to prepare. These data are "planning costs" and reasonable estimates should provide adequate cost tradeoffs. For the sample plan, the unit cost of the regular first shift production was zero dollars. This is the least-cost manufacturing alternative (commodity produced at standard cost). The model will seek to minimize the production plan costs. In general, the production plan costs are related to the amount of seasonal variation in the commodity demand.

Model options can improve the fit for specific applications.

- Twelve months or thirteen four-week periods
- One, two or three shifts
- Workforce data
- Separable programming (piecewise approximation) for inventory, overtime, and workforce change variables
- Goal programming for shift plans

A SAMPLE MASTER PRODUCTION SCHEDULE

The MPS model calculates the lot sizing and timing of production (units produced) for products. The solution will minimize the sum of family set-up costs and product inventory carrying costs.

FAMILY--A group of products sharing a common setup or changeover in the MPS.

The sample plan has families E, D and L with 3 products in family E, 3 products in family D, and 2 products in family L. Beginning inventory data and ending inventory data are specified for each product. These data contain anticipation inventory and cycle inventory. The product changeover sequence in the MPS can be affected by the beginning/ending inventory data and program constraints.

Capacity units in the sample MPS are taken directly from the commodity A production plan. These data appear in capacity constraints in the MPS model. This is an "average" total production rate, each period, for the critical resource or bottleneck operation. In the MPS model, a different production rate can be used for each individual product (capacity units required per unit of product). Thus, more units of fast products can be produced per period--fewer units for the slower running products. This is reflected in Figure 3 in the difference between data for "units produced" and "capacity units."

Figure 3

THE MASTER PRODUCTION SCHEDULE
FOR COMMODITY A

PRODUCT	BEG. INV.	1	2	3	4	5	6	7	8	9	10	11	12	13	END. INV.
		SCHEDULE PERIOD — UNITS PRODUCED													
E1	514			280	544	6	352	826		460	440	500	460	932	494
E2	460		44	310	552		378	914		294	474	504	442	871	436
E3	354			12	128	116	174	404		240	220	230	260	550	344
D1	124	695			233					180	188	192	466		114
D2	168	800								170	511	161	212		158
D3	96	78			342					107	93	95	271		92
L1	582			299		188	745		1795	525	197			16	550
L2	696		1540	876		735	183								656
	2994 TOTAL	CAPACITY UNITS FROM THE PRODUCTION PLAN													2844 TOTAL
		1733	1734	1834	1734	1125	1834	1930	1975	1974	2109	1603	2090	2135	

NOTES: No constraints were used for product minimum run quantity.
Zero units were backordered.
This program was prepared using 15 specified family setups.
The capacity units required per unit of product were: family E: .9 units; family D: 1.1 units; and family L: 1.1 units.

THE COMMODITY STRUCTURE

A production plan and a master production schedule are prepared for each commodity.

COMMODITY -- A group of products sharing a single facility.

SINGLE FACILITY -- One or more manufacturing units that can produce similar products.

You essentially master schedule a single facility with one commodity. If you have four assembly lines which can run any of the families--you have a single facility and single commodity. If you have four assembly lines and each line is dedicated to a different product group--you have a four-commodity example. A two-commodity example would be a plant which can manufacture both passenger car tires and heavy service tires. Each of the two types of tire production has its own dedicated building and curing equipment.

THE INVENTORY STRUCTURE

The inventory structure is presented to show the common framework and to show how planned changes are handled within the framework.

ANTICIPATION--Inventory accumulated to provide for later seasonal demand and to cover vacation/holiday allowances.

CYCLE--Inventory that results from batch production in the MPS.

PIPELINE AND SAFETY STOCK--Inventory that remains relatively constant in the plan. Part of the pipeline inventory would be stock-in-transit.

Anticipation inventory assists in the leveling of production and workforce. A typical production plan would have zero units of beginning and ending commodity anticipation inventory. The 150 units of beginning inventory in the sample production plan may be the result of a demand forecast error during the previous year. A negative amount (shortage) could also have been specified.

Master production schedule inventory data are the sum of anticipation inventory and cycle inventory. The specified beginning inventory in the MPS was 2,994 units and the specified ending inventory was 2,844 units (refer to Figure 3).

150 Beginning Anticipation	+	2,884 Beginning Cycle	=	2,994 Beginning Inventory
0 Ending Anticipation	+	2,884 Planned Ending Cycle	=	2,884 Ending Inventory

Please note that the sample MPS had no planned change for the cycle inventory. If there is a planned change in the cycle inventory (beginning/end of year), the following adjustments can be made in the anticipation inventory of the production plan. For a planned 100 unit increase, set the ending anticipation inventory at 100 units. For a planned 100 unit decrease, set the beginning anticipation inventory at 100 units. This plan will then be passed on to the MPS.

For any planned changes in pipeline and safety stock inventory, adjustments are made in the demand forecast data. The demand adjustments are made for both the period commodity demand in the production plan and product demand in the MPS.

Original Demand Forecast	+ -	Planned Change in Pipeline/Safety Stock	=	Adjusted Demand

USING THE PRODUCTION PLAN

The model for the production plan is an excellent tool for evaluating alternative manufacturing strategies. An important test would be: level versus variable workforce plans. The planning model will calculate a level production plan if the annual upper limit for total workforce change is set to zero.

Other typical tests are:

- Modify capacity data
- Shift sizing and timing
- Inventory strategies
- Vacation timing
- Overtime strategies

- Idle workforce
- Forecast error
- Subcontract, backorder, and lost demand.

For most applications, the master schedule will "follow" the production plan and standard hierarchical planning procedures are used. In some planning environments, the production plan would follow the MPS. For example, when the MPS production rate is level for several periods--perhaps in connection with a KANBAN type component plan. This MPS could be passed to the production plan to create the final capacity and inventory plan.

THE FINANCIAL PLAN

A one-year financial plan is based on the production plan. The linear programming output file of the production plan is the primary input data file. Cycle and anticipation inventory data are an important link between the two plans. Data for the beginning inventory (2,994 units) plus the annual anticipation inventory plan are the basis for the finished goods inventory valuation. Direct labor cost data can be based on the workforce plan or an optional data entry.

Prepared financial statements are:

- Inventory valuation
- Manufacturing cost summary
- Pro forma income statement
- Pro forma cash summary.

One mode of operation is to use only manufacturing cost data. Use production plan data and data for planned changes in material inventory and work-in-process inventory. Negative values will appear in the cash summary. This financial plan can be used as the manufacturing budget. Another mode of operation is to prepare an "enriched revenue" plan with sales revenue and general ledger data.

Options available in the off-the-shelf financial software include:

- 12 or 13 periods
- Absorption or direct costing
- Time phasing of sales receipts and material expenses
- Distribution to ledger accounts

- LP data file or manually prepared production plan
- Manufacturing costs or enriched revenue plan
- Consolidation financial software for multiple commodities and/or plants.

The basic concept for the financial framework is: total manufacturing costs for a period equal the sum of standard commodity costs and production plan costs.

Total Manufacturing Costs per Period	=	Standard Costs	+	Production Plan Costs

where:

Standard Cost = Units Produced x Standard Cost per Unit
= Workforce Units x Standard Cost per Period

Production Plan Costs = Sum of Shift Premiums, Overtime Premiums, Anticipation Inventory Costs, etc.

CREATING THE MPS/RCRP MODEL

The application generator for the linear programming model consists of two stages. Data for the application parameters are used to generate a data file with labels. Data for demand, capacity, inventory, cost, unit loads, etc., are entered in the data file. The BLP87/88 linear programming file is generated using the MPS/RCRP data file.

Sample parameters for the data file generator:

NUMBER OF MPS PERIODS	13
NUMBER OF PRODUCTS	8
NUMBER OF FAMILIES	3
FAMILY 1 ID	E
FAMILY 2 ID	D
FAMILY 3 ID	L
NUMBER OF PRODUCTS IN FAMILY 1	3
NUMBER OF PRODUCTS IN FAMILY 2	3
NUMBER OF PRODUCTS IN FAMILY 3	2
SINGLE LINE MPS	NO
FORWARD LOAD	YES
BACKWARD LOAD	YES
NUMBER OF RESOURCES	11
NUMBER OF PERIODS IN RESOURCE OFFSETS	13

If more than one product is produced at any instant (e.g., two assembly lines), additional constraints are generated which limit the maximum family production per period.

The optional RCRP program contains both forward and backward loading. Forward resource loads related to the MPS production are commonly found in the process industry where finishing operations follow the main process facility. Backward loading occurs in assembly settings where fabrication work and vendor activities precede assembly. Most manufacturing planning can be enhanced using both types of load data. The unit load data are prepared by family. The unit load profile states the quantity of each resource required, by period, for one unit of family production in the MPS. The same resource can appear in both forward and backward loading.

USING THE MPS/RCRP PROGRAM AS A CORPORATE PLANNING MODEL

A corporate planning model can be created using the MPS/RCRP generation program. The load data can represent higher planning level inputs and outputs. "Family" production of "standard business units" could appear in the MPS instead of commodity products. Data-driven, off-the-shelf software can drastically reduce the corporate modelling development time. The optimization capability of linear programming is present and manufacturing strategies can be created within specified resource limits.

Resource data for a corporate planning model are presented in Figure 4. (These data are not related to the sample MPS presented in Figure 3.)

Figure 4

A CORPORATE PLANNING MODEL PREPARED
USING THE MPS/RCRP GENERATION PROGRAM

BACKWARD LOADS

	PERIODS						
	−6	−5	−4	−3	−2	−1	Current
Energy	318	286	282	259	103	280	367
Preparation	416	162	165	310	305	438	453
Material 1	159	163	160	331	170	140	409
Material 2	353	162	165	360	364	399	480

MASTER PRODUCTION SCHEDULE

FORWARD LOADS

PERIODS							
Current	+1	+2	+3	+4	+5	+6	
197	170	225		71	229	122	Finishing 1
272	266	230	796	239	95	552	Finishing 2
258	303	374	426	485	72	106	Transport
258	202	517	544	619	416	71	Converting Plant 1
	303	278	59	338	408	249	Converting Plant 2

Alternatives for MRP

KAREL V. DONSELAAR

1 Introduction

Material coördination is the activity, which coördinates the order release decisions for successive stages in a multi-echelon production/inventory system. A well-known concept for material coördination is MRP I. See Orlicky (1975). Unfortunately, MRP I is not a very suitable concept in a stochastic environment.

In this paper two alternative concepts for material coördination are discussed. These concepts are Line Requirements Planning (LRP) and Flexible Integral Requirements planning STrategy (FiRST). In Sections 2 and 3 LRP and FiRST will be introduced. In these sections only the main characteristics of LRP and FiRST will be discussed. For a more detailed description of LRP and FiRST the reader is referred to Van Donselaar (1989).

2 LRP

LRP is a straightforward extension of BSC, which stands for Base Stock Control (see Magee (1958) and Timmer et al. (1984)). Both BSC and LRP are planning concepts for material coördination, which are based on the echelon-concept. According to the echelon-concept the reordering decision in a multi-stage system should not be based on the inventory in one local stage, but rather on the inventory in a so called echelon. An echelon for product A is a set of consecutive stages in a multi-stage system, starting with product A and including all downstream stages.

To explain the logic of BSC and LRP, a simple two-stage system is introduced. This system consists of a final product B and its component A. It takes one unit of A to produce one unit of B. See Figure 1.

The echelon reorder levels are denoted by r(i), i=A,B. The inventories on hand plus on order for products A and B are denoted by I(A) resp. I(B). The system is reviewed once every period.

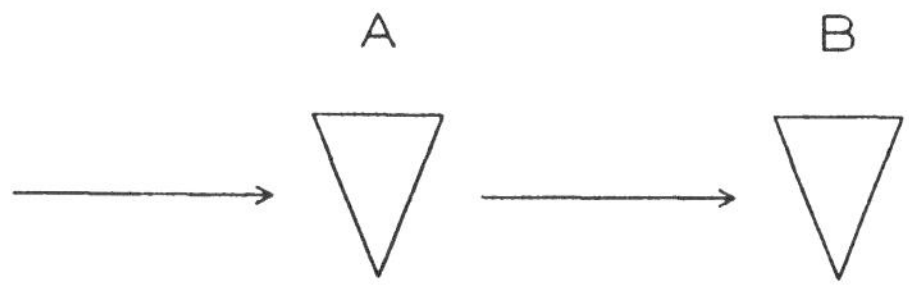

Figure 1 The product structure of B.

In case of BSC, the reorder policies for A and B can be formulated as follows:

for product B: reorder if $I(B) < r(B)$
for product A: reorder if $I(A) + I(B) < r(A)$

This shows that in case of BSC the reorder policy for the final product is equivalent with a local reorder policy. For the component however, BSC takes into account the inventory in the entire system and therefore BSC is no longer equivalent with a local reorder policy. The advantages of BSC compared with local reorder policies are discussed in e.g Timmer et al. (1984).

The logic of BSC has been slightly extended by Van Doncolaar, Jenniskens and Timmer (1987) to incorporate a time-phased element. They call their planning logic Line Requirements Planning (LRP). LRP is very similar to planning concepts proposed by e.g. Stommels (1979), Büchel (1982), Luyten (1987) and Kwikkers (1987).

The Time Phased Order Point logic is an extension of the Statistical Inventory Control logic. In exactly the same way the LRP logic is an extension of the BSC logic. So the LRP logic differs from the BSC logic in two ways:

1. The echelon reorder points r(.) are turned into dynamic reorder points, based on the demand forecasts for the final product during the echelon lead time.
2. Apart from the order release suggestion for the current period, also planned orders for future periods are determined. In contradistinction to MRP, these planned orders are <u>not</u> used for the coördination of higher level stages.

To illustrate the logic of LRP, a numerical example will be used. This example is based on the following parameters for products A and B:

	A	B
(local) lead time	1 period	1 period
lot-size	30 units	30 units
(local) inventory	20 units	20 units

The planning according to LRP is depicted in Figure 2. For this example planning tables are used instead of mathematical formulae. Since LRP is based on echelon inventories and requirements during the echelon lead time, two new concepts are introduced in these tables: integral requirements and integral inventories.

The integral requirements for a product are defined as the requirements for the final product, off-setted with the sum of all downstream lead times.

The integral inventory for a product is defined as the local inventory on hand plus the sum of all downstream inventories (on hand plus on order).

These definitions apply for linear systems. Analogous definitions can be given for non-linear systems.

In Figure 2 it is illustrated that the integral requirements for product A are equal to the off-setted integral requirements for product B. The off-setting is equal to the lead time of product B, i.e. one period. The integral inventory for product A is equal to 40 units. This is the sum of the local inventory of product B (20 units), the scheduled receipts of product B (0 units) and the local inventory of product A (20 units).

Based on the integral requirements, the integral safety stock norm, the integral inventory and the local scheduled receipts, the available balance is determined. (Note that in the example with products A and B there are no scheduled receipts and no safety stock norms). Starting from the available balance, the regular Time Phased Order Point logic is applied in order to determine the planned orders.

Product B

period	0	1	2	3	4	5	6	7
Integral requirements		10	10	10	10	10	10	10
Integral inventory	20							
Available balance	20	10	0	-10	-20	-30	-40	-50
Planned order due				30			30	
Planned order release			30			30		

Product A

period	0	1	2	3	4	5	6
Integral requirements	10	10	10	10	10	10	10
Integral inventory	40						
Available balance	30	20	10	0	-10	-20	-30
Planned order due					30		
Planned order release				30			

Figure 2 An example to illustrate the planning logic of LRP.

The main characteristic of MRP is the explosion of planned orders. From Figure 2 it will be clear, that LRP explodes requirements instead of planned orders. The main advantage of this explosion logic is the fact, that the information is less distorted. With LRP the requirements for the final product and the downstream inventories are still known separately at the component level. This gives planners a better view on the flexibility in their system. With the MRP explosion logic, the only information available at the component level are the gross (component) requirements.

One drawback of LRP is evident from the example in Figure 2. LRP notices that the inventory in the entire system (equal to 40 units) is sufficient to meet demand in the next four periods. So in period 5 the first replenishment should be due. Since the echelon lead time for product A is 2 periods, the first planned order for A should be released in period 3.

In this logic the lot-size restriction of product B is not taken into account. In fact, the planning logic of LRP is based on the assumption that lot-sizes are 100% flexible. If the lot-size restriction of product B is extremely rigid, the local inventory of 20 units A (which is less than the lot-size of B) cannot be used to satisfy the demand for product B. In such a situation the first 'planned order release' for product A should have been planned in period 1.

MRP automatically assumes that the lot-size of product B is 100% inflexible. This leads to a correct order release suggestion in case the lot-size of product B is inflexible. However, if in reality this lot-size is flexible, the MRP advise will be wrong.

Two brief comments should be made here:

- The drawback of LRP mentioned above is limited in linear and convergent systems in case appropriate lot-sizes can be chosen (e.g. in case the component's lot-size is an integer multiple of the final product's lot-size, see Wijngaard and Wortmann (1985)).
- In divergent systems with imbalance, LRP may yield incorrect planned orders. On the other hand, simulation experiments show that the impact of this imbalance on the system's service level is often negligible.

A more detailed discussion on these subjects can be found in Van Donselaar (1989).

In the next section the first impetus for a planning concept for material coördination is described, which is based on the strong features of both MRP and LRP. This planning concept is meant to be applicable in deterministic as well as in stochastic environments and in environments with flexible as well as with inflexible lot-sizes.

3 FiRST

In this paper the planning logic of FiRST will be explained for 2-stage linear systems in a deterministic environment. The planning logic for multi-stage non-linear systems and/or stochastic environments is analogous and is described in Van Donselaar (1989).

* The logic of FiRST

FiRST is primarily based on LRP, due to LRP's number one strong feature: the transparency of the planning. At the same time, as opposed to LRP, FiRST also takes into account the (minimal) restrictions imposed by lot-sizes. To take into account the lot-size inflexibility, two new variables are introduced:
- the Lot-size Inflexibility Factor (LIF)
- the minimal planned orders

The LIF is defined as the ratio of the minimal lot-size over the normal lot-size. Suppose for example that the normal lot-size for product B is equal to 30, but in reality the lot-size is allowed to be any number larger than or equal to 24 units.
In this case the LIF is equal to 24/30 * 100% = 80%.

The logic of FiRST is rather straightforward. First of all a planning is made for every product by applying the LRP logic. This results in the 'Planned orders LRP'. For the final product, these planned orders are correct. Therefore for a final product the 'Planned orders FiRST' are simply equal to the 'Planned orders LRP'.

For the component however, the Planned orders LRP may be wrong (as indicated at the end of Section 2). Therefore they have to be evaluated and possibly be adjusted.

The evaluation of the Planned orders LRP for a component is done by means of a Local Check. This check is based on the observation, that the cumulative inventory for the component should cover at least:
- the cumulative minimal planned order releases of the parent product
- the net requirements of the parent product, off-setted with the parent product's lead time; if the final products are required in period t, the components have to be ready in period t minus the final product's lead time.

To check whether both these restrictions are met, it is sufficient to check whether the maximum of both restrictions is met. Therefore the variable 'Cumulative minimal requirements' is defined for the component:

Cumulative minimal requirements :=
max(cum. minimal planned order releases, cum. net requirements)

with:

minimal planned orders := LIF * 'Planned orders LRP'

Here the 'Planned orders LRP' are defined as the planned orders, which are determined by applying the LRP logic.

In the Local Check it is investigated whether the cumulative minimal requirements are covered by the component's cumulative inventory. If they are not covered, the cumulative inventory has to be raised by adjusting the release date of the Planned orders LRP. (Note that the size of the planned orders is not adjusted). The revised planned orders are called the 'Planned orders FiRST'.

The planning logic of FiRST for a 2-stage linear system in a deterministic environment can be summarised as follows:

For the final product level:
1. Determine the planned orders according to the regular LRP logic.

For the component level:
1. Determine the planned orders according to the regular LRP logic.
   These planned orders are called 'Planned orders LRP'.

2. Check the Planned orders LRP:
   a. Determine the minimal planned orders and the net requirements of the final product.
   b. Determine the cumulative minimal requirements for the component.
   c. Determine the cumulative inventory for the component. This is the time-phased expected local inventory for the component (including the component's Planned orders LRP).
   d. Check whether the cumulative inventory is sufficient to cover the component's cumulative minimal requirements. In case the minimal requirements are not covered, it is obvious that the component's Planned orders LRP are not correct. In this case the Planned orders LRP are replanned with respect to their timing in such a way, that the minimal requirements are covered. The resulting planned orders are called 'Planned orders FiRST'.

* An example to illustrate FiRST

To illustrate how FiRST works, the example from Section 2 is used again.

** The final product level

step 1: Determination of the planned orders for B according to the regular LRP logic. This leads to the planning table for product B from Figure 2:

Product B

period	0	1	2	3	4	5	6	7
Integral requirements		10	10	10	10	10	10	10
Integral inventory	20							
Available balance	20	10	0	-10	-20	-30	-40	-50
Planned order due				30			30	
Planned order release			30			30		

** The component level

step 1: Determination of the planned orders for A according to the regular LRP logic. This leads to the planning table for product A from Figure 2:

Product A

period	0	1	2	3	4	5	6
Integral requirements	10	10	10	10	10	10	10
Integral inventory	40						
Available balance	30	20	10	0	-10	-20	-30
Planned orders LRP due					30		
Planned order release				30			

step 2: The local check.

a. Determination of the final product's (cumulative) minimal planned orders and the (cumulative) net requirements for B:
Minimal planned orders = Planned orders LRP * LIF.
In this example the LIF is equal to 80%.

Product B

period	0	1	2	3	4	5	6	7
Planned order rel. LRP			30			30		
Minimal planned orders			24			24		
Cum. min. planned orders			24	24	24	48	48	

Product B

period	0	1	2	3	4	5	6	7
Integral requirements		10	10	10	10	10	10	10
Integral inventory	20							
Available balance	20	10	0	-10	-20	-30	-40	-50
Cum. net requirements		0	0	10	20	30	40	50

b. Determination of the component's cumulative minimal requirements. (Before explosion to product A, the cumulative net requirements of product B are off-setted with the lead time of product B.)

Product A

period	0	1	2	3	4	5	6
Cum. min. planned orders			24	24	24	48	48
Cum. net requirements		0	10	20	30	40	50
Cum. min. requirements		0	24	24	30	48	50

c. Determination of the cumulative inventory for the component. This is achieved by adding the Planned orders LRP for A to its local inventory.

Product A

period	0	1	2	3	4	5	6
Local inventory	20						
Pl. orders LRP (due)					30		
Cum. inventory	20	20	20	20	50	50	50

d. The component's Planned orders LRP are replanned in such a way, that the cumulative inventory covers the cumulative minimal requirements. In this case in period 2 the cumulative minimal requirements (24 units) are not covered by the cumulative inventory (20 units). This can be solved by moving the Planned order LRP for product A from period 4 to period 2. The replanned order constitutes the Planned order FiRST.

Product A (before replanning)

period	0	1	2	3	4	5	6
Local inventory	20						
Pl. orders LRP (due)					30		
Cum. inventory	20	20	20	20	50	50	50
Cum. min. requirements		0	24	24	30	48	50

Product A (after replanning)

period	0	1	2	3	4	5	6
Local inventory	20						
Pl. orders FiRST (due)			30				
Cum. inventory	20	20	50	50	50	50	50
Cum. min. requirements		0	24	24	30	48	50

This example shows, that FiRST is able to both recognise the lot-size inflexibility (as opposed to LRP) and to give an integral view to the planner (as opposed to MRP).

If in the example the minimal lot-size for B would have been equal to 20 (which corresponds with a LIF equal to 20/30 * 100%), FiRST would have recognised that the first planned order for A should be due in period 4. To see this, note that the entire planning is the same as before, except for the minimal planned orders for B and the local check for A.

The minimal planned orders for B are then equal to:

Product B

period	0	1	2	3	4	5	6	7
Minimal planned orders			20			20		

The altered minimal planned orders lead to the following cumulative minimal requirements:

Product A

period	0	1	2	3	4	5	6
Cum. min. pl. orders			20	20	20	40	40
Cum. net requirements		0	10	20	30	40	50
Cum. min. requirements			20	20	30	40	50

This in its turn leads to the following Local Check:

Product A

period	0	1	2	3	4	5	6
Local inventory	20						
Pl. orders LRP (due)					30		
Cum. inventory	20	20	20	20	50	50	50
Cum. min. requirements			20	20	30	40	50

Since the cumulative inventory is always larger than or equal to the cumulative minimal requirements, the Planned orders FiRST are now simply equal to the Planned orders LRP.

So FiRST recognises that in case the minimal lot-size for B is equal to 20 units, it is not necessary to adapt the Planned orders LRP. The examples above show that FiRST is able to handle both flexible and inflexible lot-sizes.

4 Concluding remarks

For material coördination several concepts are available, e.g. MRP, BSC, LRP and FiRST.

- MRP is very suitable in a deterministic and in a 'frozen' environment.
- BSC is applicable in production or inventory systems with stationary stochastic demand processes in case imbalance is not dominant.
- LRP is suitable for systems with non-stationary stochastic demand processes as long as imbalance is not dominant.
- FiRST should be regarded as a first impetus for a material coördination concept, which can be used in dynamic stochastic environments in which imbalance can not be ignored.

FiRST is the most complicated concept. Simplifications of FiRST, dealing only with the most drastic drawbacks of LRP, may turn out to be useful in practice. An example of such a simplification is to only consider the first (minimal) planned order in the Local Check. Further research in coöperation with advanced users and companies developing standard software is needed here.

Literature.

Büchel, A. (1982), An overview of possible procedures for stochastic MRP, Engineering Costs and Production Economics, Vol. 6, pp.43-51.

Donselaar, K. van, F. Jenniskens and J. Timmer (1987), Line Requirements Planning alternatief voor MRP? (in Dutch), Tijdschrift voor Inkoop & Logistiek, Vol. 3, No. 11 and 12.

Donselaar, K. van (1989), Material coördination under uncertainty, Dissertation, Eindhoven University of Technology, Department of Industrial Engineering and Management Sciences, The Netherlands.

Kwikkers, R. (1987), Models of forecast uncertainty and flexibility inventory in requirements planning systems, Dissertation, University of Michigan.

Luyten, R. (1987), A systems-based approach for multi-echelon production/inventory systems, Dissertation, Leuven Catholic University - T.E.W., Belgium.

Magee, J.F. (1958), Production planning and inventory control, McGraw-Hill, New York.

Orlicky, J. (1975), Material requirements planning, McGraw-Hill.

Stommels, F.J. (1979), Evaluatie van een produktieplanningssysteem voor een keten met grote onzekerheden (samen leren leveren), Dissertation (in Dutch), Eindhoven University of Technology, Department of Industrial Engineering and Management Sciences.

Timmer, J.P.J., W. Monhemius and J.W.M. Bertrand (1984), Production and inventory control with the Base-Stock System, EUT/BDK/12, University of Technology Eindhoven, The Netherlands.

Wijngaard, J. and J.C. Wortmann (1985), MRP and inventories, European Journal of Operations Research, Vol. 20, pp. 281-293.

A Widely Acclaimed Method of Load-Oriented Job Release and its Conceptual Deficiencies

GERHARD KNOLMAYER

1. Introduction

In the last decade, a heuristic method of load-oriented job release, commonly referred to as "Belastungsorientierte Auftragsfreigabe" (BORA), has been developed (cf. e.g. (4); (53)) and very successfully promoted throughout industry. Several standalone software packages with BORA-implementations appeared in the market and BORA-modules have been added to widely used MRP II-systems like COPICS, RM-PPS, PSK2000 and INTEPS (cf. (14), p.24; (59), p. 259). The main goals pursued in connection with BORA are:

- Reduction of lead time and lead time variability by reducing work-in-process
- Reduction of inventory to reduce holding cost
- Reduction of lot sizes to smooth production flow, to prevent unpredictable machine idleness, and to avoid changing bottlenecks in the manufacturing system.

Most literature on BORA is available only in German; therefore, a short description will be given in section 2. After a comparison with related concepts in chapter 3, some empirical results of applying BORA are summarized in section 4. The advantages of BORA have been presented in a large body of literature whereas the few critical remarks on BORA appeared in a rather dispersed way. Therefore, section 5 systematically presents deficiencies of BORA from the perspective of scheduling theory.

2. Description of BORA

The main idea behind BORA is usually illustrated by the analogy of a funnel (cf. e.g. (4), pp. 14; (53), p. 58). The flow through a funnel is independent of the amount of liquid waiting for running through it; for a given funnel cross-section however, flow time is proportional to the funnel's load. Analogously, for a given machine capacity, throughput times can only be reduced by lowering work-in-process (WIP). Simulations based on actual job shop data indicated that a moderate reduction of WIP has only a negligible impact on the system's output but throughput times tend to shrink considerably; however, with further WIP-reductions output falls sharply but throughput time is improved only insignificantly (cf. e.g. (49), p. 225 and Figure 1). Lowering WIP without seriously impairing performance, is achieved by a delicate job-release heuristic which prevents a job to be released from a pool of waiting orders if at least one of its operations may have problems in obtaining the capacities needed during the planning period; thus, the build-up of undesirable queues and inventories is limited.

BORA is a *one*-period-oriented release heuristic because a reliable forecast can allegedly be given only for a single period ((60), p. 329). Therefore, data generated by a (multi-period-oriented) MRP-system must be aggregated to apply BORA. This is accomplished by a two-step procedure: First, the capacity available at each workstation during the planning period is multiplied by a load percentage (LDP) to yield a load-limit for that machine. For ease of presentation, LDP is assumed to be identical for all machines, but in practice it may be determined individually; often figures between 200 and 300 % are applied ((69), p. 131). Second, planned operation times are discounted by using a discount percentage $DP \geq 100$; the operation times are multiplied with $(100/DP)^n$ if n operations have to be performed before the operation regarded can be executed. This is predicated on the idea that the probability of executing an operation in the planning period is the smaller the larger the number of operations preceeding it. It is usually recommended to set the discount percentage DP equal to the load percentage LDP (cf. e.g. (4), pp. 76; (53), pp. 208).

After charging all jobs released in earlier planning cycles, the "candidate jobs" are regarded in increasing order of slack-time. Only those jobs are candidates whose estimated starting date is not behind a predetermined time-horizon to prevent the release of jobs which are not urgent. It has been recommended to use a time-horizon of 3 to 4 planning periods ((61), p. 38). For each candidate job, all workstations where operations have to be performed are checked; the job is released if the load-limits of all workstations have not been exceeded yet. In this case, all discounted operation times are charged to the individual load accounts of the workstations. Otherwise, the job regarded remains in the pool of waiting jobs. All discounted operation times of released and postponed jobs at a certain workstation represent its "chargeable load".

A release interval of one week is usually employed in the machine-tool industry, but shorter intervals are used in other cases ((53), p. 210). Up-to-the-minute status information on all jobs and operations running in the shop is a mandatory requirement for applying BORA. Thus, it can only be employed if a monitoring system provides an accurate collection of real-time factory data; this requirement is often violated in practice ((9), p. 376).

Similar to other modern production management concepts, BORA recommends small lot sizes. A simulation study in which "large" lots were split revealed that output as well as throughput time can be improved considerably by smaller lot sizes (cf. e.g. (49), p. 228

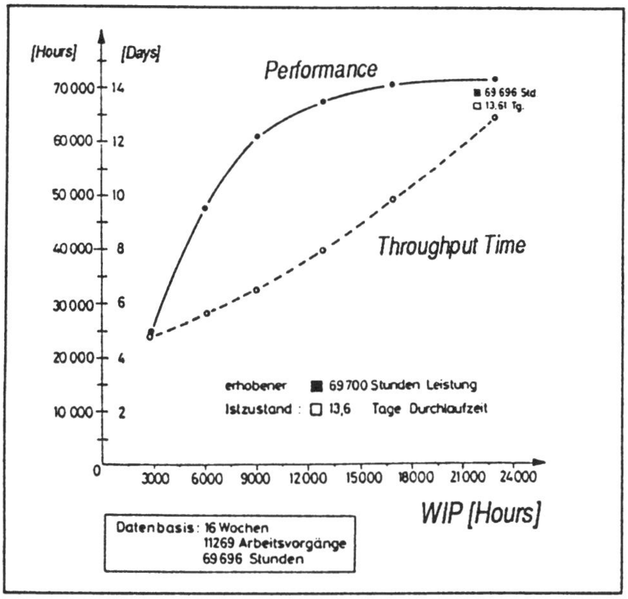

Fig. 1: Relationships beween Work-in-Process, Performance and Throughput-Time (Wiendahl, 1983, p. 225)

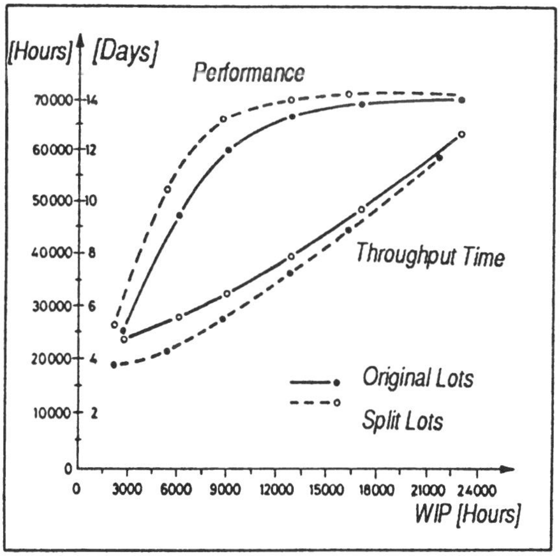

Fig. 2: Effects of Lot-Splitting on Performance and Throughput-Time (Wiendahl, 1983, p. 228)

and Figure 2). Prima-facie, the output effect is surprising because smaller lot sizes require additional setups which may be detrimental in connection with high capacity utilization rates; yet, the phenomenon may be explained by a better smoothed production which reduces the extent of temporary machine idleness due to unbalanced material flow.

3. The History of BORA and Its Connection to Other Job Release Concepts

BORA is based on ideas which originated at IFA (Institut für Fabrikanlagen), University of Hanover, at the end of the 70s. Heinemeyer ((20), pp. 14) and Wegner ((47), pp. 43) stressed the importance of job release times; Jendralski (26) and Bechte (4) developed the basic BORA-prodecure in their dissertations and performed extensive simulations, e.g. on the effects of employing different LDPs. The first publication in journals appeared in 1981 by Kettner and Bechte (27). Since then, the new chairman of IFA, H.-P. Wiendahl, and his associates promoted the BORA-concept successfully in German speaking countries by numerous publications, seminars, consultancy projects, and software development.

At the time of BORA's inception some of its basic ideas already had some tradition; references may be found in Tatsiopoulos/Kingsman (45). Plossl/Wight ((37), p. 31) presented a bathtub in which load is symbolized by the level of water and capacity by the rate at which the water is passing through; with reference to this picture, the funnel analogy is often called *Wight's funnel* in the American literature ((7), 1989, p. 7). Already Bakker (3) stressed the importance of job release times, which were also emphasized by Irastorza/Deane ((25), p. 304): "In general, the objective of the loading algorithm is the selective releasing of jobs to the shop such that job due dates can be met while a balance of work is maintained among the machines". Their mixed-integer algorithm attempts to minimize the difference between the desired aggregate load in front of a machine and its actual aggregate load. A certain provision in the objective function allows to make candidate jobs increasingly attractive for release as their due dates approach.

Input-Output Control is a method which also stresses the reduction of deviations between planned and actual input ((6); (37); (63)). It forecasts the arrival of jobs at all workstations for several periods and tries to stabilize queues according to planned lead times. The procedure does not provide decision rules for job release but merely presents What-if-results to a scheduler ((66), p. 70).

The developers of BORA didn't always pay tribute to earlier publications but their concept is also sometimes neglected by current research outside of Germany. One may refer to Srihari/Greene (43) who compare release rules closely related to BORA and to

Fuse ((16), p. 147) who presents a throughput diagram very similar to those used in connection with BORA.

4. Empirical Results of Applying BORA

Several papers report practical effects attributed to BORA-implementations, the key data of which are summarized in Figure 3. The results described are quite impressive even if some "Hawthorne-effects" may have occured. However, impressive advantages are also frequently reported for other production planning and scheduling procedures as well.

5. Deficiencies of BORA

5.1. BORA's Funnel Model May Be an Oversimplification

The analogy between the flow of liquids and material flow has fascinated many authors: The literature abounds with basins, tubes and faucets which are supposed to illustrate by analogy the intricacies of material flow systems. Referring to a model of a single isolated funnel, BORA proponents argue that WIP and throughput time are directly proportional (cf. e.g. (53), pp. 57). In other statements it is claimed that the linear relationship only holds beyond a critical point ((53), p. 207). Sometimes, two straight lines with a kink are presented; however, there is a considerable deviation between these theoretical, funnel-model based curves and their empirical counterparts ((36), pp. 421).

Some presentations of the curves (cf. (58), p. 67 and Figure 4) are fairly idealized compared with the original simulation results (cf. Figure 1): e.g. it is assumed that more than the "adequate inventory" will not improve the performance of the production system ((54), p. 31; (58), pp. 66), whereas earlier papers reported a 5 % difference ((60), p. 332). Furthermore, the original presentations show nonlinear relationships without accentuating a critical point. Other studies show that the change of WIP and throughput times is very much contingent upon situational factors ((31), pp. 432). All these results contradict the strict proportionality implied by the funnel model which may therefore be regarded as an oversimplification.

5.2. BORA's FCFS-Recommendation May Not Be Reasonable and Cannot Be Traced in Throughput Diagram Illustrations

To obtain a highly predictable material flow, a first-come-first-served (FCFS) sequencing rule is often recommended in connection with BORA. Simulation studies are presented as evidence that in a BORA-environment no improvements can be obtained by more sophisticated sequencing rules (62) since only a small number of jobs should be waiting

Source	Change of Throughput Time	Change of Work-in-process	Change of Tardiness
Sainis (1985, p. 565)	- 50 %	- 50 %	improved
Ulfers (1986, p. 161)	- 62 %		
Knecht (1986, pp. 144)	- 23 %	- 25 %	- 50 %
Büttner (1987, p. 82)	- 46 %	- 32 %	- 81 %
Heiler (1987, p. 381)	- 75 %	- 43 %	
Herrmann (1988, p. 50)	- 70 %	- 20 %	
Herrmann (1988, p. 51)		- 50 %	
Müller (1988, p. 146)	- 45 to - 65%	- 35 to - 50%	
Wiendahl (1988b, p. 86)	- 30 %	- 30 %	- 50 to - 80%
Gottschalk (1989, p. 270)	- 80 %		
Holzhüter (1989, p. 289)			
after 9 months	- 31 %	- 31 %	- 67 %
after 21 months	- 25 %	- 26 %	- 74 %
Lemke (1989, p. 321)	- 70 %	- 50 %	

Fig. 3: Empirical Results Attributed to Practical BORA-Implementations

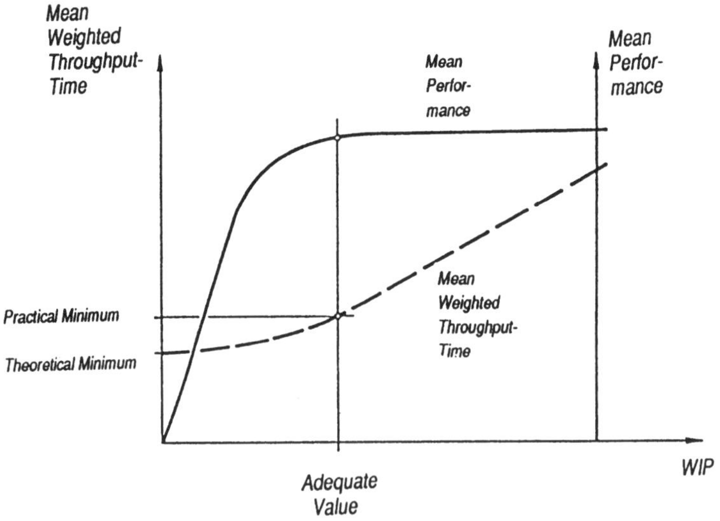

Fig. 4: Idealistic View of Figure 1 (Wiendahl, 1988b, p. 67)

in the queues. However, other simulation studies seem to indicate that a slack-time rule applied instead of FCFS may lead to considerable improvements in BORA-applications ((1), pp. 35). These contradictory results cannot be explained without further examining unpublished details of the simulated systems.

Some basics of BORA are visualized by "throughput diagrams" which depict the cumulated operation times of arriving and completed jobs at a certain workstation as input and output curves (cf. e.g. (49), p. 227 and Figure 5). Assuming identity between planned and actual operation times and no machine idleness, the output curve should be a 45°-line (cf. (2), p. 101). If jobs are executed on a FCFS-basis, the vertical segments should appear in both graphs in the same sequence. Independent from the priority rule applied, vertical segments of both curves should perfectly match. However, these obvious features are frequently missing in presentations of throughput diagrams.

5.3. BORA Does Not Coordinate Dependent Jobs

Jobs are said to be dependent if their outputs are (at least partially) needed for some (sub-)assembly operation. Such a dependency is not taken into account by BORA ((69), p. 21): If due to exceeded load-limits, some but not all dependent jobs were released, WIP will obviously build-up in front of the (sub-)assembly station. Under such circumstances, BORA must be regarded as myopic.

5.4. The Importance of BORA's Planning Period Has Largely Been Ignored

BORA has been criticized for considering only one period because the conditions after the first period should also be taken into account ((28), p. 277; (33), pp. 88; (65), p. 228; (67), p. 128). The short planning periods usually employed may lead to situations where load-limits do not prevent the build-up of queues ((33), pp. 89). Furthermore, the load-limit has been misinterpreted by Adam ((1), p. 24) as a "theoretical peak load" during the planning horizon. This shows that the transformation of multi-period to single-period data may be difficult to understand and lead to acceptance problems.

Often only load-limits and the time-horizon are regarded relevant parameters of a BORA-implementation (cf. e.g. (4), p. 71; (10), p. 46; (29), p. 135; (35), p. 19); rarely it is mentioned that the length of the planning period is relevant also ((2), p. 98; (15), p. 93; (22), p. 324). Even worse, several papers give recommendations on numerical values of LDP without specifying the length of the planning period (e.g. (42); (54), p. 33; (68), p. 92). The recommendation of a certain LDP is futile unless the length of the planning period is specified as well ((2), p. 114): Dramatically different numbers of jobs will be released for a given LDP and varying lengthes of planning periods. The problem is recognized e.g. in ((53), p. 210) where the relationship between the planning period length and LDP is discussed.

5.5. BORA's Discount Factors Cannot Realistically Be Interpreted As Probabilities

Often $(100/DP)^{n-1}$ is interpreted as probability that the nth operation of a candidate job will be finished in the planning period (e.g. (2), p. 99; (53), pp. 214; (69), pp. 131). For this reason BORA has also been qualified as a "statistics-oriented" procedure ((5), p. 59; (24); (51); (57), p. 1226). It has been argued that the interpretation as probabilities only holds if each operation in the queue has the same probability for being executed ((33), p. 86) - which contradicts the FCFS-recommendation, and if the planning period is sufficiently long and operation times are comparatively short ((69), p. 133). The discounting procedure is said to be valid only if operation times are infinitely small and the production system is perfectly balanced ((2), p. 106). Suppose, for example, the planning period is one week and the first five operations of a job each need one day execution time. In this case, the probability that the remaining operations can be performed in the planning period is obviously zero whereas BORA's discount procedure will give non-zero "probabilities" for executing these operations in the planning period because the discount factors do not take operation times into account. Therefore, realistic manufacturing conditions render the probability view of discount factors invalid.

5.6. BORA's Use of Load Percentages to Determine Discount Factors Results in Nonlinear Relationships

The load percentages LDP are usually applied to determine both the load limits and the discount factors. For reasons given above, the discounting procedure can only be regarded as heuristic. Consequentially, it should have been shown that setting LDP=DP provides better results than other methods. From a practical point of view, BORA does not necessarily require equality between LDP and DP ((30), p. 163). Identity may even be unplausible since counterveiling effects occur when these parameters are modified: Suppose, the release of a certain number of jobs is deemed "adequate" by the scheduler. If DP=100 %, unmodified operation times are charged to the workstation accounts; in this case, the LDP must be set sufficiently high to compensate for the low DP ((18), p. 3). Conversely, if a steep discounting is realized e.g. by setting DP=1000 %, a low LDP value should be used.

If both parameters are kept equal, any change in the level of the parameters will result in nonlinear effects. This becomes evident if the ratio between the load chargeable and the load limit is considered for different niveaus of LDP=DP. The effect is illustrated in Figure 6 for a set of arbitrarily chosen candidate jobs. If changes of LDP are considered,

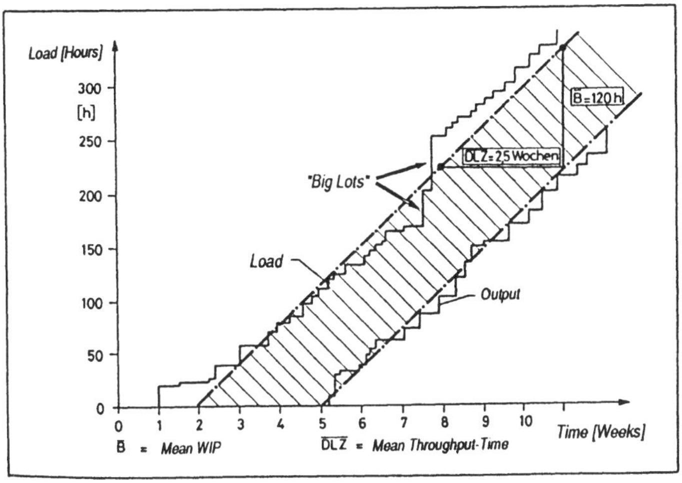

Fig. 5: Throughput Diagram (Wiendahl, 1983, p. 227)

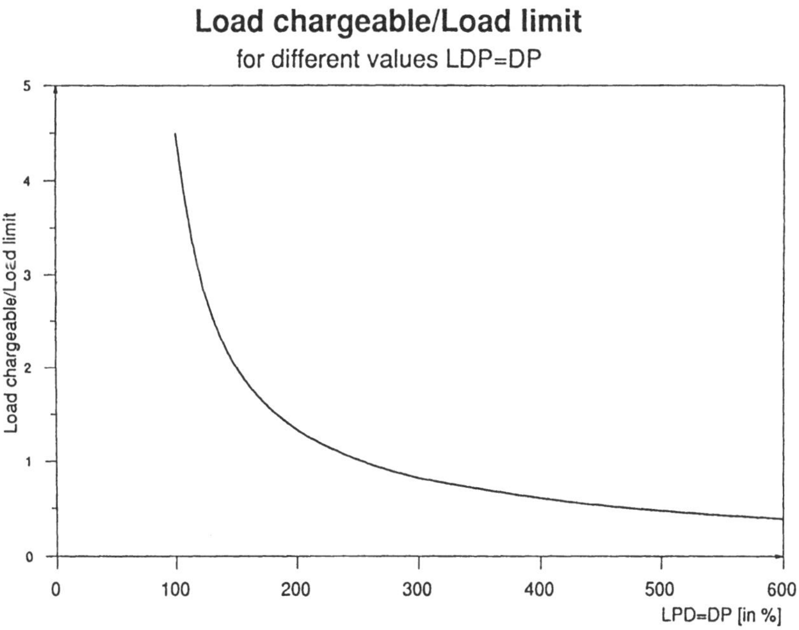

Fig. 6: Nonlinear Relationship, Resulting from Using the Load Percentage as Discount Percentage

the nonlinearity may prevent economically reasonable adjustments because users are accustomed to think in linear changes between independent and dependent variables. Therefore, applying BORA may be facilitated by holding DP constant and varying only LDP.

5.7. BORA's Release Mechanism is Based on a 'Hard' Constraint

In modelling complex systems, the application of 'hard' systems thinking to 'soft' problems has been criticized ((12), pp. 141). In practical problems, the border between feasible and infeasible solutions is rarely very well defined; e.g. in capital budgeting problems, modelled as mixed-integer programs, the introduction of a borrowing activity is used to soften a hard budget constraint (64), pp. 6).

BORA rejects the release of jobs if any one of the workstations involved has already exceeded its load-limit; conceptually, BORA does not provide a mechanism for releasing jobs because e.g. of their high external priority or because the workstation preventing the release is needed only in the distant future. A softer rejection procedure which also employs the scheduler's subjective judgements could be more acceptable because the vague, heuristic determination of LDP and DP does not justify a hard rejection constraint. Therefore, some BORA-implementations provide the planner with a listing of rejected jobs and allow him to overrule BORA's decisions ((33), p. 97). In other implementations the scheduler is asked whether he wants to modify a capacity whose load-limit is exceeded ((5), pp. 60; (46), pp. 159); to make this decision, the scheduler should not only be provided with information on the limit exceeded but also with data concerning which jobs are being affected.

5.8. BORA Recommends a Lot Size Reduction Without Reference to Capacity Utilization

In conjunction with BORA, small lot sizes are frequently recommended; this advice is derived from a simulation of data taken from a single shop in which lots with operations needing more than 40 hours to complete were split into two sublots. Both system performance and throughput time were improved by this decision (Figure 2). The problem with evaluating this result is that no information is available about how the original lot sizes were obtained, the percentage of lots split, and the shop's capacity utilization level. Furthermore, a recent remark by Wiendahl ((53), p. 281) leads to the presumption that the gain in performance may result (at least partially) from additional setups induced by lot splitting; if setup times were actually counted as output, the conclusions drawn would be rather dubious.

Generally speaking, it may be unwarranted to transfer such idiosyncratic findings to situations with e.g. very different degrees of capacity utilization or extreme lot size reductions. The above mentioned result which was obtained for a rather moderate lot splitting policy has provoked some authors to claim that a lot size of 1 will achieve maximum throughput ((8), p. 657; (42); (46), p. 166). This extrapolation is by no means acceptable: A more comprehensive simulation study shows that splitting lots may have minor advantages if capacity utilization is low but may trigger extremly unfavourable effects in high load situations ((31), pp. 431).

The results of lot splitting are usually visualized by plotting WIP against throughput time and performance (Figure 2). The points obtained by using discrete values of LDP are connected, suggesting continuous nonlinear functions. However, comparing values of these graphs for identical positions on the X-axis is questionable since the ceteris-paribus-assumption is violated. Some graphical renditions (42) generously ignore the important feature that the WIP-values of the support points in the graphs differ. In a comparison of two lot sizing procedures ((36), p. 426), the two graphs are based on 5 and 8 support points, respectively; it is by no means clear whether and, if so, which points can be compared in a meaningful way. To illustrate this problem, Figure 7 shows throughput times based on two *fictitious* observations each for original and split lot sizes. By coincidence, the first observation may give identical values for WIP and throughput time (point A); the second observation, obtained at high capacity utilization, may show that splitting lots is inferior with respect to both criteria (cf. points B and C). However, comparing points on the lines AB and AC for a given WIP erroneously suggests superiority of lot splitting (cf. points D and E). Therefore, the standard method of presenting the advantages attributed to splitting lots may be confusing ((31), p. 432).

Finally, the usual presentation conveys the impression that average WIP is an independent variable. However, WIP, performance and throughput time are contingent e.g. upon type, size and frequency of accepted orders, on capacity utilization levels and lot sizes. Thus independent and dependent variables are not always clearly separated in presentations of BORA ((70), p. 334).

5.9. BORA Focusses on Work-in-Process and Neglects Material Stocks

BORA postpones the release of a job if at least one of the workstations necessary for processing the job is overbooked; this delay will reduce WIP and subsequently holding cost. However, this postponement leads to additional material stocks (cf. (18), p. 4; (22), p. 90; (40), p. 70; (41), p. 227; (69), p. 20; (70), p. 336) because material availability is necessary for making a job a release candidate ((27), p. 463; (61), p. 38). BORA does not determine the planned starting point of single operations ((1), pp. 20); therefore, it is

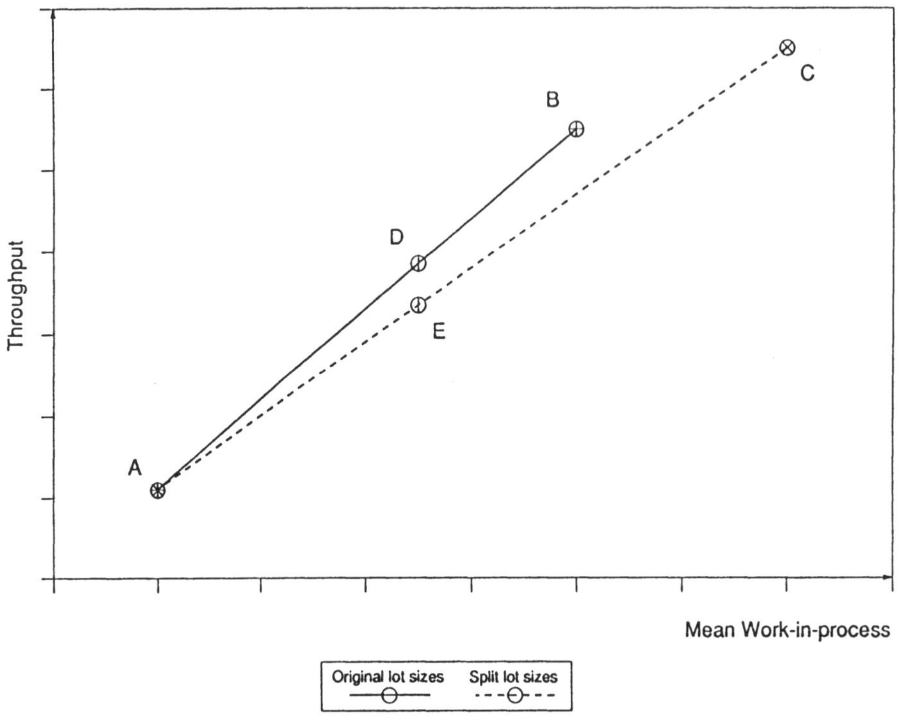

Fig.7: Deficiencies in Illustrations of Lot-Splitting Effects

usually required that *all* materials are physically available to release the job. This neglects the option of modern MRP-systems to coordinate the procurement of materials with the consuming operation (cf. (14), p. 20; (41), p. 176).

Material often accounts for the main portion of a product's cost; therefore, the net reduction of capital tied up in inventory may be small if WIP is substituted by material inventories. However, the strict application of a load-based job release may, in the long run, result in substantially reduced and less volatile throughput times; this may allow for later procurement of materials and thus reduce capital tied up in inventories.

5.10. BORA Implicitly Dictates and Ranks Goals

Proponents of BORA stress the minimization of throughput times and WIP as well as meeting due dates as the main scheduling goals whereas high capacity utilization is regarded as a secondary objective ((48), p. 1; (50), p. 11; (51), p. 291; (53), p. 18; (54), pp. 28; (60), p. 327); more recently, also the WIP-reduction is disqualified ((59), pp. 222). Greiner ((18), pp. 4) contends that the application of BORA implies the following ranking of goals (in decreasing order of importance):

1. Minimization of throughput times
2. Minimization of inventory
3. Maximization of capacity utilization
4. Minimization of tardiness.

From a methodological point of view, research should not impose value systems on decision makers. The subgoals mentioned above may influence the strategic goals of the company in different ways. A small reduction of capacity utilization may reduce profitability quite remarkably ((1), pp. 19). In general, the compromise between throughput time, inventory reduction and capacity utilization depends on many factors which are difficult to quantify ((66), p. 71).

BORA implicitly gives higher weights to goals of the production department than to those of the sales department ((30), p. 163): The release of a job may be postponed due to excessive load at one workstation although it should be released with respect to its due date and assumed throughput times. Simulations show that BORA does not improve on tardiness compared with a simple, lead-time-dependent release procedure (30). Inferior results of BORA with respect to meeting due dates also appeared in other simulation studies ((2), p. 113), and also Greiner ((18), p. 4) mentions this feature as a deficiency of BORA.

In Figure 4, the (idealized) critical point has been called the "adequate value" or "optimal point" ((13), p. 460; (15), p. 19, p. 104; (50), p. 17; (53), p. 207; (54), p. 31; (55), p. 83; (56), Fig. 23; (62), p. 39); however, whether a certain value for inventories, performance and throughput time is adequate or not depends on the goals of decision makers and should not be dictated by scientists. At least the more cautious formulation "adequate work-in-process *zone*" ((52), pp. 41; (55), p. 83; (58), p. 67; (59), p. 256) should be used instead. In earlier publications the possibility of adapting the LDP to different goal weights has been emphasized ((51), p. 296); this idea is inherently in conflict with abstractly defined optimal values or zones.

5.11. BORA Disregards the Need for Dynamically Updating Lead Times

Descriptions of BORA often do not comment on how historical throughput times are adjusted to actual occurrences. An exponential smoothing procedure may be helpful in updating throughput times ((22), p. 93; (30), p. 165; (44), p. 210). Bechte ((4), p. 83) applies a smoothing factor 0.5 for weekly relase and another paper mentions that IFA recommends a smoothing factor 0.2 for updating on a daily basis ((44), p. 210). However, Holdhof ((22), pp. 266) has shown that such high weights attached to recent data may lead to inferior results than a less hectic adjustment.

5.12. BORA Depends on a Powerful Medium-Term Production Planning System

BORA was developed as a conceptually rather simple procedure for automating short-term job release decisions. Due to its simplicity, it will not function properly without carefully coordinating orders accepted in terms of due dates, capacity availabilities and procurement of materials (cf. (33), pp. 95; (38), p. 63; (67), p. 128). Medium-term planning is certainly more important than short-term job-release-decisions for which even a theoretically questionable procedure may not cause too much harm.

6. Conclusion

BORA is a widely acclaimed release heuristic and very favourable effects have been reported in practical applications. To partially compensate for the vast amount of literature that uncritically supports this concept, some of its theoretical deficiencies and presentation inaccuracies have been described in this paper. Individuals, responsible for scheduling, must decide for themselves whether they can derive benefits from a theoretically questionable procedure such as BORA; in order to support this decision process, they should at least be provided with a list of issues to be considered when discussing a possible BORA-introduction.

References:

(1) Adam, D., Ansätze zu einem integrierten Konzept der Fertigungssteuerung bei Werkstattfertigung, in: D. Adam (Ed.), Neuere Entwicklungen in der Produktions- und Investitionspolitik, Wiesbaden: Gabler 1987, p. 17 - 52.

(2) Adam, D., Die Eignung der belastungsorientierten Auftragsfreigabe für die Steuerung von Fertigungsprozessen mit diskontinuierlichem Materialfluß, in: Zeitschrift für Betriebswirtschaft 58 (1988), p. 98 - 115.

(3) Bakker, W., De Levertijd in Machinefabrieken, Een orienterend onderzoek, Diss. TH Delft 1965.

(4) Bechte, W., Steuerung der Durchlaufzeit durch belastungsorientierte Auftragsfreigabe bei Werkstattfertigung, Diss. Univ. Hannover 1980.

(5) Bechte, W., Kontroll- und Planungssystem zur Belastungsorientierten Fertigungssteuerung im Dialog - Konzept und Realisierung, in: H.-P. Wiendahl (Ed.), Belastungsorientierte Fertigungssteuerung BOS-F, Praxis und Weiterentwicklung, Hannover: IFA 1987, p. 39 - 70.

(6) Belt, B., Integrating Capacity Planning and Capacity Control, in: Production & Inventory Management 17 (1976), Nr. 1, S. 9 - 25.

(7) Blackstone, J.H., Jr., Capacity Management, Cincinnati et al.: South-Western 1989.

(8) Börnecke, G., Flexibilisieren heißt Automatisieren, in: VDI-Z. 127 (1985), p. 657 - 661.

(9) Buchmann, W., Konzeption eines Programmsystems zur belastungsorientierten Fertigungssteuerung, in: Zeitschrift für wirtschaftliche Fertigung 77 (1982), p. 374 - 377.

(10) Buchmann, W., Belastungsorientierte Auftragsfreigabe - Steuerungsparameter, Wirkung im Rahmen der Fertigungssteuerung, Einbettung in ein EDV-System -, in: (24), p. 46 - 61.

(11) Büttner, R., Planungs- und Kontrollsystem zur Belastungsorientierten Fertigungssteuerung im Dialog - Anwendererfahrungen, in: H.-P. Wiendahl (Ed.), Belastungsorientierte Fertigungssteuerung BOS-F, Praxis und Weiterentwicklung, Hannover: IFA 1987, p. 71 - 90.

(12) Checkland, P., Systems Thinking, Systems Practice, Chichester et al.: Wiley 1981.

(13) Dietrich, W., Anforderungen an CIM-fähige Produktionsplanungs- und -steuerungssysteme, in: Zeitschrift für wirtschaftliche Fertigung 83 (1988), p. 459 - 463.

(14) EDV Studio Ploenzke (Ed.), PPS-Studie, 3rd ed., Wiesbaden et al: Ploenzke 1989, Vol. 1, Chapter 2.

(15) Erdlenbruch, B., Grundlagen neuer Auftragssteuerungsverfahren für die Werkstattfertigung, Düsseldorf: VDI 1984.

(16) Fuse, M., Stockless Production, A Manufacturing Strategy for Knowledge Society, in: C.A. Voss (Ed.), Just-In-Time Manufacture, Berlin et al.: IFS/Springer n.d., p. 133 - 157.

(17) Gottschalk, E., Erfahrungen mit der Belastungsorientierten Fertigungssteuerung als Planungs- und Steuerungsverfahren in einem Traktoren- und Dieselmotorenwerk, in: H.-P. Wiendahl (Ed.), Belastungsorientierte Fertigungssteuerung, Praxis und Weiterentwicklung, Hannover: IFA 1989, p. 261 - 277.

(18) Greiner, T., BOA, nicht immer die beste Lösung, Paper presented at the JIT-Conference, Frankfurt 1988.

(19) Heiler, M., Erfahrungsbericht über Auswahl und Einsatz des PPS-Systems PSK 2000, in: H.-P. Wiendahl (Ed.), Belastungsorientierte Fertigungssteuerung BOS-F, Praxis und Weiterentwicklung, Hannover: IFA 1987, p. 353 - 382.

(20) Heinemeyer, W., Die Analyse der Fertigungsdurchlaufzeit im Industriebetrieb, Diss. TU Hannover 1974.

(21) Herrmann, N., Werkstatt ohne Wartezeit, in: Siemens-Magazin COM 1988, Nr. 1, p. 48 - 51.

(22) Holdhof, J., Terminplanungssysteme für Werkstätten mit heterogenen Produktionsbedingungen, Düsseldorf: VDI 1986.

(23) Holzhüter, E., Erfahrungen mit dem System KPSF zur belastungsorientierten Fertigungssteuerung in einem Maschinenbaubetrieb mittlerer Größe, in: H.-P. Wiendahl (Ed.), Belastungsorientierte Fertigungssteuerung, Praxis und Weiterentwicklung, Hannover: IFA 1989, p. 279 - 299.

(24) IFA (Ed.), Statistisch orientierte Fertigungssteuerung, Hannover: IFA 1984.

(25) Irastorza, J.C., Deane, R.H., A Loading and Balancing Methodology for the Job Shop Control, in: AIIE Transactions 6 (1974), p. 302 - 307.

(26) Jendralski, J., Kapazitätsterminierung zur Bestandsregelung in der Werkstattfertigung, Diss. TU Hannover 1978.

(27) Kettner, M., Bechte, W., Neue Wege der Fertigungssteuerung durch belastungsorientierte Auftragsfreigabe, in: VDI-Z. 123 (1981), p. 459 - 466.

(28) Kistner, K.-P., Steven, M., Produktionsplanung, Heidelberg: Physica 1990.

(29) Knecht, R., Zwei Jahre belastungsorientierte Auftragsfreigabe bei Fa. Hilti, in: H.-P. Wiendahl (Ed.), Praxis der belastungsorientierten Fertigungssteuerung, Han-nover: IFA 1986, p. 119 - 148.

(30) Knolmayer, G., Holdhof, J., BORA versus DORA: Ein Vergleich zweier Terminplanungsverfahren, in: Die Arbeitsvorbereitung 23 (1986), p. 163 - 165.

(31) Knolmayer, G., Lemke, F., Auswirkungen von Losgrößenreduktionen auf die Erreichung produktionswirtschaftlicher Ziele, in: Zeitschrift für Betriebswirtschaft 60 (1990), p. 423 - 442.

(32) Lemke, J., Praktische Erfahrungen und daraus resultierende Weiterentwicklungen in der dezentralen Werkstattsteuerung, dargestellt am Beispiel der Siemens AG, in: H.-P. Wiendahl (Ed.), Belastungsorientierte Fertigungssteuerung, Praxis und Weiterentwicklung, Hannover: IFA 1989, p. 301 - 361.

(33) Missbauer, H., Optimale Werkstattbeauftragung unter dem Aspekt der Bestandsregelung, Linz: Trauner 1987.

(34) Müller, J., Erfahrungen beim Einsatz der Methode "Auftragsfreigabe mit Belastungsschranke (ABS)" in der Werkstattsteuerung, in: Schriften zur Unternehmensführung 39 (1988), p. 131 - 147.

(35) Nyhuis, F., Belastungsorientierte Fertigungssteuerung, Bewährungsprobe bestanden, in: Industrie-Anzeiger 106 (1984), Nr. 29, p. 19 - 20.

(36) Nyhuis, F., Einfluß der Losgröße auf das Durchlaufverhalten von Fertigungsaufträgen - Ein Denkansatz zur Verbesserung der Losgrößenbildung, in: H.-P. Wiendahl (Ed.), Belastungsorientierte Fertigungssteuerung, Praxis und Weiterentwicklung, Hannover: IFA 1989, p. 401 - 427.

(37) Plossl, G.W., Wight, O.W., Capacity Planning and Control, in: Production & Inventory Management 14 (1973), Nr. 3, p. 31 - 67.

(38) Ritter, K.-H., Voraussetzungen zum Einsatz der Belastungsorientierten Auf-tragsfreigabe, in: H.-P. Wiendahl (Ed.), Praxis der belastungsorientierten Fertigungssteuerung, Hannover: IFA 1986, p. 53 - 73.

(39) Sainis, P., Die neuesten Tendenzen in der Fertigungssteuerung und ihre Anwendung in der Praxis, in: Zeitschrift für wirtschaftliche Fertigung 80 (1985), p. 561 - 566.

(40) Scheer, A.-W., PPS-Systeme im Wandel, in: Industrie-Anzeiger 106 (1984), Nr. 21, p. 68 - 71.

(41) Scheer, A.-W., Wirtschaftsinformatik, Informationssysteme im Industriebetrieb, Berlin et al: Springer 1988.

(42) Siemens (Ed.), Kürzere Durchlaufzeiten, Große Lose - wirtschaftlich?, n.l.: n.d.

(43) Srihari, K., Greene, T.J., Job Release Rules Based Upon Facility Attributes, in: A. Mital (Ed.), Proceedings IXth International Conference on Production Research, Volume II, Cincinnati 1987, p. 1764 - 1770.

(44) Storfinger, R., Die belastungsorientierte Fertigungssteuerung für mechanische Fertigungen in der BMW AG (Projekt BORA), in: H.-P. Wiendahl (Ed.), Praxis der belastungsorientierten Fertigungssteuerung, Hannover: IFA 1986, p. 197 - 218.

(45) Tatsiopoulos, I.P., Kingsman, B.G., Lead time management, in: European Journal of Operational Research 14 (1983), p. 351 - 358.

(46) Ulfers, H.-A., Erfahrungen mit der belastungsorientierten Auftragssteuerung ABS bei Siemens, in: H.-P. Wiendahl (Ed.), Praxis der belastungsorientierten Fertigungssteuerung, Hannover: IFA 1986, p. 149 - 169.

(47) Wegner, N., Simulation von Einplanungs- und Abfertigungsstrategien bei Werkstattfertigung, Diss. TU Hannover 1978.

(48) Wiendahl, H.-P., Grundsätzliche Zusammenhänge zwischen Leistung, Durchlaufzeit und Beständen in der Werkstattfertigung, in: VDI-Berichte 1982, Nr. 463, p. 1 - 4.

(49) Wiendahl, H.-P., Betriebsorganisation für Ingenieure, München-Wien: Hanser 1983.

(50) Wiendahl, H.-P., Grundlagen neuer Verfahren der Fertigungssteuerung, in: (24), p. 1 - 19.

(51) Wiendahl, H.-P., Grundlagen und Anwendungsbeispiel eines statistisch orientierten neuen Verfahrens der Fertigungssteuerung, in: Fertigungstechnik und Betrieb 35 (1985), p. 291 - 296.

(52) Wiendahl, H.-P., Von der belastungsorientierten Auftragsfreigabe zur durchlauforientierten Fertigungssteuerung, in: H.-P. Wiendahl (Ed.), Praxis der belastungsorientierten Fertigungssteuerung, Hannover: IFA 1986, p. 17 - 52.

(53) Wiendahl, H.-P., Belastungsorientierte Fertigungssteuerung, München-Wien: Hanser 1987a.

(54) Wiendahl, H.-P., Belastungsorientierte Fertigungssteuerung, in: Technologie & Management 36 (1987b), Nr. 2, p. 28 - 34.

(55) Wiendahl, H.-P., Ein Modell für die CIM-gerechte Fertigungssteuerung, in: CIM Management 3 (1987c), Nr. 2, p. 77 - 84.

(56) Wiendahl, H.-P., Das Durchlaufdiagramm - Ein universelles Modell zur Abbildung, Steuerung und Kontrolle logistischer Prozesse, Paper presented at the International Software-Congress, Karlsruhe 1987d, Section L7.

(57) Wiendahl, H.-P., Erwiderung: Probleme der belastungsorientierten Auftragsfreigabe, in: Zeitschrift für Betriebswirtschaft 58 (1988a), p. 1224 - 1227.

(58) Wiendahl, H.-P., Die belastungsorientierte Fertigungssteuerung, in: Schriften zur Unternehmensführung 39 (1988b), p. 51 - 87.

(59) Wiendahl, H.-P., Grundlagen und Entwicklungsstand der Belastungsorientierten Fertigungssteuerung, in: H.-P. Wiendahl (Ed.), Belastungsorientierte Fertigungssteuerung, Praxis und Weiterentwicklung, Hannover: IFA 1989, p. 219 - 260.

(60) Wiendahl, H.-P., Buchmann, W., Realisierung alternativer Fertigungssteuerungsstrategien, in: VDI-Z. 126 (1984), S. 327 - 334.

(61) Wiendahl H.-P., Lorenz, W., Holzkämper, R., Sichere Fertigungstermine durch einen neuen Ansatz der Fertigungssteuerung, in: Management-Zeitschrift io 52 (1983), Nr. 1, p. 36 - 40.

(62) Wiendahl, H.-P., Lüssenhop, T., Wirkung von Prioritätsregeln, Eine kritische Betrachtung, in: VDI-Z. 131 (1989), p. 36 - 41.

(63) Wight, O.W., Input - Output Control: New way to cut work-in-process queues, in: Modern Materials Handling 25 (1970), Nr. 9, p. 38 - 45.

(64) Williams, A.C., Some Modeling Principles for MIP's, Paper presented at the 7th International Symposium on Mathematical Programming, Stanford Univ., 1973.

(65) Zäpfel, G., Strategisches Produktions-Management, Berlin-New York: de Gruyter 1989.

(66) Zäpfel, G., Missbauer, H., PPS-Systeme mit Bestandsregelung zur Senkung von Durchlaufzeiten und Beständen, in: Information Management 1987, Nr. 1, p. 65 - 73.

(67) Zäpfel, G., Missbauer, H., Neuere Konzepte der Produktionsplanung und -steuerung in der Fertigungsindustrie, in: Wirtschaftswissenschaftliches Studium 17 (1988), p. 127 - 131.

(68) Zenner, D., Flexible Werkstattsteuerung mit ABS-X, Unnötige Wartezeiten abbauen, in: PC Magazin 1985, Nr. 44, p. 91 - 94.

(69) Zimmermann, G., PPS-Methoden auf dem Prüfstand - was leisten sie, wann versagen sie?, Landsberg am Lech: Moderne Industrie 1987.

(70) Zimmermann, G., Produktionsplanung variantenreicher Erzeugnisse mit EDV, Berlin et al.: Springer 1988.

JIT in a Low Volume, Hi-Tech Production Environment

JOHN CARLSON, WILLIAM OSGOOD and NICK KONOVALOV

PRELIMINARY REMARKS

The implementation of JIT in medical instrument production presents a significant challenge because the devices manufactured are used in life support and critical care environments. Thus quality and reliability cannot be compromised by any change in manufacturing strategy. An additional challenge is brought about because the volume of the hi-tech, hi-cost instruments is relatively low and the cycle time per operator is relatively long, averaging approximately 30 minutes.

The company, like so many other companies, has looked to JIT to improve quality, reduce work, reduce space, improve the work environment, improve customer service, be more responsive to customer needs, reduce lead time for product enhancements, reduce production cycle time and regain market share. The success of the program is indicated by the reduction in throughput time from 75 days to 3 days in addition to a 34% decrease in rework and a 60% reduction in required space.

INTRODUCTION

Baxter Healthcare, Inc. is one of the largest healthcare corporations with sales in excess of $7 billion annually. The company has 46 operating divisions worldwide and manufactures approximately 60% of the approximately 70,000 products it markets. The products range from hi-tech microprocessor-based devices, implantables and pharmaceuticals to commodity products such as surgical gowns, gloves and other sterile hospital products.

The Edwards division produces a limited range of medical instruments with options and a variety of accessories. It is one of four instrumentation divisions within Baxter and has been producing medical equipment since the early 1970s. It has been able to hold a significant share of the market for several decades.

The products are infusion pumps, plasmapheresis instruments which separate plasma from whole blood and highly sophisticated monitoring equipment for the critical care/intensive care units of hospitals. All three of the product groups were targeted for Just-In-Time manufacture.

MANUFACTURING BEFORE ADOPTING JIT

Before JIT, production:

- was strictly batch-type, work order controlled
- was supported by a large staff group

- had many inspection points throughout the assembly process
- had high raw material and WIP inventories

- did minimal people development and training
- incurred excessive rework
- had high labor content in the products produced
- had long throughput times of over two months

- dealt with slow new product introduction
- lived with high emphasis on R&D, little on processes

- operated with excessive manufacturing overhead
- made minimal capital investment in process equipment
- made minimal investment in process/methods improvement

Change was essential for the division to avoid losing market share. This industry is no different from others in that the driving force for change stems from competition. The desire for change at the operations level is driven by a) the increased effort required to maintain present antiquated systems; b) general dissatisfaction with the present production methods and quality; and c) a vision of what could be accomplished by changing. At some point, the 'cost' of changing to a 'better way' becomes less than the 'cost' of status quo.

THE JIT PROGRAM

The approach used was to attack the in-house, non-value adding manufacturing activities first and then those involved in procurement. Among the activities changed were:

Storage	- workstations became mini-storerooms, replenished via kanbans as items reached marked bin levels.
Materials	- assembly kits were retrieved only as needed and handled by the line personnel.
Inspection	- operators were taught what to observe and were given responsibility for quality.
Re-work	- was performed before units were passed on by instituting self-inspection training and methods.
Testing	- test results were communicated directly back to operators, engineers and procurement.

Starting with a stable product, a primary ingredient for success, JIT was implemented first on the simpler, infusion pump assembly line activities. The training program was traditional and employed consultants, managers, video tapes, case studies and open discussions to 'tell the story.' Problem-solving techniques were illustrated with examples from the floor. For example, a fishbone diagram was used to elicit possible causes for "SKID PAD DOESN'T STICK" and is shown as follows:

```
              MATERIALS                               METHODS
                     \                    Instruction Sheet incorrect \
                      \                         Parts not clean enough \
  Paint prevents bond \                     Glue not applied properly \
  Texture prevents bond \                    Pads not held long enough \
         Incorrect glue \                    Pads not held tight enough \
---------------------------------------------------------------------> SKID PAD NOT STICKING
          Holding tool /                     Assembler not following /
          doesn't work /                         Instruction Sheet  /
                      /                                            /
                     /                                            /
           EQUIPMENT                               PEOPLE
```

The operators became managers of components via replenishment cards (kanbans) placed in the parts bins at each workstation. Formerly there was a great deal of miscounting of 'floor stock' which distorted the MRP reporting and destroyed confidence in MRP reports.

With the disciplines of JIT, the chaotic 'end-of-the month' syndrome was rejected as an operating mode. A program of daily progress recording and display was instituted to replace the monthly production goals which often degenerated into fire fights in order to meet them.

The results from Phase I, the pump line, and Phase II, the autophoresis line were so impressive that Phase III, the critical care instrument assembly activity was next converted to a mixed model JIT assembly and test line within a year. The results of Phase II are summarized as follows:

RESULTS (Autophoresis Line - Some data disguised)

Item	Before	After	% Reduction
Distance traveled	5436 Ft.	338 Ft.	94%
Production space	4633 Sq. Ft.	1855 Sq. Ft.	60%
WIP	$1278 K	$ 868 K	32%
Throughput (Days)	75	3	95%
Std. Hrs./Assy.	13.6 Hrs.	10.5 Hrs.	22%
Std. Hrs./Test	8.0 Hrs.	5.0 Hrs.	37%
Std. Hrs.Insp.	5.0 Hrs.	2.0 Hrs.	60%

LESSONS LEARNED

It became apparent that JIT cannot be taught solely from video tapes, manuals, simulations and presentations. As the saying goes, "there comes a time to put aside the book and video tapes on how to ride a bicycle and go out and 'ride it.'"

A few psychological and procedural errors were made in the beginning. Over enthusiasm with rework reduction caused one glaring faux pas by asking individual operators to post problems on a board entitled "My Mistakes" displayed at their own workstations. This was intimidating, demeaning and misunderstood and was summarily discontinued.

Management's role in the continuing improvement program is not always understood either by them or the operators. Enthusiasm for the project and the significant investment in consultants and training can cause top management to ignore some fundamental JIT premises required for long-term success. Because of the criteria by which they are measured, management often looks for some immediate results. Thus the integration of the major functions of material planning, procurement and control can easily be thrust aside.

In addition, top management often does not have sufficient knowledge of MRP, its required disciplines and its role in conserving resources. For example, instead of returning unsatisfactory components to vendors, these parts could introduced into production and used 'as is' in order to meet a schedule. A tolerance for poor incoming quality destroys the ability of MRP to perform its planning function and sends a very contradictory message to the operators. It was estimated that labor cost is at least 25% higher because of rework and overtime caused by shortages and components which should have been rejected.

Other flaws in the system contribute to quality problems such as obsolete engineering design and drawing practices, incorrect specifications issued to suppliers and inappropriate measures of buyer performance. The largest contributor to quality problems is over-emphasis on obtaining the "best price" instead of minimizing "total cost."

Another difficult barrier to overcome is that of jurisdiction and job classification. Operators traditionally are reluctant to perform tasks above or beneath their classification. Cross-training and the team approach can minimize this disruptive practice. Stagnation can also take over at some point as the people are besieged with 'new' approaches to doing their job. Quality circles help alleviate these problems but they must be supported by management and encouraged to hold frequent, short meetings to expose problems and recommend improvements.

Workstation layouts often suggest individual control over activities rather than team effort and cooperation. Interstation communication is absolutely essential in JIT environments. Shorter distances between workstations and a close proximity to floor stock and the stockroom enhance this communication. The workstations should be regularly reviewed to assure visual as well as oral contact among operators.

A MIXED MODEL ASSEMBLY LINE

The Phase III JIT project involved a family of electronic monitoring instruments. These instruments are in their 8th generation of production but the assembly methods were antiquated. The objective was to change from isolated workstations to a mixed model line. This was to be accomplished with the help of operator suggestions, standard data and methods analysis, video tape analysis, line balancing, simulation and JIT techniques.

A further objective was to have synchronized production with as much standardization of methods and product as possible such that changeovers could be introduced on a rolling basis. That is, a change of operation of station #1 can be started when station #5 is still working on a prior product. The set-up time reduction must leave virtually no lost time between changeovers.

New instrument designs are expected approximately every 4 months. Thus there is a need to have a flexible mixed product assembly activity so that products can be phased in and out without major disruptions to the line. For example, rather than produce a thirty- day requirement for each product, assembly should be done on a demand basis to reduce F/G inventory and the risk of obsolescence or rework. The instruments take only a few hours to assemble given that the materials and components are available. On a demand basis, the company can produce any of the products within a few days by building small quantities each of Model C, Model M, etc. on the same line.

The Bill of Materials was reduced from seven levels to one level. This alarmed Data Processing as they could not understand why nor how you could build an instrument containing in excess of 1000 components without breaking it down into subassemblies. The answer lies in the single level assembly line. The bulk of the components are fasteners and other small parts which can be kept as "line stock" at the workstations and replenished by kanbans submitted by the operator when the stock reaches a prescribed, visual level.

The cycle time per work station is about 30 minutes, very long for a JIT environment. With normal variability of cycle times at the work stations, there could be significant delays if there were no buffer units in the WIP inventory. Thus the single kanban square concept at each workstation was sacrificed for one unit buffer.

The assembly operations are not complex but new methods are constantly sought to improve quality and productivity. Continual training of employees is required as products, methods and procedures change.

To help design the work stations, a discrete element simulation software program was developed. PCMODEL was selected because of its superb animation capability, its realistic representation of the real layout and its flexibility for changing the parameter values. (PCMODEL was developed by David White of SIMSOFT, Brookings, Oregon)

Several versions of the Mixed-Product Assembly line were simulated as follows:

1) A single continuous mixed product assembly and test line
2) Two dedicated parallel assembly lines
3) Two parallel mixed product assembly lines

Figures 1, 2 and 3 are snapshots of these configurations. The results of these simulations showed that the two parallel, mixed product configurations yielded the highest output given the variable operation times and selected rework percentages.

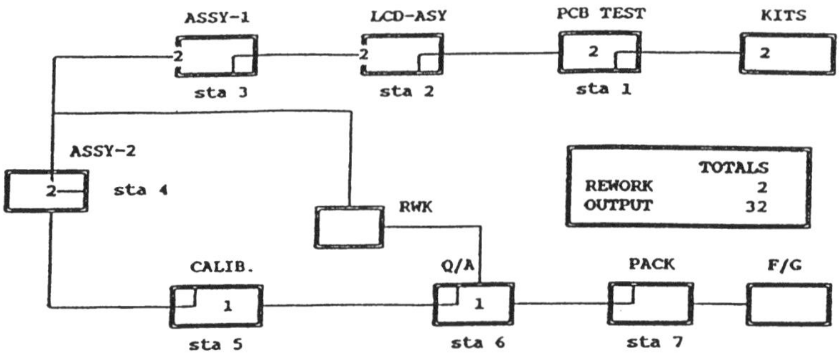

Figure 1 Single, Mixed-Product Assembly Line

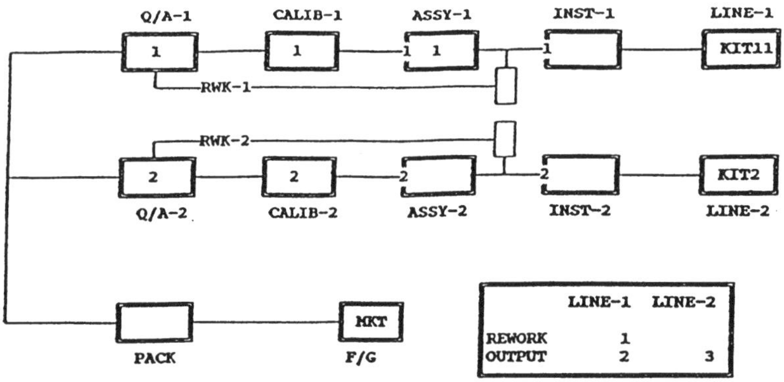

Figure 2 Pull System-Two Dedicated Lines

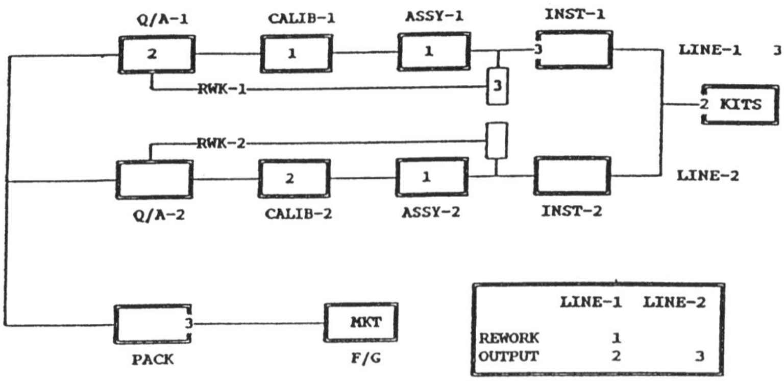

Figure 3 Pull System-Two Mixed-Product Lines

INFORMATION SYSTEM

Production problems that inhibit implementation of JIT are the same as those that prevent effective use of MRP and other systems. Highest on the list would be lack of data integrity. However, in JIT environments the speed of production can present a problem in that there may not be a satisfactory procedure or sufficient time to collect traditional accounting data. Most cost accounting systems require Work Orders and details on assembly hours to determine time per unit. With JIT, however, most accounting details are not important and costly data collection systems which have little or no use to manufacturing must be avoided.

For example, the 'accounting standards' are stated in dollars ($) which are changed every year to reflect changes in overhead rates. Another problem is that outside processing is considered as a purchase. Man-hours/unit and defects/unit data is not available in a manner useful to manufacturing.

Although information systems are essential in order to furnish the accurate data required to develop sound bases for decisions, we are reminded that:

1. People respond almost exclusively to things on which they are measured and rewarded - but measurements in JIT environments is operationally different.

2. Company objectives in output and quality must be clearly stated and understood by the employees and how they are to be rewarded for achieving those objectives.

3. Most cost accounting systems cause managers to try to optimize <u>local</u> costs at the expense of total manufacturing cost.
4. Computers can do significant harm if they generate erroneous data such as incorrect inventory status (MRP).

Focusing on better data integrity and timeliness, greater production flexibility and better planning the inventory and lead time reduction objectives are being realized. Thus instead of monthly batch requirements, systems were developed to provide weekly and daily requirements to maintain market coordination.

QUALITY MANAGEMENT

The Quality Management Program was aimed at early identification of critical quality characteristics during the design, manufacturing, engineering and vendor-sourcing processes. This set the stage for Just-In-Time production.

The division uses about 480 suppliers who provide approximately 10,000 active items required in the manufacture of the three product groups. Of these suppliers, 90% are distributors who, therefore, perform no functional tests on the component items they sell. This complicates the purchasing and quality control activity.

The few statistical tools used have been able to help in several ways including separating problem operations from acceptable ones and provide a warning to the operator that something is going wrong with the process. Defects/unit data are manually collected and plotted daily on a C-chart for everyone to see and discuss.

A quality improvement program continually verifies that there is a basic requirement to manage people differently if you expect them to contribute suggestions and creative solutions to the problems. The program not only can solve many problems but it opens the door between employees and management. Changes in their workplace or layout along with equipment purchases are now reviewed with the operators before the decisions are made.

IMPACT ON MRP

JIT almost always requires some changes in MRP. It requires good front-end planning to try to smooth the independent demands on the factory to balance the load and work flow.

The bills of material must change to reflect the levels as they exist on the floor. The planning, monitoring and control of the bills of material changes must be closely coordinated with engineering, production and the data base managers.

Activities on the floor change rapidly and these changes must be reflected at least on a daily basis. It is difficult and unnecessary for a MRP systems to provide this currency. An auto replenishment (kanban) system can perform this function but not in the traditional manner of stock status reports.

CONCLUSIONS

Without a quality commitment up-front by both management and the workers, the advantages of JIT--flexibility, constant improvement, accountability and responsibility--cannot be realized. Synchronized manufacturing cannot take place unless the barriers to quality improvement and methods improvement are removed, and only management can remove these barriers.

BIBLIOGRAPHY

Bose, G.J. and A. Rao, "Implementing JIT with MRP II Creates Hybrid Manufacturing Environment," Industrial Engineering, September 1988.

Cadley, J.A., "Insights From Simulating JIT Manufacturing," Interfaces, Vol. 19, March-April 1989.

Carlson, J.G., J.J. Lee, and G.S. Lawson, "JIT in Quick Time," Just-In-Time Manufacturing, Proceedings of 2nd International Conference, London, October 1987.

Cate, G.P., "JIT Implementation in a Small Electronics Company," Just-In-Time Manufacturing, Proceedings of 2nd International Conference, London, October 1987.

Collins, P.E., "People Really Do Count," Just-In-Time Manufacturing, Proceedings of 2nd International Conference, London, October 1987.

Galloway, D.G., and D. Waldron, "You Get What You Measure," Just-In-Time Manufacturing, Proceedings of 2nd International Conference, London, October 1987.

Hickman, R.I.F., "Managing Just-In-Time Materials," Just-In-Time Manufacturing, Proceedings of 2nd International Conference, London, October 1987.

Huang, P.Y., L.P. Rees and B.W. Taylor III, "Simulation Analysis of JIT for Multiline, Multistage Production Systems," Decision Sciences, Vol. 14, 1983.

Jewitt, R.H., "MRP vs. JIT - Is There a Conflict?" Just-In-Time Manufacturing, Proceedings of 2nd International Conference, London, October 1987.

Karmarker, U., "Getting Control of Just-In-Time," Harvard Business Review, September-October 1989.

Marsden, A.W., "Changing People's Behavior," Just-In-Time Manufacturing, Proceedings of 2nd International Conference, London, October 1987.

Sonnenburg, R.D., "The MRP and JIT Marriage," APICS, Conference Proceedings, 1983.

A Study of JIT Application under Stochastic Demand and Supply Arrival

MOHAMMAD JALALI, RAFAEL MORAS and RICHARD DUDEK

PRELIMINARY REMARKS

A simulation and analysis of variance study was conducted of a production line system operating in a near Just-In-Time (JIT) mode. The objective was to investigate the effect of four system parameters on the production line idle time and the carrying time of inventory. The system consisted of a production line producing one unit of product each production cycle to a JIT demand, each cycle was defined as one unit of time. The demand arrived stochastically during each cycle. The system was supplied by a supplier delivering materials with a stochastic arrival time of up to two cycles. The four system parameters were (1) operating inventory order quantity, representing a policy of ordering which permitted supply arrival for more than one cycle of production requirement, (2) stochastic demand arrival of up to one cycle, (3) safety stock of up to 3 units (Kanbans) of production, and (4) stochastic arrival of supply of up to three cycles.

The study is described and results reported. Predictive models were obtained for each of the dependent variables of the system, namely the production line idle time, carrying time of regular inventory and carrying time of the safety stock. The models can be utilized to determine the optimal operating policy of a near JIT production system balancing the costs of idle production capacity against the costs of inventory.

INTRODUCTION

An ever increasing need for productivity in manufacturing has been identified in recent years by American industry. Western producers have experienced the need to adopt strategies that will improve their industrial and marketing positions in competing with foreign competitors. The reasons for the lack of productivity of labor and capital investments are being investigated. Among the procedures being developed and used are technological advances in production operations such as cellular manufacturing, flexible manufacturing, or CAD/CAM; and technological advances in production planning techniques such as material requirements planning (MRP), manufacturing resources planning (MRP-II), Optimized Production Technology (OPT), and Just-in-Time (JIT). Usually there are two alternatives available to improve productivity. The first consists of using technology with increased automation, changes in product design, and advanced manufacturing processes. The second approaches the improvement of control of production resources. The latter is essential regardless of the level of technology used in the production operation.

The Just-in-Time system has been proposed as an alternative for U.S. manufacturers who are confronted with productivity problems. Major changes are necessary to implement JIT, which is considered by some to be a philosophy for achieving productivity results by continuous improvement. Some of the objectives of JIT are: (1) to identify and solve fundamental problems; (2) to reduce waste by reducing or eliminating all processes that do not add value to the product (i.e. setups, transportation, inspection, storage, delays); and (3) to devise appropriate systems that identify problems as they occur (Schonberger 1984).

In the traditional production system, the main goal is generally to optimize throughput by way of increasing labor and equipment utilization (Krajewski et al. 1987). This goal may be achieved by implementing inspection schemes and by increasing inventory to compensate for problems such as machine breakdowns, unstable demand, unreliable deliveries, etc. A reorder point is set to account for long and variable lead times. Normally, this tends to increase inventory levels, which in turn create inefficiencies such as large amounts of money tied up, the risk of obsolescence, a low inventory turnover, and also the possibility of under- or over-estimating demand.

JIT enthusiasts claim that even though the solutions obtained by methods as the EOQ and its variations are considered "optimal," these solutions may give problem solvers a false sense of reassurance in that high setup costs and other inefficiencies are taken for granted. Therefore, no efforts are made to find or eliminate the real source of the problems. An analogy that has become a classic in the JIT literature is that of viewing inventory as water that covers up problems that are like rocks. If inventory levels are systematically reduced, then the problems may be exposed and an effort can be made to solve them. Under JIT, industrial engineering principles are methodically applied to improve operations and eliminate waste. Processes are made smaller and simpler instead of bigger and more complex.

Another major difference between traditional inventory systems and JIT is the idea of a pull production system (as opposed to "push" strategies followed by the traditional production system). In JIT, a supplying work center will not produce a part until a "pull" signal is issued by the work center down the line. Therefore, the supplying work center waits for the using work center to pull these parts before working on additional parts.

A successful JIT implementation requires the establishment of appropriate relationships with vendors and customers. Vendors must be carefully chosen so that they deliver quality products, in small lot sizes and on time. Having a few selected suppliers with these characteristics is recommended over having multiple sources.

Ideally, the suppliers should be located close to the production system to achieve the on time delivery. This may not be the situation for many production systems so the system experiences some variability in supply arrival. Understanding the effects on the production system from variable supply arrival would be of benefit to any organization in the planning stages of adopting a JIT system.

This paper describes a study conducted to discern effects on the production system utilizing a JIT system under conditions of stochastic demand and supply arrivals. A simulation and analysis of variance study was conducted of a production line system operating in a near JIT mode. The objective was to investigate the effect of four system parameters on the production line idle time and the carrying time of inventory. The system consisted of a production line producing one unit of product each production cycle to a JIT demand, each cycle was defined as one unit of time.

The demand, equal to one unit, arrived stochastically during each cycle. The system was supplied by a supplier delivering materials with a stochastic arrival time of up to three cycles. The four system parameters were (1) operating inventory order quantity, representing a policy of ordering which permitted supply arrival for more than one cycle of production requirement, (2) stochastic demand arrival of up to one cycle, (3) safety stock of up to 3 units (Kanbans) of production, and (4) stochastic arrival of supply of up to three cycles.

BACKGROUND

The production strategy for multistage production and inventory systems may be classified into two categories; push and pull type production ordering systems (Kimura and Terada 1981). The push type production ordering system, known as a Just-In-Case (JIC), is a system in which orders are launched and pushed through the system to meet some established due date and/or some planned finished goods inventory. Furthermore, this manufacturing technique is based on a production schedule that is generated from forecasted demand.

A pull type production system is one in which orders are pulled through the system to satisfy the demand for end items. Furthermore, this manufacturing technique is based on the frozen master production schedule that is generated from customer orders. A pull system is one important element of Just-in-Time.

To achieve ideal JIT manufacturing, the following prerequisites need to be satisfied (Shingo 1989), (Hernandez, 1989): (1) reduction in changeovers and set up times; (2) a logical preventive maintenance program.; (3) highly skilled and flexible workforce; (4) raw material defect free; (5) a proficient plant layout for JIT production; (6) JIT delivery of materials for production; and (7) a uniform demand rate.

JIT production is based on the hypothesis that significant savings can be obtained by reducing inventory levels to an absolute minimum, i.e., one unit. JIT production is attainable if the following assumptions hold: (1) supplies arrive on time; (2) process is highly reliable; (3) constant or small variation in processing times; and (4) there is no or very small fluctuation in demand. Within American industry these factors are not always present for various reasons, such as: greater distance between supplier and buyer, which creates fluctuation in "on-time" delivery performance, reliance on multiple sources of supplies, variable master production schedule which results in fluctuation in demand, and variable processing time due to frequent employee turnover.

This research focuses on the operational control problem affected by factors from both ends of the production system; demand side and supply side of the system which are subjected to variability. The factor of interest on the demand side of the system is stochastic demand arrival time. The factors of interest on the supply side of the system are:
(1) stochastic supply arrival time; (2) variable levels of safety stock; and (3) order quantity which determines the frequency of raw material delivery in a certain planning horizon time period.

The system measures of performance are considered to be average production line idle time, average safety stock and operational inventory carrying times per kanban; each expressed in time units.

This study aims at developing models for average production line idle time, average safety stock and operational inventory carrying time per kanban as a function of operational inventory order quantity (OIOQ), range of demand and supply arrival times (RDAT and RSAT) and safety stock quantity (SS) when JIT production concepts are used. The order of supply is pulled when the last container of raw material is removed from the first stage of production line for processing which is directly related to demand arrival; i.e. a JIT supply ordering pull production strategy.

The number of pulls per time period within the planning horizon is fixed and assumed to be one pull per time period and is expected to occur within each time period. The arrival of this demand pull triggers the order of supplies which are defect free. The production line is assumed to function as a unit which is assumed to be internally failure free. The system is composed of a market that "pulls" the finished goods from the production line which in turn "pulls" the raw material from the storage which "pulls" the supplies in a "single pull," i.e., OIOQ per replenishment cycle is equal to one; or "batch pull," i.e., OIOQ for N production cycles.

Both supply arrival time and demand arrival time are assumed to be stochastic following a uniform distribution, and therefore, some level of safety stock is allowed in the system. The maximum value of RDAT is chosen as one time unit, because it has been indicated that JIT will not be cost effective if the variability in pull rate is too large. RSAT was allowed to vary up to a maximum of three time units, one period or cycle of lead time and up to two production cycles.

The purpose of this research is to develop predictive production system models of production line idle time and inventory (operational and safety stock) carrying time resulting from varying levels of the independent variables of range of demand and supply arrival times (RDAT and RSAT), safety stock (SS) and operational inventory order quantity (OIOQ). These models could aid a manufacturer to plan in advance the likely economic implications of adapting a JIT system.

LITERATURE REVIEW

As described in the survey of JIT literature by Moras, et al., (1990) of the 266 articles considered only 56 were technical papers. None were reported having made a study of the effects of stochastic supply arrival on the production system. The study reported here considered the effects of both stochastic supply arrival time and demand arrival time on the production line idle time and carried inventory.

Some analytical papers consider the implementation of JIT in a production environment which does not possess the characteristics for implementation of ideal JIT, such as: single pull, stable and steady demand, constant processing time and balance station times, reliable equipment and machinery, negligible lead time and changeover, and finally on time raw material delivery.

Kim (1985) considered deviation from a single pull, specifically a batch pull. He analyzed an operating policy for JIT production called periodic pull system (PPS). The essence of PPS was to review the status of material flow at all production stages at regular intervals. Deviation from constant operation was considered by Huang, et al. (1983). They investigate the effects of variable processing times, variable master production schedules, and imbalances between stages on a multi-line, multi-stage JIT production system. Departure from steady and stable demand in a JIT production system is considered by Ebrahimpour and Fathi (1985). They utilized a simulation to investigate WIP inventory behavior under three environmental conditions: cyclical demand, steady growth, and gradual reduction of kanban cards during stable demand. Villeda, et. al (1988) departed from the JIT requirement of having a balanced station time. They studied the effect of imbalance in a JIT production system via simulation and modeled the performance of JITPS under different unbalancing strategies. Spence and Porteus (1987) studied the relationship between setup time and capacity. Departures from operating with reliable equipment in a JIT production system are considered by Lulu and Black (1987).

METHODS AND PROCEDURES

The objectives of the study were to develop functional relationships between the dependent variables of production line idle time, average operational inventory carrying time per kanban, and average safety stock inventory carrying time per kanban, and the independent variables of operational inventory order quantity (OIOQ), range of demand arrival time (RDAT), range of supply arrival time (RSAT), and safety stock quantity (SS) within a production system utilizing JIT. Experimental design, simulation modeling, and regression analysis were utilized to achieve the above objectives.

The average production line system idle time (APLSIT), average safety stock inventory carrying time (ASICT), and average operational inventory carrying time per kanban (AOICT) were three dependent variables under the conditions of the system.

The production line system idle time (PLSIT) was the amount of time which the production line was idle due to late raw material arrival time, lack of safety stock in storage and/or lateness in demand occurrence. It was measured in terms of average idle time per time unit of production.

The average operational inventory carrying time per kanban (AOICT) was the amount of time which one kanban of operational inventory was carried before it was pulled by the first stage of the production line for processing.

The average safety stock inventory carrying time per kanban (ASICT) was the amount of time which one kanban safety stock was carried before being utilized by the production line system in the case of shortage in operational inventory.

The operational inventory order quantity, the range of demand arrival time, the range of supply arrival time, and safety stock were the four independent variables which were selected in this study.

The operational inventory order quantity specifies the order quantity for the number of cycles for which inventory is ordered and carried. For example, an order quantity of one indicates order frequency of every cycle of the production period, while an order quantity of four indicates an order frequency of every fourth cycle of the production period. The order quantity was set at three levels: 1, 4, and 7 cycles.

The range of demand arrival time was the amount of variability in demand arrival time from the beginning of each cycle within the period. The levels of RDAT were set at 0.25, 0.5, 0.75, and 1 time units.

The range of supply arrival time was the amount of variability in supply arrival time from minimum lead time of one time unit. The levels of RSAT were set at 0.5, 1, 1.5, and 2 time units (recall, each cycle of production is one time unit).

Safety stock, the amount of raw material that could be required to compensate for lack of operating inventory, set at three levels: 1, 2, and 3 kanbans.

The levels of each factor were selected considering the following JIT principles: (1) a JIT production system does not permit high variability in the range of supply or demand arrival times, (2) the order of supply is placed in small quantity and frequent delivery, and (3) existence of inventory in any form, operational inventory and safety stock, is considered the root of several problems and should be minimized.

In a pure Just-In-Time production system in which the pull rate is constant and supply arrives exactly on time, the arrival of demand triggers all stages of the production line almost simultaneously to the first stage which triggers the order of supplies and subsequently the arrival of supplies. However, in this study, due to the stochastic nature of demand and supply arrival times, triggering the order of supply may not take place simultaneously to arrival of demand, because, when demand arrives, the production line might still be busy. As soon as production completes that cycle and the last container of raw material is removed from the input stocking point of the first station, the order of supply is triggered. If the supplier is available, the order is taken right away, otherwise the order of supply (withdrawal kanban) goes on the scheduling board until the supplier arrives.

Some portion of the production line idle time is due to variability in the range of demand arrival and some will be due to the variability in the range of supply arrival. When the raw material arrives and the production line is busy or demand time passes the supply arrival time, raw material needs to wait for the production line to be available or for a customer demand to arrive. The amount of time that supply needs to wait for availability of the production line or customer demand will be the raw material carrying time.

Design of Experiment

After defining the dependent and independent variables and selecting their levels, a four-factor complete randomized factorial design of experiment with fixed level of each factor was utilized for this experiment. With four factors, two at three levels and two at four levels, a total of 144 treatments were made. A simulation model was used to collect the desired data. Each treatment was replicated 5 times, therefore, a total of 720 observation was collected for each dependent variable. Null hypotheses were tested for each of the independent variables effect on the dependent variables.

Regression Analysis

After testing the hypotheses and detecting the main and interaction effects of those variables which were significant at the one or five percent level, a graphical analysis of the main and interaction effects of variables was also performed to more clearly understand the influence of the variables on the dependent variables. Also Duncan's multiple range test was performed to detect the significant effect of different levels of a factor at the one or five percent level of significance. Finally regression models were utilized to determine the best prediction model for predicting the dependent variables.

Simulation Model Description

For this study, GPSS (General Purpose Simulation System) version H which is widely used to build computer models for discrete-event simulation was used to construct computer models for the proposed production scenarios. The model was composed of several segments.

(1) Initialization segment: performs the initialization of the variables, be executed only once in each run.

(2) Customer waiting time segment : computes the amount of time a customer needs to wait for the availability of finished goods if finished goods are not available upon arrival.

(3) Production line segment: maintains the finished goods inventory. It also sends signals to various segments including finished goods carrying time segment and the production line idle time segments.
(4) Finished goods carrying time segment: computes the amount of time that finished goods waits for a customer arrival.
(5) Raw material carrying time segment: calculates the raw material carrying time when raw material arrives and the production line is busy or demand arrival time is later than the supply arrival time. Raw material carrying time is composed of two elements; safety stock inventory carrying time and operational inventory carrying time.
(6) Production line idle time due to the variability in the range of supply arrival time: computes the production line idle time when supply arrival time exceeds the demand arrival time and finished goods completion time.
(7) Production line idle time due to variability in demand arrival time segment: calculates the production line idle time if demand arrival time exceeds the finished goods completion time.
(8) Timer segment: performs the function of a timer. This model segment terminates the simulation after five weeks steady state simulation run.

RESULTS

Two prediction models, utilizing multiple linear regression and nonlinear regression procedures of the Statistical Analysis Systems (SAS) computer package, were developed for each dependent variables; APLSIT, ASICT, and AOICT for the 'near' JIT production system investigated. From the above two models the one with the higher R^2 was selected as a prediction model for the dependent variables.

Simulation Results and APLSIT Data Analysis

As stated earlier, the levels of independent variables; the operational inventory order quantity and safety stock were at three levels; (1, 4, 7), (1, 2, 3) kanban(s), respectively. The levels of the range of supply and demand arrival times were set at four levels; (0.5, 1, 1.5, 2), (0.25, 0.5, 0.75, 1) time units, respectively.

The general linear model procedure (PROC GLM) of the (SAS) software package available at the Academic Computer Services, Texas Tech University, Lubbock, Texas, was used to perform all statistical analysis.

Main Effects

The result of analysis of variance in Table 1, showed that all the main effects except for the range of demand arrival time were statistically significant at the one percent level. The simulated APLSIT data showed that as the level of order quantity increases, the production line system idle time decreases. In other words, decreasing the frequency of delivery resulted in reduction of production line idle time.

Figure 1 illustrates how APLSIT was affected by the order quantity. It appears that APLSIT follows an exponential decay pattern. Therefore, it can be concluded that there is a point beyond which increasing the order quantity will have slight affect on reducing the production line idle time. It should be noted that choosing a high value of order quantity could result not only in higher inventory carrying time, but it also may result in more scrap and rework due to less

Table 1: Analysis of Variance Using Average Production Line Idle Time as the Dependent Variable

Source	df	SS	F Value	Pr>F
OIOQ	2	0.0994	3807.08	0.0**
SS	2	0.0277	1062.49	0.0**
RSAT	3	0.2017	5151.35	0.0**
RDAT	3	0.0001	2.17	0.09
OIOQ*SS	4	0.0137	262.73	0.0**
OIOQ*RSAT	6	0.1328	1695.99	0.0**
OIOQ*RDAT	6	0.0000	1.14	0.3401
SS*RSAT	6	0.0405	517.03	0.0**
SS*RDAT	6	0.0000	0.61	0.7224
RSAT*RDAT	9	0.0003	2.97	0.0018**
OIOQ*SS*RSAT	12	0.0179	114.20	0.0**
OIOQ*SS*RDAT	12	0.0001	0.64	0.8084
OIOQ*RSAT*RDAT	18	0.0004	1.77	0.0257*
SS*RSAT*RDAT	18	0.0001	0.59	0.9048
OIOQ*SS*RDAT*RSAT	36	0.0006	1.37	0.0786

** alpha = 0.01, * alpha = 0.05

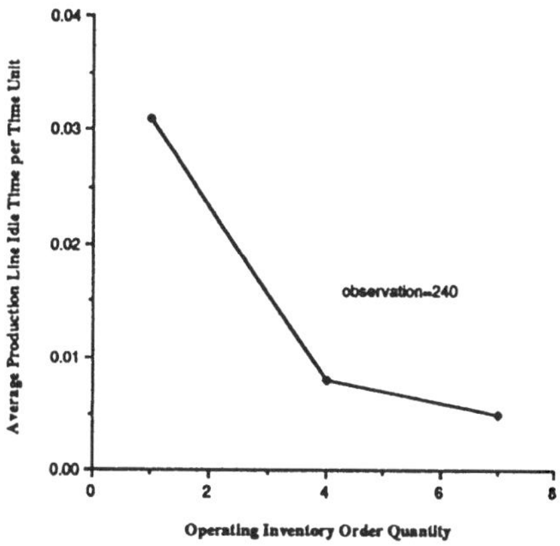

Figure 1: Effect of Operating Inventory Order Quantity on Average Production Line Idle Time.

frequent feed back which the supplier receives from the system.

As shown in Figure 2, production line idle time also decreased by increasing the levels of safety stock. Average production line idle time decreased markedly as the level of safety stock increases from one to two kanbans.

The result of the simulation indicates that as long as the range of supply arrival time did not exceed the production unit time the production line idle time was equal to zero. This is due to the fact that minimum safety stock was chosen to be one kanban and the production line produces one kanban per time unit or cycle. Therefore, even if operational supply arrives one time unit late, the safety stock will compensate for that tardiness and consequently the production line idle time will be equal to zero. As the range of supply arrival time exceeded the production time unit, the average production line idle time greatly increased. Figure 3 shows these effects of the range of supply arrival time on production line idle time.

Interaction Effects

The results of analysis of variance in Table 1 indicate that four second order and two third order interactions were statistically significant at the one or five percent levels of significance. Graphical analysis of these interactions generally reflected the same patterns and trends of the main effects. From the graphical analysis it could be concluded that the most and the least influencing factors on production line idle time were range of supply arrival time and the range of demand arrival time, respectively. This is reflected in the prediction model developed.

APLSIT Model Selection

Of the two models developed for production line idle time, the linear model had an $R^2 = 0.87$ and the non-linear model an $R^2 = 0.95$. Therefore, the non-linear model was selected as the predictive model and is given below.

$$APLSIT = 0.003858\ e^{(-0.4011\ OIOQ\ -\ 0.5318\ SS\ +\ 2.264\ RSAT\ -\ 0.0552\ RDAT)}. \qquad (1)$$

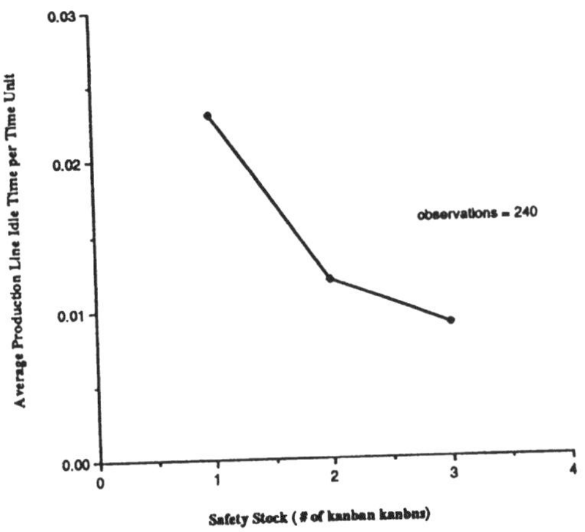

Figure 2: Effect of Safety Stock on Average Production Line Idle Time.

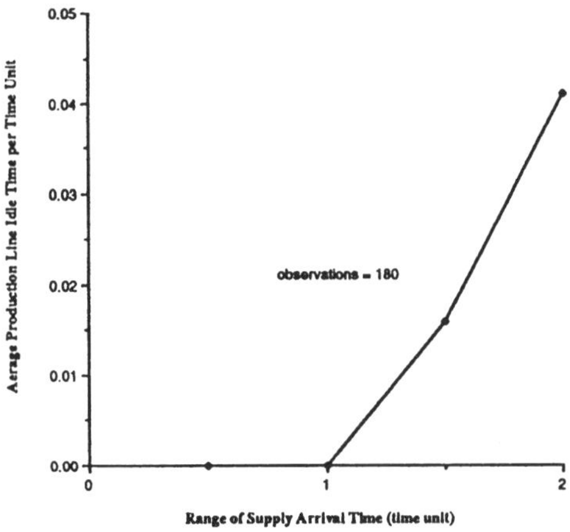

Figure 3: Effect of Range of Supply Arrival Time on Average Production Line Idle Time.

Simulation Results and ASICT/kanban Data Analysis

As in the analysis for APLSIT, the independent variables and their levels were the same. The results of the analysis of variance are depicted in Table 2, all main effects as well as most of the interactions were statistically significant at the one or five percent levels.

Main Effects

As would be expected, as the operating inventory order quantity increased, the average safety stock inventory carrying time per kanban also increased linearly. This relationship is shown in Figure 4. As depicted in Figure 5, an increase in the the level of safety stock resulted in an increase in ASICT/kanban. As expected, an increase in the range of supply arrival time resulted in a decrease in the ASICT/kanban as shown in Figure 6. The almost negligible effect of the range of demand arrival time on average safety stock inventory carrying time per kanban is illustrated in Figure 7.

Interaction Effects

The results of analysis of variance in Table 2 showed that a majority of the interactions were statistically significant. Graphical analysis of these interactions generally reflected the same patterns and trends of the main effects. From the graphical analysis could be concluded that besides the quantity of safety stock, the range of supply arrival time and the operational inventory order quantity were the most influencing variables on average safety stock inventory carrying time.

ASICT Model Development

The same statistical analysis and procedure which was used with APLSIT model development, was also utilized to develop prediction models for ASICT. The linear model displayed an $R^2 = 0.99$ while that of the non-linear model was 0.95. Therefore, the linear model was selected for prediction of ASICT.

$$\begin{aligned} ASICT = & -1.2746 + (0.4801)OIOQ + (0.4890)SS + 0.7637RSAT - \\ & (0.3622)(RSAT)^2 + 0.6364(OIOQ)(SS) + 0.0450(OIOQ)(RSAT) - \\ & 0.1722(OIOQ)(SS)(RSAT) + 0.0297(SS)(RSAT)(RDAT) \end{aligned} \quad (2)$$

Table 2: Analysis of Variance using Average Safety Stock Inventory Carrying Time per Kanban as the Dependent Variable

Source	df	SS	F Value	Pr>F
OIOQ	2	8209.76	57036.82	0.0**
SS	2	2316.57	16094.25	0.0**
RSAT	3	412.02	1908.33	0.0**
RDAT	3	0.634	2.94	0.03*
OIOQ*SS	4	513.19	1782.69	0.0**
OIOQ*RSAT	6	145.03	335.85	0.0**
OIOQ*RDAT	6	0.617	1.43	0.2008
SS*RSAT	6	84.33	195.30	0.0**
SS*RDAT	6	0.548	1.27	0.2702
RSAT*RDAT	9	1.04	1.61	0.1103
OIOQ*SS*RSAT	12	40.506	40.90	0.0**
OIOQ*SS*RDAT	12	0.831	0.96	0.4837
OIOQ*RSAT*RDAT	18	1.423	1.10	0.3494
SS*RSAT*RDAT	18	2.8407	2.19	0.0031**
OIOQ*SS*RDAT*RSAT	36	3.8411	1.48	0.0370*

** = alpha = 0.01, * = alpha = 0.05

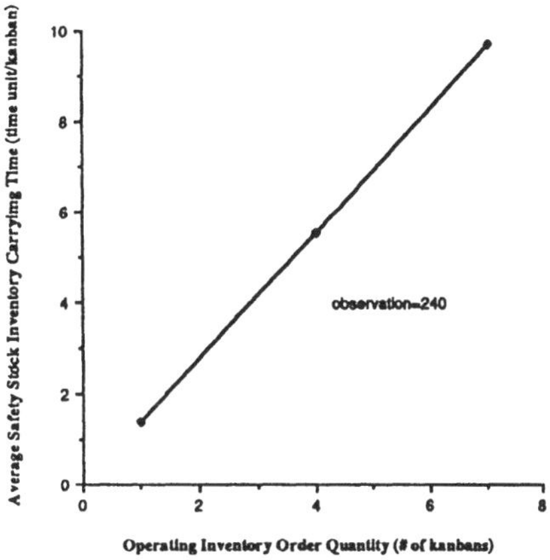

Figure 4: Effect of Operating Inventory Order Quantity on Average Safety Stock Inventory Carrying Time.

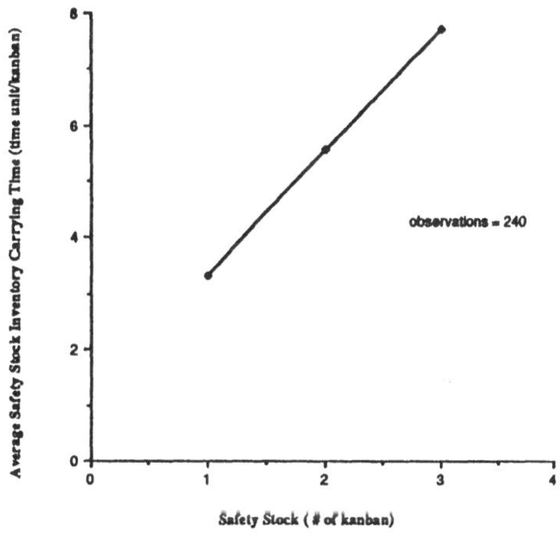

Figure 5: Effect of Safety Stock on Average Safety Stock Inventory Carrying Time per Kanban.

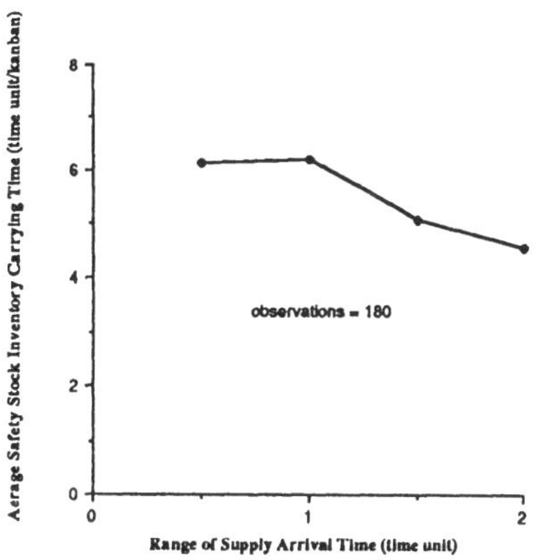

Figure 6: Effect of Range of Supply Arrival Time on Average Safety Stock Inventory Carrying Time per kanban.

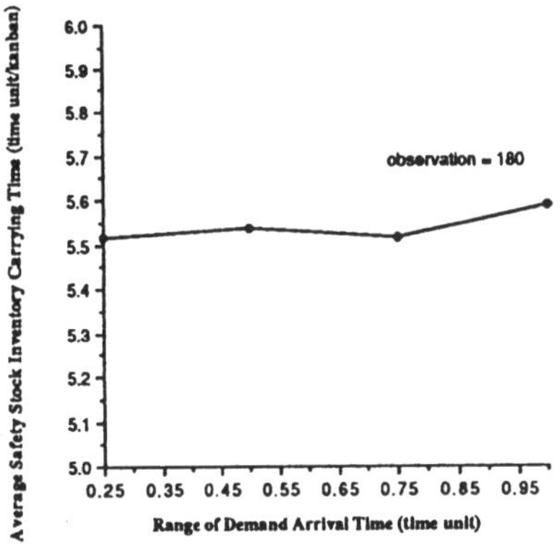

Figure 7: Effect of Range of Demand arrival time on Average Safety Stock Inventory Carrying Time per Kanban.

Simulation Results and AOICT/Kanban Data Analysis

The results of the analysis of variance for the dependent variable of average operational inventory carrying time per kanban in time units are given in Table 3.

Main Effects

All main effects with the exception of range of demand arrival time were statistically significant at the one percent level. Figure 8 shows that as the operational inventory order quantity was increased, the average operational inventory carrying time per kanban also increased. An increase in the level of safety stock resulted in an increase in AOICT as shown in Figure 9. Figure 10 shows the influence of the range of supply arrival time on the average operational inventory carrying time. The AOICT had its maximum value when the range of supply arrival time was minimum. As the range of supply arrival time was increased, the AOICT per kanban was decreased.

Interaction Effects

The results of analysis of variance in Table 3 shows that among all possible interactions only the SS*RSAT and RSAT*RDAT interactions were significant at one percent level. Graphical analysis of these interactions reflected the same patterns and trends of the main effects.

AOICT Model Development

In the AOICT model development, the linear model possessed an $R^2 = 0.997$ while the non-linear had 0.95. Therefore, the linear model was selected.

$$\text{ACICT} = 0.565 + 0.508\ \text{OIOQ} - 0.780\ \text{RSAT} - 0.001\ (\text{OIOQ})^2 + 0.187\ (\text{RSAT})^2 - 0.014(\text{SS})^2 + 0.042\ (\text{SS})(\text{RSAT}) \qquad (3)$$

SUMMARY AND CONCLUSIONS

Mathematical predictive models were developed to demonstrate how the JIT concept may be applied in production environments which do not possess the unique physical characteristics for ideal JIT production, such as: 'on time' raw material delivery and stable demand. It was concluded that when the variability in the range of supply arrival time and variability in the range of demand arrival time do not exceed the production unit time, the production line idle time is equal to zero.

Table 3: Analysis of Variance Using Average Operating Inventory Order Quantity Carrying Time as the Dependent Variable

Source	df	SS	F Value	Pr>F
OIOQ	2	1065.9359	99999.99	0.0**
SS	2	0.3856	53.17	0.0**
RSAT	3	13.694	1258.78	0.0**
RDAT	3	0.0113	1.04	0.3751
OIOQ*SS	4	0.007	0.48	0.7485
OIOQ*RSAT	6	0.0245	1.13	0.3445
OIOQ*RDAT	6	0.0164	0.75	0.6071
SS*RSAT	6	0.3832	17.61	0.0**
SS*RDAT	6	0.0106	0.49	0.8184
RSAT*RDAT	9	0.0774	2.37	0.01**
OIOQ*SS*RSAT	12	0.0720	1.66	0.0730
OIOQ*SS*RDAT	12	0.0267	0.61	0.8311
OIOQ*RSAT*RDAT	18	0.0583	0.89	0.5881
SS*RSAT*RDAT	18	0.1037	1.59	0.0576
OIOQ*SS*RDAT*RSAT	36	0.1432	1.10	0.3243

** alpha = 0.01, * alpha = 0.05

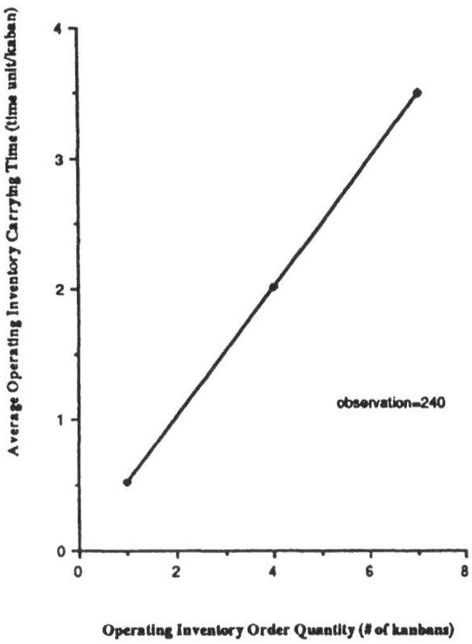

Figure 8: Effect of Operating Inventory Order Quantity on Average Operating Inventory Carrying Time per Kanban.

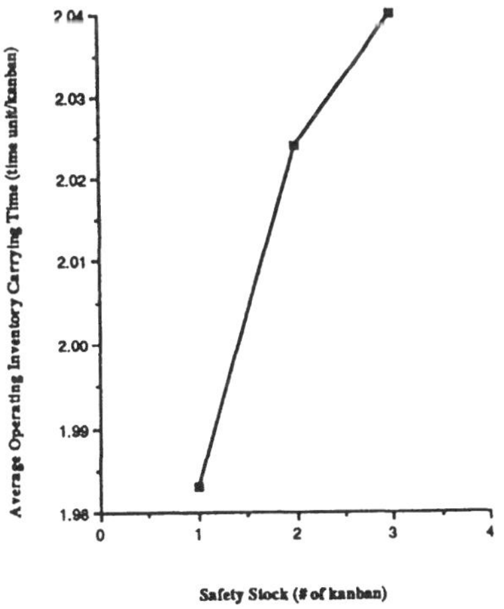

Figure 9: Effect of Safety Stock on Average Operating Inventory Carrying Time per Kanban.

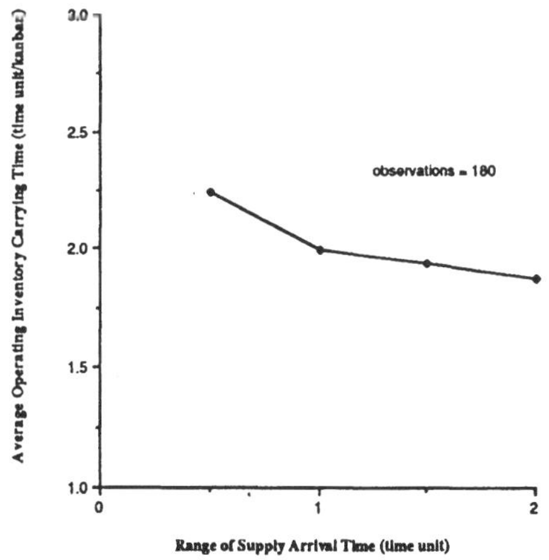

Figure 10: Effect of Range of Supply Arrival Time on Average Operating Inventory Carrying Time per kanban.

Due to the sensitivity of the production line to the range of supply arrival time, the geographical distance is of more concern in a JIT system than in a JIC system.

If a production line is subjected to random variation in both ends of the system, increases in the level of safety stock and/or operating inventory order quantity decreases the production line idle time, However, for a given level of safety stock and/or operating inventory order quantity, a threshold exists beyond which increases in the safety stock and/or operating inventory do not result in corresponding decreases in production line idle time.

The prediction models developed in this research, demonstrate how advantages of JIT manufacturing can be obtained by those companies that are in the process of implementing JIT principles but experience some variability in both demand pull rate and supply arrival time.

To explain the application of the models, any manufacturer usually has some prior information relative to range of supply and demand arrival times. With this information, arbitrarily assign a low value of operating inventory and safety stock,since the major objective of JIT is to decrease inventory, then calculate the average production line idle time and overall inventory carrying time per kanban. If the cost of production line idle time is higher than the cost of overall inventory carrying time, then increase the level of operating inventory and the safety stock until inventory carrying cost balances with the cost of production line idle time.

REFERENCES

Ebrahimpour, M. and B. F. Modarres, "Dynamic Simulation of a Kanban Production Inventory System," International Journal of Operations and Production Management, Vol. 5, No. 1, 1985, 5-14.

Hernandez, A., "Just-In-Time Manufacturing," Prentice Hall, Englewood Cliffs, NJ, 1989.

Huang, P. Y., L. P. Rees, and B. W. Taylor III, "A Simulation Analysis of The Japanese Just-In-Time Technique (With Kanbans) for a Multiline, Multistage Production System," Decision Sciences, July 1983, 326-338.

Kim, T. M., "Just-in-Time Manufacturing System: A Periodic Pull System," International Journal of Production Research, Vol. 23, No. 3, 1985, 553-562.

Kimura, O. and H. Terada, "Design and Analysis of Pull System, a Method of Multi-stage Production Control," International Journal of Production Research, Vol. 19, No. 3, 1981, 241-253.

Krajewski, L, J., B. K. King, L. P. Ritzman, and D. S. Wong, "Kanban, MRP, and Shaping the Manufacturing Environment," Management Science, Vol. 33, No. 1, January 1987, 39-57.

Lulu, M. and J. Black, "Effect of Process Unreliability on Integrated Manufacturing/Production Systems," Journal of Manufacturing Systems, Volume 6, No. 1, 1987, 15-22.

Schonberger, R. J., "Just-In-Time Production Systems: Replacing Complexity With Simplicity in Manufacturing Management," Industrial Engineering, October 1984a, 52-63

Shingo, S., "A Revolution in Manufacturing, The SMED System," Productivity press, Cambrige, MASS, 1989.

Spence A. M. and E. L. Porteus, Setup Reduction and Increased Effective Capacity," Management Science, Vol. 33, No. 10, October 1987, 1291-1301.

Villeda, R., R. Dudek, and M. L. Smith, "Increasing the Production Rate of a Just-in-Time Production System with Variable Operation Times," International Journal of Production Research, Vol. 26, No. 11, 1988, 1749-1768.

A Comprehensive Study of Quality Management Practices in JIT and Non-JIT Firms

MALING EBRAHIMPOUR

INTRODUCTION

In recent years, there have been numerous books and articles on the subject of JIT production system. These works cover many aspects of JIT such as its philosophy, definition, benefits, and requirement for successful implementation. These publications vary from descriptive, to survey, to simulation, and to actual case studies. The concept of just-in-time (JIT) production system requires continuous improvement via waste elimination throughout the organization. The concept of continuous quality improvement in a JIT environment covers all aspects of production from raw materials to finished goods and after sales services. The quality factor and its importance as an essential element for the success of JIT manufacturing system has been discussed in several books and numerous published articles. However, most studies related to JIT system treat the concept of quality lightly and do not examine it in great detail. This paper provides a detailed examination of the role of quality in the JIT environment by identifying a set of characteristics which are necessary to implement a continuous quality improvement system in an organization. This quality system should, in turn, pave the way for the successful implementation of the JIT system. To identify these characteristics comparisons are made between companies which are JIT users against those that are not currently using a JIT system. Furthermore, implementation problems encountered by JIT users are delineated and the benefits of the quality system within a JIT environment are explored. Based on the above discussion the following questions are proposed:

1- What are the major differences/similarities between firms using JIT production system and those non-JIT users in their use of continuous quality improvement?

2- What are the barriers that JIT users encounter when implementing the continuous quality improvement concept in their organizations?

3- What are the critical organizational characteristics necessary to successfully implement the continuous quality improvement concept in an organization?

Answers to the above research questions can be critical for those organizations that are contemplating to introduce JIT manufacturing systems. The result will help management to develop an effective plan to implement the continuous quality improvement successfully.

METHODOLOGY

The research design for this study consisted of field study of manufacturing firms operating in the United States. Data was collected using two different methods: in-depth case studies of selected manufacturing firms and survey questionnaires. A multiple methods approach was used in order to further assure the validity and authenticity of the results. The first method of collecting data consisted of in-depth personal observations and interviews with plant managers, quality managers, manufacturing managers, corporate vice presidents, and presidents. Six firms were selected for the on-site visits. Three firms were JIT users and three were not. The mail survey was the second method of data collection. The questionnaires were sent, by names, to the presidents, vice presidents, general managers (plant managers), quality managers, and operations managers (manufacturing managers).

The population for this study was all manufacturing firms operating in the United States which had Standard Industrial Classification (SIC) code numbers from 3500 thru 3870 which contain machinery, electrical and electronic, transportation equipment, and measuring and analyzing devices manufacturers. Questionnaires were sent to the managers in 647 randomly selected companies with the pre-specified SIC numbers. Of the total sample (n=647), only 170 completed and usable questionnaires were returned resulting in an overall response rate of 26.3%. The distribution of the usable questionnaires by types of firms was 35.9% JIT users and 64.1% non-JIT users. However, firm types did not differ significantly in terms of annual sales nor by number of employees in the plant. The distribution of firm types (i.e., JIT users and non-JIT users) was not significantly different across industries.

Analyses were performed with regard to the firm's type, i.e., JIT users and non-JIT users. Classifications of various questions in the mailed survey were: 1) top management and quality, 2) quality control/quality assurance, 3) implementation of quality improvement programs, 4) vendor evaluation, and 5) production operations.

RESULTS

Based upon seven distinct elements, Tables 1 and 2 summarize the findings from the case studies.

The following tables (3 thru 8) indicate the result of the statistical analysis of the questionnaires for the five categories which where defined in the methodology section. Each table shows mean and standard deviation for each firm type and indicate the relevant F-value. The existence of any significant difference between groups is shown by the asterisks above the F- ratio. Detail analysis of these significant differences for all categories are discussed in the following section.

DISCUSSION OF THE RESEARCH QUESTIONS

The purpose of this section is to provide detailed explanations and specific answers to the research questions which have been proposed in the Introduction.

Differences/Similarities Between the JIT Users and Non-JIT Users

Examination of all critical factors of product quality (Table 3) indicates that in the non-JIT users firms top management's

Elements	Company 1	Company 2	Company 3
1. Goal	Satisfying Customers	Meeting Competition's Challenge	Meeting Customers Specifications
2. Status of Quality	Not a Strategic Planning Tool	Strategic Planning Tool	Strategic Planning Tool
3. Problems	Lack of Top Management Support, Lack of Communication	Lack of Top Management Support, Lack of Workers Trust, Lack of true Understanding of Quality	Lack of Top Management Support, Existence of Double Standards to Measure Performance
4. Management Actions	No Specific Action	Implementation of QIP, Change of Corporate Culture	Implementation of QIP
5. Features and Effects of QIP Implementation (Expected)	Change of Management Attitude Toward Quality	Vendor Relation	Process Design
6. Responsibility of Quality Department	Inspection of Incoming Parts and Finished Products	Analysis of Causes of Defects, Defect Prevention, Analysis of Costs of Quality	Promoting Quality Ideas, Analysis of Causes of Defects, Customer Interface
7. Vendor Management	Formal Evaluation System (Being Developed), Major Criteria: Quality of Incoming Parts, On-Time Delivery	Formal Evaluation System, Major Criteria: Quality of Incoming Parts, On-Time Delivery	Formal Evaluation System, Major Criteria: Quality of Incoming Parts, On-Time Delivery

TABLE 1: Summary of the Case Studies of Non-JIT Users

Elements	Company 1	Company 2	Company 3
1. Goal	Increase Profit Via Gaining Market-Share	Continuous Quality Improvement, Increase Market-Share	Meet Competition's Challenge via Improving Quality and Lowering Cost
2. Status of Quality	Strategic Planning Tool, Highest Priority	Strategic Planning Tool,	Strategic Planning Tool, Highest Priority
3. Problems	Employee Resistence, Lack of Quality Knowledge, Lack of Communication	Lach of Top Management Support, Lack of Quality Knowledge, Lack of Feedback	Lack of Vendor Support, Lack of Training
4. Management Actions	Change Corporate Culture, Management Visible Supportand Commitmentto Quality, QualityTraining and Education, Implementation of QIP, Use of SQC	Management Visible Support and Commitment to Quality, Quality Training in All Levels, Participative Management, Management Committee to Review Suggestions	Management Support and Understanding of Quality, Management Leadership for Quality, Quality Training in All Levels, Measuring Performance by Quality not by Production Rate
5. Feauters and Effects of QIP Implementation	Plant Efficiency, Vendor Relation, Worker Participation Less Inspection	Vendor Relation, Productivity, Less Time on Inspecting Parts,	Plant Efficiency, Productivity, Worker Participation,
6. Responsibility of Quality Department	Teaching/training QC tools and Techniques	Training for Quality, Analysis of Causes of Defects	Setting up Quality Policies, Quality Taining in All Levels, Promoting QC Ideas
7. Vendor Management	Formal Vendor Selection System, Major Criteria: Vendor Use of SQC, Quality of Delivered Parts, On-Time Delivery	Team of Especialists, Vendor Certification Program, Major Criteria: Quality of Delivered Parts, On-Time Delivery, Mutual Relation	Formal Vendor Selection System, Team of Especialists, Major Criteria; Quality of Delivered Parts, On-Time Delivery

<u>TABLE 2:</u> Summary of the Case Studies of JIT Users

Factors	JIT Users Mean	JIT Users STD	Non-JIT Users Mean	Non-JIT Users STD	F ratio
Top Management Accessibility[1]	4.42	0.93	3.98	1.26	2.85*
Top Management Commitment[2]	4.21	0.92	3.55	1.13	5.66***
Quality Imprvement as a Part of Strategic Planning[3]	1.25	0.51	1.44	0.54	1.48
Process Design	4.42	0.87	3.92	1.14	3.39*
Workforce Training	4.18	1.03	3.83	1.13	1.54
Product Design	4.56	0.76	4.24	0.99	2.04
Top Management Performance	4.19	0.94	3.63	1.13	2.83*
Equipment Maintenance	4.20	0.83	3.69	1.14	2.43*
Purchased Material	4.41	0.74	4.08	0.70	2.10
Workmanship	4.42	0.70	4.23	0.96	0.98

Notes:

1 Not accessible at all=1, Slightly accessible=2, Moderately accessible=3, Reasonably accessible=4, Completely accessible=5.

2 Not Committed at all = 1, Slightly committed = 2, Moderately committed = 3, Reasonably committed = 4, Completely committed =

3 Yes = 1, No and Don't know = 2

4 Not critical at all = 1, Slightly critical = 2, Moderately critical = 3, Reasonably critical = 4, Extremely critical = 5

* $p<0.05$
*** $p<0.001$

<u>TABLE 3:</u> ANOVA Results of All Critical Factors of Product Quality

Factors #	JIT Users Mean	JIT Users STD	Non-JIT Users Mean	Non-JIT Users STD	F ratio
Promoting QC Ideas	4.12	0.98	3.80	1.22	1.77
Setting up Quality Policies	4.34	0.92	3.94	1.13	1.75
Inspection Planning	4.02	1.02	3.89	1.10	1.56
Defect Prevention Planning	4.19	0.96	3.72	1.21	2.29
Analysis of Causes of Defects	4.29	0.80	3.72	1.22	3.66**
Vendor Surveillence	3.75	0.99	3.46	1.08	0.92
Quality Costs Analysis	3.46	1.14	2.93	1.26	3.72**
Preparing Executive Reports	3.36	1.10	2.97	1.07	2.80*
Training for Quality	3.88	1.08	3.31	1.39	3.60**
Keeping Track of Quality	3.66	1.06	3.27	1.10	1.37
Monitoring Operations for	3.88	1.07	3.49	1.00	2.53*
Helping Purchasing Department to Select Vendors	3.89	1.22	3.07	1.24	4.36**
Helping Design the Product	3.26	1.26	2.83	1.42	1.97
Inspecting Incoming Parts	3.77	1.13	3.77	1.10	1.84
Inspecting Finished Products	3.95	1.10	4.04	0.99	1.75
Inspecting Production Processes	3.95	1.03	3.44	1.27	2.37
Inspecting Product Shipment	3.23	1.18	2.73	1.24	2.84*
Customer Interface	4.28	1.01	3.69	1.26	3.04*

Not important at all = 1, Slightly important = 2, Moderately important = 3, Reasonably important = 4, Extremely important = 5

* p<0.05, ** p<0.01

TABLE 4: ANOVA Results of Responsibilities of QC/QA Departments

---

Factors #	JIT Users Mean	JIT Users STD	Non-JIT Users Mean	Non-JIT Users STD	F ratio
Lack of Top Management Support	3.75	1.48	3.92	1.43	1.27
Lack of Manufacturing Support	3.61	1.27	3.83	1.27	1.07
Lack of Marketing Support	3.39	1.22	3.11	1.32	1.20
Lack of Training for Quality	3.80	1.24	3.72	1.19	1.01
Lack of Purchasing Support	3.46	1.19	3.24	1.26	2.25
Lack of Engineering Support	3.67	1.17	3.58	1.23	1.65
Lack of Communication	3.95	1.07	3.96	1.18	1.90
Lack of Customer Feedback	3.71	1.24	3.51	1.32	1.46

Not important at all = 1, Slightly important = 2, Moderately important = 3, Reasonably important = 4, Extremely important = 5

TABLE 5: ANOVA Results of BarrierstotheImplementationof Qulity Improvement Programs

Factors #	JIT Users Mean	JIT Users STD	Non-JIT Users Mean	Non-JIT Users STD	F ratio
Market Share	3.14	1.29	2.90	1.31	0.86
Productivity	3.64	1.14	3.31	1.21	1.78
Product Design	3.13	1.22	3.04	1.28	0.77
Vendor Relation	3.76	1.01	3.18	1.22	3.50**
Plant Efficiency	3.41	1.10	2.96	1.14	2.27
Process Design	3.39	1.16	2.89	1.11	2.73*
Worker Participation	3.75	0.99	3.19	1.20	2.65*
Delivery Schedule	3.37	1.10	2.81	1.20	2.47*
Distribution System	3.02	1.25	2.36	1.09	3.52**
Worker Skill	3.34	0.90	3.07	1.07	1.53
Inspection Time for:					
Incoming Parts	3.35	1.35	3.09	1.28	0.73
Finished Products	3.30	1.34	3.05	1.28	1.12

Not improved at all = 1, Slightly improved = 2, Moderately improved = 3, Reasonably improved = 4, Extremely improved = 5

* p<0.05
** p<0.01

TABLE 6: ANOVA Results of Factors Improved as a Resultof Implementation of Quality Improvement Programs

Factors #	JIT Users Mean	JIT Users STD	Non-JIT Users Mean	Non-JIT Users STD	F ratio
Price	3.95	1.06	3.95	0.98	0.48
Quality of Delivered Parts	4.56	0.87	4.26	0.90	1.89
On-Time Delivery	4.36	0.80	4.20	0.92	0.51
Design Capability	3.66	1.06	3.29	1.16	2.42
Technological Progress	3.58	1.05	3.15	1.09	2.59*
Mutual Relationship	4.11	1.13	3.87	1.03	1.28
Vendor Use of SQC	3.73	1.21	2.78	1.42	6.45***
Vendors Housekeeping	3.57	1.03	2.95	1.20	3.20*
Quality Management	3.97	1.00	3.54	1.25	2.10

Not important at all = 1, Slightly important = 2, Moderately important = 3, Reasonably important = 4, Extremely important = 5

* p<0.05
*** p<0.001

TABLE 7: ANOVA Results of the Vendor Evaluation Criteria

Factors	JIT Users Mean	JIT Users STD	Non-JIT Users Mean	Non-JIT Users STD	F ratio
Authority to Stop Production Line (1)	1.83	0.76	2.26	0.80	4.98**
Machine Operator Perform Daily Check (1)	1.77	0.69	2.24	0.89	3.56**
Workers Responsible for Quality (2)	1.84	0.84	2.07	0.83	1.38
When Corrective Action Taken (2)	1.79	0.85	2.01	0.92	1.26
Who Performs Corrective Action (3)	1.06	0.75	2.05	0.85	3.45*
Production Objectives of the Plant: (4)					
a. Low Inventory	4.17	0.84	3.49	1.17	5.46***
b. Continuous Improvement of Product Quality	4.25	0.90	3.72	1.27	2.62*
c. Meetig Production Schedules	4.59	0.56	4.63	0.62	1.21
d. Training Workers	3.62	1.02	3.27	1.17	1.68
e. Reducing Set-up Time	3.78	1.01	3.13	1.18	4.34**
f. Productivity Improvement	4.41	0.77	3.69	1.19	5.36***

Notes:

1 Yes: in all areas = 1, Yes: in some areas = 2, No and Don't know = 3.

2 When discovered = 1, At a later time = 2, No action taken and Don't know = 3.

3 Person who made it = 1, A separate team = 2, No action taken and Don't know = 3.

4 Not important at all = 1, Slightly important = 2, Moderately important = 3, Reasonably important = 4, Extremely important = 5.

* $p<0.05$
** $p<0.01$
*** $p<0.001$

TABLE 8: ANOVA Results of the Production Operations

performance, accessibility and their commitment to improve quality are significantly lower than those of the JIT users firms. Although in interviews with non- JIT users firms a few managers stated their support of quality and regarded quality planning as a strategic tool, top management's lack of leadership contradicted their statements as they assign the responsibility for quality leadership to lower level management. In addition, JIT firms perform equipment maintenance on a regular (daily) basis and improve process design in order to improve product quality whereas non-JIT firms do not.

The responsibility of the quality department in the JIT and non-JIT firms is examined in Table 4. This group consisted of eighteen different factors. As Table 4 shows JIT and non-JIT firms had significantly different opinions about eight of these factors. They include factors such as analysis of causes of defects, quality cost analysis, preparing reports for executives, training for quality, helping purchasing department in vendor selection, and customer interface. These findings agree with the personal observations and a general consensus among practitioners and academics that the most important responsibilities of quality departments in JIT firms are to analyze causes of defects, interfacing with customers to discover the existence of problems in the field, and training all employees in quality related subjects.

There were eight factors listed as the barriers to the implementation of quality improvement programs, Table 5. It appears that all these factors are important to the successful implementation of quality improvement programs, regardless of the types of firm. However, a closer examination of this finding and findings of the interviews reveals some important differences among the various firm types. The most important barriers for these firms were the lack of top management support and the existence of double standard to measure performance. The critical barriers for the JIT firms were the lack of quality training and proper communication systems. In addition, some JIT firms encountered problems such as employee resistance and the lack of training.

Table 6 shows a list of twelve factors as the benefits gained by the implementation of quality improvement programs. Although JIT and non-JIT firms vary significantly on five of these factors, all firms perceived the greatest improvements in productivity and vendor relations. In addition, some non-JIT firms expected positive changes in the management's attitudes toward quality if a quality improvement program was to be implemented. This is an important factor since in many non-JIT firms, lack of management support was indicated as the major barrier to the implementation of any major quality improvement programs.

With regard to vendor selection criteria, JIT and non-JIT firms demonstrated similar reactions to such factors as price, quality of delivered parts, on-time delivery, and vendor quality management, Table 7. However, JIT firms were significantly different from the non-JIT firms with regard to the vendor's use of statistical quality control, vendor's technological progress, and the vendor's housekeeping. One explanation is that a JIT manufacturing system requires the use of various SQC techniques on all levels of the firms. These firms urge their suppliers to use SQC tools and expect to receive a copy of various SQC charts with the delivered parts.

The last group to be elaborated upon is the production operations activities in the plant. Table 8 reveals that JIT and non-JIT firms differ significantly with regard to the responsibilities of their production workers. These results agree with the findings of the interviews, the review of the literature, and the widespread notion among practitioners as well as academics that workers in JIT firms are responsible for the quality of what they make. In JIT firms more authority is given to these workers to improve product quality. Table 8 indicates that workers in JIT firms are assigned the authority to stop the production line to rectify quality problems and to perform daily checks on their machines and equipment. Another important difference is the existence of a separate rework team in non-JIT firms to work on defective items, while in the JIT firms workers who make the bad parts are responsible for taking immediate corrective actions. In addition, production objectives of the plant indicates several significant differences exist between JIT and non-JIT firms. Lowering inventory, reducing set-up time, continuous improvement of product quality, and productivity improvement are factors that are significantly different with regard to the types of firms.

<u>Barriers to the Implementation</u>

The first barrier mentioned by almost all firms is the lack of top management understanding and support of quality. Although top management initiated the implementation of total quality management, many firms did not succeed in their initial implementation steps. The main reason behind this failure was top management's lack of leadership for quality and its lack of proper understanding of important quality issues and its direct impact on the success of JIT manufacturing system. In some cases, top management approved the programs and left lower level management in charge. Quality was not given the highest priority and was not considered part of strategic planning.

The second barrier was the lack of proper communication channels. One- way (top-down) communication discouraged employees' interaction with one another and with their supervisors. In addition, top-down communication created invisible walls between departments which hindered the communication and the smooth flow of information throughout the organization. This problem existed until top management introduced some type of participative management and sought workers' input in the decision making process.

The lack of manufacturing support in the implementation of quality improvement programs was the third barrier. In many non-JIT firms, the focus of manufacturing is on the rate of output and the quality of the product is not considered a determining factor. The output oriented manufacturing firms created huge quantities of WIP buffer which caused problems for the proper implementation of the system. The existence of a buffer and pressure to meet delivery schedules caused the manufacturing department to either hide quality problems or to ignore them. JIT firms gave the highest priority to the quality of the products. In addition, the traditional adversarial relations between the quality and manufacturing departments is eliminated by assigning the main responsibility for quality to production workers.

The fourth barrier was the lack of training for quality. The main reason for the lack of training was higher management limited understanding of the value of training. Although a few managers indicated the importance of quality training and taught various quality control techniques to their employees, their programs did not succeed until all employees (from top to bottom) were trained how to use these tools.

The next important barrier was the existence of double standards to measure employee performance. In many firms, employee performance is based on the quantity of output (rate of production) for the lower level employees and return on investment (ROI) or some other types of financial criteria for higher level employees. For those firms which attempted to implement the JIT manufacturing system, these methods of evaluation created a stumbling block, since in the JIT system, quality has the highest priority. This barrier was eliminated in many firms by introducing two strategies. The first strategy was to measure employee performance in terms of the quality of the products. The second method was to convert quality performance into some type of monetary value and add it to the total performance evaluation.

The last barrier to implementation was the resistance of the employees to the change. The employee resistance seems to be a natural reaction if they do not understand the system. In several firms, top management introduced and tried to implement the new system within a short period of time. Under such time pressure, employees were required to comply without actually comprehending the system or their role in the system. Too much pressure to implement and too little time to understand, as well as lack of training and employees skepticism of top management, created resistance in many levels of the organization. Based on the findings of this study, resistance to change subsided with top management's persistence in showing faith in its employees, slowing the implementation process, and using extensive training programs.

Organizational Characteristics

If any manufacturing firm intend to implement the total quality management and JIT manufacturing system, it must start from the very top. Top management must set up objectives and goals for the organization with regard to improving product quality. The major goal of top management should be growth of the company, not by acquisition, but by continuously improving product quality and reducing cost which will increase productivity and market share. In addition, top management must be willing to take necessary risks. In fact, the willingness to take risks was one of the important traits of the top management in JIT firms.

Organizational culture is the second characteristic important for a successful implementation of the total quality management. A change in organizational culture was mentioned by several managers as one of the most important characteristics. The new organizational culture is quality- oriented and encourages quality improvement activities. It creates an environment which nurtures pride in quality workmanship. Organizational culture helped many JIT firms create or improve trust between management and workers. The new culture is quality-intensive rather than quantity-intensive and requires the highest priority to be assigned to quality improvement on all levels of the organization.

The third organizational characteristic is planning horizon. The JIT firms have usually expressed their planning horizon in long terms, about ten years or more, whereas such a lengthy planning horizon is not common in non- JIT firms. The important point about long term planning is that it should incorporate JIT philosophy in the long term strategic planning of the organization. In addition, clear-cut quality policies are set and attempts are made to achieve them within the planning horizon.

The role of the quality department is the fourth organizational characteristic. Those U.S. firms that successfully implemented JIT required the integration of the quality department's activities into other departments such as production and design. Thus, the role of the quality department changed from one of inspection-intensive in all phases to improvement-intensive on all levels by all employees. Instead of inspecting products and processes, the quality department must teach quality control tools, educate all employees, and provide support and assistance to other departments for the overall improvement of product quality.

The next organizational characteristic is training and education. Training as an organizational characteristic is one of the most important elements for successful implementation of a JIT system and total quality management. Training and education are considered long term investments. The important role of quality as a crucial tool necessary for the survival of the organization is taught to all personnel. Training employees in several jobs coupled with teaching various quality control techniques provides a universal language for employees to effectively communicate with each other. Training and education help employees on all levels of the organization to recognize how their job relates to the total production process and how individual performance contributes to the overall well-being of all employees and the firm.

The sixth organizational characteristic is the role of production workers. The findings of this study suggests that the successful implementation of the JIT production system and total quality management requires the assignment of quality responsibility to the production workers. They will have the freedom to stop the production line in order to make necessary adjustments or corrections in order to eliminate quality problems. The quality staff will provide support, if necessary.

The seventh organizational characteristic is communication. The smooth flow of information and the lack of unnecessary delay in feedback is a trait of JIT firms. The findings of this study support the widespread notion that in JIT firms participative management is practiced in order to break the traditional boundaries among personnel and/or departments. The creation of two-way communication helps remove or reduce employee misunderstanding and resistance to any necessary changes. Participative management allows management to use employee input in the decision making process. Additionally, informal group meetings, such as quality control circles, worker involvement teams, zero defect teams, and so on, are established.

Close relations and mutual trust with suppliers is another characteristic of those firms that have successfully implemented the JIT manufacturing system. This study's findings indicated that in JIT firms, buyers select few suppliers and make efforts to establish long-term relations. Buyers usually adopt some type of formal vendor selection system and certify the selected suppliers. In addition, JIT firms encourage their suppliers to use SQC techniques and provide assistance to their suppliers to upgrade their quality.

Performance evaluation is the last organizational characteristic to be discussed. The findings of this study demonstrated that one of the major reasons for the failure of management's effort to improve quality is the existence of double standards to measure performance. However, one of the characteristics of JIT firms was that they developed or were developing devices that gave quality performance higher priority than other measuring factors such as meeting delivery schedules. Another strategy was to translate the quality performance of employees into monetary value and add it to the total performance evaluation picture.

SUMMARY AND CONCLUSIONS

This study was concern with the effect of the implementation of total quality management as a critical factor for the success of JIT manufacturing system in manufacturing firms. Firms in the survey were divided into two distinct categories of JIT users and non-JIT users. Six manufacturing firms were visited and about six hundreds and fifty questionnaires were sent to various firms.

The discussion of the research propositions revealed that there are important differences between the JIT user firms and non-JIT user firms. The most important differences exist in the areas of : 1) management commitment to quality, 2) status of quality, 3) process design as a critical factor to improve quality, 4) responsibility of the quality department, 5) improvement of market share, 6) improvement of productivity, 7) improvement of vendor relations, 8) improvement of process design, 9) increase of plant efficiency, 10) enhancement of worker participation, 11) reduction of inspection time for finished products, 12) vendor use of SQC, and 13) employees' responsibilities in the workplace. In all of the above factors, non-JIT firms consistently had significantly lower mean than their JIT users counterparts.

The discussion of research propositions demonstrated that there are important implementation barriers which included: 1) lack of top management understanding of quality, 2) lack of leadership for quality from top management, 3) lack of proper communication, 4) lack of support from manufacturing, 5) lack of proper training, 6) existence of double standards to measure employee performance, and 7) employees' resistance to change. These barriers could be eliminated if firms can develop several specific organizational characteristics. These characteristics were found in JIT firms. Non-JIT firms lacked several or all of these characteristics. These organizational characteristics were: 1) top management purpose, growth of the company through continuous improvement on quality; 2) organizational culture, quality-oriented not quantity-oriented; 3) planning horizon, long term planning for quality; 4) quality staff, teachers and facilitators of quality ideas; 5) training for quality, company-wide training and training on SQC techniques; 6) employee role, responsible for quality of products; 7) communication, use of various employee involvement teams and participative management; 8) vendor management, fewer vendors and closer relations with them; and 9) performance evaluation, measuring performance based on quality improvement accomplishments and converting them into monetary value.

It seemed that firms that have these characteristics will probably succeed in properly implementing the total quality management which should, in turn, lead to the success of JIT manufacturing system. However, examination of these characteristics shows that many of them are necessary for successful implementation of any type of system.

Trade-off between Production and Inventory Costs with Respect to a Given Demand Situation

JOACHIM REESE

I. Problem

Modern concepts in production management argue that it would be possibly preferable to instal a production rate which is fully synchronized with those rates of other production stages in the same firm or other firms. Thus, between the supplier and the customer of a finished or semi-finished product no inventory costs have to be taken into account, if the coordination will be perfect and JIT(Just-In-Time)-production takes place. Within a firm the synchronization may be a matter of planning and organization, and the kanban-principle is one possible result, though inventory is not really zero, but there exists always a buffer stock. As soon as two different firms are concerned, there are also different planning processes, and the advantage of the customer may become simultaneously the disadvantage of the supplier.

The general difficulty is that the supplier has to take care of precise delivery times given by the customer. Shall he accept the corresponding production rate and the equivalent lot size, respectively, if the minimization of total cost subject to a given demand situation is the problem to solve?

FANDEL, FRANCOIS and MAY (1988) show that to avoid inventory costs often means to split the economic lot size at the price of additional setup costs. This restriction becomes more and more doubtful, when the customer reduces his call forward intervals and, as a consequence, the order quantities at each call forward.

Additionally, if the supplier wishes to produce his goods as late as possible, the deviation from the optimal production rate can increase unit costs of production because of overtimes, extra shifts and more rejects, for example. This trade-off between the production costs and the inventory costs as a result of starting the production earlier, but with the optimum rate, is examined here.

The discussion refers to different types of problem structures: Deliveries can take place immediately after each unit of production or lot for lot. The demand can occur in parts, e.g. once a period, or altogether at the end of production. Information of call forward deliveries can be available "early enough" or "too late" to optimize the production process. The last case is often verified in context with JIT-production.

In the analysis we mainly restrict to an underemployed firm, so that there is at least enough capacity to shift the production lots to the desired periods. Besides of the presentations in chapter V the optimum lot size is predetermined. Transportation costs are due to the customer, i.e. the problem solution for the supplier does not depend on the frequency of transports. All demands are deterministic. The production process is single staged: One product is produced on only one machine.

II. The Starting Point: JIT-Production at the Optimum Rate

In many firms the planning procedure seems to be utmost easy: To minimize costs, first of all the optimum production rate p^* is determined by minimizing the unit costs of production $c(p^*)$. If the unit costs increase monotonously, the optimum rate equals the minimum rate p_{min}, but as a rule the unit costs will decrease, when the minimum rate is increased, e.g. until the optimal number of revolutions of a motor is reached (cf. GUTENBERG 1951). Furthermore, economic considerations are as well relevant for such a cost structure as technical terms. If one takes the rate of fixed costs and possible opportunity costs into consideration, which becomes most striking at any slow production, the decrease of the cost function $c(p)$ becomes even more evident. But economies of scale often tend to turn round after some while. When a motor runs beyond optimality or work has to be done on an extra shift, production costs will usually arise again to a higher amount per unit. At any rate, the global minimum exists and is finite, if the interval of feasible production rates is confined by a minimum rate p_{min} and a maximum rate p_{max}.

Given a lot x^* the optimal total production costs will be in general

$$(1)\ C_1=c(p^*)\cdot x^*,$$

provided the production rate is not altered during the whole manufacturing process.

If the lot is delivered at the end of the manufacturing process in one batch, there exists a stock only from the beginning to the end of the process. The production interval can be calculated from the lot and the production rate as $t=x^*/p^*$. Thus, the total inventory costs sum up to

$$(2)\ C_2=(x^*/2)\cdot h\cdot(x^*/p^*)$$

with $x^*/2$ denoting the average stock and h the holding costs per unit of time and quantity.

Supposed the production has to start earlier as a result of the total production planning in the firm, as there are no other free capacities, the inventory costs raise correspondingly, whereas the production costs remain unchanged, when choosing the optimum rate.

On the contrary the inventory costs disappear, as soon as the deliveries to the customer take place unit for unit instead of lot for lot. This is true for any strict JIT-production. Transportation times between supplier and customer raise considerably and usually restrict this structure to some extraordinary circumstances (cf. FANDEL and REESE 1989). The advantages of JIT-production have been discussed in the literature at some length, especially those which are not quantifiable (cf. SUZAKI 1985, WILDEMANN 1987, 1990 and many others).

III. Minimization of Production and Inventory Costs subject to a Discrete Demand Date

There is no doubt that taking the optimum production rate p^* cannot always be the dominant strategy for a firm, which wants to minimize total costs of the production process. It has to be examined inasfar an increase of the production costs caused by a higher rate is more than only equalized through sunken inventories. The first step in this direction is to minimize the cost function

$$(3)\ C(p)=C_1(p)+C_2(p).$$

The optimality condition developed from this problem with a given convex cost function c(p) runs as follows:

$$(4)\ dc/dp=h\cdot(x^*/p^2)/2.$$

The right side of equation (4) is positive. Therefore the derived production rate p^0 exceeds p^*, i.e.

$$(5)\ p^0 > p^*.$$

Inequation (5) states that the production should be accelerated to save inventory costs $C_2(p^0)$, while the amount of production costs $C_1(p^0)$ increases just moderately. A comparison of both cost terms on the basis of (4) says:

$$(6)\ C_1 = C_2 \text{ if and only if } (dc/dp)/(c/p) = 1.$$

Production costs should thereby equal inventory costs, if the cost elasticity with regard to the production rate is one. The elasticity characterizes the percentage variation of the unit costs of production, when the production rate is altered to one per cent. (6) will be e.g. fulfilled in the case of a strict linear cost function c(p). Substituting the equation symbols in (6) by inequation symbols the corresponding propositions can also be proved.

A concave cost function c(p) may perhaps lead to an optimum $p^0 = p_{max}$. This happens, if (4) describes a maximum condition instead of a minimum one. Inventory costs shrink to an absolute minimum for the case of a lot for lot delivery. JIT-production is approximated as its best and is even optimal. Whether the presupposition of a concave cost function c(p) is realistic or not, is not to be discussed here. Individual experiences will prove or disprove it.

As stated earlier the claimed production time results as $t = x^*/p$. High production rates lead to corresponding time reductions. Informations with respect to forthcoming call forwards or new orders must be available early enough to plan the production process optimally. This necessary lead time diminishes with an increasing rate of holding costs h, because these costs force high production rates (cf. (4)). On the other hand increasing marginal costs of production dc/dp slow down the optimal production per unit of time. This causes again a longer lead time for the preceding information procedure. So it is evident, that information and production structure closely depend on each other. Late call forwards can be accepted more easily, if there exist favourable marginal production costs and stock-keeping is rather expensive.

Informational delays give rise to a further acceleration of the production beyond the optimal degree. As there is only a limited time $t < t^0 = x^*/p^0$ left, the production rate grows to at least $p = x^*/t > x^*/t^0 = p^0$. Thus, production costs raise to such an amount, that they cannot be compensated any more by lower inventory costs. The increase of total cost must be interpreted as the transactional costs due to a lack of information. It puts an additional burden on the supplier, but does not bring any utility for the customer. A better decision support, e.g. a more sensitive information system, can therefore avoid such sunken costs (cf. also FANDEL and REESE 1989).

Former examinations have shown, that any unforced alteration of the production rate within the same production process can only be suboptimal, if the function of unit costs is convex with an increasing area after a decreasing one and the rates can be adapted continuously (cf. DELLMANN and NASTANSKY 1969). This is an immediate result from convex optimization theory. Nevertheless, the problem of different production rates in a process may not be neglected in case of discrete adaptations only. For feasible rates p_i, $i=1,...,I$, identical time intervals $t=1,...,T$ can be assigned in that way, that every t permits another rate. When the length of a time interval is chosen for the basic unit of the production rate, the periodical output is just

$$(7)\ x_t=p_i\cdot y_{it}.$$

y_{it} symbolizes a binary variable, which becomes one, if and only if production in period t is done with the rate p_i.

Modelling the whole problem means:

$$(8)\ \text{Minimize } C_1(y_{it})+C_2(y_{it})$$

subject to

$$(9)\ C_1(y_{it})=\Sigma_{i=1,...I;t=1,...,T}[c(p_i)\cdot p_i\cdot y_{it}],$$
$$(10)\ C_2(y_{it})=h\cdot\Sigma_{i=1,...I;t=1,...,T}\{[(2T+1)/2-t]\cdot p_i\cdot y_{it}\},$$
$$(11)\ \Sigma_{i=1,...,I;t=1,...T}(p_i\cdot y_{it})=x^*,$$
$$(12)\ \Sigma_{i=1,...,I}y_{it}=1 \text{ for } t=1,...,T, \text{ and}$$
$$(13)\ y_{it}\in\{0,1\}.$$

While (9) and (10) define the production and inventory costs, (11) assures, that the production will be fulfilled until the given planning horizon T. (12) and (13) guarantee a unique assignment of production rates to the different time periods.

The problem can be solved by a dynamic programming approach in the following way: Starting with t=1 the cost functions

$$(14)\ F^t(x)=\min_{x_t}\{F^{t-1}(x-x_t)+C^t(x_t)\}$$

are to be determined subsequently for $t=1,...,T$. $F^t(x)$ denotes the minimum costs accumulated over all periods till t, $C^t(x_t)$ describes the total costs of period t, if x_t is produced during this period. $F^T(x^*)$ is the minimum required.

The results of this discrete optimization correspond to those derived theoretically for the continuous case. An acceleration of the production process will take place first of all in the last periods, whereas the optimum rate p^0 should be chosen in the first periods, if production is not limited to some few late periods anyhow. This strategy is demonstrated in an example in chapter VI.

IV. Minimization of Production and Inventory Costs subject to a Continuous Demand Rate

As yet mentioned, customers are often very engaged in receiving their materials in small portions to save inventory costs. The extreme situation of JIT-supply ends in a continuous demand over the whole production period. It seems to be a wide-spread opinion, that this tendency brings discredit upon the suppliers, because they have to reduce their economic lot sizes. But if the emerging transportation problem is solved satisfactorily, there remains a new problem structure on the suppliers' side, which offers some revealing facts.

Let us assume that the total demand is still x^*, which shall be now splitted in continuous call forward quantities at each time unit during the planning horizon T. The corresponding demand rate shall be given constantly as $d^*=x^*/T$. If the production rate is $p(\geq d^*)$, there exists a maximum stock

$$(15)\ x^* \cdot (p-d^*)/p$$

at time

$$(16)\ t=x^*/p,$$

which is a result of the extended EOQ-model with constant demand rates.

The calculation of production costs has to be done now as before, i.e.

$$(1a)\ C_1=c(p) \cdot x^*,$$

whereas inventory costs alter to

$$(2a)\ C_2=(x^*/2) \cdot h \cdot (x^*/p) \cdot [(p-d^*)/d^*].$$

Compared to the amount in (2), C_2 in (2a) is lower, if

$$(17)\ (p-d^*)/d^*<1.$$

Inequation (17) is true particularly for strict JIT-production, i.e. where $p=d^*$ holds. This is not unexpected, as the savings from inventory costs must serve to compensate increasing production costs, which can result from disadvantageous demand rates.

Minimizing total cost for a given concave or "weakly" convex function of unit costs c(p) leads indeed to the optimum

$$(18)\ p^+=d^*.$$

But if c(p) is "strongly" convex, the optimum condition becomes

$$(4a)\ dc/dp=-h\cdot(x^*/p^2)/2\ (<0),$$

supposed that d^* can be accomplished with the resulting optimum rate p^+. Otherwise (18) holds again. In comparison with (4) equation (4a) denotes just the negative amount of marginal unit costs for optimality. This means $p^+<p^*$ and deceleration of production, respectively.

It seems remarkable, that (4a) does not depend on the actual demand rate d^*, but only on the total demand x^*. Thus, it cannot be stated generally, whether (17) is fulfilled or not. At any rate there exists a cost-dependent lower bound d_{min}, where the supplier is no more able to compensate his steadily growing production costs by means of lower inventory costs, though a slow production is optimal under the given circumstances. JIT-supply becomes infavourable, as production is too far from an optimal coordination of the process activities. The same is true for an upper bound, which involves JIT-production in spite of (4a).

As concerns the composition of total costs, the equality of production and inventory costs is optimal in the case of (4a), if and only if the cost elasticity equals $d^*/(p-d^*)$, i.e.

$$(6a)\ C_1=C_2 \text{ if and only if } -(dc/dp)/(c/p)=d^*/(p-d^*).$$

The corresponding inequations are also true. So, if the demand rate d^* is further reduced by the customer, the ceteris paribus consequences are increases of the inventory costs as well as of the total costs C, supposed (17) is yet fulfilled. The optimum production strategy should not be altered.

In the case of the optimum (18) there are obviously no inventory costs at all. JIT-supply implies strict JIT-production, and cost evaluation can be restricted to production costs. Best of all would be a call forward rate $d^*=p^*$, so that the minimization of supplier's costs can be realized perfectly. But continuous demands d^* are also preferable to a batch delivery at the end of the planning horizon, as long as the additional unit costs of production are lower than the saved inventory unit costs over the average storage time $(x^*/p^0)/2$. This condition expressed analytically runs:

$$(19)\ c(d^*)-c(p^0)<h\cdot(x^*/p^0)/2.$$

For any cost function c(p), which is increasing monotonously, the validity of inequation (19) is obvious, as long as $d^* \leq p^0$ holds.

It must be stated analogously to chapter III, that the choice between discrete production rates for a finite planning horizon usually involves different rates in single periods. The modelling approach has to be modified as follows: Instead of (10) the inventory costs become

$$(10a)\ C_2(y_{it})=h\cdot\Sigma_{i=1,...I;t=1,...,T}\{[(2T+1)/2-t]\cdot p_i\cdot y_{it}-(T-t)\cdot d_t\}.$$

d_t symbolizes the accurate demand at the end of period t. Generalizing the above presumptions, it is now even possible to alter the call forward quantity from period to period. Moreover, the condition (11), which guarantees the fulfilment of production in time, has to be supplemented by

$$(11a)\ \Sigma_{i=1,...,I;t=1,...\tau}(p_i\cdot y_{it})=\Sigma_{t=1,...,\tau}d_t \text{ for } \tau=1,...,T-1,$$

and

$$(11b)\ \Sigma_{t=1,...T}d_t-x^*.$$

The conditions (9), (12), and (13) remain unchanged.

The dynamic programming solution procedure is the same as already described. An example is given in chapter VI.

V. The Optimal Lot Size

In the preceding chapters discussions always ran under a constantly given lot size x^*. Nevertheless the solution will not be optimal, if modifications of x^* and influences on the total costs were not examined. Because any x^* is possible, we can restrict to analysing, whether a splitting of x^* in smaller lots is favourable or not (for further discussions on possible setup time reductions cf. e.g. ESROCK 1985, PORTEUS 1985, KARMARKAR 1987, SPENCE and PORTEUS 1987, KNOLMAYER and LEMKE 1990).

Presuming that each lot effects setup costs s, there have to be taken into consideration total setup costs

$$(20)\ C_3=s\cdot(x^*/x),$$

if x denotes the lot in question and (x^*/x) therefore the number of necessary setups. C_3 has to be added to the production costs C_1 (cf. (1a)) and the inventory costs, which are now dropped to

$$(2b)\ C_2=(x/2)\cdot h\cdot(x^*/p)\cdot[(p-d^*)/d^*]$$

due to the reduced average stock $(x/2)\cdot(p-d^*)/d^*$. The minimization of the total costs then results in the optimal strategy, which is characterized by the optimum parameters p_{opt} and x_{opt}, i.e. the best production rate as well as the appropriate lot splitting. The optimum conditions are (validity restrictions as discussed in chapter IV):

$$(4b)\ dc/dp=-h\cdot(x/p^2)/2,$$

and

$$(21)\ x^2=2s\cdot p/h\cdot[d^*/(p-d^*)].$$

Equation (21) corresponds to the well-known classical formula (cf. HARRIS 1915). It is perhaps noteworthy that (21) implies a decrease of the optimal lot size, if the demand rate d^* is reduced. The opposite would be true for the production rate, if there had been a choice: Any further acceleration from the optimum rate would diminish the optimal lot size, but altogether to higher costs. E.g. there may be an increase of setup costs s, which causes a corresponding increase of the optimal lot, but leads simultaneously to a decrease of the optimal production rate. In the whole, from (21) it follows immediately that

$$(22)\ C_2=C_3,$$

whereas $C_1=C_2$ is only true with the presupposition (6a).

If $p_{opt}=d^*$ holds instead of (4b) for concave or only "weakly" convex cost functions c(p) (cf. chapter IV), there should be of course no lot splitting, i.e. $x=x^*$ is optimal instead of

(21). Setup costs as well as inventory costs are at their absolute minimum in this case of JIT-production.

As earlier discussed for different discrete production rates within a finite planning horizon, it can just as little be excluded from the beginning of a planning procedure, that there exist no different lot sizes. First of all the optimal splitting depends on the individual call forward quantities during the time regarded. To examine this, a dynamic programming approach is well suited as before, though it has to be modified as against the approaches developed in the chapters III and IV. It is not intended to plunge into details here.

VI. Some Numerical Examples

In this chapter the theoretical reflections are illustrated by examples for the different demand situations as well as production and cost structures.

A first function of unit costs shall be given as

$$(23)\ c(p)=(9-2.5p)^2+4.$$

The holding cost rate is assumed as h=2, and the demand at the end of the planning horizon as $x^*=80$. Minimizing c(p) leads to $p^*=3.6$. This causes production costs $C_1=320$ and inventory costs $C_2=1778$ according to (1) and (2) respectively. Total costs become C=2098. The values noted here and afterwards are sometimes merely approximations.

In the case of a global optimization with respect to (4) an acceleration of production as to $p^0=4$ is preferable. Thus, production costs rise to $C_1=400$, but inventory costs drop to $C_2=1600$, and total costs to C=2000, consequently.

If the demand x^* is partitioned into call forward rates $d^*=1$ ($d^*=1.5$; $d^*=1.767$; $d^*=2$; $d^*=3.6$, $d^*=5.433$), the optimum production rate will be always equal to the demand rate, supposed there is no minimum rate $p_{min}>d^*$. For $p^+=d^*$, production costs are $C_1=3700$ ($C_1=2525$; $C_1=2000$; $C_1=1600$; $C_1=320$; $C_1=2000$), whereas inventory costs do not occur. Thus, permanent deliveries are profitable for supplier and customer as well, if the demand rate d^* lies between 1.767 and 5.433. For demand rates outside this interval permanent deliveries are disadvantageous for the supplier.

The optimality condition (4a) is not relevant for the chosen cost structure, as the convexity of the unit cost function is still below the limit, which guarantees the optimality of any acceleration of production. But if the function alters to

$$(23a)\ c(p)=(12-5p)^2+4,$$

the optimum production rate will be indeed $p^+=2$, i.e. exceed the demand rate for $d^*\leq 2$. The inventory costs increase now as compared to the former structure, but the production costs diminish more. For $d^*=1$ ($d^*=1.5$; $d^*=1.767$; $d^*=2$) total costs become C=3560 (C=1427; C=782; C=360). As soon as $d^*>2$ holds, $p^+=d^*$ is inevitable. Nevertheless, further cost decreases occur for $2<d^*\leq p^*$.

Setup costs have been disregarded by now, because the lot $x^*=80$ was fixed. With a known cost rate s per lot it should be examined yet further, whether any lot splitting is favourable or not. For s=100, $d^*=1$, and the cost structure (23a), the optimality conditions (4b) and (21) lead to the optimum lot size $x_{opt}=13.2$ and the optimum production rate $p_{opt}=2.35$. Total costs are reduced to C=1539, supposed that each lot is produced as late as possible. The clear advantage is, that the lotwise production can be speeded up to approximately $p^*(=2.4)$, because breaks become possible and the inventory is thereby not permanently filled up. The production of a single lot x=80 at a rate p=2 would be optimal, if and only if the setup cost rate rises to s=3200.

If the unit cost function is monotonously increasing, results differ widely. Be

$$(23b)\ c(p)=3.2p+4,$$

h=2, and $p_{min}=2$, then obviously $p^*=p_{min}$ holds. Costs become C=4032. According to (4) $p^0=5$ leads to an acceleration of production and decreases costs to C=2880. A constant demand rate is met optimally by $p^+=p_{min}$ again. If $d^*=1$, total costs equal once more C=4032, because the lower average inventory and the longer storage time are balanced. But costs diminish for increasing rates $d^*\geq 2$. This effect is reinforced by an optimal lot splitting. With $x_{opt}=14.14$ and $p_{opt}=2$ the absolute minimum C=1963 is attained.

Finally, the dynamic programming procedure shall be exemplified in two cases. At first the demand structure requires a total delivery of $x^*=8$ after four periods. The internal cost structure is as follows:

p	0 ($=p_{min}$)	1	2	3 ($=p_{max}$)
c(p)	0	6	4	5

The holding cost rate is h=4 per piece and period. Within a period the production rate shall be invariable. Between two successive periods it can be changed. No setup costs occur. The planning tableau is shown below:

x	x_1	C^1	x_2	C^2	x_3	C^3	x_4	C^4
1	1	8						
2	2	12	2	12				
3	3	21	3	21				
4			2	32				
5			3	41	3	41		
6			3	54	3	54		
7					3	69		
8					3	82	3	82

C^t denotes the total costs until period t, x_t denotes that quantity, which is produced optimally in period t. If there is no entry in the tableau, the corresponding production till this period is not feasible because of the maximum production rate and delivery restrictions, respectively. For $x^*=8$ total costs are C=82, and the periodical productions are: $x_4=3$, $x_3=3$, $x_2=2$, and $x_1=0$. The optimum rate should be practised in period 2 only. In the periods 3 and 4 the production should be speeded up. Period 1 is free for other productions.

If quantities are ordered at the end of every period in smaller portions, the optimal solution must be adapted. For call forwards $x_1^*=1$, $x_2^*=2$, $x_3^*=2$, and $x_4^*=3$, the modified tableau looks as follows:

x	x_1	C^1	x_2	C^2	x_3	C^3	x_4	C^4
1	1	8						
2	2	12						
3	3	21	2	20				
4			2	28				
5			3	37	2	32		
6			3	50	3	41		
7					3	53		
8					3	66	3	53

As a whole, costs are obviously below those of the preceding case, because the intermediate call forwards empty the stock at the end of each period at least partially. The optimal production plan is: $x_4=3$, $x_3=2$, $x_2=2$, and $x_1=1$, i.e. JIT-production is superior to any kind of storage. Costs drop to C=53. Holding a stock and producing over the whole time horizon at the optimum rate $d^*=2$ becomes favourable, if the holding cost rate h can be lowered considerably.

VII. Main Results and Concluding Remarks

In the course of all JIT-discussions suppliers' problems have been described accurately. Costs explode and raise serious difficulties for the firm. Reasons have been found in the necessity of smaller production lots and accelerated production at all, if the storage capacity is limited. Any fundamental analysis must be done on combining all these important arguments with the individual cost and demand situation. Some hints are given here by means of a marginal analysis and discrete optimization. It becomes evident that there exist a lot of structures, in which JIT-supply can be managed to the suppliers' satisfaction, too. Furthermore, JIT-production is occasionally well suited to support JIT-supply. An essential precondition for the acceptance of the inherent organizational changes will be an optimal cooperation between the suppliers and the customers, especially as concerns the information flow. Beyond that, the novel requirements for the suppliers must be met by an appropriate planning process. Instead of stepwise optimization, e.g. isolated calculation of optimal lot sizes and optimal production speeds, an integrated procedure within an extensive production planning and information system is necessary. The utility of these systems has to be extended also to small firms.

Literature

Dellmann, K., and L. Nastansky: Kostenminimale Produktionsplanung bei rein intensitätsmäßiger Anpassung mit differenzierten Intensitätsgraden, in: Zeitschrift für Betriebswirtschaft, 1969, pp. 239-268

Esrock, Y. P.: The Impact of Reduced Setup Time, in: Production and Inventory Management, 1985, Vol. 4, pp. 94-101

Fandel, G., and J. Reese: "Just-in-Time"-Logistik am Beispiel eines Zulieferbetriebs in der Automobilindustrie, in: Zeitschrift für Betriebswirtschaft, 1989, pp. 55-69

Fandel, G., P. Francois, and E. May: Effects of Call-Forward Delivery Systems on Suppliers' Serial per Unit Costs, in: G. Fandel, H. Dyckhoff, and J. Reese (eds.): Essays on Production Theory and Planning, Berlin et al. 1988, pp. 66-84

Gutenberg, E.: Grundlagen der Betriebswirtschaftslehre, Vol. I: Die Produktion, Berlin et al. 1951

Harris, F.: Operations and Costs, in: Factory Management Series, Chicago 1915, pp. 48-52

Karmarkar, U. S.: Lot Sizes, Lead Times and In-Process Inventories, in: Management Science, 1987, pp. 409-418

Knolmayer, G., and F. Lemke: Auswirkungen von Losgrößenreduktionen auf die Erreichung produktionswirtschaftlicher Ziele, in: Zeitschrift für Betriebswirtschaft, 1990, pp. 423-442

Porteus, E. L.: Investing in Reduced Setups in the EOQ Model, in: Management Science, 1985, pp. 998-1010

Spence, A. M., and E. L. Porteus: Setup Reduction and Increased Effective Capacity, in: Management Science, pp. 1291-1301

Suzaki, K.: Just-in-Time and Quality Control, Los Angeles 1985

Wildemann, H.: Das Just-in-Time-Konzept, Produktion und Beschaffung auf Abruf, Frankfurt 1987

Wildemann, H.: Kundennahe Produktion und Zulieferung: Eine empirische Bestandsaufnahme, in: Die Betriebswirtschaft, 1990, pp. 309-331

Combined Optimization of Safety Stocks and Processing Lead Times in Multi-Stage Production Systems

KARL INDERFURTH

1. INTRODUCTION

Production planning in a multi-stage environment usually is performed with MRP systems using inputs in form of known and constant processing lead times, yields, and product demands. In practical problems one often is faced with the problem of appropriate integration of demand uncertainty and lead time variability in a MRP planning process.

Risks from uncertain demands generally are covered by interstage safety stocks. Mainly due to implementation advantages procedures like hedging (i.e. overestimating of demands) or setting up safety times sometimes are preferred to using safety stock norms [10]. The former methods, however, suffer from a serious drawback, in that they are not able to generate optimal safety stocks under cost and service level aspects. In early MRP literature (e.g. [6]) building up safety stocks was recommended only for end-items of the production process which are affected directly by uncertainties of demand. While under these conditions the correct size of safety stocks reflecting cost impacts and desired service levels can easily be evaluated, the total cost minimum in general will not be reached because a distribution of safety stocks also among intermediate-stage products may be less costly. This effect stems from the fact that products at upstream stages represent less value added and that these products offer opportunities of risk pooling with respect to final-stage demands when component commonality or divergent production structures are given [9].

Optimal policies for simultaneous safety stock and lot-size determination only can be derived for multi-stage problems with specific production and cost structures [1], [11]. Therefore, in order to simplify the decision problem in multi-stage manufacturing processes often a separation of lot-sizing and safety stock calculation is performed by solving the safety stock optimization problem under the condition of a lot-for-lot production with periodic control [2]. This assumption is just valid if production is ruled by a base-stock control policy like it is appropriate for JIT manufacturing. Even for this kind of production control a cost minimizing allocation of safety stocks under the constraints of reaching desired service levels is very difficult to obtain because of the very complex interactions of single stage delivery service and final customer service levels. Only for two-stage systems these dependencies have been described and employed analytically [8], [9].

Now, for many practical problems these interactions may be of low importance, because even in cases of excessive demand situations intermediate production stages can fulfil their service commitments by utilizing emergency actions like accelerating internal processing or initiating express deliveries from outside supplies if safety stock protection is not sufficient. Assuming total reliability of internal production stages by this kind of operating flexibility an effective safety stock optimization procedure has been developed for serial systems in [7] which has been extended to divergent production structures in [3].

Multi-stage divergent structures can especially be found in chemical production processes like production of active ingredients in the pharmaceutical industry where a specific raw material is processed to an enlarging set of intermediates over several consecutive steps until the final ingredients for pharmaceuticals are produced. Under the assumptions of stationary demands and unrestricted processing capacities in [4] a powerful safety stock optimization procedere has been shown to exist. This approach will be extended to a situation where a limited capacity for final-stage operations is given which has to be allocated to the different processes (e.g. by specifying priority rules), thus having impact on their lead times. In pharmaceutical production this problem arises when decisions on utilization of manpower for compulsary assays of active ingredients at the final production stage have to be met. In this case the respective processing times have to be treated as variable in a certain range whereas their total amount is fixed according to the given capacity. Safety stock optimization now must be performed jointly with processing time determination.

2. OPTIMIZATION PROBLEM AND SOLUTION PROPERTIES

The multi-stage production system underlying the safety stock and processing time optimization problem is characterized by a general divergent structure as shown in Figure 1.

In Figure 1 each node describes a single product as well as an inventory in which this item may be stocked. Each link means a specific production process for manufacturing the succeeding item. The production stages starting with procure-

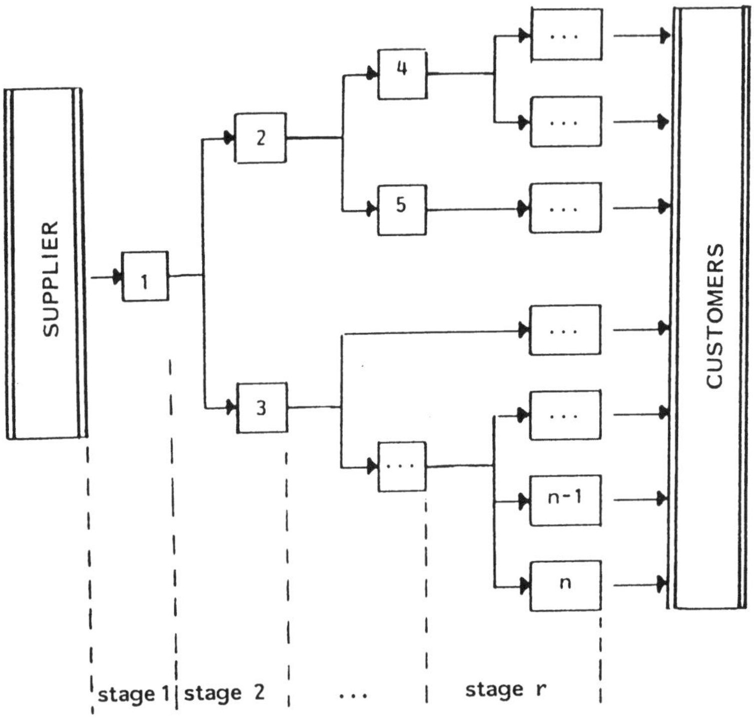

Figure 1: Divergent production system

ment of raw material and then built up by succeeding production processes are numbered from 1 to r. The single items, running from 1 to n, are assigned to the highest feasible stage in the system. So all finished goods which are sold to customers belong to the final stage.

In addition, we use the following notations for sets of products:

E :	final-stage items	
V :	intermediate-stage items	
N(i) :	immediate successors of item i	$(i \in V)$
E(i) :	final-stage items succeeding i	$(i \in V)$
V(i) :	predecessors of item i	$(i \in V \cup E)$
v(i) :	immediate predecessor of item i	$(i \in V \cup E)$
P(s) :	items assigned to stage s	$(s = 1,2,\ldots,r)$

The following data of production structure and demands are fixed and given:

σ_k^2 :	demand variance of final-stage item k	$(k \in E)$
$\rho_{k\ell}$:	correlation coefficient of demands for items k and ℓ	$(k \in E,\ \ell \in E,\ k \neq \ell)$

a_{ik} : multi-stage production coefficients $(i \in V,\ k \in E)$

t_i : lead times of intermediate processes $(i \in V)$

Resulting from the divergent production structure the intermediate-stage demand variances (for $i \in V$) are

$$\sigma_i^2 = \sum_{k \in E(i)} a_{ik}^2 \sigma_k^2 + \sum_{k \in E(i)} \sum_{\substack{\ell \in E(i) \\ \ell \neq k}} a_{ik} a_{i\ell} \sigma_k \sigma_\ell \rho_{k\ell} \tag{1}$$

Data for costs and desired delivery service are given as follows:

h_i : inventory holding costs for product i $(i \in V \cup E)$

α_i : internal service level for intermediate-stage product i $(i \in V)$

α_k : external service level for final-stage product k $(k \in E)$,

with service levels as proportion of periods in which the current demand is fully satisfied from running production and safety stocks.

If we assume that the end-item demands of succeeding periods are independent, identically distributed random variables with a normal distribution the safety stock I_j for each item $i \in V \cup E$ results as

$$I_j = q_j \sigma_j \sqrt{RTL_j} \tag{2}$$

where RTL_j means the respective replenishment lead time for product j and q_j is a safety factor which is computed as α_j-percentile of the standard normal distribution. The RTL_j - except for the first-stage item - are depending on lead times and safety stocks of preceding items in the production structure. That means, by assigning safety stocks to a specific product j and all its predecessors a special <u>service time</u> S_j is generated which defines the maximum time that may elapse before the orders placed against process j - by the successors of j - will be fulfilled within the framework of the desired delivery service. For the intermediate stages $i \in V$ the following relationship holds

$$RTL_i = S_{v(i)} + t_i - S_i \tag{3}$$

At the final stage the RTL_k also depend on the decision on the allocation of a fixed total lead time T^+ to the different processes $k \in E$ yielding processing times T_k. Additionally, it must be reflected that customer service times have to be equal to zero ($S_k = 0 \ \ \forall \ k \in E$) and that - along with the assumption of periodic control -

the final-stage lead time has to be enlarged by the review period. So for $k \in E$ we have:

$$RTL_k = S_{v(k)} + T_k + 1 \tag{4}$$

Using formula (2) and defining $c_j = h_j q_j \sigma_j$ $(j \in V \cup E)$ the expected inventory holding costs for safety stocks in the whole system are given as

$$C = \sum_{j \in V \cup E} c_j \sqrt{RTL_j} \tag{5}$$

Now, using the safety stock approach with service times S_i $(i \in V)$ as decision variables and taking into account the variability of final-stage processing times T_k $(k \in E)$ and their restriction by given minimum levels T_k^- the combined <u>optimization problem (P.C)</u> of determining safety stocks and lead times in multi-stage production systems can be formulated as follows:

Minimize

$$C = \sum_{i \in V} c_i \sqrt{S_{v(i)} + t_i - S_i} + \sum_{k \in E} c_k \sqrt{S_{v(k)} + T_k + 1}$$

subject to

$$0 \leq S_i \leq t_i + S_{v(i)} \qquad (i \in V)$$

$$T_k \geq T_k^- \qquad (k \in E)$$

$$\sum_{k \in E} T_k = T^+$$

$$\text{with} \quad \sum_{k \in E} T_k^- \leq T^+ \quad \text{and} \quad S_{v(1)} = 0 \, .$$

Caused by the specific problem structure along with the minimization of a special concave function over a convex polyhedron an extreme point property of the optimal solution can be shown to exist (see [3]):

$$\begin{aligned} &S_i^* \in \{0;\ S_{v(i)}^* + t_i\} && (i \in V) \\ &T_k^* \in \{T_k^-;\ T_k^+\} && (k \in E) \end{aligned} \tag{6}$$

$$\text{with} \quad T_k^+ = T^+ - \sum_{\substack{\ell \in E \\ \ell \neq k}} T_\ell^-$$

This solution property of choosing either its minimum or its maximum feasible value for each product's service and processing time results in an "all-or-nothing" policy for determining safety stocks all over the production system:

$$I_i^* \in \{ q_i \sigma_i \sqrt{S_{v(i)}^* + t_i}\ ;\ 0 \} \qquad (i \in V)$$

$$I_k^* \in \{ q_k \sigma_k \sqrt{S_{v(k)}^* + T_k^- + 1}\ ;\ q_k \sigma_k \sqrt{S_{v(k)}^* + T_k^+ + 1} \} \qquad (k \in E)$$

So the interrelationship between setting safety stocks at various production stages and fixing processing times becomes apparent.

3. ITERATIVE SOLUTION PROCEDURE

3.1 PROBLEM SEPARATION AND ITERATIVE SOLUTION

Taking into account that for practical problems production structures with more than 10 stages, 100 products, and 50 final-stage items can occur, the numerical solution of the joint optimization problem is a very cumbersome task. Even by using the specific extreme point property in formula (6) there remain $b\ 2^a$ possible service and processing time combinations (with $a = |V|$ and $b = |E|$) which yield the set of feasible solutions.

Fortunately an efficient iterative solution procedure can be applied making use of a separation method which allows the reduction of the combined optimization problem to a sequence of far more simple problems of service time optimization for given processing times (S-optimization) on the one hand and lead time optimization for given service times (T-optimization) on the other. For each of the separate optimization problems a powerful dynamic programming algorithm exists which generates numerical solutions even for very large problems within a few seconds of PC computing time.

The complete solution can be evaluated by an iterative procedure of solving separate problems of T-optimization and S-optimization as presented in Figure 2 which, however, will not in general converge to the global optimum. Thus, besides of certain cases, where such convergence properties exist, this iterative method has to be viewed as a heuristic approach (see [5]). Generally the optimal combined poliy can be determined by performing S-optimization for each of the p extreme points of the T-optimization problem and applying complete enumeration.

3.2 SEPARATE SERVICE TIME AND LEAD TIME OPTIMIZATION

By fixing processing lead times at any iteration ($T_k = T_k^{(m)}\ \forall\ k \in E$) the optimization problem (P.C) is reduced to the problem of just determining the service times $S_i (i \in V)$ which minimize inventory holding costs and guarantee desired service levels:

Minimize

$$C_s = \sum_{i \in V} c_i \sqrt{S_{v(i)} + t_i - S_i} + \sum_{k \in E} c_k \sqrt{S_{v(k)} + T_k^{(m)} + 1}$$

subject to $(i \in V)$

$$0 \leq S_i \leq t_i + S_{v(i)} \quad \text{and} \quad S_{v(i)} = 0\ .$$

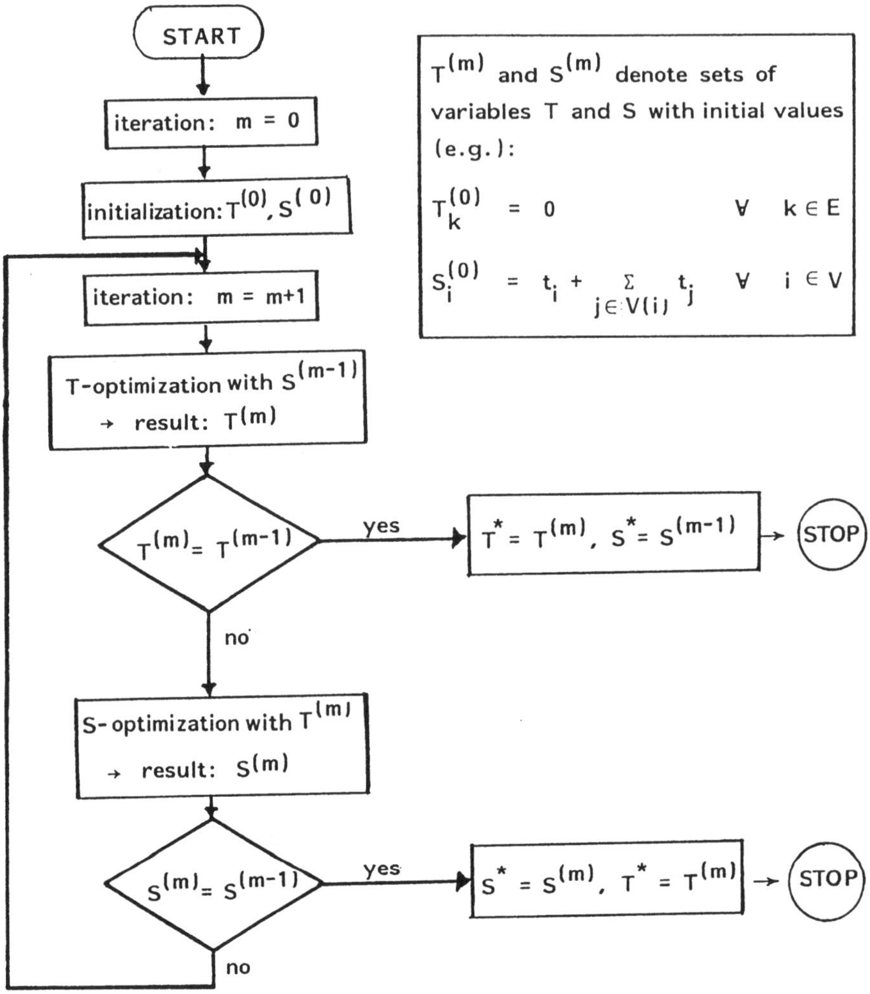

Figure 2: Flow chart of the iterative (T,S)-optimization procedure

This optimization problem (P.S) can be interpreted as a multidimensional allocation problem of distributing the total replenishment lead time $T_k^{(m)} + \sum_{i \in V(k)} t_i$ of all final-stage items $k \in E$ all over the various steps of the multi-stage process. It can be solved by a very efficient dynamic programming algorithm utilizing the fact that the recursive relations of the dynamic optimization procedure at each stage s ($s \in \{1,2,\dots,r\}$) can be separated into $|P(s)|$ isolated optimizations for each product $i \in P(s)$. Moreover, along with the extreme point property of the solution state space and decision space are drastically reducable to s and 2 elements respectively for each optimization at stage s (see [4]).

Given predetermined service times $S_i = S_i^{(m)}$ ($i \in V$) optimization problem (P.C) is transfered to a simple processing time optimization problem (P.T) with decision variables T_k ($k \in E$):

Minimize

$$C_T = \sum_{k \in E} c_k \sqrt{S_{v(k)}^{(m)} + T_k + 1}$$

subject to

$$T_k \geq T_k^- \qquad (k \in E)$$

and

$$\sum_{k \in E} T_k = T^+$$

Problem (P.T) is a very simple one-dimensional allocation problem of cost minimizing distribution of total processing time T^+ to the different final-stage processes $k \in E$. As each optimal allocation for a specific item $k \in E$ is restricted to only two values T_k^- and T_k^+ - again due to the extreme point property - dynamic programming can be used here too as a very favourable solution procedure (see [5]).

4. NUMERICAL RESULTS

4.1 AN EXAMPLE

For a simple example of a three-stage divergent production structure as depicted in Figure 3 results of combined safety stock and processing time optimization shall be demonstrated.

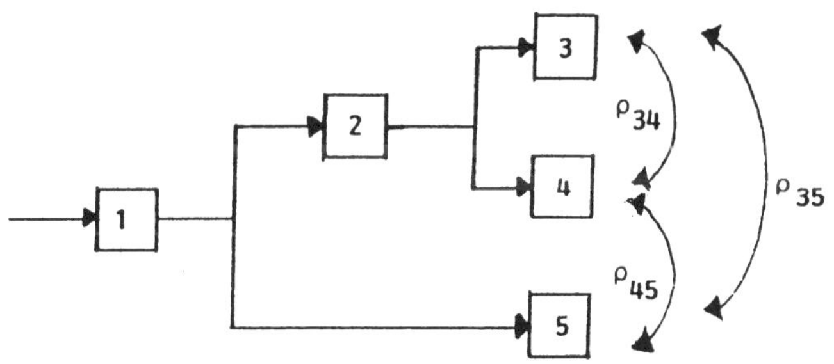

Figure 3: Three-stage production system

The various problem parameters are shown in Table 1:

	intermediate-stage items		final-stage items	
		i = 1 , 2	k =	3 , 4 , 5
demand variances σ_k^2				100 , 400 , 625
correlation coeffients $\rho_{k\ell}$			$\forall\, \ell \neq k$	0 , 0 , 0
production coefficients a_{ik}	$\forall\, k \in E(i)$	1 , 1		
holding costs h_i, h_k		1 , 2		3 , 3 , 2
service levels α_i, α_k		99% , 99%		99% , 99% , 99%
fixed processing times t_i		1 , 2		
min.processing times T_k^-				0.5 , 0.5 , 1.5
total processing time				$T^+ = 4.0$

Table 1: Parameters of the three-stage production example

Starting the iterative solution procedure as proposed in Figure 2 the optimal solution is reached after the first iteration resulting in

$$S_1^* = 0,\ S_2^* = 2,\ T_3^* = 2.0,\ T_4^* = 0.5,\ T_5^* = 1.5\ .$$

This solution leads to optimal safety stocks amounting to

$$I_1^* = 78,\ I_2^* = 0,\ I_3^* = 52,\ I_4^* = 87,\ I_5^* = 92\ .$$

In this example the most effective distribution of safety stocks is characterized by establishing buffer inventories only at the first (i.e. raw material procurement) stage besides concentrating on the final stage where safety stocks for reasons of guaranteeing a desired customer service have to be held by all means. This result goes hand in hand with fixing processing time priorities for final-stage products 4 and 5 - mainly due to their high demand variances - to the debit of product 3.

4.2 SENSITIVITY ANALYSIS

For a specified production structure like in Figure 3 distribution, size, and value of the optimal safety stocks especially depend on two characteristics: the flexibility in allocating processing times to the single final-stage processes and the correlation between end-item demands which determines the degree of risk-pooling opportunities by establishing intermediate-stage safety stocks.

The flexibility in fixing lead times is mainly characterized by the difference between total processing time T^+ and the sum of minimum lead times $\sum_{k \in E} T_k^-$. For $\sum_{k \in E} T_k^- = T^+$ no flexibility exists. This situation is given e.g. for $T_3^- = T_4^- = 1.0$ and $T_5^- = 2.0$ in our example. In this case the distribution of buffer inventories does not change, but size (I^*-size) and value (I^*-value) of total final-stage safety stocks becomes larger yielding a 3 % increase of total costs. Raising flexibility by decreasing the minimum lead times T_k^- ($\forall\, k \in E$) by an equal percentage p for constant total processing time ($T^+ = 4.0$) results in decreasing amounts of total safety stock size and value as shown in Figure 4.

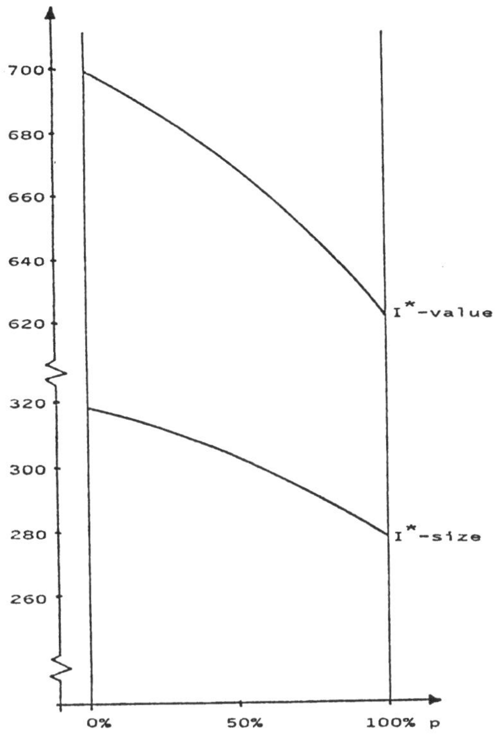

Figure 4: Safety stock results for varying flexibility p

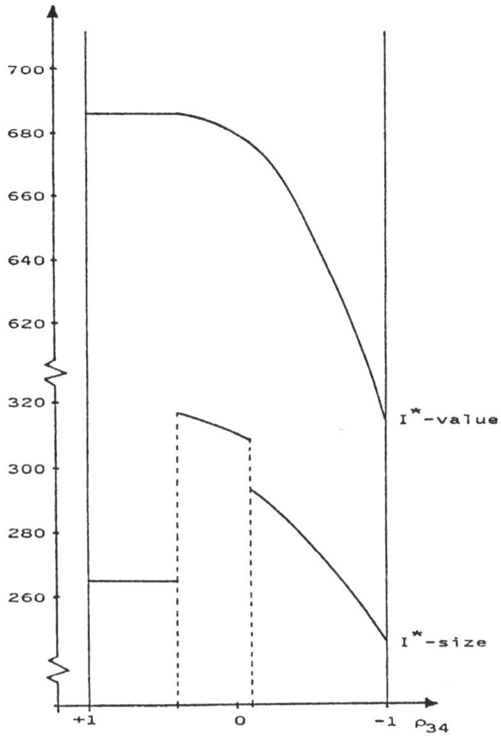

Figure 5: Safety stock results for varying correlation ρ_{34}

Compared with processing time flexibility the sensitivity of the safety stock allocation seems to be higher with respect to parameters of the stochastic demand. Especially the correlation of demands is very significant for distribution and size of safety buffers. In Figure 5 the dependence of safety stock size and value on correlation of demands ρ_{34} between products 3 and 4 (holding $\rho_{35} = \rho_{45} = 0$) is demonstrated for the original example in chapter 4.1. It can be seen that the total safety stock value is monotonously decreasing with a declining correlation coefficient, especially rapidly falling within the region of negative correlation. This is mainly due to the fact that demands which are not completely positive correlated offer chances for risk-pooling in intermediate-stage inventories thus allowing to decrease the amount of necessary stocks. This effect is especially significant if the correlation tends to be very small or even negative.

These results are also indicated by the graph of the total safety stock size in Figure 5. The discontinuities of the I^*-size function result from changes in the optimal safety stock allocation. While for $\rho_{34} < -0,1$ safety stocks are held for all items but product 1 - especially utilizing risk-pooling opportunities by product 2 -

for ρ_{34} > + 0,4 only the final-stage products are buffered. For the intermediate region of correlation the safety stock distribution as described in chapter 4.1 is optimal. Utilizing risk-pooling chances for small or negative correlation allows a remarkable reduction in safety stock holding costs of up to 14 % for ρ_{34} = -1 .

In the simple example we considered the optimal allocation of final-stage processing lead times did not change with varying correlation coefficients ρ_{34}. In general, demand correlations will have an impact on the processing time decisions.

5. CONCLUSIONS

The procedure for combined determination of safety stocks and final stage production lead times presented in this paper gives an insight in the importance of interrelationships between both tasks in a multi-stage production environment and offers a powerful approach for the numerical solution of this complex decision problem.

This approach is limited to problems of deciding upon operation times for a single - here the final - processing stage. Optimizing the employment of scarce resources for operations in more than one production stage can be formulated as a decision problem just like in chapter 3.1, but the very efficient solution procedure presented in chapter 3.2 can no longer be applied.

If lot-sizing has to be considered for some processes and production is not totally performed lot-by-lot as presumed the safety stock procedure integrated in our approach will overestimate the size of necessary buffer inventories. In this case the amount of the respective service levels should be reduced somewhat alike it should be increased if the assumption of total reliability of internal production processes does not seem to be completely realistic.

The approach described is developed just for divergent production structures. Nevertheless, it is quite general since multi-stage divergent systems are not only met in a lot of practical problems - especially in chemical production and physical distribution - but also can be found as subsystems in more general structures or can be identified to be related to the problem of dealing with component commonality in convergent systems. Last, but not least, the well-known serial production structure is nothing else but a special case of a divergent system.

REFERENCES

[1] CLARK, A.J., SCARF, H., "Optimal Policies for a Multi-Echelon Inventory Problem", Management Science 6(196o): 475-49o

[2] GRAVES, S.C., "Safety Stocks in Manufacturing Systems", Journal of Manufacturing and Operations Management 1(1988): 67-101

[3] INDERFURTH, K., "Zur Festlegung von Sicherheitsbeständen in mehrstufigen Lagerhaltungssytemen mit divergierender Struktur", Discussion paper No. 2o3, Department of Economics, University of Bielefeld, Bielefeld 1989

[4] INDERFURTH, K., "Safety Stock Optimization in Multi-Stage Inventory System", Discussion paper No. 209, Department of Economics, University of Bielefeld, Bielefeld 1990

[5] INDERFURTH, K., "Ein Ansatz zur simultanen Planung von Sicherheitsbeständen und Durchlaufzeiten in mehrstufigen Produktionsprozessen", Discussion paper No. 218, Department of Economics, University of Bielefeld, Bielefeld 1990

[6] ORLICKY, J.A., "Material Requirements Planning", McGraw-Hill, New York 1975

[7] SIMPSON, K.F., "In-Process Inventories", Operations Research 6(1958): 863-873

[8] VAN DONSELAAR, K., WIJNGAARD, J., "Practical Application of the Echelon Approach in a System with Divergent Production Structures", in: AXSÄTER, S., SCHNEEWEISS, Ch., SILVER, E.A. (eds.), Multi-Stage Production Planning and Inventory Control, Springer, Berlin-Heidelberg - New York 1986 : 182-196

[9] VAN DONSELAAR, K., WIJNGAARD, J., "Commonality and Safety Stocks", Engineering Costs and Production Economics 12(1987): 197-204

[10] WIJNGAARD, J., WORTMANN, J.C., "MRP and Inventories", European Journal of Operations Research 20(1985): 281-293

[11] ZIJM, W.H.M., LANGENHOFF, L.J.G., "An Analytical Theory of Multi-Echelon Production/Distribution Systems", Note No. 75 of Centre for Quantitative Methods, Nederlandse Philips Bedrijven, B.V., Eindhoven 1989

On the Commonality Problem in Multi-Stage Inventory Control Systems

HENRIK JÖNSSON and EDWARD SILVER

0. PRELIMINARY REMARKS

This paper addresses a so-called assemble-to-order environment. There are a number of end items each of which requires an assembly of several components. Some of the components are unique to specific end items while others are common to two or more final products. Components must, due to their long leadtimes, be ordered prior to knowing the end item demand levels, but final assemblies can be quickly completed after knowing the demands.

The objective is to maximize the expected profit contribution subject to a budget constraint on the component inventory investment. The associated two-stage stochastic programming problem is very difficult, if not impossible, to solve.

For the special case of a very simple commonality structure, with normally distributed end item demand levels, the optimal allocation is developed. A much simpler heuristic allocation procedure is shown to give excellent performance over a wide range of parameter values.

Due to the solution difficulties for more realistic sized problems, heuristics and bounding procedures, that give good results, are suggested. The proposed heuristics can easily be extended to problems with continuous demand distributions.

A new solution procedure, the so-called scenario aggregation principle, for stochastic programming problems in general and for the commonality problem in particular, is discussed. The main idea of this approach is to solve a deterministic problem, given each of a number of scenarios, and then iteratively adjust the solutions obtained so as to identify a unique solution that maximizes the objective for the set of scenarios used.

1. INTRODUCTION

The so-called assemble-to-order environment, discussed by Wemmerlöv [1984], including commonality among the components to be assembled, poses many challenging research problems. In the product structure there are *n* final products (end items), which are assembled from a set of *m* components. Some of the components are unique to specific end items while other components are common to two or more end items. The name *commonality (inventory) problem* is due to the latter fact. See the illustration in Figure 1. The label associated with each arc in Figure 1 indicates the number of units of the particular component needed for each unit of the different end items.

Production/acquisition of the components takes much longer than the final assembly operations, hence it is assumed that, for a specific demand period, the demands for the end items are known *after* the production/acquisition of the components but *before* the final assembly operations. The problem addressed is how to allocate a fixed budget among the various components so as to maximize the expected profit from selling the final products in such a single demand period. The rationale for dealing with a single period has been discussed elsewhere [Jönsson and Silver, 1989b]. In the latter article a list of

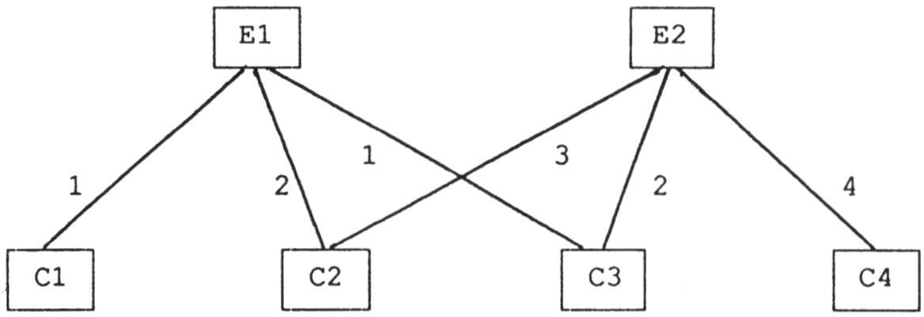

Figure 1. Product structure example.

references related to inventory management of common components is presented. The importance of this problem area has been emphasized by CONDOR [1988] as follows:

> "One fundamental issue that cuts across many industries is the commonality of parts. Increasingly, it has been recognized that fewer parts overall, along with modular designs, can result in dramatic savings. Better models are needed to help quantify these savings and to identify improvements to be achieved."

Collier [1982] initiated an interest in taking advantage of the commonality situation. He derived an approximate model of the effect of commonality on required safety stocks, but some of his implicit assumptions were questioned by McClain et al [1984].

Baker [1985] discusses the problem of determining the end item stockout probabilities on the basis of the probability distributions for the components. Numerical examples are based on a product structure with 2 end items and 3 components of which one is common to both end items. A rather unusual service measure and choice of end item demand distributions (uniform probability density functions), were used.

In a subsequent study by Baker et al [1986] the following 3 properties for some commonality problems are identified:

Property 1: Introduction of commonality reduces the total inventory required to meet a specified service level.

Property 2: The optimal stock of the common component is lower than the combined stock it replaces.

Property 3: The combined optimal stocks of product-specific components are higher with commonality than without.

Gerchak et al [1988] have studied in what problem instances the abovementioned properties are valid. In Table 1 below the results are summarized, and we see that all of the properties are valid in at least one of the three problem types.

	Problem		
	Unconstrained profit maximization	Cost minimization subject to service level constraints	
Property		Arbitrary costs	Equal costs
1	Yes	Yes	Yes
2	No	No	Yes
3	Yes	No	Yes

Table 1. Situations in which the 3 properties are valid.

Gerchak and Henig [1986, 1989] have dealt with the unconstrained profit maximization problem. In a situation with negligible fixed costs for setups, stationary demand and a forward carrying of unused stock, they have shown that the optimal solution to a single period problem also is optimal in a multi-period problem, i e the solution is myopic in nature. Furthermore, they argue that when adding a storage capacity constraint to the basic problem the optimal solution is still myopic, but this is, in general, not the case when a production or purchase budget capacity constraint is included. In an example they also demonstrate that the optimal stocks of some product-specific components in the constrained case can, contrary to intuition, be *lower* in an assemble-to-order system than in a corresponding make-to-stock system.

Gutierrez and Hausman [1989] discuss multiple supply decisions and rationing of common components during an order cycle (corresponds to a single period). The rationing depends on the shortage costs for different end products and the remaining time until the next resupply point.

The current paper is an overview of the research carried out by the authors on the commonality inventory problem. In Section 2 the problem formulation and notation used are introduced, and then different aspects of the research are summarized in each of the following three sections.

2. PROBLEM FORMULATION

We wish to allocate resources for acquisition of the m components. The overall objective is to maximize the expected net profit contribution, determined as expected sales revenue minus expected costs for shortages, component acquisition/production, assembly-to-order and inventory holding. An alternative objective used is to minimize the expected number of units short, which is a special case of the presented overall objective.

The constraints on the solutions involve:

1. The budget constraint for the components.
2. The component requirements for the assembly of final products must be less than or equal to the number of components acquired beforehand.
3. The number of units sold of an end item cannot exceed its demand.

In order to formulate the problem mathematically the notation below is introduced (the scenarios listed represent a subset of the various possible demand realizations and will be discussed further in Section 5):

m = number of distinct components

n = number of distinct final products

Z_j = stochastic variable denoting the demand of final product j during a single period (or order cycle), $j=1,2,\ldots,n$

$\phi(u)$ = unit normal probability density function (p.d.f.)

$\Phi(k)$ = cumulative unit normal distribution function = $\int_{-\infty}^{k} \phi(u)\,du$

$G_u(k)$ = unit normal loss function = $\int_{-\infty}^{k} (u-k)\phi(u)\,du$

$p(z_j)$ = probability mass function (p.m.f.) for the discrete stochastic variable Z_j

z^l = an n-vector for the lth demand realization, $l=1,2,\ldots,L$

p_l = probability of the lth demand realization, $l=1,2,\ldots,L$

$p_l = \prod_{j=1}^{n} p(z_j^l)$ (assuming independent demand distributions)

$A = \{a_{ij}\}$ = m by n matrix describing the product structure where a_{ij} represents the number of units of component i required for each unit of end item j

y_i = number of units of component i that are produced/acquired

h_i = inventory holding cost per unit of component i (\$/unit/period)

v_i = unit value of component i (\$/unit)

V = available budget for acquisition of components (\$)

s_k = scenario k that is identical with one of the possible demand realizations z^l, $l=1,2,\ldots,L$

S = a set of scenarios chosen to represent the full problem. $S = \{s_1, s_2, \ldots, s_K\}$ where, in general, $K << L$

x_j = number of final assemblies of end item j that are sold

x^s = an n-vector with the x-values corresponding to scenario s where $s \in S$

r_j = revenue per unit sale of end item j (\$/unit)

b_j = shortage cost per unit of unsatisfied demand of end item j (\$/unit)

The inventory allocation problem described can be formulated mathematically according to eq (1) below (which is a two-stage stochastic optimization problem with recourse, cf Kall [1976]):

$$\max_{y} f = E_Z [Q(y,Z)] - h^T y$$
$$\text{s t} \begin{cases} v^T y \le V \\ y \ge 0 \end{cases} \tag{1}$$

where

$$Q(y,Z) = \max_{x} \{ r^T x - b^T (Z-x) \mid Ax \le y;\ x \le Z;\ x \ge 0 \}$$

3. A SIMPLE CASE OF COMPONENT COMMONALITY

Jönsson and Silver [1989a] deal with a problem with the same product structure as in Baker [1985], i e two end items where each assembly requires one unique component (denoted by *1* and *2* respectively) and one common component (denoted by *3*). However, in contrast with Baker's analysis, the demands of the end items are assumed to be normally distributed (with mean μ_j and standard deviation σ_j for $j=1,2$) and the expected number of units short is minimized. The sum of the allocated stocks of the three components must be less than *V* units (i e $v_i=1$ for all *i*). In Figure 2 the shortage situations are depicted.

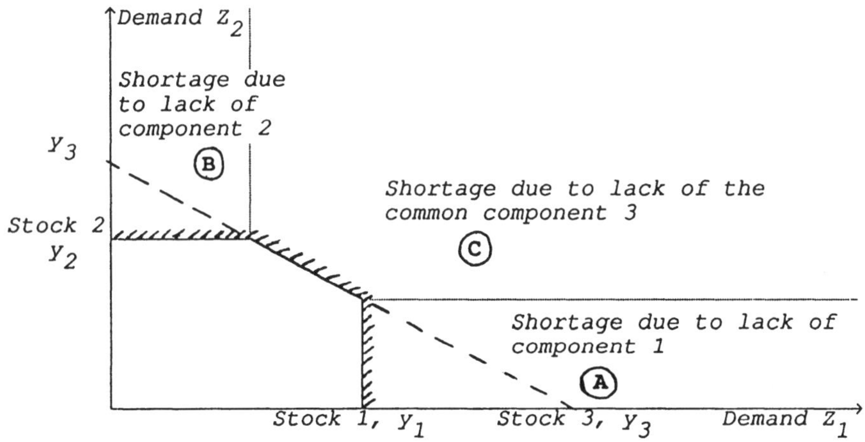

Figure 2. Demand situations causing shortages given the component inventory levels y_1, y_2 and y_3.

It is easy to realize that the assembly and sales of the end items must satisfy the conditions $x_1 \leq \min\{y_1, y_3\}$ and $x_2 \leq \min\{y_2, y_3\}$, and that the sum of the inventories for the unique components must not be less than that of the common component, i e $y_1+y_2 \geq y_3$. Shortages occur when demand for the end items falls into any of the areas A-C in Figure 2. Through lengthy calculus and algebra the expected number of units short, *EUS*, can be derived and expressed according to:

$$EUS = \sigma_1 G_u(k_1) + \sigma_2 G_u(k_2) + \sigma_2\left[[1-\Phi(k_1)]G_u(c_2) - [1-\Phi(c_1)]G_u(k_2) + \int_{c_1}^{k_1} G_u(c_0-\beta u)\phi(u)\,du\right] \tag{2}$$

where

$$k_i = (y_i-\mu_i)/\sigma_i, \qquad i=1,2$$
$$c_0 = (V-y_1-y_2-\mu_1-\mu_2)/\sigma_2$$
$$c_1 = (V-y_1-2y_2-\mu_1)/\sigma_1$$
$$c_2 = (V-2y_1-y_2-\mu_2)/\sigma_2$$
$$\beta = \sigma_1/\sigma_2$$

The minimum of *EUS* was found by using Newton's method for numerical unconstrained optimization. A very simple heuristic inventory allocation rule, based on using the normal p.d.f. with mean $\mu_3=\mu_1+\mu_2$ and variance $\sigma_3^2=\sigma_1^2+\sigma_2^2$ for the common component and assuming independent demands for the three components, was derived. It gives a common safety factor k^{heu} for all three components and the corresponding allocation rule is:

$$k^{heu} = \left[V-2(\mu_1+\mu_2)\right]/(\sigma_1+\sigma_2+\sigma_3)$$
$$y_i^{heu} = \mu_i + k^{heu}\sigma_i, \qquad i=1,2,3 \tag{3}$$

It turns out that the result can be expressed in terms of only two independent parameters, namely

$$\alpha = \left[V-2(\mu_1+\mu_2)\right]/(\sigma_1+\sigma_2)$$
$$\beta = \sigma_1/\sigma_2 \tag{4}$$

In the numerical examples α and β were chosen from the intervals [-0.1, 2.0] and [0.1, 1.0] respectively. The values of *EUS* when using the simple heuristic allocation rule according to eq (3), turn out to be only between *0* and *0.762* % above the optimum, extremely good results. In most cases k^{heu} lies between k_1 and k_2. When $\alpha > 0$ (i e the budget exceeds the expected requirements) the heuristic puts too much into the common component. In cases when β is near *1* and $\alpha < 0$, k^{heu} is larger than both k_1 and k_2, i e too little is put into the common component.

4. HEURISTICS AND UPPER BOUNDS FOR MORE GENERAL COMMON COMPONENT INVENTORY PROBLEMS

In the paper by Jönsson and Silver [1989b], commonality problems with more complex product structures are studied. The problem formulation is as given in Section 2 including the compact formulation of eq (1). Since the evaluation of $Q(y,Z)$ involves a maximization over x for each demand realization Z, an exact evaluation of the expected value, $E_Z[Q(y,Z)]$, is impossible for continuous demand distributions. A discretization of the p.d.f.'s would permit an approximate evaluation, but the number of discrete demand vectors increase exponentially with the number of end items. Here we assume that we have a finite, discrete p.m.f. with probabilities p_l and demand vectors z^l for all $l = 1,2,\ldots,L$. Given this the problem of eq (1) can be formulated as an equivalent one-stage integer programming problem according to:

$$[P] \quad \max_{y,x^l} \sum_l p_l \left\{ r^T x^l - b^T (z^l - x^l) - h^T y \right\}$$

$$\text{s t} \quad \begin{cases} v^T y \le V & \\ A x^l \le y & \text{for } l = 1,2,\ldots,L \\ x^l \le z^l & \text{for } l = 1,2,\ldots,L \\ y \ge 0, & \text{integer} \\ x^l \ge 0, & \text{integer for } l = 1,2,\ldots,L \end{cases} \qquad (5)$$

This turns out to be a gigantic problem for almost any practical situation. Noting that the important decision variables are y-variables, relatively few in number, one could assume that the x-variables are continuous variables. Then the problem structure allows for the Benders decomposition approach, cf Benders [1962] or Lasdon [1970], which can be used to provide an upper bound for the optimal solution and successively better y-values. Although this approach may be useful for research and/or bounding purposes, it is not practical for routinely solving the complex commonality problems in practice. Thus heuristic approaches are recommended. The following heuristics are suggested in the paper by Jönsson and Silver [1989b].

Heuristic 1: Allocate the components according to an optimal policy for stockholding of end items.

Heuristic 2: Allocate the components on the basis of the derived p.m.f.'s of demand for the components.

Heuristic 3: A mixed integer programming approch (as the problem in eq (5) but with continuous *x*-variables)

Heuristic 3 can be expected to provide reasonably good solutions and so it does in the numerical examples, but it can only be applied to relatively small problems. Therefore it is used as a basis for comparison. Because of the relaxation of the integer *x*-variables it also provides an upper bound to the optimal solution.

Heuristic 1 is a straightforward appproach that ignores the inherent two-stage nature of the problem, and the associated potential benefit of the commonality among components. When one just stocks end items it follows that $y = Ax$. This leads to a non-linear knapsack problem in the *x*-variables that easily can be solved with dynamic programming. As expected this approach does not perform very well in the numerical examples.

Heuristic 2 is an extension of the simple heuristic that performed well in Section 3. It is based on the derived demand, u, of the components where $u = AZ$ $(u_i = \sum_{j=1}^{n} a_{ij} Z_j)$. In the objective function suitable weights, expressing the approximate sales revenue, w_i^r, and the approximate shortage cost, w_i^b, associated with a unit of component i, are introduced. This approach leads to a non-linear knapsack problem in the *y*-variables that again can easily be solved with dynamic programming. The problem formulation is:

$$\max g = \sum_{i=1}^{m} \left[w_i^r \sum_{0}^{y_i} u_i p(u_i) + \sum_{y_i+1}^{\infty} \left\{ w_i^r y_i - w_i^b (u_i - y_i) \right\} p(u_i) \right] - h^T y$$

$$\text{s.t.} \quad \begin{cases} v^T y \le V \\ y \ge 0, \text{ integers} \end{cases} \tag{6}$$

where

$$w_i^r = \sum_{j=1}^{n} \left[r_j E[Z_j] / a_{ij} \mid a_{ij} > 0 \right] \Big/ \sum_{j=1}^{n} \left[E[Z_j] \mid a_{ij} > 0 \right]$$

$$w_i^b = \sum_{j=1}^{n} \left[b_j E[Z_j]/a_{ij} \mid a_{ij}>0 \right] / \sum_{j=1}^{n} \left[E[Z_j] \mid a_{ij}>0 \right]$$

In the numerical examples binomial p.m.f.'s were used for all end items. The number of end items was *3* or *4*, and the number of components was *4* or *5*. Six basic cases were used (a case is a product structure with given demand distributions). Each case was divided into a number of subcases *(1, 2* or *3)* based on various budget levels *V*. The total number of resulting problems was *13*.

In *6* of the *13* subcases heuristic 2 gave the same result as the best one, heuristic 3. Through the determination of the objective function value of eq (5) with fixed *y*-values as discussed above, we found that heuristic 3 provided the optimal solution in *4* subcases.

In summary, we believe that heuristic 2 would be the best one of these *3* to use in a practical situation. Another interesting feature of the approach of heuristic 2 is that it leads to an extremely simple solution for the *continuous* version of the problem in eq (6), i e one could easily deal with problems involving continuous demand distributions (see Jönsson and Silver [1989b]).

5. THE SCENARIO AGGREGATION APPROACH TO COMMON COMPONENT INVENTORY PROBLEMS

In this section we outline a very different approach to obtaining a reasonable solution to the complex optimization problem discussed in Section 4. (Further details of the approach are available in Jönsson et al [1990]). The aforementioned combination of a sequence of combinatorial decisions (integer numbers of components, then assemblies) and stochastic demands suggests the use of scenario aggregation, an approach recently advocated by Rockafellar and Wets [1987]. The basic idea is to consider only a relatively small subset *S* of the typically extremely large number of stochastic outcomes (demand vectors). Given each of these so-called scenarios, it is relatively easy to find the optimal values of the decision variables (component and assembly quantities). However, such solutions would not be implementable since the assumed knowledge of the demand in each scenario would result in different component allocations for the various scenarios. This cannot be allowed since only one component allocation can be implemented prior to actually experiencing the demand in practice. The scenario aggregation approach involves an iterative scheme aimed at "pricing" the component allocations in such a manner that an implementable solution ultimately is ensured.

The convergence to optimal solutions by the scenario aggregation approach of Rockafellar and Wets [1987] is only guaranteed for problems with continuous variables. However, Björnestad et al [1989] have shown how the approach can be used for combinatorial problems, such as our commonality problem. Further descriptions of the scenario aggregation technique, as well as applications, are given in Mulvey and Vladimirou [1988a,b,c] and Wets [1989].

Consider a set of representative scenarios S that is a subset of all the L possible different demand realizations (see also Section 2), where a particular scenario s is identical with a corresponding demand realization z^s for all $s \in S$ (the demand realization is denoted z^s in order to simplify notation although z^s may not be identical to the s:th of the original demand realizations). Associated with each individual scenario s is a probability π_s (≥ 0), that properly reflects the importance of this scenario in the overall problem. The chosen probabilities must satisfy

$$\sum_{s \in S} \pi_s = 1 \tag{7}$$

Suppose we let y^s be the vector of y values for demand realization s. Then, in our problem the component allocation variables y^s, determined in the first stage, must be the same for all scenarios, since we must choose how to allocate the component budget before the demands for final products are known in the second stage. Thus, the implementability (or non-anticipatory) condition can be stated as

$$N = \{y^s \mid y^s = y^{s'} \text{ for all } s, s' \in S\} \tag{8}$$

The problem in eq (5) can now be approximated (in the sense that only a small subset of the possible demand realizations are considered) as follows:

$$[P] \quad \max_{y^s,x^s} \sum_s \pi_s \left\{ r^T x^s - b^T(z^s - x^s) - h^T y^s \right\}$$

$$\text{s t} \begin{cases} v^T y^s \le V & \text{for all } s \in S \\ Ax^s \le y^s & \text{for all } s \in S \\ x^s \le z^s & \text{for all } s \in S \\ y^s = y^{s'} & \text{for all } s, s' \in S \\ y^s, x^s \ge 0, \text{ integer} & \text{for all } s \in S \end{cases} \tag{9}$$

The set of feasible solutions for a particular scenario can be written in compact form as:

$$C_s = \{ (y,x) \,|\, Ax \le y;\ x \le z^s;\ v^T y \le V;\ y,x \ge 0 \text{ and integers} \} \tag{10}$$

Utilizing this and the N notation, the scenario aggregation problem of eq (9) can now be rewritten as:

$$[P] \quad \sum_{s \in S} \max_{x^s,y^s} \pi_s \{ r^T x^s - b^T(z^s - x^s) - h^T y^s \}$$

$$\text{s t} \begin{cases} (y^s,x^s) \in C_s & \text{for all } s \in S \\ y^s \in N & \text{for all } s \in S \end{cases} \tag{11}$$

By relaxing the non-anticipatory conditions, $y^s \in N$, problem P separates into K small scenario subproblems that can be solved without too much difficulty. We associate with each relaxed constraint in N a multiplier vector that, after some manipulation, results in one multiplier vector w^s for each scenario solution vector y^s. This leads to the following Lagrangean dual problem:

$$[LD] \quad \min_w L(w) \tag{12}$$

where

$$L(w) = \sum_{s \in S} \max_{(y^s,x^s) \in C_s} \left[\pi_s \left\{ r^T x^s - b^T(z^s - x^s) - h^T y^s \right\} - w^{s^T} y^s \right]$$

For each choice of multiplier vectors (the w^s's) one can solve the separate scenario subproblems in the dual problem *LD* . The resulting value of *L(w)* gives an upper bound (*UBD*) to the optimal solution of problem *P*. For reference purposes, let the solutions of the dual subproblem be denoted by $(\bar{y}^s, \bar{x}^s)$.

This is a slightly modified version of the scenario aggregation approach of Rockafellar and Wets, for problems with discrete variables which has been suggested by Björnestad et al [1989]. The Lagrangean relaxation approach for obtaining bounds in combinatorial optimization problems has been used extensively during the last several years, see e g Fisher [1981, 1985] and Geoffrion [1974].

In our inventory problem it is easy to find feasible solutions since all integer allocations of components that satisfy the budget constraint are implementable. One option is to identify the solution $\hat{y}$, among the scenario solutions $\{\bar{y}^s\}_{s\in S}$, that maximizes

$$F = \max_{\hat{y}\in\{\bar{y}^s\}_{s\in S}} \sum_{s\in S} \pi_s \left\{ \max_{x^s\in C_s} \left[r^T x^s - b^T(z^s - x^s) - h^T \hat{y} \right] \right\} \tag{13}$$

and choose $\hat{y}$ as the new, best, feasible solution, denoted by $\bar{\bar{y}}$, if the objective function value *F* improves the existing lower bound (*LBD*). If the lower bound is not improved, the best one available is kept. In the process a set of feasible solutions will be obtained with successively better objective function values.

Schematically the scenario aggregation algorithm is as follows:

Step 0: Initialize
Step 1: Solve problem *LD* with $w^s = \bar{w}^s$ for all $s \in S$. Update the bounds; $UBD = \min\{UBD, L(\bar{w})\}$ and $LBD = \max\{LBD, F\}$. *F* is obtained from (13)
Step 2: Stop if *UBD* and *LBD* are "close enough".
Step 3: Update the multipliers $\bar{w}^s$. Go to step 1.

We have done considerable preliminary experimentation with the same numerical examples as discussed in Section 4. The results are encouraging in that in all examples the best feasible solution was obtained at a reasonably early stage, often before the *10*th iteration. However, there are a number of open issues, that we are exploring, including

1) How to decide on the number of scenarios, precisely which ones to use, and how to estimate the associated probabilities (the π_s's)?
2) How to update the w^s vector multipliers in the most effective fashion?
3) How to reduce the observed rather large gap between the upper and lower bounds remaining even after extensive iterations? Should one concentrate on obtaining better upper bounds or better feasible solutions?

6. SUMMARY

In this paper we have presented an overview of recent research on the commonality problem in inventory management. The problem, involving both combinatoric and stochastic aspects, is inherently very difficult to solve in an optimal sense. Thus, we have described two very different approaches. In the first a simple heuristic decision rule allocates stock to the various components based upon their marginal distributions of demand. The second approach is that of scenario aggregation, where only a small subset of the possible demand realizations are considered and the individual scenario solutions are iteratively adjusted in an algorithm that produces better and better feasible solutions.

ACKNOWLEDGEMENTS

The research leading to this paper was partially supported by the *Natural Sciences and Engineering Research Council of Canada* under grant A1485.

REFERENCES

Baker K: Safety Stocks and Component Commonality, *Operations Management*, Vol 6, No 1, 1985, pp 13-22

Baker K, Magazine M and Nuttle H: The Effect of Commonality on Safety Stock in a Simple Inventory Model, *Management Science*, Vol 32, No 8, 1986, pp 982-988

Benders J F: Partioning Procedures for Solving Mixed-Variables Programming Problems, *Numerische Mathematik*, Vol 4, 1962, pp 238-252

Björnestad S, Hallefjord Å and Jörnsten K: Discrete Optimization under Uncertainty: The Scenario and Policy Aggregation Technique, WP 89/06, Centre for Petroleum Economics, Chr. Michelsen Institute, Bergen, 1989

Collier D A: Aggregate Safety Stock Levels and Component Part Commonality, *Management Science*, Vol 28, No 1, 1982, pp 1296-1303

Committee On the Next Decade in Operations Research (CONDOR): Operations Research: The Next Decade, *Operations Research*, Vol 36, No 4, July-August 1988, pp. 619-637

Fisher M: The Lagrangean Relaxation Method for Solving Integer Programming Problems, *Management Science*, Vol 27, No 1, 1981, pp 1-18

Fisher M: An Applications Oriented Guide to Lagrangean Relaxation, *Interfaces*, Vol 15, No 2, 1985, pp 10-21.

Geoffrion A M: Lagrange Relaxation for Integer Programming, *Mathematical Programming Study 2*, 1974, pp 82-114.

Gerchak Y and Henig M: An Inventory Model with Component Commonality, *Operations Research Letters*, Vol 5, 1986, pp 157-160

Gerchak Y and Henig M: Component Commonality in Assemble-to-Order Systems: Models and Properties, *Naval Research Logistics*, Vol 36, 1989, pp 61-68

Gerchak Y, Magazine M and Gamble B: Component Commonality with Service Level Requirements, *Management Science*, Vol 34, No 6, 1988, pp 753-760

Gutierrez G and Hausman W: Optimal and Approximate Dynamic Rationing Policies in an Assemble-to-Order System, The University of Texas at Austin and Stanford University, 1989.

Jönsson H and Silver E: Optimal and Heuristic Solutions for a Simple Common Component Inventory Problem, *Engineering Costs and Production Economics*, Vol 16, 1989a, pp 257-267

Jönsson H and Silver E: Common Component Inventory Problems with a Budget Constraint: Heuristics and Upper Bounds, *Engineering Costs and Production Economics*, Vol 18, 1989b, pp 71-81

Jönsson H, Jörnsten K and Silver E: Application of the Scenario Aggregattion Algorithm to a Two-Stage, Stochastic, Common Component, Inventory Problem with a Budget Constraint, Dept of Business Administration, Gothenburg School of Economics, Working Paper, 1990

Kall P: *Stochastic Linear Programming*, Springer-Verlag, Berlin, 1976

Lasdon L: *Optimization Theory for Large Scale Systems*, MacMillan Publishing, 1970

McClain J O, Maxwell W L, Muckstadt J A, Thomas L J and Weiss E N: Comment on 'Aggregate Safety Stock Levels and Component Part Commonality', *Management Science*, Vol 30, No 6, 1984, pp 772-773

Mulvey J and Vladimirou H: Solving Multistage Stochastic Networks: An Application of Scenario Aggregation, Report SOR-88-1, Dept of Civil Engineering and Operations Research, School of Engineering and Applied Science, Princeton University, 1988a

Mulvey J and Vladimirou H: Stochastic Network Optimization Models for Investment Planning, Report SOR-88-2, Dept of Civil Engineering and Operations Research, School of Engineering and Applied Science, Princeton University, 1988b

Mulvey J and Vladimirou H: Evaluation of a Distributed Hedging Algorithm for Stochastic Network Programming, Report SOR-88-14, Dept of Civil Engineering and Operations Research, School of Engineering and Applied Science, Princeton University, 1988c

Rockafellar R T and Wets R J-B: Scenarios and Policy Aggregation in Optimization under Uncertainty, WP-87-119, IIASA, Laxenburg, Austria, 1987

van Donselaar K and Wijngaard J: Commonality and Safety Stocks, *Engineering Costs and Production Economics*, Vol 12, 1986, pp 197-204

Wemmerlöv U: Assemble-to-Order Manufacturing: Implications for Materials Planning, *Journal of Operations Management*, Vol 4, 1984, pp 347-368

Wets R J-B: Stochastic Programming: Solution Techniques and Approximation Schemes, in Bachem A, Grötschel M and Korte B, *Mathematical Programming: The State of the Art*, Springer-Verlag, Bonn, 1982

Wets R J-B: The Aggregation Principle in Scenario Analysis and Stochastic Optimization, in Wallace S W (ed), *Algorithms and Model Formulations in Mathematical Programming*, Springer-Verlag, Berlin, 1989

Effect of the Standardization of the Hierarchy Parts Structure of a Product (SHPSP) on Production Ordering and Inventory Levels

KAZUYOSHI ISHII and TOSHIHIKO SUMINOKURA

1. INTRODUCTION

In recent years, changes in users' values have diversified and complicated users' needs. This diversity of needs has made the market more segmented and has led to an increase in the number of items produced by each manufacturer as well as to increased fluctuations in product demand. Companies must continue to improve product quality and reduce lead times and costs to succeed in severe competition for sales.

Standardization of the hierarchy parts structure of a product (SHPSP) has been proposed to improve product quality and productivity, that is, to reduce the level of safety stock and costs. However, SHPSP is problematic in that it can decrease the possibility of fulfilling the various users' needs with the result that the market share of the product group decreases.

Some previous investigators have discussed this problem. Hirano (1983) developed three kinds of indices which can evaluate the degree of the standardization of the parts structure of products according to the types of standardization on a product family tree, such as within a product, between products, and among product groups. Although these indices are effective at the product design stage, it is impossible to evaluate the effectiveness of the standardization on the productivity and reliability of production systems in the production process planning stage in which production process plans and production ordering systems are designed. Many process planning systems have been developed for mainly the machining process (Descotte and Lamobe 1981, Wolfe and Kung 1984, Phillips and Zhang 1984, Hukuda 1986). A few systems for the assembling process have been developed (Miyakawa and Ohashi 1986, Hayashi et al. 1987).

Many effective production ordering systems have been developed to satisfy various purposes in the world of multistage production and inventory systems. These ordering systems are classified into two types. One is the "push"-type production ordering system, which may be called a "centralized" production ordering system such as the MRP system (Burns and Sivazlian 1978, Van Aken 1980, Muramatsu et al. 1985). The other is the "pull"-type production ordering system, which may be called a "decentralized" production ordering system such as the Kanban system (Kimura and Terada 1981, Tabe et al. 1980, Bitran and Chang 1987, Hiraki et al. 1989). Soshiroda (1971), Muramatsu et al. (1985) and Takahashi et al. (1987) have proposed "amplifications" of production ordering and inventory levels as evaluating functions of the productivity and reliability of multistage production and inventory systems.

In this paper, the following steps are taken to propose a method of evaluating SHPSP and to demonstrate the effectiveness of SHPSP under the conditions of a process plan which is designed according to the manner of SHPSP on a product family tree, a push-type production ordering system, the production facility and market demand.

(1) The structural formulations of production ordering and inventory levels of a product are revealed according to the manner of the standardization of the component parts in which the influences of the forecasting error, the difference between the forecasted demands which were estimated at different periods, the down-time of the production facility and the safety stock of the component part planned to be standardized upon production ordering and inventory levels of a product are clarified.

(2) The influences of the coefficient of correlation between the product demands, the autocorrelation coefficient of product demand, average loading ratio, the coefficient of variation of down-time and the safety stock level of the component part to be standardized upon amplifications of production ordering and inventory levels are clarified in comparision to the manner of the standardization of the part.

2. MODEL DESCRIPTIONS

2.1 Notations

D_t^i: the demand of product i during the t-th period (i=A,B)

$\hat{D}_{t,t+k}^i$: the forecasted demand of product i for the (t+k)-th period forecasted at the end of the t-th period (i = A,B ; k = 1,2,...)

A(i,j): the required quantity of the j component part to manufacture one unit of product i (i = A,B ; j = z,z1,z2,x,y)

S^j: the safety stock level of product/component part j

S*: the factor of the safety stock level of component part j

C^j: the production capacity of the process to manufacture product/component part j

X_t^j: the down-time occurring during the process of manufacturing component part j during the t-th period, which is measured in terms of the number of component part j (j = z,z1,z2)

$O_{t,t+2}^j$: the production ordering level of product/component part j for the (t+2)-th period calculated at the end of the t-th period j

P_t^j: the actual production level of product/component part j during the t-th period

R_t^j: the backlog of the production ordering level of product/component part j which cannot be produced because of the restriction of the production capacity during the t-th period or of the inventory level of the available component part to manufacture a product

I_t^j: the inventory level of product/component part j at the end of the t-th period

$Amp(O_{t,t+2}^i)$:the amplification of the production ordering level of the product

$Amp(I_t^i)$: the amplification of the inventory level of product i

ρj: the average loading ratio of the production stage to manufacture component part j

λi: the autocorrelation coefficient of the demand of product i

r: the coefficient of correlation between product A and B demands

ηj: the coefficient of variation of the down-time at the production stage to manufacture component part j

E(*): the expectation of *

V(*): the variance of *

Cov(*,**):the covariance between * and **

$\lfloor \ \rfloor$: the symbol defining the maximum integer to be less than the value in brackets

2.2 Assumptions

The system model developed in this paper assumes the following.

(1) The production system under consideration manufactures two kinds of products, A and B, with standard specifications. The product has the family tree shown in Fig.1. In the family tree, component parts z1 and z2 could potentially become a common component part z. This manner of standardization is that of "between products" in the same level of a product family tree and the index is 2 (Hirano 1983).

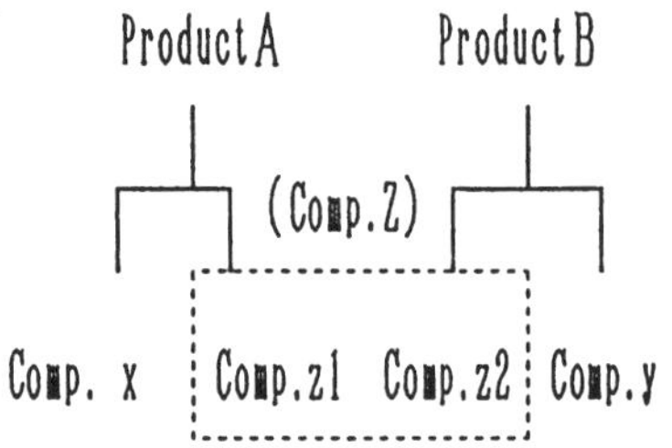

Fig. 1 The family tree of the products

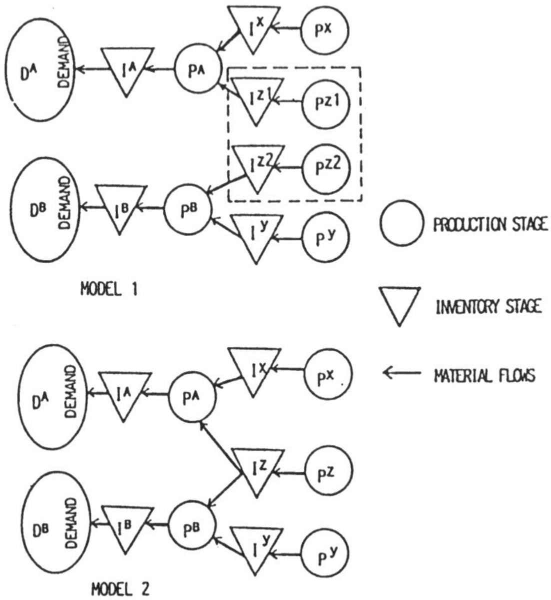

Fig. 2 The process according to the method of the standardization of the component parts

(2) The material flows in process plans, shown in Fig.2, depend on the manner of the standardization of component parts z1 and z2. That is, we call the case in which product A has component part z1 and product B, z2, model 1. In model 2, products A and B have a common component part Z.

(3) Each production stage in the system produces a single standard item.

(4) The production capacity of each stage is planned to be constant for any period.

(5) Machine breakdown occurs at the production stage of manufacturing the component parts. The available production level at this stage varies from period to period.

(6) The raw material inventory is always sufficient, and the handling time between the processes is disregarded.

(7) A production lead time in each production stage is one planning period. The production ordering level of the product/component part calculated at the end of the t-th period is produced in the stage during the (t+1)-th period, and supplied to the inventory stage at the beginning of the (t+2)-th period.

(8) The demand for product i at each period is the stationary time series with an exponential autocorrelation (Harvey 1981) whose average, variance and covariance are known, and the following relations are assumed:

$$Cov(D_t^i, D_{t+k}^i) = \lambda_i^k * V(D_t^i), \quad (i = A,B;\ k = 1,2...) \tag{1}$$

$$V(D_t^i) = V(D_{t+k}^i), \quad (i = A,B;\ k = 1,2,...) \tag{2}$$

$$E(D_t^i) = E(D_{t+k}^i), \quad (i = A,B;\ k = 1,2,..). \tag{3}$$

When $\lambda_i = 0$, the time series has a correlation between products A and B, shown in the following relation:

$$r = Cov(D_t^A, D_t^B) / \sqrt{V(D_t^A) * V(D_t^B)}. \tag{4}$$

(9) A forecasted value of the demand for product i is calculated by the autocorrelation method as follows (Soshiroda 1971):

$$\hat{D}_{t,t+k}^i = \lambda_i^k * D_t^i + (1 - \lambda_i^k) * E(D_t^i)$$

$$(i = A,B;\ -1 < \lambda_i < 1;\ k = 1,2,...). \tag{5}$$

2.3 Model formulations

(1) Production ordering level

The production ordering is based on a fixed-interval ordering system which is designed based on information such as forecasted demand considering the planning period as a production lead time and the feedback of term-end inventory at the end of each planning period (Soshiroda 1971, Muramatsu et al. 1985, Takahashi et al. 1987). The production ordering level of product i is calculated at the end of the t-th period as follows:

$$O_{t,t+2}^i = \hat{D}_{t,t+2}^i + (\hat{D}_{t,t+1}^i - P_t^i) - I_t^i + S^i, \quad (i=A,B). \tag{6}$$

The production ordering level of the component parts is calculated by eqns.(7) and (8) according to the manner of the standardization:

$$O_{t,t+2}^z = \sum_{i=A,B} A(i,z) * \hat{D}_{t,t+3}^i + \left(\sum_{i=A,B} A(i,z) * \hat{D}_{t,t+2}^i - P_t^z\right) - I_t^z + S^z \tag{7}$$

$$O_{t,t+2}^j = \hat{D}_{t,t+3}^i * A(i,j) + (\hat{D}_{t,t+2}^i * A(i,j) - P_t^j) - I_t^j + S^j,$$

$$(i,j) = (A,x), (A,z1), (B,y), (B,z2). \tag{8}$$

(2) The production level

The production ordering level of the product i is limited by the production capacity and the inventory level of the component part. Thus the actual production level of product i is determined according to the manner of the standardization as follows:

MODEL 1

$$P^i_{t+1} = \min[C^i ,\ O^i_{t,t+2} ,\ \min_j \lfloor (I^j_t + P^j_t)/A(i,j) \rfloor],$$

$$(A,j)\ ;\ j = z1,x \text{ and } (B,j)\ ;\ j = z2,y; \qquad (9)$$

MODEL 2

$$P^A_{t+1} = \min[C^A ,O^A_{t,t+2} ,\ \min\lfloor (O^A_{t,t+2}/ \sum_{i=A,B} A(i,z)*O^i_{t,t+2})(I^z_t + P^z_t)/A(A,z),$$

$$(I^x_t + P^x_t)/A(A,x) \rfloor] \qquad (10)$$

$$P^B_{t+1} = \min[C^B ,\ O^B_{t,t+2} ,\ \min\lfloor ((I^z_t + P^z_t) - P^A_{t+1} * A(A,z))/A(B,z),$$

$$(I^y_t + P^y_t)/A(B,y) \rfloor]. \qquad (11)$$

The production level of component part j is as follows:

$$P^j_{t+1} = \min(C^j - X^j_{t+1} ,\ O^j_{t,t+2}), \quad (j=x,y,z1,z2,z). \qquad (12)$$

(3) The inventory level

The inventory level at the end of the t-th period of product i and component part j is calculated as follows:

$$I^i_t = I^i_{t-1} + P^i_{t-1} - D^i_t, \quad (i=A,B) \qquad (13)$$

$$I^j_t = I^j_{t-1} + P^j_{t-1} - P^i_t * A(i,j), \quad (i,j)=(A,x),(A,z1),(B,y),(B,z2) \qquad (14)$$

$$I^z_t = I^z_{t-1} + P^z_{t-1} - (P^A_t *A(A,z) + P^B_t *A(B,z)). \qquad (15)$$

(4) The backlog of production ordering level

The production ordering level varies period by period according to changes in product demand. Restriction of production capacity, down-time and the level of the inventory of the component part occasionally prevent the fulfillment of a production ordering level in one period. The backlog of the production ordering level of the product caused by the restrictions mentioned above is calculated as follows:

MODEL 1

$$R^{i}_{t+1} = \max[0\ ,\ O^{i}_{t,t+2} - C^{i}\ ,\ O^{i}_{t,t+2} - \min_{j} \lfloor (I^{j}_{t} + P^{j}_{t})/A(i,j) \rfloor],$$

$$(i,j)=(A,z1),(A,x),(B,z2),(B,y); \tag{16}$$

MODEL 2

$$R^{A}_{t+1} = \max[0\ ,O^{A}_{t,t+2} - C^{A},$$

$$O^{A}_{t,t+2} - \min\lfloor (O^{A}_{t,t+2}/ \sum_{i=A,B} A(i,z)*O^{i}_{t,t+2})*(I^{z}_{t} + P^{z}_{t})/A(A,z),$$

$$(I^{x}_{t} + P^{x}_{t})/A(A,x) \rfloor] \tag{17}$$

$$R^{B}_{t+1} = \max[0,\ O^{B}_{t,t+2} - C^{B},\ O^{B}_{t,t+2} - \min\lfloor (I^{z}_{t} + P^{z}_{t} - P^{A}_{t+1}*A(A,z))/A(B,z),$$

$$(I^{y}_{t} + P^{y}_{t})/A(B,y) \rfloor]. \tag{18}$$

The backlog of the production ordering level of the component part is calculated by

$$R^{j}_{t+1} = \max[\ 0,O^{j}_{t,t+2} - (\ C^{j} - X^{j}_{t+1})],\ (\ j = z1,z2,z,x,y). \tag{19}$$

(5) The evaluating functions

To evaluate the effect of SHPSP on the reliability and productivity of production systems, the amplifications defined as the following equations are proposed.

(a) The amplification of the production ordering level of a product:

$$Amp(O^{i}_{t,t+2}) = V(O^{i}_{t,t+2})/V(D^{i}_{t}),\quad (i=A,B). \tag{20}$$

(b) The amplification of the inventory level of a product:

$$Amp(I^{i}_{t}) = V(I^{i}_{t})/V(D^{i}_{t}),\quad (i=A,B). \tag{21}$$

3. ANALYSIS

3.1 The structure of the production ordering and inventory levels of a product

In order to analyze the amplifications of the production ordering and the inventory levels of a product, the influence of forecasting error, the difference between forecasted demands, down-time, safety stock level and the backlog of the production ordering level upon the production ordering and inventory levels of a product are clarified in each model which is developed according to the manner of standardization.

The forecasting error between the forecasted demands of product i calculated at the end of the (t-k)-th period and the actual demand is defined as follows (Takahashi et al. 1987):

$$\varepsilon^{i}_{t-k,t} = \hat{D}^{i}_{t-k,t} - D^{i}_{t}, \quad (i = A,B ; k = 1,2,..). \qquad (22)$$

The difference between forecasted demands of product i for the t-th period calculated at the end of the (t-k)-th period and the (t-m)-th period is defined as follows:

$$\xi^{i}_{t-k,t-m}(t) = \hat{D}^{i}_{t-k,t} - \hat{D}^{i}_{t-m,t}, \quad (i = A,B ; k = 1,2,.. ; m = 1,2,.. ; k \neq m). \qquad (23)$$

(1) The production ordering level of a product

The production ordering level of product i is defined by eqn.(6). Equations (9),(10) and (11) are transformed into eqn.(24) by use of eqns.(16),(17) and (18):

$$P^{i}_{t} = O^{i}_{t-1,t+1} - R^{i}_{t}, \quad (i=A,B). \qquad (24)$$

Equation(6) is transformed into eqn.(25) by the use of eqns.(6),(13) and (24):

$$O^{i}_{t,t+2} = \hat{D}^{i}_{t,t+2} - (\hat{D}^{i}_{t-1,t+1} - \hat{D}^{i}_{t,t+1}) - (\hat{D}^{i}_{t-1,t} - D^{i}_{t}) + R^{i}_{t}, \quad (i=A,B). \qquad (25)$$

As a result, eqn.(25) is described as follows by the use of eqns.(22) and (23):

$$O^{i}_{t,t+2} = \hat{D}^{i}_{t,t+2} - \xi^{i}_{t-1,t}(t+1) - \varepsilon^{i}_{t-1,t} + R^{i}_{t}, \quad (i=A,B). \qquad (26)$$

(2) The inventory level of a product

The term-end inventory level of product i is defined by eqn.(13). Equation (13) is transformed into eqn.(27) by the use of eqns.(13),(22),(23) and (24):

$$I^{i}_{t} = \varepsilon^{i}_{t-2,t} + \varepsilon^{i}_{t-2,t-1} - R^{i}_{t} + S^{i}, \quad (i=A,B). \qquad (27)$$

(3) The backlog of production ordering level

Furthermore, in a similar manner as shown above, the backlog of production ordering level, which is composed of both the production ordering and inventory levels described by structural formulations (26) and (27), is shown in Table 1 according to the manner of standardization.

Table 1 The structural formulations of the backlog of production ordering level

R	MODEL 1	MODEL 2
R^i_{t-1}	$\max[0, \hat{D}^i_{t-2,t} - H^i_t + R^i_{t-2} - C^i,$ $\hat{D}^i_{t-2,t} - H^i_t + R^i_{t-2} -$ $\min_j L\{A(i,j)*$ $(\hat{D}^i_{t-3,t} + H^i_{t-1} + R^i_{t-2} - R^i_{t-3})$ $+ S^j - R^j_{t-2}\}/A(i,j)\rfloor]$	$R^A_{t-1} = \max[0, \hat{D}^A_{t-2,t} - H^A_t + R^A_{t-2} - C^A, \hat{D}^A_{t-2,t} - H^A_t + R^A_{t-2} - \min$ $L(D^A_{t-2,t} - H^A_t + R^A_{t-2})\{\sum_{i=A,B} A(i,z)(\hat{D}^i_{t-3,t} + H^i_{t-1} + R^i_{t-2} - R^i_{t-3}) + S^z$ $- R^z_{t-2}\}/\{\sum_{i=A,B} A(i,z)(\hat{D}^i_{t-2,t} - H^i_t + R^i_{t-2})A(A,z)\}, A(A,x)*$ $(\hat{D}^A_{t-3,t} + H^A_{t-1} + R^A_{t-2} - R^A_{t-3}) + S^x - R^x_{t-2}\}/A(A,x)\rfloor]$ $R^B_{t-1} = \max[0, \hat{D}^B_{t-2,t} - H^B_t + R^B_{t-2} - C^B, \hat{D}^B_{t-2,t} - H^B_t + R^B_{t-2} -$ $\min L\{\sum_{i=A,B} A(i,z)(\hat{D}^i_{t-3,t} + H^i_{t-1} + R^i_{t-2} - R^i_{t-3}) + S^z - R^z_{t-2} -$ $A(A,z)(\hat{D}^A_{t-2,t} + H^A_t + R^A_{t-2} - R^A_{t-1})\}/A(B,z), \{A(B,y)*$ $(\hat{D}^B_{t-3,t} + H^B_{t-1} + R^B_{t-2} - R^B_{t-3}) + S^y - R^y_{t-2}\}/A(B,y)\rfloor]$
R^j_{t-1}	$\max[0, A(i,j)(\hat{D}^i_{t-2,t+1} - \xi^i_{t-3,t-2}(t) -$ $H^i_{t-1} - R^i_{t-2} + R^i_{t-3}) + R^j_{t-2}$ $- C^j + X^j_{t-1})]$, $(j=z1,z2)$	$\max[0, \sum_{i=A,B} A(i,z)(\hat{D}^i_{t-2,t+1} - \xi^i_{t-3,t-2}(t) - H^i_{t-1} - R^i_{t-2}$ $+ R^i_{t-3}) + R^z_{t-2} - C^z + X^z_{t-1}]$.

where $H^i_t = \xi^i_{t-3,t-2}(t-1) + \varepsilon^i_{t-3,t-2}$ $(i=A,B)$

(4) Results and discussions

From eqns.(26) and (27) and Table 1, the following information is obtained.

(a) The different influence between the manner of standardization upon the production ordering and inventory levels of the product occurs in the term of the backlog which is determined by forecasting error, the difference between the forecasted demands, down-time, the restrictions of safety stock and production capacity.

(b) Although the backlog has a recursive structure in both models, the backlog of the production ordering level of a product of model 2 is influenced by the manner of the distribution of component z to be standardized among products.

3.2 Amplifications of the production ordering and inventory levels of product

This analysis aims to clarify the influences of average loading ratio, auto-correlation coefficient of demand, coefficient of correlation between the product demands, standard deviation of down-time and the safety stock level of the component part upon the amplifications of production ordering and inventory levels of the product.

(1) Preparation for the analysis

The previous analysis shows that when the production capacity and safety stock level of the component parts are sufficient, the backlog of the production ordering level does not occur. Therefore, the production ordering and inventory levels of a product are independent of the manner of the standardization.

As a result, in the case of $R^i_t = 0$ for all i and t, the amplifications of production ordering and inventory levels can be shown by the following equations by the use of eqns.(1),(2),(3),(5),(22) and (23):

$$Amp(O^i_{t,t+2}) = V(\hat{D}^i_{t,t+2} - \xi^i_{t-1,t}(t-1) - \varepsilon^i_{t-1,t})/V(D^i_t)$$

$$=[V(\hat{D}^i_{t,t+2}) + V(\xi^i_{t-1,t}(t-1)) + V(\varepsilon^i_{t-1,t}) - 2Cov(\hat{D}^i_{t,t+2}, \xi^i_{t-1,t}(t-1))$$

$$+ 2Cov(\xi^i_{t-1,t}(t-1), \varepsilon^i_{t-1,t}) - 2Cov(\hat{D}^i_{t,t+2}, \varepsilon^i_{t-1,t})]/V(D^i_t)$$

$$= 1 + 2\lambda_i + 2\lambda_i^2 - 2\lambda_i^4 - 2\lambda_i^5 \qquad (28)$$

$$Amp(I^i_t) = 2 + 2\lambda_i - \lambda_i^2 - 2\lambda_i^3 - 2\lambda_i^4 . \qquad (29)$$

(2) Numerical analysis of amplification

When the backlog occurs, it is difficult to analyze the amplification theoretically so that numerical experiment focuses on the terms of the backlog under the additional following assumptions.

Figure 3 shows the flow chart for calcuation of the amplifications of production ordering and inventory levels of the products. The flow chart was developed on the basis of the structural formulations obtained in the previous analysis.

(3) Assumptions and numerical conditions

1) $D_t^i \sim N(100,25)$, $\lambda A = \lambda B = \lambda$, $\lambda = -0.5, 0.0, 0.5$;
in the case of $\lambda = 0.0$, $r = -0.5, 0.0, 0.5$.

2) The machine breakdown occurs in the production stage of manufacturing component parts z1, z2 and z, and the down-time of each stage has the following normal distribution:

$$X_t^z \sim N(30,\sigma^2),\ Cov(X_t^j, X_{t+k}^j)=0,\ Cov(X_t^{z1}, X_t^{z2})=0,\ Cov(D_t^i, X_t^j)=0,\ E(X_t^j) = E(X_t^z)/2.$$

The coefficient of variation of down-time is defined as follows:

$$\eta = \sqrt{V(X_t^z)}/E(X_t^z) = \sqrt{V(X_t^j)}/E(X_t^j),\ (j=z1,z2) \qquad (30)$$

$\eta = 0.0, 0.1, 0.2, 0.3,$ $\sigma = 0.0, 3.0, 6.0, 9.0.$

3) The average loading ratio of the production stage which manufactures component parts z1, z2 and z is defined by eqn.(30). The numerical condition is for the cases of $\rho = 0.8, 0.9, 1.0$:

$$\rho = \sum_{i=A,B} A(i,z)*E(D_t^i)/(C^z - E(X_t^z)) = A(i,j)*E(D_t^i)/(C^j - E(X_t^j)),\ (j=z1,z2). \qquad (31)$$

4) The factor of the safety stock level of component parts z1, z2 and z is defined by eqn.(32). This factor is standardized to allow comparison with the manner of standardization. The factor is set at $S^* = 0.0, 0.25, 0{,}5, 1.0$:

$$S^* = S^z / \sum_{i=A,B} A(i,z)*E(D_t^i) = S^j / A(i,j)*E(D_t^i),\ (j=z1,z2). \qquad (32)$$

5) $A(i,j)=1$, $\sum_{i=A,B} A(i,z)=2$, $(j=z1,z2)$.

6) The production capacities C^i, C^x and C^y are sufficient (i=A,B).

7) The safety stock levels S^i, S^x and S^y are sufficient (i=A,B).

8) The number of simulation period is 500.

(3) Results and discussions

Figure 4 shows the relationships between the factor of the safety stock level of the component part and the amplifications of the production ordering

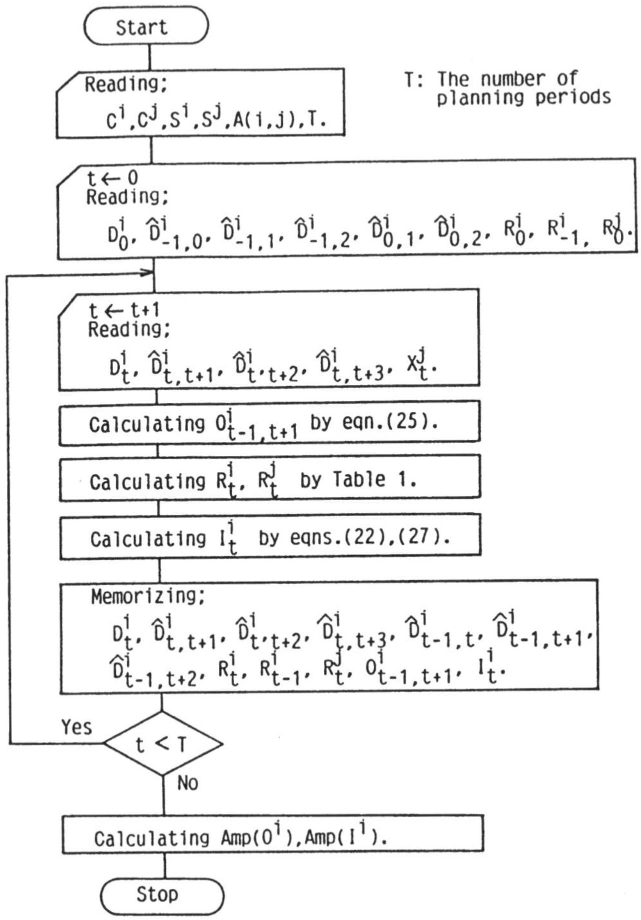

Fig.3 The flow chart for calculation of the amplifications

and inventory levels of product A under the conditions of average loading ratio and autocorrelation coefficient. Those of product B show similar results.

Figure 4 shows that

1)the amplifications of the production ordering and inventory levels of the products decrease exponentially with the increase of the factor of the safety stock of the component part;

2)the effect of reducing the amplifications decreases with the increase of the factor of the safety stock of the component part so that the amplifications are constant depending on the autocorrelation coefficient when the factor is greater than 0.5 and average loading ratio is less than 0.9;

3)the higher the average loading ratio and autocorrelation coefficient, the greater the effect of reducing the amplifications by the safety stock of the component;

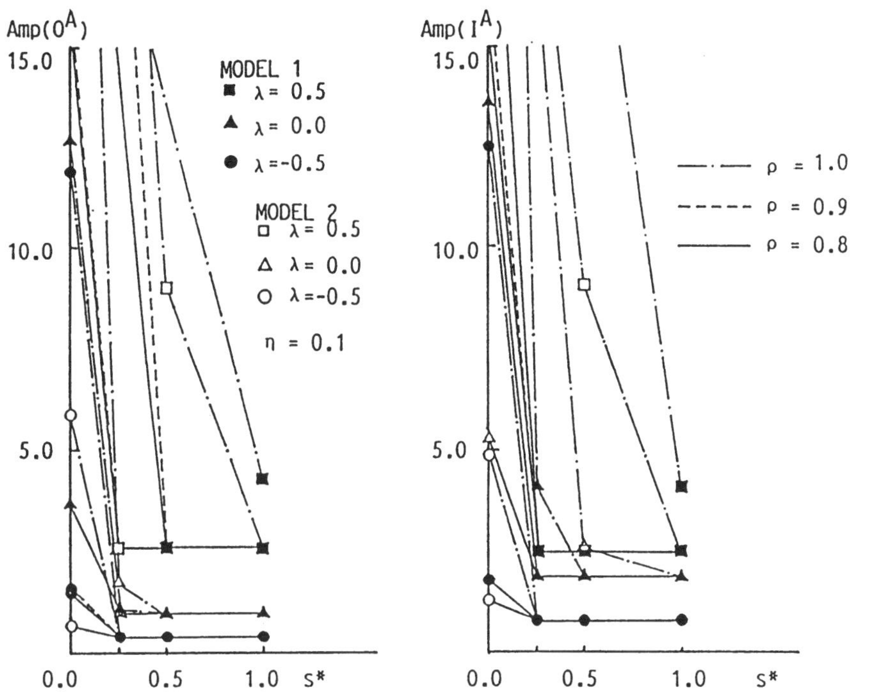

Fig.4 The amplifications of production ordering and inventory levels of a product (η= 0.1)

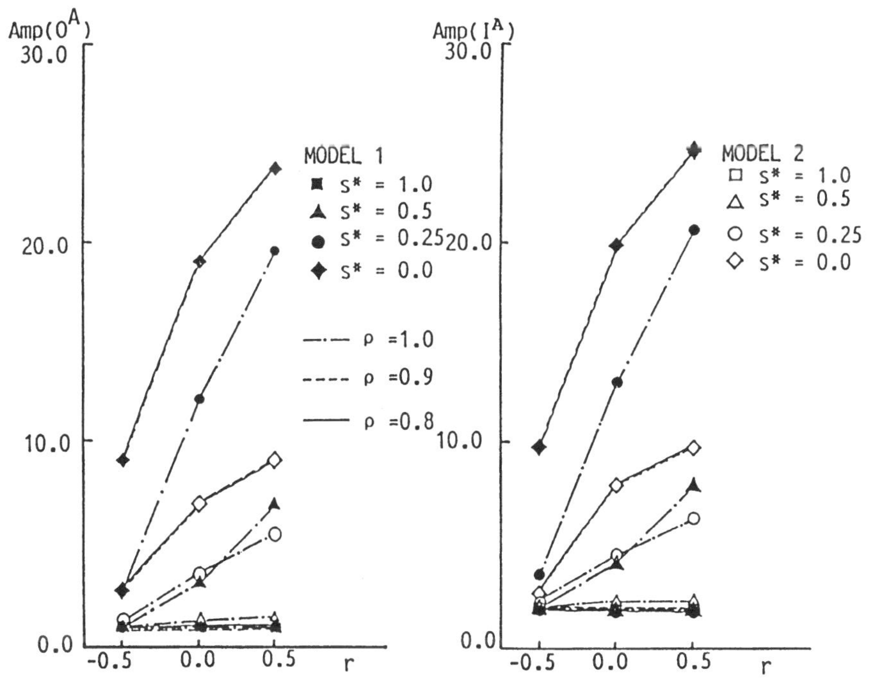

Fig.5 The relationships between the coefficient of correlations between product demands and the amplifications of production ordering and inventory levels of a product(η=0.1, λ= 0)

4)the higher the average loading ratio and autocorrelation coefficient and the lower the factor of the safety stock of the component part, the more SHPSP can reduce the amplifications.

Figure 5 shows the relationships between the coefficient of correlations between product A and B demands and the amplifications under the conditions of average loading ratio and the factor of the safety stock of the component part. Figure 6 shows the relationship between the coefficient of variation of down-time occurring in the production stage of manufacturing component parts z1, z2 and z, and the amplifications under the conditions of average loading ratio and the factor of the safety stock of the component part.

These figures indicate that

1)the higher the coefficient of correlation between the product demands and average loading ratio, the greater the effect of reducing the amplifications by the safety stock of the component part;

2)the amplifications are constant and independent of the coefficient of correlations between product demands when the average loading ratio is less than 0.9 and the factor of the safety stock is greater than 0.25;

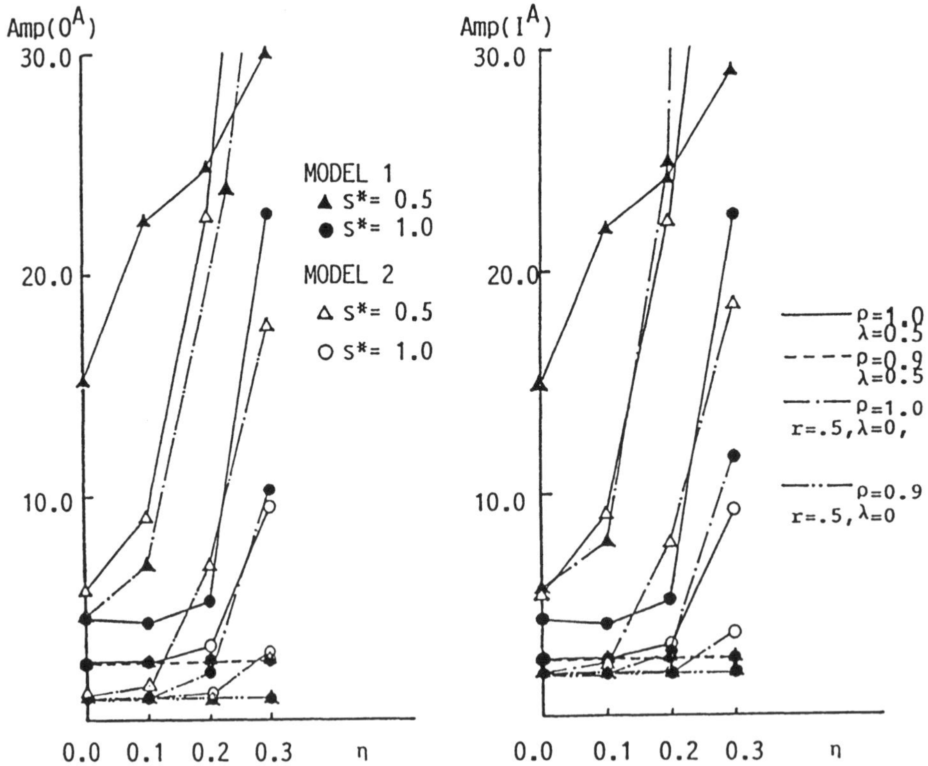

Fig.6 The relationships between the coefficient of variation of down-time and the amplifications of production ordering and inventory levels of a product(s*=0.5/1.0, ρ =1.0/0.9,[λ,r]=[.5,0]/[0,.5])

3)the higher the coefficient of correlation, the more the SHPSP can reduce the amplifications;

4)the higher the coefficient of variation of down-time, the higher the amplifications;

5)there is a case in which SHPSP increases the amplifications when the coefficient of variation of down-time, average loading ratio and autocorrelation coefficient of product demand are higher.

4. CONCLUSIONS

In order to propose an evaluation method for the standardization of the hierarchy parts structure of a product (SHPSP) in a process planning stage, the amplifications of the production ordering and inventory levels of a product were analyzed according to the manner of the standardization of the component part, and the following conclusions were reached.

(1)The influence of the backlog of the production ordering level caused by forecasting error, the difference between forecasted demands, the down-time, the production capacity and the safety stock of the component part to be standardized upon the production ordering and inventory levels of a product is different according to the manner of the standardization of the component part.

(2)The effect of SHPSP on the amplifications of the production ordering and inventory levels of a product is determined by the conditions of the level of the safety stock of the component part, average loading ratio, variation of down-time, autocorrelation coefficient and coefficient of correlation between product demand.

(3)If an opportunity loss wherein the manner of SHPSP decreases the market share of the product group is estimated, it is possible to determine the optimal manner of SHPSP by consideration of the inventory holding costs of the product and component part to be standardized, the investment cost of production capacity and the opportunity loss of a decreased market share.

5. ACKNOWLEDGMENT

This study was supported by the Takahashi Research Fund.

References

BITRAN,G.R. and CHANG,L.,1987, A mathematical programming approach to a deterministic Kanban system,Management Science,33,4,427-441.

BURNS,J.F. and SIVAZLIAN,B.D.,1978,Dynamic analysis of multi-echelon supply systems,Computer and Industrial Engineering,2,181-193.

DESCOTTE,Y. and LAMOBE,J.,1981,GARI:A Problem Solver That Plans How to Machine Mechanical Parts,Proceedings of the 7th International Joint Conference on Artificial Intelligence,Vancouver,Canada,766-772.

HARVEY,A.C.,1981,Time Series Models(Philip Allan Publishers).

HAYASHI,N.,ISHII,K. and ICHIMURA,T.,1987,A case study of production design based on the fusion model, Innovation : Adaptation and Growth, 61-72 (Elsevier Science Publishers).

HIRAKI,S.,ICHIMURA,T.,ISHII,K.,NAGAMURA,K.,ANDO,Y.,TAKAHASHI,K. and MURAMATSU, R.,1989,Low-inventory ordering systems for the international co-operative-knock-down production systems,International Journal of Production Research,25,5,831-846.

HIRANO,T.,1983,Research on the Standardization of Product Design - Especially on the Standardization of the Hierarchy Parts Structure of Product-, Doctoral Thesis of Waseda University, Tokyo,(in Japanese).

HUKUDA,Y.,1986,Automated Process Planning System,Journal of Japan Industrial Management Association,37,4,211-217(in Japanese).

KIMURA,O. and TERADA,H.,1981,Design and analysis of pull system:a method of multi-stage production control,International Journal of Production Research,19,3,241-253.

MIYAKAWA,S. and OHASHI,T.,1986,The Hitachi Assemblability Evaluation Method-(AEM),Proceedings of International Conference on Product Design For Assembly,Newport,Rhode Island,April.

MURAMATSU,R.,ISHII,K. and TAKAHASHI,K.,1985,Some ways to increase flexibility in production systems, International Journal of Production Research,23,4,691-703.

PHILLIPS,R.H. and ZHANG,Z.S., 1984, Artificial intelligence based generative process planning,Proceedings of the 12th NAMRC,414-417.

SOSHIRODA,M.,1971,A study on the variances of the ordering and inventory levels in the periodic ordering system, Ph.D.Thesis of Waseda University(in Japanese).

TABE,T.,MURAMATSU,R. and TANAKA,T.,1980,Analysis of production ordering quantities and inventory variations in multi-stage production ordering systems,International Journal of Production Research,18,245-257.

TAKAHASHI,K.,MURAMATSU,R. and ISHII,K.,1987,Feedback method of production ordering system in multi-stage production and inventory systems,International Journal of Production Research,25,6,925-941.

VAN AKEN,J.,1980,Aggregate Planning in Decentralized Organizations,OMEGA,8,3, 323-332.

WOLFE,P.M. and KUNG,H.K.,1984,Automating process planning using artificial intelligence, Proceedings of Annual International Industrial Engineering Conference,387-395.

A Stochastic Model for In-Process Inventory

SANKARAN RAMANI

PRELIMINARY REMARKS

The problems of In-process Inventory can be successfully formulated in terms of the Inventory levels which allow the maximum production or the lowest per unit cost. In this paper a reliability - theoretic approach to solve In-process Inventory problem in a Concrete Brick Manufacturing Plant is proposed. The model deals with stock out probability of the bunker of rated capacity assuming that the depletions of the stock from the bunker is linear with respect to the failure rate of the bunker.

If we consider a line consisting of two stages and a bunker for in-process storage, it is interesting to note that the results for a single stage can be extended with suitable modifications to more stages of a production line. The bunker acts as a supply source for the succeeding state, when the proceeding stage suffers from stoppages due to breakdown, tool setting, etc. "Consider a production line consisting of two machines say A and C, corresponding to stages 1 and 2. A bunker 'B' of capacity N units is located between the two machines. This bunker acts as a supply source for machine C whenever machine A breaks down or requires tool setting". Each machine can be in one of two states (i.e.) working state and non-working state. The bunker can also be in any one of two states, i.e. working and non-working state. The working state of the bunker represents the state of the bunker when it contains atleast one unit in storage. The non-working state corresponds to no unit being in the bunker.

The bunker B acts as a standby to machine A. When machine A breaks down, the bunker instantaneously takes over. However, when the bunker fails, it will be rectified at the end of the shift, i.e. for a three shift operation the rectification period may be the week-end or month-end or the time taken to procure from outside sources and fill the bunker. The bunker is made ready only for the next day's operation. The time taken to rectify the bunker is the lead-time for procurement.

The Markov renewal process is analysed to find out the first time passage distribution. The mean time to system failure can be used to optimise the In-process Inventory level, provided we know the cost functions governing both the In-process Inventory level as also the production down time. The method of treating the In-process Inventory as a standby unit can be extended to multistage production processess as well. It may be worthwhile to explore whether some of the existing models in reliability theory can be suitably modified to study multistage in-process inventory problems on the above lines.

INTRODUCTION

In-process inventories have not been treated as extensively in the literature as the more direct inventory situations, eventhough such inventories are responsible for the utilisation of expensive machinery in production processess. Experience and intuition continue to govern in-process storage and production decisions, eventhough the application of some recently developed mathematical techniques can lead to a better comprehension and efficiency of the systems.

The major problems in the design and operation of production lines have been identified by Buchan, and Koenigsberg (1) as

i) The sequencing of jobs into stages and the determination of the number of stages in the line.

ii) The location of bunkers or storage spaces for in-process inventory.

iii) The size or capacity of these pulsating stores so called because the content of each storage space increases and decreases continuously.

Buchan and Koenigsberg have suggested that since production capacity, labour and inventory can be substituted for one another over a limited range, one cannot seek to minimise either the amount of inventory or the cost of the inventory system alone. These authors feel that the problems are more successfully formulated in terms of the in-process inventory levels which allow the maximum production, or the lowest per unit cost or which minimize the effects of production stoppages due to absence of material or lack of space to store them for succeeding stages.

A detailed analysis and review of the basic problems associated with the efficient operation of production and assembly lines has been made by Koenigsberg (2). He considers that complex production systems have not been examined in detail by analytic methods, and that operations

Research workers have only made preliminary efforts in providing some understanding of production line processes and in obtaining measures of the desirability and effectiveness of in-process storage. Koenigsberg has reviewed three basic approaches to evaluate the effectiveness of internal storage. The loss transfer model, the stochastic model and the queueing model are compared. These methods are applicable to problems which, in physical appearance are different from production lines, such as chemical production (process manufacture). Koenigsberg has outlined an industrial application for a system consisting of reactors, holding tanks and stills in a chemical plant. This study aimed at determining the equipment required (the number of holding tanks) to attain a desired production level. Queueing theory was used to obtain estimates of the required holding tank capacity.

Michael Freeman (3) has outlined the operational and economic aspects of production stages and analysed a three stage production line by resorting to simulation in a digital computer. Knott (4) has derived approximate formulae to study trade-offs between inactivity, queue capacity, number of stations and servicing time distributions for the designof a particular class of series production lines, and points out that there is scope for developing more accurate formulae.

The development of more accurate formulae is relatively scarce in the literature concerning in-process inventory. However some of the reliability models based on stochastic point processes appear to be capable of solving some of the in-process inventory problems with appropriate modification. This paper is devoted to a reliability-theoretic approach to solve an in-process inventory problem in a concrete brick manufacturing plant.

DETAILS OF THE CASE-STUDY

An opportunity to evaluate and optimise the bunker capacity in a cellular concrete brick manufacturing plant of the Tamil Nadu State Housing Board was made available to the author in the course of a consultancy assignment. The brick manufacturing plant was being designed and erected at Ennore in the outskirts of Madras City. The lime kiln was subcontracted to a separate agency and the plant was erected much in advance of the lime kiln.

The problem of designing a bunker of optimum capacity was posed to the author. Figure 1. illustrates the role of the bunker, in the production process. The entire production would come to a standstill if the bunker becomes empty when the lime kiln is in a repairable state.

The models applicable to the above situation have been outlined in the following pages. The first model deals with the mean value of the first time to system down state. The model is adapted from Osaki (5). The second model deals with the stockout probability of the bunker of rated capacity, assuming the depletion of the stock from the bunker to be linear with respect to the failure rate of the bunker. This model is adapted from Ramesh (6).

The main problem is to find the optimum bunker capacity for holding lime. The kiln is the main source of supply of lime for the brick-making plant. However, shut down of the lime kiln due to breakdowns and maintenance requires an inprocess storage of lime to be fed from a bunker during the periods of shut down of the bunker.

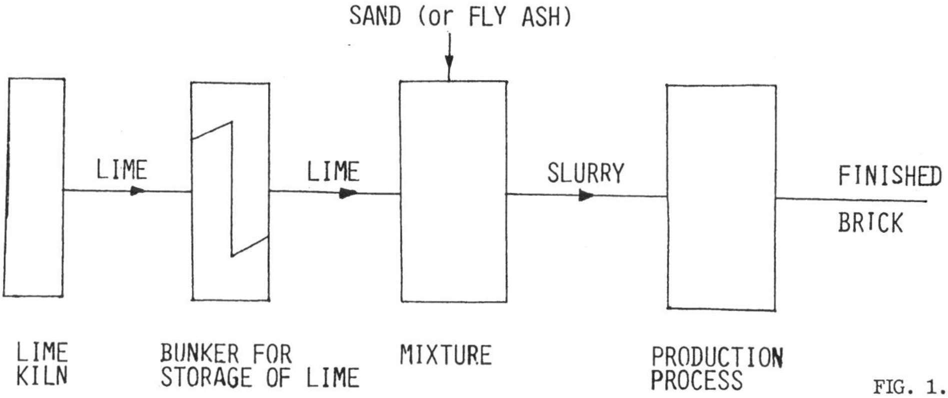

FIG. 1.

The brick-making plant requires lime, sand and gypsum as raw materials. Whenever the kiln is shut down, the bunker will be commissioned to function. Once the bunker is commissioned, it will function continuously until it becomes empty. The reason for this continuous operation of the bunker is that lime is a perishable raw material, and must be used within a few days of its being unpacked.

Since there is no other source of lime near the plant, the decision of "When to procure" and "how much to procure" must be taken in advance and the bunker capacity designed accordingly. Each time the bunker is commissioned, it feeds into the line until it is empty. Subsequently it is filled with lime procured from outside sources. In case the kiln is rectified by the time the bunker becomes empty, there is no production down-time. A knowledge of the mean time between subsequent fillings of the bunker is necessary for taking procurement decisions, and integrating the procurement-inventory problem. Thus, if the lime kiln, the bunker and the subsequent mixing process of lime and sand are viewed as a system, a knowledge of the mean time (and also the variance) for system down time is required for optimising the bunker capacity.

PROBLEM FORMULATION

In-process inventory is closely connected with problems of production. Several attempts have been made to study the variation in production due to changes in in-process inventory levels. One variation of the in-process inventory problem has been stated thus by Buchan and Koenigsberg (1).

"Given a production line process and the mean output rate of each stage, what is the output of the system if a store of N units is provided between each stage, or N_1, N_2, N_3......., N_i units are stored between i^{th} and $(i+1)^{st}$ stages ?"

If we consider a line consisting of two stages and a bunker for in-process storage, it is interesting to note that the results for a single stage can be extended with suitable modifications to more stages of a production line. The bunker acts as a supply source for the succeeding stage, when the proceeding stage suffers from stoppages due to breakdown, tool setting etc. Consider a production line as shown in Figure 2 , consisting of two machines A and C corresponding to stages 1 and 2. A bunker 'B' of capacity N units is located between the two machines. This bunker acts as a supply source for machine C whenever machine A broaks down or required tool setting. Each machine can be in one of two states (i.e.) working state and non-working state.

The bunker can also be in any one of two states - working and non-working state. The working state of the bunker represents the state of the bunker when it contains atleast one unit in storage.

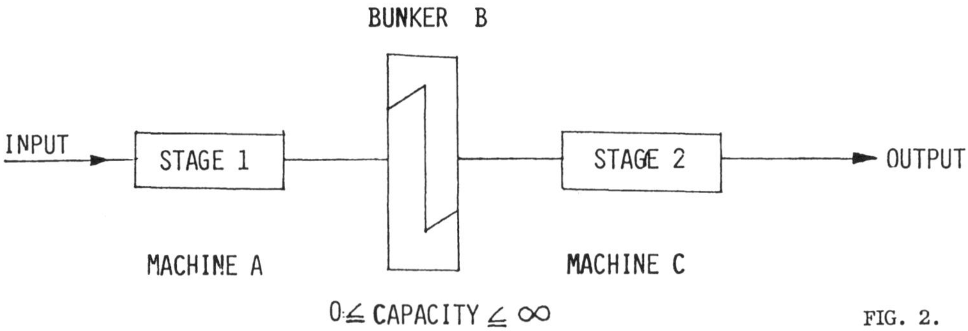

FIG. 2.

The bunker B acts as a standby to machine A. When machine A breaksdown, the bunker instantaneously takes over. However, when the bunker fails, it will be rectified at the end of the shift and made ready only for the next day's operation. The time taken to rectify the bunker is the lead-time for procuring lime.

THE MODEL

The failure time of machine A is assumed to have an exponential distribution

$$F_1(t) = 1 - \exp(-\lambda_1 t) \quad (t \geq 0)$$

the failure time of the bunker is a constant K, where K depends on the number of units N stored in the bunker. The repair time of machine A is assumed to be an arbitrary distribution $G_1(t)$ $(t \geq 0)$. The repair time of the bunker is assumed to be infinite for the single shift operation under consideration.

The switch-over times from the failure to the repair, from the repair to standby state, and from standby state to the operative state are assumed to be instantaneous. We also assume that after repair a unit recovers its function perfectly. The bunker does not fail in standby. $\bar{F}_i(t)$ and $\bar{G}_i(t)$ are survival distributions of $F_i(t)$ and $G_i(t)$ respectively.

To start with, at time t = 0, let us assume that machine A begins to be operative and at that time the bunker begins to be standby. Then the possible states of the system are as follows:

State s_o : machine A begins to be operative and the bunker begins to be standby.

State s_1 : the bunker begins to be operative and machine A begins to get repaired.

State s_2 : machine A begins to be operative and the bunker begins to get repaired.

State s_3 : the machine A and the bunker are under repair or failure. This state denotes the system down.

The problem is to find the mean value of the first time to system down (State 3) starting from state s_o at t = 0.

The State transition diagram is shown in Figure 3.

In terms of of Markov renewal process, the state s_o, s_1, s_2 and s_3 of the system denote the regeneration points of the process.

The only situation causing a change from the state s_o is when the operative machine A fails before the bunker fails. In this case, the system goes from state s_o to state s_1.

The Laplace transform of the time distribution from state s_o to state s_1 is $q_{o1}^*(s)$ given by

$$q_{o1}^*(s) = f_1^*(s) \qquad \text{-- (1)}$$

Where

$$q_{o1}(t) = f_1(t) \qquad \text{-- (2)}$$

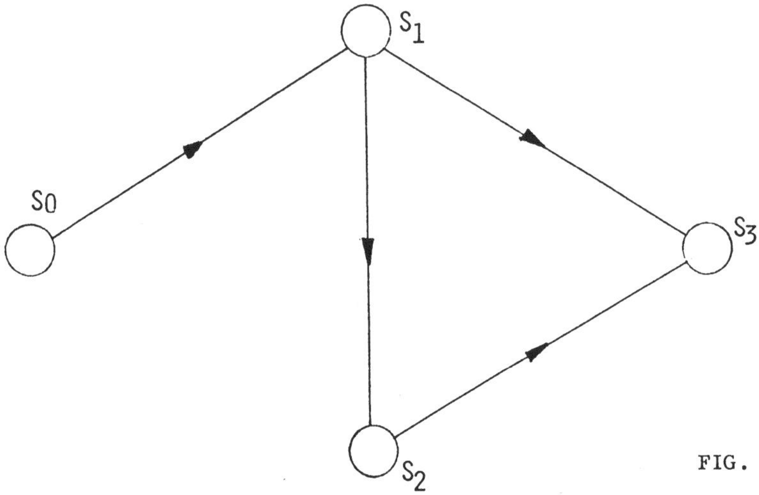

FIG. 3.

Henceforth we denote the Laplace transforms of $G_i(t)$ by $g_i(s)$. In the state s_1 we consider the following two (exclusive and exhaustive) cases:

i) The operative bunker fails before the completion of repair of machine A.

ii) After the completion of repair of machine A, this machine is kept in standby and the machine B fails before the standby machine A fails.

In the case (i) above, the system goes to state s_3 and the Laplace transform of the time distribution to state s_5 from state s_1 is

$$q_{13}^*(s) = \int_o^\infty e^{-st}\, \bar{G}_i(t)\ dF_2(t)$$

$$= \int_o^\infty \delta\ (x-a)\ f(u)\ du \qquad \text{-- (3)}$$

In case (ii) above the system goes to state s_2. The Laplace transform of the time distribution from state s_1 to state s_2 is

$$q_{12}^*(s) = \int_o^k g_1(u)\ du\ e^{-ks} \qquad \text{-- (4)}$$

where

$$q_{12}(t) = \delta(t-k) \int_o^k g_1(u)\,du \qquad \text{-- (5)}$$

From the state s_2, the system eventually goes to state s_3 and we can obtain a similar equation as follows:

$$q_{23}(t) = g_1(t)\ H\ (\lambda - t)$$

$$= \int_o^k e^{-st}\ f_1\ (t)\ dt \qquad \text{-- (6)}$$

The transition from state s_2 to s_1 is given by

$$q_{21}(t) = f_1(t)\ H(t - \lambda\ ) \qquad \text{-- (7)}$$

where

$$q_{21}{}^*(s) = \int_0^\infty e^{-st} f_1(t)\ dt$$

$$= \frac{\lambda}{\lambda + s}\ e^{-\lambda(\quad + s\ )} \qquad \text{-- (8)}$$

We now define $\emptyset_i^*$ (s) (i = 0, 1, 2), as the Laplace transform of the first time distribution from state s_1 to state s_3 respectively.

We can now find the first passage time distributions in the above Markov renewal process, where state s_3 is absorbing.

In state s_1 we have

$$\emptyset_o^*\ (s) = q_{o1}(s)\quad \emptyset_1(s) \qquad \text{-- (9)}$$

Similarly for $\emptyset_1(s)$ and $\emptyset_2(s)$

$$\emptyset_1^*\ (s) = q_{13}(s) + q_{12}(s)\quad \emptyset_2(s) \qquad \text{-- (10)}$$

$$\emptyset_2^*\ (s) = q_{23}(s) + q_{21}(s)\quad \emptyset_1(s) \qquad \text{-- (11)}$$

Solving equations (7.8), (7.9) and (7.10), we have

$$\emptyset_o^*\ (s) = q_{01}(s)\ [\bar{q}_{13}(s) + q_{12}(s)\quad q_{23}(s) + q_{21}(s)]/\ \frac{\emptyset_o^*}{q_0} \qquad \text{-- (12)}$$

This denotes the Laplace transform of the first time distribution from state s_o to state s_3.

The mean time E(T) is given by

$$E(T) = \left. \frac{d\, \emptyset_o^*(s)}{ds} \right|_{s=o} \quad \text{-- (13)}$$

The mean square value of T denoted by E (T^2) is given by

$$E(T^2) = \int_o^{\infty} t^2 \emptyset_o(t)\, dt$$

$$= -2 \left. \frac{d}{ds} \emptyset_o^*(s) \right|_{s=o} \quad \text{-- (14)}$$

The variance of T is given by

$$\text{Var}\,(T) = E(T^2) - \left\{E(T)\right\}^2$$

$$= -2 \left. \frac{d}{ds} \emptyset_o^*(s) \right|_{s=o} - \left. \left[\emptyset_o^*\,(s) \right]^2 \right|_{s=o} \quad \text{-- (15)}$$

The mean time to system failure can be used to optimise the inprocess inventory level provided we know the cost functions governing both the in-process inventory level as also the production down time. The method of treating the in-process inventory as a standby unit can be extended to multistage production processes as well. It may be worthwhile to explore whether some of the existing models in reliability theory can be suitably modified to study multistage in-process inventory problems.

REFERENCES

1. BUCHAN, J. and KOENIGSBERG, E., (1963), Scientific Inventory Management, Prentice Hall International Inc.

2. KOENIGSBERG, E., (1959), Production Lines and Internal Storage - A Review, Mgmt. Sci., 5, p.410.

3. MICHAEL FREEMAN, (1964), The Effects of Breakdowns and Interstage Storage on Production Line Capacity, J. Ind. Engg., 15, p.194.

4. KNOTT, A.D., (1970), The Inefficiency of a Series of Work Stations - A simple Formula, Int. Jnl. Prod. Res., 8, 2, p.109.

5. OSAKI SHUNJI, S., (1970), Reliability Analysis of a Two Unit Standby Redundant System with Standby Failure, Opsearch, 7, p. 13.

6. RAMESH, K.S. (1970), Point Processes in Reliability and other Related Topics, Ph.D. Thesis, Indian Institute of Technology, Madras.

An Inventory Model with Lateral Transshipments

SVEN AXSÄTER

1. Introduction

In practice it is quite common to have several parallel regional warehouses when distributing products over a large geographical area. Each warehouse is then providing service to a certain local market. The regional warehouses normally replenish their stocks from a common supplier. In the event of a stockout at a warehouse it may be advantageous, however, to replenish from a nearby parallel warehouse instead of from the common supplier. Such lateral transshipments may improve the service-performance of the total system considerably. Therefore, many companies use lateral transshipments systematically. It is, in general, quite difficult, though, to model lateral transshipments in inventory systems, and most inventory models assume that such transshipments are not allowed.

In a recent paper (Axsäter (1990)) a new approach for modelling lateral transshipments was applied to a two-echelon inventory system with repairable items. The purpose of the present paper is to illustrate how this approach can be used when dealing with a single-echelon case and consumable items. Our model assumes that each regional warehouse applies a continuous review one-for-one replenishment system ((S - 1, S)-system) and that the warehouses are facing independent Poisson demand processes. These assumptions are often relevant when dealing with expensive items with relatively low demand, for example certain types of spare parts.

Other inventory models that consider lateral transshipments between parallel stocking locations are described in Gross (1963), Das (1975), Hoadley and Heyman (1977), Karmarkar and Patel (1977), Dada (1985), Bowman (1986), Cohen, Kleindorfer and Lee (1986), Slay (1986) and Lee (1987). We refer to Lee's paper for an overview of different approaches.

The paper is organized as follows. In Sec. 2 we give a detailed problem formulation. Thereafter in Sec. 3 we present our modelling technique. A technique for numerical solution is described in Sec. 4. Finally, we give some conclusions in Sec. 5.

2. Problem formulation

We consider a group of regional warehouses facing independent Poisson demand processes. Each warehouse is applying a continuous review one-for-one ordering system.

When an item is demanded at a warehouse the item is delivered from stock provided that the warehouse has stock on hand. At the same time a new unit is ordered from the supplier. However, if there is no stock on hand at the warehouse the item is instead ordered as an emergency transshipment from another (randomly chosen) warehouse in the group that has stock on hand. This warehouse orders a new unit from the supplier. If no warehouse in the group has stock on hand the unit is backordered from the warehouse where the demand occurred.

We introduce the following notation:

m = number of regional warehouses in the group

λ_j = demand rate at warehouse j

$\lambda_0 = \sum_{j=1}^{m} \lambda_j$ = demand rate at the group of warehouses

S_j = inventory position at warehouse j

$S_0 = \sum_{j=1}^{m} S_j$ = sum of inventory positions at warehouses in the group

L = average leadtime when ordering from a warehouse in the group, the same for all warehouses

π_k = steady-state probability that the inventory on hand at the group of warehouses is k

π_k^j = steady-state probability that the inventory on hand at warehouse j is k

The behaviour of the system will depend on the inventory positions. Our purpose is to determine, for given S_j, the proportions of the demand that will be met immediately, met by lateral emergency transshipments or backordered.

β_j = probability that a demand at warehouse j is met immediately by stock on hand

α_j = probability that a demand at warehouse j is met by an emergency transshipment

θ = probability that a demand at a warehouse in the group is backordered

Obviously $\beta_j + \alpha_j + \theta = 1$. We shall assume that $S_j > 0$ since otherwise $\beta_j = 0$.

When we have derived the proportions β_j, α_j and θ it is, in general, relatively straightforward to estimate the costs of a certain policy. Note that our simple rule for emergency transshipments is, in general, not optimal.

3. Modelling approach

The group

Let us first consider the inventory position at the whole group of regional warehouses. The group is applying a standard $(S_0 - 1, S_0)$-policy. Provided that successive leadtimes when ordering from the supplier are independent it is well known (Palm (1938)) that the distribution of outstanding orders is Poisson independent of the leadtime distribution. We can therefore replace a stochastic leadtime by its mean. It is then easy to derive the exact steady-state probabilities for the inventory on hand in the considered group of warehouses.

$$\pi_{S_0-k} = \exp(-\lambda_0 L) \frac{(\lambda_0 L)^k}{k!} \qquad k = 0, 1, 2, \ldots \tag{1}$$

Identical warehouses

We shall first consider the case when the warehouses in the group are identical, i.e. have the same average demand and apply the same inventory policy. We can then suppress subscript or superscript j (referring to warehouse j) in the notations. For simplicity we shall assume that the leadtime distribution is exponential.The service rate is then 1/L. Recall that without the lateral transshipments the leadtime distribution does not affect the solution at all. Our approximation is therefore reasonable. Let us also initially assume that we know the probabilities β, α and θ that we wish to determine.

We shall model the demand at a warehouse as a Poisson process but let the intensity depend on whether the warehouse has stock on hand or not. This approximation together with the assumption of an exponentially distributed leadtime will enable us to model the inventory situation at a warehouse as a simple queuing system. A warehouse has evidently stock on hand during the proportion β of the time. While the inventory is positive the warehouse is facing the normal demand λ plus a demand from other warehouses due to lateral transshipments. The average rate of this demand is $\alpha\lambda/\beta$ (since the overall average is $\alpha\lambda$). When the warehouse has no inventory on hand the only demand is the demand that has to be backordered. This demand is λ minus demands to other bases due to emergency transshipments with the rate $\alpha\lambda/(1 - \beta)$. The demand rates are consequently:

positive inventory on hand — demand rate $= g$

$$= \lambda(1 + \alpha/\beta) = \lambda(1 - \theta)/\beta \quad (2)$$

zero or negative inventory on hand — demand rate $= h$

$$= \lambda(1 - \alpha/(1 - \beta)) = \lambda\theta/(1 - \beta) \quad (3)$$

Using (2) and (3) we can formulate the standard equations for the steady-state probabilities π_k for the inventory on hand at a warehouse.

$$\pi_S g = \pi_{S-1}/L$$

$$\pi_{S-k}(g + k/L) = \pi_{S-k+1}\, g + \pi_{S-k-1}(k + 1)/L \qquad k = 1, 2, \ldots, S - 1$$

$$\pi_0(h + S/L) = \pi_1 g + \pi_{-1}(S + 1)/L \qquad (4)$$

$$\pi_{S-k}(h + k/L) = \pi_{S-k+1}\, h + \pi_{S-k-1}(k + 1)/L \qquad k = S + 1, S + 2, \ldots$$

By solving the system (4) we obtain

$$\pi_{S-k} = \pi_0 \frac{S!}{k!\,g^{S-k}L^{S-k}} \qquad k = 0, 1, \ldots, S-1 \qquad (5)$$

$$\pi_{S-k} = \pi_0 \frac{S!\,h^{k-S}L^{k-S}}{k!} \qquad k = S, S+1, \ldots \qquad (6)$$

where

$$1/\pi_0 = \sum_{k=0}^{S-1} \frac{S!}{k!\,g^{S-k}L^{S-k}} + \sum_{k=S}^{\infty} \frac{S!\,h^{k-S}L^{k-S}}{k!} \qquad (7)$$

Equations (5)-(7) describe the inventory situation at a warehouse, but the probabilities are still expressed in terms of the unknown parameters β and θ that we wish to determine. This means that we need two additional equations.

First it is obvious that

$$\beta = \sum_{k=1}^{S} \pi_k \qquad (8)$$

Consider now a situation where the system is just about to leave or has just reached a state where no warehouse has positive stock on hand. This means that one of the warehouses has no stock on hand and no backorders. The other m - 1 warehouses may have backorders. The average number of backorders in the whole group δ can therefore be approximated as

$$\delta = (m-1)\gamma \qquad (9)$$

where γ is the average number of backorders at a warehouse given that there is no stock on hand.

$$\gamma = \left(\sum_{k=0}^{\infty} k\pi_{-k}\right)/(1-\beta) \qquad (10)$$

Since δ is normally not integer we can use the closest lower and upper integers δ' and δ'' and a linear interpolation when estimating θ:

$$\theta = (\delta''-\delta)\sum_{k=\delta'}^{\infty} \pi_{-k} + (\delta-\delta')\sum_{k=\delta''}^{\infty} \pi_{-k} \qquad (11)$$

The unknown parameters in the solution (5) - (7) can now be determined from (8) - (11).

Nonidentical warehouses

If the warehouses are nonidentical it is still possible to use essentially the same technique when modelling the demand at a warehouse. The proportions β_j and α_j will, of course, now vary between the warehouses due to different demand rates and inventory policies, while θ is the same for the whole group. We shall illustrate the extension to nonidentical warehouses by considering m = 2 warehouses. (A case with m = 3 is considered in Axsäter (1990)). If we consider warehouse 1 we can replace (2) and (3) by

positive inventory on hand $\quad$ demand rate $= g_1 = \lambda_1 + \lambda_2\alpha_2/\beta_1$ $\quad$ (12)

zero or negative inventory on hand $\quad$ demand rate $= h_1 = \lambda_1 \theta/(1 - \beta_1)$ $\quad$ (13)

Given the demand rates we can treat each warehouse separately and apply (5) - (8). Since γ_j is different for the two warehouses we must change the determination of δ. Using that the average flow to and from the zero stock state at warehouse 1 is $\pi_0^1 (h_1 + S_1/L)$ and the probability of no inventory on hand at warehouse 2 is $1 - \beta_2$ we suggest the following approximation

$$\delta = (v_1 \gamma_2 + v_2 \gamma_1)/(v_1 + v_2)$$

$$v_1 = \pi_0^1 (h_1 + S_1/L)(1 - \beta_2) \qquad (14)$$

$$v_2 = \pi_0^2 (h_2 + S_2/L)(1 - \beta_1)$$

The determination of θ is still according to (11).

4. Numerical solution

In order to determine β_j, α_j and θ we need a numerical procedure for solving the system of equations that define the solution. Let us first consider the case with identical warehouses. The following iterative procedure will converge rapidly to a solution for normal data. Start with initial values of β and θ in (2) and (3). We can then obtain a new β from (5) - (8) and a new θ from (9) - (11). The new values of β and θ are then again inserted in (2) and (3) and the process is repeated until convergence.

We can apply essentially the same numerical technique in case of m = 2 nonidentical warehouses. This means that we use (12) and (13) instead of (2) and (3). Then we have to solve (5) - (8) and (10) for each warehouse. When determining δ we use (14), while θ is still obtained from (11).

For a more detailed discussion of the numerical solution and numerical results we refer to Axsäter (1990)

5. Conclusions

We have considered a technique for modelling lateral transshipments in inventory systems. This technique was originally suggested in connection with a multi-echelon inventory system and repairable items (Axsäter (1990)). In this paper we have shown that the same approach can be used when modelling consumable items in a single-echelon inventory system with several parallel warehouses.

References

AXSÄTER, S., "Modelling Emergency Lateral Transshipments in Inventory Systems", Management Sci. (to appear), 1990.

BOWMAN, J.P., "A Multi-Echelon Inventory Stocking Algorithm for Repairable Items with Lateral Resupply", Working Paper, 1986.

COHEN, M.A., P.R. KLEINDORFER and H.L. LEE, "Optimal Stocking Policies for Low Usage Items in Multi-Echelon Inventory Systems", Naval Res. Logist. Quart., 33 (1986), 17-38.

DADA, M., "A Two-Echelon Inventory System with Priority Shipments", Working Paper, 1985.

DAS, C., "Supply and Redistribution Rules for Two-Location Inventory Systems: One Period Analysis", Management Sci., 21 (1975), 765-776.

GROSS, D., "Centralized Inventory Control in Multilocation Supply Systems", in Multistage Inventory Models and Techniques, H.E. Scharf, D.M. Gilford and M.W. Shelly (Eds.), Stanford University Press, Stanford, CA, 1963, 47-84.

HOADLEY, B. and D.P. HEYMAN, "A Two-Echelon Inventory Model with Purchases, Dispositions, Shipments, Returns and Transshipments", Naval Res. Logist. Quart., 24 (1977), 1-19.

KARMARKAR, U.S. and N.R. PATEL, "The One-Period, N-Location Distribution Problem", Naval Res. Logist. Quart., 24 (1977), 559-575.

LEE, H.L., "A Multi-Echelon Inventory Model for Repairable Items with Emergency Lateral Transshipments", Management Sci., 33 (1987), 1302-1316.

PALM, C., "Analysis of the Erlang Traffic Formula for Busy Signal Arrangements", Ericsson Technics, 5 (1938), 39-58.

SLAY, F.M., "Lateral Resupply in a Multi-Echelon Inventory System", Working Note, Logistic Management Institute, 1986.

A Flexible Decision Support Framework for Production Scheduling

AUKE WOERLEE

1 Introduction

Momentarily decision support systems (DSSs) can roughly be categorized in two groups. DSSs in the first group are designed for a single decision situation and are mostly hard to adapt for other decision situations. DSSs in the second group can be considered as DSS generators. A broad range of decision situations can be defined by these systems but they are difficult to manage. Often these DSS generators require extensive programming experience to be able to construct a DSS for a particular decision situation. Furthermore, variations in the decision situation usually imply that a completely new DSS has to be generated. Neither of the two groups possess adequate flexibility. They either cannot be extended or are hard to adapt for a particular decision situation.

A dynamic environment requires a flexible modeling and scheduling mechanism. Not only should a DSS be able to suggest solutions for a particular decision situation, it should also be adaptable during the decision process. Consequently, the solution methods maybe need adjustments as well. A DSS for production scheduling should not only be capable to manage changes in the problem data but also modifications in the underlying model.

In this chapter we will propose an approach, powerful enough to handle a broad variety of scheduling environments which are partly determined by fuzzy restrictions and human capriciousness. A proper implementation of this shell can cope with almost any practical scheduling problem. It will increase the possibilities to adapt models but moreover it will improve the possibilities to switch between modeling and solving. However, the most important feature is that the user's expertise can be incorporated directly (interactively) in the modeling and solving process. Therefore, a shell for production scheduling can support the decision process considerably.

It is not our intention to define a meta model that is capable to solve any (mathematical) problem. The purpose of the shell is to increase the chance that the DSS will be accepted because only an opportunity to model and solve wide range of scheduling problems does not guarantee that the DSS will be used by decision makers. If it is too complicated to specify models or to generate solutions the user will give up sooner or later.

In the first section the planning environment is defined. It is explained what kind of problems is considered and how scheduling problems have to be specified. In the section the communication between the DSS and the outer world (for example the user) is considered whereas the remainder of this chapter is reserved for the components within the DSS.

2 Planning Environment

Scheduling problems can occur in many variations. Therefore, it is difficult to define a DSS shell for production scheduling that captures all aspects of practical scheduling problems.

An essential question in designing a DSS shell is which part should be predefined by the system, which part should be premodeled by an analyst and which part should be left to the user. If the flexibility of a DSS is a crucial aspect, the only thing that should be predefined in the system is the fact that it is meant to support production scheduling decisions. So, there is an amount of resources that can be utilized to process the customer orders. Either the analyst or the user has to specify the characteristics of the resources and the orders, how the resources are used, and which criteria have to be used by constructing schedules.

Should the characteristics of scheduling decision situation be embedded in the system or should they be specified in a kind of modeling language? Basically, only the modeling language differs. In case the characteristics are embedded in the DSS, a third or fourth programming language is used. If not, a modeling language is used which may be considered as a fourth generation language as well. The main advantage of a modeling language is its potential to change the model within the DSS. On the other hand, the model is to be interpreted by the DSS which makes it slower then a DSS with an embedded model. Modeling languages are designed for a broader class of mathematical problems than just scheduling problems. Therefore, it can be hard to specify a particular scheduling restriction since the system has to understand all types of restrictions of mathematical problems.

We are mainly interested in a shell for scheduling problems and not particularly in developing a modeling language. Therefore, we propose a partly embedded approach; the structure of scheduling models is incorporated in the system but the model and its characteristics are not. With respect to scheduling problems, this approach offers enough flexibility.

In this section we will present a description of such a scheduling environment. We will introduce some mathematical formulations which will be used later on to create, manipulate, and evaluate schedules. Contrary to the framework described by Hackman and Leachman [6] it is not our intention to give a complete scheduling description which can be reduced to simpler ones but we will give a basic description which can be extended to more complex ones. The advantage of this approach is that simpler problems do not contain redundant decision variables and constraints. Furthermore the approach is more flexible because problem definitions can be extended at any time.

A production scheduling decision is defined as how to process a set of activities $\mathcal{A}$ given a set of resources $\mathcal{R}$ on a time axis τ. Each decision problem can be identified by specific characteristics (entities). The entities of a production scheduling decision are activities, resources (machines, equipment, buffers, labor, material, capital, etc.) and time. An entity is characterized by a number of attributes. For example the activity entity can hold a processing time and a processing interval, and the resource entity can hold a capacity and a cost per time unit.

An entity table can be composed if all attributes have been defined. The rows in the table correspond to an element (an activity, a resource) and the columns correspond to the attributes. The individual cells of the table can both hold simple constants (the capacity of a resource) or complex functions (processing time of an activity which depends on the resources used). A new row is appended to the table whenever a new element is specified. A new column is added whenever a new attribute is specified. Since rows and columns can be added at any time, the DSS should be able to maintain dynamic entity tables. Clearly, this dynamic approach is extremely flexible since the problem specification can be modified at any time. Table 1 shows an sample entity table of resources.

Here, the cost attributes of the resources 'machine x' and 'machine y' are assumed to be functions of the set of activities $\mathcal{A}$. Naturally, the costs of the resources can also be dependent on the set of resources $\mathcal{R}$. A similar remark can be made about the capacities of the resources which are assumed to be constants in this example. The purpose of this example is to illustrate the flexibility of an entity table.

The decision variables of a production scheduling problem can be divided in two types. One type determines the time at which the activities are processed whereas the other type determines how and when the capacity of the resources is used. It is clear that the two types are closely interrelated;

	Resource attributes			
	Name	Capacity	Cost	
Resource elements	machine x	c_x	$f_x(\mathcal{A})$	$\cdots$
	machine y	c_y	$f_y(\mathcal{A})$	$\cdots$
	tool a	$f_a(\mathcal{R})$	$\cdot$	$\cdots$
	$\vdots$	$\vdots$	$\vdots$	

Table 1: Resource entity table

if activities are processed in certain intervals then some capacity of the resources is consumed in the same interval.

The decision variables are subject to constraints. Constraints are defined on single elements (activity A may not start before its release date) or on groups of elements (activity A_3 may not be processed before activities A_1 and A_2 have been completed). Obviously, constraints for a particular element can affect other elements. If a constraints says that no more than q units of resource R may be used at the same time this automatically implies that all activities en masse may not use more than q units of resource R at a time.

The previous definition of scheduling problems only determines the scheduling problem class. Specific problem characteristics are not included at this time. Below we will give some examples of regularly occurring attributes and derived entities. If an attribute is not defined for a particular element, the corresponding cell in the entity table will simply be left open.

First of all we can define the order entity $\mathcal{J}$ which is derived from the activities. An order J is a set of activities $\mathcal{A}(J)$ with a specific processing order. We assume that each order J has a release date $r(J)$, a due date $d(J)$ and a deadline $b(J)$. We also assume that an order J has a size $u(J) \in \mathbf{R}^+$ which denotes how much of order J has to be processed.

Now the set of activities can be given by $\mathcal{A} = \bigcup_{J\in\mathcal{J}} \mathcal{A}(J)$. An activity is expected to hold a processing time per unit $v(A)$. Activities may be split in tasks. If $\mathcal{T}(A)$ denotes the set of all possible combinations in which activity A can be split, then the function $H : A \rightarrow \mathcal{T}(A)$, $H(A) = \mathcal{T}_A$ denotes how activity A is split into a set of tasks $\mathcal{T}_A \in \mathcal{T}(A)$ and $\mathcal{T} = \bigcup_{A\in\mathcal{A}} \mathcal{T}_A$ denotes the set of all tasks. Each task can be performed by a combination of the resources. Let the $\mathcal{F}(T)$ denote the class of sets which can be used to process task $T \in \mathcal{T}$. Note that if resources can be substituted continuously, the class $\mathcal{F}(T)$ is infinite.

A resource $R \in \mathcal{R}$ is expected to have a capacity $c(R)$ (in each period there are $c(R)$ units of resource R available) and a quota $q(R)$ (there are $q(R)$ units of resource R available in the entire planning period).

Typical constraints are precedence relations between the activities of an order, capacity restrictions on the resources, completion of orders before their due date, etc. The description of orders, activities and resources will serve as a basis for any scheduling problem. The user can add and drop attributes and constraints for each of the entities.

Now we will give the definition of a schedule. Let $\mathcal{S}$ be the set of all partial schedules. Then a schedule $S \in \mathcal{S}$ is determined by a pair of functions (F, G) on $\mathcal{T}$. A task $T \in \mathcal{T}$ in schedule S is processed by a combination of resources $F(T) \in \mathcal{F}(T)$ and starts at time $S(T)$. The completion time is given by $C(T) = S(T) + p(T, F(T))$ where $p(T, F(T))$ is the time needed to process task T using the combination $F(T)$ of resources.

How does a user actually make decisions during a scheduling session? The vector of p objective functions $\vec{Z} : \mathcal{S} \rightarrow \mathbb{R}^p$ provides an indication of the quality of a schedule. In order to determine the best schedule, the user should be able to arrange the function values of the p criteria in some kind of order. A scheduling decision will only be made if the objective function values $\vec{Z}(S) \geq \alpha$ ($\alpha \in \mathbb{R}^p$); each criterion should be at least equal to a certain aspiration level. An insertion or a deletion of a task changes the values of objective criteria, some will improve others will get worse. It is up to the user or the algorithm to accept or reject a particular alteration.

3 Communication

How can a mathematical model be specified in a shell? Before we answer this question we have to clarify the difference between user and analyst. We will assume that an analyst is able and a user is not necessarily able to use a formal language to construct models. A shell for production scheduling should therefore include separate communication methods for both analyst and user. Both communication methods express the information in high level descriptions which cannot be managed by the system directly. Hence, besides the communication methods for analyst and user we also consider a communication method on system level.

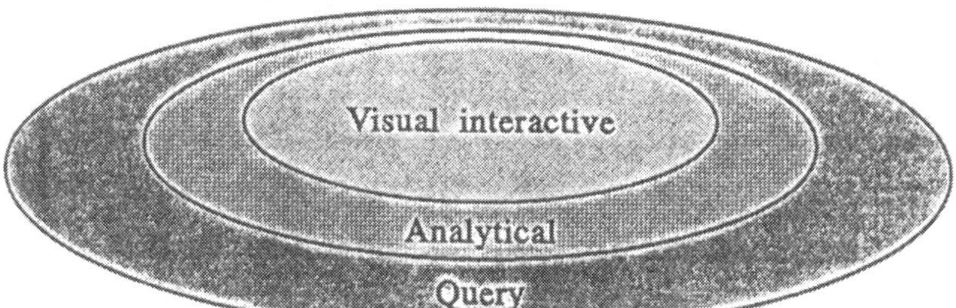

Figure 1: Communication methods

A user does not need the whole potential of the analytical and query language. For that reason the possibilities of the visual interactive language are only a subset of the capabilities of these languages. Similarly, an analyst does not need the full potential of the query language. Accordingly, the functionality of the analytical language is in turn a subset of those of the query language. In figure 1 the different communication levels in the shell are depicted.

3.1 Query language

The most elementary way to communicate with a decision support system is by specifying model structure and problem instances through data properly arranged in tables. Information entered through a formal language is necessarily converted into tables since this is the only way in which it can be analyzed. Though rudimentary, its existence in a shell is essential since it can easily exchange data with the outside environment. For example, a monitoring system registers the start and completion times of tasks and sends this information to the shell by means of the system level communication method. Now suppose a job J has to be scheduled in the interval $[t_1, t_2]$ on machine M. Before job J can be scheduled, a check should be performed to certify that machine M is not already allocated in the interval $[t_1, t_2]$. The query

```
SELECT *
FROM schedule
WHERE start_time < t2
AND end_time > t1
AND machine = 'M';
```

investigates all records in the table 'schedule' with columns 'start_time', 'end_time' and 'machine' and returns the group of records which (partly) overlap the interval $[t_1, t_2]$. If no records are returned, job J can be scheduled in the indicated interval.

3.2 Analytical language

Primarily the analyst is responsible to define the underlying model of the scheduling situation. Of course he can develop the system by use of the tabular method· but it is time consuming and information expressed in tables is not clear. As a result the model is difficult to adapt and, even worse, to maintain. Hence, there is a need for a formal language that offers the analyst an easy way to develop models and at the same time increases the maintainability of the DSS.

The main question is how specific problems can be captured into a general framework. Modeling languages are a possibility to describe problems easier and to reduce the amount of errors in the model. The design of one the first modeling languages such as GAMS [1,2] was initiated to improve and simplify the model construction. Problems are represented in a compact and well-organized form, changes in the model can be made easily, and the model is specified independently from solution algorithms. The ideas of languages like GAMS originate from both relational data base theory and mathematical programming. The data base theory is used for data organization and the description of the problems is inspired by mathematical programming.

The structured modeling approach of Geoffrion [4] is more than a language. It not only improves the productivity of modeling but its most important feature is that it can be used for a broad class of optimization problems contrary to for example GAMS which aims at a particular decision situation.

Modeling languages improve the flexibility and surveyability of the modeling process but they can mainly deal with deterministic data. Since, an objective of a shell should be that users have to participate actively in decision making, it should also be possible to deal with user knowledge. Logic programming languages are especially developed to manage knowledge (see for example Lloyd [8], Gallier [3]). The logic programming approach can further improve the flexibility since it is easy to integrate dynamic knowledge about the decision situation quickly.

The last modeling approach we will mention here is the functional language approach (see for example Glaser et al. [5]). Functional languages represent models according to the functional relations between the individual elements in the model.

Now we consider the example of job J again which has to be scheduled in the interval $[t_1, t_2]$. In a formal language an expression like

$$[S_j, C_j] \cap [t_1, t_2] = \emptyset, \quad \forall j \tag{1}$$

indicates that all scheduled jobs with starting time S_j and completion time C_j are intersected with the interval $[t_1, t_2]$. If equation 1 evaluates to 'true' for all scheduled jobs then job J can be scheduled in the interval.

3.3 Visual interactive language

All languages suffer from a major drawback. They may be useful for analysts to develop models but they can hardly be understand by a regular user. However, should a regular user have an opportunity to enter information into the system? Users have specific knowledge about the decision situation. They have experience and intuition which can be employed to control the automatic components of the DSS. Besides background knowledge, there is also a continuous flow of actual information about the decision situation; machines can break down, people can become sick, etc. It has to be the user who is going to enter all this information into the DSS. An easier method should therefore be made available for the users of the DSS; a method that is easy to use but at the same time is powerful enough to specify a broad range of modeling elements.

Visual interactive modeling is such an approach. The user has an opportunity to model and solve problems with visual interactive tools. What does this mean for the previous example? If a job J has to be scheduled in interval $[t_1, t_2]$, the only thing that should be done is to check in the Gantt chart on the screen whether or not machine M is allocated in the interval.

A user oriented modeling language should be easy to use but flexible enough to build complicated models. Unfortunately these objectives mostly contradict. The more modeling methods are presented to the user the more it will confuse him. It becomes unclear which method should be used for a particular task.

One of the problems with ordinary modeling methods is that the specification and the justification of the model are completely separated. The user formulates the model of a decision situation and this model is validated by a model interpreter. The interpreter reports specification errors which should be corrected by the user. Even if the two components are combined, there is not a true interaction between user and system. The purpose of interactive modeling is to make the modeling and problem solving process simpler and more effective. Additionally, interaction should also have a positive effect on the integration of modeling and problem solving.

How should the interaction between user and system look like? In many circumstances the visualization of a model increases the comprehension of the user and makes it easier to make justifications. For example, suppose that figure 2 is an exact replication of a production system.

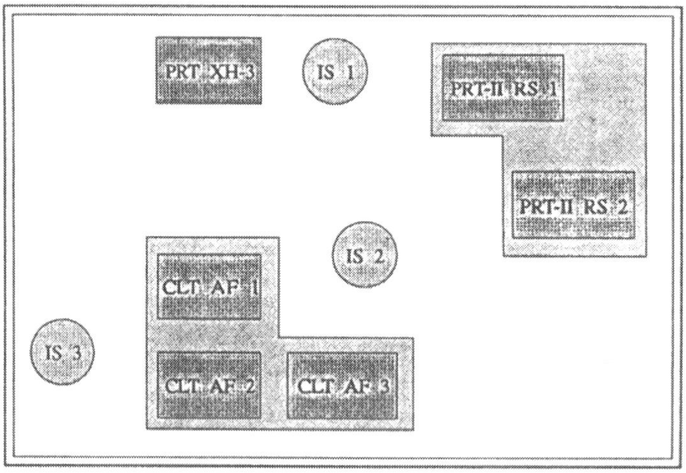

Figure 2: Map of production system

These representations have at least two advantages. First of all the user can easily identify the individual components of the production system on the screen. Secondly, modeling functions can be performed simply by pointing to the individual components on the screen. For example, a precedence constraint between an activity on 'PRT XH-3' and an activity on 'IS 1' can be accomplished by respectively pointing to the two resources. In this way the user is able to specify the route of a job through the system. Figure 2 recognizes two types of resources: the resources denoted by an ellipse and resources denoted by a rectangle. The representation can be further improved in case the resources are denoted by icons, each clearly characterizing the properties of the resource. The closer the representation on the screen corresponds to the real world, the more the user feels confident of using the modeling environment.

A similar statement can be made for problem solving. Visual techniques presented in every step of the solution process will increase the insight the planner has about the solution process. Interaction will allow him to react immediately if the results of a particular action are not satisfactory. For example, suppose that the Gantt chart in figure 3 represents a schedule for the production system

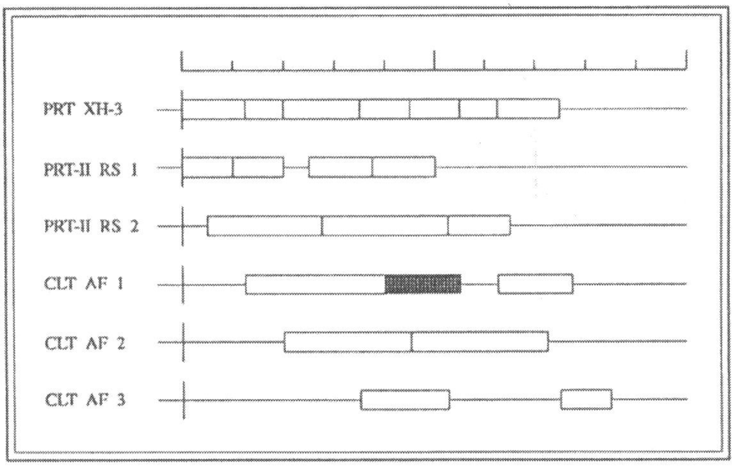

Figure 3: Gantt chart of schedule

represented in figure 2. The marked job in the schedule depicts an infeasibility which occurred after the last scheduling operation. The visualization of the infeasibility gives the user a possibility to respond directly.

Hence, visualization improves the user's perception of both modeling and problem solving. However, visual interactive techniques also reduce the number of mistakes. Consider the production system in figure 2 again. Obviously, a precedence constraint can be specified less erroneously by pointing to two resources than by writing that the activity on resource 'IS 1' can only be started if the activity on resource 'PRT XH-3' is completed in some formal language.

Visual methods also offer substantial support to the user in the problem solving process. Consider the Gantt chart in figure 3. We already observed an infeasibility in the schedule. Infeasibilities occur when a user utilizes the system interactively. We allow a user to make small mistakes; it is permitted to process a task at any time in a schedule as long as the restrictions upon the resources are not violated. Therefore it can happen that a task is scheduled at an illegal position. Since the system directly repports infeasibilities, the user has a possibility to correct these mistakes.

Visual methods can reduce the number of mistakes. However, we can much more exploit the interaction. For instance, only meaningful modeling and problem solving commmands are presented to the user. Interaction can also prevent illegal user actions. If a user tries to specify a precedence constraint which would generate a cycle in the job's route through the system, the system will recognize the cycle and disallow the addition of the precedence constraint. If the user attempts to schedule a task on a resource which cannot be used to process the task, the system will observe the mistake and disallow the inclusion of the task into the schedule.

We will conclude this exposition with perhaps the most important feature of visual interactive methods: the user attains control over the algorithms in the model base. The user can exploit the algorithms to compute any number of schedules. However, it will be the user who has to indicate if a proposed schedule is acceptable. At all times the user is able to adapt the schedule. He can employ his knowledge and intuition to make subtle alterations. The following example makes this more clear.

Example 1 *Suppose that a task T can be carried out by a number of resources but that the user believes that the task should be carried out by a particular resource R at time t. Now the visual interactive language shows the exact location and the only thing that remains for the user is pointing at position R, t.*

The last example again demonstrates that modeling and problem solving are intertwined. The user's instruction that a task should be performed on a particular resource is of course nothing else than a subtle adaptation of the model. The consistency of a generated schedule for a specific model is even a better illustration. Suppose that several tasks in a schedule utilize a resource and this resource is removed from the model. This implicates that the schedule becomes invalid.

3.4 Flexibility

What does flexibility mean at each of the communication levels? Flexibility is the ability of the system to adapt to altering decision situations. On each of the three communication levels flexibility has a specific meaning.

At system level the decision support system has to communicate with monitoring, sales, inventory and personnel systems. Each of these external systems exchange information with the planning system which can have direct influence on the planning process. The following two examples show what external information can mean for the operational planning.

Example 2 *Suppose that a monitoring system keeps track of all the jobs in the production system. It reports when jobs are started and when they are completed. Now suppose that the monitoring system reports that a job J is completed at a time later than the completion time in the planned* *schedule. Assume that job J makes use of a machine M which can only process a single job at a time. This implicates that jobs performed after job J which use machine M cannot be started before job J is actually finished. Consecutively, other jobs can in turn be influenced by one of the jobs on machine M.*

It is not immediately clear what should be done in these circumstances. A possible approach is to delay the jobs on machine M and preserve the schedule on other machines. However, alternatively the schedule could be reconstructed starting from the completion of job J on machine M.

Example 3 *Suppose that a personnel system reports the absence of an employee who is normally controlling a machine M. Therefore, machine M cannot be utilized by any of the jobs without the assistance of another employee.*

Again one can prefer to use a strategy which preserves the schedule but this is only possible if there is some employee who can control the machine M. Optionally, one can choose to recalculate the schedule given the unavailability of a resource.

In case of the other external systems, it is easy to come up with similar examples. For instance, the inventory system can indicate that a particular material is out of stock or the sales system can report a rush order. The information of each of these external systems can have direct consequences on the status of the planning process. Basically there are two approaches to treat this information. The first one is to preserve the schedule as much as possible whereas the other approach reconstructs the schedule completely.

It is of the utmost importance that all relevant information is available in the planning system at any time. Therefore, the communication level on system level should have a high flexibility to deal with external information. It should not only process the information quickly but it should also notice the relevancy of the information for the planning process.

An analyst should be offered a modeling environment in which he can model any mathematical property. The analytical language should not only be able to model regular model properties such as that a machine can only process a single job at a time, but also irregular properties. The following example shows the kind of relations that can occur.

Example 4 *Consider the jobs J_1 and J_2, and a buffer B. On the one hand, if job J_1 is already in buffer B it is not allowed to add the job J_2. On the other hand, if job J_2 enters the buffer before job J_1 it is allowed to add job J_1.*

The analytical language should be able to model this and other awkward relations. Therefore, it should possess the flexibility to model any relevant characteristic of the planning process but simultaneously it should be defined in a comprehensible and consistent manner and should not contain any superfluity.

The tripartition of the communication methods within the decision support system is mainly based on the activities which have to be performed. Low level information transactions are considered to be dealt with on system level, modeling is expected to be dealt with on analyst level whereas the actual planning is assumed to be carried out by the user. However, on user level the system should offer some possibilities to adjust the model. If the actual planning process is considered more closely, one will notice that the user is really altering the underlying model by using his intuition and experience. The following example will illustrate how the user's experience is incorporated in the planning process.

Example 5 *Suppose that the model allows the processing of job J on machine M_1 or on machine M_2. The user knows from past experience that machine M_1 sometimes gives problems while processing job J. Therefore, the user concludes that job J should be processed on machine M_2. In doing this however, he is in fact adding a new restriction to the model.*

Note that by specifying a particular machine for a job, the user acts as an analyst. The second example shows that a user also plays an important role in incorporating external irregularities in the planning process.

Example 6 *Suppose that a power failure shuts down a group of machines. Since there is no electricity, the monitoring system cannot report the unavailability of the machines. The user can however, respond immediately by blocking the machines in question.*

Both examples indicate that the human involvement in the planning process is extremely important. Therefore, the user should be offered a communication method that is flexible, that is, a method which offers the user the possibility to anticipate adequately to arbitrary events in the planning process.

4 Model Management

The intention of a shell for production scheduling is to improve the flexibility of the decision situation. On the one hand the shell allows a user to model a wide range of scheduling problems. On the other hand a shell should be able to solve all models since a DSS for production scheduling has a supporting function. A consequence is that we cannot suffice with a built-in algorithm which is capable to solve all problems simply because such an algorithm does not exist. How can we possibly determine which algorithm should be applied to solve a particular problem?

Each manipulation will affect some jobs and resources. Therefore, the constraints and functional relations of these jobs and resources have to be reevaluated. As a result other jobs and resources can be affected. For example, consider a partial schedule in which the user removes an activity. A constraint expresses that a succeeding activity can only start when this activity is completed. This implicates that the restriction is violated and that the succeeding activity is affected and should be reevaluated as well.

What is an algorithm? The purpose of an algorithm is to solve stringently defined computational problems under which scheduling problems. Before an algorithm is to be applied, it has to fulfill some properties. Does it obtain an optimal solution? What is its computational time? How well did it behave in empirical studies? These (and other) properties determine which algorithm can best be chosen for a particular scheduling problem.

What is a scheduling problem? Each scheduling problem can be characterized by a particular utilization of the resources and precedence relations between activities, its class. Furthermore, the choice of an algorithm is influenced by the size of the problem parameters. Though structurally identical, a problem in which the processing requirements are about the same for all activities is not identical to a problem in which the processing requirements differ significantly. Consequently, it is reasonable to assume that algorithms (in case available) will be chosen differently. The number of structural differing scheduling problems is enormous. Since the number of scheduling classes is large, it is likely that there is not a custom-made algorithm available for all problems in the model base. How can an algorithm be obtained if there is not any available? The simplest solution is to look for an algorithm which is intended to solve a closely resembling problem. However, there it does not guarantee that this algorithm will perform equally well.

We will use templates to represent algorithms (for example see Sneller [9]). Each template comprises two parts. The first part describes the characteristics a problem should have to be solvable by the algorithm. Evidently, the number of characteristics in the template tends to increase if the scheduling problems become more complex. The second part contains some specific attributes of the algorithm such as its objective function, the quality of the solution, empirical results, etc. The following example illustrates how templates can be used to investigate if a particular problem can be solved by a proposed algorithm.

Consider a single machine scheduling problem. The second column of table 2 shows a template of an algorithm for this problem, the branch and bound method of Carlier. The structure of problem P (third column) fits the template of the algorithm exactly. Since the algorithm generates optimal solutions, problem P will be solved optimally. The structure of problem Q (fourth column) does not exactly match the template of the algorithm since there may occur off-line times on the machines. Hence there is no certainty that the algorithm will generate an optimal solution for problem Q. Even worse, there is no certainty that the algorithm will construct a feasible solution. Generally, algorithm templates that do not exactly meet the structure of a problem do not guarantee feasible solutions.

Which algorithm is selected to solve a scheduling problem? Before an algorithm can be chosen,

Characteristic	Algorithm	Problem P	Problem Q
Release dates	yes	yes	yes
Due dates	yes	yes	yes
Job priorities	no	no	no
Off-line times	no	no	yes

Description	Branch and bound method
Objective function	Minimize $L_{\max}$
Solution qualification	Optimal
Running time	$O(2^n)$

Table 2: Algorithm template of a single machine problem

the structure of the problem has to be determined and an objective function has to be chosen. We will assume that such a function exists although it is doubtful if it fully meets the requirements of the user. Once the structure and the objective function are known, the problem template can be compared with the algorithm templates in the model base. In principle all algorithms can be applied of which the first part of the template completely correspond to the problem template. The second part of the algorithm template contains information about the objective function. This function may conflict with the chosen objective function. In this case either a selection rule or the user have to decide whether or not to use the algorithm.

However, even if there is an exact match between algorithm and problem in the top part of the table, it is doubtful if the objective function in the bottom part is appropriate. Mostly, a single objective criterion from the vector $\vec{Z}$ of all objective functions is chosen. If the objective function is the kth component of $\vec{Z}$ than surely for the constructed schedule $S \in \mathcal{S}$, $Z_k(S) \geq \alpha_k$ but it is likely that for some objective criterion $l \neq k$, $Z_l(S) < \alpha_l$ for the reason that the criterion was not considered during the generation of the schedule S.

What if there is not an algorithm in the model base which exactly suites the structure of a problem? For example, the template of problem Q in table 2 does not match the algorithm template. However, the problem and algorithm template are almost identical. An inference mechanism should conclude if an algorithm is still applicable for a particular problem even though it does not match the problem structure. A more or less identical decision situation is the selection of the best algorithm among two algorithms with identical template. The problem parameters should also play an important role in this case. The best approach to decide if an algorithm suits to solve a particular problem instance is a knowledge base system but this is beyond the scope of this thesis. However, some simple rules can be extremely fruitful. For instance, it is reasonable to assume that some of the characteristics are more important than the other ones. Accordingly, algorithms will be considered if they meet the important characteristics but fail to meet one or more of the other ones. More formally, suppose that a scheduling problem is represented by the characteristics C_i

($i = 1 \cdots p$) and assume that they are sorted in order of importance. Then an algorithm can be applied for a particular problem if and only if the characteristics $C_1 \cdots C_k$ ($k \leq p$) of the algorithm correspond to those of the problem. An alternative approach is to quantify the characteristics and to arrange them in order of their importance. Then algorithms will be considered if the sum of the concordant characteristics exceeds a certain value. More formally, suppose that $V(C_i)$ is the quantification of the importance of characteristic C_i. Furthermore, if element w_i of the vector $\vec{w}$ be 1 if the characteristic C_i of the algorithm corresponds to the one of a problem and be 0 otherwise. Then the algorithm can be applied if and only if $\sum_{i=1}^{p} w_i V(C_i) \geq c$, where c is a user determined value.

We realize the simplicity of the rules above but we think that these rule more or less reflect the selection of an algorithm by a user if there is not an exact shaped algorithm at hand.

5 Dialog Management

Visual interactive techniques are an important contribution to the functional requirements of a DSS. In section 3 it was shown that visual interactive techniques can simplify the modeling and solution process. However, specific visual interactive techniques cannot be employed in all circumstances.

Individual users have different tastes and consequently will prefer different techniques. For example, a Gantt chart is often used to represent the machine utilization in a planning period. An alternative method is to represent the progression of the jobs in the planning period. In order to combine both methods Jones [7] proposed a 3-dimensional Gantt chart. The advantage is that all information is available at the same time. However, in many cases the perception of a user is not sufficient to understand everything.

The structure of scheduling problems can exclude certain techniques. What if a job can be processed by two machines simultaneously? The Gantt chart in the previous example cannot be used anymore since it is only able to show a single task of a job at a time. The machine utilization representation cannot be used if machines have a capacity which can be shared by two or more jobs at a time. If machines can be used by more than one job and tasks of jobs can be processed simultaneously than even the 3-dimensional Gantt chart cannot be used.

Another aspect of the dialog is how it should react to changes in the decision situation. The system can be configured to react immediately whenever a particular situation occurs or only to observe the change. For example, suppose that a user assigns a job to a machine which is not available for that job. The most suitable action of the system in this case is to reject the assignment and to report the mistake to the user. If the user assigns a job at a time where a precedence relation is violated then it is not directly clear how the system should react. It can simply reject the assignment or it can report the violation, either explicitly or not. Each user has a personal opinion about the system's reaction..

Decision situations are not exactly alike. The dialog method which is good in one situation is not necessarily good in another. A Gantt chart is an appropriate way to communicate with a user during a scheduling session but is almost unusable if several schedules have to be evaluated and compared. A table of the objective criteria and the various scenarios gives the user better understanding.

The above examples illustrate that there is a need for a dialog management mechanism that selects an appropriate dialog method for particular problems. The manager should be able to select dialogs which correspond to both problem structure and user preferences. Furthermore, the manager should be able to respond directly whenever the decision situation gives rise to an alternative dialog method.

6 Knowledge Management

The scheduling environment described in this chapter supports the decision maker with visual interactive techniques. Both model and dialog management can be improved if available knowledge of user and past experience can be incorporated in the decision process. We will confine ourself to mention possible application areas.

Solution algorithms have a good performance in one situation but behave badly in another. If the system is able to recognize these circumstance, this information could be used to select a solution procedure.

One of the objectives of a decision support system for production scheduling is to make the decision process less dependent of a human planner. If the activities of a user during a planning session can be captured and if the way of thinking can be recognized and stored in a knowledge base, this information can be utilized by other (less experienced) users.

Dialog methods offer a proper insight in some decision situations but nothing to speak of in others. A system with some knowledge about the quality of the cognitive properties of dialog methods can help to select appropriate dialog methods in any circumstance thereby improving the decision environment.

References

[1] J. Bisschop and A. Meeraus. On the development of a general algebraic modeling system in a strategic planning environment. In *Math. Programming Stud. 20*, pages 1–29. North-Holland, Amsterdam, 1982.

[2] A. Brooke, D. Kendrick, and A. Meeraus. *GAMS: a user's guide.* The World Bank, 1988.

[3] J.H. Gallier. *Logic for computer science: foundations of automatic theorem proving.* Harper & Row, 1986.

[4] A.M. Geoffrion. An introduction to structured modelling. *Management Science*, 33(5):547–588, 1987.

[5] H. Glaser, C. Hankin, and D. Till. *Principles of functional programming.* Prentice-Hall, 1984.

[6] S.T. Hackman and R.C. Leachman. A general framework for modeling production. *Management Science*, 35(4):478–495, 1989.

[7] C.V. Jones. The three-dimensional Gantt chart. *Operations Research*, 36:891–903, 1988.

[8] J.W. Lloyd. *Foundations of logic programming.* Springer-Verlag, second edition, 1987.

[9] A.C.W. Sneller. Classificatie van schedulingproblemen. Master's thesis, Erasmus University Rotterdam, Rotterdam, The Netherlands, 1989.

Supported by the Netherlands Organization for Scientific Research

The Leitstand - A New Tool for Decentral Production Control

AUGUST-WILHELM SCHEER and ALEXANDER HARS

I. NEW REQUIREMENTS FOR PRODUCTION CONTROL

Current applications for production planning and control (PPC) lack adequate support for shop-floor control functionalities. In an environment of changing markets and fierce competition this becomes increasingly disturbant as demands for short delivery times, low batch sizes in production down to a batch size of one, holding of promised delivery dates and flexibility in production increase significantly.

A. Deficits of traditional PPC systems

Traditional PPC systems offer only low support for master requirements planning which is oriented more towards a longer term and for the short term shop-floor control functions [SCHE 1986]. In contrast, a well-balanced PPC-system would have to give at least as much weight to the long-term planning issues, whose decisions are binding for the subsequent planning levels, and to short term planning and control as to medium term material requirements and capacity planning.

Traditional electronic data processing (EDP) systems for production planning and control base upon a uniform planning philosophy. All the functions - beginning with master planning, material requirements planning, capacity planning and ending with detailed scheduling and monitoring - traditionally are supported by one uniform system. In a complex production environment, objectives and planning philosophies, however can easily be different depending on the different production types. One production site might have the goal of minimising the inventories while another might aim to maximize the loading of certain high-cost machines. This can not be accomplished simultaneously by a central uniform PPC system.

A central approach to production control also means a uniform approach for all the production sites. This affects both the production control philosophy and the handling of objects as order and operation. Algorithms for capacity smoothing and short term scheduling however have proven not to be successful for complex, centralized applications. In addition, hardware and software lack the capabilities for efficient man-machine interaction required for making the most of the human's intelligence. Graphical man machine interfaces do not exist for host-based systems today.

Furthermore, in a production environment on the level of shop-floor control, the flow of information is so large and complex that a tradional host-oriented system is not capable of handling the huge amount of data in the short time spans necessary. For production control applications the traditional long- and medium term planning systems have to be equipped with real-time capabilities. This generally exceeds the potentials of the hard- and system software employed for traditional PPC systems. The only solution offered today is to utilize separate tools for short-termed production control which are closely linked to the medium and long term planning of a traditional PPC system. Such systems are already available. They transfer, however, the centralised, automation-oriented approach of traditional PPC systems to the shop floor. Their main intention is to control the production, not to support the short term planning.

Traditional PPC systems also present large problems when they have to blend with the heterogeneous data processing equipment typically encountered on the shop-floor. In contrast, integration with the hardware zoo which frequently exists in today's factories requires an open software conception. The production control software has to be designed to link with other information systems on the shop floor regardless of their operating system, database connection, network approach etc. This is not the case for PPC systems.

B. New conceptions for production organization

It is not, however, only the deficits of current PPC installations which lead to the specification of additional requirements for new conceptions within production control. There also is a trend to migrate to new types of production organisation.

While traditionally the work was organised along similar activities resulting in an efficient utilization of equipment capacity, today the conception of organising the production along the work flow is gaining in importance. Production then is organized around product groups. For selected parts, "factories within the factory are established" wich perform the complete production process for these parts. Current trends lead to the development of flexible manufacturing systems (FMS) and production islands. The organisationally most radical approach is presented by the production island conception.

According to the German AWF (Ausschuß für wirtschaftliche Fertigung = Committee for economical production) a production island has the task of producing final products as complete as possible starting from a given set of components. The required equipment is spatially and organisationally assigned to the production island. The activities of the group working within the production island bear the following characteristics: largely autonomous management of planning, decision and control functions within given limits and dispensing with the strong tayloristic principle of the division of labour in favour of functional re-integration. [AWF 1984, p. 5].

The organisational structure of production which centers upon the group technology principle is fundamentally different from that within a centralized job-shop environment. Figure 1 shows the structure of a decentral production control system. Much stronger are the similarities between a production island organisation and a decentral job-shop environment. A decentral job-shop environment requires basically the same functionalities and in particular the coordination between different job-shops equally requires a separate functionality.

The key functionalities of such a decentralised control system are:

The Coordination function is not proprietary to each decentral control area but coordinates the work between all control areas. This functionality comprises the link to the PPC system for the exchange of order and operation data. It will be discussed more in detail in section II.C.

Data exchange and management comprises the link to the coordination system for the exchange of order and operation data. It provides the adaption of the orders and operations transmitted from the MRP system via the coordination function to the format required for scheduling. In addition, this function allows the creation, update and deletion of orders within the decentral control area and operations to support the reaction to failures in production and for the management of rush orders.

Short term scheduling is the most striking function of a leitstand. Thereby resources are allocated to operations and orders observing capacity and availability constraints.

The scheduling is followed by the release. Thereby a part of the schedule - corresponding to a very short time horizon - is freezed for a certain time span and thus transferred from the planning to the realization state. The release function transfers planned actions to reality.

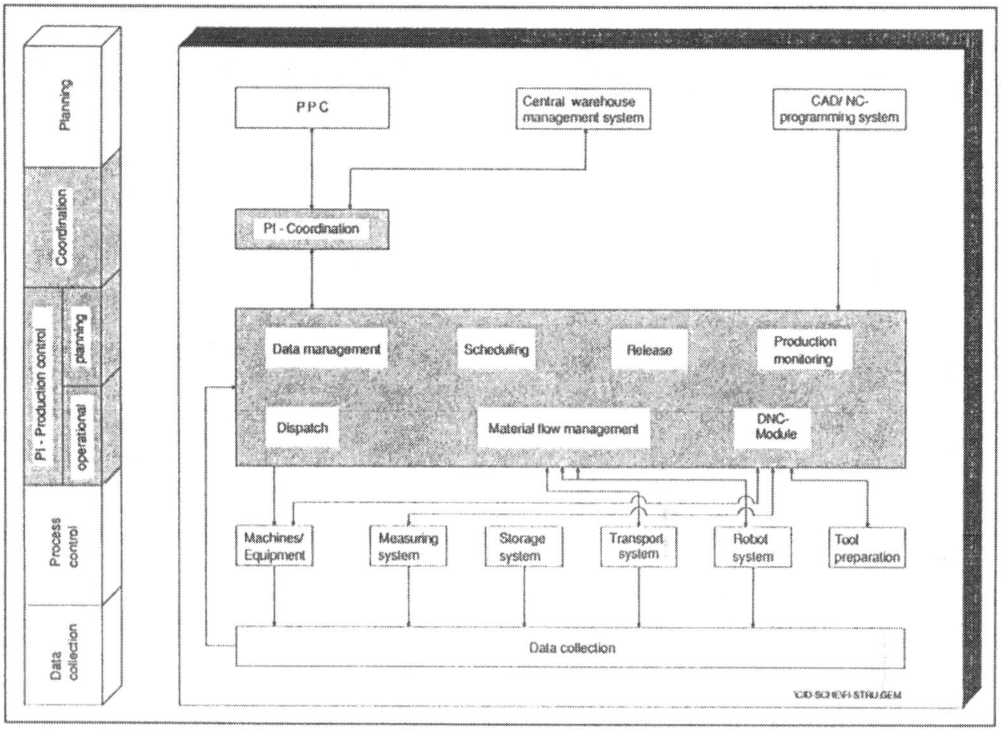

Figure 1: Structure of a decentral production control system (from [KeKe 1990])

The goal of dispatch is the distribution of orders on the basis of the released operations. The recipients of orders are either the workers or subordinated information systems e.g. dynamic numeric control systems (DNC) systems or material handling systems.

Production monitoring evaluates the deviations between planned and actual data. It allows fast reaction on failures and is split into order progress monitoring, failure identification and the evaluation of quality control data.

Material flow management is relevant in highly automated decentral environments e.g. flexible manufacturing systems (FMS) or flexible assembly systems (FAS). It creates, releases, monitors and coordinates material flow processes [KeKe 1990].

A DNC-module is the prerequisite for NC-programming within for example a production island. Compared to conventional central NC programming, this increases the autonomy and in many cases velocity and reliability of the NC programms designed.

II. CHARACTERISTICS OF A LEITSTAND

A solution to the requirements identified in section one are EDP systems which are called "Leitstand" or graphical control systems. The term "Leitstand" originates from the German systems, but has recently been introduced into the English scientific literature [ADEL 1989]. A Leitstand is an EDP-system which is characterized by its user friendly man machine interface showing a GANTT chart and bases on an interaction style which is oriented to support, not to automate the thougt processes of the human required for scheduling and shop-floor control applications. Leitstands are autonomous systems which operate on their own database. They are implemented on a workstation or on a personal computer. The term Leitstand refers to the classic control center within a manufacturing company whose most striking element has been the planning board.

Supply and demand for leitstands has dramatically increased in the last years. The number of leitstand suppliers approaches the twenties. According to a recent survey the number of leitstands installed in Germany industry could top 320 by the end of this year [Koeh 1990]. For a survey of leitstands see [HaSc 1990], [FrGr 1989].

In the following paragraphs, the most important characteristics of leitstands are discussed with a view to utilizing them within a decentral environment.

A. Interactive scheduling

The type of computer support for scheduling is dependent on the complexity of the scheduling problem. To decide, whether the computer can schedule by itself or if it merely is capable of supporting the human scheduler, the nature of the scheduling problem has to be investigated. The goal of scheduling is to provide the "best possible schedule". However, already the definition of the properties of the best schedule is a very significant problem.

Typically, an industrial enterprise discerns the different scheduling goals shown in figure 2. The goals in many cases compete with one another. The optimum utilisation of work stations for example will result in optimum machine load but increase inventories and throughput times. To be certain to reach customer satisfaction by meeting due-dates, an enterprise has take into account that inventories will consequentially rise. Reality is not that simple that each company could identify exactly one of these goals as being its primary goal. A company will always have a multiplicity of goals within production and the goals can not easily be related to each other in a realistic manner. This problem can not be solved by weighing the different goals or by defining a priority sequence. An additional problem is that the goals of a company are not stable. Furthermore the goals as shown in figure 2 are not yet measurable. Although the goal to minimise the througput time might be analysed by averaging the througput times of all work in process, the reduction of some orders (e.g rush orders) might have a higher priority than the reduction of througput time for other orders (standard orders). So simple averaging is not the answer even while evaluating only one goal. Another problem for automatic optimization is how to generally define certain decisions which the scheduler takes in non-standard cases. How should the cases where splitting is to be carried out be defined? Can the decision to increase the available capacity by overtime be automated? For decisions of this kind, the interaction with the scheduler is indespensable.

The complexity of problems encountered for scheduling is a very important reason why automatic scheduling is not the right approach. Even the best mathematic algorithms could not produce optimum schedules because it is not possible to define what exact properties a good schedule has to meet. This is only aggravated by the fact that until today mathematical algorithms that solve precisely defined model scheduling cases (e.g. five machines, five orders with five operations, maximise average load) are not available.

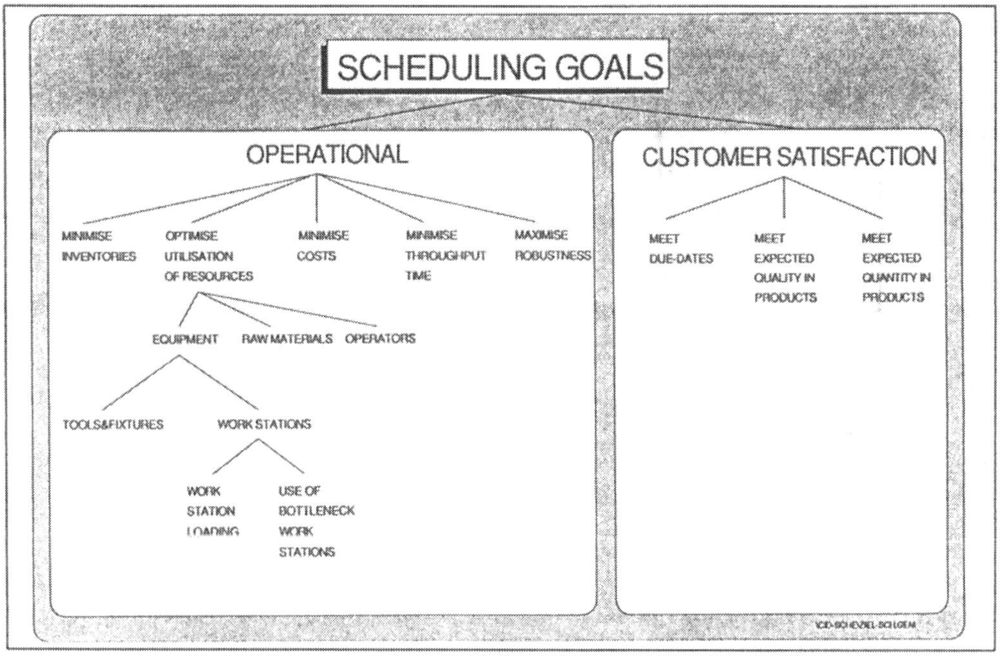

Fig. 2: Goals that need to be satisfied by the scheduling process (from [Farh 1990, p.93])

Current Leitstand algorithms for automatic scheduling are primitve heuristics. Heuristics are approximative procedures to solve a certain problem, which don't necessarily reach the optimum solution [Müll 1971]. The decision component of the current heuristics is based upon simple priority rules. The allocation of operations to machines is either based upon externally determined priorities, e.g. the importance of the order for the enterprise, or on priority values depending on characteristics of the orders and operations considered. Priority rules could be based for example upon the

- shortest operation time
- longest operation time
- lowest deviation from due date

The efficiency of these heuristics is low [WiLü 1989].

Therefore leitstands focus on the principle of interactive scheduling. The leitstand is utilized as electronic version of a scheduling board where the object-oriented logic of the board can be imitated by positioning the operations with the mouse on screen. Using a leitstand it is easier, to execute procedures as splitting or combining of operations and to test changes to a schedule than with the traditional scheduling board.

B. Open and flexible structure

As a consequence of the high complexity of the planning problem in industry, leitstands have to be capable of satisfying a multiplicity of different requirements. To avoid costly adaption of leitstand software, they have to be so flexible in their structure as to adapt to the need of scheduling within every enterprise.

Problems of leitstand scheduling procedures can be divided into the problem of selecting between alternative machines which is required for the allocation of an operation to one of several similar machines. In this context, leitstands have to support the splitting of an operation into smaller batches which are to be processed on different machines. The second problem is to determine the sequence of operations on a single machine. Hence leitstands have to support combining operations to be processed one after the other on one machine with the aim of reducing set-up times and costs. For this specific functionality, efficient algorithms can be employed to derive suggestions for the order of operations on a machine. In addition overlapped processing of operations have to be supported. This is particularly important for the application in a flexible manufacturing system environment.

Furthermore, leitstands have to provide the capability to support different activation principles. Activation principles correspond to the application of the push or the pull principle for the scheduling process. This is characterized by determining the direction of planning. For forward scheduling the scheduling process starts with the earliest start date of the first operation of the earliest order and continues working its way forward in time with the successor operations and orders. The advantage of forward scheduling is a high machine load factor at present and the creation of buffer times for the latest finish date. In contrast, during backward scheduling, the creation of the schedule starts out from the latest end date of an order and then continues backward in time. Backward scheduling always considers the last operation of an order first. The advantages of backward scheduling are a reduction of the stock on hand associated with a reduction of capital costs. Some leitstands offer a combination of both activation principles which is termed bottleneck scheduling. Starting from a bottleneck machine, the operations which are predecessors of the operations to be processed on the bottleneck machine undergo backward scheduling, while the successors of the operations on the bottleneck machine undergo forward scheduling. The advantage of bottleneck scheduling is the reduction of lead times of orders passing the bottleneck. Figure 3 shows the utilization of the different scheduling directions for the scheduling of an order which consists of three operations. The first and the third operation are to be processed on a milling machine, the second operation is to be processed upon a boring machine. The operations of this order are vertically hatched. The boring machine is a bottleneck machine which already is loaded with other operations represented by the diagonal hatching.

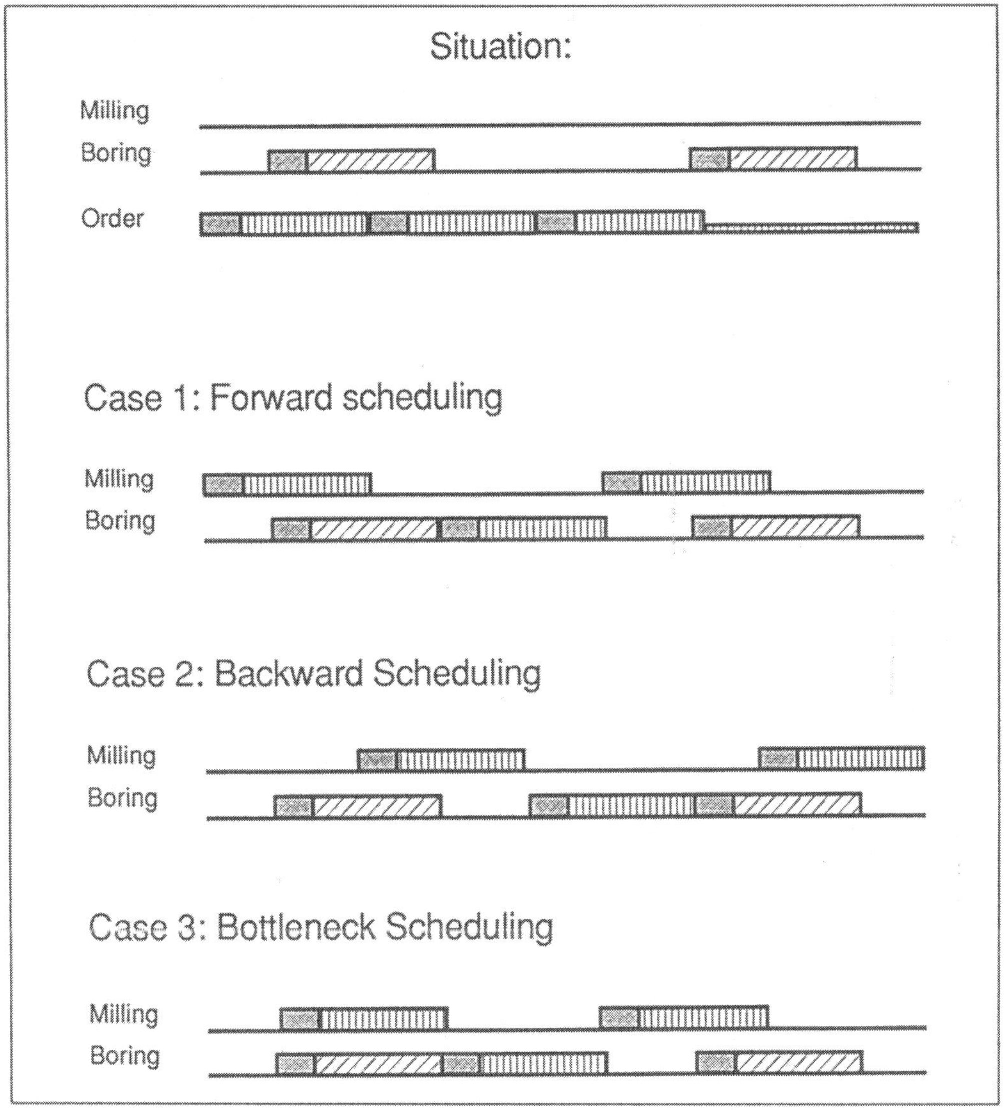

Figure 3: Application of activation principles

C. Coordination functionality

In a decentral production environment, scheduling has to ensure the correspondance between different control areas in which products are manufactured. A decentral production environment can be thought of as consisting of several more or less independent control areas. In a production island environment for example, each production island would be regarded as one control area.

The different control areas can schedule according to different planning principles and objectives. However, each order which arrives from the production planning and control system has to be assigned to a certain control area. Frequently, it will not be possible to manfuacture one order within one control area. The order will have to pass several control areas. This is evident within a job-shop environment. But in a production island environment orders frequently will have to leave the production island for processing or pass through different islands. Therefore a functionality is required which coordinates the different decentral control areas. The leitstands present functionalities to support this requirement. Figure 4 shows the conception of Leitstand FI-2 to realize the coordination of several production island leitstands.

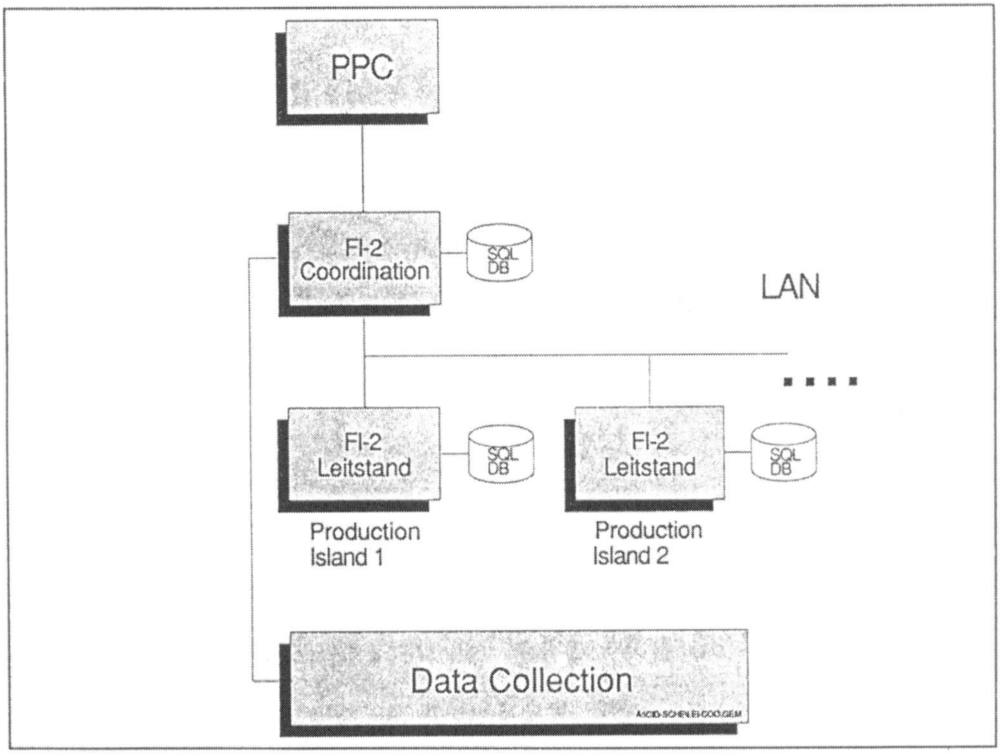

Fig. 4: The coordination functionality provided by the Leitstand FI-2

The coordination module is linked to the different leitstands belonging to the decentral control areas. In addition, the data collection system is linked directly to the coordination module. All data collection and shop floor monitoring acitivities therefore pass through the coordination module. The coordination module receives the orders from the PPC systems and sends completion information back to the production planning and control system. The data which is managed by the coordination module is stored within a relational database. This database contains information which is duplicated within the proprietary databases of each control area leitstand. The orders received from the PPC system are evaluated, completed as necessary with shop-floor specific information and then allocated to different control areas. According to the decisions of the foreman and to the specific properties of the order, the operations are allocated sequentially to the control areas. Where several succeeding operations are allocated to one control area, there these operations are grouped within the coordination module as independent planning unit. Each independent planning unit is equipped with a latest completion date and sent to the corresponding control area. All monitoring by the coordination module corresponds to the planning units, not to specific operations. Figure 5 shows the breakdown of planning units of one order into the sequential operations within two control areas. Operations 10 to 50 are grouped within planning unit 1, operations 60 to 80 belong to planning unit 2 within the coordination module. If within one control area a breakdown or any other failure delays certain operations, this is not reported to the coordination module as long as the completion of the last operation of this planning unit within the control area is not endangered. Only then is the coordination module activated. The module subsequently informs the succeeding control areas of the delay enabling them to plan accordingly. The coordination module also enables the foremen of the different control areas to gather information about the status of work within preceeding or succeeding control areas thus providing transparency by a multi-area data view. Furthermore it supplies the management of the interface to the PPC and data collection sytems.

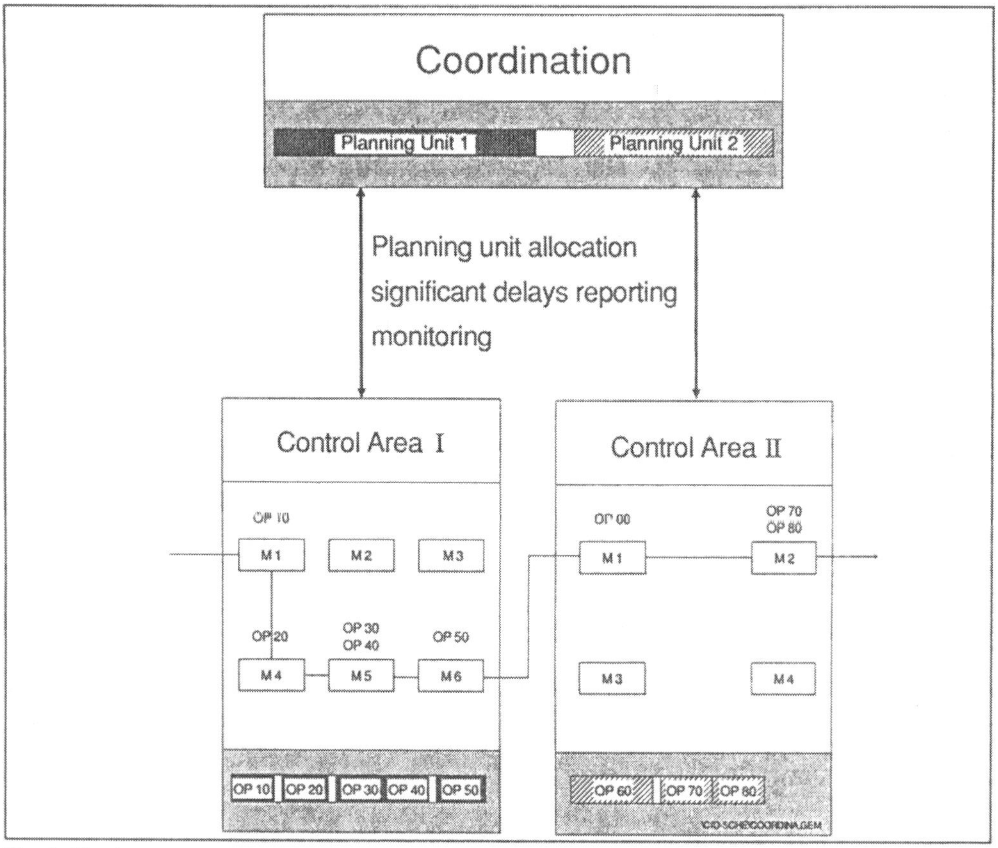

Fig. 5: The breakdown of planning units into operations within of different control areas

D. Graphical User Interface

The main output device is a graphical screen. It is the essential prerequisite for the visualisation of a GANTT chart. Some leitstands have a separate alphanumeric screen which displays textual information. The most important input device is the mouse. Leitstands with a highly developed man machine interface require only a very small amount of interaction via keyboard. This interaction is only necessary for those inputs where a selection between predetermined options is not possible. To improve the visualisation of the schedule, leitstand provide different types of diagrams. All leitstands show the scheduled operations depending upon time. For such a GANTT chart, the operations are shown as bars whose length expresses the sum of processing and set-up time upon the machines and wich are consecutively shown in a row. Each row represents a machine. Operations are shown in different colours to show the pertinence to a certain type of order or to identify properties such as "in work", "in delay", "not feasible". Too much colouring, though, reduces the clarity of the schedule. Besides the GANTT chart showing the machine load, which is represented in the upper half of figure 6, some leitstands provide a diagram showing the orders not yet allocated to the machines. This representation - which varies greatly from leitstand to leitstand - is called order pool. It contains for example in each row an order with the operations that have not yet been allocated to the machines. A small bar shows the buffer time during which the order can be delayed without affecting its latest completion date.

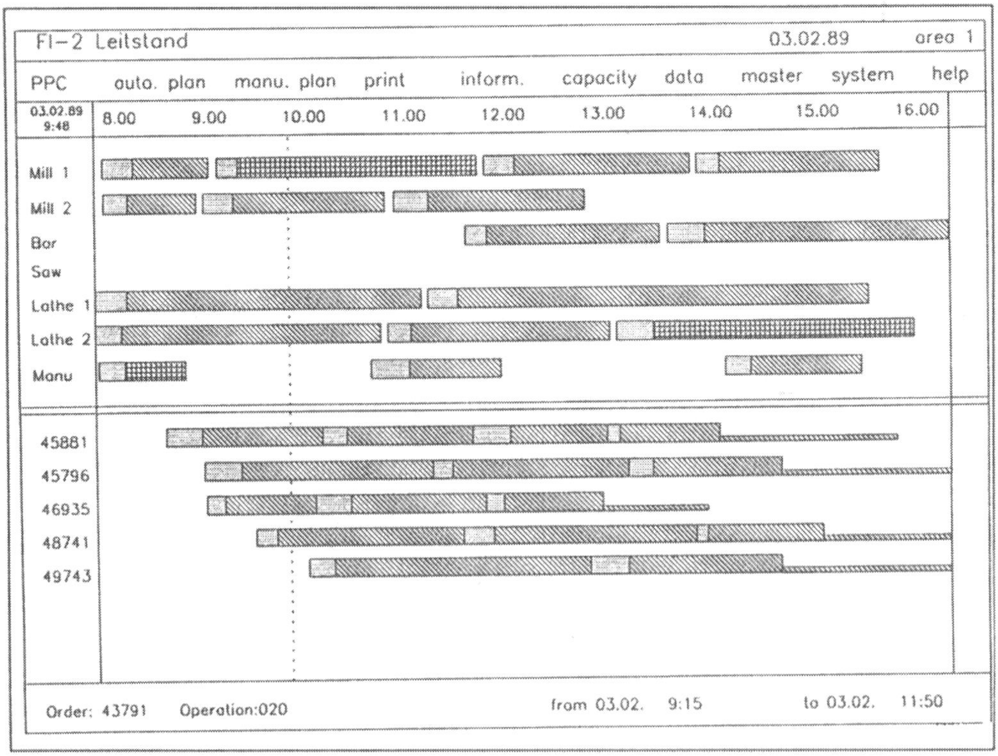

Figure 6: Schematic user interface of the leitstand FI-2

The lower half of figure 6 shows such an order pool. Besides these diagrams, the majority of the leitstands provide load diagrams which show the capacity and total load of certain machines or machine groups over a certain period including the capacity required for the operations within the order pool. A schematic representation of such a capacity diagramm is shown in figure 7. Various leitstands utilize a horizontal axis representing time. This time ruler can be utilized to modify the time slice shown on screen using the mouse. In addition the scale of the time ruler can be modified to expand or compress the part of the GANTT chart shown on screen. This allows e.g. to vary the scale from half a shift per screen to several months per screen. In a separate time window the actual time for a certain position on the GANTT chart is displayed. The time ruler functions are of high importance for the user as they reduce the disadvantage stemming from the smaller screen of the leitstand in comparison to the much larger traditional scheduling board. The timer ruler reduces the danger that the user loses the orientation on the screen. The user friendlyness of leitstands is very high resulting from the grafical representation, the interaction logic which corresponds to the human logic, the set-up of preferably soft plausibility checks, low requirements for human memorizing capabilities and a low complexity of interaction. The user friendlyness is the highest with those leitstands that utilize a standard graphical user interface. Therefore leitstands do not require specially qualified personnel with high EDP-expertise.

E. Integration nucleus

Generally, leitstands can operate without links to other EDP-systems. This is an advantage for the implementation in companies which currently have but a low level of EDP-support and for testing different leitstands in a real production environment.

(1) Interface to the PPC-system

All leitstands provide interfaces to specific production planning and control systems and propose the development of interfaces to PPC systems not yet supported. A PPC system sends orders and operations roughly scheduled to the leitstand where each order, in some cases each operation, is equipped with an earliest start and a latest end date. The leitstand then has to schedule the order and operations within the corresponding interval.

By linking the leitstand via the coordination functionality to the PPC system the amount of paper for order and operations vouchers in production is significantly reduced. This leads to a reduction of transfer times which otherwise would be necessary for centralized printing and transportation of centralized order vouchers and reduces the quota of misplaced order vouchers. The flow of orders released by the PPC system becomes transparent only through the leitstand. The link between PPC and leitstand is not unidirectional. The leitstand reports orders completed, mile stones reached and machine breakdown via the coordination functionality to the PPC system. Via mile stones the PPC system synchronizes its own planning with the progress in production and thus increases the quality of planning.

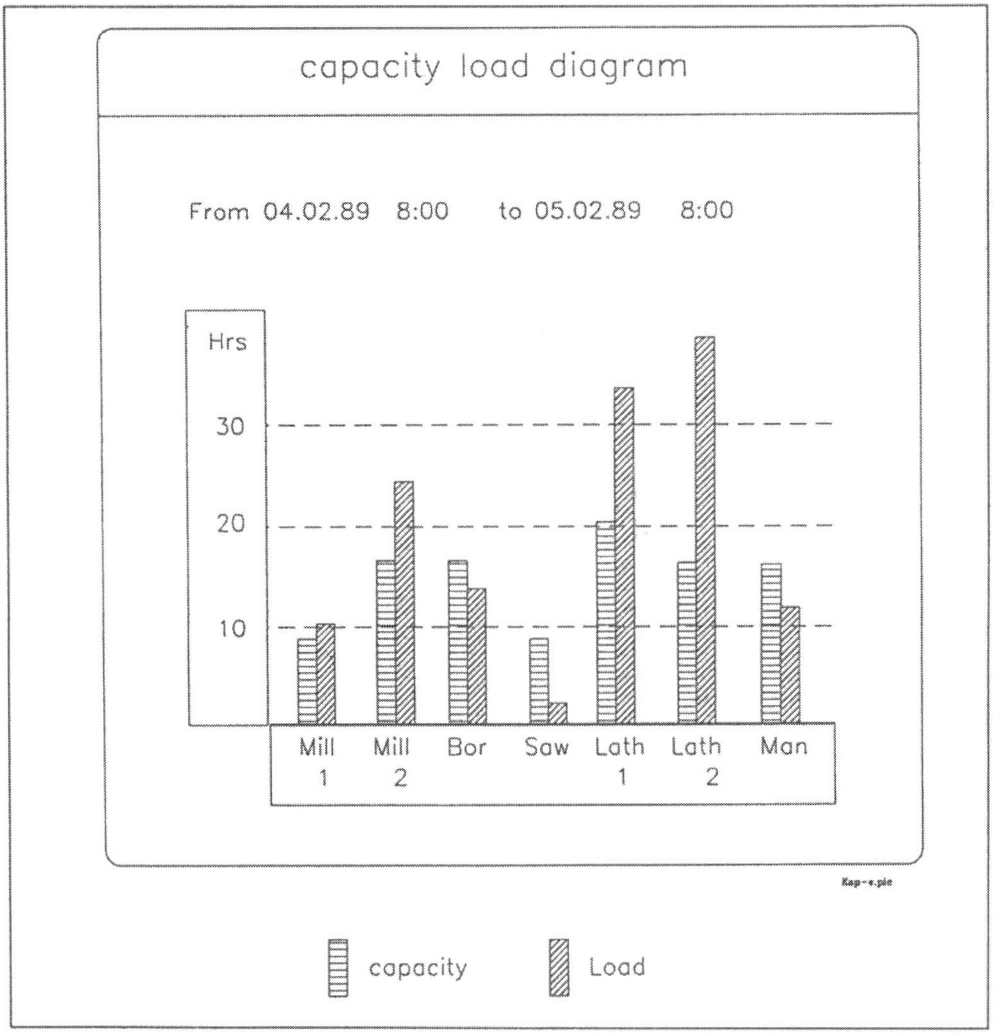

Figure 7: Schematic capacity load diagram

(2) Interface to the Data Collection System

Linking the leitstand with the data collection system is the second direction of integration. The data collection system can monitor the order progress. This can be accomplished by reporting order ompletion via data collection terminals. Order completion, however, can also be reported directly to the leitstand. Therefore order vouchers are stacked and entered into the leitstand for example via bar code reader. This has to take place prior to every scheduling session. A second functionality of data collection is to report machine breakdown to the leitstand. Breakdown often is highlighted using a red bar. Some leitstands generate an acoustic signal to report such an event. A leitstand in addition can be utilized to store production data and support various information and statistical evaluation functions. A leitstand can be enhanced to store all production data centrally and distribute it to other systems. Currently much effort is put into improving the leitstand's data collection facility.

(3) Interfaces to other EDP systems

Besides the production planning and control and the data collection interface a link to the dynamic numeric control systems is possible. This interface functionality is only supported in some cases and always is directed to the needs of the specific enterprise. Normally, the leitstand does not operate on more than the number and the status of a NC-program. During dispatch, the corresponding program number is transferred to the NC system. Some leitstands offer an availability check functionality for NC programs. Links to material handling systems and to tool management systems are not standardly available.

III. FUTURE TRENDS

A. Simulation and animation

A possibility to increase the efficiency of scheduling is to provide leitstands with additional simulation functionality. The application of simulation would allow to generate different schedules according to different scheduling strategies and to a variation of the parameters that drive the strategies. Today the vast majority of leitstands does not support the analysis of the schedules generated interactively or automatically. This would require the evaluation of different characteristics of a generated schedule as e.g. the average lead time, the amount of delays, the capacity load factor, the average amount of started but not yet completed orders, the sum of set-up cost etc. The user would thus be enabled to compare different schedules using the criteria which correspond best to his objectives. To a large extent this is a problem of presenting the output of a simulation run in a way that is suitable for the human. Cifers and tables are only a crude way of reaching the human mind. A better approach are graphs and figures. Hence there remains much room for improving the analysis and presentation of the properties of the schedules generated.

Animation is another solution that allows to better visualize the results of the generation of a schedule and for their comparison. Animation is a means by which the process of production can dynamically be monitored in advance by the scheduler and by which he can identify critical parts of a schedule. [ScHZ 1990, p.152] Animation will also help to reduce the level of qualification required for working with the leitstand or of interpreting its results. Therefore this will increase the possibility of being accepted or even utilized by every member of a work group within a partially autonomous decentral production island.

That simulation and animation today generally are not supported by leitstands is a consequence of the high computing power such functionalities require. In contrast to the personal computers on which the leitstands originated, today's workstations, however, provide the potential of carrying out simulation and animation.

B. Re-allocation of functionalities to production

Currently the trend can be identified that the spectre of leitstand functionalities is increased. This is one reason why it becomes more and more difficult to distinguish unambiguously between leitstands and other - more traditional - shop floor control systems. Leitstands provide more and more functions of operational control. The utilization of leitstands within production islands could pave the way for allocating further EDP-supported functionalities back to the production level. Functions such as NC-programming, computer-aided planning (CAP) or quality assurance and -control would be likely candidates [Ruff 1990].

C. Expert systems

Leitstand functionalities often have been proposed for support by expert systems. The applicability of expert systems in this area, however, has to be carefully analysed. They certainly will not solve the dilemma of automatic scheduling, which even advanced mathematics has not been able to solve. In some areas, an application of expert systems, however, could be useful. Expert systems could provide help in searching for strategies which reflect the goal structure of the scheduler. They could suggest alternative approaches to solve a certain production problem. Expert systems might also be useful to solve well-specified problems as the calculation of set-up times depending upon the order of operations at a certain machine. Expert systems might also be utilized to automatically search for ways of increasing the quality of a given schedule. They would have to evaluate the schedule and apply rules which reflect the knowledge of the scheduler and the specific situation of the company. The sequence of operations at a certain machine for example always leaves room for improvement which depends very specifically upon the characteristics of the operations. In this area leitstand functionality still can be enhanced significantly.

(® LEI-HAGN.TXT ®)

[AWF 1984] AWF (Ed.): Flexible Fertigungsorganisation am Beispiel von Fertigungsinseln, 1984, p.5.

[Farh 1990] Farhoodi, F.: A knowledge-based approach to dynamic scheduling. In: International Journal of Computer Integrated Manufacturing, Volume 2 No. 3 (March-April 1990), pp.84-95.

[FrGr 1989] Fertigungsleitsystem. In: VDI-Z 131 Nr.8, August 1989, pp.97-105.

[HaSc 1990] Hars, A.; Scheer, A.-W.: Leitstände - ein neues Instrumentarium zur Fertigungssteuerung. In: v.Behr, M.; Köhler, Ch. (Eds.): Werkstattoffene CIM-Konzepte, Projekkträgerschaft Fertigungstechnik Forschungsbericht KfK-PFT 157, 1990, pp.51-78.

[HeZe 1990] Herterich, R.; Zell, M.: Dezentrale Fertigungssteuerung. In: VDI-Z 131 No. 5, May 1990, pp.19-25.

[KeKe 1990] Keller, G.; Kern, S.: Das Fertigungsinselprinzip als Bestandteil von CIM. In: CIM-Management 1/90, 1990, p.44-49.

[Müll 1971] Müller-Merbach, H.: Operations Research. 2. Auflage, München, 1971, p.290.

[Ruff 1988] Ruffing, T.: Die integrierte Auftragsabwicklung bei Fertigungsinseln - Grobplanung, Feinplanung, Überwachung. In: Proceedings of the AWF Fachtagung Fertigungsinseln, 6-7 December 1988, pp.281-309.

[Ruff 1990] Ruffing, T.: Fertigungssteuerung bei Fertigungsinseln - Funktionale und Datentechnische Informationsarchitektur. PhD thesis manuscript (in print), Saarbrücken 1990.

[Sche 1986] Scheer, A.-W.: Neue Architektur für EDV-Systeme zur Produktionsplanung und -steuerung. Veröffentlichungen des Instituts für Wirtschaftsinformatik, Heft 53, 1986.

[Sche 1988] Scheer, A.-W.: Computer Integrated Manufacturing - Computer Steered Industry. Berlin et. al. 1988.

[ScHZ 1990 Scheer, A.-W.; Herterich, R.; Zell, M.: Final DFG project report Verteilte Produktionsplanungs- und -steuerungssysteme unter Einsatz von Mikrocomputern. Saarbrücken, 1990.

[WiLü 1989] Wiendahl, H.-P.; Lüssenhop, T.: Wirkung von Prioritätsregeln. In: VDI-Z 131 (1989) No. 1, pp36-41.

Scheduling with Alternative Process Plans

JAEKYOUNG AHN and ANDREW KUSIAK

1. INTRODUCTION

Automated manufacturing systems may require a considerable capital investment. The key to success in implementation of automated manufacturing systems is effective utilization of manufacturing resources through application of efficient scheduling algorithms.

Kusiak [1990] presented a knowledge-based system for scheduling in automated manufacturing. Kusiak and Ahn [1990] proposed the Most Dissimilar Resources (MDR) dispatching rule that has been designed for scheduling operations in an automated manufacturing environment. The latter paper shows that the MDR dispatching rule provides schedules of better quality and consistency than a number of the exiting dispatching rules. In areas other than automated manufacturing, extensive research has been done for the scheduling problem with multiple resources [Pritsker et al. 1969, Davis 1973, Graham et al. 1979, Lloyd 1981, Monma 1982, Blazewicz, Lenstra, and Rinnooy Kan 1983, Dobson and Karmarkar 1986, 1989].

A process plan in an automated manufacturing system specifies requirement for machines, tools, auxiliary devices (eg., fixtures, pallets, and feeders) as well as operations to be performed. Many of the scheduling models for automated manufacturing systems are based on the assumption that for each part, one process plan only is available [Kusiak and Finke, 1988]. A manufacturing system with alternative process plans appears to have a potential for increasing throughput rate by eliminating bottlenecks and handling resource breakdowns.

Srihari and Greene [1988] discussed process planning strategies that allow to avoid bottlenecks, reduce in-process inventory, balance machine utilization, and reduce flowtime. The system provides alternative process plans for each job. Chang et al. [1989] developed a bottleneck-based beam search method for job scheduling in a flexible manufacturing system with alternative process plans. They focused on bottleneck operations in the critical path. Sherali et al. [1989] presented a model for selecting a set of process plans and scheduling operations on machines as to minimize the makespan. Kim [1990] analyzed characteristics of the job shop scheduling problem with multiple identical jobs and alternative process plans. His computational results were differed from those obtained for the classical job shop problem in which order sizes and alternative process plans were not explicitly considered.

In this paper, the effect of alternative process plans on scheduling performance measures in an automated manufacturing system is analyzed. Salient features of the manufacturing environment, i.e., concurrent use of multiple resources and the existence of the alternative process plans are considered.

Two formulations of the scheduling problem are described in Section 2. In Section 3, scheduling models in automated manufacturing systems are presented. In Section 4, computational experience is discussed. Conclusions are drawn in Section 5.

2. SCHEDULING PROBLEM IN AUTOMATED MANUFACTURING SYSTEMS

In the model ((1)-(10)) presented in this section, the traveling time of parts between machines is not considered, while in the model ((11)-(14)), the traveling time is incorporated.

2-1. Scheduling Model Without Traveling Times

To present the scheduling models, the following notation is introduced :

i : part index, $i=1,..,n$
k : operation index, $k=1,..,m$
n : number of parts
m : number of operations
c : resource type index, $c=1,...,s$
s : number of resource types
r : resource index, $r=1,...,r_c$
r_c : number of resources r of type c
k_q : alternative q of operation k
q_{ik} : number of alternatives of operation k of part i
O_i : set of operations of part i
R_k : set of resources used by operation k, $R_k = \{c_1^{r_1}, c_2^{r_2},..., c_s^{r_s}\}$
P_i : set of pairs of operations [k,l] of part i, where k precedes l
Q_i : set of ordered pairs of operations [k,l] of part i, where k and l can be performed in any order
c^r: the rth resource of type c
N_c^r : set of operations to be performed by the resource r of type c, $c=1,..,s$ $r=1,..,r_c$
A_{ik} : set of alternatives of operation k of part i
t_{ik_q} : processing time of the operation k_q of part i
f_{ik} : completion time of operation k of part i
T : makespan
M : arbitrary large number

$$z_{ik_q} = \begin{cases} 1, & \text{if alternative operation q of operation k belonging to part i is selected} \\ 0, & \text{otherwise} \end{cases}$$

$$x_{kl} = \begin{cases} 1, & \text{if operation k precedes operation l} \\ 0, & \text{otherwise} \end{cases}$$

The objective of the scheduling model is to minimize the makespan.

$$\text{Min } T \tag{1}$$

s.t. $f_{il} - t_{ilq} + M(1-z_{ilq}) \geq f_{ik}$

$[k,l] \in P_i,\ q \in A_{il}$ for all i (2)

$f_{il} - t_{ilq} + M(1-z_{ilq}) \geq f_{ik} - Mx_{lk}$

$[k,l] \in Q_i,\ q \in A_{il}$ for all i (3)

$f_{il} - t_{ilq} + M(1-z_{ilq}) \geq f_{jk} - Mx_{lk} - M(1-z_{jkq'})$

$l \in O_i,\ k \in O_j,\ R_{l_q} \cap R_{k_{q'}} \neq \phi,\ q \in A_{il},\ q' \in A_{jk}$ for all ordered pairs (i,j), $i \neq j$ (4)

$x_{kl} + x_{lk} = 1$

for all k, l such that $[k,l] \in Q_i,\ q \in A_{il}$ for all i, or $l \in O_i,\ k \in O_j,\ R_{l_q} \cap R_{k_{q'}} \neq \phi,\ q \in A_{il}$, $q' \in A_{jk}$ for all ordered pairs (i,j), $i \neq j$ (5)

$f_{ik} \leq T$ for all i, k (6)

$f_{ik} - t_{ikq} + M(1-z_{ikq}) \geq 0$

$q \in A_{ik}$ for all i, k (7)

$\sum_{q \in A_{ik}} z_{ikq} = 1$ for all i, k (8)

$x_{kl} = 0,1$ for all k, l (9)

$z_{ikq} = 0,1$ for all i, k, q (10)

Constraint (2) ensures that the operations of each part are processed according to the precedences required. Constraint (3) implies that any two operations of each part cannot be processed at the same time. Constraint (4) implies that a machine can process only one operation at a time. Constraint (5) imposes a precedence relation between a pair of operations. Constraint (6) implies that the completion time of each operation is not greater than the makespan. Constraint (7) states that the completion time of an operation is not smaller than its processing time. Constraint (8) ensures that for each operation only one alternative operation is selected. Constraint (9) and (10) impose integrality.

2-2. Scheduling Model With Traveling Times

In addition to the previously introduced notation, define matrix $[\tau_{r'r}]$ of travel times between machines r' and r. Note that $[\tau_{r'r}] = \tau[k^1_{q1}, k^2_{q2}]$ where operations k^1_{q1} and k^2_{q2} are processed on machines r' and r, respectively.

$$\text{Min } T \tag{11}$$

s.t. $f_{il} - t_{ilq} - \tau_{r'r} + M(1-z_{ilq}) \geq f_{ik} - M(1-z_{ikq'})$

$[k,l] \in P_i,\ q \in A_{il},\ q' \in A_{ik}$ for all i and $lq \in N_1^r$ and $kq' \in N_1^{r'}$ (12)

$f_{il} - t_{ilq} - \tau_{r'r} + M(1-z_{ilq}) \geq f_{ik} - Mx_{lk} - M(1-z_{ikq'})$

$[k,l] \in Q_i,\ q \in A_{il},\ q' \in A_{ik}$ for all i and $lq \in N_1^r$ and $kq' \in N_1^{r'}$ (13)

(4), (5), (6), (7), (8), (9), and (10) (14)

Constraint (12) ensures that the operations in a part are processed according to the precedences required, allowing for travel of the part between operations. Constraint (13) implies that any two operations in a part cannot be processed at the same time, and travel time of the part is considered.

3. SCHEDULING ALGORITHMS

It is not difficult to conclude that the two scheduling models described in Section 2 are NP-hard. Therefore, the effort for solving these problems optimally in an acceptable time is futile. In order to solve the scheduling problem discussed, in this section, two heuristic algorithms are developed. Before the algorithms will be presented, the following notation and definitions are introduced:

t : current time

f_k : completion time of operation k

rt_k : remaining processing time of operation k

Operation k is schedulable at time t, if :

(1) no other operation which belongs to the same part is being processed at time t,

(2) all operations preceding operation k have been completed before time t,

(3) all resources required by the process plan to perform operation k are available at time t.

s_k : status of operation k ;

$$s_k = \begin{cases} 1, & \text{if operation k is schedulable,} \\ 2, & \text{if operation k is nonschedulable,} \\ 3, & \text{if operation k is being processed,} \\ 4, & \text{if operation k has been completed,} \\ 5, & \text{if operation k satifies the first two conditions in the definition of schedulability} \end{cases}$$

sr_{cr} : status of resource r of type c ;

$$sr_{c^r} = \begin{cases} 1, & \text{if resource } c^r \text{ is avaliable,} \\ 0, & \text{otherwise} \end{cases}$$

S_j : set of operations with $s_k = j$, $j = 1,...,5$

C_j : temporal set of operations

3-1. Solving the Model ((1)-(10))

To solve the model ((1)-(10)), an algorithm similar to the heuristic incorporated into the knowledge-based system discussed in Kusiak [1990] is presented. The algorithm presented incorporates the MDR dispatching rule presented in Kusiak and Ahn [1990].

<u>Algorithm 1</u>

Step 0. Initialization :

- current time $t = 0$

- resource status, $sr_{cr} = 1$, for all r and c
- operation status and set of operations with the operation status are initialized as follows :
  (i) Set $S_3=S_4=\phi$
  (ii) For each part, if operation k has no preceding operations, set $s_k = 1$; otherwise, $s_k = 2$
  (iii) Construct S_1 and S_2 for operations in (ii)

Step 1. If all the operations have been completed, STOP; otherwise go to Step 2.

Step 2. Set $S_5= \phi$, and if $S_1 =\phi$ then go to Step 8; otherwise, go to Step 3.

Step 3. In the set S_1, select an operation k* with a dispatching rule specified by the user (part i* corresponding to operation k* is automatically selected).

Step 4. Set :

- the remaining processing time of operation k*, $rt_{k*} = t_{ik*}$
- resource status $sr_{cr} = 0$ for $c^r \in R_{k*}$

Construct :

- $C_1 = \{k \mid k \neq k^*, k \in Qi^*, k \in S_1\}$
- $C_2 = \{k \mid \{k^*, k\} \in N_{cr}$ for all c and r, $k \in S_1\}$

Update :

- set of schedulable operations $S_1 = S_1 - \{k^*\} - \{C_1 \cup C_2\}$
- set of nonschedulable operations $S_2 = S_2 \cup C_1 \cup C_2$
- set of processing operations $S_3 = S_3 \cup \{k^*\}$
- set of partially schedulable operations $S_5 = S_5 \cup C_2 - C_1$

If $S_1 \neq \phi$, go to Step 3.

Step 5. If $S_1 =\phi$ and $S_5 \neq \phi$, then go to Step 6; otherwise, go to Step 8.

Step 6. If the sum of the waiting time and processing time for operation $k \in S_5$ in the basic process plan is longer than the one in the alternative process plan, then $S_1=S_1 \cup \{k\}$ and replace the basic operation with the corresponding alternative operation. Set $S_5 =\phi$.

Step 7. Go to step 3.

Step 8. Set :

- completion time $f_k = rt_k + t$, $k \in S_3$
- current time $t = \min_{k\#} \{f_k\}$, $k \in S_3$
- resource status $sr_{cr} = 1$, $c^r \in R_{k\#}$
- remaining time $rt_k = f_k - t$, $k \in S_3$

Step 9. For operations in the set S_2, construct :

- $C_1 = \{k \mid$ all the preceding operations of operation k have been completed $\}$
- $C_2 = \{k \mid sr_{cr} = 1, c^r \in R_k\}$
- $C_3 = S_2 - C_2$

Update :

- set of schedulable operations $S_1 = S_1 \cup \{C_1 \cap C_2\}$

- set of nonschedulable operations $S_2 = S_2 - \{C_1 \cap C_2\}$
- set of processing operations $S_3 = S_3 - \{k^\#\}$
- set of completed operations $S_4 = S_4 + \{k^\#\}$
- set of partially schedulable operations $S_5 = S_5 \cup \{C_1 \cap C_3\}$

Step 10. Repeat Step 8 and 9 until $\{k \mid rt_k = 0, k \in S_3\} = \phi$. Go to Step 11.

Step 11. If $S_5 \neq \phi$, do Step 6. Go to Step 1.

Example 1

Consider the problem of scheduling 8 operations belonging to 3 parts (Figure 1). The resources required by each operation and the corresponding processing times are shown in Table 1. Assume that each operation has 2 alternative process plans, and the MDR dispatching rule is used in Step 3 of the algorithm.

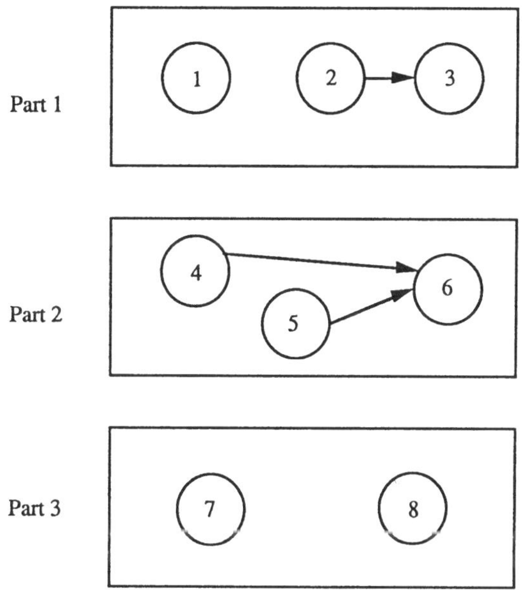

Figure 1. Three parts with eight operations

Table 1. Process plans for three parts

Part No.	Operation No.	Process Plan	Processing Time	Machine	Tool	Fixture
1	1	1	4	m_2	t_3	f_4
		2	5	m_1	t_1	f_3
	2	1	3	m_3	t_4	f_2
		2	4	m_3	t_2	f_4
	3	1	2	m_2	t_5	f_4
		2	3	m_1	t_3	f_2
2	4	1	1	m_3	t_6	f_1
		2	2	m_1	t_4	f_3
	5	1	2	m_1	t_5	f_3
		2	3	m_2	t_6	f_2
	6	1	8	m_2	t_4	f_3
		2	11	m_2	t_1	f_4
3	7	1	5	m_1	t_3	f_1
		2	8	m_3	t_2	f_4
	8	1	5	m_3	t_1	f_2
		2	7	m_1	t_5	f_3

Step 0. Initialize the variables :

$t = 0$

resource status is set to 1 for all operations

$S_1=\{1,2,4,5,7,8\}, S_2=\{3,6\}$, and $S_3=S_4=\phi$.

Step 1. Since $S_4=\phi$, go to Step 2.

Step 2. Set $S_5=\phi$. Since $S_1\neq\phi$, go to Step 3.

Step 3. The MDR dispatching rule selects operations $\{1,2,5,7,8\}$. Operation 1 is chosen arbitrarily.

Step 4. Set :

- remaining processing time of operation 1, $rt_1 = 4$,
- resource status $sr_{m2} = sr_{t3} = sr_{f4} = 0$.

Construct :

- $C_1 = \{2\}, C_2 = \{7\}$.

Update :

- set of schedulable operations $S_1 = S_1 - \{1\} - \{2,7\}=\{4,5,8\}$,
- set of nonschedulable operations $S_2 = S_2 \cup \{2\} \cup \{7\}=\{2,3,6,7\}$,
- set of processing operations $S_3 = S_3 \cup \{1\}=\{1\}$,
- set of partially schedulable operations $S_5 = S_5 \cup \{7\}=\{7\}$.

Since $S_1\neq \phi$, go to Step 3.

Step 3. The MDR dispatching rule selects operations 5 and 8. Operation 5 is selected arbitrarily.

Step 4. Set :

- remaining processing time of operation 1, $rt_5 = 2$,
- resource status $sr_{m1} = sr_{t5} = sr_{f3} = 0$.

Construct :

- $C_1 = \{4\}$, $C_2 = \phi$.

Update :

- set of schedulable operations $S_1 = S_1 - \{5\} - \{4\} = \{8\}$,
- set of nonschedulable operations $S_2 = S_2 \cup \{4\} = \{2,3,4,6,7\}$,
- set of processing operations $S_3 = S_3 \cup \{5\} = \{1,5\}$,
- set of partially schedulable operations $S_5 = \{7\}$.

Since $S_1 \neq \phi$, go to Step 3.

Step 3.Operation 8 has been selected.

Step 4. Set :

- remaining processing time of operation 6, $rt_8 = 5$,
- resource status $sr_{m3} = sr_{t1} = sr_{f2} = 0$.

Construct :

- $C_1 = \phi$, $C_2 = \phi$.

Update :

- set of schedulable operations $S_1 = S_1 - \{8\} = \phi$,
- set of nonschedulable operations $S_2 = \{2,3,4,6,7\}$,
- set of processing operations $S_3 = S_3 \cup \{8\} = \{1,5,8\}$,
- set of partially schedulable operations $S_5 = \{7\}$.

Since $S_1 = \phi$, go to Step 5.

Step 5. Since $S_1 - \phi$ and $S_3 \neq \phi$, go to Step 6.

Step 6. Since $sr_{m1} = sr_{m2} = sr_{m3} = 0$, go to Step 8.

Step 8. Set :

- completion time $f_1 = 4$, $f_5 = 2$, and $f_8 = 5$,
- current time $t = \min\{4,2,5\} = 2$, and $k^{\#} = 5$,
- resource status $sr_{m1} = 1$, $sr_{t5} = 1$, $sr_{f3} = 1$,
- remaining time $rt_1 = 2$, $rt_5 = 0$, $rt_8 = 3$.

Step 9. For operations in the set S_2, construct :

- $C_1 = \{2,4,7\}$, $C_2 = \phi$, and $C_3 = \{2,3,4,6,7\}$.

Update :

- set of schedulable operations $S_1 = \phi$,
- set of nonschedulable operations $S_2 = \{2,3,4,6,7\}$,
- set of processing operations $S_3 = S_3 - \{5\} = \{1,8\}$,

- set of completed operations $S_4 = \{5\}$,
- set of partially schedulable operations $S_5 = \{2,4,7\}$.

Step 10. Since $\{k \mid rt_k = 0, k \in S_3\} = \phi$, go to Step 11.

Step 11. Since $S_5 \neq \phi$, do Step 6.

The Gantt chart of the final schedule generated in 7 iterations of Algorithm 1 is shown in Figure 2.

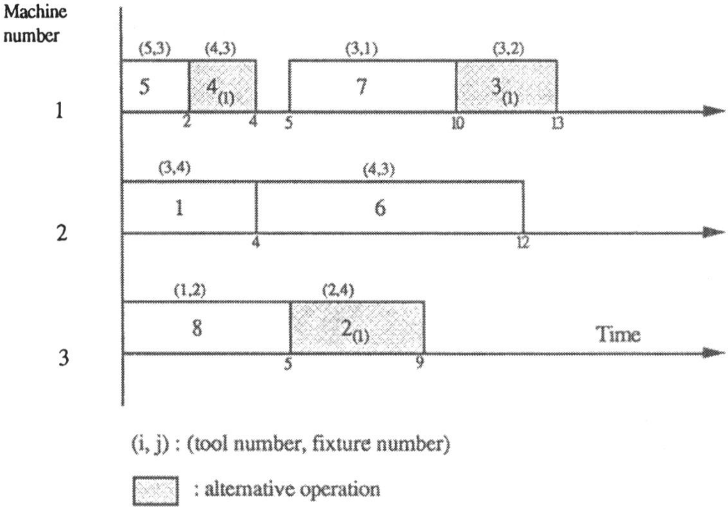

Figure 2. Gantt chart of the schedule in Example 1

3-2. Solving the Model ((11)-(14))

To solve the model ((11)-(14)), the previously provided definition of schedulability is extended as follows :

(1), (2), (3) See the previous definition of schedulability,

(4) The part with operation k has been moved to a specified machine by the material handling system.

<u>Algorithm 2</u>

Step 0. Initialize the variables.

- current time $t = 0$
- resource status, $sr_{cr} = 1$, for all r and c
- operation status and set of operations with the operation status are initialized as follows :
  - (i) Set $S_1=S_2=S_3=S_4=\phi$
  - (ii) For each part, if operation k has no predecessor operations then $s_k = 1$; otherwise, $s_k = 2$
  - (iii) Construct S_1 and S_2 for operations in (ii)

Step 1. If all the operations have been completed, STOP; otherwise go to Step 2.

Step 2. If $S_1 = \phi$, go to Step 5; otherwise, go to Step 3.

Step 3. Select an operation $k^*_{q\#}$ in the set S_1, based on a dispatching rule provided (part i* corresponding to operation $k^*_{q\#}$ is automatically selected).

Step 4. Set :

- remaining processing time of operation k*, $rt_{k^*} = t_{ik^*_{q\#}}$
- resource status $sr_{cr} = 0$ for $c^r \in R_{k^*_{q\#}}$

Construct :

- $C_1 = \{q \mid q \in A_{i^*k^*} \setminus q^\#\}$
- $C_2 = \{k_q \mid k_q \neq k^*_{q\#}, k_q \in Q_i^*, k_q \in S_1\}$
- $C_3 = \{k_q \mid \{k^*_{q\#}, k_q\} \in N_{cr}$ for all c and r, $k_q \in S_1\}$

Update :

- set of schedulable operations $S_1 = S_1 - \{k^*_{q\#}\} - \{C_1 \cup C_2 \cup C_3\}$
- set of nonschedulable operations $S_2 = S_2 \cup C_2 \cup C_3$
- set of processing operations $S_3 = S_3 \cup \{k^*_{q\#}\}$

If $S_1 \neq \phi$, go to Step 3 ; otherwise, go to Step 5.

Step 5. Set :

- completion time $f_k = rt_k + t$, $k \in S_3$
- current time $t = t + 1$
- remaining time $rt_k = rt_k - 1$

If $rt_{k^*} = 0$ in $k \in S_3$, then resource status $sr_{cr} = 1$ for $c^r \in R_{k^*_{q\#}}$

Step 6. In set S_2, Construct :

- $C_1 = \{k_q \mid$ all the preceding operations of operation k_q have been completed $\}$
- $C_2 = \{k_q \mid sr_{cr} = 1, c^r \in R_{k_q}\}$
- $C_3 = \{k_q \mid$ current time - completion time of the immediate predecessor operation of operation $k_q >$ τ[immediately preceding operation of operation k_q , k_q]$\}$

Update :

- set of schedulable operations $S_1 = S_1 \cup \{C_1 \cap C_2 \cap C_3\}$
- set of nonschedulable operations $S_2 = S_2 - \{C_1 \cap C_2 \cap C_3\}$
- set of processing operations $S_3 = S_3 - \{k^\#\}$
- set of completed operations $S_4 = S_4 \cup \{k^\#\}$

Step 7. Go to Step 1.

Algorithm 2 is illustrated in Example 2.

Example 2

Use the data in Example 1, and assume that the MDR dispatching rule has been specified in Step 3. The matrix of travel times is given in (15).

$$[\tau_{rr'}] = \begin{array}{c} \\ m_1 \\ m_2 \\ m_3 \end{array} \begin{array}{c} \begin{array}{ccc} m_1 & m_2 & m_3 \end{array} \\ \begin{bmatrix} 0 & 1 & 3 \\ 2 & 0 & 1 \\ 3 & 2 & 0 \end{bmatrix} \end{array} \qquad (15)$$

In the first iteration of the Algorithm 2, the MDR rule generates the set of schedulable operations, $S_1=\{1,4,8_{(1)}\}$. Thus, at time t=1, $S_1=\phi$, $S_2=\{2,3,5,6,7\}$, $S_3=\{1,8_{(1)}\}$, $S_4=\{4\}$.

Step 5. At time t=1, $rt_4=0$, and $k^{\#}=4$.

Step 6. For the set S_2, construct :

- $C_1 = \{5\}$, $C_2 = \phi$, $C_3 = \phi$.

Update :

- set of schedulable operations $S_1 = \phi$
- set of nonschedulable operations $S_2 = \{2,3,5,6,7\}$
- set of processing operations $S_3 = \{1,8_{(1)}\}$
- set of completed operations $S_4 = \{4\}$

Step 7. Go to Step 1.

After 15 iterations, the schedule illustrated with the Gantt chart shown in Figure 3 has been generated.

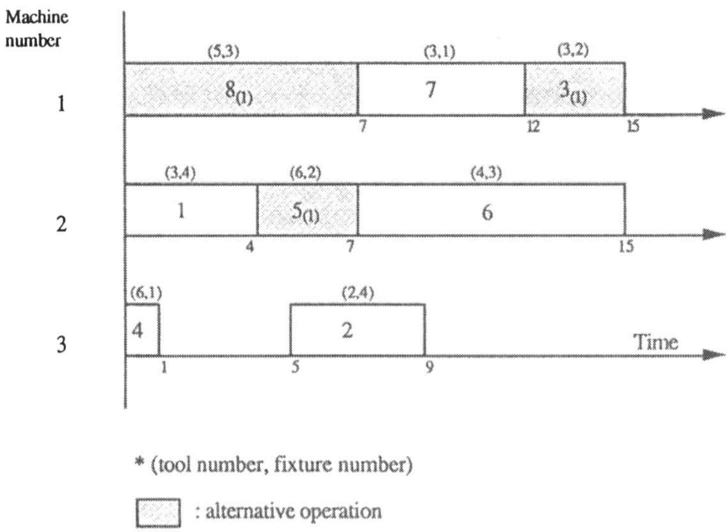

Figure 3. Gantt chart of the schedule in Example 2.

4. COMPUTATIONAL EXPERIENCE

In this section, each of the five dispatching rules incorporated in the two heuristic algorithms is discussed.

- dispatching rule R1 : the MDR dispatching rule.
- dispatching rule R2 : selects an operation with the smallest number of remaining operations.
- dispatching rule R3 : selects an operation belonging to a part with the minimum number of currently schedulable operations.
- dispatching rule R4 : selects an operation with the shortest processing time.
- dispatching rule R5 : selects an operation randomly.

Characteristics of the problems considered in this section are shown in Table 2.

Table 2. Characteristics of the test problems

Problem No.	Problem Size	No. of Parts	No. of Opn's	Number of Resources				No. of Runs
				Machines	Fixtures	Pallets	Tools	
1	Small	5	60	3	3	5	5	10
2				3	3	20	15	10
3				3	15	20	15	10
4				15	15	20	15	10
5	Medium	8	160	5	5	8	8	10
6				5	5	60	20	10
7				5	16	60	20	10
8				20	20	60	20	10
9	Large	10	250	8	8	10	10	10
10				8	8	30	20	10
11				8	15	30	20	10
12				20	20	60	40	10

Note that for the data in Table 2, the following holds :

(i) As the problem number increases, the number of operations per part, number of parts in the problem, and number of available resources increase.

(ii) Each instance of the problem has the same data (e.g., precedence relations, processing times, etc.) except of the alternative process plans and traveling times.

The processing times for operations in the basic process plans were generated according to the discrete uniform distribution with parameters [1,10]. Given the basic processing time generated for each operation, processing times for each of the three alternative process plans were calculated as 120%, 130%, and 150% of the basic processing time, respectively. Thus, processing times of operations in an alternative process plan are assumed to have greater values than that of the basic process plan. Traveling times for parts were imposed according to the basic processing times generated, i.e., the traveling time from one machine to another was generated using the discrete uniform distribution with 10%, 30%, and 50% of the basic processing time. For simplicity, T1 denotes the scheduling problems with traveling time of 10% of the basic processing time, T2, 20%, and T3, 30%, respectively.

To evaluate the quality of schedules, two performance measures were considered :

- makespan (T)

$$T = \max (C_j), \; j = 1,...,n$$

where : C_j is completion time of operation j n is the number of operations

- average machine utilization rate ($\bar{u}$)

$$\bar{u} = \frac{\sum_{i=1}^{m} u_i}{m}$$, where : m is the number of machines

$$u_i = \frac{\sum_{r \in N(i)} t_r}{T}$$

t_r is the processing time of operation r on machine i

N(i) is the set of operations processed on machine i

In order to show the relationship between the makespan and average machine utilization rate, the problems in Table 2 were solved. To evaluate the quality of the solutions generated by each of the five dispatching rules incorporated in the two algorithms, the frequency of the best solution (F_{R_i}) generated by rule R_i, i=1,..,5 was used. The total of the frequency of the best solution generated by each of the two algorithms are shown in Table 3.

Table 3. Summary of the frequency of the best solution

Rule No.	Frequency of the best solution								
	Algorithm 1		Algorithm 2						
	Scheduling without alternative process plans	Scheduling with alternative process plans	Scheduling without Alternative process plans			Scheduling with Alternative process plans			
			T1	T2	T3	T1	T2	T3
R1	56	73	74	75	74	68	69	80
R2	7	13	7	4	8	8	9	4
R3	32	23	9	22	18	10	8	11
R4	21	13	30	14	12	32	32	18
R5	18	18	22	21	14	9	13	16

The results in Table 3 show that the MDR dispatching rule considerably produces schedules of better quality than the other dispatching rules.

The effect of alternative process plans on the quality of schedules is measured with the followings :

(i) Percentage of improvement of the makespan

$$\Delta T_{R_i} = \frac{T'_{R_i} - T_{R_i}}{T'_{R_i}} \; 100\ \%,$$ where: T_{R_i} is the makespan of the schedule generated by dispatching

rule Ri with alternative process plans, T'_{R_i} is the makespan of the schedule generated by dispatching rule Ri without alternative process plans,

(ii) Percentage improvement of the average machine utilization

$\Delta U_{R_i} = \frac{U_{R_i} - U'_{R_i}}{U'_{R_i}}$ 100 %, where: U_{R_i} is the average machine utilization of the schedule generated by dispatching rule Ri with alternative process plans,U'_{R_i} is the average machine utilization of the schedule generated by dispatching rule Ri without alternative process plans.

(iii) Percentage change of CPU time

$\Delta t_{R_i} = \frac{t_{R_i} - t'_{R_i}}{t'_{R_i}}$ 100 %, where: t_{R_i} is the CPU time of the schedule generated by dispatching rule Ri with alternative process plans, t'_{R_i} is the CPU time of the schedule generated by dispatching rule Ri without alternative process plans.

Table 4 through 6 summarize the effect of alternative process plans on the quality of the schedules generated with Algorithm 1.

Table 4. The percentage improvement of the makespan of schedules generated with Algorithm 1

Rule No. \ Problem No.	1	2	3	4	5	6	7	8	9	10	11	12
R1	15.0	16.0	11.3	7.7	11.3	15.1	16.3	7.4	11.0	12.2	15.3	8.9
R2	16.2	11.9	7.1	11.0	10.9	11.8	11.2	7.7	10.5	11.1	13.6	6.4
R3	11.9	17.4	6.9	13.0	10.0	4.5	9.1	7.6	10.9	10.8	11.3	5.1
R4	14.8	12.0	11.1	9.2	8.9	11.9	10.2	6.6	8.6	10.1	13.0	7.1
R5	14.7	15.6	14.2	7.6	11.5	12.5	15.0	6.4	9.8	11.0	13.1	4.8

Table 5. The percentage improvement of the average machine utilization of schedules generated with Algorithm 1

Rule No. \ Problem No.	1	2	3	4	5	6	7	8	9	10	11	12
R1	23.0	26.2	14.4	12.8	18.0	22.6	21.5	12.5	17.6	19.3	22.2	22.7
R2	26.3	18.8	11.7	17.1	18.5	18.9	15.4	12.2	18.0	18.4	20.7	19.6
R3	20.2	28.6	10.2	18.8	16.3	22.7	13.0	11.1	18.7	17.9	17.9	5.8
R4	25.2	21.3	15.7	17.9	16.3	20.7	15.0	12.9	17.3	18.7	20.9	36.4
R5	22.3	23.0	19.3	13.0	18.0	19.5	21.2	9.7	16.8	17.9	20.0	29.6

Table 6. The percentage change of CPU time of Algorithm 1

Rule No. \ Problem No.	1	2	3	4	5	6	7	8	9	10	11	12
R1	14.2	11.4	45.4	15.4	1.6	8.5	23.2	5.9	2.6	1.3	3.5	9.5
R2	24.6	30.5	54.9	31.5	5.4	12.8	31.9	15.7	0.4	- 3.0	- 2.8	0.3
R3	22.4	27.1	48.0	24.7	3.9	9.1	23.5	13.1	- 3.5	- 1.1	- 7.2	9.3
R4	18.9	29.9	48.6	30.8	4.5	11.0	30.9	13.5	- 4.3	- 6.0	- 2.1	3.5
R5	32.7	37.7	52.6	30.7	8.0	15.1	32.2	17.1	- 0.7	- 2.4	- 5.2	1.8

The makespan and the average machine utilization rate of the scheduling problems with alternative process plans improve, while the CPU time increases. Table 7 through 9 summarize the effect of alternative process plans on the quality of schedules generated with Algorithm 2.

Table 7. The percentage improvement of the makespan for Algorithm 2

Rule No.		1	2	3	4	5	6	7	8	9	10	11	12
T1	R1	11.4	10.9	9.6	- 1.8	9.1	8.9	1.4	1.9	9.5	10.1	8.9	1.5
	R2	11.6	10.1	- 1.0	4.9	2.6	6.6	- 6.8	6.1	8.3	5.1	4.2	2.1
	R3	13.9	10.9	- 8.3	1.1	6.6	7.4	- 3.4	- 0.1	4.4	8.4	8.2	- 1.4
	R4	5.6	14.2	6.5	2.4	10.6	11.1	9.0	4.3	10.0	9.1	9.4	4.1
	R5	15.0	8.4	- 0.7	2.7	7.3	7.5	- 1.0	0.3	6.8	7.9	8.7	- 0.2
T2	R1	12.4	10.2	6.8	- 0.8	9.6	10.4	8.7	6.6	10.3	8.4	8.6	1.5
	R2	8.6	10.6	- 4.4	5.0	4.1	7.1	- 4.8	6.1	5.2	9.8	4.3	7.0
	R3	11.1	4.2	- 8.2	7.6	4.5	7.7	- 5.3	0.2	6.4	7.6	8.1	1.3
	R4	14.4	10.5	6.5	6.8	9.9	12.6	10.2	4.7	8.0	10.8	12.6	3.6
	R5	13.6	10.3	- 0.5	7.4	7.2	7.7	2.6	3.8	7.6	7.5	6.4	3.1
T3	R1	13.5	10.9	8.3	5.1	8.7	10.5	8.2	5.1	7.7	9.6	9.2	6.0
	R2	12.6	9.6	- 3.7	10.6	7.1	8.1	- 1.6	3.6	8.5	10.3	7.6	8.6
	R3	13.2	7.1	- 3.6	5.8	8.2	5.4	0.1	4.8	7.0	8.4	10.2	4.8
	R4	12.3	9.9	8.5	9.9	9.7	9.3	9.5	6.9	9.8	12.1	11.0	6.9
	R5	10.7	12.9	- 0.5	9.2	8.3	10.3	5.6	4.7	8.5	6.9	9.0	6.0

Table 8. The percentage improvement of the average machine utilization for Algorithm 2

Rule No.	Problem	1	2	3	4	5	6	7	8	9	10	11	12
T1	R1	31.0	28.5	21.0	12.5	29.2	28.8	21.3	15.3	29.3	30.3	28.3	15.8
	R2	29.5	27.7	9.6	18.8	19.3	24.0	5.9	19.8	27.2	22.8	21.3	15.5
	R3	34.2	30.2	6.9	16.9	25.2	26.7	13.7	16.7	24.1	28.8	29.5	16.0
	R4	25.8	30.0	14.6	16.3	27.3	27.7	17.8	16.8	28.4	26.1	24.7	16.7
	R5	34.0	25.5	11.5	16.2	26.6	26.8	15.2	16.1	27.2	27.0	28.1	15.9
T2	R1	32.5	29.4	18.2	14.6	29.0	30.6	22.3	24.2	31.7	27.8	27.6	17.3
	R2	25.6	30.3	8.0	19.3	22.3	25.4	8.8	22.2	22.7	30.4	22.2	22.5
	R3	31.4	21.5	7.7	26.3	24.8	28.3	11.2	17.6	26.6	27.4	29.3	18.4
	R4	32.1	26.9	16.2	23.1	26.7	30.0	20.4	20.1	25.6	28.7	30.5	19.2
	R5	34.8	29.6	13.8	24.7	27.2	28.0	19.1	21.1	27.2	26.9	25.5	19.6
T3	R1	33.8	28.7	20.0	21.6	29.9	30.0	23.0	22.5	26.8	29.8	29.4	24.7
	R2	31.1	28.3	10.1	29.9	26.0	26.8	12.8	20.4	28.0	30.0	26.2	26.2
	R3	34.4	25.6	11.6	25.6	28.2	26.0	18.2	23.0	27.4	28.9	31.6	23.2
	R4	28.8	26.6	20.6	25.9	26.6	25.8	22.1	24.3	26.8	30.7	27.9	24.0
	R5	27.6	31.2	13.0	26.8	28.0	32.0	22.4	22.4	28.4	26.3	28.7	24.5

Table 9. The percentage change of CPU time for Algorithm 2

Rule No.	Problem No.	1	2	3	4	5	6	7	8	9	10	11	12
T1	R1	94.6	181.1	255.0	334.4	135.9	190.1	328.8	467.1	177.1	229.7	290.2	523.1
	R2	- 2.4	21.2	73.8	45.5	- 7.1	1.2	55.7	31.8	- 14.3	- 12.5	- 11.4	18.2
	R3	0.3	9.1	58.4	39.6	- 12.1	- 7.4	35.5	30.0	- 14.1	- 18.3	- 20.4	15.3
	R4	- 2.1	17.8	63.1	40.5	- 3.6	3.6	45.5	30.2	- 11.5	- 9.0	- 8.8	14.8
	R5	- 1.5	12.7	42.3	35.4	- 10.2	- 8.3	30.3	24.6	- 16.7	- 21.1	- 20.3	10.3
T2	R1	82.4	158.8	224.1	292.8	127.9	181.9	298.0	400.6	163.2	216.8	273.9	458.3
	R2	5.2	19.2	63.8	39.3	- 5.3	1.3	48.7	22.8	- 11.9	- 16.1	- 12.2	8.0
	R3	- 4.5	16.1	62.3	33.1	- 8.4	- 6.7	33.5	22.5	- 16.3	- 17.4	- 18.3	6.4
	R4	0.6	10.0	59.5	37.7	- 7.0	2.4	35.8	26.0	- 9.2	- 11.5	- 10.1	7.7
	R5	5.4	8.6	44.4	27.0	- 11.6	- 3.2	21.6	19.8	- 13.8	- 19.6	- 17.3	4.5
T3	R1	87.2	161.5	198.9	260.8	120.2	165.0	259.2	363.3	157.6	195.3	240.9	394.9
	R2	0.0	16.9	53.4	25.6	- 6.2	1.3	35.3	24.0	- 12.3	- 15.5	- 14.6	5.1
	R3	- 6.5	6.2	44.2	26.9	- 8.0	- 7.5	26.6	19.2	- 14.5	- 17.0	- 16.6	3.6
	R4	0.1	17.0	51.4	26.1	- 7.7	7.1	26.9	20.1	- 14.7	- 13.6	- 13.6	5.2
	R5	- 11.2	10.9	53.5	28.1	- 12.5	- 6.8	16.1	20.3	- 12.8	- 20.4	- 15.6	0.6

For all the problems with traveling times (T1, T2, and T3), the introduction of the alternative process plans improves the quality of schedules in most of the cases solved.

The two scheduling algorithms were coded in Pascal and ran on the Prime 9955 computer. The CPU times of the scheduling problems with alternative process plans for Algorithm 1 are shown in Table 10. The results shown in Table 10 are representative of all other problems.

Table 10. Average CPU time (sec) for 12 problems with alternative process plans for Algorithm 1

Rule No. \ Problem No.	1	2	3	4	5	6	7	8	9	10	11	12
R1	1.2	1.3	1.7	1.3	12.5	13.8	16.0	12.0	40.5	39.3	38.5	37.7
R2	0.9	1.0	1.3	1.0	10.1	11.3	12.0	8.6	32.1	29.6	27.6	22.4
R3	1.0	1.1	1.4	0.9	11.1	11.9	12.4	8.4	33.2	31.7	27.9	23.0
R4	1.0	1.1	1.3	1.0	10.7	11.8	12.6	8.3	32.0	30.3	27.9	22.1
R5	1.0	1.1	1.3	0.9	10.8	12.1	12.6	8.7	33.6	31.9	28.7	22.7

5. CONCLUSIONS

In the paper, two algorithms were developed for scheduling problems with and without alternative process plans in automated manufacturing systems. The effect of alternative process plans on the performance of the schedules generated was studied. The computational studies show that the quality of schedules (makespan, utilization of resources) improves when alternative process plans are used.

The testing effort involved a set of 9600 scheduling problems obtained for randomly generated data. The results show that each of the two algorithms can be used in practice for solving scheduling problems with alternative process plans.

REFERENCES

Blazewicz, J., Lenstra, J.K., and Rinnooy Kan, A.H.G. (1983) "Scheduling subject to resource constraints : Classification and complexity", Discrete Applied Mathematics, Vol. 5, pp. 11-24.

Chang, Y-L., Matsuo, H., and Sullivan, R. (1989) "A bottleneck-based beam search for job scheduling in a flexible manufacturing system", International Journal of Production Research", Vol. 27, No. 11, pp. 1949-1961.

Davis, E.W. (1973) "Project scheduling under resource constraints - Historical review and categorization of procedures", AIIE Transactions, Vol. 5, No. 4, pp. 297-313.

Dobson, G. and Karmarkar, U.S. (1986) "Large-scale shop scheduling: Formulations and Decomposition", Working Paper No. QM8631, Graduate School of Management, University of Rochester, N.Y.

Dobson, G. and Karmarkar, U.S. (1989) "Simultaneous resource scheduling to minimize weight flow times", Operations Research, Vol. 37, No. 4, pp. 592-600.

Graham, R.L., Lawler, E.L., Lenstra, J.K., and Rinnooy Kan, A.H.G. (1979) "Optimization and approximation in deterministic sequencing and scheduling : A survey", Annals of Discrete Mathematics, Vol. 5, pp. 287-326.

Kim, Y-D. (1990) "A comparison of dispatching rules for job shops with multiple identical jobs and alternative routeings", International Journal of Production Research, Vol. 28, No. 5, pp. 953-962.

Kusiak, A (1990) Intelligent Manufacturing Systems, Prentice Hall, Englewood Cliffs, N. J.

Kusiak, A and Ahn, J. K. (1990) "A Dispatching Rule for the Scheduling Problem in Automated Manufacturing", Working paper No. 90-03, Department of Industrial Engineering, University of Iowa, Iowa, IA.

Kusiak, A and Finke, G. (1988) "Selection of Process Plans in Automated Manufacturing Systems", IEEE Journal of Robotics and Automation, Vol. 4, No. 4, pp. 397-402.

Lloyd, E.L. (1981) "Concurrent task systems", Operations Research, Vol. 29, No. 1, pp. 189-201.

Monma, C.L. (1982) "Linear-time algorithm for scheduling on parallel processors", Operations Research, Vol. 30, No. 1, pp. 116-124.

Pritsker, A.A.B., Watters, L.J., and Wolfe, P.M. (1969) "Multiproject scheduling with limited resources : A zero-one programming approach", Management Science, Vol. 16, No. 1, pp. 93-108.

Schrage, L. (1984) Linear, Integer, and Quadratic Programming with LINDO, Scientific Press, Palo Alto, CA.

Sherali, H.D., Sarin, S.C., and Desai, R. (1989) "Models and algorithms for job selection, routing, and scheduling in a flexible manufacturing system", Working Paper, Dept. of Industrial Engineering and Operation Research", Virginia Politechnic Institute and State University, Blacksburg, Virginia

Srihari, K. and Greene, T.J. (1988) "Alternate routings in CAPP implementation in a FMS", Computers and Industrial Engineering, Vol. 15, Nos 1-4, pp. 41-50.

Scheduling and Resource Allocation Problems in Some Flow Type Manufacturing Processes

ADAM JANIAK

1 INTRODUCTION. EXAMPLES OF MANUFACTURING PROCESSES WITH PRODUCTION SCHEDULING AND RESOURCE ALLOCATION

Many manufacturing processes characterize by a flow of works (jobs) through successive machines (processors, devices) in a given technological order. The classical multi-machine flow-shop problems have traditionally been considered in many papers (e.g. in [11]) under the assumption that work processing times are-constant parameters. However, in many manufacturing processes work processing times may vary over the ranges of durations bounded by the maximum (normal) and minimum (crash) value of duration. In other words in such situations a scheduler (an algorithm) may have at its disposal certain continuously - divisible limited resources, such as financial outlay, energy, gas, fuel, catalyzer, raw materials, power, which may be applied to reduce effectively the processing times of individual works. Such problems require a more general approach to scheduling, one that determines not only the order in which works should be processed, but also the distribution of limited resources to individual works or to machines on which they are processed.

Now we very shortly describe three manufacturing systems in which the situation mentioned above exists.

I. The first process exists between converter plants and blooming mill in steel mills [13]. After the termination of a cyclic process of converter heat in one of several converters, hot liquid metal is poured into steel ladles and next into ingot moulds where it solidifies. Then after the ingot stripper process, ingots are appriopriately segregated into batches and next transported to the suitable batery of soaking pits, where the process consists in preheating ingot batches up to the requested temperature. Each ingot batch consists of several ingots, and is preheated in a separate soaking pit. Ingots are preheated by gas. For each soaking pit, the maximum and minimum gas flow intensity is limited. Individual soaking pits may be available for preheating at diffirent moments (cyclic process). After the preheating process has been completed, ingots from each batch are hot-rolled by a blooming mill. There is usually only one blooming mill in the whole procuction line. A blooming mill is very expensive (more than all the batteries of soaking pits) and so it may be treated as a critical machine in this production line. Soaking pits have practically no bottlenecks in the process. Preheating time of each ingot batch in each soaking pit (up to an appriopriate temperature-about 1400°C) is inversely proportional to the gas flow intensity. (This time also depends on the number, format and type of pig iron of ingots in batch and on the technical state of the employed soaking pit. These influences are usualy included in the constant parameters of the preheating time models). Each batch of ingots preheating in a separate soaking pit creates a work. The rolling time of each ingot batch may vary for different ingot batches, because it depends on the number, format, temperature and type of pig iron of ingots in batches, however, for each ingot batch it is fixed and given. It may be given a constraint on the gas flow intensity carried to the whole battery of soaking pits or it may be given a deadline on the completion time of the hot-rolling of ingots on the blooming mill.

It is easy to notice, that the above described process may be modelled by means of a single machine work scheduling task with release dates (i.e. moments at which works are available for processing) being positive, strictly decreasing continuous functions of resource amounts. This is due to the fact that the preheating time of each ingot batch may be treated as a release date for the work of ingot batch hot-rolling on the blooming mill (single machine).

The solution manners of the above task for various criteria are shortly presented in Section 2.

II. Very similar processes (however much more complicated) appear in the forge shop department of steel plants. This department consists of such technological devices as: various forge furnaces, forging presses and transport facilities (like heater cars and crane facilities). Works are forging stocks and forgings (e.g. working rolls, crankshaft of marine engine, turbine wheel) which may flow many times among different forge furnaces and forging presses (with different forging pressures). Forgings are preheated (up to an appriopriate temperature) by gas in forge furnaces. Their preheating times are dependent on gas flow intensity (which is limited for each forge furnace and may be limited for the whole battery of forge furnaces).

It is easy to notice that the above described process may be modelled by means of the flow shop scheduling problem with parallel machines in each work centre and with work processing times dependent on resource.

III. Similar problem appears during the copper electrorefining process in copper industry [10] where the problem of scheduling of electrocopper production and allocation of electric power among groups of electrolytic tanks may be also reduced to the flow shop problem with identical parallel machines (groups of electrolytic tanks) in the first work centre and (usually two) identicall parallel machines (overhead cranes) in the second work centre. At the same time, production times of copper cathodes in each of two phases of anodic cycle are inversely proportional (functions) to the amount of alloted electric power (positive strictly decreasing, nonlinear, convex, contrinuous functions). Global amount of electric power may be limited. Analogical problem occurs during the aluminium electrolysis process in aluminium industry.

The solution manner of the flow-shop problem with work processing times dependent on resource on some machines are presented in Section 3 under the assumption that each work centre contains one machine. (The algorithm presented there may be easily generalized on the case with several parallel machines in each work centre). Section 4 reports on some computational experience with the algorithm presented in Section 3 and contains some concluding remarks.

2 SINGLE MACHINE SCHEDULING CASE

This section is devoted to the solution of the following three problems which may be considered for the first manufacturing process I, described in the previous section.

Ia. For the case when the global gas flow intensity (carried to the whole battery of soaking pits) is limited, find a feasible control, i.e. a processing order of ingot batches on the blooming mill and an allocation of the gas flow intensity (locally and globally limited) into the individual soaking pits, such that the standstills of the blooming mill are minimized. It may be easily proved that the minimization of the standstills is equivalent to the minimization of the maximum job completion time.

Ib. For the case when a deadline on the completion time of the hot-rolling of ingots on the blooming mill is given, find a feasible control, i.e. an ingot batches processing order on the blooming mill and gas (locally limited) allocation, such that the total gas consumption is minimized, subject to a given deadline.

Ic. Bicriterial approach. Find the entive set of Pareto optimum efficient controls each consisting of an ingot batches processing order and gas allocation, i.e. compromise points between the blooming mill standstills (representing in a certain sense the quality of the control) and the total gas consumption (representing in a certain sense cost of the control).

It is easy to notice that the problem Ia may be reduced to the maximum completion time minimization problem for the single machine scheduling with release dates dependent on continously - divisible resource amounts. This problem for different models of release dates is strongly NP-hard [9], whereas for identical models of release dates (being certain positive strictly decreasing continuous functions) it may be solved in polynomial time [4]. In [9] several approximate algorithms (with worst case analysis) is given for the NP-hard case with models of release dates being various positive strictly decreasing linear functions.

In turn, the second problem Ib may be transformed to the problem of minimization of the total gas consumption for the single machine scheduling with release dates dependent on resources subject to a given common deadline. Similarly as for the Ia problem, this problem is strongly NP-hard [7] for different models of release dates and it is polynomially solved for identical models of release dates [8]. Analogical properties hold for the third problem Ic, i.e. for the bicriterial approach (see [7,8]).

3 THE OPTIMAL CONTROL IN MULTI-MACHINE FLOW TYPE MANUFACTURING PROCESSES

3.1 PROBLEM STATEMENT

It is mentioned in Section 1 that the second II and the third III manufacturing processes (described in Section 1) may be reduced to the flow-shop problem with work processing times dependent on continuously-divisible resource amounts. The present section is devoted to the statement and solution of this problem, which may be formulated as follows. A set of n multiple-activity works $W_1,\ldots,W_i,\ldots,W_n$ is available for processing at time zero. There may exist technological precedence constraints between works that can be conveniently represented by sets $A_i \hat{=} \{d | W_i < W_d\}$ (or equivalently by $B_i = \{d | W_d < W_i\}$) $i=1,2,\ldots,n$, where $W_i < W_d$ means that the processing of W_d cannot start before W_i has been completed. The shop contains m different machines M_j, $j \in M \hat{=} \{1,2,\ldots,m\}$, and each work W_i consists of m activities $O_{i1},\ldots, O_{ij},\ldots, O_{im}$, each of which requires a different machine. The flow of work is unidirectional from the first machine M_1 to the last one M_m.

For the cases when fewer activities exist in a work, the corresponding processing times can be taken to be zero. Moreover, it is assumed that machines are continuously available and each of them can process only one activity at a time. Individual activities are not preemtable. Furthermore, we shall assume that for the activities performed on machines with indices from the set $M^1 \subset M$ the processing times t_{ij} of activities are constant, i.e. for $i=1,2,\ldots,n$, $j \in M^1$. Whereas, for the activities performed on machines with indices from the set $M^2 \hat{=} M \setminus M^1$ the processing time t_{ij} of the activity O_{ij} depends linearly on an amount of resource r_{ij} alloted to realization of this activity, i.e.

$$t_{ij} \hat{=} b_{ij} - a_{ij} \cdot r_{ij}, \quad i=1,2,\ldots,n, \quad j \in M^2,$$

where $b_{ij} > 0$, $a_{ij} > 0$ are known parameters.

A resource allocation $r \hat{=} [r_{1j_1}, r_{2j_1},\ldots,r_{nj_1},r_{1j_2},\ldots,r_{nj_2},\ldots, r_{ij_k},\ldots,r_{nj_c}]$ (where $j_k \in M^2$, $k=1,2,\ldots,c$, and c is the cardinality of the set M^2) will be called admissible for the production process under consideration if the following constraints are satisfied:

$$\underline{r}_{ij} \leq r_{ij} \leq \bar{r}_{ij}, \quad i=1,2,\ldots,n, \quad j \in M^2, \qquad \sum_{j \in M^2} \sum_{i=1}^{n} r_{ij} \leq R',$$

where $\underline{r}_{ij}$, $\bar{r}_{ij}$ are known technological constraints on the amount of resource alloted to realization of the activity O_{ij}, $0 \leq \underline{r}_{ij} \leq \bar{r}_{ij} < b_{ij}/a_{ij}$, and R' is the global amount of resource alloted to realization of the set of works $W_1,\ldots,W_n$, $R' \geq \sum_{j \in M^2} \sum_{i=1}^{n} r_{ij}$. The set of all admissible allocations of resource will be denoted by R.

A processing order of works in the considered production process may be represented by the vector $p \hat{=} [p^1, p^2,\ldots, p^j,\ldots,p^m]$, where $p^j = [p^j(1), p^j(2),\ldots, p^j(n)]$ is the permutation of work indices defining the succession of works (activities) on the machine M_j. Let P be the familly of all work processing orders p satisfying given precedence constraints.

The production process may be evaluated according to various optimality criteria. In the presented paper we will minimize the length of the time T required to complete all works: $T \triangleq T(p,r) \triangleq C_{pm(n)}(p,r)$ where, for the processing order of work $p \in P$ and resource allocation $r \in R$, $C_{pm(n)}(p,r)$ is the moment at which the last work $W_{pm(n)}$ on the last machine M_m is completed. In other words, the problem is to find such a control, i.e. such a processing order of works $p^* \in P$ on successive machines and such a resource allocation $r^* \in R$ that the time required to complete all works T(p,r) is minimized, $T(p^*,r^*) = \min_{p \in P} \min_{r \in R} T(p,r)$.

The problem under consideration is a generalization of the very difficult classical flow-shop problem, i.e. with constant processing times of activities, which was considered e.g. in [11].

In [3] it was proved that the two machine case of the problem under consideration is NP-hard even for identical slopes of models of works on the first machines (i.e. for $a_{ij} = a_j$, $i=1,2,\ldots,n$) and constant work processing times on the second one.

The first attempt to solve the problem under consideration was made in [2], where some branch and bound algorithm was presented. In each step of that algorithm only one work (from the set of candidates which can be moved without violating the fixed precedence constraints) was moved before or after some work of the permutation obtained in the previous step. The candidates for movement was evaluated by means of some lower bounds. Now we will outline another branch and bound algorithm in which candidates for movement are evaluated by means of special expressions which were constructed analogically as for the algorithm solving the job-shop problem with resource constraints, presented in [1]. Subsection 3.2 is devoted to basic notions and notations. Some properties of the considered problem are also given in subsection 3.2. Outline of the mentioned algorithm is given in subsection 3.3.

3.2 BASIC NOTIONS AND NOTATIONS. PROBLEM PROPERTIES.

The following notions and notations will be used.

For a given processing order of works $p \in P$, a sequence of activities such that the first and the last activity of this sequence is an activity performed as the first one on the machine M_1 and the last one on M_m, respectively, and the direct descendant of any activity O_{ij} (except the last one) in this sequence is either the activity O_{ij+1} (i.e. the next one in the work W_i) or the activity performed directly after O_{ij}

on M_j in p, is called a path in p.

By $r_p^* \in R$ will be denoted the optimal resource allocation in $p \in P$, i.e. a resource allocation minimizing the length of the time required to complete all works for the fixed p, i.e. T(p,).

For the optimal resource allocation r_p^* in $p \in P$, the sum of the processing times of activities creating the given path will be called the length of this path.

A path with the maximal length in $p \in P$ will be called the critical path in p.

It can be easily proved that the length of the critical path in $p \in P$ is equal to $T(p, r_p^*)$ and there is no idle time inserted between any two succeeding activities of the critical path. It is easy to notice that there may be several critical paths in $p \in P$ under the $r_p^* \in R$.

A subsequence of succeeding activities $\langle O_{i_1 j}, O_{i_2 j}, \ldots, O_{i_{f-1} j}, O_{i_f j} \rangle$ of a certain critical path in $p \in P$ containing the maximal number of activities performed on the same machine (M_j) such that all its activities belong to the same critical paths, at the same time $O_{i_1 j}$ or $O_{i_f j}$ may also belong to other critical paths (different for $O_{i_1 j}$ and $O_{i_f j}$), will be called a section in p.

It is easy to notice that there may be several sections on each machine M_j in $p \in P$ under $r_p^* \in R$.

By S_k and k_p will be denoted the set of activities of the k-th section and the number of sections in p, respectively.

The following property can be proved [2]:

<u>Property 1</u>

For each $p \in P$, if the processing order of works $p' \in P$ has been obtained by an interchange of the succession of activities in p and if $T(p', r_{p'}^*) < T(p, r_p^*)$, then in p′ at least one activity in at least one section of p is moved before its first or after its last activity.

The above property is the crucial fact on which the branch and bound algorithm, presented in the next section, is based.

3.3 BRANCH AND BOUND PROCEDURE

3.3.1 Lower bound

In the branch and bound method starting with any initial processing order of works $p_1 \in P$ a sequence of processing orders $p_r \in P$ is generated. For each p_r a lower bound LB_r on the value of time required to complete all jobs $T(p_s, r_{p_s})$ for every $p_s \in P$ which can be generated from p_r is calculated. The lower bound is obtained (see [2]) by the relaxation of the capacity constraints on all the machines except for the j-th one. In this order for each activity O_{ij}, $i=1,2,\ldots,n$, we calculate a head h_{ij} of the activity O_{ij}, i.e. the length of the longest path from the activity processed as the first one on the first machine in p_r to the considered activity; the length of each of these paths being calculated for the optimal allocation of the resource amount $R' - \sum_{j \in M^2} \sum_{i=1}^{n} r_{ij}$ among the activities creating the path considered. For each activity O_{ij}, $i=1,2,\ldots,n$ we also calculate a tail q_{ij} of the activity O_{ij}, i.e. the length of the longest path from the considered activity to the activity processed as the last one on the last machine in p_r. The length of each of these paths is calculated analogically to h_{ij}. The processing time of each activity O_{ij}, $i=1,2,\ldots,n$, performed on the machine M_j is equal to t_{ij}, at the same time $t_{ij} = b_{ij} - a_{ij} r_{ij}$ for $j \in M^2$ and $t_{ij} = \text{const}$ for $j \in M^1$.

Analogically to the algorithm of solving the flow-shop problem in [2] for each p_r generated under the presented algorithm the set F_{p_r} of some precedence constraints between activities is fixed. (In [2] the set of, a certain disjunctive arcs was fixed in each considered graph representing given processing order of jobs.). Thus the lower bound may be provided by the minimized value of the time required to complete activities performed on M_j under the conditions that each activity O_{ij}, $i=1,2,\ldots,n$ (a) becomes available at h_{ij}, (b) is processed during t_{ij} units without interruption (if $j \in M^2$, then $t_{ij} = b_{ij} - a_{ij} r_{ij}$), (c) is completly finished q_{ij} units after leaving M_j and (d) the given technological precedence constraints and fixed F_{p_r} ones in p_r are respected. It may be proved that the above problem is NP-hard and solution time of it is too excesive to apply it as a lower bound. In the implemented algorithm lower bound is obtained from the above lower bound by the relaxation of the values h_{ij} (or q_{ij}), i.e. by assumption that all h_{ij} (or q_{ij}) are equal to $h^* = \min_{1 \le i \le n} h_{ij}$ (or $q^* = \min_{1 \le i \le n} q_{ij}$).

Anallogically to Theorem 1 in [5] it can be proved that the optimal solution value LB_j^q of the above relaxed problem (for $q_{ij} := q^*$) may be obtained by ordering p^j the activities on the machine M_j according to nondecreasing h_{ij} with simultaneous regard to the precedence constraints beetwen activities and, in the case when $j \in M^2$, by the allocation, of the resource amount $R' - \sum_{j \in M^2} \sum_{i=1}^{n} r_{ij}$ among the activities O_{ij}, $i=1,2,\ldots,n$, obtained due to the following procedure:

Procedure 1

Step 1. Set $r^*_{ij} = \underline{r}_{ij}$ for any $i=1,2,\ldots,n$, $R':=R' - \sum_{j\in M^2}\sum_{i=1}^{n} r_{ij}$ and $c:=1$. Find the starting times of activities on M_j using the following recursive formulaes:

$S_{p^j(1)} := h_{p^j(1),j}$, $S_{p^j(i)} := \max[h_{p^j(i)}, S_{p^j(i-1)} + b_{p^j(i-1),j} - a_{p^j(i-1),j} \cdot \underline{r}_{p^j(i-1),j}$, $i=2,3,\ldots,n$ and go to Step 2;

Step 2. Find the greatest index k, $c \leq k \leq n$ that satisfies the following equation $h_{p^j(k),j} = S_{p^j(i)}$. Then find the set $B:=\{p^j(i) \mid k \leq i \leq n,\ r^*_{p^j(i),j} < \bar{r}_{p^j(i),j}\}$. If the set B is empty or $R' = 0$, then Stop - r^*_{ij}, $i = 1,2,\ldots,n$, is the optimal resource allocation for p^j, otherwise go to Step 3;

Step 3. Find an index t, that satisfies $a_{p^j(t),j} := \max_{p^j(i)\in B} a_{p^j(i),j}$ and then set $d:= \min_{t<i\leq n} (S_{p^j(i)} - h_{p^j(i),j})$, $z:=\min\{\bar{r}_{p^j(t),j} - \underline{r}_{p^j(t),j},\ R',\ d/a_{p^j(t),j}\}$. Next bring up to date resource allocation and starting times by setting $r^*_{p^j(t),j} := r^*_{p^j(t),j} + z$, $R' := R' - z$, $S_{p^j(i),j} := S_{p^j(i),j} - z \cdot a_{p^j(t),j}$ for $t < i \leq n$. Finally set $c:=k$ and go to Step 2.

The value LB^h_j, for the case when $h^* = \min_{1\leq i\leq n} h_{ij}$, may be obtained in a similar way.

Consequently, in the problem under consideration, for each processing order of works $p_r \in P$, the lower bound is given by:

$$LB_r = \max_{1\leq j\leq m} [\max(LB^h_j, LB^q_j)].$$

3.3.2 Branch and bound algorithm

In branch and bound procedure a sequence of work processing orders $p_r \in P$ is generated. An initial processing order $p_1 \in P$ may be obtained by a certain heuristic method (see e.g. [2]). For each p_r from this sequence the lower bound LB_r and the optimal allocation of resource $r^*_{p_r} \in R$ is computed, next all the sections, that contain at least two operations are identified and enumerated $k=1,2,\ldots,k_{p_r}$. Each new processing order of works $p_s \in P$ is obtained from an immediately preceding processing order $p_r \in P$ of the sequence by moving one activity of some section in p_r before its first (or after its last) activity (see Property 1).

By C^b_k (C^a_k) will be denoted the set of activities which can be moved before the first (after the last) activity of the k-th section in p_r

without violating the given technological precedence constraints and fixed in p_r F_{p_r} ones. From the activities of the sets C_k^a and C_k^b, $k=1,2,\ldots,k_{p_r}$ we want to choose an activity which generates a descendant p_s of p_r with the smallest possible $T(p_s, r_{p_s}^*)$ in order to obtain quickly a good upper bound.

For evaluation of all the descendants $p_s \in P$, which can be generated from $p_r \in P$, special expressions $\Delta_k^b(O_{ij})$ and $\Delta_k^a(O_{ij})$ are useful. These expressions are analogical to the expressions $\Delta_k^b(j)$ and $\Delta_k^a(j)$ used in the same purpose in the job-shop problem in [1]. The expression $\Delta_k^b(O_{ij})$ ($\Delta_k^a(O_{ij})$) evaluates the descendant $p_s \in P$ which can be obtained from p_r by moving the activity O_{ij} before the first (after the last) activity in the k-th section in p_r.

Algorithm 1

Step 1. Compute the lower bound LB_r for p_r. If $LB_r \geqslant T(*,*)$, where $T(*,*)$ is the current upper bound, then go to Step 4, otherwise, go to Step 2.

Step 2. Compute $r_{p_r}^* \in R$ and $T(p_r, r_{p_r}^*)$ for p_r. If $T(p_r, r_{p_r}^*) < T(*,*)$, then set $T(*,*):=T(p_r, r_{p_r}^*)$. Identify the C_k^a and C_k^b - sets of candidates in p_r. If $C_k^a = \emptyset$ and $C_k^b = \emptyset$, for $k=1,2,\ldots,k_r$, then go to Step 4. Otherwise, for each activity O_{ij} of these sets compute expressions $\Delta_k^b(O_{ij})$ and $\Delta_k^a(O_{ij})$ [1] and go to Step 3.

Step 3. If $C_k^a = \emptyset$ and $C_k^b = \emptyset$, then go to Step 4. Otherwise, from among the candidates in p_r, choose an activity O_{ij} with the smallest value $\Delta_k^e(O_{ij})$, $e \in \{a,b\}$. If $\Delta_k^b(O_{ij})$ (or $\Delta_k^a(O_{ij})$ is chosen, then generate a new processing order of works $p_s \in P$ by moving the activity O_{ij} before the first (or after the last) activity in the k-th section in p_r, fixing the precedence constraints F_{p_s} and setting $C_k^a:=C_k^a-\{O_{ij}\}$ (or $C_k^b:=C_k^b-\{O_{ij}\}$. Then set $p_r:=p_s$ and go to Step 1.

Step 4. Backtrack to the direct predecessor p_p of p_r. If p_r has no predecessor, then the algorithm terminates: the processing order of works p^* and the resource allocation r^* associated with the current upper bound $T(*,*)$ are optimal. Otherwise, eliminate the processing order p_r and all its data, set $p_r:=p_p$ and go to Step 3.

4. COMPUTATIONAL EXAMPLE. CONCLUDING REMARKS.

The algorithm described in the previous section was tested using a problem constructed on the basis of the known job-shop problem with six works (each containing six activities) performed on six machines, which is one of the three benchmark problems discussed in the literature [12, p. 236].

The tested problem was constructed in the following way. There were changed the technological precedence constraints in each work in such a way, that an activity of a work W_i performed on the j-th machine in the benchmark problem became the O_{ij} activity in W_i. In this way the flow-shop problem was obtained from the job-shop one. Furthermore, it was assumed that on the j-th machine the processing times of activities are linearly dependent on resource amounts in the following way: $t_{ij} = b_{ij} - a\,r_{ij}$, $0 \leqslant r_{ij} \leqslant \bar{r}$, i=1,2,...,6 and the processing times of the activities performed on the other machines are equal to that of the benchmark problem. This assumption is considered for successive six machines from the problem, i.e. for v=1,2,...,6. It is arbitrarily assumed that a=10, r=15 for all investigated examples of the problem under consideration and $R' := 0.1\cdot(6\,\bar{r})$, $0.5\cdot(6\,\bar{r})$, $0.9\cdot(6\,\bar{r})$. It is supposed that $b_{ij} = d_{ij} - a\cdot\bar{r}$ for i=1,2,...,6, $j \in M^2$, were d_{ij} is the processing time of the activity of the work W_i performed on the j-th machine in the benchmark problem. Table I shows the computational results for $R' = 0.5(6\,\bar{r})$.

j	Initial solution $T(p_1, p^*_{p_1})$	Optimal solution $T(p^*, r^*)$	Initial lower bound LB_1	Solution time in seconds (Algol on computer ODRA 1325)	Number of generated nodes
1	497	492	492	69	2
2	483	483	483	12	1
3	482	476	469	146	4
4	488	475	473	329	9
5	513	508	501	220	6
6	506	502	499	287	8

Table I. Computational results.

It is obvious that the addition of resource constraints to the classical flow-shop problem [11] makes more difficult the solution of it. It is easy to notice that the change of the machines with resource constraints have some influence on the solution time and the number of generated nodes. An increase in the number of generated nodes for which the optimal resource allocation is calculated does increase the solution time. An increase in the number of candidates, for which expressions evaluating them are calculated, involves an increase of the solution time as well.

The same example was also used for testing the previous algorithm, given in [2]. Comparing the results obtained in [2] and here it is difficult to state univocally which algorithm is better. There are some examples for which the algorithm presented above is better, however there are ones for which the previous algorithm is better.

Analogically as for the job-shop problem (see [6]), the presented algorithm may be generalized to the case with several machines in each work centre. In this case the problem must be modelled by means of a disjunctive graph.

The method proposed may be used for solving other production problems in which the critical path concept can be used.

5. REFERENCES

|1| J. Grabowski, A. Janiak, Job-shop scheduling with resource-time models of operations, Europ. J. Oper. Res., 28, 58-73, 1987.

|2| A. Janiak, General flow-shop scheduling with resource constraints, Int. J. Prod. Res. 26, 1089-1103, 1988.

|3| A. Janiak, Minimization of resource consumption under a given deadline in the two-processor flow-shop scheduling problem, Inf. Proc. Letters, 32, 101-112, 1989.

|4| A. Janiak, Time-optimal control in a single machine problem with resource constraints, Automatica, 22, 745-747, 1986.

|5| A. Janiak, One machine scheduling problems with resource constraints, in: Proc. of the 12th IFIP Conference System Modelling and Control, Budapest, Springer-Verlag, 84, 358-364, 1986.

|6| A. Janiak, Job-shop problem with parallel machines subject to resource constraints, Scientific bulletins of the University of Mining and Metallurgy, 928, 339-350, 1982 (in Polish).

|7| A. Janiak, Single machine scheduling problem with a common deadline and resource dependent release dates, Europ. J. Oper. Res. (to appear).

|8| A. Janiak, Resource-optimum control in a single machine problem with completion time constraint Report No 63/89 Inst. Eng. Cyber., Wrocław Techn. Univ.

|9| A. Janiak, Minimization of the blooming mill standstills - Mathematical model - Suboptimal algorithms, Scientific bulletins of the University of Mining and Metallurgy 8/2, 37-49, 1989.

|10| A. Janiak, Optimum and approximation algorithms of sequencing jobs and resources allocation in discrete and discrete-continuous manufacturing processes. Monograph of Technical University of Wrocław 1990/91 (to appear, in Polish).

|11| B.J. Lageweg, J.K. Lenstra, A.H.G. Rinnooy Kan, A general bounding scheme for the permutation flow-shop problem Oper. Res. 28, 53-67, 1978.

|12| J.F. Muth, G.L. Thompson (eds), Industrial Scheduling Prentice Hall, Engelwood Cliffs, 1963.

|13| T.J. Williams(ed), Analysis and design of hierarchical control systems: with special reference to steel plant operations, North-Holland, 1986.

The paper was partially supported by R.P.I.02 "Theory of control and optimization of continuous dynamic systems and discrete processes".

Intelligent Control of Flexible Manufacturing Systems

SANJAY JOSHI and RICHARD WYSK

1. Background

Computer Integrated Manufacturing (CIM) has existed, in concept, for several years. While it is still generally believed that CIM can have a major positive impact on productivity, it is far from commonplace in today's factories. The high costs of software development, maintenance, and integration are among the most prominent reasons for our slow evolution to CIM. More specifically, many computer packages have been developed to address specific manufacturing tasks like scheduling or process planning. Unfortunately, these software systems fail to provide the "hooks" needed for total Manufacturing System integration. This happens because these programs are: 1) developed by different vendors, on different hardware platforms, 2) not meant to be integrated with programs from other vendors, and 3) not part of a overall CIM architecture.

This paper addresses these topics as they relate to a shop floor control system (SFCS) for CIM. The objectives of the paper are to show and demonstrate a generic, flexible, cost effective architecture to plan, schedule, and control shop floor actions. This will eventually lead to a reduction in the time required to convert a manufacturing system design into an operational manufacturing system.

2. The Proposed System Architecture

A prototype hierarchical shop floor control system (SFCS) consisting of several levels is described. The detail of the cell, workstation and equipment levels for a CIM system is also presented. System equipment includes NC machining centers, robot systems, AGVs, or any other computer controlled manufacturing equipment and related tooling resources. We assume that the cell controller periodically receives a list of jobs, with associated due dates from a Master Production Scheduler. For each job, a process plan will be retrieved either from a database or as a file transfer. The SFCS will determine the sequence of activities necessary to execute that process plan. This involves the selection of a specific process routing and activities, scheduling of all activities at both the cell and workstation levels, coordinating those activities across the equipment, monitoring activities, job tracking, and some error recovery. To our knowledge, no system exists which distributes and integrates all of these functions within and across levels of a control hierarchy.

2.1 The Architecture

A "systems approach" to this problem is essential to achieve the level of integration that is required for CIM. A primary component of any system theory is the definition of the state of the system. For any complex system like manufacturing, the definition of the system state could involve thousands of variables, some discrete, some continuous, many stochastic. As the system evolves in time, those variables take on many different values. The goal is to define a set of decision rules which govern the way in which that evolution takes place. These rules can be generated "off-line" or "on-line", and are invoked to "control" how and when the system changes. Since the complete problem is too large and with complex interactions, a decomposition of the problem is necessary to create a series of well defined solvable subproblems. The decomposition should result in creating components that perform tasks that are relatively independent within the context of the larger system. The need for decomposition, creates the necessity of providing clear and well defined "hooks" to integrate the various decomposed problems into an integrated whole. The ability to effectively decompose and define the necessary "functional hooks" provides scaleability to the architecture.

A 3-level hierarchical decomposition of the SFCS is shown in Figure 1. Each hierarchical controller module performs 3 main functions - planning, scheduling and control. The equipment level corresponds to the physical devices such as Numerically Controlled (NC) machines, robots, measuring and inspection machines, material handling devices, programmable fixtures, etc. The equipment is usually controlled by a device controller which determine the various capabilities of the equipment. The equipment may perform a single or variety of operations on a part. The workstation level as defined in the hierarchy, corresponds to several integrated pieces of equipment. At the simplest level, a workstation could be a robot tending a single machining center, along with the requisite fixtures, buffers, and sensors. A more complex workstation could be a single robot tending several machines. As an example, a rotational workstation could consist of a robot and several machines (NC turning centers, NC grinders, etc.) performing a various rotational operations. The definition of what constitutes a unit operation determines the configuration, size and individual characteristic of a workstation. The notion of a unit operations is vital to the definition of a workstation. The minimum requirement of a workstation is that a part be coupled and uncoupled with at least one piece of processing equipment. A cell is viewed as several integrated workstations, coupled by material transport workstations. A cell can be considered a virtual entity and exists to couple and uncouple the unit operations of the workstation level. The primary activity of the cell is to provide integrate workstation activities into a broader system. Typical architectural characteristics for each level are shown in Table 1.

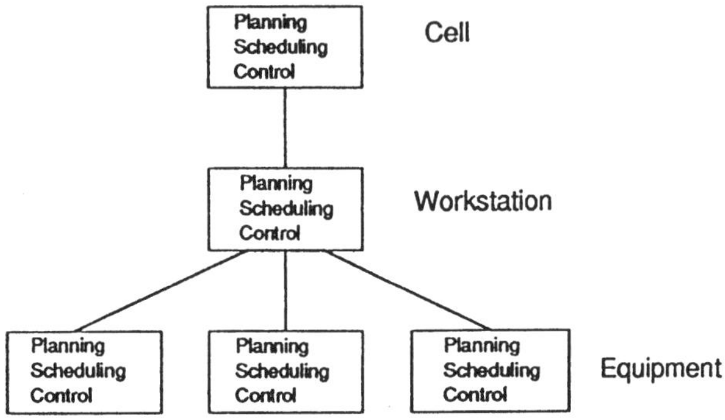

Figure 1. Three-level control hierarchical decomposition.

A key element of this architecture is that each level controller module performs the same functions--planning, scheduling and control. For the proposed SFCS, *planning* is the activity responsible for selecting/updating the process plan or strategy to be used in executing assigned jobs and generating revised completion times for those jobs. Cell level process plans contain routing summaries for each job. Each workstation plan contains the operations to be done and the information necessary to generate part programs or their equivalent for each processing step. At each level, *scheduling* is responsible for evaluating candidate process plans and generating/updating a sequence of expected start and finish times (the schedule) for the activities in the selected plan. *Control* is responsible for interfacing with all subordinates and other applications, and executing scheduler commands. It initiate start-up and shutdown, issues commands to perform assigned activities, uses feedback to monitor the execution of those activities, and oversees error recovery. Note that these functions are executed at different frequencies within the same level and across separate levels. Table 2 shows a functional breakdown of the activities at each level in the hierarchy.

Execution of commands by each controller occurs from a top- down execution. In case of failure and error recovery, the information and requests flow in a "bottom-up" manner, both with each controller and between the different levels as shown in Figure 2.

There are several issues which must be addressed before this approach can be implemented in a real system: state definition at each level; robust schedules; statistical analysis and rule selection; and error recovery. At this point, we expect to use some type of AI knowledge representation scheme. This

Table 1. Typical Architectural Characteristics

Item	Equipment	Workstation	Cell
EXAMPLES Hardware	Lathe, Mill, T-10 Bridgeport Series I IBM 7545 Robot	Robot tended Machine Center, Cartrac Material Handling System	Variable Mission System, Several Integrated Workstations
Controller Hardware	Mark Century 2000, Accuramatic 9000, Custom-single-board system	Allen-Bradley PLC-4, IBM-PC/AT, etc.	VAX 11/750, SUN Workstation, etc.
Type Controller	Single-board processors, Machine tool controller, Servo-Controller, etc	PLC, PC Minicomputer	PC, Microcomputer Super-Minicomputer
Language Application	Assembler, Part programming, Robot programming, etc.	C, Lodder logic, Pascal and other sententual languages	C, LISP, FORTRAN and other high level languages
Memory/Size Requirements	8k - 128k RAM plus custom ROM, EPROM, etc.	256k - 1 Meg RAM 1 Meg - 80 Meg Hard drive	512k - 4 Meg RAM 10 Meg - 1Gigabyte Hard drive
Response time	< 10exp(-3) sec	< 1 sec	< 20 sec
Machines/ Interconnects	1 - 1 connect	1 - many 1 - {1,8] Machine tools 1 - {1,50] Material handling	1 - many 1 - {1,15] Workstations

Table 2. Functional Breakdown of Control Architecture

Level / Functions	Equipment	Workstation	Cell
Planning	Tool selection, parameter specification, tool path refinement, G,M,T code, tool assignment to slots, job setup planning	-Resource allocation jobs -Batch splitting and equipment load balancing	Batching, Workload balancing between workstations, Requirements Planning -Task allocation to workstations
Planning Horizon	Millisecs - Minutes	Minutes - Hours/Days	Hours - Days/Weeks
Scheduling	-Operation sequencing at individual equipment	-Sequence equipment level subsystems -Deadlock detection and avoidance -Gantt chart or E. S. based scheduling -Buffer management	-Assignment of due dates to individual workstations -Look ahead ES/ simulation based scheduling -Optimization based tech -Batch sequencing
Control	-Interface to workstation controller -Physical control (motion control at NC and robot pick and place level) -Execution of control programs (APT, AML, etc.)	-Monitor equipment states and execute part and information flow actions based on states -Synchronize actions between equipment (e.g. robot and machine while loading/unloading parts) -Ladder logic execution	Organizational control of workstations, interface with MPS, generation of reports, etc.

allows us to capture the dependencies that exist between the two and to have efficient updating procedures.

The key to developing this type of distributed system lies in the generation of "robust schedules". Robust means that the schedule remains close to optimal (relative to some performance measures) over a realistic range of uncertainties. This means that the schedule is rigorous enough to absorb minor perturbations. This will limit the number of reschedulings that must be done when something goes wrong [Storer,1990].

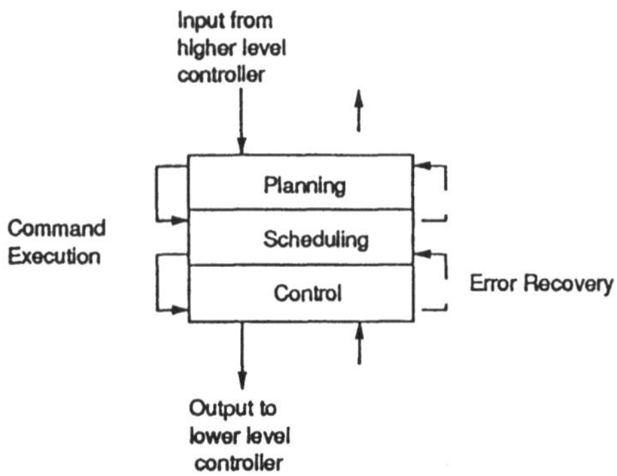

Figure 2. Execution schema.

Error recovery in this context means dealing with the impact of delays. This is a two step process: determine the impact on the current schedule and update the schedule. The first step requires a robust data structure for the schedule which allows one to determine the "ripple effect" of a delay on the current start and finish times. Based on this analysis, the recovery can be a quick fix (in which a few times get revised), a major rescheduling, and/or a major replanning.

2.2. Process Plan Representation

Process plans play an important role in the definition of the control structure, as well as in defining alternative sequences of operations that must be known by the planning and scheduling modules. The need for an alternative manifests itself in two important ways:

(i) it provides dynamic schedules with various alternatives available at any instant

(ii) the process of creating control software to implement the scheduling actions requires that alternatives be known priori, so that control software can provide for the necessary links in the control architecture.

A process plan representation is required that is capable of representing all multi-level interactions and possible precedence that occur among the planning and processing decisions. An AND/OR precedence graph based representation is proposed as a compact representation of the process plan [Mettala and Joshi, 1989]. The hierarchical nature of CIM control makes the graph representation even more attractive. Figure 3 shows the use of hierarchical AND/OR graphs to represent different levels of process planning required. The cell level process plan represents that the part must visit four workstations, and two alternative sequences exist ({W1-W2-W3-W4}, {W1- W3-W2-W4}) based on the precedence requirements. Within each workstation there may be alternate routes through machines, and this is represented by the expanded graph of the cell level node. In the example, if the part is processed at workstation, W1, then the two alternative machine sequences that exist are ({M1-M2-M3}, {M1-M2-M4}). The individual machine nodes at the workstation level can be further expanded to represent the alternate tools and tool sequences. In the example, the part can be processed at machine M1 using the two tool alternatives ({T1-T2}, {T3}).

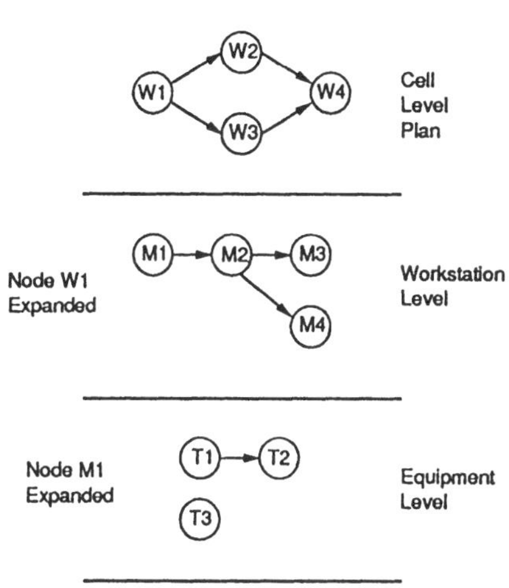

Figure 3. Process planning graphs.

The process plan decomposition parallels that of the control architecture, and thus ensures that the routing alternatives are provided for decision making at an appropriate level of detail suitable for each level in the architecture.

2.3. Detailed Architecture of a Controller Module

Efficient control of shop floor control system is accomplished using a Multipass Expert Control System philosophy. This philosophy is illustrated in Figure 4. The philosophy is based on the premise that a good set of manufacturing knowledge can be collected and stored in an expert system that can be used for a manufacturing control module. The major problem associated with planning, scheduling and control is the intractable size of most problems; however with the architecture presented, the problem can be decomposed into smaller solvable subproblems.

The workstation controller module is illustrated in Figure 5. As can be seen in the figure, it receives commands from the cell controller, and plans the activities of the various equipment to successfully insure completion of activities in the desired time frame. It should be noted that the expert scheduling system as well as a deadlock detection system operate at both the cell and workstation levels. The expert scheduling system for the workstation control module maintains the necessary knowledge base for this level's activity. It uses a simulation model coupled with expert system logic to determine future courses of action. Thus the effects of the decision can be previewed before execution. The simulation is based on actual shop status maintained in the database, incorporates flexible process plans, and utilizes deadlock detection/avoidance (see Section 3.2) modules to determine "best" part movement strategy. This enables dynamic routing/rerouting of parts in cases of machine failure or other unforeseen circumstances.

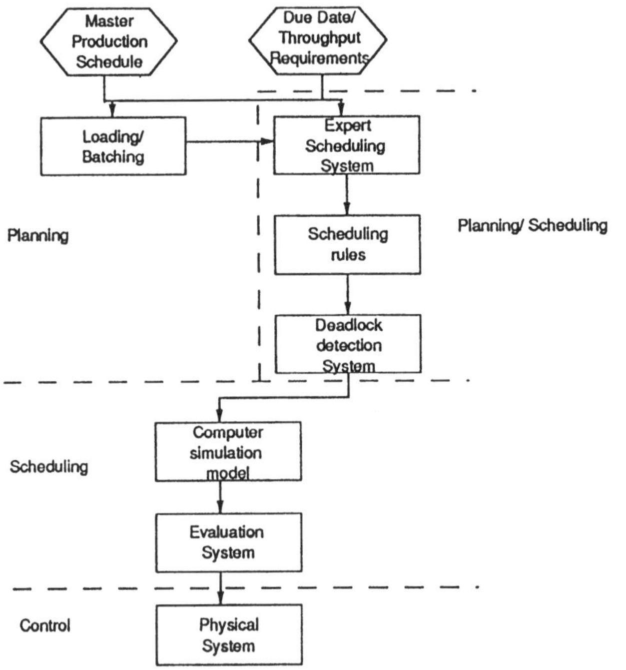

Figure 4. Functional breakdown of controller modules.

The error recovery module, houses requisite error recovery algorithms/procedures and is invoked when commands cannot be completed as requested. The error recovery module determines the type of error, viable actions for recovery, interface with scheduler to determine efficiency/performance of the recovery strategy, initiate actions to effect the requisite part and information flow required to complete the recovery process.

The dynamic deadlock and detection avoidance and recovery provides algorithms and procedures required to maintain the system in a deadlock free state where possible, and to initiate recovery from an unavoidable situation. It should be noted that many of the same functions and activities will be performed at both the Workstation and Cell levels. The concept of unit operations is the discriminator between these levels.

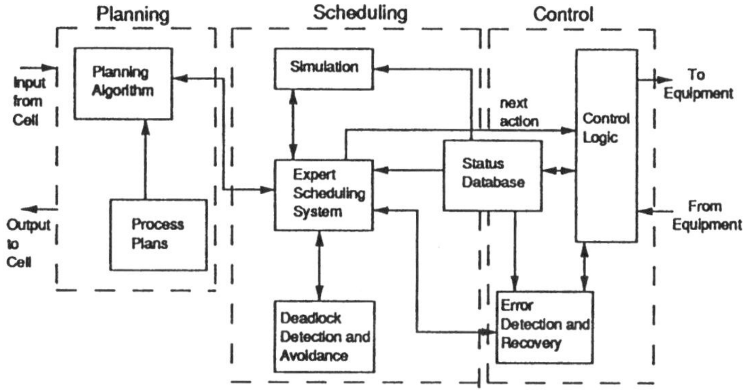

Figure 5. Workstation controller module.

3.0 Software Elements for Intelligent Control

The software required in a shop floor control system consists of several components:
(i) control software to affect part and information flow
(ii) scheduling software to affect part flow and timing
(iii) software for contingency planning (deadlock detection and recovery, error recovery, etc.)
(iv) data management and data collection software
In the following section, issues and algorithms relating to the development of software for intelligent control as pertaining to 1) Expert Scheduling System, 2) On line procedure to detect and avoid system deadlocks, and 3) automatic software generation of control software are presented.

3.1. Automatic Generation of Control Software

The development of software for control systems is characterized by the need to carefully model: machines on the factory floor, the interaction of parts on those machines as specified by a process plan, specific data updates that occur within computers constituting the control system, and, finally, to specify the interaction between the control commands and models that characterize the computer control system. To focus on automatic software generation, equivalence of formal models of the manufacturing system and of software must be achieved. Loosely speaking, analogous techniques have been used in compiler construction [Aho, 1985], but in order to apply those techniques to manufacturing control systems, strict equivalence of the properties of manufacturing systems, and the corresponding messages and formal models used to control the system must be created.

The software generation system significantly increases the likelihood of producing correct control software with greatly reduced software development time. Since the control elements of the software will be generated from a formal model of the system and the process plans that execute on the system, automatic upgrades will be possible when changes to the system layout or product mix occur. Additionally, the use of a formal model to generate run time software can be used to generate the pre-and post-conditions required for theoretical proofs of correctness associated with each action in the manufacturing system.

Several authors [Naylor and Volz 1987, Biemans and Blonk, 1986, Naylor and Maletz, 1986, Baxter 1987] have developed formal descriptive models of the factory floor with the intent of supporting software generation, but have not shown methods for systematically extending their formal models into running software. Other authors have developed rapid software prototyping systems [Maimon and Fisher, 1988] that provide only for the structural specification of cell level control software, still demanding enormous development of hand crafted code to realize effective systems. In order to overcome the current impediments to the adoption of FMS technologies by small to medium firms, we believe that reliable, affordable factory floor level control software generators deserve consideration.

The idea of controlling an FMS as a by-product of grammar recognition is developed. Each (cell, workstation, and equipment) controller is modelled as a layered parser (Figure 6) capable of satisfying the varying levels of command and processing required. Within each control module the highest level of control is executed by a control parser which is an automatically constructed push down automata. The parser recognizes a control grammar which is generated by searching a system state space graph. The control parser is responsible for recognizing control elements of the input, and activating the appropriate control action, which are typically associated with the movement of parts in the system,

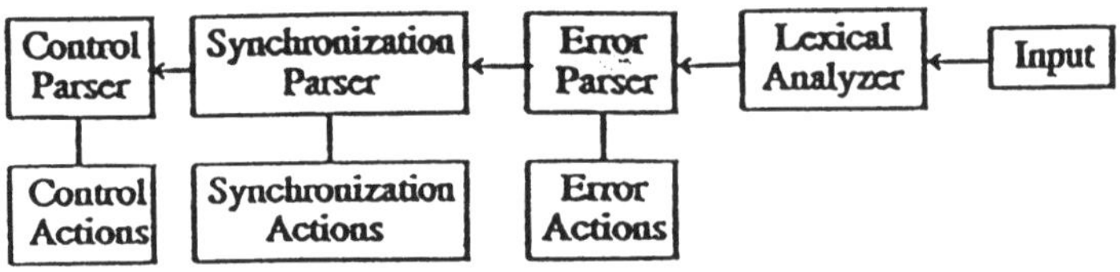

Figure 6. Layered parsers for control.

consultation of schedulers, or preparation of reports. The next layer of actions is the Synchronization Actions which is activated upon recognition of grammatical constructs reserved for synchronization of machine interaction (e.g., when a robot holding a part in a machine fixture requests that the machine grasp the part). Synchronization is defined in a separate parser layer since the states in the underlying model are not affected by synchronization actions. Error actions are executed upon recognition of error conditions and implemented via the use of error parser. The parsers and associated context free control grammars are generated automatically from the formal models used to describe the system.

The process involves the following steps:

(i) Development of formal models of manufacturing systems.

This stage of the research is concerned with the development of a formal model capable of modeling the physical part flow and system level information flow. This is achieved through the use of physical and system level graph models of the manufacturing system. The physical graph model which models physical relationship between machines is developed using machines as nodes and relations on arcs that specify feasible movements of parts.

From the physical graph a system model is created which models the part-in-contact relationships (system states) as the nodes, and system relations which signify actions which may begin and end in a given state. These system relations are created systematically using a mapping from the physical graph to system graph.

The formal model is then created based on the preconditions for movement. For example, a physical relation "pickup' where a robot grasps a part form a machine, or input/output port, can be mathematically stated as:

$$St_i \text{ pickup } St_j \text{ iff } (V_{MH}(i) \wedge V_{MP}(j)) \quad (V_{MH}(i) \wedge V_{port}(j))$$

where St_i and St_j represent initial and final states in physical graph and $V_{MH}(i)$ specifies that the value of machine in state is of type material handling, and $V_{MP}(j)$ and $V_{port}(j)$ are processing machines and input/output ports, respectively.

(ii) Creating context free grammars representing control actions for each controller.

After the state space models for the various controllers are created, the next step is to reduce the graph models to grammars. The graph models can be proved equivalent to push down automata and hence to context free grammars. This is of fundamental importance since it establishes the fact that there must exist a set of grammars equivalent to the manufacturing system. It is the recognition of these grammars that provides the requisite control actions.

(iii) Parsers for recognizing context free grammars

Programs that recognize the languages are called parsers. Methods of automatically creating parsers from context free grammars are well known and understood [Tremblay and Sorenson, 1985], and are used to produce the parsers which then control the manufacturing operation as a by product of recognizing the commands.

(iv) Associating semantic actions with control grammars.

Semantic actions have to be associated with control, synchronization and error parsers in each controller. In a manufacturing environment, a typical semantic action might be to cause a robot to move, or cause a message to be transmitted to another computer. The semantic actions are embedded in the control grammars and produce the actions to be executed as a result of reduction of specific command recognition production in the CFG's.

(v) Automatic control software generation

The final step in the process is to create actual software that will reside on each computer in the control architecture. Figure 7 shows the structure of the automatic control software generator. The grammars are used to create source code files which are table driven parsers that carefully integrate the finite state

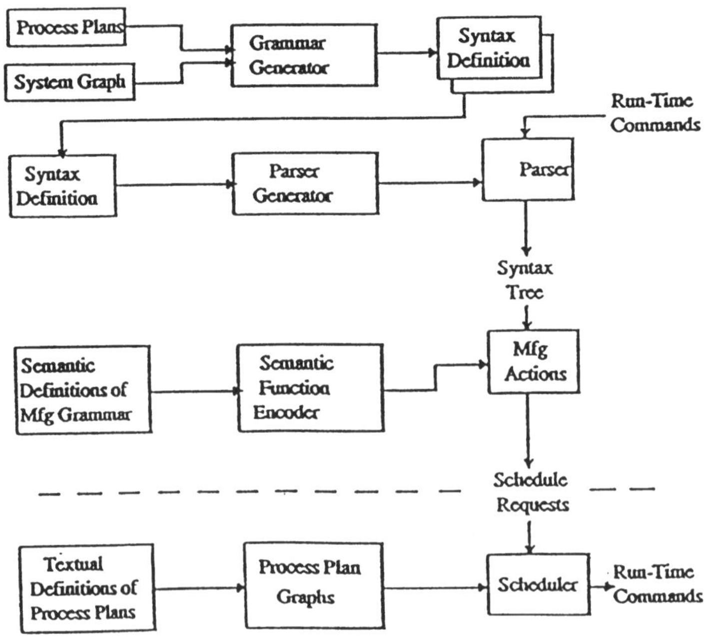

Figure 7. Automatic control software generator.

recognizers associated with normal context free grammar processing with certain states of particular interest in the context of a manufacturing system. A lexical analyzer is also created for each grammar automatically. The lexical analyzers tokenize the input stream and return values to the parser. The source code, lexical analyzer, and utility files for communication handling are then compiled to produce an executable file for each machine. The resultant code is very much like a compiler, where the generated systems do not process an input file but rather, an input command string obtained from an interaction communication system, such as network interface, or serial data communication.

3.2. System Deadlock:Detection and Avoidance

In order to achieve true flexibility in an FMS, the parts must be capable of being routed through the cell in a manner that does not inhibit part flow. System deadlock is a situation where machines in a manufacturing system have been allocated parts in such a way that part flow is impossible. Deadlocking can arise even in the simplest of automated manufacturing systems. Consider a two machine cell consisting of machine A and B, processing two parts P1, and P2, and serviced by a robot (Fig. 8a). The routing for part P1 is A-->B and for part P2 is B-->A. Suppose part P1 arrives and is loaded onto machine A, and then part P2 arrives and is loaded on machine B. Any further part flow is now inhibited resulting in a system deadlock. A deadlock may involve more than one machine as shown in Figure 8b. Although a system deadlock can eventually be resolved if there is a storage to buffer at least one individual part, a deadlock can still occur if the part storage is not properly used as illustrated in Figure 8c. Once a deadlock occurs, automated control for the manufacturing system may have to evoke manual recovery procedures, resulting in degrading performance of automated manufacturing systems. Hence the detection and avoidance of system deadlocks should be an important integral part in any automated manufacturing system.

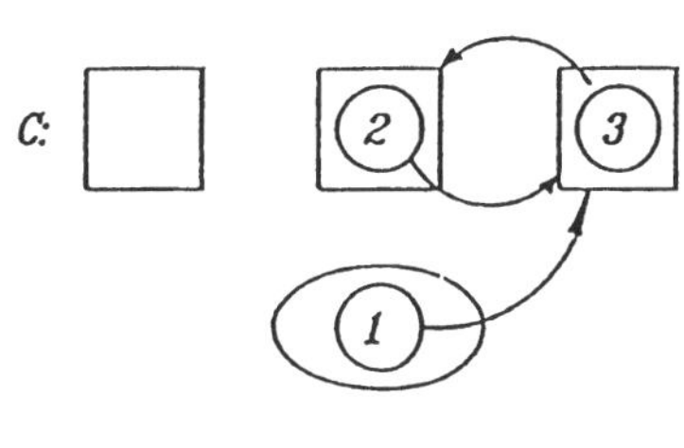

Figure 8. Examples of system deadlocks.

Two factors that effect system deadlock are complexity of part routing and availability of part storage. A flow line FMS will not have a deadlock problem, but on the other hand, if each part has a different routing through the FMS the frequent deadlocks can be expected. The complexity of the problem can be illustrated by the following example. A manufacturing system with 6 machines and parts capable of moving from any one machine to another can have a total of 57 deadlock possibilities ($C^6_2+C^6_3+C^6_4+C^6_5+C^6_6$), where C^6_2 corresponds to the number of 2 machine deadlocks possible in a 6 machine system, C^6_3 corresponds the number of 3 machine deadlocks, and so on). Several deadlocks could exists simultaneously. Availability of storage space reserved for deadlock situation can make recovery possible if properly used. However some manufacturing systems may not be equipped with any storage at all. In such a case deadlock must be avoided. In this proposal two approaches are presented: (i) a deadlock detection and recovery approach when special storage is available, and (ii) an avoidance approach for a system with no storage.

System deadlocks can have an adverse effect on performance of automated systems, and have been ignored in most scheduling studies, which typically assume infinite capacities and few limits on physical constraints imposed by material handling equipment. Conventional implementations of FMS systems allocate large amounts of storage space to avoid deadlocking, and rely on creating batches of parts using linear flow characteristics [Co 1984] which results in inefficient use of resources and major loss in flexibility of systems.

Deadlocking can be prevented in three ways: 1) Providing large in-process queues and storage areas *(Planning)*, 2) Restricting flow of parts, such that they only have a positive unidirectional flow direction *(Planning and scheduling)*, and 3) Controlling parts flow to inhibit deadlocking *(Scheduling)*. Co and Wysk[1986] showed that if the maximum number of parts in a system is n and there are $n - 1$ queuing stations, then deadlocking can not occur. In practice, several queueing stations are installed in an FMS to plan for worse case operation.

Deadlocking problems have been studied by researchers in computer science in the context of multiple coprocesser operating systems [Haberman 1969, Holt 1971, Tsutsui & Fujimoto 1987] and distributed systems [Gray 1978, Menasce & Muntz 1979]. These studies cannot be directly applied to manufacturing systems but provide insight into deadlock resolution in manufacturing systems.

The deadlock detection and avoidance process interfaces with the scheduling system, to assist in sequencing part flow in a manner that does not lead to a deadlock situation. The approach presented here implements the detection and avoidance in three steps:

Step 1: Development of Detection Procedure

This step is required to isolate combinations of parts and machines which could lead to a potential deadlock in the context of a running system.

A graph theoretic approach is proposed for the identification of system deadlocks, where a graph G=(V,E) is created such that V={set of nodes corresponding to the machines in the system}, and E={set of directed arcs corresponding to the part routing}. Based on the definition of a system deadlock, a circuit is a necessary condition for a system deadlock. Hence the next step is to create a procedure for detecting circuits in the graph.

A string multiplication technique developed by [Boffey 1982] is used to identify all circuits in the graph. However, it is well known that identification of all circuits is a NP hard problem, but for our application the following observation will make the problem size manageable by providing an upper bound on the search procedure for closed circuits. A system deadlock in a manufacturing system composed of M machines and N parts, cannot possibly involve more than M machines or N parts. Hence the string multiplication technique for identifying circuits will terminate after x iterations where x=MIN{M,N}.

Not all circuits correspond to system deadlocks. Attributes and characteristics of system deadlocks are correlated with properties of the graph to further narrow down the search for circuits that have potential for deadlocks. Further details on deadlock detection are presented in [Wysk, Yang and Joshi, 1990].

Step 2: Avoidance of Deadlocks

The avoidance problem deals with problems of detecting and preventing a system deadlock from materializing in a running system. The complete route of each newly arriving part as well as the remaining routes of all parts already in the system are converted into a graph, and the graph is examined for all circuits which could potentially lead to deadlocks. These circuits are stored in a database which provides the scheduling system the capability of directing part flow without causing a system deadlock, and are updated dynamically as parts flow through the system.

This process can be further complicated by the interaction of two or more circuits to create a deadlock situation. Consider Figure 9, which shows 4 machines A,B,C,D and 3 part routes marked 1,2,3. Part 1 is at machine A, part 2 at machine B, and part 3 is at machine D. Each of the circuits by itself cannot cause a system deadlock, but as soon as either part 2 or part 3 moves, one of the circuits will result in a deadlock. Hence the avoidance system must also be capable of preventing such situations from arising. Such circuit interactions are identified by identifying higher level closed circuits created by collapsing the circuits into single higher level node, and retaining the arcs emanating from/to the collapsed circuit as arcs in the higher level graph [Yang, 1989]. The higher level circuits identification procedure is the same as the one discussed in step 1.

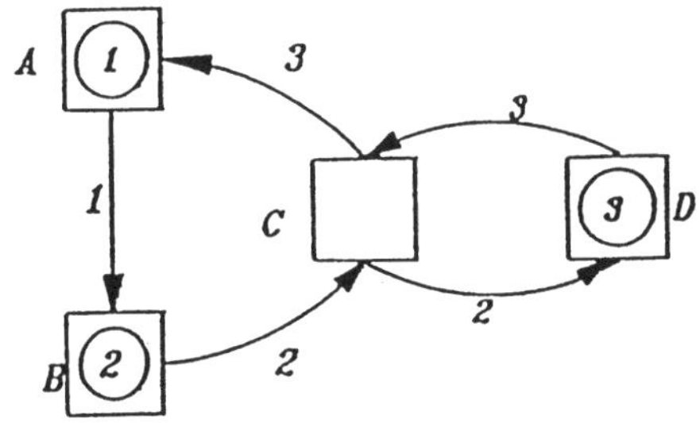

Figure 9. Interaction between circuits.

Step 3: Recovery from Deadlock

This approach is applied in a manufacturing system that reserves special storage for a deadlocking situation. Here deadlocks are allowed to occur and the recovery procedure activated to resolve it. In this case, the circuits provide both the necessary and sufficient conditions for system deadlocks. Furthermore, circuits do not interact with each other, and all circuits interact independently.

The first step in the correction procedure is to move one of the deadlocked parts to the specially reserved storage. Interaction with the scheduling module is used to determine which part should be moved to storage. The remaining parts are processed on requested machines in an order to prevent subsequent deadlocks (if only one storage location is provided). The last step is to reinsert the part from storage back to the machines for processing. It can be seen that if the recovery approach is not carefully planed it can result in unproductive time spent in moving parts between storage and machines.

3.3. Multipass Expert Scheduling System

All decisions on part movement will be made by the scheduler, and the control system merely serves to execute the commands from the scheduler. The multi-pass expert scheduling integrates with the structure as shown in Figure 5.

Multipass scheduling is an approach to resolving factory scheduling problems and has shown promise recently. Wu and Wysk [1990] showed significant throughput improvements can be made using a simulation model to determine the future course for a manufacturing system. Essentially the procedure works as follows. A simulation model of the system is resident on the system control computer. At each decision point, a deterministic simulation is run to see what control policy (from a series of rule based policies) impacts the current system most favorably. This control policy is then chosen and the appropriate control response is signaled for execution on the system.

One unfortunate drawback of this procedure is that FMS's are not static (the types of parts, demand, tooling, etc. change over time). Each time the specifics change, the simulation model must be rewritten or updated which has to be done manually. Although the benefits can be significant, the implementation costs can be quite expensive.

The basic idea behind the multipass expert scheduling system is to use deterministic simulation as a short-term predictive tool for alternate control strategies in a manufacturing cell. There are many controllable (endogenous variables that must be affected). There are also many uncontrollable (exogenous) factors that will impact a system. A deterministic simulation can be run as a Gantt chart like scheduler to analyze short-term effects. Decisions based on interference, machine utilization, and other system performance measures can be assessed for short-term sequencing and operational decisions. This allows for use of a reasonably simple set of rules that are evaluated to determine performance.

The simulation does not examine exogenous factors. The system will respond to exogenous factors by creating a new simulation experiment/model when unforeseeable events occur. However, the scheduler need only respond to these events rather than try to minimize conflict from them. The system integrates with cell databases on line deadlock detection, avoidance and recovery module, and error recovery modules.

The multipass expert scheduling system consists of three major components:

(i) an intelligent scheduling module,
(ii) a simulator,
(iii) interfaces to other modules, databases and cell controllers.

3.3.1. Intelligent Scheduling Module (ISM)

Scheduling is the principle vehicle for utilizing the resources efficiently, responding to managerial objectives rapidly, and satisfying the system constraints effectively. The scheduling module enables generation of various schedules based on input from the factory control system along with the actual status at the shop floor. ISM utilizes expert methodologies to assist in shop scheduling.

The ISM consists of the following components:

(i) The acquisition module for transferring the knowledge and expertise of scheduling into facts and rules. This module is constructed so that ISM can be continuously expanded since scheduling knowledge is domain specific and will vary based on scheduling environments of different machine part configurations with different levels of complexity. In addition, the acquisition module employs machine techniques to learn from feedback from the shop floor.

(ii) The knowledge base consists of primary and meta knowledge. The primary knowledge comprises current cell status, deadlock circuits, information, dynamic dispatching rules and scheduling heuristics. The meta knowledge consists of knowledge to apply scheduling rules and heuristics to previously unexperienced circumstances, and consist of general criteria for selecting scheduling rules, and learned heuristics.

(iii) Inference engine to manipulate the rules and facts in the database to generate an alternate space consisting of several "good" alternative scheduling rules which are simulated to evaluate their performance potential.

3.3.2. Simulation Model

The major function of the simulation model is to evaluate alternative control policies. The simulation model queries the current databases and determines the future system status by making a pass of deterministic simulation according to a rule(s) defined in the rule module. The system performance predicted by each pass of simulation is a measure of closeness to an objective function from the higher level factory control system. Thus, at the end of all passes of the simulation, the best schedule is then applied to the physical manufacturing system. The simulation model will operate on parts only currently available to the system (exogenous arrivals will not be permitted). This creates a deterministic system that can be easily analyzed.

An important issue that must be considered is the length of the simulation window, that determines how long a simulation should look ahead. Intuitively, this value is critical to overall performance of the multipass scheduling mechanism. Further research will determine the most desirable simulation window based on its effect on performance of the scheduling mechanism, computational burden on the system and other factors that will arise.

4. A Prototype System

Figure 10 contains a schematic of an operation Flexible Machining Cell (FMC) at The Pennsylvania State University. The FMC currently consists of five Workstations: a Rotation Workstation, a Prismatic Workstation, an Assembly Workstation, a Material Handling Workstation, and a Warehousing Workstation. It should be noted that the various Workstations couple and uncouple from the Material

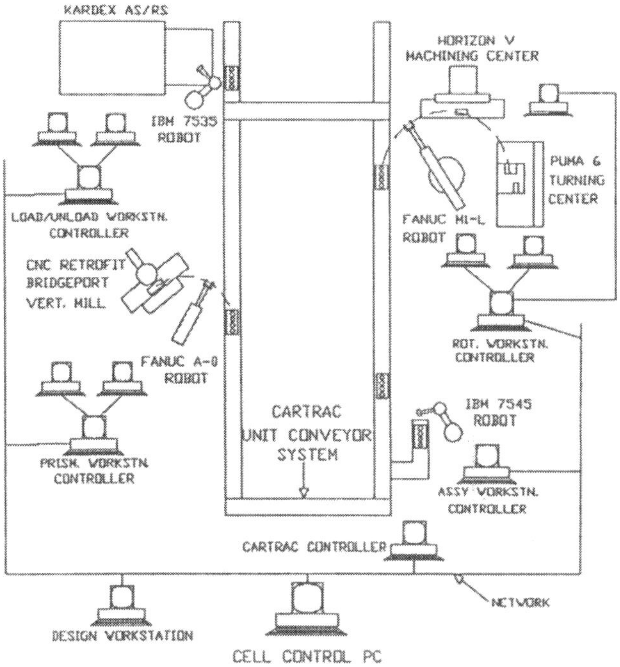

Figure 10. Schematic of Penn State FMS.

Handling Workstation through predefined coupling/uncoupling points. It should be further noted that the Workstations contain from one (1) to four (4) Equipment components, with the Rotation Workstation containing four (4) equipment elements: a Fannc M-1, a Daewoo Puma and Turning Center, a Pratt & Whitney Horizon V Machining Center and parts invertor/queue.

Control of the system is initiated by the FMC Controller - an 80286 PC based controller. The FMC controller communicates to the Workstation controllers (also 80286 based PC's) via a DECNET LAN using a common set of grammars. Vocabulary from the FMC controller consists of instructions such as (to the Rotational Workstation):

MANUFACTURE (batch)

The Rotational Workstation then begins to plan, schedule and control the necessary equipment required to produce the parts in the batch identified by the pt_no. The batch quantity is variable between 1 - 4 (constrained by the material handling device in use). Instructions are also sent to the material handling station to deliver the raw material to rotational workstations. The batching task is performed at the cell controller. Table 3 lists a sample of cell-workstation commands. If the part pt_no has not been

Table 3

Cell - Workstation Commands	
Command	Returned Values
GET STATUS - Requests the current status of the workstation.	Block of information detailing the status of the workstation. Includes information on the current batch, expected completion time, and inoperative machines.
SEQUENCE (batch) - Requests the workstation to sequence the parts in batch. This function will be used by the cell controller when determining a viable batch configuration.	Runtime of the batch INVALID PART TYPE INVALID BATCH ERROR
MANUFACTURE (batch) - Requests the workstation to start processing batch.	Expected completion time INVALID PART TYPE INVALID BATCH ERROR

Table 4

Workstation -Equipment Commands	
Control Function	Response
Machine Tools	
GET_STATUS - Requests the operational status of the machine.	Block of data including information on the currently loaded part (if any), currently loaded NC file name, and currently loaded tools.
GRASP (part) - Instructs the machine to close the fixture or part chuck. The parameter part gives the part information required for flexible fixtures	FIXTURE CLOSED PART TYPE UNKNOWN ERROR
RELEASE (part) - Instructs the machine to open the fixture.	FIXTURE OPEN PART TYPE UNKNOWN ERROR
PROCESS (part) - Instructs the machine to start processing the part.	PROCESSING COMPLETE PART TYPE UNKNOWN MACHINE DOWN ERROR
STOP - Stops processing immediately	MACHINE STOPPED
Load/Unload Robots	
MOVE (part, loc1, loc2) - Instructs the robot to move part from location loc1 to location loc2. Part is defined for flexible grippers.	MOVE COMPLETED LOCATION INVALID PART TYPE UNKNOWN ERROR
MOVE (loc) - Instructs the robot to move to the location loc.	MOVE COMPLETED LOCATION INVALID ERROR
GRASP (part) - Instructs the robot to close the gripper.	GRIPPER CLOSED PART TYPE UNKNOWN ERROR
RELEASE (part) - Instructs the robot to open the gripper.	GRIPPER OPEN PART TYPE UNKNOWN ERROR
STOP - Stops the robot immediately	ROBOT STOPPED
Automated Storage and Retrieval System (AS/RS)	
RETRIEVE (part) - Instructs the system to locate part and bring it to the load/unload station. Part could represent finished goods or raw materials.	PART RETRIEVED PART NOT FOUND PART TYPE UNKNOWN ERROR
STORE (part) - Instructs the system to find part's storage location and either bring the location to the load/unload station or take the part to the storage location from the load/unload station.	PART STORED SPACE NOT AVAILABLE PART TYPE UNKNOWN ERROR
LOCATE (part) - Requests the current storage location for part.	PART NOT FOUND PART TYPE UNKNOWN ERROR
ALLOCATE (part) - Instructs the system to allocate space for part.	SPACE ALLOCATED SPACE NOT AVAILABLE PART TYPE UNKNOWN

produced in the workstation previously, process planning information will be exchanged. If the part has been produced, a resource requirements list and part program will remain resident in memory. If the Rotation Workstation is currently idle and empty, the coupler (Material Handling Workstation - Rotation Workstation) will be inspected until a part pt_no arrives. The workstation controller then coordinates the actions of the various equipment using the set of commands (Table 4) for communicating with the various equipment controllers [Smith, 1990]. The management of the interactions of the equipment may or may not require scheduling (for a single machine) or deadlock detection. If these functions are required, they are included in the workstation controller.

The above FMC is operational and produces several parts belonging to two part families (rotational prismatic). Early experience with the architecture looks promising from both a control and expendability viewpoint. There are however many control problems requiring significant research and development before generic procedures will exist.

5. Summary

In this paper, a scaleable architecture for FMS control has been presented. The key elements of this architecture are that each controller performs the same functions (planning, scheduling and control), and the degree of complexity of the controllers can be scaled without affecting the control elements. Various elements of the control architecture have already been implemented at the Penn State FMS Lab, and further research is underway towards implementation of the complete architecture. Early experience has been promising, but further refinements will have to be made to adjust for some implementation difficulties that may be encountered.

BIBLIOGRAPHY

Aho, A.V., Sethi, R., and Ullman, J.D., *Compilers - Principles, Techniques and Tools*, Addison Wesley, Reading, Massachusetts, 1985.

Baxter, R.D., "A Brief Discussion of Formally Based Approaches to Rapid Prototyping," British Aerospace Technical Report BAE-WAA-RP-RES-SWE-440 (Unclassified), 5 March 1987.

Biemans, F.P.M., and Blonk, P., "On for the Formal Specification and Verification of CIM Architectures Using LOTOS," *Computers in Industry* 7, 491-504, 1986.

Boffey, T.B., *Graph Theory in Operations Research*, McMillan Computer Science Series, 1982.

Co, H.C., "Design and Implementation of FMS -- Some Analysis Concepts," Ph.D. Thesis, VPI-SU, 1984.

Chang, P., Wu, S., and Storer, R., "A One-Machine Rescheduling Heuristic with Efficiency and Stability as Criteria", Lehigh University working paper.

Davis, W. and Jones, A., "Issues in Real-Time Simulation for Flexible Manufacturing Systems", Proceedings of the European Simulation Multiconference, Rome, Italy, June 7-9, 1989.

Davis, W. and Jones, A., "A Functional Approach to Designing Architectures for Computer Integrated Manufacturing", Systems, Man, and Cybernetics special issue on Manufacturing, Vol. 19, No. 2, 164-174, March, 1989.

Davis, W. and Jones, A., "A Real-Time Production Scheduler for a Stochastic Manufacturing Environment", International Journal of Computer Integrated Manufacturing, Vol. 1, No. 2, 101-112, 1988.

Dar -El, E.M. and Wysk, R.A., "Job Shop Scheduling - A Systematic Approach," *Journal of Manufacturing Systems*, Vol. 1, no. 1, 1982.

Gurecki, R. and Nof, S.Y., "Decision Support for a Computerized Manufacturing System," The Optimum Planning of Computerized Manufacturing Systems, NSF Report No. APR74-15256, No. 14, 1979.

Habermann, A.N., "Prevention of System Deadlocks," *Communications of the ACM*, p. 373, Vol. 12, 1969.

Holt, R.C., "Comments on Prevention of System Deadlocks," *Communication of the ACM*, Vol. 14, 1971.

Jones, A., Barkmeyer, E., and Davis, W., "Issues in the Design and Implementation of a System Architecture for Computer Integrated Manufacturing", International Journal of Computer Integrated Manufacturing special issue on CIM architecture, Vol 2, No. 3, 1989.

Joshi, S., Wysk, R.A, and Jones, A, "A Scaleable Architecture for Control of FMS", paper *CIMCON'90*, Washington, May 1990.

Maimon, O.Z., and Fisher, E.L., "An Object-Based Representation Method for a Manufacturing Cell Controller," *Artificial Intelligence in Engineering*, 3, 1, 1988.

McLean, C.R., Mitchell, M., and Barkmeyer, E., "A Computer Architecture for Small Batch Manufacturing," *IEEE Spectrum*, 59-64, May 1983.

Menasce, D.A. and Muntz, R.R., "Locking and Deadlock Detection in Distributed Data Bases," *IEEE Transactions on Software Engineering*, Vol. SE-5, No. 3, May 1979.

Mettala, E., Joshi, S., and Wysk, R., "CIMGEN - A Case Tool for CIM Development", Proceedings of the FMS Conference, Boston, MA, August, 1989.

Mettala, E.G. and Joshi, S., "A Compact Representation of Process Plans for FMS Control Activities", *Penn State University Industrial Engineering Working Paper Series*, 1990.

Naylor, A.W., and Maletz, M.C., "The Manufacturing Game: A Formal Approach to Manufacturing Software," *IEEE Transactions on Systems, Man, and Cybernetics*, SMC-16, 321-334, May/June 1986.

Naylor, A.W., and Volz, R.A., "Design of Integrated Manufacturing System Control Software," *IEEE Transactions on Systems, Man, and Cybenetics, SMC-17, 6, 1987.*

Smith, J.S., Development of Hierarchical Control Model for a Flexible Manufacturing System, Master's Thesis, Industrial and Management Systems Engineering, Pennsylvania State University, 1990.

Tremblay, J.P., and Sorenson, P.G., *The Theory and Practice of Compiler Writing*, McGraw-Hill Book Company, new York, 1985.

Tsutsui, S. and Fujimoto, Y., "Deadlock Prevention in Process Computer Systems," *The Computer Journal*, Vol. 30, No. 1, 1987.

Wysk, R.A., Wu, S.D., and Yang, N.S., "A Multi-Pass Expert Control System (MPECS) for Flexible Manufacturing Systems," *Bound Volume of the Symposium on Integrated and Intelligent Manufacturing (ASME)*, 1986.

Wu, S.D. and Wysk, R.A., "An Application of Discrete-Event Simulation to On-Line Control and Scheduling in Flexible Manufacturing Systems, *International Journal of Production Engineering*, 1990.

Wysk, R.A., Yang, N.S, and Joshi, S., "Detection of System Deadlocks in Flexible Manufacturing Cells", *Penn State University Industrial Engineering Working Paper Series, 1990.*

Yang, N. S., "Resolution of System Deadlocks in Real Time Control of Flexible Manufacturing Systems," Ph.D. Dissertation., The Pennsylvania State University, 1989.

Event Graphs for Modeling and Evaluating Modern Production Systems

GEORGE HARHALAKIS, SAID LAFTIT and JEAN-MARIE PROTH

1. INTRODUCTION

In this paper, the performance problem of job-shop and assembly systems with deterministic manufacturing times under cyclic production process is addressed. In both types of systems, the ratios of the various product types to manufacture are given. The set of the ratios is known as the **product mix**. The control of a production system in order to meet the given ratios is obtained by associating a machine sequence to each machine. Such a sequence defines the sequencing of the product types (or the sequencing of the components of the products types, in the case of an assembly system) on a machine. The proportion of a product type (or a component) in a machine sequence tally with the related ratio. If one of the machine sequences does not reflect the corresponding ratio, it has been proved in [1] that the system blocks after a finite period of time.

The modeling of job-shop and assembly systems is based on Event Graphs ([1], [2] and [5]), a special class of Petri Nets. The model obtained is a strongly connected event graph, as shown in [4] for the job-shop systems.

The first goal of this paper is to explain how to model job-shop and assembly systems using event graphs. The second goal is to show that, for both types of production systems, it is always possible to fully utilize the bottleneck machine, and thus to reach the maximal productivity. Finally, the machine sequences being given, we show that the optimal solution to the problem (i.e. the solution which fully utilizes the bottleneck machine with a minimal work-in-process (WIP)) is the solution to an integer linear programming problem. Note that, in real-world production systems and in particular Flexible Manufacturing Systems (FMS), minimizing WIP is the same than minimizing transportation resources such as pallets or carts, which are often very expensive.

The paper is organized as follows. Section 2 is devoted to the presentation of the event graphs and their properties. In section 3, we show how to use event graphs to model job-shop as well as assembly systems. In section 4, we derive the performance evaluation of the production systems from the properties of the event graphs. Finally, we explain how to reach the optimal solution to the system by solving an integer linear programming problem.

2. PETRI NETS AND EVENT GRAPHS

A Petri net is a bipartite directed graph with two types of nodes: the **places** (represented by circles) and the **transitions** (represented by bars). These nodes are joined by **directed arcs**. These arcs can be weighted. In the following, we assume that every arc has a weight equal to 1. The two ends of a given arc are of different types. **Tokens** evolve in the Petri net and represent the dynamics of the system. They are represented by **black dots**. A Petri net is given in figure 1. In this example, transitions are denoted by t_i (i = 1, ..., 6) and places by p_j (j = 1, ..., 6). Furthermore, p_1 contains two tokens, p_3 one and the other places are empty.

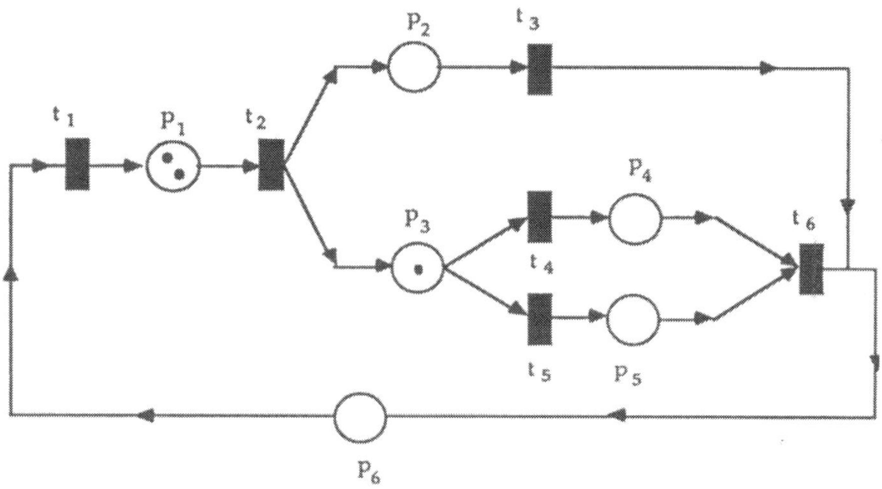

Fig. 1: A Petri net example

Let P and T be respectively the set of places and the set of transitions. A **marking** of a Petri net is a function

$$M: P \rightarrow \{0, 1, 2, \ldots\}$$

which assigns a non-negative number to each place. Such a number represents the number of tokens in the place. A transition t is **enabled** by a marking M if and only if each of its input places contains at least one token (assuming that the weight of each arc is equal to 1). For instance, in figure 1, t_2, t_4 and t_5 are enabled. The **firing** of a transition consists in removing one token from each input place and adding one token to each output place. In a Petri net, it may arise that several transitions are enabled, but only some of them can be fired. For instance, in figure 1, t_4 and t_5 are enabled, but only one of them can be fired because p_3 contains only one token. In that case, the tokens can evolve in the net only if we add some decision-making processes to make this kind of decision.

It is possible to assign a time to each transition. In that case, the Petri net is called **timed Petri net**. Such a time represents the time taken by the related transition to fire. In a timed Petri net, a firing is initiated by removing one token from each of the input places.

The firing terminates after a period of time equals to the time assigned to the transition, and one token appears in each output place at the instant when the firing terminates.

An **event graph** is a Petri net such that each place has exactly one input transition and one output transition and such that every arc is 1-weighted. Figure 2 gives an example of event graph.

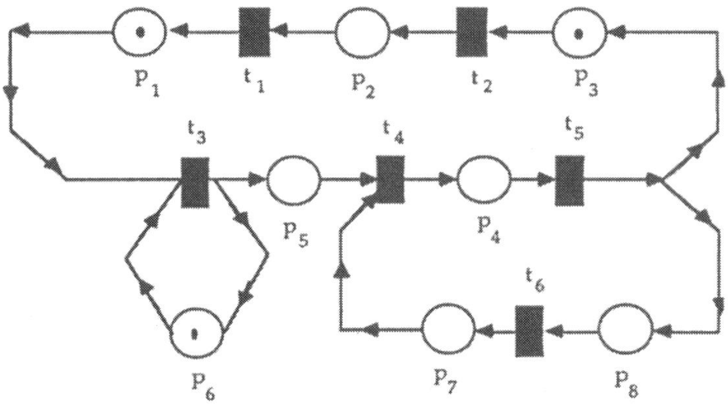

Fig. 2: An event graph

In such a Petri net, we never have to make a choice between several enabled transitions to decide which one to fire. In other words, an event graph is a **decision-free Petri net.** In the following, we consider only timed event graphs.

An **elementary circuit** is a directed path which goes from one transition back to the same transition, and which has at most one time assigned to each transition. For instance, in the event graph given in figure 2, the following circuits are elementary:

$\gamma_1 = \{t_1, p_1, t_3, p_5, t_4, p_4, t_5, p_3, t_2, p_2, t_1\}$

$\gamma_2 = \{t_3, p_6, t_3\}$

$\gamma_3 = \{t_4, p_4, t_5, p_8, t_6, p_7, t_4\}$

The following property has been demonstrated by Commoner and al. [2]:

Property 1:

The total number of tokens in any elementary circuit is invariant by transition firing.

We define the **cycle time** of an elementary circuit as follows:

$$C(\gamma) = \mu(\gamma) / M(\gamma) \qquad (1)$$

where:

$\mu(\gamma)$ is the sum of the times assigned to the transitions of γ
$M(\gamma)$ is the number of tokens in γ

According to property 1, $C(\gamma)$ remains constant by transition firing and $C(\gamma)$ can be computed using $M^{\circ}(\gamma)$, initial marking of γ.

Γ being the set of elementary circuits of a strongly connected event graph, the maximum cycle time over all the elementary circuits is:

$$C^{*} = \mathcal{M}ax_{\gamma \in \Gamma} \; C(\gamma) \qquad (2)$$

If $\gamma \in \Gamma$ is such that:

$$C(\gamma) = C^{*}$$

γ **is called critical circuit.**

The following property is due to Chretienne and al. [1]:

Property 2:

Assuming that the transitions fire as soon as they are enabled (i.e. ready to fire) and that the event graph considered is strongly connected, then the firing rate of all transitions in steady state is given by:

$$\lambda = 1 / C^{*} \qquad (3)$$

Broadly speaking, λ represents the speed at which the strongly connected event graph evolves. As a consequence, it is possible to increase the "speed" of the event graph by decreasing C^{*}, i.e. by increasing the number of tokens in the critical circuits.

Finally, property 3 also holds:

Property 3:

If one of the elementary circuits does not contain any token, then the system blocks after a finite period of time.

In the next section, we show how to model job-shop and assembly systems using event graphs.

3. MODELING JOB-SHOP AND ASSEMBLY SYSTEMS

For the sake of clarity, we explain separately how to model job-shop and assembly systems using event graphs.

3.1 Modeling job-shop systems

In the following, we explain how to model a job-shop system using the example shown in figure 3.

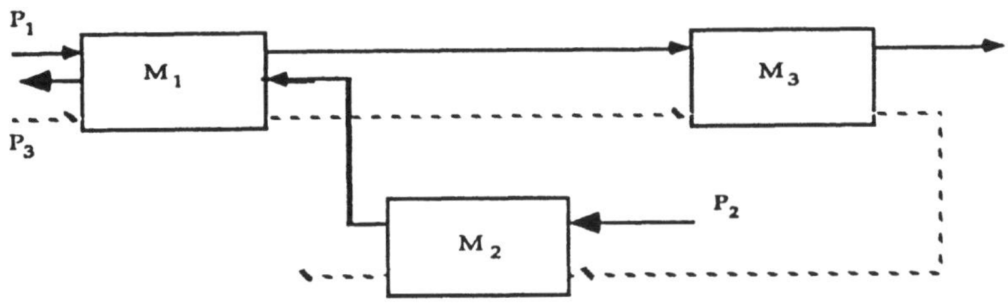

Fig. 3: A job-shop system

The job-shop under consideration is composed of three machines denoted by M_1, M_2 and M_3. Three types of products, denoted by P_1, P_2 and P_3, can be manufactured using this job-shop system. Let us assume that the product mix is (0.25 ; 0.50 ; 0.25), which means that the goal is to manufacture 25% of products of type P_1, 50% of products of type P_2 and 25% of products of type P_3. A sequence which reflects the product mix is P_1, P_2, P_2, P_3. We assign to each element of this sequence a **process circuit** as presented in figure 4.

Each process circuit is an elementary circuit and models the routing of the related product type. Each transition firing corresponds to the execution of an operation, according to the order specified by the manufacturing process. The time assigned to each transition is the time corresponding to the related operation. Each token in a process circuit is a work-in-process (WIP). A place corresponds to a storage buffer and we assume that a new part is launched in the system as soon as a part is completed. Thus, in the process circuits, tokens recirculate indefinitely.

We then model the sequencing of the part types in the machines by means of **command circuits**. A command circuit is an elementary circuit which connects all the transitions that correspond to operations performed on the same machine. In figure 5, we present the command circuit corresponding to machine M_1.

The order of the transitions in command circuits is determined by the input sequence assigned to the corresponding machines.

Note that a command circuit contains only one token, and a token in a command circuit does not represent a part.

We can see that the model is a strongly connected event graph which contains three types of elementary circuits:

- the command circuits
- the process circuits
- the mixed circuits, that include nodes of both processing and command circuits.

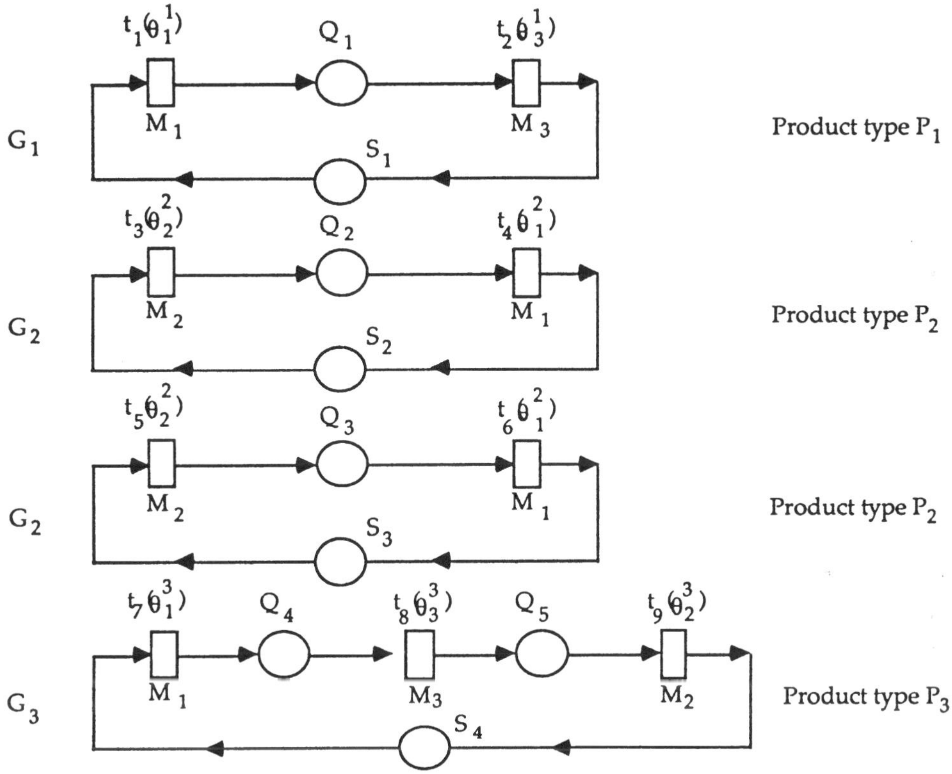

Fig. 4: The process circuits

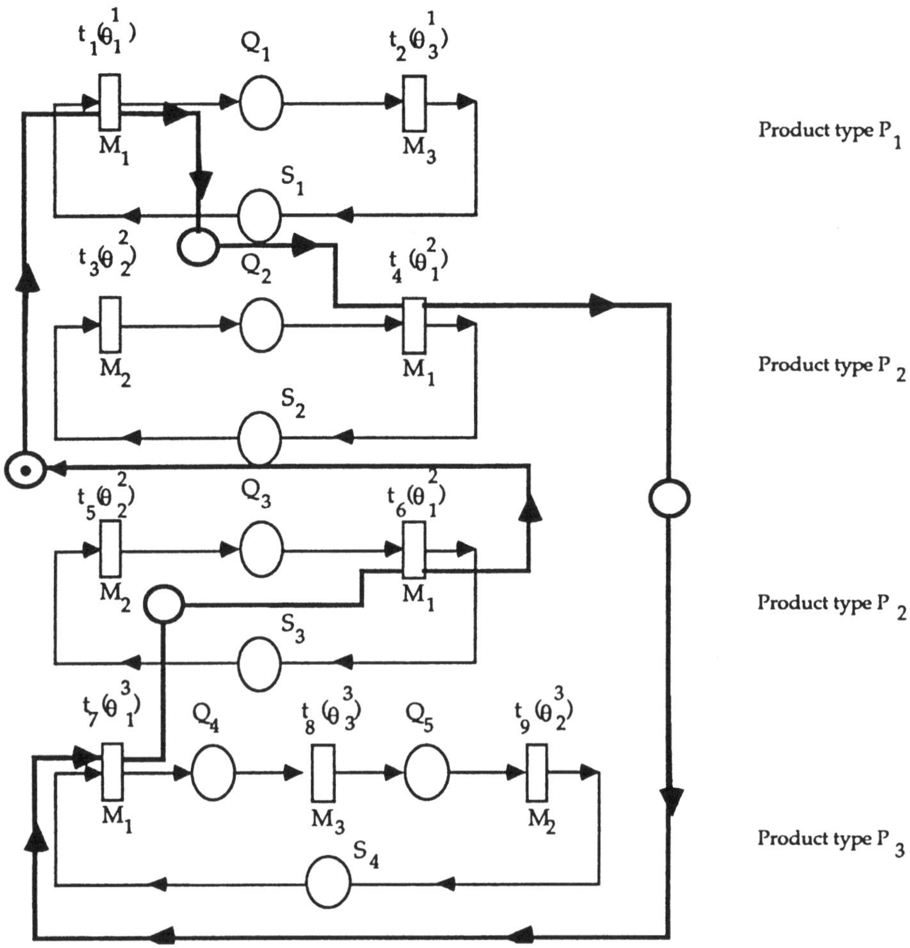

Fig. 5: The job-shop model

3.2. Modeling assembly systems

In this section, we use again a small example to explain how to model an assembly system. Let us consider for instance two product types P_1 and P_2 whose manufacturing processes are presented in figure 6.

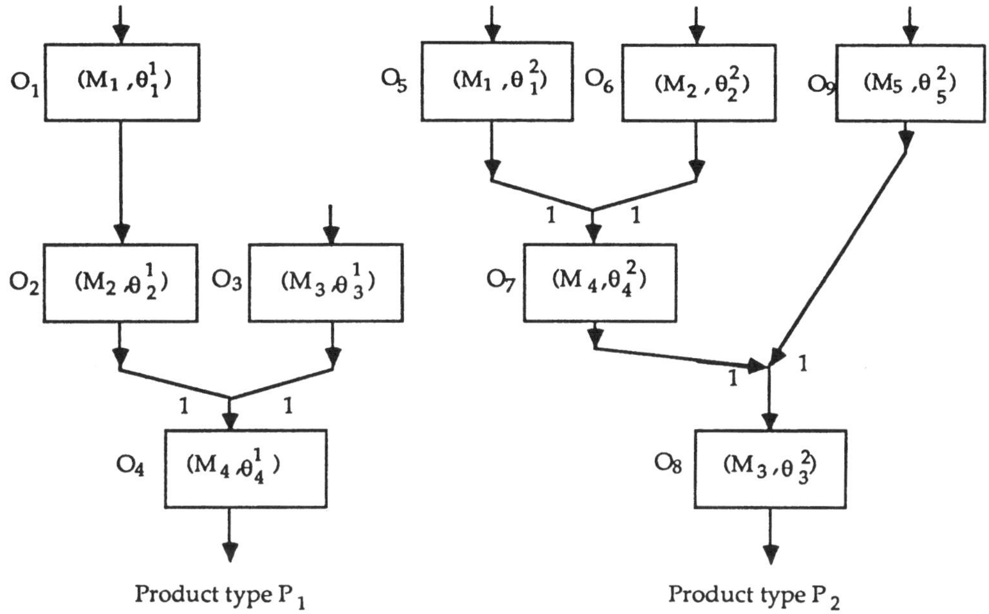

Fig. 6: Product Types P_1 and P_2

In figure 6, M_i (i = 1, 2, 3, 4, 5) are the machines, O_i (i = 1, ..., 8) represent operations and θ_i^j represent the manufacturing times. Two types of operations are involved in the manufacturing process. Operations such as O_1, O_2 just transform a component of the final product. In the following, we refer to these operations as **regular operations**. Another type of operations are the **assembly operations**. For instance, O_4, O_7 and O_8 are such types of operations. An assembly operation puts together components to obtain a more complex component of a final product or the final product itself. In figure 6, the integer values assigned to the arrows represent the number of components of each type of product required to obtain one unit of the next component.

In figure 7, we present the models related to the previous manufacturing processes.

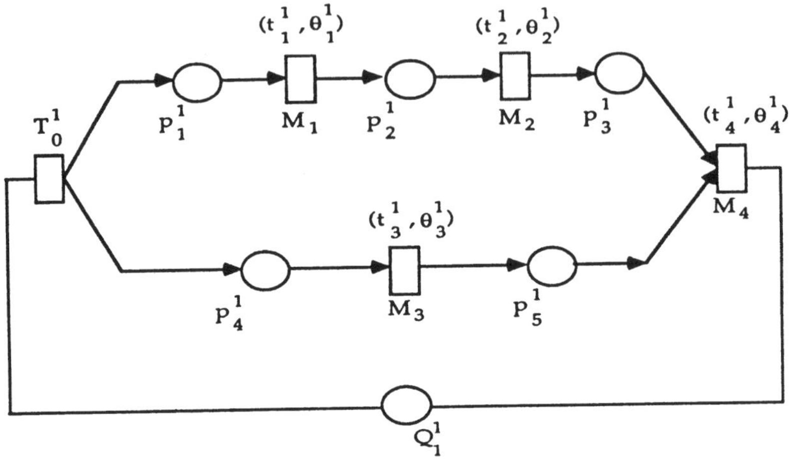

Fig. 7.1: Model related to manufacturing process P_1

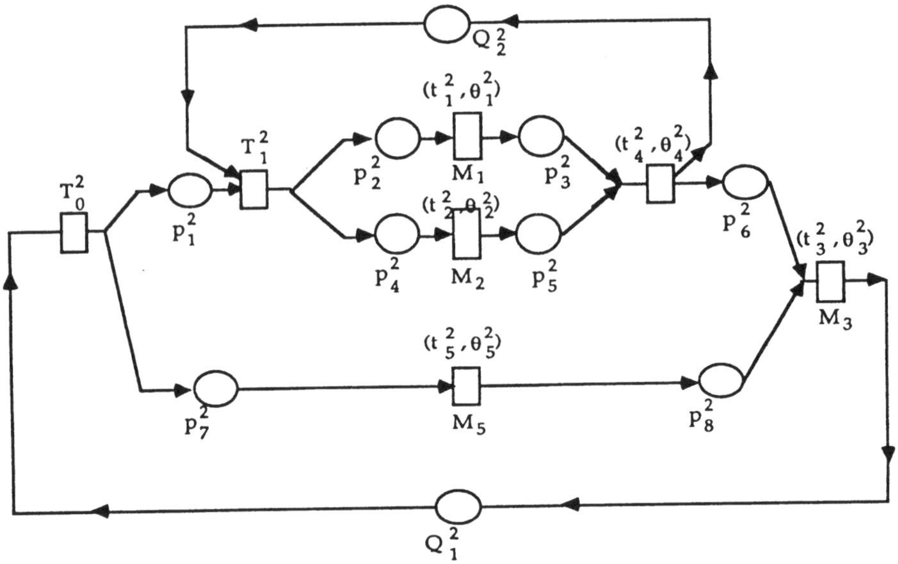

Fig. 7.2: Model related to manufacturing process P_2

Note that these models make use of transitions which do not represent operations but the beginning of the manufacturing of components which are required to assemble a more sophisticated component or the final product. In figure 7.1, it is the case of T_0^1. It is also the case of T_0^2 and T_1^2 in figure 7.2. We call these transitions **launching transitions**. A directed path connects each transition representing an assembly operation to the related launching transition. In practice, it means that the transportation resources used to carry the components at this level of complexity recirculate as soon as the assembly operation is completed. Thus, transportation resources are assigned to each level of manufacturing.

The manufacturing process models correspond to the process circuits in the case of job-shops. These models are completed by command circuits which model the sequencing of the part types in the machines. In figure 8, we represent the command circuit related to machine M_1, assuming that the sequencing related to M_1 is (P_1, P_2, P_2).

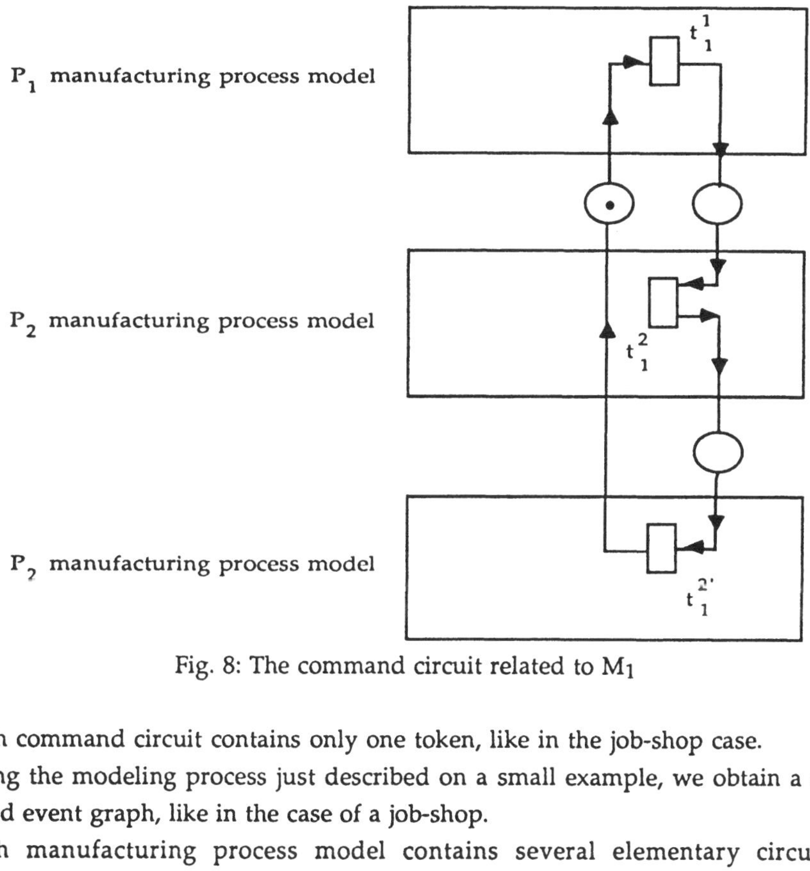

Fig. 8: The command circuit related to M_1

Each command circuit contains only one token, like in the job-shop case.

Using the modeling process just described on a small example, we obtain a strongly connected event graph, like in the case of a job-shop.

Each manufacturing process model contains several elementary circuits. For instance, the model related to the manufacturing process P_1 (see figure 7.1) contains two elementary circuits:

$$\gamma_1 = \{p_4^1, t_3^1, p_5^1, t_4^1, Q_1^1, T_0^1, p_4^1\}$$

$$\gamma_2 = \{p_1^1, t_1^1, p_2^1, t_2^1, p_3^1, t_4^1, Q_1^1, T_0^1, p_1^1\}$$

The model related to the manufacturing process P_2 (see figure 7.2) contains the following elementary circuits:

$$\gamma_3 = \{T_1^2, p_2^2, t_1^2, p_3^2, t_4^2, Q_2^2, T_1^2\}$$

$$\gamma_4 = \{T_1^2, p_4^2, t_2^2, p_5^2, t_4^2, Q_2^2, T_1^2\}$$

$$\gamma_5 = \{T_0^2, p_1^2, T_1^2, p_4^2, t_2^2, p_5^2, t_4^2, p_6^2, t_3^2, Q_1^2, T_0^2\}$$

$$\gamma_6 = \{T_0^2, p_1^2, T_1^2, p_2^2, t_1^2, p_3^2, t_4^2, p_6^2, t_3^2, Q_1^2, T_0^2\}$$

$$\gamma_7 = \{T_0^2, p_7^2, t_5^2, p_8^2, t_3^2, Q_1^2, T_0^2\}$$

Others elementary circuits are the command circuits and the mixed circuits, like in the job-shop case.

4. PERFORMANCE EVALUATION OF THE PRODUCTION SYSTEMS

Let us consider an event graph model of a job-shop or an assembly system. Such a system blocks after a finite period of time if one of the elementary circuits does not contain any token (see property 3). If C^* is the cycle time related to the critical circuit, $\lambda = 1 / C^*$ is a parameter defining the average productivity of the system: on the average, the system produces one sequence of products reflecting the product mix every λ–period (see property 2).

Finally, if the initial marking of the event graph is given, it is possible to compute the productivity of the related manufacturing system or to see if it will block after a finite period of time, providing that the elementary circuits have been identified. Note that the initial marking is known as soon as the initial work-in-processing and the machine sequences are known.

Algorithms to compute the elementary circuits are given in [5] and [6].

5. OPTIMIZATION OF THE PRODUCTIVITY

The productivity of a manufacturing system is maximal if one of the command circuits of the related model is a critical circuit. Thus, optimizing the productivity consists in adding as few tokens as possible in the elementary circuits (except in the command circuits which contain only one token) in order to make critical a command circuit.

Let $\Gamma = \{\gamma_1, \gamma_2, \ldots, \gamma_n\}$ be the set of elementary circuits *except the command circuits* and $P = \{p_1, p_2, \ldots, p_q\}$ the set of places in the event graph model.

γ being an elementary circuit, we denote by $\mu(\gamma)$ the sum of the times assigned to the transitions of γ.

If Γ_c is the set of command circuits, let:

$$M = \mathcal{M}ax_{\gamma \in \Gamma_c} \ \mu(\gamma)$$

and, for i = 1, 2, ..., n:

$$n_i = \lceil \mu(\gamma_i) / M \rceil$$

where $\lceil x \rceil$ is the smallest integer greater than or equal to x.

n_i is then the minimal number of tokens to be put in the elementary circuit $\gamma_i \in \Gamma$ in order that γ_i be not critical.

We define:

- the matrix $A = [a_{ij}]$; i = 1, ..., n ; j = 1, ..., q

where:

$$a_{ij} = \begin{cases} 0 \text{ if } p_j \text{ does not belong to } \gamma_i \\ 1 \text{ otherwise} \end{cases}$$

- the matrix $X = [x_1, \ldots, x_q]^t$, where x_i is the number of tokens in p_i
- the matrix $N = [n_1, n_2, \ldots, n_n]^t$

The problem to solve is as follows:

$$\text{Minimize} \sum_{j=1}^{q} x_j$$

s.t.

$$\begin{cases} A\,X \geq N \\ x_j \geq 0 \text{ and integer for } j = 1, \ldots, q \\ \text{the } x_j \text{ corresponding to places belonging to command circuits are known} \end{cases}$$

In [3] an efficient heuristic algorithm has been proposed to solve this integer LP-problem.

The optimal solution of the previous integer LP-problem is the initial marking of the strongly connected event graph. This initial marking corresponds to the initial WIP of the production system. Some of these work-in-process may not actually exists. In that case, we introduce artificial WIP, and the system reaches its steady state when all these artificial WIP are removed from the system.

The problem with the previous algorithm is that the number of elementary circuits (i.e. the value of the parameter n) grows very fast with the size of the model. As a consequence, only quite small systems can be handled using this approach.

Another approach is currently under development. This approach also leads to an integer LP program, but the number of constraints is equal to the number of places. A step by step heuristic algorithm is under development around this new approach. It will allow to handle models with thousands of places.

6. CONCLUSION

In this paper, we have explained how to model job-shop and assembly systems using event graphs. In the case when the demand is expressed as a product mix, we show that models are always strongly connected event graphs. Thus, the initial state of the system being given, we can derive the behavior of the manufacturing system from the properties of the strongly connected event graph. We also explain how to reach the maximal productivity with a minimal number of parts in process, the machine sequences being given.

The critical point in the previous work is the computation of the elementary circuits whose number increases very fast with the size of the model. A research currently in progress investigates how to avoid the computation of these elementary circuits.

BIBLIOGRAPHY

[1] CHRETIENNE P.,
"Les réseaux de Petri temporisés", Université Paris VI, Thèse d'Etat, 1983.

[2] COMMONER F., HOLT A., EVEN S. and PNUELI A.,
"Marked Directed Graphs", Journal of Computer and System Science, vol. 5, n° 5, 1971.

[3] HILLION H.P. and PROTH J.M.,
"Analyse de fabrications non linéaires et répétitives à l'aide de graphes d'événements temporisés", RAIRO, vol. 22, n° 2, Septembre 1988.

[4] HILLION H.P. and PROTH J.M.,
"Performance Evaluation of Job-Shop systems Using Timed Event Graphs", IEEE Transactions on Automatic Control, vol. 34, n° 1, Janvier 1989.

[5] RAMAMOORTY C.V. and HO G.S.,
"Performance Evaluation of asynchronous concurrent systems using Petri nets", IEEE Trans. Software Eng., vol. SE-6, n° 5, pp. 440-449, 1980.

[6] RAMCHANDANI C.,
"Analysis of asynchronous concurrent systems by timed Petri Nets", Lab. Computer Science, MIT, Cambridge, MA, Tech. Rep. 120, 1974.

Scheduling a Work Conserving Queue with Deadlines: Minimizing the Cost of Getting the Work Done on Time

RANDOLPH HALL

INTRODUCTION

The literature on job-shop scheduling with deadlines has been shaped by this paradigm:

Paradigm 1

(1) Jobs are processed over a fixed number of hours per day, and overtime is unavailable;

(2) Some jobs will inevitably be tardy; and

(3) The performance of a scheduling system should be measured on the basis of job tardiness (or earliness).

In virtually every textbook on production and operations management, and in virtually every paper on job-shop scheduling with deadlines, Paradigm 1 prevails (see, for example, the reviews by Baker and Scudder [1990], Blackstone et al [1982], French [1982], Graves [1981] and Gupta and Sen [1984]).

In light of the worldwide emphasis on just-in-time manufacturing, the inevitability of tardiness should surely be questioned. Any firm that wishes to retain or attract customers must provide reliable service. A single late job may cause the customer to take his or her business elsewhere. Persistent tardiness can permanently damage a firm's reputation and possibly cause bankruptcy. Yet any student who opens a textbook on job-shop scheduling will be told that tardiness is inherent to production systems. The message that tardiness is the product of poor planning, inadequate production control, and indifference to customer service is not communicated.

In today's environment, a more relevant paradigm seems to be the following:

Paradigm 2

(1) All jobs must be delivered on time;

(2) A published due date policy must be followed; and

(3) The objective is to minimize the cost of delivering jobs on time.

Under Paradigm 2, random variations in workload are accommodated through undertime and overtime, not through late deliveries.

Paradigm 2 is the subject of this paper. It will be investigated with respect to scheduling rule, capacity optimization and due date policy. The paper begins by stating assumptions in the following section.

ASSUMPTIONS

Jobs are processed in a work-conserving single-stage queueing system without sequence dependent setup times. The queueing system operates intermittently within cycles (e.g., a day or week). Each cycle contains a service period, when jobs are received and delivered, and a regular production period, when jobs are processed. These periods may or may not coincide. Overtime production occurs outside the regular production period, as needed.

The amount of work that can be performed within a regular production period will, by definition, equal the value one. The amount of work performed during overtime will be measured in the same units as regular work. For example, .5 units of overtime is equivalent to one-half the regular production capacity per cycle. Job size will also be measured in units of work (e.g., a job of size .3 would take 30% of a cycle's regular production capacity).

Definitions

$\Lambda(t)$ = cumulative units of work arrived by time t

$\Omega(t)$ = cumulative units of work completed by time t

$O(t)$ = cumulative units of work completed by time t, through use of overtime production

$R(t)$ = cumulative capacity for processing work by time t, through use of regular time production

$D(t)$ = cumulative units of work that must be delivered by time t

T = time that the last job is completed.

By assumption, sufficient overtime will be available between the times jobs are submitted and received to allow all work to be completed on time. In reality, very large jobs may occasionally violate this assumption. Nevertheless, most firms could foresee an impending shortfall and circumvent late deliveries by turning down orders.

The objective will be to minimize production cost, which will comprise overtime and regular time cost. However, once a regular production capacity is specified, overtime will be the only controllable cost component, and the objective will be to minimize expected overtime per cycle. This will be the objective in the following section.

The system resembles a queue with a finite buffer, except that when the buffer fills, work is diverted to overtime production rather than lost. A key difference is that finite buffer models ordinarily place a limit on the number of customers in the system, not the quantity of work in the system. Further, finite buffer models do not account for the periodic nature of production and service.

THE OPTIMALITY OF EDD WITH RESPECT TO MINIMIZING OVERTIME COST

The optimality of EDD is first demonstrated for static scheduling with deterministic (but variable) processing times, and then extended to dynamic scheduling. Stochastic processing times and the shortest-processing-time rule are also discussed.

Static Schedule

Under static scheduling, regular production should be used to capacity from the start of production until the last job is completed. Because all work must be completed on time, a lower bound on the cumulative work completed by time t is:

$$\Omega(t) = O(t) + R(t) \geq D(t), \qquad T \geq t \geq 0 . \tag{1}$$

Therefore:

$$O(t) \geq D(t) - R(t) , \qquad T \geq t \geq 0 . \tag{2}$$

Because O(t) is a non-decreasing function, it must also be true that:

$$O(t) \geq \max_{\tau \leq t} \{D(\tau)-R(\tau)\} , \qquad T \geq t \geq 0 . \tag{3}$$

For certain queue disciplines, overtime may exceed the lower bound in Eq. 3. Let:

A(t) = amount of advance work completed by time t that possesses a due date after time t.

Ω(t) must be sufficient to produce all of the work due by time t, plus all of the advance work completed before time t. Therefore:

$$O(t) \geq \max_{\tau \leq t} \{D(\tau) + A(\tau) - R(\tau)\} , \quad T \geq t \geq 0 . \tag{4}$$

The lower bound of Eq. 3 cannot be realized unless A(t) equals zero at the time when the deviation between D(t) and R(t) is largest. Suppose that EDD is implemented as follows:

(1) Jobs are sequenced in ascending order according to due date
(2) Full regular production capacity is used whenever jobs are waiting, and
(3) Overtime is only used when a job would otherwise be late.

Then jobs are only completed in advance of their deadline if they do not add to overtime, and regular production is used to capacity until the queue vanishes. Therefore:

$$A(t) = \begin{cases} R(t)-D(t), & D(t) \leq R(t),\ 0 \leq t \leq T \\ 0, & \text{otherwise} \end{cases} . \tag{5}$$

Whenever D(t) > R(t), A(t) = 0. Hence, Eq. 3 is an equality.

As a demonstration, Figure 1 shows an example D(t) curve, along with R(t) and Ω(t). For clarity, it has been assumed that the service and regular production periods coincide, and that the time scale only measures events that occur during these periods. Because overtime occurs between these periods, it is represented by a vertical step in Ω(t). R(t) is a line of slope one through the origin. In the figure, overtime is scheduled at the first possible instant (though it could have been spread over the planning horizon). Therefore, Ω(t) vertically translates R(t) by the quantity of overtime work, which is the maximum of D(t)-R(t).

Dynamic Schedules/Deterministic Processing Times

With dynamic scheduling, jobs arrive at different times, as defined by A(t), and cannot be scheduled simultaneously. There are two important consequences: (1) the server may go idle as it waits for jobs to arrive, and (2) some work may be processed out of the static EDD order. Both consequences can lead to increases in overtime.

Let DEDD denote the dynamic version of the earliest-due-date discipline. That is, the job being processed at any time must have the earliest due date among all jobs currently in the system. DEDD will be differentiated from the static EDD, which specifies that the entire set of jobs received over a planning period is sequenced by due date in ascending order. DEDD assumes preemption as higher priority work

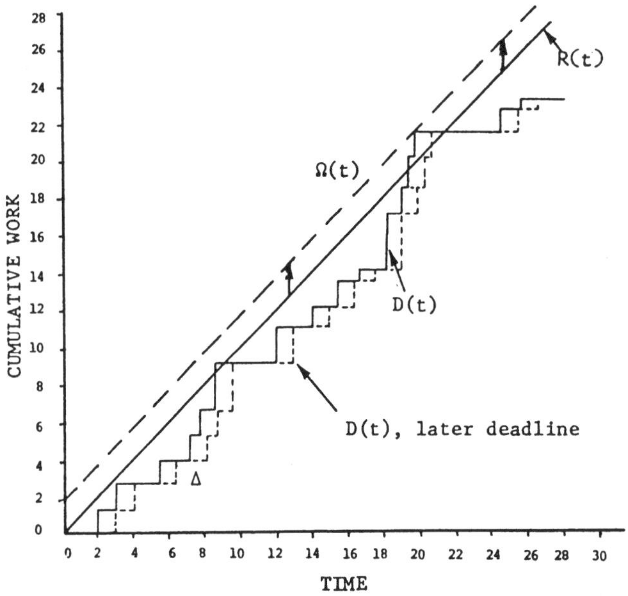

Figure 1. Cumulative arrival and departure diagram for static EDD rule.

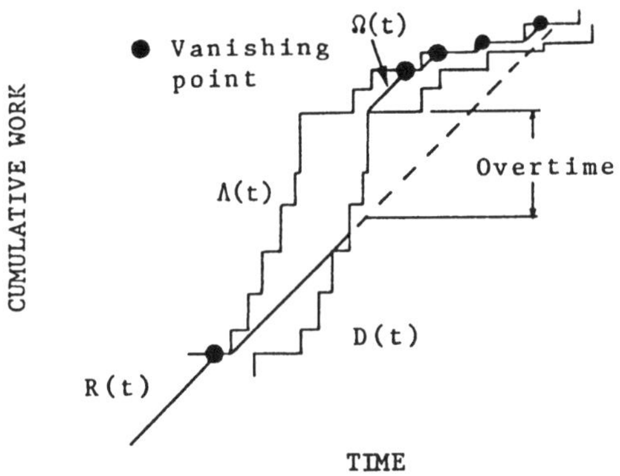

Figure 2. For dynamic arrivals, overtime depends on maximum separation between R(t) and D(t)

arrives. Resumption will be allowed with no change in total processing time. As before, full regular production capacity will be used whenever jobs are waiting.

For any cycle, overtime is only scheduled when essential to complete jobs on time for the next service period. If the unfinished set of jobs due in the next service period can be completed through regular production, no overtime is scheduled. That is, the amount of overtime scheduled in any cycle is defined by the maximum value of $D(t)+A(t)-R(t)$ over the subsequent service period. Under these conditions, DEDD will be shown to minimize overtime for any realization of $A(t)$ and $D(t)$.

Let a vanishing point be a time when the quantity of work in the system drops to zero and let a queueing incident be the period from when the first job arrives after a vanishing point until the next time that $A(t) = \Omega(t)$ (or, if the queue does not vanish, until the end of a planning period). If $R(t)$ exceeds $D(t)$ throughout the incident, and jobs can be sequenced according to static EDD, then all jobs can be delivered on time without overtime. Otherwise, overtime may be needed to complete some of the work on time.

Let the starting time of a queueing incident equal time 0, and let $O(0)=D(0)=A(0)=R(0)=0$. Then the cumulative overtime required up to any time t must be:

$$O(t) = \max_{\tau \leq t} \{D(\tau)+A(\tau)-R(\tau)\} , \qquad T \geq t \geq 0 . \tag{6}$$

Eq. 6 guarantees that the cumulative production equals or exceeds cumulative work due plus the quantity of work completed early. Because processing times are deterministic, and because overtime is assumed to be available without limit, there is no need to schedule more overtime than specified by Eq. 6.

Unlike the static case, DEDD does not guarantee that $A(t) = 0$ whenever $D(t) > R(t)$: the entire set of jobs cannot be scheduled simultaneously. However, DEDD does minimize $A(t)$ for all t and, therefore, does minimize overtime during a queueing incident. Under DEDD, none of the arrived work due after any time t can take priority over arrived work due on or before time t. This guarantees that regular production capacity is never diverted to jobs with later due dates, and that jobs due on or before any time t are processed at the maximum possible rate through regular production. For any realization of $A(t)$ and $D(t)$ (such as Figure 2) DEDD minimizes $O(t)$ throughout the queueing incident.

DEDD continues to minimize $O(t)$ until the start of the next queueing incident, during which time no work arrives and $O(t)$ must be constant. DEDD will then minimize any additional overtime in the next queueing incident, because the process is repeated. Therefore, DEDD minimizes total overtime for any realization of $A(t)$ and $D(t)$, given the assumed flexibility in scheduling overtime.

Stochastic Processing Times

The main impact of stochastic processing times is in the scheduling of overtime. The overtime scheduled in any cycle must be sufficient to insure that R(t)+O(t) exceeds D(t)+A(t) throughout the subsequent service period. Because these values are not known with certainty in advance, a safety margin of supplemental overtime is needed to ensure that jobs are completed on time. The size of this safety margin depends on the standard deviation of the processing time for the work remaining to be completed in the following regular production period and due in the next service period.

To reduce the size of the safety margin, and the amount of overtime worked, it may be advantageous to give priority to jobs with uncertain processing times over those with more predictable processing times. If the uncertain jobs are processed in overtime, and the predictable jobs are left over for the following regular production period, the size of the safety margin can be reduced. Therefore, DEDD is not necessarily optimal. Nevertheless, priority should not be extended to jobs due later than the next service period. As with the deterministic case, completing jobs far in advance would increase A(t) without the benefit of reduced safety margin. Therefore, DEDD should be close to optimal. (If no regular production can occur between the time overtime is scheduled and the end of the subsequent service period, then no safety margin is needed: all remaining work that is due in the next service period must be processed in overtime.)

Scheduling Rules Based on Processing Time

Prior research on job shop scheduling has emphasized incorporation of processing times in the scheduling rule. The shortest-processing-time discipline, for instance, can minimize the expected number of waiting jobs under a variety of conditions. Under Paradigm 1, this translates into a small number of jobs being late and a small value for the expected tardiness. While these are important accomplishments under Paradigm 1, they offer no benefit when overtime prevents jobs from being delivered late. Nothing is gained by reducing the number of jobs in the system if the quantity of outstanding work remains unchanged. In fact, SPT adds to overtime because it causes A(t) to increase. Paradigm 2 clearly demands different scheduling rules than Paradigm 1.

DUE DATE POLICY AND REGULAR PRODUCTION CAPACITY

Given that the EDD (or DEDD) discipline is followed, the principle ways to control system cost are through the selection of a regular production capacity and through the setting of the due date policy. Under Paradigm 1, these issues demand subjective judgments. The optimal regular production capacity is determined by a tradeoff between capacity cost and job tardiness, an especially fuzzy measure of performance. (In the job-shop literature, the number of ways to measure tardiness is virtually limitless.) The due date policy is determined by a tradeoff between two service measures: time until due and tardiness.

Paradigm 2 makes the tradeoffs tangible. The production capacity depends on a tradeoff between two cost measures: regular cost and overtime cost. The due date policy is defined by a tradeoff between cost and time until due, a more concrete measure of performance than tardiness. Therefore, quite independent of the merits of utilizing overtime to eliminate tardiness, Paradigm 2 offers a more definitive framework for making key decisions.

This section evaluates the cost impacts of changing the due date policy and regular production capacity. To keep the analysis manageable, production periods and service periods will not be allowed to overlap. For example, work may be received during a daytime 8:00-5:00 period and performed overnight. The effects of overlapping periods are discussed in the appendix.

Because production and service periods do not overlap, it can be assumed that deadlines occur at the start of service periods. That is, there is no benefit in specifying a deadline in the middle of a service period if jobs cannot be processed in the intervening time. For simplicity, all work that arrives in a service period will be assigned the same deadline, which is represented by the number of cycles allowed between receipt and delivery. Because service and production periods do not overlap, there is no need to specify when an arrival occurs within a service period or when a job is completed within a production period. The values of interest are the total amount of work to arrive and to be produced during a period.

Equal Delivery Times

Suppose that a firm advertises the policy that all work will be delivered at the start of the d'th service period after the one received. For a given arrival process of jobs, let:

ρ = expected quantity of work received per cycle

$E(O|d,\rho)$ = expected overtime per cycle, with d cycle delivery and a work per cycle of ρ.

The expected overtime per cycle is bounded from below by the expected work received per cycle in excess of regular production capacity:

$$E(O|d,\rho) \geq \begin{cases} \rho-1, & \rho \geq 1 \\ 0, & \rho < 1 \end{cases} . \tag{7}$$

These bounds would be achieved if the arrival process is deterministic and stationary. In other cases, fluctuations in arrivals can create overtime.

For any realization of an arrival process, there exist alternating periods during which any small scalar increase in job sizes would either cause an increase in overtime or not cause an increase in overtime. During periods following the imposition of overtime, up to the next vanishing point, or during queueing incidents that do not require overtime, small increases in job size can be absorbed through an equivalent reduction in idle time. These will be called idle periods. For periods preceding the imposition of overtime, beginning when the queue first forms, increases in job size must be compensated through increased overtime. These will be called overtime periods. In no realization can an increase in job size cause overtime to increase by more than an equal amount. Therefore, for scalar increases in job size:

$$\frac{\partial E(O|d,\rho)}{\partial \rho} \leq 1 \ . \tag{8}$$

As job sizes increase, overtime periods must become longer and idle periods must become shorter. Therefore, the probability that any job falls in an overtime period must increase as ρ increases. For a scalar increase in job sizes, this probability represents the derivative of $E(O|d,\rho)$ with respect to ρ, and, furthermore:

$$\frac{\partial^2 E(O|d,\rho)}{\partial \rho^2} \geq 0 \ . \tag{9}$$

When jobs are very small, the probability that a job falls in an overtime period must be small, and the derivative of $E(O|d,\rho)$ with respect to ρ must be close to zero. When ρ is very large, the probability that a job falls in an overtime period must be close to one, and the derivative of $E(O|d,\rho)$ must be close to one. In these extremes, $E(O|d,\rho)$ is accurately approximated by Eq. 7, the lower bound on expected overtime per cycle.

Let the excess overtime be the expected overtime per cycle in excess of the lower bound in Eq. 7. As ρ approaches one from below, the excess overtime must increase at the same rate that $E(O|d,\rho)$ increases. For $\rho > 1$, the excess overtime must decrease at the rate $1-\partial E(O|d,\rho)/\partial\rho$. The excess overtime, which represents the potential reduction in overtime through the alteration of the due date policy, is largest when $\rho = 1$.

Increasing the lead-time, d, clearly has the effect of reducing the expected overtime per cycle. This can be visualized through Figure 1. Within any queueing incident, delaying all of the deadlines by a small value Δ decreases the maximum of $D(t)-R(t)$ by a like amount, which translates into reduced overtime if $\max\{D(t)-R(t)\}\geq d$. (In cases where the due date policy is not uniform, $A(t)$ can also decline.) Further, the change in expected overtime per cycle due to a small change Δ must equal the number of overtime periods occurring per cycle, multiplied by Δ. Therefore:

$$\frac{\partial E(O|d,\rho)}{\partial d} = - \text{[expected frequency of overtime periods per cycle]}$$

The frequency of overtime periods tends to be largest when $\rho \approx 1$. For large values of ρ, altering the due date policy provides little benefit because incidents tend to be long-lasting and infrequent. For small values of ρ, altering the due date policy also provides little benefit, because excess capacity keeps queues from growing large.

For a given arrival process, ρ can be optimized to minimize the production cost per unit of work received. Because multiplying all job sizes by a scalar value γ is equivalent to multiplying the regular production capacity by the scalar $(1/\gamma)$, optimizing ρ has the effect of optimizing regular production capacity. The relevant objective function is expressed as:

$$C = \frac{E(\text{overtime cost per cycle}) + E(\text{regular production cost per cycle})}{E(\text{work per cycle})}$$

$$= \left[\frac{\beta E(O|d,\rho)}{\rho} + \frac{1}{\rho}\right]\alpha, \tag{10}$$

where:

α = cost per unit of regular production capacity

β = premium paid for overtime production, as a ratio to the cost of regular production capacity.

Through simulation and linear regression, $E(O|d,\rho)$ was estimated for the independent uniform $[0,a]$ ($\rho = a/2$) and independent exponential ($\lambda = 1/\rho$) distributions for work per cycle. In each case, the arrival and departure processes were simulated over 500 day periods, and replicated for various values of ρ. For each replication, the average work received per cycle and the average overtime per cycle were calculated and plotted (Figure 3 for uniform and Figure 4 for exponential). Exact values were calculated for d = 1, in which case, $E(O|1,\rho) = \int_1^{\infty}(x-1)f(x)dx$, where $f(x)$ is the probability density function for the work received per cycle.

In all cases, for values of ρ close to one, the relationship between average overtime per cycle and ρ was accurately approximated by the relationship:

$$E(O|d,\rho) \approx (\rho-1) + \frac{a}{\rho^b}, \tag{11}$$

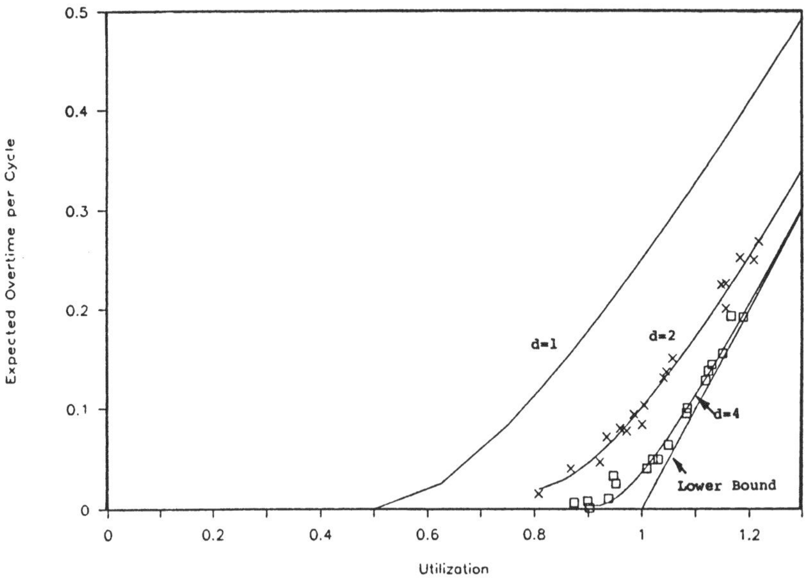

Figure 3. Expected overtime versus utilization for uniform work distribution.

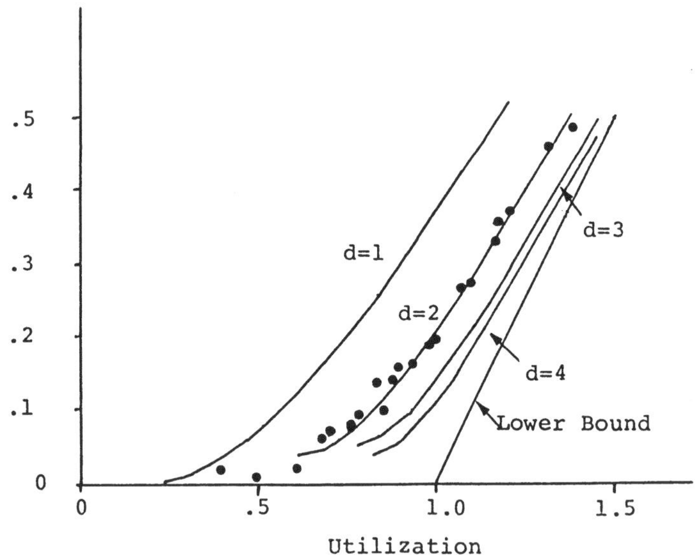

Figure 4. Expected overtime versus utilization for exponential work distribution

where a and b are parameters specific to the probability distribution and deadline policy. Equation 11 generalizes the exact result for the uniform/d=1 case, for which a=.25 and b=1. For other cases, a and b were estimated through least-squares linear regression on the log of the function:

	d=	1	2	3	4
Uniform	a=	.25	.101	.055	.036
	b=	1	3.36	6.28	9.91
Exponential	a=	---	.206	.139	.111
	b=	---	1.50	2.65	3.50

For d=1 and exponential, the exact result for $E(O|d,\rho)$ is $\rho \cdot e^{-1/\rho}$. For both distributions, increasing d has the effect of increasing the power to which $1/\rho$ is raised. This causes $E(O|d,\rho)$ to more closely approximate the lower bound of Eq. 7. For all values of ρ, $E(O|d,\rho)$ is larger for the exponential distribution than the uniform distribution, owing to its larger coefficient of variation (1 versus .29). Equation 11 is not accurate for small values of ρ because it begins to increase as ρ declines. In this range $E(O|d,\rho)$ should be approximately zero.

Using the uniform distribution approximation for $E(O|d,\rho)$, the objective function of Eq. 10 is plotted as a function of ρ for varying values of d and β in Figs. 5-7. Clearly, there is a tendency for ρ^* to decrease as β increases: if overtime is expensive, then it is important to have more regular capacity on hand. Increasing d has the effect of smoothing out variations in cyclic workload. This means that the optimal value of ρ tends to be closer to one than it would be for small values of d.

The optimal value of ρ can be found as a function of a,b and β by taking the derivative of Eq. 10 and setting to zero:

$$\rho^* \approx \left[\frac{a(b+1)\beta}{1-\beta}\right]^{1/b} \tag{12}$$

Results are shown below for four due date policies with β=2:

ρ^*	Days Until Due			
	1+	2	3	4
Uniform	1.0	.96	.97	.98
Exponential	.69	1.02	1.01	1.00

\+ Exact Result

In all but one case, $\rho^* \approx 1$, meaning that the expected overtime per cycle equals the expected idle time per cycle, a logical result considering that the opportunity costs for idle time and overtime are identical when β=2. For the special case d=1,

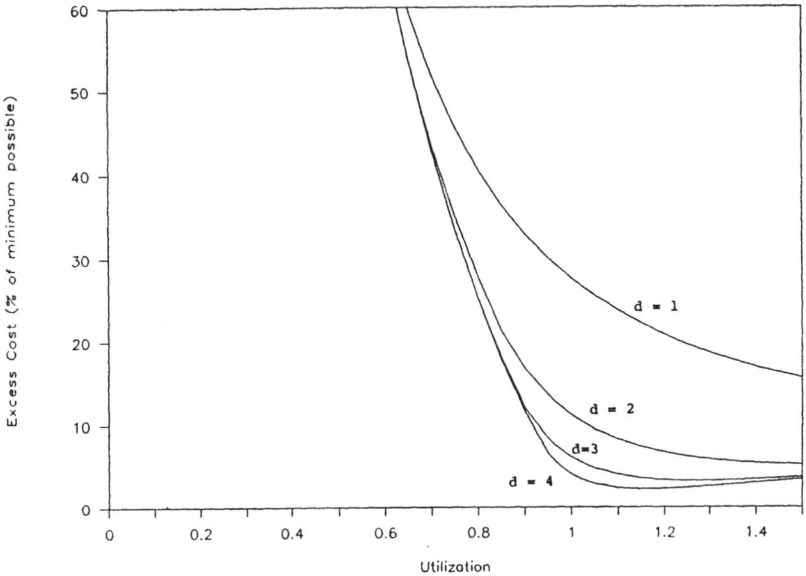

Figure 5. Excess cost versus utilization, β= 1.1

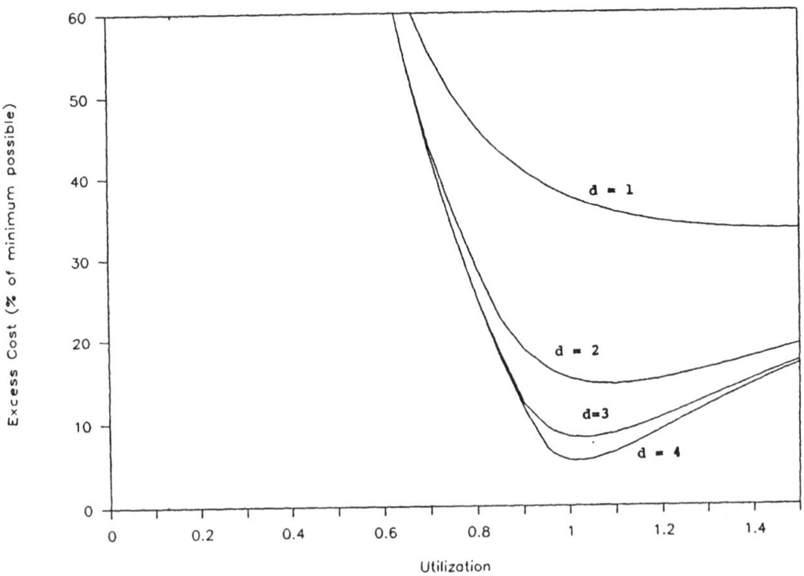

Figure 6. Excess cost versus utilization, β = 1.5

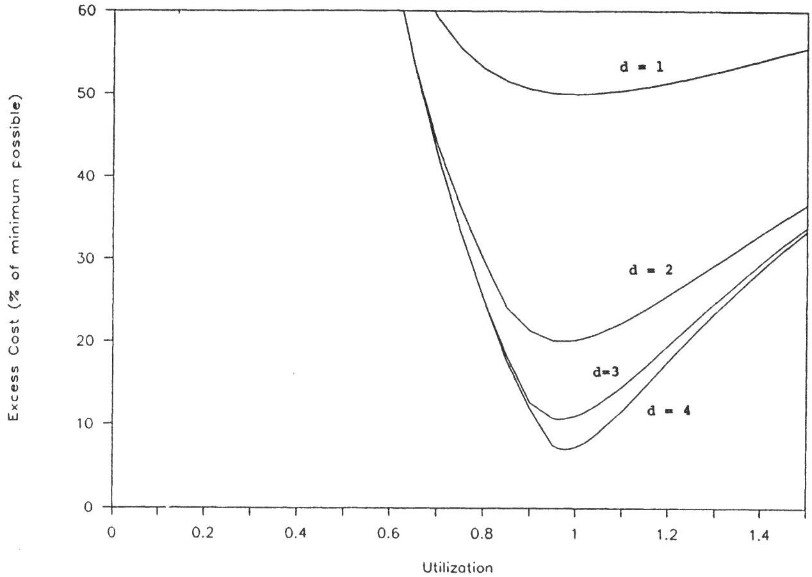

Figure 7. Excess cost versus utilization, $\beta = 2$

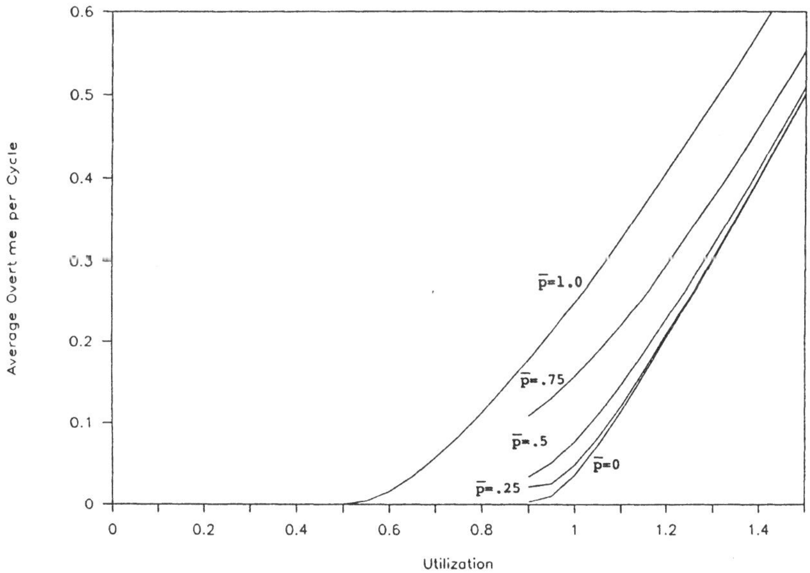

Figure 8. Average overtime per cycle for mixed due-date policy. Figures based on regression to simulated data.

optimizing ρ is equivalent to optimizing the order quantity in the classic newsboy problem. This means that ρ^* is the median of the work per cycle distribution. Because the exponential distribution is not symmetric, $\rho^* \neq 1$ when d=1.

Mixed Due Dates

It may be advantageous to offer two levels of service: a high price rush service and a low price ordinary service. Within any queueing incident, adding a rush service can cause D(t) to increase as well as force some work to be processed out of the static EDD order.

Suppose that work can be due either after one cycle or after four cycles. Let $\bar{p}$ represent the mean proportion of work due after one cycle. For $\bar{p}=0$, E(O) is identical to $E(O|d=4,\rho)$ and when $\bar{p}=1$, E(O) is identical to $E(O|d=1,\rho)$. For other values of $\bar{p}$, E(O) falls between these extremes. Figure 8 gives E(O) for different values of $\bar{p}$, where the proportion in any cycle is uniform over $[0,2\bar{p}]$ and the work received per cycle is uniform over $[0,2\rho]$. The least-squares values of a and b for Eq. 11 are provided below:

$\bar{p}$ =	1	.75	.5	.25	0
a =	.25	.157	.077	.048	.036
b =	1.00	2.69	5.26	8.79	9.91

When $\bar{p}$ is decreased even slightly below 1, E(O) declines appreciably. Allowing just a small fraction of work to be due after four cycles greatly increases the flexibility in scheduling work. Once $\bar{p}$ has declined as far as .25, E(O) is nearly at its minimum. Overall, it seems that a split policy can provide substantial benefits to customers who demand fast service without causing cost to increase appreciably.

FURTHER ISSUES

The dichotomy between Paradigm 1 and Paradigm 2 extends to most of the issues in the scheduling literature, such as:

(1) Sequence dependent setup times
(2) Interactions between queues in a network of servers.
(3) Non-stationarities in arrival rate
(4) Arrival rate dependencies within customer classes
(5) Cutoff times and deadlines within periods
(6) Preemption with increase in processing time.

In all cases, one would expect that due date should play a larger role, and service time a smaller role, under Paradigm 2 than under Paradigm 1. Little is gained by reducing the expected number of jobs in the system. The key to minimizing overtime is to minimize the quantity of outstanding work due by any given time. This is accomplished by insuring that the system utilizes regular production capacity to its maximum and by giving priority to jobs with early due dates over those with later due dates.

In addition, Paradigm 2 provides a more definitive framework for optimizing the regular production capacity and for selecting a due date policy. It eliminates much of the ambiguity inherent to tardiness based performance measures. It makes the tradeoffs tangible.

REFERENCES

Baker, K.R. and G.D. Scudder (1990) "Sequencing with Earliness and Tardiness Penalties: A Review," Operations Research, V. 38, pp. 22-36.

Blackstone, J.H., D.T. Phillips and G.L. Hogg (1982) "A State-of-the-art Survey of Dispatching Rules for Manufacturing Job Shop Operations," International Journal of Production Research, V. 20, pp. 27-45.

French, S. (1982) Sequencing and Scheduling: An Introduction to the Mathematics of the Job-Shop, Ellis Horwood, Chichester, U.K.

Graves, S.C. (1981) "A Review of Production Scheduling," Operations Research, V. 29, pp. 646-675.

Sen, T. and S.K. Gupta (1984) "A State-of-Art Survey of Static Scheduling Research Involving Due Dates," Omega, V. 12, pp. 63-76.

APPENDIX-OVERLAPPING PERIODS

If service and production periods overlap, it is possible for the system to go idle during cycles in which the amount of work available exceeds the production capacity. This is especially true when d=1. If service and production periods coincide, and work is due at the start of the service period, each cycle would begin idle with no work in the system. Nevertheless, even when d=1, some overlap can be tolerated without appreciably adding to overtime or idle time. The issue of permissible overlap is examined here.

For simplicity of exposition, suppose that the service and production periods have identical lengths and that the production period begins c units of time after the start of the service period and ends c units of time after the end of the service period. Also, assume that all work is due at the start of the service period following the period received (i.e., d=1). Larger leadtimes would moderate the effects of overlapping periods, so d=1 is an extreme case. The arrival process will be assumed to be stationary, and processing times will be independent. Now consider three regimes:

Small ρ (.6 or less): In this case, queues will vanish and reform during the overlapped portion of the regular production period, and the system may approach steady-state by the time when the service period ends. Then the work remaining at the end of the service period is approximated by the steady-state queue size, and overtime will be small if c is adequate to accommodate the steady-state queue:

$$c \geq k \cdot E(\text{steady-state queue size, measured in work}) \ . \tag{13}$$

k should be approximately twice the coefficient of variation for the steady-state queue size, plus one.

Medium ρ (.6 to 1) In this range, overtime will occasionally be needed whether or not service and production periods overlap. A reasonable objective would be to insure that the overlap does not create idle time in those cycles in which overtime is needed. Hence, overtime is kept at a minimum.

Suppose that the amount of work received in a cycle equals exactly 1 (i.e., $\Lambda(1)=1$), the maximum amount of work that can be processed without overtime. For such a cycle, it would be desirable for c to be sufficiently large to keep the system from going idle during the regular production period.

The expected work in the system at any time t is the difference between the expected work received and the expected work completed:

$$E(\Lambda(t)-\Omega(t)\,|\,\Lambda(1)=1) \;=\; t-(t-c) \;=\; c, \qquad c \leq t \leq 1 \ . \tag{14}$$

If c is sufficiently large, random fluctuations in the receipt of work will not cause the system to go idle.

The amount of work received in any cycle is a random sum of random variables. If these random variables are independent, and the arrival process of jobs is stationary Poisson, then:

$$V[\Lambda(t)] \;=\; tV(X) + tE^2(X) \;=\; tE^2(X)\,[C^2(X)+1], \tag{15}$$

where E(X),C(X), and V(X) are the expectation, coefficient of variation and variance of a job's processing time. Given the condition that the amount of work received in the cycle equals one, independence is not provided and Eq. 15 only gives an upper bound on the variance.

Let $\tilde{\Lambda}(t)$ equal the amount of work received between time t and the end of the service period. Given the condition that $\Lambda(1)=1$, $\Lambda(t)+\tilde{\Lambda}(t) = 1$ for $0 \leq t \leq 1$. Further, $V[\Lambda(t)]=V[\tilde{\Lambda}(t)]$ over this same domain. An upper bound on $V[\tilde{\Lambda}(t)]$ is defined by Eq. 15, with (1-t) replacing t. Combining this upper bound on $V[\tilde{\Lambda}(t)]$ with the upper bound on $V[\Lambda(t)]$ provides the following:

$$V[\Lambda(t)] = V[\tilde{\Lambda}(t)] \leq E^2(X)[C^2(X)+1]\cdot\min\{t,1-t\} = \sigma^2 . \qquad (16)$$

To prevent idle time in a cycle where $\Lambda(1)=1$, a conservative constraint on c would be:

$$c \geq 2\cdot E(X)\sqrt{[C^2(X)+1]\cdot\min\{t,1-t\}} , \qquad c \leq t \leq 1 . \qquad (17)$$

Unless c is very large, the right-side of Eq. 17 is largest when t=.5, in which case the expression becomes:

$$c \geq \sqrt{2}\cdot E(X)\sqrt{C^2(X)+1} . \qquad (18)$$

c should be largest when the expected work per job is large, and when the variation in work per job is large.

Large ρ ($\rho > 1$) In this regime, it would be sufficient to prevent idle time during the overlap on an average day. As with the previous regime, $E[\Lambda(t)] - 2\sigma$ should be greater than the expected work completed by time t:

$$\rho\cdot t - 2\cdot E(X)\sqrt{[C^2(X)+1]\max\{\rho t,\rho(1-t)\}} - [t-c] \geq 0, \quad c \leq t \leq 1. \qquad (19)$$

The minimum of Eq. 19 would occur sometime between the start of the production period (when the expected work in the system is smallest, over [c,1]) and the middle of the service period (when the standard deviation of the work received is the largest). Because $\rho>1$, the expected work in the system grows with time, meaning that c does not have to be as large as for the previous regime.

Overall, it appears that small values of c are tolerable in the following cases: (1) when ρ, and the steady-state queue size, are small; (2) when ρ is large, and the system rarely goes idle; and (3) when ρ is comparable to one, and job sizes are small and constant. Small values of c are least tolerable when ρ is comparable to one and job sizes are large and variable.

These issues can be studied in greater detail through an analysis of transient queueing conditions, either analytically or by simulation. Analysis could also be extended to specifying a deadline within a cycle, the effects of which are similar to specifying the overlap.

Analytical and Simulation Techniques Used to Gain Insight into Multi Product Machine Shop Control

PREETINDER CHEEMA and DENIS TOWILL

1. PRELIMINARY REMARKS.

Traditional classical control theory has been well established in the design techniques developed in hardware applications such as machine tool drivers, aircraft autoland systems [1] etc. Equivalent modern control theory which yields similar results has been applied to manufacturing systems [2]. In both cases, we can therefore apply the principles learnt in "hard" systems to "soft" and organisational systems [3].

This paper illustrates some of the useful theoretical concepts of Industrial Dynamics (ID), and by way of an example, investigates the dynamics of a live 'To Make' order system for a multi-product machine shop.

The manufacturing system is modelled using modern ID and classical control theory (Laplace Transformations), the system is analysed by looking at deterministic and stochastic inputs. The scope of this paper (part 1) describes the model and illustrates some of the results. Part 2 of this paper will fully investigate the results from this, and subsequently improved 'To Make' models.

2. NOMENCLATURE.

Industrial Dynamics

PIC RATE	Statistical function for parts to be on shelves.
COMRATE	Rate of completion for orders.
FRST	Freestock or inventory in finished goods.
TOMK	Rate at which orders are placed.

BUF	Buffer stocks.
LTQTY	Lead time quantity estimation for the completion of parts.
OSORD	Outstanding orders still to be filled.
LDTM	Lead time.
AWDMD	Average Weekly demand for parts.
WIP	Work In Progress.
TORD	Total Order Rate, or sales.
TAI	Time to adjust inventory deficit to zero for no further changes to TORD.
TAC	Time over which total orders are averaged.

Transfer function techniques

S	Laplace operator.
τ_{pa}	Actual production lag time constant.
τ_{pb}	Theoretical production lag time constant.
τ_a	Total orders averaging time constant.
τ_i	Time over which inventory error is adjusted.
K_w	WIP adaptive time constant.
K_2	Lead time gain factor ($k_2 * \tau_{pb} S$).
K_3	Pick Rate gain factor ($1+k_3*\sigma$).
σ	Variant for Pick Rate.
M	Multiplying Factor.
θ_{oss}	Steady State Value Of FRST.

3. INTRODUCTION.

In recent years, the increasingly complex problem of analysing Manufacturing Systems has led to a combination of simulation and analytical techniques based upon a Control Engineering methodology.

Tustin [4] in 1953 suggested that control theory could have a major part in the analysis of "soft systems". He had already made a considerable reputation due to his contributions to knowledge in radar control systems and man-machine systems, so was well qualified to appreciate the benefits of modelling which would lead to the concept of design by analogy.

Towill [5] in 1982 bridged the gap between "hard" and "soft" systems when he identified particular manufacturing systems operations with a family of equivalent "hard" systems including satellite trackers, machine tool drivers, and aircraft autoland systems.

The use of control theory in manufacture includes the specification, modelling, simulation, and subsequent optimisation of systems which are characterised by orders; material flows; information flows; and decision processes which operate to produce effective output from a sequence of machines and manufacturing processes.

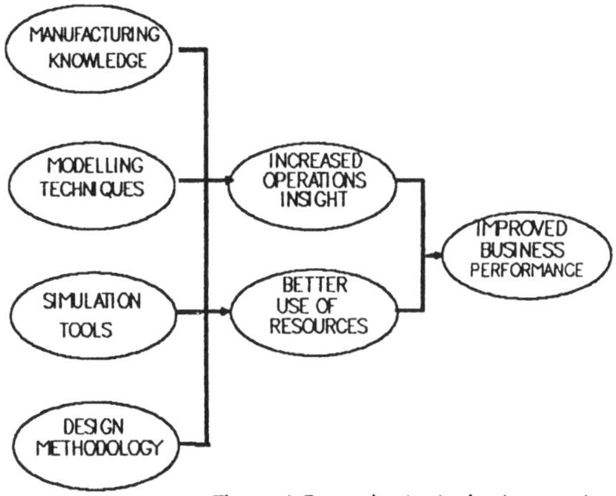

Figure 1 Dynamic Analysis Approach To Manufacturing Systems Design [6]

Figure 1 shows the objectives required for a dynamic analysis approach to manufacturing systems design.
The basic requirements are, manufacturing knowledge, modelling techniques, and simulation tools integrated within a design methodology.Much work has been undertaken by the Industrial Systems Group at Cardiff [7,8,9] in the field of production control systems. Much of the work is in collaboration with South Wales industry, and many grants have been awarded by research councils in pursuit of gaining insight into industrial systems.

The current situation being investigated is the 'To Make' model of Biomet Ltd, a leading manufacturer of orthopedic joints and supporting devices, who are based in Bridgend, South Wales. The model represents the production interaction of sales with operating variables such as inventory, work in progress, production order and completion rates, lead time and 'pic rate' estimations.

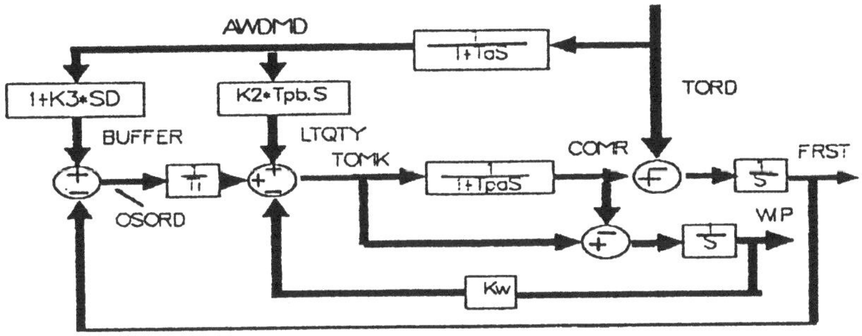

Figure 2. The Biomet 'To Make' Model.

A series of transfer functions will be derived to analyse the model. Exponential smoothing or EXSMO [10] is used to simulate the model for different types of input sales demand and variance of the time constants. The simulated results are then cross checked with the analytical results to test the accuracy and validity of the model.

4. MODEL BUILDING.

The essential steps in modelling and design of manufacturing systems can be summarised as shown in Figure 3.

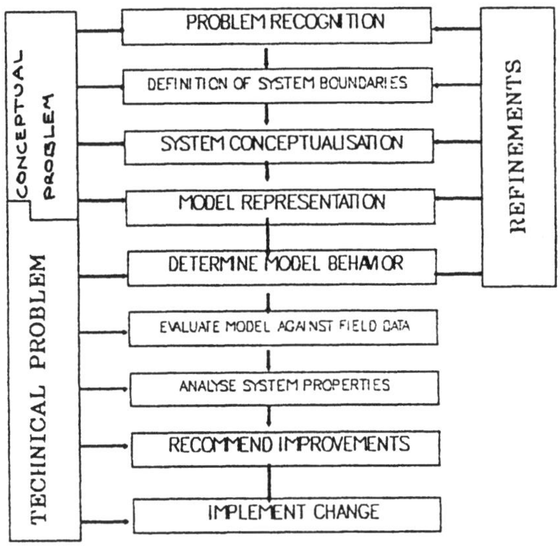

Figure 3. Essential Steps In The Modelling And Design Of A Manufacturing System.

The problem is identified from existing operating records and the first hand knowledge of production managers, manufacturing engineers, and shop floor operators. Much data is usually collected in the form of operation times; machine down times and other delays, but this, more usually than not, is not in the form required for model building. It is probable that the system modeller will have to conduct activity sampling, interviews, and interpolation/extrapolation of existing data to put in a useful form for the model. [11]

The definition of the system boundaries must represent the current problem. The original 'To Make' model at Biomet has undergone changes to reflect the current computerised production system and improvements as designed by the system engineer. Cross checking and validation of the model are essential steps needed to test how closely the design resembles what actually exists. Any scepticism on model validity may seriously undermine any future results.

Potential improvements are tested via computer simulations for different types of inputs, may they be deterministic or stochastic. A sensitivity analysis of the effects on varying key parameters is also undertaken to determine system robustness.

The new proposed system is then subject to a cost-benefit analysis against the business plan of the user, for example, the trade off between customer service levels and stock turnovers.

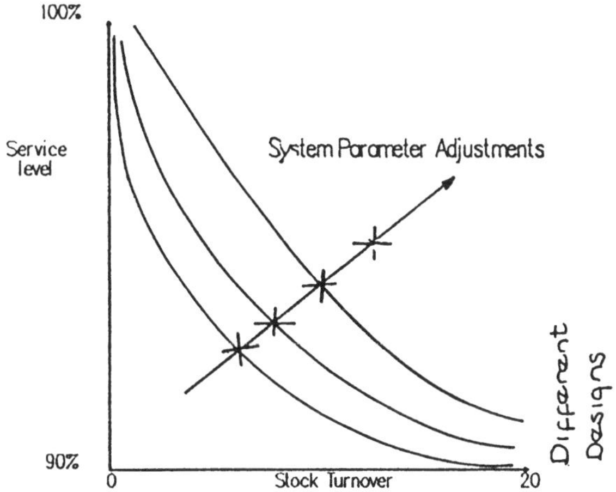

Figure 4. Trade Off Between Service Level And Stock Turnover.

Figure 4 shows the trade off between customer service levels and potential stock turnovers for various designs. Therefore if customers are to be satisfied totaly, this implies an infinite amount of freestock must be held, thus limiting the number of stock turnovers. Whereas if the service level were to fall to say 90%, then less freestock would be held (but this may risk stockouts), and turnover would increase. Therefore the critical design would have to be a compromise.

5. THE DEVELOPMENT OF THE 'TO MAKE' MODEL.

The product range in the host company is in excess of 6000 parts ranging from replacement joints (hip, knee etc) to a range of fixation devices (bone screws, pins, nails), to operating tools and various supporting devices. Of these 6000 parts, 500 are in regular production.

Marketing policy promises the customer a service level such that the goods will be delivered within 48 hours of orders received. This implies that there must be sufficient products which can be picked off the shelf to satisfy customer requirements, hence a minimum inventory level is required. A statistical quantity - the 'Pic Rate' is calculated to give this minimum requirement.

Figure 2 shows the block diagram of the 'To Make' model. It shows that there are two linear feedback loops governed by FRST and WIP, and two major feedforward loops which smooths out the average sales demand.

5.1 Lead Time and Lead Time Quantity.

The lead time quantity refers to the amount of orders expected to be in the production pipeline in order to ensure the average weekly demand for that product is met. In practice, this expected lead time quantity (like the averaging demand) is automatically adjusted by the 'To Make' model in real time to represent the last observed process time. This is done using the date a part order is released onto the shop floor, and the date it has been manufactured and then booked into stores. This data is entered into the computer system as an integral part of the product management system.

The time that the batch has spent on the shop floor gives the latest available production lead time for that product. The lead time used in the LTQTY calculation is updated by this latest figure. Therefore, the new lead time estimate becomes:

LDTM=1/3*(latest lead time)+2/3*(previous lead time) (1)

and LTQTY=AWDMD*LDTM (2)

5.2 The 'Pic Rate'.

The quantity of buffer stock reflects the company policy for safety stock. This buffer level is set for each product depending on its ranking on the Pareto scale (and hence the company determines customer service level).

This quantity is based on an A B C analysis by sales value, wherein all products are classed within 5 categories. For example, the Pic Rate for the top 200 parts, which account for 86% of the total sales, should be 97.5%.

The Pic Rate is interpreted via an extra buffer stock quantity on the shelf additional to one weeks average demand. This is equal to a constant K_3 times the standard deviation (σ) of the average weekly demand from its mean value. The constant K_3 for the first (A) Pareto category is a 97.5%

probability of occurrence averaging the random variance in a normal distribution of standard deviation (σ) . The Pic Rate and constant values can then be summarised as in Table 1.

CATEGORY	PICK RATE	PARETO RANGE	CONSTANT K_3
A	97.5%	1 TO 200	1.96
B	95.0%	201 TO 500	1.65
C	85.0%	501 TO 1000	1.04
D	70.0%	1001 TO 2000	0.53
E	-	2001 TO 6000	0.00

Table 1 Summary Of Pic Rates

As can be seen from Table 1, Category A has a greater priority weighting than B which in turn has a greater weighting than category C. Using steady state analysis ,it can be deduced then that the theoretical lead times increase as the Pic Rate value for a category reduces. This is as expected and can lead to higher stock turnover for lower demand at the expense of a lower service level.

Thus, the Pic Rate is a compromise that has to satisfy both the customer and the director of production. The customer wants his parts on time with the specified quality requirements. The production director wants to hold the minimum stock quantity to satisfy the customer without risking stockouts.

5.3 Work In Progress.

At the time the computer program at the factory is run to estimate current lead times, the work in progress is the amount of work that has been authorised onto the shop floor by the works director, but has not yet been completed and booked into stores as finished goods.

6. SYSTEM PARAMETER ESTIMATIONS.

Extensive work has been carried out by Coleman [12] on the Lead time estimation on categories A to E.

From Figure 5 it can be seen that there is no real correlation between the theoretical and actual lead times. Also, the actual lead time is itself a variable. Nevertheless, one of the tasks of the present work is to see if it is possible to improve the system design and develop an algorithm to improve the service level - stock turnover trade off.

From Figure 5 the time constants τ_{pb} and τ_{pa} have been evaluated. The existing model in operation has the values of τ_i and K_w deliberately set to unity. We are presently investigating whether these parameters can later be tuned to improve the overall system performance. The averaging demand parameter, τ_a, is an exponentially moving average built up over 12 months. Therefore it can be thought of as a smoothed (AWDMD) 'roll on roll off' value.

7. THE LAPLACE AND INVERSE LAPLACE TRANSFORMS

The Laplace and inverse Laplace transform solutions of the transfer functions was evaluated for the 'To Make' model [13]. The following is a summary of the major transfer functions.

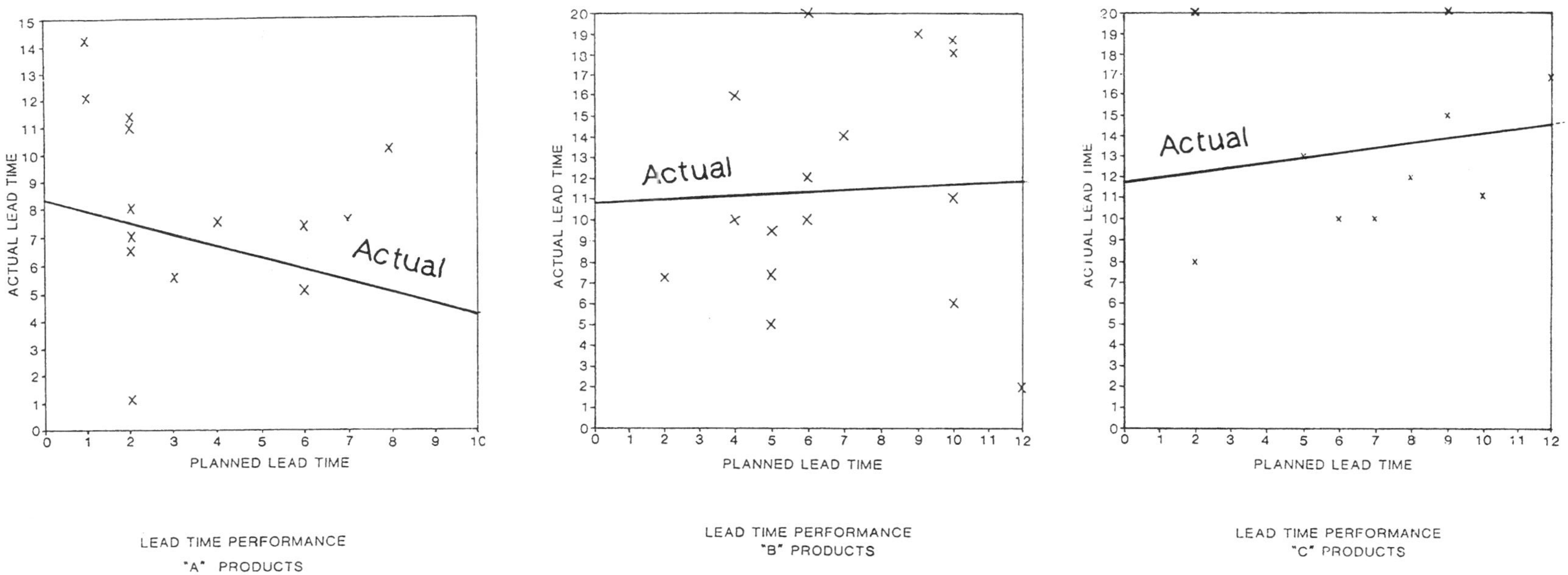

Figure 5. Lead Time Estimation For Categories
A, B, and C

$$\text{If} \quad \frac{1}{[(\tau_i S+1)+\tau_{pa}S(\tau_i S+K_w\tau_i)]} = \{.\}$$

$$\text{Then} \quad \frac{FRST}{TORD} = \{.\}*[\frac{K_2}{(1+\tau_a S)} + \frac{\tau_i K_3}{(1+\tau_a S)} - \tau_i(1+\tau_{pa}S)-\tau_i K_w \tau_{pa}] \quad \ldots(3)$$

$$\frac{COMRATE}{TORD} = \{.\}*[\frac{\tau_i S(K_2/\tau_i + K_3) + (1+\tau_a S)}{(1+\tau_a S)}] \quad \ldots(4)$$

$$\frac{WIP}{TORD} = \tau_{pa}*\{.\} *[\frac{\tau_i S(K_2/\tau_i + K_3) + (1+\tau_a S)}{(1+\tau_a S)}] \quad \ldots(5)$$

$$\frac{OSORD}{TORD} = \frac{K_2}{1+\tau_a S} - \{.\}*[\frac{K_2}{(1+\tau_a S)} + \frac{\tau_i K_3}{(1+\tau_a S)} - \tau_i(1+\tau_{pa}S)-\tau_i K_w \tau_{pa}] \quad \ldots(6)$$

If TORD is a unit step input of value 1/S, the transfer function inverse for FRST and OSORD is in the form of;

$$\{FRST/OSORD\} = M*\frac{S^2 + a_1 S + a_0}{S(S+\alpha)(S+\beta)(S+\mu)} \quad \ldots(7)$$

and the inverse Laplace solution is;

$$\{FRST/OSORD\} = \ldots$$

$$M*[\frac{a_0}{\alpha\beta\mu} + \frac{(\alpha^2-a_1\alpha+a_0)e^{-\alpha t}}{\alpha(\alpha-\beta)(\mu-\alpha)} + \frac{(\beta^2-a_1\beta+a_0)e^{-\beta t}}{\beta(\alpha-\beta)(\beta-\mu)} + \frac{(\mu^2-a_1\mu+a_0)e^{\mu t}}{\mu(\alpha-\mu)(\mu-\beta)}] \quad \ldots(8)$$

The transfer function for COMRATE and WIP is in the form;

$$\{COMRATE/WIP\} = M* \frac{S + a_0}{S(S+\alpha)(S+\beta)(S+\mu)} \quad \ldots(9)$$

and the inverse Laplace solution is;

{COMRATE/WIP}=...

$$M*\left[\frac{a_0}{\alpha\beta\mu} + \frac{(\alpha-a_0)e^{-\alpha t}}{\alpha(\beta-\alpha)(\mu-\alpha)} + \frac{(\beta-a_0)e^{-\beta t}}{\beta(\alpha-\beta)(\mu-\beta)} + \frac{(\mu-a_0)e^{-\mu t}}{\mu(\alpha-\mu)(\beta-\mu)} \right] \quad \ldots(10)$$

The coefficients for α, β, μ (the poles) are common for all four transfer functions, and are as follows;

$$\alpha = 1/\tau_a$$

$$\beta = \frac{1}{2\tau_{pa}} * [(\tau_{pa}K_w+1) - \sqrt{[(\tau_{pa}K_w+1)^2 - 4\tau_{pa}/\tau_i]}]$$

$$\mu = \frac{1}{2\tau_{pa}} * [(\tau_{pa}K_w+1) + \sqrt{[(\tau_{pa}K_w+1)^2 - 4\tau_{pa}/\tau_i]}]$$

Table 2 summarises the coefficients M, a1, and a0 for all four functions.

	M	a_0	a_1
FRST	-1	$-(\tau_i(K_3-1-K_w\tau_{pa})+K_2)/(\tau_i\tau_{pa}\tau_a)$	$(\tau_a+\tau_{pa}+K_w\tau_{pa}\tau_a)/(\tau_a\tau_{pa})$
OSORD	$\frac{(K_2 + \tau_a)}{\tau_a}$	$\frac{K_w\tau_{pa} - (K_3-1)}{\tau_{pa}(K_2 + \tau_a)}$	$\frac{(K_2(\tau_{pa}K_w+1) + \tau_a + \tau_{pa} + K_w\tau_{pa}\tau_a)}{\tau_{pa}(K_2 + \tau_a)}$
COMRATE	$\frac{K_2 + \tau_iK_3 + \tau_a}{\tau_a\tau_i\tau_{pa}}$	$\frac{1}{K_2 + \tau_iK_3 + \tau_a}$	
WIP	$\frac{K_2 + \tau_iK_3 + \tau_a}{\tau_a\tau_i}$	$\frac{1}{K_2 + \tau_iK_3 + \tau_a}$	

Table 2. Summary Of Inverse Laplace Coefficients.

8. SYSTEMS ANALYSIS

Much work has been carried out on validating the model against real life data and testing the simulation equations against the Inverse Laplace solution [13]. The simulation responses were found to agree closely with the analytical results.

The analysis to date has been undertaken in both the time and frequency domains. It is the belief of the authors that to get a true picture of a system's response it is helpful for both these domains to be explored.

Interesting cross checks can be found if we apply the Final Value Theorem (FVT). This simple theorem gives considerable insight without using demanding mathematics.

The FVT states that;

$$\lim_{t \to \infty} [F(t)] = \lim_{S \to 0} [S*F(S)]$$

If we consider the existing 'To Make' model, (that is with $K_w=1$, and $\tau_i=1$), then FRST becomes;

$$\frac{FRST}{TORD} = \ldots$$

$$\frac{1}{[(1+\tau_{pa}S)(1+S)]} * \left[\frac{K_2}{(1+\tau_a S)} + \frac{K_3}{(1+\tau_a S)} - (1+\tau_{pa}S) - \tau_{pa}\right] \quad \ldots(11)$$

If TORD is a step input, θ_{oss} is;

$$\theta_{oss} = S*\left[\frac{1}{S}\left[\frac{1}{[(1+\tau_{pa}S)(1+S)]} * \left[\frac{K_2}{(1+\tau_a S)} + \frac{K_3}{(1+\tau_a S)} - (1+\tau_{pa}S) - \tau_{pa}\right]\right]\right] \quad \ldots(12)$$

Substituting for K_2 and K_3, the FVT becomes;

$$\theta_{oss} = k_2\tau_{pb} + (1+k_3\delta) - (1+\tau_{pa}) \quad (13)$$

In an ideal world, in "steady state", the theoretical lead time would equal the actual lead time,

Therefore $\tau_{pb} = \tau_{pa}$

and if $k_2 = 1$ (weighting function),

$$\underline{\theta_{oss} = k_3\delta} \quad (14)$$

The significance of this is quite enlightening. It states that if ideal conditions occur, (that is the theoretical lead times equal actual lead times, and all relative weightings are equal), the steady state value of Freestock would effectively meet the target Pic Rate. This is the basis of the original "To Make" model.

At steady state, all orders that are released onto the shop floor are (on completion) immediately taken from the freestock levels to the customer. The only inventory that is left is the 'extra quantity' (derived from the Pic rate) needed to service customers in case of random fluctuations about the steady state.

The studies were carried out by varying the following parameters;

τ_{pa}: 2,4,8,16

τ_{pb}: 2,4,8,16

τ_a: 4,8,16,24,48

$k_w = \tau_i = 1$, $\sigma = 0$.

As can be seen from Figure 6, the optimum design is a trade off between long lead times and inventory holdings. The shorter lead times have quick rise times (taken as 90% of final value), but at the expense of high stock levels. Whereas longer lead times tend to give much slower rise times, low stock levels, and have the danger of stock outs. Note that in the responses of Figure 6, the initial stock levels were considered as zero. In actual fact, a stock level depending on its Pic Rate would suffice for initial conditions, and hence Figure 6 would have a normalised stock level.

All the responses show an initial inventory deficit. This is because stock is taken from the existing inventory level to satisfy customer demands, and because there is a production delay, there has been no time to replenish this stock level. By looking at equations (3) and (8), the Laplace and Inverse Laplace Transformations for FRST, it can be seen that there is negative value in the zeros, hence it can be established that there is an initial negative response.

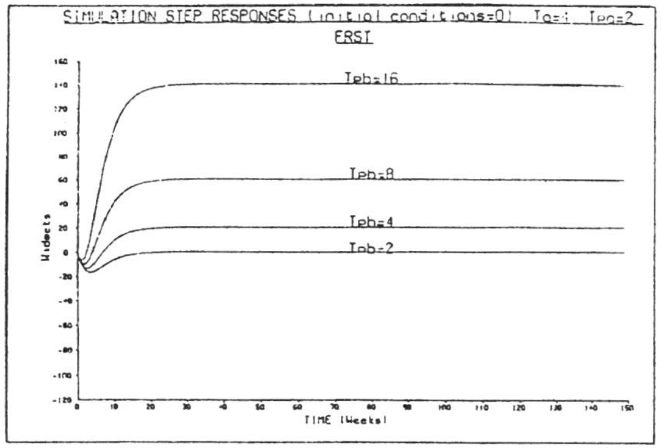

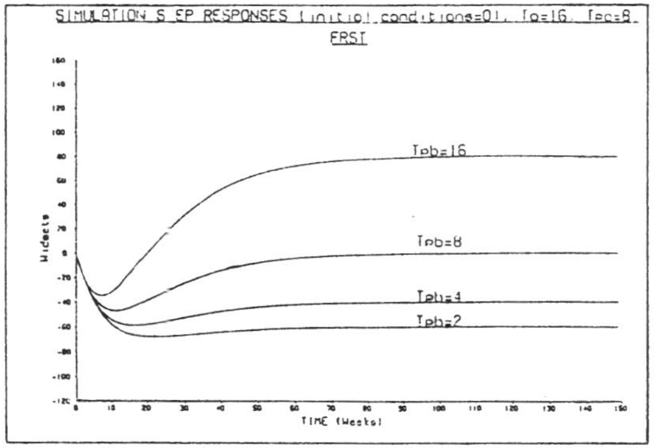

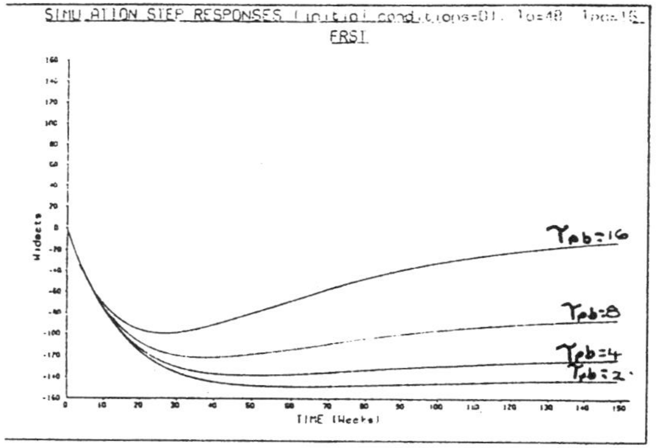

τ_{pb}	Rise time (weeks)	Final Value* (widgets)
2	12	0
4	12	20
8	12	60
16	12	140

τ_{pb}	Rise time (weeks)	Final Value* (widgets)
2	30	-60
4	40	-40
8	50	0
16	60	80

τ_{pb}	Rise time (weeks)	Final Value* (widgets)
2	65	-140
4	75	-120
8	125	-80
16	>150	0

* Values not normalised to take into account initial stock levels.

Figure 6. Sample Of Step Responses For FRST
(for a step increase of 10 widgets)

9. CONCLUSIONS

The beginning of the paper has shown the 'tools' needed to bridge the gap between "hard" and "soft" control systems. The theory has been applied to a real life study, where the production and decision making has been modelled into a production control system.

The model has then been "brought to life" by the use of EXSMO simulations and via the Laplace transform, where the dynamics can be analysed. It has been shown how mathematical modelling can be used to gain insight into manufacturing systems.

Simulation studies are currently being conducted to improve upon the present model, and indeed, the paper is to be considered as one part, as future work showing comprehensive results is intended to be published.

10 ACKNOWLEDGEMENTS.

The authors would like to acknowledge the co-operation and support of Brian Bishop and the staff of Biomet Ltd, Bridgend, for their help in the research activities of P.S.Cheema.

11 REFERENCES.

(1) D.R.Towill., 1977, Low Order Modelling Of Systems And Processes, Jnl Automatic Control of the Benelux National Member Organisations Of IFAC Journal A, 18 pp 200-207.

(2) J.L.Christensen, W.L.Brogran.,1971, Modelling And Optimal Control Of A Production Process, Int.J.Syst.Sci. 1(3), pp 247-255.

(3) J.Parnaby.,1979, Concept Of A Manufacturing System, Int.J.Prod.Res, 17(2), pp 123-134.

(4) A.Tustin., 1953, The Mechanism Of Economic Systems, Harvard University Press, Cambridge, Mass.

(5) D.R.Towill., 1982, Dynamic Analysis Of An Inventory And Order Based Production Control System, IJPR, Vol 20, No 6, pp 671-687.

(6) D.R.Towill., 1990, The Practical Use Of Control Theory In Manufacturing Systems Modelling And Design, SERC ACME Workshop On "CAPM" Reference Models Derived From Control Theory", Castle Donnington.

(7) A.L.del Vecchio., D.R.Towill., 1989, Conceptual Modelling Of Industrial Production/Distribution Systems via Standard Form, Seventh International Conference On Mathematical And Computer Modelling, 2-5 August 1989, Chicago, USA.

(8) D.R.Towill., 1988, The Dynamic Analysis Approach To Manufacturing Systems Design, The University of Wales Review, No 3.

(9) J.S.Edghill, C.Olsmats, D.R.Towill., 1988, Industrial Case Study Of The Dynamics And Sensitivity Of A Close Coupled Production Distribution System, IJPR, Vol.26, No 10, pp 1681-1693.

(10) D.R.Towill., 1977, Exponential Smoothing Of Learning Curve Data, IJPR, Vol 15, No 1, pp 1-15.

(11) P.S.Cheema., 1987, The Performance And Evaluation Of SMT Machines At AB Electronic Plc, University Of Wales Institute Of Science And Technology, MSc dissertation, Appendix A.

(12) J.L.Coleman., 1988, Lead Time Reduction By System Simplification, University Of Wales College Cardiff Msc dissertation, pp 55-63.

(13) P.S.Cheema., (In Preparation) PhD Thesis, University of Wales College Cardiff.

Part II: Production as a Competitive Weapon

World Class Manufacturing in the 1990s: Integrating JIT, TQC, FA, and TPM with Worker Participation

PHILIP HUANG

INTRODUCTION

In the 1980s, we have witnessed revolutionary changes in the concepts and practices of the management of manufacturing systems. Numerous unconventional production management techniques, mostly tributed to such well respected Japanese production management pioneers as Taiichi Ohno and Shigeo Shingo, have been not only taught but also applied in many American firms. Technical terms originated in Japan such as JIT (Just-in-Time), Kanban, SMED (Single-Minute Exchange of Die), and Poka-yoke have already become terms frequently used in the United States by practitioners in industries and faculty and students in academic institutes as well. There is no doubt that Japanese production management experts and their innovative concepts and techniques have greatly facilitated productivity improvements in many American plants. However, the most important lesson we can learn from Japan, in our opinion, is their attitude of "never satisfying with the status quo". We believe that "continuous improvement" is the real key to gain competitiveness in the 1990s. American companies should not simply dwell on what they have learned from their Japanese competitors and be content with the productivity improvements they have managed to gain so far. Facing an increasingly dynamic international markets, American firms cannot afford to overlook any potential means for "continuous improvement" [21].

The purpose of this paper is to address the need to integrate several recent but important developments in production management to form a cohesive manufacturing strategy. This integrated manufacturing strategy should include the following basic cornerstones: Just-in-Time (JIT), Total Quality Control (TQC), Factory Automation (FA), and Total Productive Maintenance (TPM). In this paper, a brief introduction to each of the four cornerstones will be provided first. A discussion on the importance of each cornerstone and its interrelationships with others will then be followed. Moreover, we believe that the vital medium to effectively glue these cornerstones together is a group of well trained, dedicated, and loyal employees who are never satisfied with their current achievements and who are continuously seeking means to better themselves. A company taking this integrated approach should find themselves in much better competitive position in the 1990s.

THE NEED FOR AN INTEGRATED MANUFACTURING STRATEGY

International competition in the 1990s is expected to be more fierce than ever because of the following factors: 1) rapid advancement in technology, 2) revolutionary changes in the management of production systems, 3) aggravated international trade protectionism, and 4) more sophisticated and demanding consumers. To survive and to thrive in this increasingly competitive environment, a manufacturing company needs to cultivate the spirit of KAIZEN (continuous improvement) in order to become a world class manufacturer. "In search of excellence" now becomes a process of continuous improvement rather than a mere attainment of standards. In fact, "achieving and maintaining excellence," as viewed by many Japanese firms, is a journey, not a destination.

Changing Business Environment

The first energy crisis in the early 1970s signaled the end of the mass-production era. Consumers, who have played a much more crucial role in today's marketplace, have become not only more sophisticated but also more demanding. To satisfy the ever changing consumer tastes and to respond to the dramatically shortened product life cycles, a company will have to possess the capability of frequently and rapidly designing new products and improving existing products. The flexibility of the manufacturing system has consequently become a critical concern. More manufacturers have realized the importance of adopting a manufacturing strategy which emphasizes more on the small-lot production targeting toward "the economy of scope", rather than the mass-production targeting toward "the economy of scale."

To manage the much shortened product life cycle effectively, every phase of the life cycle including the identification of consumer needs, preliminary product design, final product design, process design and selection, production, distribution, customer support, maintenance, and product phase-out, will have to be considered concurrently with designing the product, capacity, and the support and maintenance capability.

Cost Reduction and Productivity Improvement

Being able to produce products with good quality and reasonable lead time is a necessary but not sufficient condition for a company to compete successfully in the marketplace. Today, price is still an important competitive weapon. However, the traditional pricing strategy, which determines the product selling price by adding profit margin to the the production cost, has been seriously challenged. A relatively new concept, which has been variably named as "non-cost," "target costing," or "cost planning," has become increasingly popular among Japanese manufacturers [15]. The advocaters of this new concept believe that the price of a product is dictated by the marketplace, not the manufacturing shop floor. Consequently, the difference between the product's competitive market price and its planned profit margin becomes a "target cost" for the manufacturer. To ensure reasonable profit and to sustain the company's healthy growth, it is critical that the production cost can be continuously reduced to reach the "target." Understandably, Toyota considers its "non-cost" principle as the real motivation for cost reduction through waste elimination [22].

From the standpoint of manufacturing, "waste" can be defined as anything, including materials, manpower, machine, space, time, and capital, spent for the purpose of producing a product but failed to directly add value to the product. Using this rather loose definition, wastes can be observed in various forms such as quality rejects, machine downtimes, inventories, setup times, unnecessary motions or double handling, etc. All these different forms of waste will merely add "cost," not "value," to the product.

Consequently, the main theme of the well-known Toyota production system, which has been regarded as one of the most productive ones, is "absolute elimination of waste." Toyota has relentlessly attacked all forms of waste by closely examining every component and every operation in their manufacturing process. The identified wastes will then be reduced or entirely eliminated. Toyota's enviable profit records during the slow growth period of the first energy crisis in 1973 have proved that the most effective approach to improve productivity is waste elimination. It is therefore evident that for a company to compete favorably in the 1990s, it is necessary to alter the traditional pricing strategy and to focus on cost reduction through waste elimination and process rationalization.

THE FOUR CORNERSTONES OF WORLD CLASS MANUFACTURING

Product quality, price, availability, and after-sales service are powerful means to win customer orders in the market. However, to gain the above features, a company will have to possess a flexible and productive manufacturing system. To build such a manufacturing system, recent developments in production management provide us with the essential cornerstones.

Just-in-Time production concept, which has been gaining an enormous popularity in the United States, stresses on the importance of producing the necessary amount of the necessary items at the necessary time. It promotes "waste free" operations on the shop floor. Total Quality Control, also known as Company-Wide Quality Control (CWQC) in Japan, dedicates the responsibility of quality assurance to the entire work force, not just the QC inspectors. Factory automation employes the current technology to improve the manufacturing operations both in terms of efficiency and quality consistency. Total Productive Maintenance is a program which focuses on improving the facility utilization rate by reducing machine breakdowns through an integration of preventive maintenance and worker participation.

The above four productivity improvement programs are proven methods for effective cost reduction, quality improvement, and leadtime shortening. With a group of dedicated workers who are trained to handle multiple tasks and to identify and solve operational problems, JIT, TQC, FA, and TPM can be effectively integrated and implemented. In the following sections, each cornerstone and its interrelationship with others will be addressed.

Total Quality Control

There is no doubt that "quality" has played a crucial role in Japan's dramatic economic recovery since the end of World War II. Japanese companies have proved that the traditional trade-off analysis of product quality versus quality control cost is inadequate. Efforts in improving product quality can lead to not only producing attractive products but also reducing wastes caused by poor product quality or inappropriate manufacturing process. Total Quality Control is perhaps the best approach to ensure product quality.

According to Feigenbaum [2], the originator of TQC concept, TQC is defined as:

> "... an effective system for integrating the quality development, quality maintenance, and quality improvement efforts of the various groups in an organization so as to enable production and service at the most economical levels which allow for full customer satisfaction."

While believing that "quality is everybody's job," Feigenbaum advocates Western-style TQC conducted essentially by QC specialists for fearing that quality may become "nobody's job." Japanese, however, took a different approach to TQC. Their concept of TQC involves all divisions as well as all employees in the organization. It is therefore also called company-wide quality control.

Evolutionary quality efforts have taken many faces over the century: quality inspection, Statistical Quality Control (SQC), Quality Assurance (QA), Statistical Process Control (SPC), Zero Quality Control, Zero Defects, TQC, Strategic Quality Management, and Total Quality Management (TQM). Quality gurus such as Deming, Juran, Crosby, Feigenbaum, and Ishikawa have contributed in shaping and implementing today's quality management concept. Recently, Shingo [19] introduced the concept of "judgmental inspection, informative inspection, and source inspection" in an effort to categorize QC activities. He believes that the adoption of source inspection and Poka-yoke methods can eliminate defects at the source and achieve Zero Quality Control.

TQC involves more than the traditional quality control of ensuring product quality by QC testing, process control, etc. It basically involves the understanding of customers' needs and making effective use of company's resources to satisfy these needs [23]. Customers can be defined as anyone, within or outside the company, affected by the product or service a company delivers. Thus, successful TQC efforts could be realized through company-wide participation and customer consciousness of all employees. Interestingly, Chase and Garvin [1] recently advocated the practice of the "service factory," where the shop floor employees who actually built the product may have to communicate directly with their customers to ensure customer satisfaction or to assist product services.

Ishikawa [8] lists the advantages of TQC as follows:

1. TQC provides true quality assurance. It is possible to build at every step in every process and attain 100 percent defect-free production. It is not enough just to find defects and flaws and to correct them. What one must do is determine the causes which create these defects and flaws. TQC and process control help workers to identify and then remove these causes.

2. TQC opens up channels of communication within a company, filling it with a breath of fresh air. It allows companies to discover a failure before it turns into a disaster, because everyone is accustomed to speaking to one another truthfully, frankly, and in a helpful manner.

3. TQC makes it possible for the product design and manufacturing divisions to follow the changing tastes and attitudes of customers efficiently and accurately so that products can be manufactured to meet customer preference consistently.

4. TQC fosters probing minds that can detect false data. It can help companies avoid relying on false sales figures and false production figures.

In short, TQC involves every division and every individual in the organization to ensure total customer satisfaction. Its philosophy promotes the concept of designing and producing products right at the first time. It is indispensable for a company to elbow its way through the competitive business environment in the 1990s. To succeed in implementing a TQC program, however, requires: 1) management commitment and active involvement, 2) company-wide education and training, 3) design and implement effective quality planning, measurement, and reward system.

Just-in-Time Production

In principle, JIT means producing the necessary items including parts, sub-assemblies, assemblies, and finished products, in the necessary quantities at the necessary time or just in time [3,5,11]. JIT philosophy considers inventory as the root of all evils because it ties up capital resources, occupies space, creates double handling, slows down the process flow, and, worst of all, hides problems such as defective parts, faulty manufacturing process, careless workmanship, ill-maintained machines, disarrayed tools, and inadequate scheduling. Lowering the inventory level, those hidden problems obviously will surface. The production process may have to stop. However, this will also provide an opportunity to examine the problem and find out permanent solutions.

When the current problem is resolved, inventory level can be further reduced to surface additional hidden problems. As indicated in the famous "river and rocks" diagram, the practice of continuously lowering the inventory level, exposing the hidden problems, examining the source of the problem, and subsequent removing those operational problems can lead to a smoother and waste-free operation in the long run.

Among those hidden rocks in the river, overproduction is a prominent one. JIT production tributes the excess inventory as a result of overproduction. In JIT philosophy, a company will be better off by letting machines and operators idle rather than using the excess production capacity to manufacture unwanted products and thus create excess inventory. The popular "pull" system, in which each operation, starting from the shipping dock and working back through the process, pulls the necessary items from the previous operations as needed [4] was designed to prevent such overproduction.

The first step towards achieving an ideal JIT production is, according to Shingo [22], through thorough examination of the production processes, and identification of waste components. Improvements in the production processes occur as a result of eliminating the identified waste components. Taiichi Ohno classified wastes into the following seven categories [14]:

- Waste of overproduction
- Waste of time on hand (waiting)
- Waste in transportation
- Waste of processing itself
- Waste of stock on hand (inventory)
- Waste of movement
- Waste of making defective products.

Numerous unconventional techniques have been developed to eliminate those wastes arising in the production settings. These techniques include: 1) quick exchange of dies, e.g., SMED and OTED [19], 2) source inspection and the Poka-yoke system [13,20], 3) small lot sizes, 4) level production plan, and 5) synchronization, 6) one-piece flow operation, or, non-stock production [3,21], 7) multi-machine handling and multi-process handling, Nagara system [22], 8) labor cost reduction [9,14], and others.

Based on Toyota's market-driven strategy, namely, non-cost principle and order-based production, JIT production emphasizes heavily on two major concepts: continuous improvement and waste elimination. JIT success stories of numerous American, European, and Japanese companies are reported extensively in the literature [16,17,24].

Factory Automation

The rapid development in technology has made available a whole host of automated devices with diverse applications in manufacturing. Functional areas such as product design, product testing, process design, fabrication, spray painting, assembly, inspection, material handling, warehousing, and production planning and control have all been dramatically affected by automation.

Essentially, automated equipment are mechanical devices controlled by a host computer which can be programmed and reprogrammed to provide various operational instructions. A general term for employing those automated equipment for manufacturing is Factory Automation. FA, a term which has gained enormous popularity in Japan, typically includes Computer-Aided Design/Computer-Aided Manufacturing (CAD/CAM), Computer-Aided Testing (CAT), Computer-Aided Process Planning (CAPP), robotics, Numerical Control (NC) machines, Computer Numerical Control (CNC) machines, automated material handling system, automated warehousing, Group Technology (GT), Manufacturing Resources Planning (MRP), bar coding, and Office Automation (OA) [6,10].

With a loaded computer program, the automated production facilities can tirelessly repeat the same manufacturing process without deviations. When a different job is scheduled on the same machine, it only requires a reloading or reprogramming of the controller. Quality consistency and flexibility consequently are the two distinguished benefits of automation. In addition, reductions in labor cost, setup time, inventory, and leadtime have all been reported as the significant benefits of automation [6,7]. Those reported benefits indicated clearly that FA is a natural extension of TQC and JIT. FA provides a promising environment for further waste elimination and quality assurance.

Total Productive Maintenance

Despite those attractive benefits, there are obstacles in pursuing automation. For example, the high cost of automated facilities has significantly hindered small and medium size companies from aggressively automating their factory floor. Those reluctant FA users believe that the idled automated equipment will constitute an unacceptable form of wastes. Therefore, effective use of expensive automated equipment becomes a critical concern. Moreover, as an unavoidable result of automation, manual operations will be gradually replaced by computer controlled machine processing. Eventually, the machine operator's job will be limited to loading and unloading only. Excess labor will soon become a significant form of waste on the shop floor.

In Japan, an innovative approach to solve the above dual problem is called Total Productive Maintenance [12]. Basically, TPM advocates the idea of extending the machine operator's job from tending the machine to maintaining the automated equipment. Workers are organized in small groups and trained to conduct preventive maintenance. Moreover, to ensure a smooth operation without unexpected interruptions or breakdowns, the automated factory requires a clean and orderly work environment. The TPM teams therefore are also responsible for maintaining a clean and organized work area.

AN INTEGRATED MANUFACTURING SYSTEM

JIT, TQC, FA, and TPM have been identified as the four cornerstones for building a world class manufacturing system for the 1990s. It is important, however, to realize that these four modules are closely linked by the common goal of improving a company's competitiveness in the marketplace through quality assurance, waste elimination, cost reduction, and leadtime compression. Moreover, these cornerstones should firmly planted on the solid common ground of worker participation.

JIT and TQC

Quality reject is perhaps the worst form of waste on the shop floor because there are only two ways to handle it: rework or discard. Apparently, both remedies will only add cost, not value, to the product. TQC's focus on quality control at the source certainly conforms to the underlying goal of JIT: waste elimination.

However, in the traditional manufacturing environment where jobs are processed in large lots and where machines are laid out according to their functions, TQC would be very difficult, if not impossible, to implement. In an JIT environment, on the other hand, where setup times are drastically shortened, production lot sizes are significantly reduced, machines are arranged according to the concept of group technology, production process are streamlined, workers are trained to handle multiple tasks, and plant loads are smoothed, the source of quality problem can easily be detected and swiftly removed. Consequently, JIT creates a better environment for a company to implement TQC.

JIT and FA

Factory automation when properly implemented should be capable of significantly reducing product development time, machine setup time, order processing time, material handling time, quality inspection time, and eventually the total market response time. The flexibility of the automated production facilities will certainly enhance the realization of just-in-time production. It is therefore natural for a JIT plant to actively pursue automation. This is, in fact, the exact path which many Japanese manufacturers are currently taking.

On the other hand, JIT's philosophy of waste elimination requires close examination of every manufacturing process in order to identify and to solve operational problems. This practice can actually create the ideal environment for implementing automation. Many Japanese practitioners have indicated that a company should not automate just for the sake of automation. Automation can only succeed if the most needed process can be selected and automated first. JIT production consequently provides an excellent environment for successful implementation of FA.

JIT and TPM

A TPM program typically enlarges the shop floor workers' responsibility from tending machines to such areas as detecting machine failures, basic maintenance, and keeping work area clean and organized. A successful TPM program can directly contribute toward more efficient utilization of facilities and better utilization of labor force. Moreover, the practice of TPM can also eliminate wastes arising from a messy work area where workers have to spend unnecessary time to find the right tools or right parts, where parts can easily get dirty or damaged, where machines can easily collect dusts or unwanted particles which can cause breakdowns. From this stand point, TPM should be an integral part of JIT production.

On the other hand, JIT production can also pave the way for implementing a TPM program successfully. In a JIT environment, machines are laid out in U-shape using the concept of cellular manufacturing, and multifunction workers are developed to handle different number of machines according to demand. In this flexible JIT environment, both workers and work areas are adaptable to TPM implementation. Moreover, the JIT concept of uniform plant load creates a stable environment which in turn can facilitate the implementation of a TPM program, because preventive maintenance can be conducted more easily in a plant with a stable production schedule.

TQC and FA

The traditional practice of quality inspection has now been viewed as a rather backward approach to achieve quality assurance. Currently, the emphases of quality activities are centered on product design and actual manufacturing process. In both areas, factory automation offers a great opportunity for quality enhancement. For example, in the area of product design, CAD and CAT can rapidly and efficiently develop and then evaluate numerous design ideas. As a result, products with better quality, in terms of performance and physical appearance, can be designed. In the area of manufacturing process, CAPP can help to design a better manufacturing procedure to ensure better product quality. Moreover, current technology enables the automated production equipment to process products with greater precision and higher consistency which are two important aspects of product quality. Last, but certainly not the least, is the fact that automation can take away the tedious and boring tasks from machine operators so that their time can be freed to participate in various TQC activities.

Quality control circle has been proven as a very effective vehicle to implement total quality control [8]. QC circle activities, however, are not limited to quality control and quality improvement. Examining the existing process and developing better production process are also frequently conducted by QC circles. Consequently, the promotion of TQC will provide better opportunity to identify the needed areas to automate and therefore enhance the odds for a successful FA implementation.

TQC and TPM

Conceptually, TPM can be viewed as an extension of TQC. These two programs share extensive similarities. For example, both are built on small group activities; both programs have explored the possibility of extending machine operators' responsibility; and both programs have focused on productivity improvement. With an active TQC program, a company should find progressing toward TPM a rather smooth and simple task.

More specifically, workers participating in a TQC program are usually trained quite extensively for effective problem identification and problem solving. TPM employs a similar tactics but focuses on the identification of machine operating problems. Workers in a TPM program are typically trained to maintain the machine and to detect the mechanical problems in advance to prevent costly breakdowns. Consequently, a TQC program should be able to provide a company with a good foundation to establish a TPM program.

On the other hand, an effective TPM program can also facilitate the realization of "total" quality control. When machines are well maintained, tools are logically arranged, parts are carefully and conveniently stored, work areas are spotlessly cleaned, many sources of quality problems can be removed. Obviously, a TPM program can create a much better work environment for achieving total quality control.

FA and TPM

TPM is a direct by-product of Japanese companies' zealous pursuit of factory automation. Since most Japanese companies are still complaining about the high price of automated equipment [7,8], keeping those expensive equipment in running condition understandably holds the key to a successful implementation of factory automation. TPM addresses not only the issues of maintaining machines but also the importance of creating a clean and organized environment to avoid conditions which may cause machine breakdowns. Moreover, TPM encourages participants to closely study the production process and to suggest or design the adequate facilities to carry out that process. With this practice, a company can better allocate its limited resources to more promising automation projects.

As indicated earlier, FA actually has created the need for developing a TPM program. Skeptics of TPM may fear that the sophistication of automated facilities may pose serious technical problems for machine operators. However, viewing this from a different angle, FA applies a proper pressure to the company. Facing the rapid development in automation technology, a world class manufacturer needs to have a group of highly competent workers who have neither fears nor resistance to using advanced manufacturing systems. TPM is an effective program to cultivate such type of workers. As the pace of automation speeds up, a company will soon be compelled to adopt programs similar to TPM in order to make their automation investment cost effective.

WORKER PARTICIPATION

JIT, TQC, FA, and TPM are all prominent tools to develop a world class manufacturing system. However, none of these tools can reach its potential without worker participation. In other words, the core of this competitive world class manufacturing system should be a group of dedicated, well-disciplined, extensively-trained, innovative, and well-respected workers who are actively participating in various productivity improvement activities. In fact, workers, because of their daily responsibility, are the ones who have the thorough knowledge about manufacturing process which they perform daily. Productivity improvement ideas proposed by these workers are usually more feasible and more effective. Furthermore, many productivity improvement programs have failed because of lack of workers' support. Improvement suggestions originated from workers will certainly obtain the necessary support. Therefore, tapping the wisdom of the workers is essential for a company to engage in continuous improvement.

On the other hand, participation fosters motivation, job satisfactions, and attitude of accepting changes. Participation also stimulates innovative ideas, improves communications, and enhances workers' self-esteem [20]. Confronted with an increasingly competitive environment where changes become norms, a company should find that worker participation can provide the vigor for its continuous prospering. For the 1990s, a world class manufacturer should employ worker participation as the medium to effectively integrate JIT, TQC, FA, and TPM into a cohesive manufacturing strategy which emphasizes heavily on continuous productivity improvement.

CONCLUSION

The admirable success of Japanese economy has shed new lights on the strategic importance of manufacturing. Since the competition in the world markets will get even more intense in the 1990s, manufacturing becomes the focal point of establishing a winning business strategy. How to develop a world class manufacturing system which is capable of producing quality products with low cost and short leadtime has therefore become a critical issue to address. In this paper, an integrated approach is proposed to build a manufacturing system which continuously pursues productivity improvement. The basic cornerstones of this integrated manufacturing system include: JIT, TQC, FA, and TPM. However, it would be a strategic mistake to treat these tools as independent productivity improvement programs which are pursued with great enthusiasm for a certain period of time and then soon be discarded. A world class manufacturer should employ worker participation as the vehicle to integrate these individual tools together to form a winning manufacturing system which vigorously pursues "perfection." This manufacturing system will equip a company with a formidable weapon to successfully compete in the 1990s' world markets.

REFERENCES

1. Chase, Richard B. and David A. Garvin, "The Service Factory," Harvard Business Review, (July-August 1989), pp.61-69.

2. Feigenbaum, A. V., Total Quality Control, New York: McGraw-Hill, 1986.

3. Hall, Robert W., Zero Inventories, Homewood, Illinois: Dow Jones-Irwin, 1983.

4. Hay, Edward J., The Just-In-Time Breakthrough, John Wiley & Sons, 1988.

5. Hirano, Hiroyuki, JIT Factory Revolution: A Pictorial Guide to Factory Design of the Future, Cambridge, MA.: Productivity Press, Inc., 1989.

6. Huang, Philip Y. and Michiharu Sakurai, "An Assessment of Factory Automation in Japan: A General Mail Survey," Manufacturing Review, Vol.2, No.3 (September 1989), pp.162-169.

7. Huang, Philip Y. and Michiharu Sakurai, "Factory Automation: The Japanese Experience," accepted for publication in IEEE Transactions on Engineering Management. (forthcoming)

8. Ishikawa, Kaoru, What is Total Quality Control? The Japanese Way, Englewood Cliffs, New Jersey: Prentice-Hall, Inc., 1985.

9. Japan Management Association, KANBAN: Just-In-Time at Toyota, Cambridge, MA.: Productivity Press, Inc., 1986.

10. Kochan, A. and D. Cowan, Implementing CIM, IFS (Publications) Ltd., UK: Springer-Verlag, 1986.

11. Monden, Yasuhiro, Toyota Production System, Norcross, Georgia: Industrial Engineering and Management Press, 1983.

12. Nakajima, Seiichi, TPM: Introduction to TPM, Cambridge, MA.: Productivity Press, Inc., 1988.

13. Nikkan Kogyo Shimbun, Ltd., Poka-yoke: Improving Product Quality by Preventing Defects, Cambridge, MA.: Productivity Press, Inc., 1988.

14. Ohno, Taiichi, Toyota Production System: Beyond Large-Scale Production, Cambridge, MA.: Productivity Press, Inc., 1988.

15. Sakurai, Michiharu and Philip Y. Huang, "Factory Automation and Its Impact on Management Control Systems: A Japanese Survey," in Yasuhiro Monden and Michiharu Sakurai (eds.), Japanese Management Accounting, American Productivity Center (forthcoming).

16. Schonberger, R. J., World Class Manufacturing Casebook, New York: Free Press, 1987.

17. Sepehri, M., Just in Time, not Just in Japan: Case Studies of American Pioneers in JIT Implementation, Falls Church, Virginia: APICS, 1986.

18. Shin, S. and D. S. Sink, "White Collar Just-In-Time: A Theoretical Framework," in Sumanth, D. J., J. A. Edosomwan, D. S. Sink, and W. B. Werther, Jr. (eds.), Productivity Management Frontiers-II, Geneva, Switzerland: Interscience Enterprises Limited, 1988.

19. Shingo, Shigeo, A Revolution in Manufacturing: The SMED System, Cambridge, MA.: Productivity Press, Inc., 1985.

20. Shingo, Shigeo, Zero Quality Control: Source Inspection and the Poka-yoke System, Cambridge, MA.: Productivity Press, Inc., 1986.

21. Shingo, Shigeo, Non-stock Production: The Shingo System for Continuous Improvement, Cambridge, MA.: Productivity Press, Inc., 1988.

22. Shingo, Shigeo, A Study of the Toyota Production System from an Industrial Engineering Viewpoint, (revised edition), Cambridge, MA.: Productivity Press, Inc., 1989.

23. Snee, R. D., Implementing Total Quality, notebook for ASQC Radford-Roanoke section workshop, Blacksburg, Virginia, October 13, 1989.

24. Vollum, R. B. and T. J. O'Malley, "Improving on JIT Success with Constraint Management and Synchronous Flow: A Case Study," Proceedings of the 32nd Annual International Conference of APICS, Orlando, Florida, 1989.

Time-Based Competition: Speeding New Product Development

JOSEPH BLACKBURN

0. Introduction

The emergence in an industry of a firm with a significant time advantage in delivering a product to market fundamentally changes the nature of competition from a cost basis to response time. How to compress time in all phases of the product delivery cycle and turn it into a competitive weapon is the essence of time-based competition.

In this paper we investigate methods of time compressing the new product introduction process. Speed in manufacturing must be coupled with rapid deployment of new products to stay abreast of the competition. Heated competition hastens obsolescence, and the edge in the marketplace shifts to the firm with the new-product that has the technology and features that consumers crave.

To isolate the keys to new-product development (NPD) speed, it is important to note which firms are the global leaders. Barry Bebb of Xerox contends that most of the leading firms are Japanese. He states that the Japanese dominate U.S. competitors on two key dimensions: (1) high-quality, low-cost, timely products through their Just-in-Time manufacturing systems; and (2) the rapid commercialization of technology in the form of new products [2]. To investigate this further, a group of manufacturing executives recently was asked to perform a simple experiment: make a list of the world-class leaders in NPD and, then, make a second list of the fastest manufacturers, those that have mastered JIT. To the surprise of no one, most of the firms on both lists were Japanese. A sprinkling of North American firms were on both lists--firms such as Hewlett-Packard, Motorola, Northern Telecom, and GE appeared frequently; however, the predominant names were Honda, Toyota, Sony, and Canon. The startling observation was that the two lists of most of the executives were virtually identical! In their view, the firms with time-compressed manufacturing were also the world's fastest new-product developers, thus extending Bebb's hypothesis that the same Japanese firms dominate the twin dimensions of manufacturing and product development.

The theme of this paper is that the high correlation observed between speed in manufacturing and NPD is not coincidental. The skills needed to win the race are the same in both cases. These two processes have more than speed in common, for the steps that a manufacturing operation must take to move to JIT are remarkably similar to the path to achieve rapid new-product deployment.

The market potential of speed in NPD is examined in the next section. Subsequently, the new-product introduction process is dissected and examined. The linkages to JIT are established and sharp comparisons are drawn between conventional American practices and those of Japanese competitors. The concluding section develops a set of guidelines based on JIT principles for accelerating the speed of new-product introductions.

1. New Products: The Essential Competitive Weapon

"Better never than late!" For new-product development, the old saying must be inverted to understand today's markets. Late to market means lost profits; profits foregone during the part of the product cycle in which the firm can least afford to lose them--that is, before competitors have brought their technology on-stream. Several significant examples of the market advantages of fast NPD are described below. A more extensive analysis is given in Chapter 5 of [3].

A 1987 study in the automotive sector by Clark, Chew, and Fujimoto indicates that, for the case of a $10,000 car (a small one by today's standards), each day of delay in introducing a new car model into the market represents, conservatively, $1 million in lost profits [4]. .Another study by Clark [5] indicates that the Japanese can, on average, complete a new-car development project about 18 months faster than either their U.S. or Western European competitors. So, if a U.S. firm and a Japanese firm each started on a new-car project at the same time, then, with average performance, coming to market a year and a half behind the Japanese firm could cost the lagging U.S. firm on the order of a half billion dollars in lost profits.

Late to market can, in fact, be fatal. In the global video-cassette recorder (VCR) market, the first widely available, affordable video cassette recorder was introduced in 1972 by Philips, the Dutch electronics conglomerate. Japanese firms did not become significant players in the market until the mid-1970s. By the end of the decade, Philips was no longer the market leader and, today, its VCR has vanished from the market. When Philips introduced its second generation VCR near the end of the 1970s, Japanese competitors such as Panasonic, Sony, and Hitachi had already *introduced and retired* three generations of VCRs. Philips was blasted out of the market by a ponderous new-product development process [9].

In the expanding market for electronic pagers ("beepers") and mobile phones, Motorola has shown that U.S. manufacturers can use "first-to-market" capability to turn the tables on Japanese competition. Motorola designed a new pager as well as the factory to produce it in 18 months, half the usual 3 years [6]. This swiftness enabled the firm to regain a position as market leader in the pager market. With its new Micro-Tac phone, Motorola extended this marketing capability to the mobile phone market.

2. A Closeup View of the New Product Introduction Process

Although U.S. and Japanese firms follow a similar sequence of stages in NPD, they do not complete them at the same speed. Figure 1 presents the results of a comparison by George Stalk of the Boston Consulting Group of the efforts made by one U.S. and one Japanese firm to develop a new marine gear. Although the technology being introduced was similar, the Japanese firm was faster at every stage except the design review process. This outcome confirms a common observation about the Japanese style of management: they devote a lot of time in the decision process to make sure that everyone agrees with the decision; once consensus is reached, new-product projects move more quickly than those of their Western counterparts.

Steps such as detail design and prototype production, Stalk found, were four times slower at the U.S. firm. Typically, the U.S. designers must produce more designs than a Japanese team before obtaining an approved design. Delays in prototype production also can be traced to engineering changes and problems deeply rooted in manufacturing. Observe in Figure 1 that the Japanese project was finished in about half the time taken by the U.S. project. Assuming that the U.S. firm had made the technological advances needed to begin the project at the same time as the Japanese firm, the latter would have been in the market with the new product a full year and a half ahead. When the U.S. firm enters the market with its product, it may discover, to its dismay, that the Japanese competitor can respond by cutting prices because the months of lead time has allowed the firm to trim fat from its manufacturing process.

More ominous is the fact that by replicating the speed of its first product development project, the Japanese firm could have completed a *second* project while the U.S. competitor was developing its first product. After 18 months of being alone in the market with new technology, the Japanese firm, conceivably, could bring a second new product to market to coincide with the first U.S. product introduction and, thereby, render the U.S. entry obsolete the day it hits the market.

3. Why Japan Leads the West in New Product Development

To examine why the West lags, conventional Western practices in both new-product development and manufacturing were

FIGURE 1

Improving Response Time In New Product Development

Marine Gears in 1987

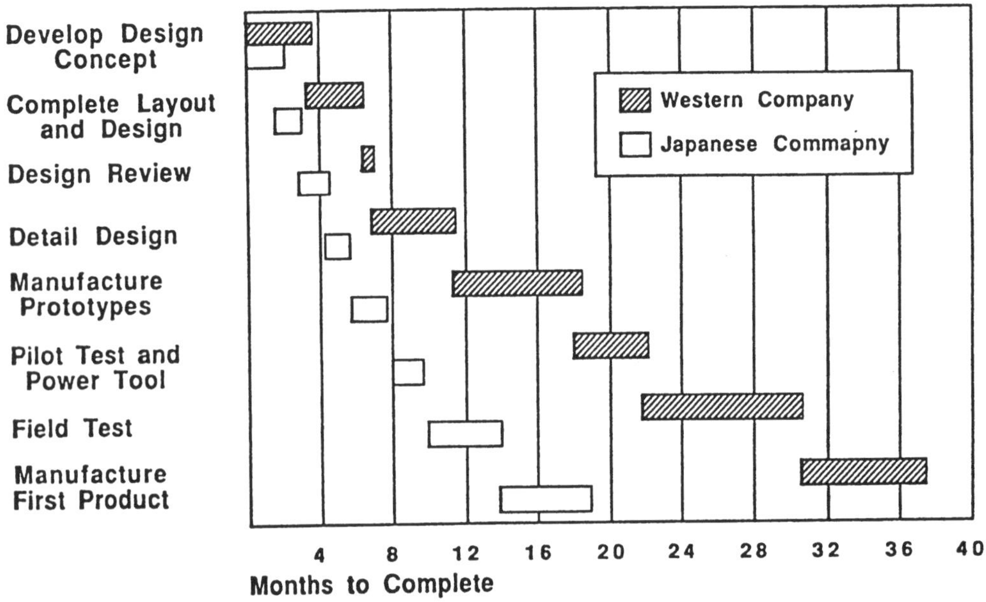

Source: Boston Consulting Group

compared with their Japanese counterparts. The results of this comparison are striking and insightful: Western practices in new-product development and manufacturing bear a strong resemblance, but differ sharply in significant ways, from their Japanese counterparts. The typical U.S. NPD process seems to have been cut from the same cloth as conventional batch manufacturing. The key features of NPD that have been emerging from Japan are virtually identical to those of JIT manufacturing. First, a comparison is drawn between conventional Western practices in manufacturing and new-product development. Then the similarities for these two functions that have emerged from Japan are examined. These two sets of comparisons reveal a clear path that U.S. firms must follow to achieve world-class speed in new-product development.

The table below summarizes the similarities between conventional practices typically found in manufacturing and new-product development in U.S. firms. Each component of these practices is subsequently discussed.

Comparison of Western Practices in Manufacturing and New-Product Development

Process Parameters	Conventional Batch Manufacturing	Conventional (Western) New Product Development
Batch sizes	Large batch production; transferred in large batches	Information processed and transferred in large batches
Layouts	Job shop	Functional by department
Process flow	Sequential activities	Sequential
Scheduling	Centralized scheduling	Centralized control
Employee involvement	Low	Low

Supply relations	Little coordination with suppliers	Low involvement with suppliers in design of components
Quality	High defect ratios; rework	Numerous engineering changes rework
Automation	Islands of automation; isolated robots or transfer devices	Isolated systems (e.g., CAD systems with negligible integration)
Lead times	Long	Long

Batch Sizes

Long changeover times are the economic driver for the large batch sizes that pervade U.S. manufacturing. The long lead times found in conventional manufacturing are causally linked to the time lags and work-in-process (WIP) inventory created by large-lot production. Studies of new-product design in the United States have shown that, as in manufacturing, information is commonly processed in large batches--that is, a "phased approach" to problem solving[4]. The analog of the changeover in new-product design is the change from working on one design problem to another. In a study of the use of CAD/CAM in product design, Adler found that "drawings often sat idle, waiting to be worked on, just as long in electronic form as in paper form[1]." Adler's findings suggest that Western approaches to product design are disturbingly similar to large-batch manufacturing.

Layouts and Process Flow

In conventional batch manufacturing the job shop layout, in which machines are grouped by function, prevails. Machining centers are spaced far apart to accommodate WIP inventory. In new-product development, the layout is analogous to the job shop: Groups tend to be located in offices according to their function. Product engineers occupy one office, process engineers reside in another, and designers are located elsewhere.In both cases the partitioned layout encourages sequential performance of development activities. A large batch of work is finished and then transferred "over the wall" to the next department where it waits in a queue for further processing.

Management and Employee Involvement

Conventional manufacturing offers few incentives for shop floor employee involvement in planning and control functions. Schedules come from "above," and process control tends to be centralized. In the U. S., new-product management also tends to be centrally controlled and scheduled. Because functions are compartmentalized, the personnel assigned to the project team are dispersed and rarely are in direct communication. This functional detachment often is a formidable obstacle to teamwork and coordination.

Supply Relations

Traditional supplier relationships in the United States tend to be at arms-length and, at times, adversarial rather than cooperative. In a comparative study of U.S. and Japanese supply chains, Welch, Blackburn, and Jueptner found that U.S. manufacturers were less likely to form long-term, cooperative relationships with their parts suppliers than were their Japanese counterparts[11]. Traditionally, Western manufacturers rarely involve their suppliers in new-product development. Suppliers first see a component after it has been designed, when they are asked to bid on a supply contract. No attempt is made to take advantage of the supplier's technology or design expertise.

Quality

Both conventional batch manufacturing and conventional product design are beset by quality problems for reasons due to large-lot production. This approach, as most of the gurus of quality have pointed out, inhibits the awareness of quality needed on the shop floor if defects and rework are to be eliminated. Moreover, large batches also increase the time between occurrence of the quality problem and detection. Quick detection of defects is necessary to determine the cause of the quality problem. In product design, quality means getting it right the first time. Not getting it right means rework (or reworked designs)--a major cause of long lead times in development cycles.

Automation

Automation in conventional U.S. manufacturing has been applied sporadically and inconsistently. Often, "islands of automation" have been created when a robot was inserted, like a surgical implant, into the production layout to relieve some bottleneck or deal with a quality problem. No effort was made to prepare the process for automation; in many cases, firms simply succeeded in "hardwiring" an inefficient, obsolete process[7].

In product design, similarly, automation has been applied to isolated segments of the process: CAD systems, for example, are widely used to automate the design function and in plant relay out tasks. In his studies, Adler found that "in most cases, CAD/CAM procedures had merely automated the paper system without any redesign of the actual information flow path or even any change in the allowed processing times. The need to reconfigure engineering procedures to improve . . . engineering operations efficiency is urgent in most of the sample businesses[1]."

To complement the analysis of Western practices, JIT manufacturing practices were compared with the emerging Japanese procedures for new-product development. The table below summarizes this comparison in which the emerging Japanese procedures for new-product development are remarkably similar to JIT manufacturing processes.

Comparison of Japanese Practices in Manufacturing and New Product Development

Process Parameters	JIT Manufacturing	Japanese New Product Development
Batch sizes	Small-batch production; transferred in small lots	Information processed and transferred in small batches
Layouts	Product layout	Grouped by Project Team
Process flow	Coordinated activities	Overlapping activities; simultaneous engineering
Management	Localized control; high level of employee involvement	Local control; management by project team
Supply relations	Close coordination with suppliers	High involvement with suppliers in design of components; technology and information exchange
Quality	Low defect rates	Few engineering changes
Automation	Integrated systems; automation follows process simplification	Automation of information flows; integrated CAD/CAE/CAM
Lead times	Short	Short

Batch Sizes

Manufacturing by JIT is launched by small-lot production, made possible by reduced changeover times. As the firm refines the production process and moves toward the goal of a unit lot size, it finds that product quality improves, inventory diminishes, and, significantly, lead time shrinks.

In product design, the batch size is measured by the amount of information to be processed or the scope of the problem to be solved. Clark's research on product development in the auto industry suggests that Japanese firms tend to bite off small chunks of the design problem at successive stages; that is, they reduce the information batch size[5]. Clark compares the phased and overlapping approaches to problem solving in terms of their efficacy in compressing a new-product development cycle. In the phased approach, common to U.S. new-product development, information about an aspect of product design is passed along only after the batch of work is complete at the prior stage. The procedure resembles the way material in conventional manufacturing is transferred in large lots from one work station to the next.

As the phased approach is to conventional manufacturing, the overlapping approach (used by leading Japanese firms) is to JIT. In the overlapping approach, information is released in incremental units--the batch size for information transfer from one design stage to the next is reduced to the smallest unit possible. The downstream stage begins design work before all work from predecessor stages is complete. To do this, the transition, or setup time, must be reduced because each team member must learn to pick up the work whenever new units of information are transferred instead of waiting for the entire batch to arrive.

Layout and Process Flow

In JIT, small-batch production forces changes in layouts, material flows, and communication between work stations. Compact process flow layouts are formed to ensure a smooth flow of materials between work centers. Unlike conventional manufacturing, the flow in JIT operations goes in two directions: Material is pulled forward as needed, but information flows back upstream to provide feedback on material requirements and quality. In new-product development, overlapping among design activities and transfer of design information in smaller batches fosters similar layout changes in the more progressive Japanese firms. Instead of organizing by function or department, the project team is pulled together in one location, thus creating a "project layout." A project layout fosters frequent, two-way communication between team members, and this interaction helps break down the tendency to attack tasks sequentially, rather than in parallel.

Management and Employee Involvement

In JIT, scheduling of production is localized. Management encourages work groups to assume responsibility for short-term scheduling and product flow. Greater employee involvement and responsibility for the process promotes a cycle of quality improvement.In new-product development in Japan, scheduling responsibility is pushed down to the lowest level. With a team approach, designers and engineers move closer together into a more interactive environment. They have a shared vision of the overall project goal and shared responsibility for the project's success.

Supply Relations

The use of JIT demands supplier involvement not only in frequent deliveries, but also in obtaining cooperative solutions to quality problems. A problem for a manufacturer or a supplier also became the partner's problem; joint responsibility for problem solving was an integral part of the supply-chain relationship. Japanese companies extend that relationship to new-product development by fostering extensive supplier involvement in the design and engineering of new components. According to Clark, "Supplier involvement (and strong supplier relationships) accounts for about one-third of the manhours advantage and contributes four to five months of the lead-time advantage (held by Japanese automakers)[5]."

Quality

During most JIT implementations, quality, as measured by the proportion of defective parts produced and the amount of rework, improves steadily as setup times and batch sizes are reduced. The actions taken to reduce setup times simultaneously attack the causes of defective parts. Consequently, problem solving and information sharing becomes a cooperative, team effort. Reaching the goal of synchronized flow in JIT depends on eliminating quality problems.

In new-product development, high quality means minimal rework or redesign. Design quality problems are rarely due to sloppy, incompetent work by designers. Instead, they typically arise because of isolation of the designer and incomplete information or misunderstandings on the part of the designer concerning product requirements. From a quality standpoint, the overlapping approach to design helps prevent rework by transmitting design information bit by bit, instead of waiting until the entire activity is completed. With a two-way information flow, problems tend to be detected earlier and solved before they contaminate the rest of the project.

Automation

The proper role of automation in a JIT system can be summed up in the phrase "First simplify, then automate." Shigeo Shingo, who helped develop some of Japan's first JIT systems, claims that automation should be used in setup-time reduction *only* after the process has been studied and simplified to remove all the extraneous steps[8]. Automation must not be viewed as a substitute, or shortcut, to process improvement.

Japanese firms use automation in new-product development similarly to the way it is used in JIT. They recognize that automation alone is relatively powerless to turn around a slow, ineffective product development team. CAD systems can speed up parts of the design function, but if the project is managed in a sequential, departmentalized way, the advantages of automation are likely to be dissipated by frequent engineering changes and redesigns. In the U.S., CAD is often viewed as a tool that encourages frequent design revisions. The firms best positioned to integrate these computer design activities and accelerate product development cycles with automation, however, are the firms that have already developed a smoothly functioning design team with overlapping activities.

4. Managing New-Product Development with the JIT Model

Firms tend to manage new-product development the same way that they manage manufacturing. The traditional Western way of managing new-product development is sequential batch processing; the emerging Japanese model is JIT, and the empirical evidence on performance weighs heavily in favor of the Japanese model for new-product development.

The JIT mode is appealing for several reasons? First, the basic objectives of JIT--eliminating waste in the organization, simplicity, total quality, and speed--are desirable attributes for all of the firm's processes. Second, the JIT model promotes organizational learning: knowledge attained through JIT implementation is recyclable for use throughout the firm. Third, JIT embodies the principle of *Kaizen*, or incremental, but continuous, improvement. Firms that adopt the philosophy of *Kaizen* tend to attack new-product development in small, manageable chunks--singles, instead of home runs.

The model outlined for managing new-product development is built upon three underlying principles of JIT.

Principle 1: To reduce throughput time, process and transfer smaller batch sizes.

Principle 2: To process smaller batches efficiently, reduce changeover times.

Principle 3: Solve the problems exposed by applying principles 1 and 2.

Satisfying principles 1 and 2 unleashes a raft of other problems--quality, layout, information and material flow, and other areas. These problems arise at a similar stage in the evolution of JIT as inventory is reduced and throughput rates increase, and no firms have solved all of them completely. The pursuit of speed, like excellence, is unending.

<u>Principle 1: Reduced Batches of Information</u>

New-product development is essentially an information-processing function in which problem solving is the fundamental activity. In NPD, the analog of the manufacturing lot size is the batch size of information. Packets of information are processed and transmitted between groups.

What are the implications of small-batch information processing for product development? First of all, there must be the early release of information from a product designer to a process engineer. Instead of waiting until a final design is complete, partial information is sent downstream so that other members of the project team can begin their work. An obvious advantage of this activity is design quality: early detection of problems. Even more serious problems can be avoided by early transmission of information to marketing and manufacturing. Many of the delays in product development projects can be traced to rework or redesign caused by a misunderstanding of product requirements.

Principle 2: Changeover Time Reduction

Changeover time reduction is an apt metaphor for the NPD process: A new-product introduction is essentially a changeover from an old product to a new one, and the goal is to make the transition as quickly and efficiently as possible. The setup reduction process is also an analog for the set of management activities that must be accomplished in new-product development to accommodate the overlapping activities. The four basic activities that apply to changeover time reduction in new-product development:

1 Internal versus external activities,

2 Eliminate and simplify tasks,

3 Teamwork,

4 Automation.

These four activities are described in greater detail in Chapter 5 of [3]. *Internal versus External Activities* in NPD refers to moving much of the design work off the critical path: engineers begin working on phases of the problem while design is still being fine-tuned. *Eliminate and Simplify Tasks* means to eliminate certain activities, such as prototyping, and simplify the design. One of the simplest, cost-effective techniques in design is to use off-the-shelf components. Many advanced computer manufacturers, such as Cray and DEC, have designed their new machines to work with standard, off-the-shelf parts[10]. *Teamwork* suggests a move by the firm toward overlapping activities in which information flows become more complex, and, therefore, teamwork becomes increasingly important. In an effective, cross-functional team, information processing, instead of being a one-way flow, is more like distillation: Design information is churned and refined through feedback from the other functions. *Automation* should be viewed as the culminating, not the initiating, activity in a firm's attempt to crash time in product development. Automation has a seductive, quick-fix appeal to impatient managers with short-term goals. In their rush to come up to speed in new-product development, U.S. managers should approach automation warily. Automating a sequential, batch process of design can, admittedly, quicken certain tasks, but it is unlikely to bring development time up to the level of global competition.

Principle 3: Solve Other Problems that Arise

In product development, as with JIT, smaller batch sizes spark the outbreak of other problems, like small brush fires, that management must solve if cycle times are to be reduced. The problems are analogous to those that arise in manufacturing--design quality, layout and information flow, worker involvement, and supplier relations--and, as in JIT, these are problems that predated the process change but were obscured by large-batch information transmission. Several of these problems were addressed in the activities to reduce changeover times.

If new-product development is restructured to follow the JIT model, quality problems may actually increase in frequency, but not in severity. As the lot size of information shrinks and the transfer rate between designers and other team members accelerates, small flaws are detected early in the process. In the long run this improves the process because the team can quickly determine that the emerging product design is veering slightly off specifications. Corrections are made before the problem magnifies.

Teamwork is essential to effective new-product design. Barry Bebb states that "The Japanese recognize that manufacturing quality, cost, and schedule are largely determined by how well engineers execute design engineering. Manufacturing cannot produce competitive products unless engineering joins with manufacturing to design competitive products[2]."

AT&T and others are finding that the fastest companies form multifunctional teams--AT&T is using it with success in product development. AT&T applied the team concept to the development of the new cordless 4200 phone. Typically, the development time for a new phone product at AT&T was about two years. The projects were also carried out like a relay race--or the "over the wall" concept. They formed teams and gave them greater decision-making and spending authority. The results: development time was one year instead of the usual two, with lower costs and increased quality[6].

5. The Tide is Turning

There are encouraging signs that some U.S. firms are catching up in the global new-product development race. Significantly, and not coincidentally, the fast-closing firms are the ones that have already successfully implemented JIT manufacturing. They, like the Japanese, have transferred the skills, honed in manufacturing, to new product development. Manufacturers in the United States must continue to cut development time and cost. Just-in-Time provides an ideal template for managing a new production process for two important reasons. First, the focus of JIT is on time and quality. Second, management does not have to learn a new set of tricks; the techniques can be learned first in manufacturing and then transferred to new-product development. This provides additional impetus for undertaking a JIT campaign in manufacturing: There are ancillary benefits for the other functions of the organization.

For firms restrained by dilatory new-product development processes, the writing on the wall is clear: the race is being won by the swift. The greatest cost, by far, is the cost of being late to market.

REFERENCES

1. Paul S. Adler, "CAD/CAM: Managerial Challenges and Research Issues," IEEE Transactions on Engineering Management, vol. 36, no. 3 (August, 1989), pp. 202-215.

2. H. Barry Bebb, "Quality Design Engineering: The Missing Link to U.S. Competitiveness," Key-note Address to the NSF Engineering Design Research Conference, Amherst, MA, June 11, 1989.

3. Joseph D. Blackburn (ed.), Time-Based Competition: The Next Battleground in American Manufacturing, Business One Irwin, New York, 1991.

4. Kim B. Clark,B. Chew, and T. Fujimoto, "Product Development in the World Auto Industry," Brookings Papers on Economic Activity, vol. 3, 1987, pp. 729-771.

5. Kim B. Clark, "Project Scope and Project Performance: The Effect of Parts Strategy and Supplier Involvement on Product Development," *Management Science*, vol. 35, no. 10, October, 1989, pp. 1247-1263.

6. Brian Dumaine, "How Managers Can Succeed Through Speed," *Fortune*, February 13, 1989.

7. Robert A. Millen, Joseph D. Blackburn and Edward Popper, "is the Company of Today Ready for the Factory of the Future," Owen Graduate School of Management Working Paper No. 86-36, Vanderbilt University, 1986.

8. Shigeo Shingo, *A Revolution in Manufacturing: The SMED System* (Cambridge, Mass: Productivity, Inc., 1985).

9. Bro Uttal, "Speeding New Ideas to Market," *Fortune*, March 2, 1987.

10. Paul Wallich, "How and When to Make Tradeoffs," Part II of "Optimizing the Design, *IEEE Spectrum*, May 1987, pp. 33-39.

11. James Welch, Joseph D. Blackburn and Peter Jueptner, "Buyer/Supplier Relations: A Contrast between Japan and the United States," Operations Roundtable Working Paper, Owen Graduate School of Management, Vanderbilt University, 1989.

Concurrent Life-Cycle Engineering for the Optimization of Products, Systems, and Structures

WOLTER FABRYCKY

0. INTRODUCTION

Production is not an end in itself. It is a means for delivering superior product attributes to meet customer expectations. Superior product attributes are quality "effects" resulting from design "causes". This paper addresses concurrent life-cycle engineering methodology for bringing quality products, systems, and structures into being.

Product optimization is the desideratum with design dependent parameters as the key. When partitioned from design independent parameters and design variables a comparison can be made of product, process, and support alternatives concurrently. A life-cycle based decision evaluation function provides the needed vehicle for optimization.

I. PREREQUISITE PHILOSOPHY AND CONCEPTS

Opportunities for product optimization identified in this paper derive philosophically from the poorly understood process by which products, systems, and structures are brought into being. Although somewhat abstract, this philosophy provides a foundation for the fundamental technological activities needed to improve the people-made world [5].

Natural and People-Made Systems

The world in which we live may be divided into the natural world and the people-made world. Included in the former are all elements of the world which came into being by natural processes. The people-made world encompasses all products and systems made by people for the use of people.

All people-made systems are embedded into the natural world. Important interfaces often exist between people-made systems and natural systems and each affects the other in some way as is indicated by Figure 1. The effect of people-made systems on the natural world and upon people has only recently become a keen subject for study by concerned individuals, especially in those instances where the effect has turned out to be undesirable.

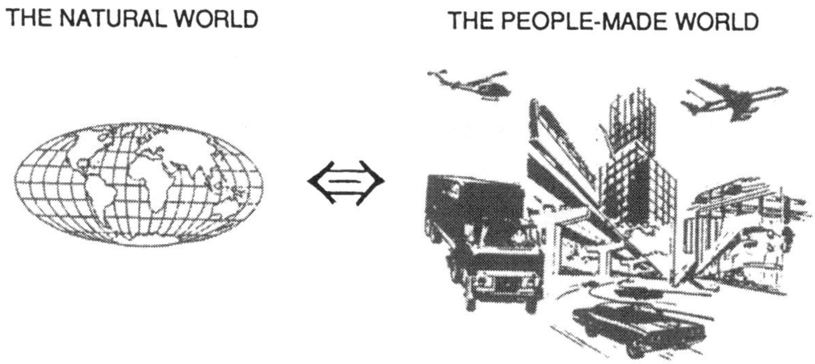

Figure 1. The Natural and People-Made World.

Significant people-made systems are of relatively recent origin; their chief engineer being human. Primitive beings had little impact on the natural world, but were impacted significantly by it. With the development of pervasive technologies, natural impacts on people have been largely mitigated. The greater concern now is with the impact of the people-made world upon the natural world, upon economic systems, and upon people themselves.

The Creation of Utility

All products, systems, and structures which have utility (the power to satisfy wants) are physically manifested. This is obvious with regard to an automobile, a loaf of bread, or a school building. It is also true for intangeable things; music is enjoyed because sound waves strike the ear, pictures are manifested by light waves, and even friendship is realized physically through the senses.

It follows that utilities are created by altering physical factors. The purpose of engineering design effort is to determine how the physical environment may be altered to create the most utility for the least cost, both in terms of product cost and social cost. Engineering is a scientifically based activity that specifies means for altering physical factors to create utility in the economic environment.

As stated in the definition adopted by the Accreditation Board for Engineering and Technology:

> Engineering is the profession in which a knowledge of the mathematical and natural sciences gained by study, experience, and practice is applied with judgement to develop ways to utilize, economically, the materials and forces of nature for the benefit of mankind.

Since engineering is largely responsible for bringing the people-made world into being, it is incumbant upon this profession to discover ways of meeting people's needs in more appropriate and economically viable ways. This must be the highest goal sought as are sought means, through engineering, for enhancing economic competitiveness.

Engineering activities of analysis and design are not an end in themselves, but are a means for satisfying human wants. Thus, engineering has two aspects. One aspect concerns itself with the materials and forces of nature. The other is concerned with the needs of people. Both aspects are illustrated in Figure 2.

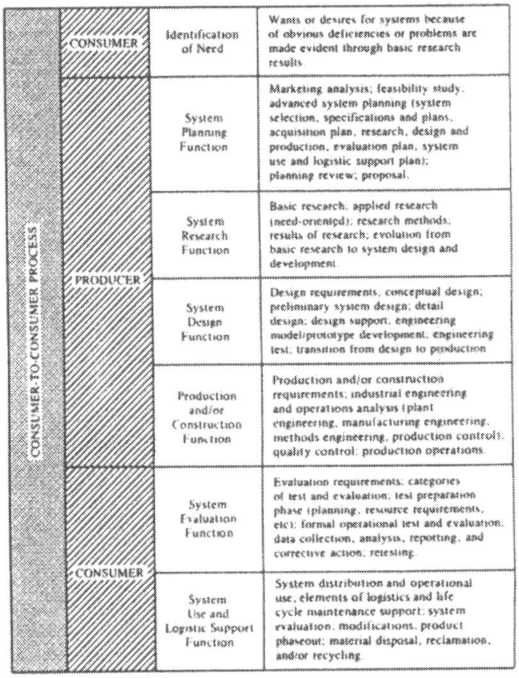

Figure 2. The Consumer-to-Consumer Process [2].

All other factors being equal, people will meet their needs by procuring products and systems which offer the highest value-cost ratio, subjectively evaluated. This ratio can be measured by giving more attention to the resource constrained world in which producers exist. Competitive pressure is increasing on companies to manufacture innovative products at low cost with high quality and outstanding customer service.

The producers response to this challenge must be to adopt a winning product strategy within an effective business environment. Product characteristics are the key; they must serve both the producer and the consumer. Consumers satisfy their needs through product attributes such as functionality, availability, reliability, affordability, and service. Producers require positive revenue flows to maintain market position and to finance corporate renewal. The degree to which desirable product attributes are identified and realized is essential in linking corporate objectives with product strategy.

Appropriate product attributes must be the focus for improvements in design, manufacturing, and product support. Concurrency in the development process is the key. It should encompass the following areas:

1) A conceptual process that defines the product attributes which will satisfy customer values and the producer's corporate objectives.
2) An advanced product design morphology which embraces more functions concurrently and considers more alternatives in an accelerated development cycle.
3) Infrastructure improvements that are based on target product attributes as the basis for justification of investment in new technologies.

The contest for market leadership will be won by those producers who can make effective use of both design methodology and technology. The causal linkage between product attributes and business strategy is indirect and uncertain. It is not easy to manage methodology and technology to achieve the desired end. The effectiveness of corporate decisions will be enhanced provided the focus is the product and its attributes in service to the consumer in a timely manner [1][3].

II. CONCURRENT LIFE-CYCLE ENGINEERING

Products, systems, and structures are designed and developed in accordance with a process not as well understood as it might be. Concurrent life-cycle engineering is suggested as a integration approach for bringing competitive products into being in such a way as to minimize their deficiencies and life-cycle cost [2].

Concurrent Engineering Definitions

Concurrent engineering is an emerging methodology which has great promise for assuring the competitiveness of new products, structures, and systems. In the defense sector, competitiveness means deploying weapons systems which are competitive in both performance and life-cycle cost. In the commercial sector, this means bringing products into being that compete successfully in the world marketplace.

As defined in a recent report by the Institute for Defense Analysis,

> "Concurrent engineering is a systematic approach to the integrated, concurrent design of products and their related processes, including manufacture and support. This approach is intended to cause the developers, from the onset, to consider all elements of the product life cycle from conception through disposal, including quality, cost, schedule, and user requirements" [12].

A recent book on concurrent design makes the following statement in its preface,

"We are particularly concerned with coordinated design of products and processes so that effective manufacturing will be possible. The term 'concurrent engineering' is used in this book to refer to this process" [9].

As used in this paper, concurrent engineering is defined as "a systematic approach to creating a product design that considers all elements of the product life cycle from conception through disposal to include consideration of manufacturing processes, transportation processes, maintenance processes, and so on" [2].

The System Life-Cycle Concept

Fundamental to the practice of concurrent life-cycle engineering is an understanding of the system life cycle. The life cycle of a product, system, or structure begins with the identification of a need and extends through conceptual and preliminary design, detail design and development, production and/or construction, distribution, customer use, support, and then phaseout and disposal.

The life-cycle concept is universal in its applicability. It originates with the perception of a need and terminates with product phaseout and disposal. Between these end points there are two major life-cycle phases. The first is acquisition, including the several interative steps necessary to define the need, perform the design, test and evaluate, and finally produce and distribute the product. The utilization phase follows and involves activities required to deal with the product in being. These include operating, maintaining, modifying, retiring, and product disposal.

A life-cycle approach for producing competitive products and systems must go beyond considerations of the life cycle of the product itself. It must simultaneously embrace the life cycle of the manufacturing system as well as the life cycle of the product service system. Accordingly, there are three concurrent life cycles progressing in parallel as is illustrated in Figure 3.

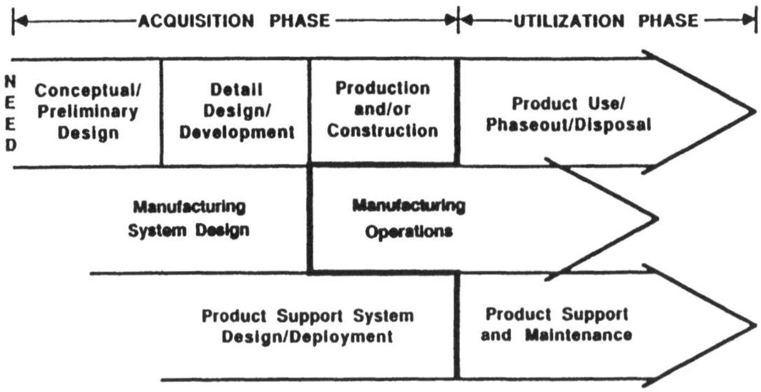

Figure 3. Product, Process, and Support Life Cycles [2].

The need for a new product comes into focus first, initiating the product life cycle. Conceptual and preliminary product design follows need determination, with simultaneous consideration for manufacturing system design; activities best pursued as a parallel life cycle before detail product design. Product support system design, too often omitted as a design imperative, is the third life cycle. It should be synchronized with product design and manufacturing system design [6].

The Life-Cycle Design Process

A detailed presentation of the elaborate technological interactions which must be integrated over the coordinated life cycles is shown in Figure 4. The progression is iterative from left to right and not serial in nature, as might have been implied by Figure 3. This progression embraces all of the life-cycle activities which must be considered in the concurrent engineering approach. The paragraphs which follow briefly describe the life-cycle phases of conceptual design and others upon which product optimization depends.

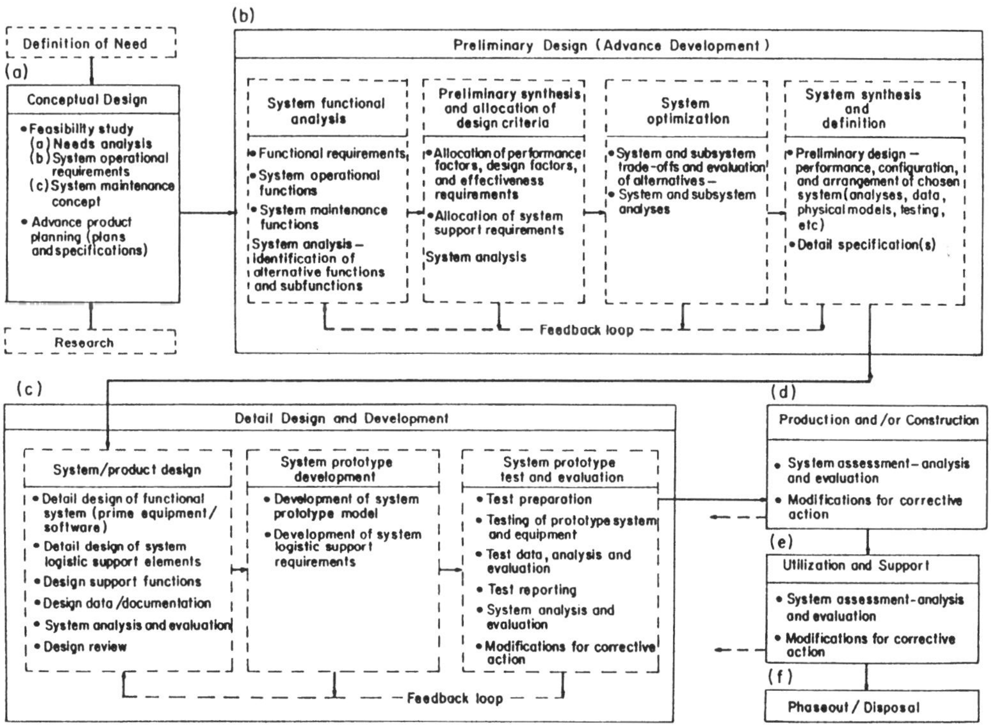

Figure 4. Life-Cycle Design Activities and Interactions [2].

Conceptual Design (Figure 4, Block a). Regardless of the product or system, design begins with an identified need arising out of a perceived deficiency. An individual and/or organization identifies the need or function to be performed, and a new (or modified) product is conceived and produced to meet the need.

Preliminary Design (Figure 4, Block b). Preliminary design begins with the technical baseline as defined in a feasibility analysis, and proceeds through the translation of system-level requirements into detailed qualitative and quantitative design requirements. This includes functional analysis and requirements allocation, the accomplishment of trade-off studies and optimization, system synthesis, and configuration definition in the form of detailed specifications.

Detail Design and Development (Figure 4, Block c). Detail design begins with the concept and configuration derived from the primary design activities identified in Block b. When an overall design configuration has been established, it is necessary to progress through further definition leading to the realization of hardware, software, personnel, data, and support capability.

The Life-Cycle Approach

The life-cycle approach is applicable to both small and large-scale products and systems. Further, the process is applicable to many different categories of systems, (e.g., an electronic system, an airplane or missile system, a manufacturing or production system, a structure or facility, etc.). Although requirements may vary from one system to the next, the process is essentially the same.

Traditional engineering design has focused mainly on the acquisition phase of the life cycle. But, recent experience indicates that a properly coordinated and functioning product or system, which is competitive in the marketplace, cannot be achieved through efforts applied largely after it comes into being. Accordingly, it is essential that engineers be sensitive to operational feasibility during the early stages of product development and that they assume the responsibility for life-cycle design which has been largely neglected in the past [6].

Good design for a product's primary function often produces side effects in the form of operational problems. This is due to exclusive consideration of the primary function, rather than to the more challenging problem of designing in the face of the several "ilities." Enough specialized knowledge exists to solve this problem. The impediment to its solution is the integrated use of what is known in a systematic manner. This integration capability is inherent to the engineering profession.

Product development embracing life-cycle methodology has an excellent chance of enhancing consumer satisfaction, corporate identity, and company profitability through the integration of important design considerations (appearance, cost, performance, and quality). But these desiderata may not be attained unless the completeness inherent in the methodology is nurtured and promulgated.

Promulgating Life-Cycle Completeness

Several research and publication activities with significant impact are now underway. These can be easily altered at little cost to accelerate progress toward completeness in concurrent engineering design. Completeness is interpreted here to be in agreement with the IDA definition [12]. Some activities with which the author is familiar are mentioned in the paragraphs which follow.

A very significant initiative by the Defense Advanced Research Projects Agency (DARPA) is addressing concurrent engineering. Entitled the DARPA Initiative on Concurrent Engineering (DICE), this project involves a consortium of industry and universities, principally General Electric and West Virginia University. In its initial year, the DICE project has focused almost entirely on the acquisition phase of the life cycle. Fortunately, the information system architecture being fashioned will permit integration of engineering design tools addressing usability factors such as reliability and maintainability. Because of the magnitude of this initiative, its demonstration status, and the influence of the sponsor, it is essential that this project move quickly toward life-cycle completeness.

The National Science Foundation (NSF) has an innovative research program underway in design theory and methodology. At the present time, the focus of this program is largely on mechanical artifact design and manufacture. A review of nearly fifty papers presented by grantees in June 1989 at Amherst reveals that little attention is being given to usability factors. Reliability was mentioned less than a half dozen times, and maintainability even less. NSF has a great impact on the thinking of academia. Accordingly, the Foundation can contribute significantly to a more complete understanding by setting goals for its design research programs which encourage consideration for all life-cycle outcomes.

DOD finds itself so resource constrained by the high cost of maintaining and operating defense systems that it now funds some research and development programs specifically focused on design for supportability. Incompleteness regarding life-cycle coverage is often tolerated. A recent example is the USAF RAMCAD program (Reliability and Maintainability in Computer Aided Design) where incompleteness is due to a lack of attention to producability and manufacturability. Fortunately, the DOD CALS initiative (Computer-aided Acquisition and Logistics Support) is defined completely regarding the life cycle. A concurrent engineering task force under the CALS initiative has a mandate to address design in the context of the entire life cycle.

The area of publications can be improved too. Regretably, the only book on concurrent engineering design does not address and explain design impacts on the utilization phase of the life cycle [9]. The index does not reveal any reference to reliability, maintainability, availability, usability, or even durability. Quality is mentioned, but in a limited way. This is unfortunate because this book is based on work at the respected and influential Draper Laboratory. Readers are likely to come to believe that concurrent engineering as discussed is all that really matters.

Finally, the new journal entitled Research in Engineering Design is a welcome publication from Springer International. However, the first issues have nothing in them dealing with design effects beyond manufacture. This can be corrected in future issues if appropriate papers are prepared and submitted.

For concurrent engineering design to make a meaningful impact on product competitiveness, no phase of the life cycle can be omitted. The competitiveness of products and systems in the world marketplace depends upon designs which yield desirable outcomes during utilization. These outcomes are best addressed during design.

III. OPPORTUNITIES FOR PRODUCT OPTIMIZATION

Greater contributions to competitive production can be made by getting involved earlier in the life cycle, promulgating concurrent engineering, developing computer-aided capability, evaluating engineering designs, extending engineering economics, and seeking quality outcomes. These opportunities are discussed in the paragraphs which follow.

Getting Involved Earlier

Engineers are most at ease in detail design and development (Figure 4, Block c) due to good technical preparation. However, the opportunities in conceptual and preliminary design (Figure 4, Blocks a and b) are excellent. Great benefit can be derived from the application of knowledge earlier in the life cycle, as is illustrated in Figure 5. Fully two-thirds of the life-cycle cost of a system is committed by the time conceptual and preliminary design is completed. This commitment reaches about 80% upon completion of detail design and development.

Systems engineers, because of their broad training, can contribute successfully to decision making in conceptual and preliminary design. The objective is to narrow the "gap" (shown in Figure 5) between available knowledge and life-cycle cost committed.

Promulgating Concurrent Engineering

Concurrent engineering is an emerging methodology based on the system life-cycle approach. It has great promise for assuring the competitiveness of new products and systems.

The concurrent engineering design process emphasizes functional analysis, the allocation of requirements, synthesis, trade-off and optimization, and so on. These activities (by themselves) are not new to engineers, nor is the approach. However, in many instances, designers will start identifying

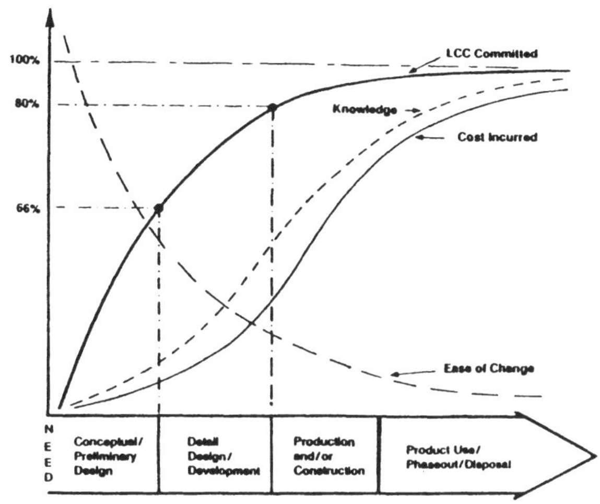

Figure 5. LCC Committed and Incurred.

"hardware" for a given system without first fully identifying essential manufacturing requirements [9]. Further, the designer will often ignore reliability, maintainability, and life-cycle cost factors. The emphasis must be on a concurrent process discipline necessary for the orderly development of the product, the manufacturing process, and the support system in a coordinated manner [2].

It is the commercial understanding of concurrent engineering that is the most incomplete. Unlike the Department of Defense, commercial firms do not assume life-cycle responsibility for the products they produce (except for warranties). Although desirable outcomes during use are the ultimate proof of product quality, short term thinking often overshadows the longer term benefits in prospect from an acquired product reputation. Engineering professionals can be of great help in fostering longer term thinking by practicing concurrent engineering.

Developing Computer-Aided Capability

The communication and coordination needed to bring the product, the process, and the service system along in a coordinated manner is not easy to achieve. Progress in this endeavor will likely be facilitated by new technologies making possible the more timely acquisition and utilization of design information. CAD/CAM is only one of these technologies. Others need developing which can integrate relevant activities of the enterprise over the spectrum of life cycles. The most promising of these is Computer-Aided Concurrent Engineering Design (CACED), illustrated schematically in Figure 6.

Opportunities for progress exist in Block 3 (estimation and prediction), Block 4 (design evaluation), Block 5 (economic and physical data base development), and in developing the HCI (human computer interface). Networking considerations for CACED workstations within the organization can

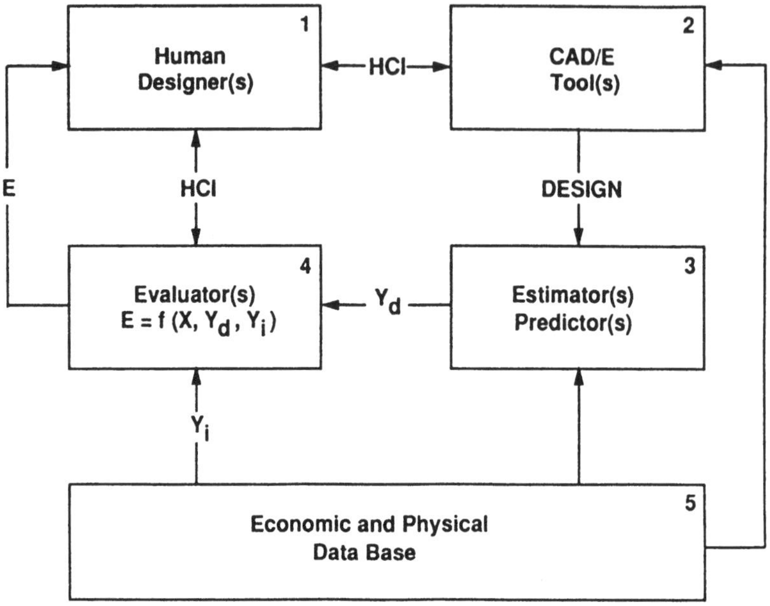

Figure 6. A Schematic for CACED [2].

also be the concern of communication engineers. With the advent of information system networking, and the increased availability of computer-aided X (where X = design, cost estimating, process planning, etc.), "facts" become more accessible and hence of diminishing relative value. The ability to apply facts (knowledge) in a creative, and more rapid manner, has become more important.

Evaluating Engineering Designs

Design evaluation in terms of life-cycle cost and system effectiveness is an activity open to those individuals with the appropriate systems analysis capabilities. This evaluation includes numerous facts, some of which are illustrated in Figure 7.

Block 4 in Figure 6 shows the placement of evaluation with CACED, but evaluation is often needed outside the domain of a workstation. In either case, the design dependent parameter approach is proving to be of considerable value. The evolution of this approach is shown in Figure 8.

The design dependent parameter approach is a mathematical way to link design actions with operational outcomes. It utilizes the design evaluation function in Figure 8. From the following definitions of terms in the function, the opportunities for engineers (particularly those with good operations analysis backgrounds) should be evident. Figure 9 is an illustration of the role of the design evaluation

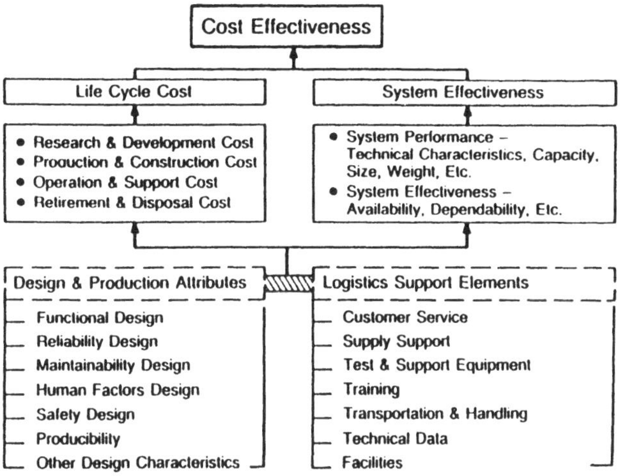

Figure 7. Factors to be Evaluated in Systems Design.

References	Functional Form	Application
Churchman, Ackoff, & Arnoff (1957)	$E = f(x_i, y_j)$	Operations
	E = system effectivenss x_i = variables under control y_j = variables not subject to direct control	Operations
Fabrycky, Ghare, & Torgorcon (1984)	$E = f(X,Y);\ g(X,Y) \lesseqgtr B$ E = systems effectiveness X = controllable variables Y = uncontrollable variables	Operations
Banks & Fabrycky (1987)	$E = f(X,Y_d,Y_i);\ g(X,Y_d,Y_i) \lesseqgtr C$ E = system effectiveness X = procurement level and procurement quantity Y_d = source dependent parameters Y_i = source independent parameters	Procurement Operations
Blanchard & Fabrycky (1981, 1990)	$E = f(X,Y_d,Y_i);\ g(X,Y_d,Y_i) \lesseqgtr C$ E = evaluation measure X = design variables Y_d = design dependent parameters Y_i = design independent parameters	Design Optimization

Figure 8. Design Evaluation Function Evolution.

function in optimizing acquisition/utilization. An example of its application, in constrained form, is given in Section IV.

To Optimally Achieve
ACQUISITION / UTILIZATION
Mathematically Link
DESIGN / OPERATIONS
Utilizing The
• GENERAL DESIGN EVALUATION FUNCTION •
For Choosing Among Design Alternatives

$$E = f(X, Y_d, Y_i)$$

$$\text{Subject to } g(X, Y_d, Y_i) \lesseqgtr C$$

Figure 9. Optimizing Acquisition/Utilization.

Extending Engineering Economics

Opportunities for the more effective application of engineering economics derive from the life-cycle process. The life-cycle approach holds great promise for helping engineering economics reach its full potential to evaluate design, manufacturing, and operations concurrently [7].

Engineering economics has always been associated with time; the time value of money, receipts and disbursements over time, etc. The central "model" in engineering economics is the money flow diagram depicting estimates of income and outlay over time [11]. Accordingly, engineering economics and the product life cycle are of the same "dimension." Missing from many applications of engineering economics is a complete view of the life cycle.

Life-cycle engineering economics is a good opportunity area for development by engineers. It can be maximally effective during conceptual and preliminary design. Economic outcomes cannot be influenced much after the design is completed. Furthermore, economic considerations with life-cycle connections must be made during manufacturing system design and support system design in a concurrent manner.

Seeking Quality Outcomes

Quality outcomes are desirable "effects" derived from design "causes." Progress in gaining the needed understanding of cause and effect is not progressing rapidly enough. More must be done to discover the essential links between design and usability. Design engineers can and should contribute insight and provide the needed leadership.

It is well known that quality cannot be inspected into a product. Producers have turned to statistical process control as a better approach. However, quality must first be designed in. This opens a new area for investigation: that of understanding and applying parameter design ideas due to Taguchi [10]. An opportunity exists to bring experimental designs and indirect experimentation into product design and development as illustrated schematically in Figure 10.

Quality outcomes have been achieved by production workers concurrently producing and "inspecting;" this in lieu of leaving quality concerns to others. In an analogous way, quality can be even more greatly enhanced when engineering managers find organizational and informational ways of assisting designers to concurrently consider factors upon which quality depends (reliability, maintainability, operating efficiency, etc.). The goal is to facilitate the embodiment of quality responsibility during the design person.

Production workers cannot impart quality beyond that intended and specified by design engineers. Design engineers can influence quality from the very beginning of the product life cycle. When design is done in a truly concurrent manner, benefits far exceeding those obtained by production workers can be expected [2][10].

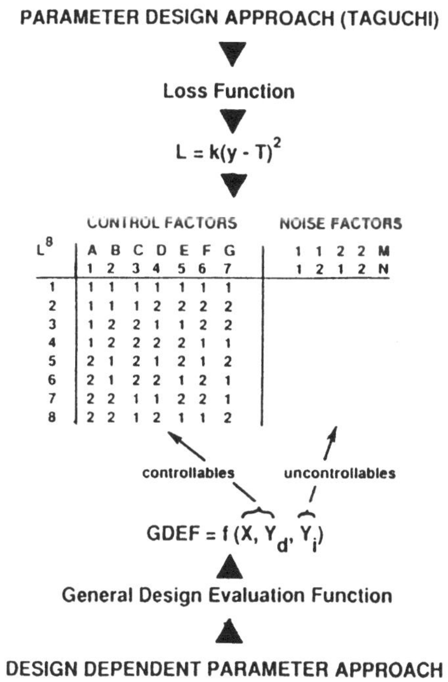

L^8	A 1	B 2	C 3	D 4	E 5	F 6	G 7	1 1	1 2	2 1	2 2	M N
1	1	1	1	1	1	1	1					
2	1	1	1	2	2	2	2					
3	1	2	2	1	1	2	2					
4	1	2	2	2	2	1	1					
5	2	1	2	1	2	1	2					
6	2	1	2	2	1	2	1					
7	2	2	1	1	2	2	1					
8	2	2	1	2	1	1	2					

Figure 10. Taguchi and GDEF Approaches [2].

IV. THE DESIGN DEPENDENT PARAMETER APPROACH: AN EXAMPLE

The design dependent parameter approach for the optimization of products, systems, and structures was presented conceptually in Section III. Due to the abstract nature of this approach, a simple example from the classical area of bridge design is presented in this section.

Consider the problem of specifying a configuration for the superstructure of a bridge. Two alternative superstructure configurations are shown in Figure 11. Regardless of the final configuration specified, the weight of the superstructure per foot will be a function of the span between piers, S. Let

L = bridge length in feet

W = superstructure weight in pounds per foot

C_s = erected cost of superstructure in dollars per pound

C_p = installed cost of piers in dollars per pier.

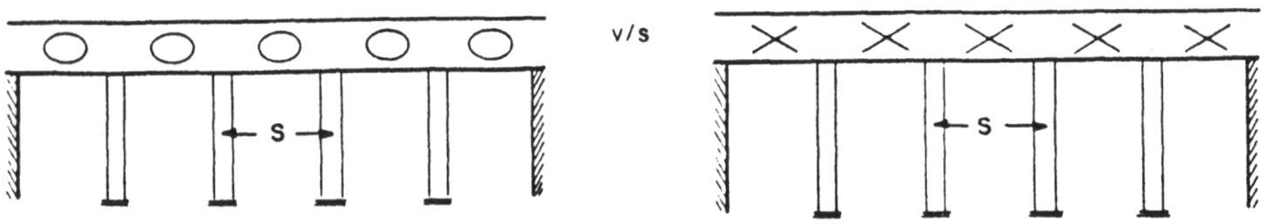

Figure 11. Bridge Design Alternatives.

The design evaluation function expressed as

$$DEF = f(X, Y_d, Y_i)$$

establishes S as the design variable (X), W as the design dependent parameter (Y_d), and C_s and C_p as design independent parameters (Y_i). Design optimization requires a search for the superstructure configuration which will result in the lowest possible total cost for the bridge. Accordingly, a total cost function must be derived.

Total cost for the bridge is the sum of the superstructure cost and the pier cost as follows:

$$TC = SC + PC$$

The superstructure cost may be expressed as

$$SC = (W)(L)(C_s)$$

where W = AS + B. And the pier cost as

$$PC = (L/S + 1)(C_p)$$

where each abutment is costed a pier. Total cost is then

$$TC = (AS + B)(L)(C_s) + (L/S + 1)(C_p)$$

from which $S^* = \sqrt{C_p/AC_s}$ and

$$TC^* = 2\sqrt{AC_p L^2 C_s} + BLC_s + C_p$$

In TC* above, W does not appear. Rather, W is given in terms of the parameters of the weight estimating relationship; A and B. These are derived by curve fitting to empirical data.

Consider a example with design independent parameters as follows:

$$L = 1{,}000$$

$$C_s = \$0.65$$

$$C_p = \$220{,}000$$

For Design A, the weight of the superstructure is found to be

$$W = 16S + 600$$

per the curve of Figure 12. And, for Design B, the weight of the superstructure is found to be

$$W = 22S + 0$$

per the curve of Figure 12.

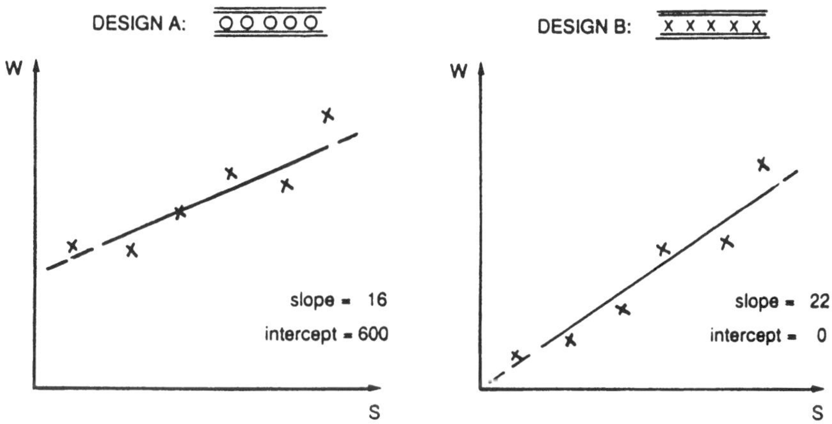

Figure 12. Weight Estimating Relationships.

From the parameters A and B for each superstructure design configuration, and the design independent parameters, the TC* can be found. It is

$$TC^* = \$2{,}219{,}158$$

for Design A with $S^* = 74.8$ feet. For Design B, it is

$$TC^* = \$2{,}295{,}454$$

with $S^* = 87.7$ feet. Figure 13 shows the total cost as a function of the span between piers.

Design B should be chosen for the unconstrained case. However, if the span between piers must be at least 100', so that barge traffic may pass, then Design A would be chosen. Note the design decision reversal due to the presence of a constraint.

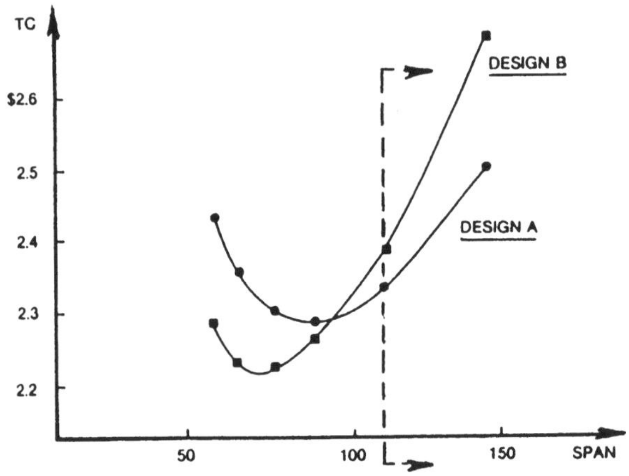

Figure 13. Bridge TC as a Function of S.

V. SUMMARY AND CONCLUSIONS

In these times of intensifying global competition, commercial firms are searching for ways to gain a sustainable competitive advantage in the marketplace. Advertising alone is not sufficient. One promising strategy is to adopt a life-cycle development methodology for proposed new products, their required manufacturing activities, and product support systems.

Product development embracing concurrent life-cycle methodology has excellent chance of enhancing consumer satisfaction, corporate identity, and company profitability through the integration of important design considerations (appearance, cost, performance, and quality). But these desiderata may not be attainable unless the completeness inherent in the methodology is recognized and implemented.

Optimization is formalized by the identification of design dependent parameters and the formulation of a general design evaluation function. By incorporating life-cycle factors into this function, design alternatives can be compared so that improvements in the design can be recognized and adopted.

REFERENCES

1. Adam, J. A., "Competing in a Global Economy," IEEE Spectrum, Vol. 27, No. 4, April, 1990.
2. Blanchard, B. S. and W. J. Fabrycky, Systems Engineering and Analysis, Second Edition, Prentice-Hall, Inc., 1990.
3. CAM-I, Inc., "Product Optimization Prospectus," September, 1989.
4. Churchman, C. W., R. L. Ackoff, and E. L. Arnoff, Introduction to Operations Research, John Wiley, Inc., 1957.
5. Fabrycky, W. J., "System Life-Cycle Engineering: Bringing Competitive Products Into Being," NSF Workshop on Emerging Engineering Opportunities, Washington, D.C., March, 1985.
6. Fabrycky, W. J., "Designing for the Life Cycle". Mechanical Engineering, Vol. 109, No, 1, January 1987.
7. Fabrycky, W. J. and B. S. Blanchard, Life-Cycle Cost and Economic Analysis, Prentice-Hall, Inc., 1991.
8. Fabrycky, W. J., P. M. Ghare, and P. E. Torgersen, Applied Operations Research and Management Science, Prentice-Hall, Inc., 1984.
9. Nevins, J. A. and D. E. Whitney, Editors, Concurrent Design of Products and Processes, McGraw-Hill, Inc., 1989.
10. Taguchi, S. and D. Byrne, "The Taguchi Approach to Parameter Design", Proceedings, ASQC Quality Congress, 1986.
11. Thuesen, G. J. and W. J. Fabrycky, Engineering Economy, Seventh Edition, Prentice-Hall, Inc., 1989.
12. Winner, R. I., J. P. Pennell, H. E. Bertrand, and M. M. G. Slusarczuk, The Role of Concurrent Engineering in Weapons Systems Acquisition, Report R-338, Institute for Defense Analysis, December, 1988.

Strategic Formation of Manufacturing Cells

CHWEN SHEU and LEE KRAJEWSKI

1. INTRODUCTION

Achieving the status of a world-class manufacturer requires a direct linkage between the needs of the marketplace and the manufacturing process needed to produce the products. Companies seek market niches by emphasizing certain characteristics of their products or manufacturing systems such as low cost, high quality, fast delivery speed, or volume flexibility. We call these characteristics ***competitive priorities***. However, many firms find themselves in the position of producing products with differing subsets of these competitive priorities on the same manufacturing processes. This can happen when new products are introduced or the firm enters new markets over time. This often causes difficulties for managers because of the conflicting manufacturing tasks that are implied [10]. For example, achieving low costs might require large batch runs or dedicated equipment utilizing low-skilled workers, while providing product flexibility might require low-volume runs on general purpose machinery utilizing skilled workers. In addition to differences in resource needs, there may be differences in infrastructure needs such as reward/incentive systems, inventory control systems, or vendor contracts [11].

The key to linking market requirements to manufacturing process seems to be the concept of manufacturing ***focus***. The concept of focused facilities, first proposed by Skinner [17], relates to limiting the set of products, technologies, volumes and markets any given plant is responsible for. The guiding principle in achieving focus is to assign products and resources to plants such that there is a homogeneous manufacturing task that is consistent with the company's overall strategy. Manufacturing focus is an intuitively appealing principle supported by many others [7], [8], [9], [10], [11], [12], [13], [15], and [21]. Most research has been empirical, based on relatively small sample sizes, but has generally supported the proposition that achieving manufacturing focus improves that firm's competitive capability and profits.

It is not always possible to devote one plant to a single market with limited products and technologies. Some plants must compete in multiple markets with conflicting competitive priorities. For these plants, Skinner and others propose the concept of plant-within-a-plant (PWP). Different manufacturing groups within the same plant having different focuses should be separated as much as possible, each with its own staff, resources and products, to form quasi-autonomous units. We call these units ***focus units***. There is empirical evidence that the PWP concept is beneficial. For example, Schonberger [15], defining manufacturing focus from the engineering perspective of process requirements only, observed that PWP made plants easier to control and simpler to maintain at Hewlett Packard, Kodak, Control Data and Westinghouse.

In this paper we concentrate on plants which must compete in multiple markets. We seek to link market requirements to manufacturing process design. Burbidge [2] [3] and Dale, Burbidge, and Cottam [6] conceptualized the procedure for arriving at such a linkage. First, large firms should be divided into separate semi-autonomous factory units using the focused factory concept. Second, the organization of departments inside the factories should be changed from a process organization to a product organization. This is essentially the PWP concept. Finally, at the shop floor level, Group Technology is used to complete the change from a process to a product organization. Here is where the manufacturing cells are actually created. This paper addresses the second step, the organization of departments into focus units. No methodology addressing this problem has been reported in the literature, yet is is a critical step in the design of manufacturing cells.

In the next section we present a nonlinear mixed-integer programming model for this problem. Because optimally solving this model poses practical difficulties, in Section 2 we propose a two-stage heuristic approach to generate solutions for management to evaluate. Section 3 contains a numerical application.

2. MODEL

Objective Function

There are two objectives to consider in this problem. The primary objective is to maximize the average degree of focus for the focus units. We first define the "strategic similarity" between two products i and j, $0 \leq SS_{ij} \leq 1$, as:

$$SS_{ij} = \alpha \left\{ 1 - \left[\sum_{p \in P_i \cup P_j} \frac{|W_{ip} - W_{jp}|}{2} \right] \right\} + (1-\alpha) \left\{ 1 - \frac{1}{n[M(i,j)]} \left[\sum_{m \in M(i,j)} \frac{|B_{im} - B_{jm}|}{Max(B_{im}, B_{jm})} \right] \right\} \qquad (1)$$

where:

P_i = set of competitive priorities for product i.

W_{ip} = weight of competitive priority p assigned to product i; $\sum_{p \in P_i} W_{ip} = 1.$

B_{im} = average batch volume of product i on machine type m, expressed in machine hours.

α = managerial parameter which sets the degree of strategic similarity to assign to the concordance of competitive priorities, $0 \leq \alpha \leq 1$.

$n[M(i,j)]$ = number of machine types visited by both items i and j.

The first part of (1) measures the degree of similarity between products i and j relative to the weights management has assigned to their competitive priorities. The second part of (1) measures batch volume similarity. Batch volumes can be an important consideration in the design of focus units when products with large batch volumes are contemplated for the same machinery as products with small batch volumes. It is likely that scheduling conflicts will arise that will hamper management's ability to achieve the competitive priorities set down for those respective products. The parameter α adjusts the relative importance of the batch volume consideration.

Once a focus unit has been designed, the degree of focus within that unit is calculated as the average degree of commonality with regard to the strategic similarity measure defined in (1). The within-unit strategic similarity for unit c, $0 \leq WSS_c \leq 1$, is given by

$$WSS_c = \frac{2\sum_{i \neq j} SS_{ij} X_{ic} X_{jc}}{N_c(N_c - 1)} \tag{2}$$

where:

X_{ic} = 1 if product i is assigned to focus unit c;
0 otherwise

N_c = number of products assigned to focus unit c

If there is only one product assigned to a focus unit, WSS_c is specified to equal 1. The average degree of focus for a given number of focus units, C, is

$$f_1 = \frac{\sum_c WSS_c}{n(C)} \tag{3}$$

where n(C) is the number of focus units where $N_c > 1$. We assume management is primarily concerned with those units where product grouping takes place because it is those units that will have the most potential for divergent manufacturing tasks.

The second objective is to economize the budget allocation. Forming focus units by only considering the strategic similarity of the products assigned to them will usually result in duplication of equipment and other resources. Forming units with considerations for the process requirements of each product will tend to result in lower capital investment and material movement between focus units because of the pooling of resources. Consequently, given an annual budget allocation for the purpose of forming focus units, we seek to maximize the difference between annual budget underruns, d^- and annual budget overruns, d^+, as a proportion of the annual budget allocation, A:

$$f_2 = \frac{(d^- - d^+)}{A} \tag{4}$$

The objective function of the model is a weighted average of f_1 and f_2:

$$\text{MAX} \quad \lambda f_1 + (1 - \lambda) f_2 \tag{5}$$

where , $0 \leq \lambda \leq 1$, is a managerial parameter that can either be used to generate focus unit alternatives exhibiting a variety of characteristics or to reflect management's utility for tradeoff in degree of focus versus budget utilization.

The Constraints

The constraints in the model are given below.

$$\sum_c X_{ic} = 1 \qquad i = 1,2,\ldots, I \tag{6}$$

$$\sum_i X_{ic} = N_c \qquad c = 1,2,\ldots, U_c \tag{7}$$

$$F_c \leq N_c \qquad c = 1,2,\ldots, U_c \tag{8}$$

$$\sum_c F_c = C \tag{9}$$

$$L_c \leq C \leq U_c \tag{10}$$

$$\sum_n \sum_{i \in V_n} Q_{mn} Z_{imnc} X_{ic} + \sum_{c' \neq c} \sum_{i \in V_m} Z_{immc} X_{ic'} \leq R_m Y_{mc} \qquad m = 1,2,\ldots, M; \; c = 1,2,\ldots, U_c \tag{11}$$

$$\sum_n Q_{nm} Z_{inmc} X_{ic} + \sum_{c' \neq c} Z_{immc'} X_{ic} \geq D_{im} X_{ic} \qquad i \in V_m; \; m = 1,2,\ldots, M; \; c = 1,2,\ldots, U_c \tag{12}$$

$$\sum_{c' \neq c} \sum_m C_m Y_{mc} + \sum_{c' \neq c} \sum_m \sum_{i \in V_m} \frac{h_i q_i}{g_{im}} Z_{immc'} X_{ic} + d^- - d^+ = A \tag{13}$$

$$Y_{mc} \geq 0, \text{ and integer} \qquad m = 1,2,\ldots, M; \; c = 1,2,\ldots, U_c \tag{14}$$

$$Z_{imnc} \geq 0 \qquad i = 1,2,\ldots, I; \; m, n = 1,2,\ldots, M; \; c = 1,2,\ldots, U_c \tag{15}$$

$$X_{ic} = (0,1) \qquad i = 1,2,\ldots, I; \; c = 1,2,\ldots, U_c \tag{16}$$

$$F_c = (0,1) \qquad c = 1,2,\ldots \ldots, U_c \tag{17}$$

$N_c \geq 0$, and integer $\qquad c = 1,2,\ldots \ldots, U_c \qquad (18)$

The indices, decision variables and parameters not already defined are given below:

m, n = indices for machine types, $m, n = 1, 2, \ldots, M$

Y_{mc} = number of machines of type m assigned to focus unit c

Z_{inmc} = total production requirements of product i on machine type m to be carried out by machine type n in unit c

F_c = 1, if there is at least one product assigned to focus unit c
0, otherwise

D_{im} = the total estimated machine hours of machine m required for product i per year

g_{im} = average processing and setup time of one unit of product i on machine type m

h_i = cost of moving one pound of product i between two focus units

q_i = average weight (pounds) per unit of product i

L_c = lower bound on number of focus units management will consider

U_c = upper bound on number of focus units management will consider

V_m = set of products that require operations on machine type m

R_m = annual productive time for one unit of type m machine

Q_{nm} = substitutability of machine type n for machine type m; greater than 1 if n is more efficient than m; equal to 1 if n and m are equally efficient; less than 1 if n is less efficient than m. For instance, a value of 0.8 would mean that n is only 80% as efficient for substituting m, and would take more time to perform the same tasks.

C_m = Annualized equivalent total costs for a given economic life of machine m

Some of the constraints need further elaboration. Constraints (11) impose the capacity limitations of machine type m and ensure that the machine hours used do not exceed the machine hours available. In this formulation we have allowed for "inter-unit" transfers. Products actually assigned to another focus unit could be processed in focus unit c on machine type m if capacity exists. Constraint set (12) ensures that each product receives its required processing time either on machine type m (its primary source) in its assigned focus unit or another unit c', or on a suitable substitute, such as machine type n within the same focus unit. Constraint set (13) measures the actual budget request, consisting of annualized capital costs of machinery and inter-unit transfer costs, relative to the budget allocation. In this formulation we assume that the firm is starting from scratch and has no existing machinery. Situations where existing machinery exists, can be easily modelled as well. Also, if inter-unit transfers are not to be allowed, the cost parameter h_i can be made arbitrarily very large.

The major obstacle to using the model presented in Section 2 is the difficulty in solving it due to the nonlinearities present in the objective function and constraints. However, it is noted that if the values of X_{ic} are known and substituted in the objective function and constraints of the model, the model is reduced to a simple mixed-integer linear program. This insight is the basis for our two-stage heuristic. In the first stage we seek to maximize f_1 by assigning products to focus units for a given C over the range $[L_c, L_c+1, ..., U_c]$. In the second stage, for each C in the feasible range and for given values of X_{ic}, we allocate machine types and quantities to focus units so as to maximize f_2.

Product Assignment

Figure 1 provides an overview of the entire focus unit formation heuristic. In stage 1 we wish to

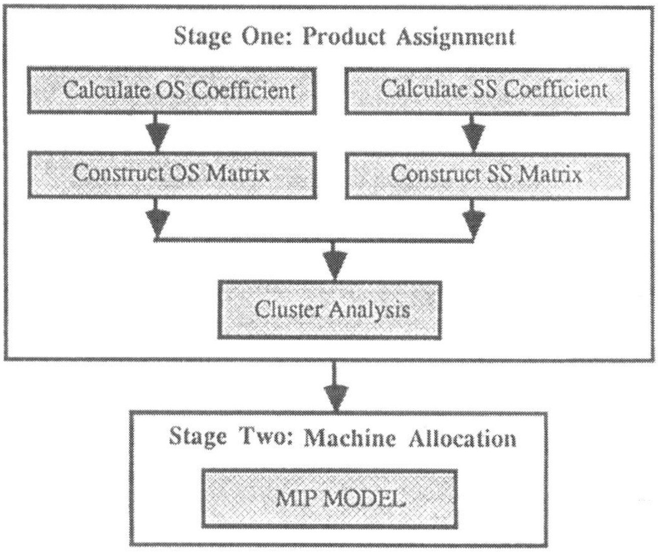

Figure 1. Focus Unit Formation Heuristic

group products that have a high degree of similarity from the standpoint of their competitive priorities. We can do this by using the strategic similarity index (1) defined earlier. However, now that we have decomposed the original problem into two subproblems, we must be careful to recognize the capital cost implications of the product assignments. Intuitively, product groupings based solely on engineering considerations for processing needs will result in more economical resource allocations than those based solely on market factors. To recognize these engineering considerations we utilize an operational similarity index. This is our surrogate for objective f_2 while in the product assignment stage. We combine the strategic similarity and the operational similarity indices into a composite similarity matrix using λ, the managerial parameter for balancing the considerations for degree of focus and budget utilization. The composite similarity matrix is then used in a clustering algorithm to produce product assignments for a given number of focus units.

Operational Similarity (OS) Index. The operational similarity between two products is a function of the number of machine types visited by both products. While a number of item-grouping indices have been proposed in the literature [4], [5], [16], [18], [19], we shall use the functional form of one described by Vakharia and Wemmerlov [18]. The OS index between products i and j, $0 \leq OS_{ij} \leq 1$, is given by

$$OS_{ij} = 0.5 \left\{ \frac{\sum_{m \in M(i,j)} \gamma_{mi}}{\sum_{n \in N(i)} \gamma_{ni}} \right\} + 0.5 \left\{ \frac{\sum_{m \in M(i,j)} \gamma_{mj}}{\sum_{n \in N(j)} \gamma_{nj}} \right\} \tag{19}$$

where

γ_{mi} = 1, if machine type m is needed for product i, or $i \in V_m$
0, otherwise
M(i,j) = set of machine types visited by both products i and j
N(i) = set of machine types visited by product i

Composite Similarity (CS) Index. Once the operational similarity index is calculated, it can be combined with the strategic similarity index (1) to form a composite similarity index, $0 \leq OS_{ij} \leq 1$, by computing a weighted average for each product pair:

$$CS_{ij} = \lambda SS_{ij} + (1-\lambda)OS_{ij} \tag{20}$$

The matrix of OS_{ij} values is the primary input to the clustering algorithm. Only the values above the diagonal are needed since the matrix is symmetric.

Clustering Analysis. Figure 2 shows the steps in the cluster analysis. After the CS matrix is computed, the algorithm begins by assigning all products to their own focus unit, and then systematically merging focus units together until the lower limit L_c is reached. For each value of C in the feasible range, the specific product assignments and value of f_1 (3) are recorded.

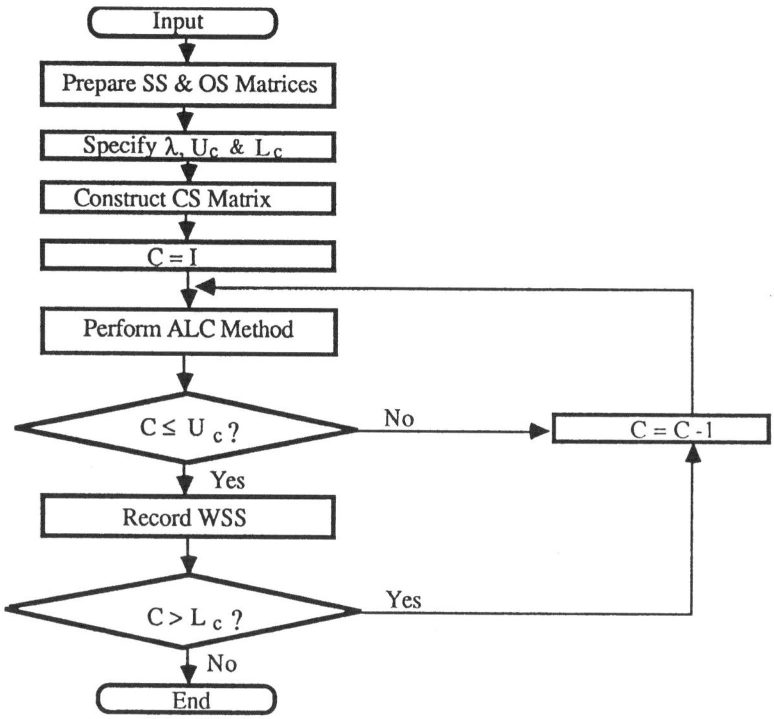

Figure 2. Cluster Analysis

We use the Average Linkage Clustering algorithm [1] to assign products to focus units. The method fuses individuals or groups of individuals with the highest value of CS_{ij}. At the end of each iteration, the composite similarity between a newly formed group and another individual or group is defined as the average of the similarities between all pairs of individuals in the two groups. The updated CS index between focus units c and c' at the end of iteration t is given by:

$$CS^{(t)}_{cc'} = \frac{\sum_{i \in c} \sum_{j \in c'} CS^{(0)}_{ij}}{N_c \, N_{c'}} \tag{21}$$

The updated index is computed with the initial values of the index to properly reflect the instances where the composite similarity between two products is zero.

The result of stage 1 is a series of product assignments for various values of C and λ. The corresponding values of CS_{ij} for these assignments are now submitted to the second stage: machine allocation.

Machine Allocation

Given C and the values of X_{ic}, the machine allocation problem is to maximize f_2 subject to constraints (11) through (15). The machine allocation model is a linear mixed-integer program where Y_{mc} is the only integer variable. The number of integer variables will be much less than MC in general because typically every focus unit will not require every machine type. Since we already have the product assignments, the machine allocation model determines the number of machines of each type to house in each focus unit, allowing for the possibility of product transfers between focus units. It all depends on the budget allocation and the cost of transfers related to capital costs.

4. NUMERICAL EXAMPLE

In this section we will demonstrate the use of the heuristic with a simple example. There are seven products and five machine types. Table 1 contains the batch volumes and total annual machine requirements for each product. Table 2 shows the weights placed on the competitive priorities of each product. The amortized machine cost is $5,000 for all machine types and there is very little machine substitutability. In particular, $Q_{ab} = 0.9$, $Q_{ce} = 0.9$, $Q_{mm} = 1$ for all m, and all other $Q_{mn} = 0$. Each product weighs 2 pounds and the transfer costs are $1 per pound for all products. In this example, the value of α in the strategic similarity index is 0.5, $L_c = 2$, $U_c = 5$, and $\lambda = 0.5$. Later we will show how the value of C and λ affect focus-unit design. Finally, the value of the budget allocation is $43,000.

Product Assignment

We begin by constructing the similarity matrices. Table 1 contains the information to determine the values of γ_{im} to compute the operational similarity (OS) matrix. For example, $\gamma_{1b} = 1$, $\gamma_{1c} = 1$ and $\gamma_{1e} = 1$ because machines b, c, and e are needed to produce product 1. All other values for γ_{1m} are zero. Equation (19) can be used to develop the OS matrix below.

Operational Similarity

Product	1	2	3	4	5	6	7
1	-	0	0.83	0	0.42	0.33	0.83
2		-	0	1.00	0.50	0.83	0
3			-	0	0	0.42	0.50
4				-	0.50	0.83	0
5					-	0.42	0.50
6						-	0.42
7							-

The SS matrix is developed from Table 1 and Table 2 with equation (1).

Strategic Similarity

Product	1	2	3	4	5	6	7
1	-	0.30	0.56	0.15	0.39	0.80	0.44
2		-	0	0.51	0.49	0.65	0.05
3			-	0.35	0.45	0.29	0.95
4				-	0.76	0.41	0.40
5					-	0.34	1.00
6						-	0.34
7							-

Finally, the OS and SS matrices are combined with equation (20) for $\lambda = 0.5$.

Composite Similarity

Product	1	2	3	4	5	6	7
1	-	0.15	0.70	0.08	0.41	0.57	0.64
2		-	0	0.76	0.50	0.74	0.03
3			-	0.18	0.23	0.36	0.73
4				-	0.63	0.62	0.20
5					-	0.38	0.75
6						-	0.38
7							-

Table 1. Batch Volumes and Annual Machine Requirements (Hours)

	Product													
	1		2		3		4		5		6		7	
Machine Type	B_{im}	D_{im}	B_{im}	D_{im}	B_{im}	D_{im}	B_{im}	D_{im}	B_{im}	D_{im}	B_{im}	D_{im}	B_{im}	D_{im}
a			8	800			5	1000	7	1400	4	200		
b	8	400			7	1400								
c	4	200			7	1400					4	200	7	1400
d			4	400			5	1000			8	400		
e	16	800							6	1200			6	1200

Table 2. Competitive Priority Weights

	Weights			
Product	Cost	Delivery Speed	Delivery Reliability	Quality
1	0.30	0.60	0.10	
2		0.70		0.30
3	0.90		0.10	
4	0.70			0.30
5	0.80		0.10	0.10
6		0.90		0.10
7	0.80		0.10	0.10

Iteration 1. In the first iteration we merge products 2 and 4 ($CS_{2,4} = 0.76$). The CS matrix is updated as follows:

$$CS^{(1)}_{1,(2,4)} = \frac{CS^{(0)}_{(1,2)} + CS^{(0)}_{(1,4)}}{2} = \frac{0.15+0.08}{2} = 0.12; \qquad CS^{(1)}_{3,(2,4)} = 0.09;$$
$$CS^{(1)}_{5,(2,4)} = 0.57; \qquad CS^{(1)}_{6,(2,4)} = 0.68; \qquad CS^{(1)}_{7,(2,4)} = 0.12.$$

The rest of the entries in the CS matrix remain unchanged.

Product	1	3	5	6	7	(2,4)
1	-	0.70	0.41	0.57	0.64	0.12
3		-	0.23	0.36	0.73	0.09
5			-	0.38	0.75	0.57
6.				-	0.38	0.68
7					-	0.12
(2,4)						-

The number of cells (C) is six, and greater than U_C. Consequently, we proceed to the next iteration.

Iteration 2. Products 5 and 7 are merged next and the updated CS matrix is shown below.

Product	1	3	6	(2,4)	(5,7)
1	-	0.70	0.57	0.12	0.53
3		-	0.36	0.09	0.48
6			-	0.68	0.38
(2,4)				-	0.35
(5,7)					-

The number of cells equals 5, and $WCS^{(5)} = \frac{0.51+1.0}{2} = 0.76$. We proceed to iteration 3.

Iteration 3, 4 and 5. Products 1 and 3 are merged in iteration 3 and $WCS^{(4)} = 0.69$. In iteration 4 we merge product 6 with cluster (2,4) and $WCS^{(3)} = 0.69$. Finally, in iteration 5, we merge cell (5,7) with cell (1,3), $WCS^{(2)} = 0.58$ and $C = L_C$. Consequently, we terminate the procedure for $\lambda = 0.5$.

We can generate a set of prospective product assignments by repeating the cluster analysis for different values of λ. Table 3 contains twelve product assignments generated from setting $\lambda = 1, 0.5$, and 0. When $\lambda = 1$ we have disregarded considerations for process similarity and have grouped products according to their competitive priorities and batch volumes only. When $\lambda = 0$, we only consider process similarities in the product assignments. As we shall see in the next stages of the procedure, the $\lambda = 0$ assignments result in the lowest budget requirements.

Table 3. Experimental Results for Various Values of λ and C

	Focus Cell Formation		
C	Strategic Focus (λ = 1)	Composite Focus (λ = .5)	Operational Focus (λ = 0)
2	(1,2,6) (3,4,5,7)	(1,3,5,7) (2,4,6)	(1,3,7) (2,4,5,6)
3	(1,6) (2) (3,4,5,7)	(1,3) (2,4,6) (5,7)	(5) (1,3,7) (2,4,6)
4	(1,6) (2) (4) (3,5,7)	(1,3) (2,4) (5,7) (6)	(1,3) (2,4,6) (5) (7)
5	(1,6) (2) (3) (4) (5,7)	(1) (2,4) (3) (5,7) (6)	(1,3) (2,4) (5) (6) (7)

Machine Allocation

Table 4 contains the budget implications resulting from the optimal allocation of machines to focus units. The results are ranked according to objective f_1 so that the tradeoffs in required funding are more easily seen.

The lowest-cost solution is C = 2, $\lambda = 0$ because of the heavy consideration for process similarities and the pooling of production requirements on each machine type, reducing the need for more machines. However, the strategic similarity of the product assignments within the two focus units is near the bottom of the twelve alternatives generated. As we increase λ, strategic similarity and funding needs also increase. In general, for any given number of focus units desired, increasing the consideration for competitive priorities and batch volumes will increase the funds required to provide the necessary resources.

Table 4 also reveals that increasing the number of focus units for a given value of $\lambda > 0$ increases the strategic similarity of products assigned to them as well as funding requirements. Authorizing more focus units allows more opportunities to match products with respect to competitive priorities, batch volumes or process requirements. However, it will also require more funds to take advantage of these more closely-matched focus units.

Table 4. Results of Machine Allocation

Rank	Design (C,λ)	f_1	Annual Funds Required	f_2
1	(5,1)	0.90	$ 56,800	-0.32
2	(4,1)	0.80	56,800	-0.32
3	(5,.5)	0.76	49,600	-0.15
4	(3,1)	0.73	52,000	-0.21
5	(3,.5)	0.69	42,400	0.01
6	(4,.5)	0.69	47,200	-0.10
7	(2,1)	0.62	45,000	-0.05
8	(2,0)	0.59	40,000	0.07
9	(3,0)	0.59	42,400	0.01
10	(2,.5)	0.58	42,400	0.01
11	(4,0)	0.55	43,840	-0.02
12	(5,0)	0.54	46,240	-0.08

Finally, the results indicate that the highest values of f_1 do not always occur for $\lambda = 1$. There is an interaction between C and λ. For example, $C = 5, \lambda = 0.5$ is superior to $C = 3, \lambda = 1.0$ for objective f_1. However it does not dominate $C = 2, \lambda = 1.0$ for the budgetary objective f_2. Consequently, f_2 must be used to arrive at a final solution. In our example, A = $43,000 and the "best" solution would be $C = 3, \lambda = 0.5$. However, if management is willing to accept a 15% budget overrun, $C = 5, \lambda = 0.5$ would be the choice.

5. DISCUSSION

The successful application of Group Technology could very well depend upon the logical formation of focus units which simplify the manufacturing tasks of managers responsible for them. We have presented a methodology for forming these focus units. This is a first step in defining manufacturing cells that support the competitive strategy of the company. Given the resource allocations and product assignments of the focus units, standard methodologies for forming group technology cells can be used to design the manufacturing cells.

The issue of how many focus units to have has not been fully addressed in this paper. We have recognized the costs that increase as C increases: amortized capital costs and transfer costs. We have neglected overhead costs which will be affected by organizational changes of the nature we are proposing. The reason we have not included these costs is that no research determining the affect of focus-unit formation on overhead has been reported in the literature. We hypothesize the relationship for overhead costs in Figure 3.

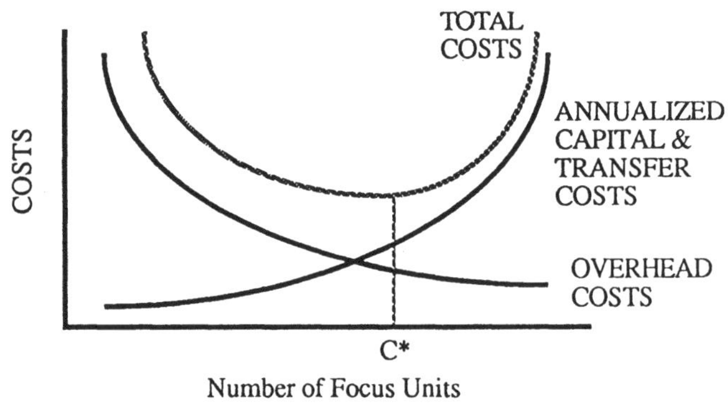

Figure 3. Cost Implications of Choosing C

There is evidence to indicate that overhead costs decrease as the size of the manufacturing unit decreases [20]. As C increases, the size of each focus unit decreases. Consequently, overhead costs within each focus unit should decrease as size decreases, however there is probably some point where the size gets too small to be efficient. Unfortunately, no research on the effect of the number of focus units (as opposed to their size) on overhead costs has been reported. We can reasonably assume that there are some overhead benefits for increasing C, but there are undoubtedly limitations to that general rule. Optimal determination of C from a cost perspective must await developments in this critical area.

REFERENCES

1. Anderberg, M.R., *Cluster Analysis for Applications,* Academic Press, New York, N.Y., 1973.

2. Burbidge, J.L., "The Simplification of Material Flow Systems," *International Journal of Production Research,* Vol. 20, No. 3, 1985, 339-347.

3. Burbidge, J.L., "New Methods of Organization to Improve Production Efficiency," *Proceedings 22nd MTDR Conference,* UMIST Manchester, 1981.

4. Carrie, A. S., "Numerical Taxonomy Applied to Group Technology and Plant Layout," *International Journal of Production Research,* Vol. 1, No. 4, 1973, 399-416.

5. Choobineh, F., "A Framework for the Design of Cellular Manufacturing Systems," *International Journal of Production Research,* Vol. 26, No. 7, 1988, 1161-1172.

6. Dale, B.G., J.L. Burbidge and M.J. Cottam, "Planning the Introduction of Group Technology," *International Journal of Operations and Production Management,* Vol. 4, No. 1, 1985, 34-47.

7. Hayes, R.H. and R.W. Schmenner, "How Should You Organize Manufacturing?" *Harvard Business Review*, January-February 1978, 105-118.

8. Hayes, R.H. and S.G. Wheelwright, "Linking Manufacturing Process and Product Life Cycles," *Harvard Business Review*, January-February 1979, 133-140.

9. Hayes, R.H. and S.C. Wheelwright, *Restoring Our Competitive Edge: Competing Through Manufacturing,* John Wiley and Sons; New York, NY., 1984.

10. Hill, T.J. and R.M.G. Duke-Woolley, "Progression Or Regression in Facilities Focus," *Strategic Management Journal*, Vol. 4, 1983, 109-121.

11. Hill, T., *Manufacturing Strategy: Text and Cases*, Richard D. Irwin, Homewood, Ill., 1989.

12. Krajewski, L.J. and L.P. Ritzman, *Operations Management: Strategy and Analysis,* Second Edition, Addision Wesley: Reading, Massachusetts, 1990.

13. Richardson, P.R., A.J. Taylor, and J. R. M. Gordon, "A Strategic Approach to Evaluating Manufacturing Performance," *Interfaces*, Vol 15, No. 6, November-December 1985, 15-27.

14. Schmenner, R.W., *Making Business Location Decisions*, Prentice-Hall; Englewood Cliffs, NJ., 1982.

15. Schonberger, R.J., "Frugal Manufacturing," *Harvard Business Review*, September-October 1987, 95-100.

16. Selvam, R. P. and K. N. Balasubramanian, "Algorithmic Grouping of Operation Sequences," *Engineering Costs and Production Economics*, Vol. 9, 1985, 125-135.

17. Skinner, W., "The Focused Factory," *Harvard Business Review*, May-June 1974, 113-121.

18. Vakharia, A.J. and U. Wemmerlov, "Designing A Cellular Manufacturing System: A Material Flow Approach Based on Operation Sequences," to appear in *IIE Transactions*, 1990a.

19. Vakharia, A.J. and U. Wemmerlov, "An Investigation of Similarity Coefficients and Hierarchical Clustering Methods Applied to the Manufacturing Cell Formation Problem," Working paper, 1990b.

20. Ward, P.T., P.D. Berger, J.G. Miller, and S.R. Rosenthal, "An Empirical Analysis of Manufacturing Process Technology and Support Staff Composition,' Working Paper, WPS 90-38, The Ohio State University, 1990.

21. Wheelwright, S.C., "Reflecting Corporate Strategy in Manufacturing Decisions," *Business Horizons*, February 1978, 57-66.

Scale, Scope or Division of Labour: Coping with Volume, Variety and Variability in Manufacturing

JOHN BUZACOTT

INTRODUCTION

We are all familiar with Adam Smith's description in "The Wealth of Nations" of the organization of the manufacture of pins based on the principle of the division of labour. Ever since, conventional wisdom has held up the desirability of using this principle of dividing up the work required in manufacture into a number of steps performed in series. It is only when demand is insufficient to keep more than one person busy that division of labour would not be used. Yet, when one looks at current developments in manufacturing, it is surprising the extent to which there is a general trend away from the organization of manufacture around simple series flow with considerable division of labour. Nor does there seem to be much of a place for the individual craftsman, capable of making a very wide variety of products, with whom Adam Smith contrasted the, to him new, form of organization around coordinated series flow. Rather there seems to be several, perhaps converging, trends in manufacturing organization. One is the development of cooperatives of crafts people who share accounting and marketing services and who agree amongst themselves about how work should be shared. This form of organization seems to be most developed in Northern Italy but it can be found wherever adequate communication facilities for both information and goods exist and perhaps also where relatively inexpensive computer controlled manufacturing equipment can be used. Another trend evident in the Eurpean and North American automobile industry is the abandonment of the use of the traditional moving belt assembly line, with its rigid work pacing and tight linking of work tasks, for those manufacturing operations that still require people to perform the production tasks (see Buzacott 1990). The new systems place the responsibility for quality on the worker, give each worker substantially more work content per vehicle and use Automated Guided Vehicles (AGVs) to move jobs from one worker to the next. The AGVs permit asynchronous work flow and also they enable work stations to be paralleled. Lastly, in other industries where there is substantial assembly work required, such as footwear, clothing or electronics, manufacture is occasionally organized around work stations to which work is dispatched from a central location where all work in process is kept unless it is actually being operated on. That is, there appears to be a trend away from strict serial organization towards cordinated parallel organization.

Another trend in manufacturing organization that is also evident relates to the *scope* or the degree of product variety which a manufacturing system is required to handle. The concept of Group Technology (GT) was developed in the USSR in the 1940s to overcome the problems in operating large job shops which make a wide variety of products (see Petrov 1969). It was found advantageous, using the same equipment, to group it into cells, each making a specific subset of the overall set of products. Provided demand remained reasonably stable so that cell utilization was acceptable, GT was found to offer significant advantages. Next Jaikumar(1986) has pointed out that Flexible Manufacturing Systems (FMS) in the USA typically make a relatively narrow range of products, on the whole substantially less than FMS in Japan. While his explanations tend to reflect adversely on the competence of US managers, it is quite possible that there are good and rational reasons and they do not only relate to such obvious explanations as excessive set-up and changeover times at the machines between different products.

There have been numerous models developed in the past to try and understand issues relating to economies of scale, diseconomies of scope and the appropriate level of subdivision of manufacturing tasks. Many of these models focus on the impact of the variability and unpredictability of demand and it is obvious that this will influence many aspects of manufacturing system design. For example, Manne(1961), Freidenfels(1980) and Buzacott and Chaouch(1988) have all looked at how the optimal plant size is influenced by demand variability and unpredictability. Economy of scope has been discussed by Goldhar and Jelinek(1983), Fine and Freund(1990) and others. Boucher(1988) has a comprehensive review of the issues relating to division of labour, specifically the impact of learning, line balance and motivation.

However, in this paper a different perspective will be taken. Our interest is on the influence of variability within the manufacturing system itself. Our perspective is that the trends in manufacturing system design have evolved out of attempts to manage and cope with this variability, which can arise from a variety of different sources, some of which are susceptible to direct control by management while others can only be dealt with by using different system designs. Some of the variability is innate in the manufacturing process and could be measured by a process capability study. Another source is machine failures and the consequent machine down time. If people are performing production tasks there is inherent variability in the time for and quality of task performance by an individual and also between individuals. Next, different jobs have different processing requirements so this will also introduce variability into the system. It seems obvious that, as the degree of this internal variability increases, it should impact on the structure of the manufacturing system. The purpose of this paper is to apply a number of stochastic models in order to provide insight into these issues. None of the models are in themselves new and most of them have been used in the past for performance evaluation of specific systems. However, their use to provide insight into the more strategic issues relating to the structure and operation of manufacturing systems appears to be new.

Basics of the Modelling Approach

The basic issues in our use of models to address the strategic aspects of manufacturing can be understood if we consider a single worker who produces a variety of different products. But suppose demand increases so that using the old methods he can no longer cope. What should he do? Let us review some of the system related options. If he does nothing then some other worker will eventually go into business making much the same set of products and it would be expected that demand would be split between them in a more or less random way. Next, our worker could encourage another worker to go into business but they would agree to assign work equitably, e.g. "I take the first job, you take the second job, I take the third job" and so on. Another possibility is to have different workers specialize in different jobs. If so, how should the jobs be allocated to the different workers? Another approach is for our worker to acquire some machinery that is capable of processing jobs faster. What are the merits of using this approach as compared to specialization? Alternatively, our worker might hire a number of other workers and subdivide the overall manufacturing task into steps with each worker performing just one step. Typically this would result in a flow line system in which all work flows from one worker to the next and all work follows the same sequence. Because this could result in large amounts of work accumulating at some workers it might be considered attractive to eliminate all or most of the inprocess work and synchronize work flow so all workers start each job at the same time. Sometimes the different steps of production need not be done in a fixed sequence so it may be possible to allow jobs to follow different paths through the system. Is there any advantage in doing this? In such a system there may not be much control over work flow and task assignment, so, to get around this, central dispatch might be used with all work in process kept centrally thus giving ready availablity of information on utilization and availability of facilities.

Model assumptions.

In order to understand the issues raised above we will use a model of a simple manufacturing system that captures many of the features of interest. Since our interest is not with the impact of disturbances in demand, we assume that demands for product j, $j = 1, \ldots, n$ arrive in accordance with a Poisson process with parameter λ_j and $\lambda = \sum_{j=1}^{n} \lambda_j$. Product j requires tasks $u = 1, \ldots, k_j$ in order to be manufactured and there are m facilities on which these tasks might be performed. The specifics of the allocation of tasks and products to facilities will depend on the system alternative that we are considering. When facility i is working on product j it will have k_{ij} tasks assigned to it. If product j has task u performed on facility i then we characterize the time to perform the task by its mean $1/\mu_{iju}$ and its variance σ_{iju}^2. We will assume independence of the different tasks so that the time to complete tasks $u = 1, \ldots, k_{ij}$ assigned to facility i on product j will have mean $1/\mu_{ij} = \sum_{u=1}^{k_{ij}} 1/\mu_{iju}$ and variance $\sigma_{ij}^2 = \sum_{u=1}^{k_{ij}} \sigma_{iju}^2$.

Another assumption that we will often make is that the time to perform task u of product j on facility i is a random variable $S_{iju} = s_{iju} X_i$ where s_{iju} is a constant dependent on i, j and u and X_i is a random variable with mean 1 and variance σ_i^2. This means that $C_{S_{iju}}^2$, the squared coefficient of variation of S_{iju}, will be such that $C_{S_{iju}}^2 = C_{X_i}^2 = \sigma_i^2$. We will call this assumption the *time scaling* assumption.

Performance measures.

In order to compare different systems it is necessary to choose performance measures that are appropriate. Two such measures are (1) *Production capacity TH*, i.e. the maximum total demand rate for which the level of work in process is finite, and (2) *Average work in process WIP* for a given total demand rate. Little's law specifies that the average flow time $\bar{F}$ and the average work in process are connected by $WIP = \lambda \bar{F}$, thus the ranking of systems on the basis of WIP will be the same as their ranking on the basis of $\bar{F}$. However, as we will see it does not always follow that the system with the higher TH will always have lower WIP for all values of $\lambda < TH$. In make-to-order systems $\bar{F}$ is an appropriate measure of the performance of the system as seen by a customer. However, in a make-to-stock system the customer would be interested in the delay Δ in filling a demand. Appropriate measures of the delay would be the probability of no delay $P\{\Delta = 0\}$ and the expected delay $E[\Delta]$. In order to estimate these delay related measures we will assume that in a make to stock system there is an output store from which customer demands are met and in which the stock level is z after a long period with no demands. Then for aggregate system analysis it is usually possible to regard the system as equivalent to a single server. So if $I(t)$ is the inventory level in the store and $B(t)$ is the level of backordered demand at time t a reasonable approximation for $P\{I(t) - B(t) = n\}$ is given by

$$\lim_{t\to\infty} P\{I(t) - B(t) = n\} = \begin{cases} 1-\rho & n = z \\ \rho(1-\hat{\rho})\hat{\rho}^{z-n-1} & -\infty < n < z \end{cases}$$

where $\hat{\rho} = \frac{WIP-\rho}{WIP}$ and WIP is the level of work in process if the same system is operated on a make to order basis with no output store. ρ is the level of utilization of the single server. If this approximation is valid then $P\{\Delta = 0\} = 1 - \rho\hat{\rho}^{z-1}$ and $E[\Delta] = \frac{\rho\hat{\rho}^z}{\lambda(1-\hat{\rho})}$. Hence it is thus reasonable to concentrate on TH and WIP as the performance measures of the system.

Single facility model.

To determine the performance of a single facility we can use results for the $M/G/1$ queue. If there is a single customer class

$$WIP = \frac{\lambda^2 E[S^2]}{2(1-\rho)} + \rho$$

where $\rho = \lambda E[S]$. If there are n customer classes with class j having generic service time S_j then $\rho = \sum_{j=1}^n \lambda_j E[S_j] = \sum_{j=1}^n \rho_j$ where $\rho_j = \lambda_j E[S_j]$ and

$$WIP = \frac{\lambda^2 \sum_{j=1}^n f_j E[S_j^2]}{2(1-\rho)} + \rho$$

where $f_j = \lambda_j/\lambda$. Now $E[S_j^2] = 1/\mu_j^2 + \sigma_j^2 = 1/\mu_j^2(1 + C_j^2)$. If we make the time scaling assumption then $C_j^2 = C_S^2$, a constant for all j and

$$WIP = \frac{(1 + C_S^2)\sum_{j=1}^n \rho_j^2/f_j}{2(1-\rho)} + \rho.$$

It is possible to regard this multi product system as equivalent to a single product system with squared coefficient of variation C'^2 such that $\rho^2(1+C'^2) = (1+C_S^2)\sum_{j=1}^n \rho_j^2/f_j = (1+C_S^2)\lambda^2 \sum_{j=1}^n f_j/\mu_j^2$. Note that $\sum_{j=1}^n f_j/\mu_j^2$ can be written in the form

$$\begin{aligned}\sum_{j=1}^n f_j/\mu_j^2 &= \sum_{j=1}^n f_j(1/\mu_j - 1/\bar{\mu})^2 + (1/\bar{\mu})^2 \\ &= \sigma_{1/\mu}^2 + (1/\bar{\mu})^2 \\ &= (1/\bar{\mu})^2(1+C_{1/\mu}^2)\end{aligned}$$

where $1/\bar{\mu} = \sum_{j=1}^n f_j/\mu_j$ and $\sigma_{1/\mu}^2$ is the weighted sum of the squared deviations of the product j mean processing time from the overall mean processing time $1/\bar{\mu}$. Hence one can write

(1)
$$WIP = \frac{\rho^2(1+C_S^2)(1+C_{1/\mu}^2)}{2(1-\rho)} + \rho$$

and also it follows that

(2)
$$1 + C'^2 = (1+C_S^2)(1+C_{1/\mu}^2).$$

Parallel Expansion

We next consider the increase of capacity by means of *parallel expansion*, i.e. each facility performs all the required tasks to complete a product and each job only visits one facility.

Independent versus Cooperative Facilities.

Suppose expansion occurs by means of *independent* facilities with each facility making the whole range of products. It would be reasonable to assume that the arrival stream at each facility would continue to be Poisson. Thus with m facilities the arrival rate of demands for type j products would be λ_j/m and hence the fraction of time facility i would be occupied in serving type j products would be $\rho_j = \lambda_j/(m\mu_j)$ if the facilities have the same processing characteristics (i.e. ignoring individual differences). Hence

(3)
$$WIP^{(i)} = m\left\{\frac{\rho^2(1+C'^2)}{2(1-\rho)} + \rho\right\}.$$

Next, suppose expansion occurs via *cooperative* facilities in which arriving demands are allocated using a cyclic allocation rule, i.e. server j gets demands $j, j+m, j+2m, \ldots$. The arrival stream at any facility is then no longer Poisson and it will have a squared coefficient of variation of $1/m$. WIP can be then found using one of the approximations for the $GI/G/1$ queue. If C_a^2 and C_S^2 are the squared coefficient of variation of arrivals and service the approximation (Marchal 1977)

$$WIP = \frac{\rho^2(1+C_S^2)(C_a^2+\rho^2 C_S^2)}{2(1-\rho)(1+\rho^2 C_S^2)} + \rho \tag{4}$$

will be used. Then for the cooperative system as a whole it follows that

$$\begin{aligned} WIP^{(c)} &= m\left\{\rho + \frac{\rho^2(1+C'^2)(1/m+\rho^2 C'^2)}{2(1-\rho)(1+\rho^2 C'^2)}\right\} \\ &= m\left\{\rho + \frac{\rho^2(1+C'^2)}{2(1-\rho)}\left(1 - \frac{1-1/m}{1+\rho^2 C'^2}\right)\right\}. \end{aligned} \tag{5}$$

Note that $WIP^{(c)} < WIP^{(i)}$, i.e. the cooperative solution will always be better than the independent solution.

Specialization.

Another possible allocation procedure would be for the facilities to agree to specialize on particular types of products. The question that then arises is what should be the basis for division of the set of products among the facilities. As in the cooperative situation we will seek "equitable" allocations so that the utilization of each facility is identical. Strict adherence to this requirement may require splitting product type j between more than one facility and it will be assumed that if this is necessary the splitting is done randomly, although, as with the cyclic allocation rule above, other allocation rules would be better.

Given the time scaling assumption applies, one can write the total WIP as

$$WIP^{(sp)} = \frac{\rho^2(1+C_S^2)}{2(1-\rho)} \sum_{i=1}^{m} (1 + C_{1/\mu_i}^2) + m\rho \tag{6}$$

where C_{1/μ_i}^2 is the squared coefficient of variation of the mean processing times of the products assigned to machine i about the overall mean processing time of the jobs assigned to that machine. It follows that the goal is thus to minimize $\sum_{i=1}^{m} C_{1/\mu_i}^2$, i.e. by reducing the diversity of the mean service times of the products assigned to each facility. This can be achieved by ranking the products according to their mean processing times and assigning them in sequence to each facility in turn until the required utilization ρ has been achieved on the facility, thus reducing the range of mean processing times of the products assigned to the facility.

It is easy to show that specialization will never be worse than random assignment and is better unless all products have the same mean processing time. However, cyclic allocation can be better than specialization if the range of mean processing times of the set of products is not too large. This can be illustrated if $n = 2$ and $m = 2$. Assume that the demand rates for the two products are such that it is always true that $\lambda_1/\mu_1 = \lambda_2/\mu_2 = \rho$ so that with specialization each facility will process only one product. Let $r = \lambda_1/\lambda_2$ and without loss of generality assume $r \geq 1$. Then it follows that with cyclic assignment $1 + C'^2 = (1 + C_S^2)\frac{(1+r)^2}{4r}$ and the condition under which cyclic assignment is preferable to specialization is then

(7) $$\frac{(1+r)^2}{4r}\left\{1 - \frac{1-1/2}{1+\rho^2 C'^2}\right\} < 1.$$

Fig. 1 shows the resulting combinations of r and ρ for which specialization is preferable to cyclic assignment for $C_S^2 = 0$ and $C_S^2 = 1$.

Scale-up.

Another possibility is to replace the m facilities by a single facility that has a processing rate of m times a single facility, hence the utilization ρ^* of this facility will be the same as the utilization of an unscaled individual facility, i.e. $\rho^* = \rho$. Assuming that the time scaling hypothesis applies then the mean processing time of product j will be $1/\mu_j^* = 1/(m\mu_j)$ and the squared coefficient of variation will be unchanged at C_S^2. It follows that

(8) $$WIP^{(sc)} = \frac{\rho^2(1 + C'^2)}{2(1-\rho)} + \rho$$

and the condition for scale-up to be better than specialization can be shown to be

$$C_{1/\mu}^2 < (m-1)\frac{2 - \rho(1 - C_S^2)}{\rho(1 + C_S^2)}.$$

For a two facility two product system with each facility processing only one product $C_{1/\mu}^2 = (r-1)^2/4r$. Hence it follows that scale-up is better than specialization if

(9) $$r < r^* = 1 + 2\frac{2 - \rho(1 - C_S^2)}{\rho(1 + C_S^2)}\left\{1 + \sqrt{1 + \frac{\rho(1 + C_S^2)}{2 - \rho(1 - C_S^2)}}\right\}.$$

Fig. 1: Specialization vs. Cyclic
Critical demand ratio (r*)

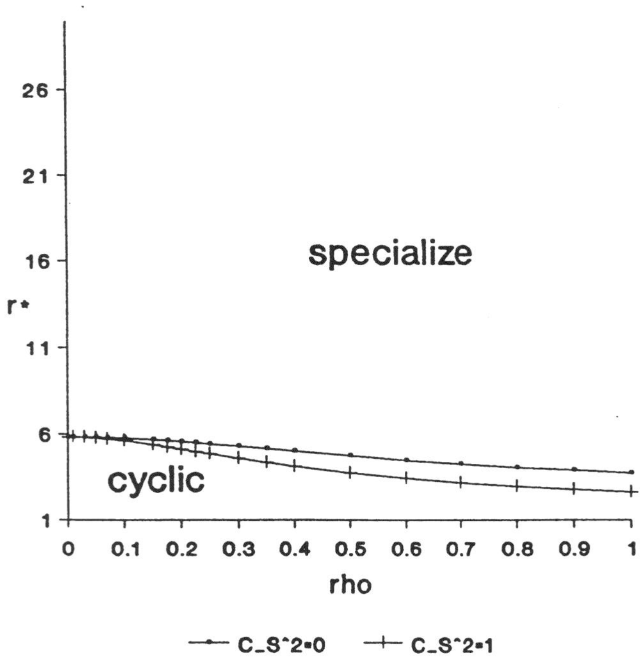

Fig. 2: Specialization vs. Scale-up
Critical demand ratio (r*)

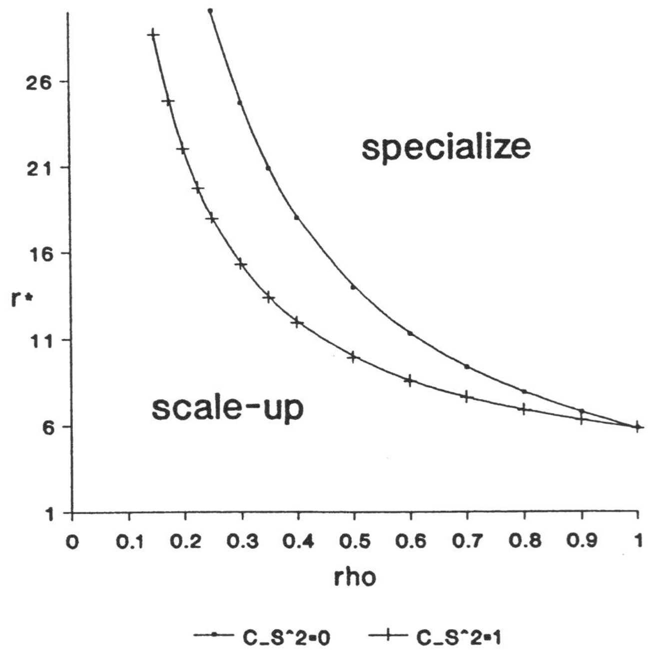

If $C_S^2 = 1$ the condition is

$$r < r^* = 1 + \frac{2}{\rho}\left(1 + \sqrt{1+\rho}\right). \tag{10}$$

Note that r^* is decreasing in ρ and as $\rho \to 1$, $r^* \to 3 + 2\sqrt{2} = 5.828$ independent of C_S^2. At this critical r^* $C_{1/\mu}^2 = 1$. Fig. 2 shows the critical r^* as a function of ρ for $C_S^2 = 0$ and $C_S^2 = 1$. Figure 3 shows WIP as a function of ρ when $r = 10$ while figures 4 and 5 show WIP as a function of r for $C_S^2 = 0$ and $C_S^2 = 1$ respectively when $\rho = 0.8$.

The scale-up results can be used as the basis of an approximation for the performance of a system consisting of m facilities but where there is a common queue of work and jobs are assigned to the facilities on the basis of the facility that completes work first gets the next job (earliest completion time assignment). We use the approximation for a $M/G/m$ queue that $\bar{N}_q$, the average number of jobs waiting in the queue is given by

$$\bar{N}_q = \frac{1 + C'^2}{2}\bar{N}_{M/M/m}$$

where $\bar{N}_{M/M/m}$ is the average number of jobs waiting in an $M/M/m$ queue. If $m = 2$ then $\bar{N}_{M/M/2}$ is given by

$$N_{M/M/2} = \frac{2\rho^3}{1-\rho^2}$$

and hence

$$WIP^{(ect)} = \frac{1 + C'^2}{2}\frac{2\rho^3}{1-\rho^2} + 2\rho. \tag{10}$$

It follows that specialization will be better if

$$C_{1/\mu}^2 > 1/\rho$$

independent of C_S^2. Note that this is the same condition as for specialization to be preferable to scale-up when $C_S^2 = 1$.

Series Expansion

We now consider approaches that require the work to be divided up into steps with each step performed on a different facility.

Fig. 3: Shared vs. Specialized
C_S^2=0, demand ratio=10:1

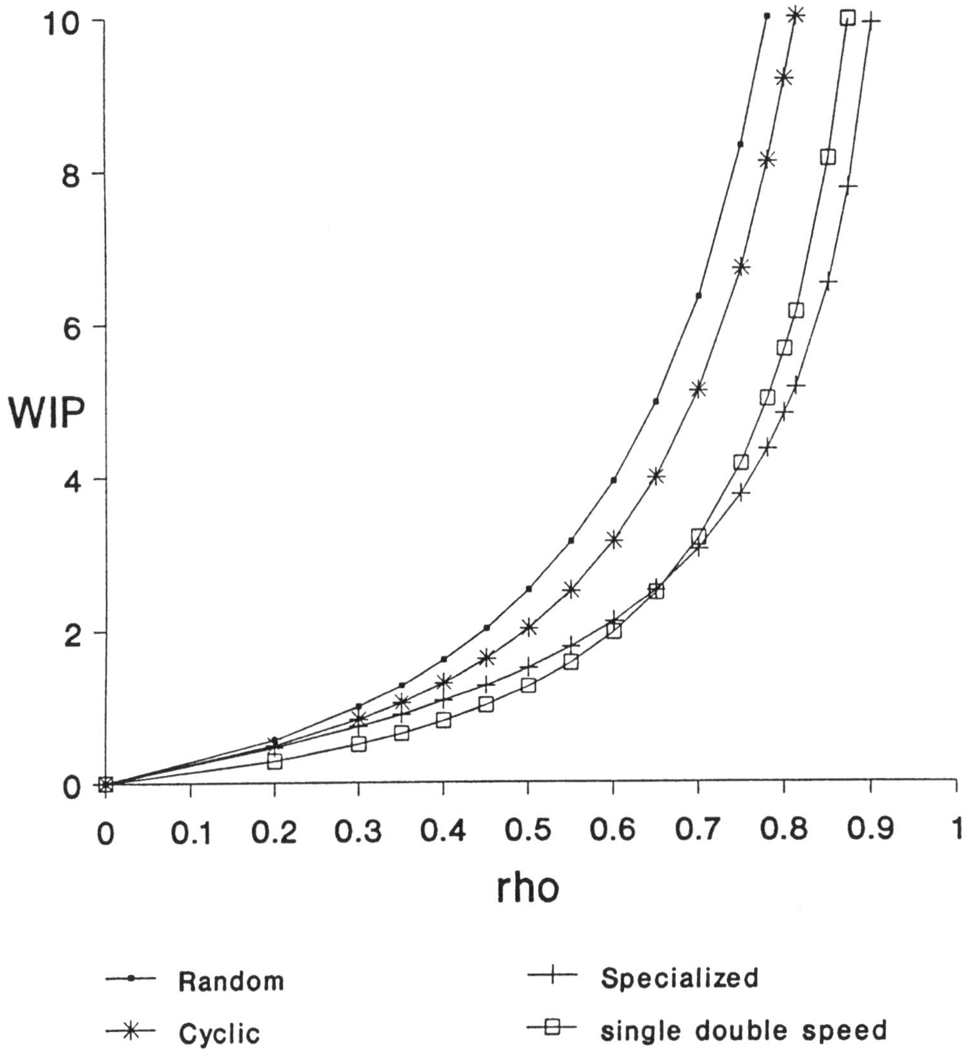

Fig. 4: Shared vs. Specialized
C_S^2=0, rho=.8

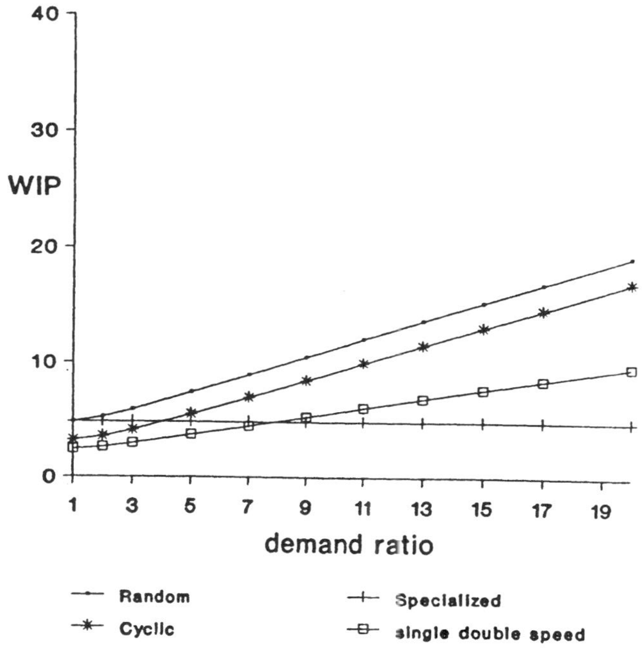

Fig. 5: Shared vs. Specialized
C_S^2=1, rho=.8

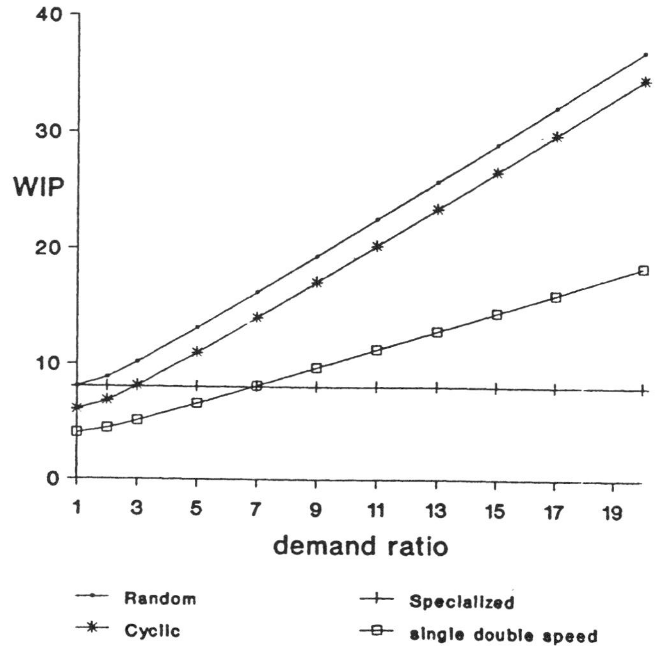

Task allocation.

If increased demand is met by subdivision of the work and allocating specific tasks to each facility then it is necessary to decide on an allocation of work. We will assume that a requirement is that each facility has the same work assignment, i.e that there is balanced load and that such a task allocation is feasible. If the overall work content on product j has mean $1/\mu_j = \sum_{i=1}^{m} 1/\mu_{ij}$ where $1/\mu_{ij}$ is the mean work content of the tasks on product j allocated to facility i, our assumption of balanced load implies that $1/\mu_{ij} = 1/(m\mu_j)$. Now under our assumption of task independence the variance of the total time to produce product j, σ_j^2, will be such that $\sigma_j^2 = \sum_{i=1}^{m} \sigma_{ij}^2$ where σ_{ij}^2 is the variance of the time required by product j on facility i. Since the means are balanced it would be reasonable to assume that $\sigma_{ij}^2 = \sigma_j^2/m$, i.e. that the variances are also balanced. Usually this is the assumption we will make but we will also mention other possible allocations of variance. Note that with balanced variance allocation the squared coefficient of variation of the processing time at facility i, $C_{Sij}^2 = \sigma_{ij}^2\mu_{ij}^2$, will be related to the overall squared coefficient of variation of the total processing time of product j, $C_{Sj}^2 = \sigma_j^2\mu_j^2$, by $C_{Sij}^2 = mC_{Sj}^2$.

Asynchronous Workflow.

We will first consider a series system where jobs flow from facility i to $i+1$ for $i = 1, 2, \ldots, m-1$ and where work flow is asynchronous. There is no limit on storage space for waiting jobs at a facility. Such a system can be modelled by queues in series with the departure process from facility i the arrival process at facility $i+1$. We will use the following two moment approximations to determine WIP_i and C_{di}^2, the squared coefficient of variation of the departure process at facility i,

$$WIP_i = \frac{\rho_i^2(1 + C_{Si}^2)(C_{ai}^2 + \rho_i^2 C_{Si}^2)}{2(1-\rho_i)(1+\rho_i^2 C_{Si}^2)} + \rho_i \tag{11}$$

and

$$C_{di}^2 = 1 + \rho_i^2(C_{Si}^2 - 1) + \frac{(1-\rho_i^2)(C_{ai}^2 - 1)}{1 + \rho_i^2 C_{Si}^2}. \tag{12}$$

With our assumption of balanced allocation of work (and variances) to each facility $C_{Si}^2 =$

mC_S^2 and $\rho_i = \rho$ for all i. Hence the total WIP is given by

$$(13) \qquad WIP^{(a)} = m\frac{\rho^2(1+mC_S^2)}{2(1-\rho)} + m\rho + \frac{\rho^2(1+mC_S^2)}{2(1-\rho)(1+\rho^2 mC_S^2)}\sum_{i=1}^{m}(C_{ai}^2 - 1)$$

and

$$(14) \qquad C_{ai}^2 - 1 = \rho^2(mC_S^2 - 1) + \frac{(1-\rho^2)}{1+\rho^2 mC_S^2}(C_{a(i-1)}^2 - 1) \quad i = 2,\ldots,m.$$

Note that if $C_{a1}^2 \leq 1$ and $mC_S^2 < 1$ that $C_{ai}^2 \leq C_{a(i-1)}^2$ for $i = 2,\ldots,m$, while if $C_{a1}^2 \geq 1$ and $mC_S^2 > 1$ then $C_{ai}^2 \geq C_{a(i-1)}^2$ for $i = 2,\ldots,m$. For small values of C_S^2 the asynchronous system will have lower WIP than the same number of facilities arranged in parallel with random allocation. However it seems that cyclic allocation is always better than the asynchronous system. If $C_S^2 = 0$ then with cyclic allocation $WIP^{(c)} = m\rho + \rho^2/(2(1-\rho))$ while in an asynchronous system $WIP^{(a)} = m\rho + (1-(1-\rho^2)^m)/(2(1-\rho))$ and it can be seen that the cyclic system is always better.

It is of interest to compare the performance with balanced allocation of variance with other allocations. Suppose all variance is allocated to one facility s so $\sigma_{sj}^2 = \sigma_j^2$ while $\sigma_{ij}^2 = 0$ for $i = 1,\ldots,m, i \neq s$. Then it is possible to show that if $C_{a1}^2 \leq 1$ that the total WIP is minimized if $s = m$. This therefore will be a lower bound on WIP for all possible allocations of variance. Note that then $C_{Sm}^2 = m^2 C_S^2$. Fig. 6 shows a comparison of series and parallel expansion for a given ρ as a function of C_S^2.

Synchronous Workflow.

With synchronous work flow all queueing occurs in front of the first facility while each facility always has only one job. With synchronous movement there are two options for determining when movement should occur: (1) *Free transfer*, i.e. wait until all facilities have completed their required tasks; (2) *Fixed transfer*, i.e. move at prespecified constant intervals with the time between successive movements defined as the cycle time. We first consider fixed transfer. Note that with fixed transfer the cycle time may not be sufficient for all tasks to be completed at each facility, thus a quality goal has to be considered in specifying the cycle time. Assuming the time required to complete all tasks at each

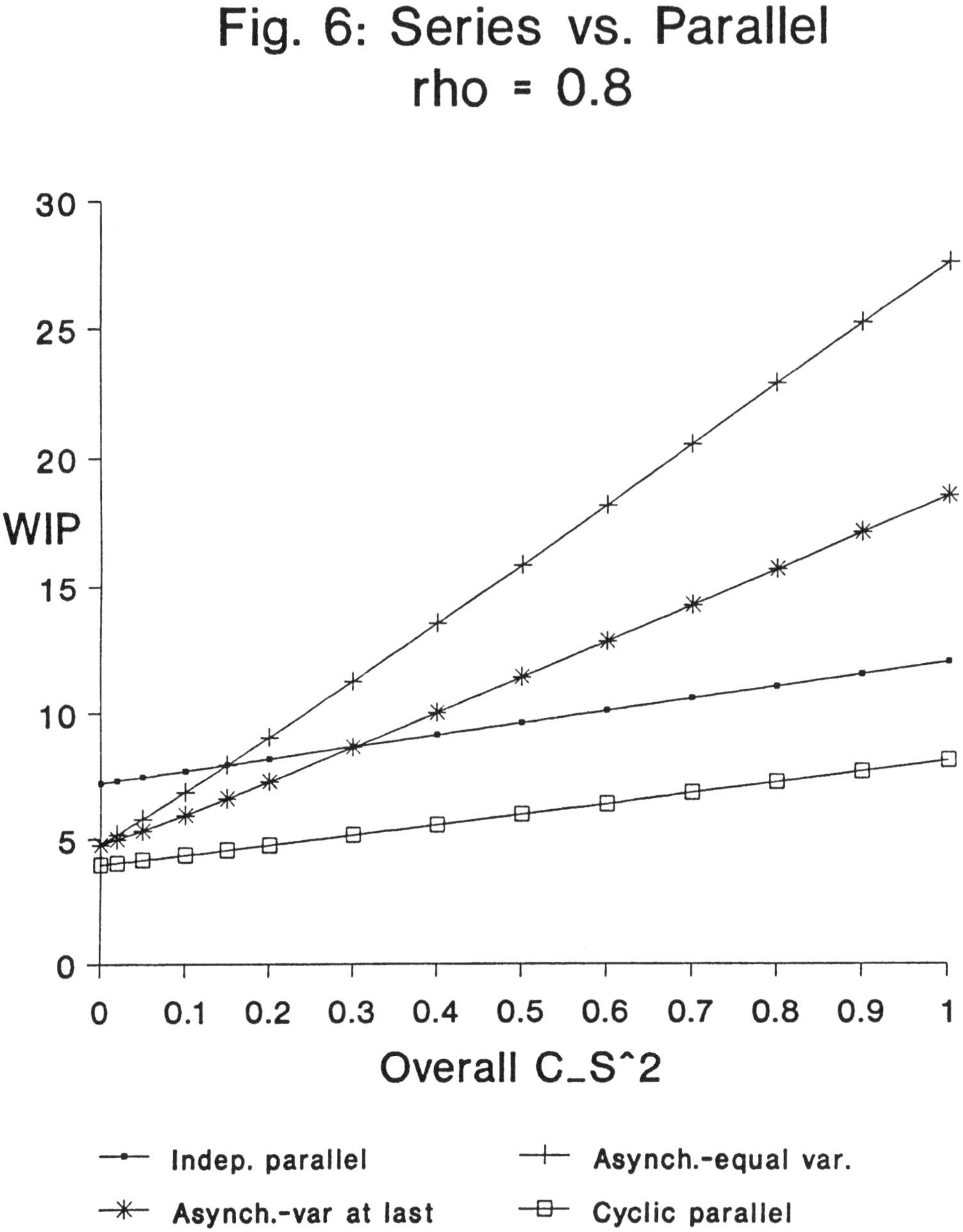
Fig. 6: Series vs. Parallel
rho = 0.8
WIP
30
25
20
15
10
5
0
0
0.1
0.2
0.3
0.4
0.5
0.6
0.7
0.8
0.9
1
Overall C_S^2
Indep. parallel
Asynch.-equal var.
Asynch.-var at last
Cyclic parallel

facility are independently and identically distributed with distribution $F_i(t) = F(t)$ for $i = 1, \ldots, m$ then for a given goal Q for the probability of completing all tasks at all work stations the cycle time τ has to be such that $\prod_{i=1}^{m} F_i(\tau) = Q$ or $F(\tau) = Q^{1/m}$. If $\bar{F} = 1 - F$ then this will be approximated by $\bar{F}(\tau) = (1 - Q)/m$. Write $\tau = (1 + k(Q,m)\sqrt{mC_S^2})/(m\mu)$. Note that $TH = Q/\tau < m\mu$ where $m\mu$ is the throughput of the asynchronous system. Because τ is constant the number of waiting jobs can be found by regarding the series facilities as equivalent to a single server with constant service time τ

$$\bar{N}_W = \frac{\rho^2(1 + k(Q,m)\sqrt{mC_S^2})^2}{2(1 - \rho(1 + k(Q,m)\sqrt{mC_S^2}))}. \tag{15}$$

If $C_S^2 = 0$ it is clear that synchronous transfer will have a lower number of waiting jobs than asynchronous transfer as it has the same initial queue and eliminates all queueing at other facilities. However, once $C_S^2 > 0$ the superiority disappears rapidly because the increase in the apparent utilization of the server due to the difference between τ and $1/m\mu$ has a greater effect than the reduction in queueing due to the reduced variability of service time.

With free transfer the system can similarly be regarded as equivalent to a single server whose service time is the distribution $(F_S(t))$ of the maximum of the times required at each facility with $F_S(t) = \prod_{i=1}^{m} F_i(t) = F^m(t)$. Because there is always sufficient time to complete the tasks at a facility $Q = 1$. The variability of the maximum will be somewhat less than the variability of the processing times at the individual facilities but the mean of the maximum will be greater than the individual means. These two effects mean that free transfer is better than asynchronous systems for small C_S^2 and ρ. The parameters of the single server equivalent to the free transfer system can be found using approximations from extreme value statisitics. If t_m is such that $F(t_m) = 1/m$ and near t_m it is possible to approximate $\bar{F}(t)$ by $\bar{F}(t) = e^{-\alpha(t-t_m)}/m$ the mean $\bar{t}$ and variance $\sigma^2(m)$ of the maximum will be given by

$$\begin{aligned} \bar{t} &= t_m + 0.577/\alpha \\ \sigma^2(m) &= (1/6)(\pi/\alpha)^2 \end{aligned} \tag{16}$$

For a normal distribution of facility processing times with variance σ^2, $\alpha = k/\sigma$ with k increasing slightly with m and is approximately 1.5 -2.0. If $m = 3$ then $\bar{t} \approx 1/(m\mu) + 0.85\sigma$

and $\sigma^2(m) \approx 0.6\sigma^2$ while when $m = 6$ then $\bar{t} \approx 1/(m\mu) + 1.3\sigma$ and $\sigma^2(m) \approx 0.5\sigma^2$. Note that, as with fixed transfer $TH < m\mu$.

Fig. 7 illustrates the performance of both fixed transfer and free transfer series systems as a function of C_S^2 for given ρ while Fig. 8 illustrates how the performance depends on ρ for given C_S^2 for $m = 3$. Note that it is only if C_S^2 is very small and ρ is near 0.6 that the synchronous systems are noticeably better than asynchronous systems. Note that the average number of waiting jobs has been used in comparing asynchronous and synchronous systems. This is because in some synchronous systems it often makes operational sense to always have a job at each facility, rather than letting the system empty out and refill.

Randomized Routing.

In a series asynchronous system, if the squared coefficient of variation of processing times at a facility is greater than one then the variability of the departure process increases along the line and this will result in average queue lengths increasing along the line. Thus if it is technologically feasible it could be preferable to introduce some randomization into the routing of the jobs. If tasks can be done in any sequence then this is easy to do by splitting up jobs into classes with each class corresponding to one of the $m!$ possible routings. Arriving jobs can be allocated to classes randomly, although there would be better allocations that take into account the queue sizes at each facility and such allocations can be implemented by a material handling system that permits central dispatch.

The system can be analysed following the approach described in Shanthikumar and Buzacott (1981). With randomized routing and all facilities balanced, the squared coefficient of variation of the arrival process at facility i will be C_{ai}^2, the same for all i. Similarly the squared coefficient of variation of the departure process, C_{di}^2, will be the same for all i. The arrival process at facility i will then be the composition of the stream of arrivals of new jobs and the streams coming from the other facilities. Since the selection from the departure stream from l of jobs to go to facility i is equivalent to Bernoulli trials, we have that the scv. of this stream will be $(m-1)/m + C_{dl}^2/m$. Since the external arrivals are allocated at random to facilities the scv. of external arrivals to i will be $(m-1)/m + C_a^2/m$. For the combination of all the streams arriving at i we will use the asymptotic approximation

Fig. 7: Asynchronous vs. Synchronous
Waiting Jobs: 3 stn. rho=.6

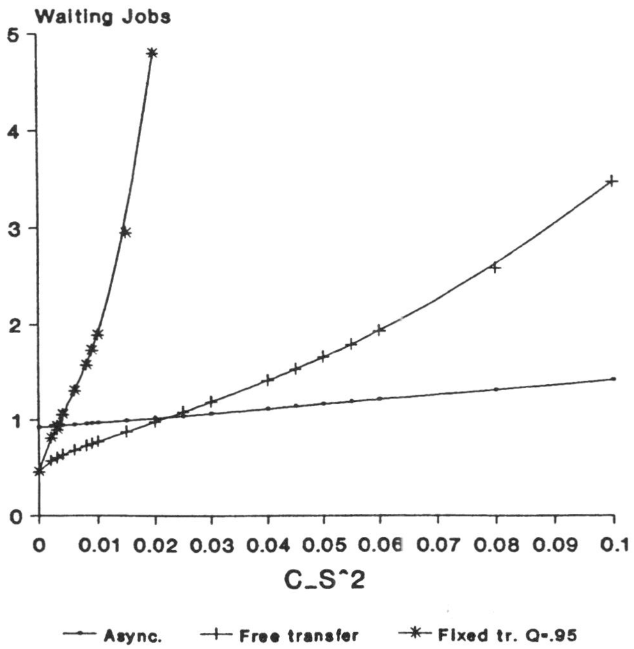

Fig. 8: Asynchronous vs. Synchronous
Waiting Jobs: 3 stn. C_S^2=.005

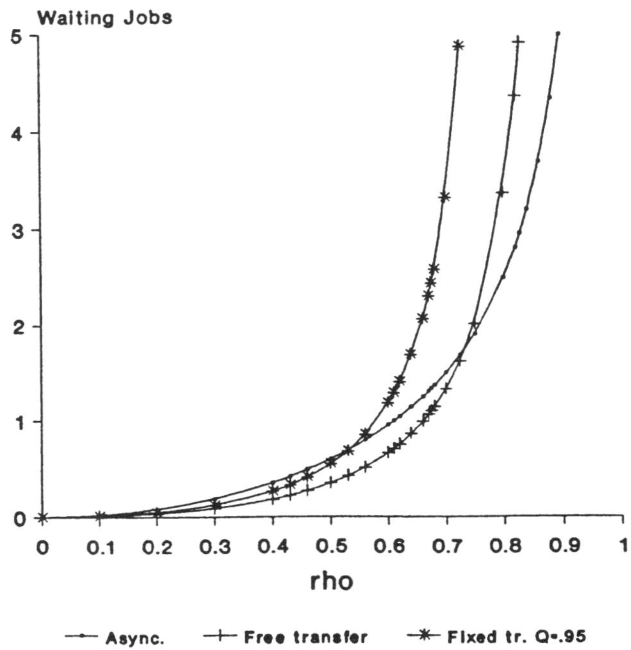

(Whitt 1982) so that its scv. is such that

$$C_{ai}^2 = \frac{m-1}{m}\left(\frac{m-1}{m} + \frac{1}{m}C_d^2\right) + \frac{1}{m}\left(\frac{m-1}{m} + \frac{1}{m}C_a^2\right)$$

where we have set $C_{dl}^2 = C_d^2$ for all $l \neq i$. For $C_a^2 = 1$ this equation can be rewritten as

$$C_{ai}^2 - 1 = \frac{m-1}{m^2}(C_d^2 - 1).$$

Now using the result for C_d^2 from above

$$C_d^2 - 1 = \rho^2(C_{Si}^2 - 1) + \frac{(1-\rho^2)}{1+\rho^2 C_{Si}^2}(C_{ai}^2 - 1)$$

it is possible to solve for C_{ai}^2 in terms of ρ^2 and C_{Si}^2 giving

$$C_{ai}^2 - 1 = \frac{(m-1)\rho^2(C_{Si}^2 - 1)/m^2}{1-(m-1)(1-\rho^2)/(m^2(1+\rho^2 C_{Si}^2))} \tag{17}$$

and hence WIP can be found using

$$WIP^{(r)} = m\left(\rho + \frac{\rho^2(1+C_{Si}^2)(C_{ai}^2 + \rho^2 C_{Si}^2)}{2(1-\rho)(1+\rho^2 C_{Si}^2)}\right). \tag{18}$$

A modification of this approach could be used to represent the cyclic assignment of arriving jobs to the facilities. One could then write

$$C_{ai}^2 = \frac{m-1}{m}\left(\frac{m-1}{m} + \frac{1}{m}C_d^2\right) + \frac{1}{m}\frac{1}{m}C_a^2$$

which when $C_a^2 = 1$ simplifies to

$$C_{ai}^2 - 1 = \frac{m-1}{m^2}(C_d^2 - 2)$$

Fig. 9 shows a comparison of randomized routing with other configurations. In a central dispatch system where jobs always return to a central dispatch point after processing at a facility and tasks can be performed in any sequence the WIP would be less than the above results. Indeed it may approach the performance of a parallel system where work is assigned to the first available facility.

Fig. 9: Random Routing vs. Parallel
3 machines rho = 0.8

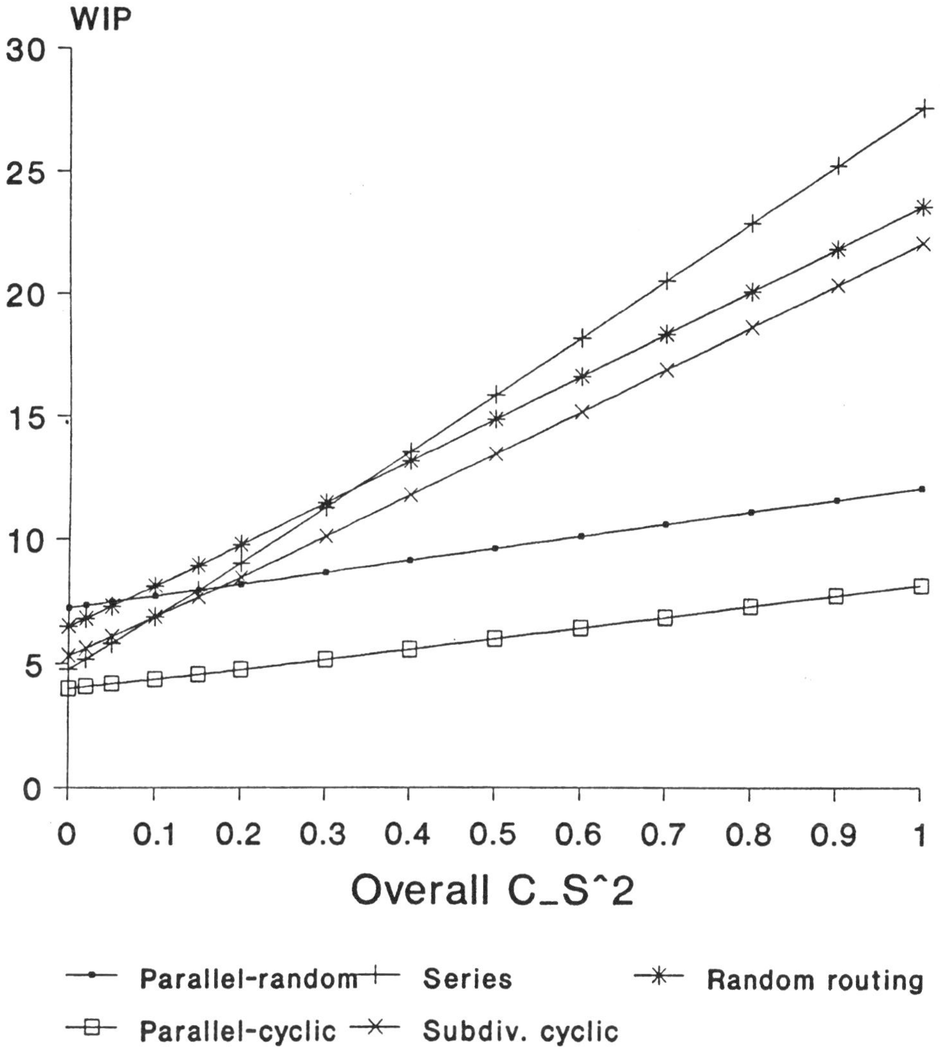

Applicability of Models.

Although only series and parallel expansion have been modelled, the results can be extended to more complex systems. For example, the comparison between a multiple product series system and specialized single product systems which may be single facility or multiple facilities in series would be expected to exhibit similar properties to the comparisons beween specialized and multiple product systems presented above. Similarly, it would also be expected that similar conclusions would be obtained if multiproduct randomly routed job shops are compared with a set of single product cells.

Product Variety.

In all these situations (single facility, flow lines, job shops) it would be expected that the multiproduct system is preferable only if the range of processing times of the different products is relatively narrow. A wide diversity of processing times creates unavoidable variance in service times and hence results in high levels of WIP. This in turn makes it difficult to provide good delivery performance. This observation "explains" why it does not make sense to contemplate manufacturing systems with a large product variety unless the product mix is very uncertain and unstable. With stable product mix, such as has charcterized many US installations of FMS, then the models support Jaikumar's observation on the relatively small product variety that is typical of these systems (Jaikumar 1986). It also supports the Soviet introduction of group technology and the division of large job shops into smaller cells with limited product variety (see Petrov 1968). On the other hand, if product variety is unavoidable, for example because product mix is unstable (this is mentioned by Petrov as one of the difficulties with introducing Group Technology ideas), then the high variability of processing times implies that a parallel system or a system with central dispatch (if there is any diversity in routings or operation sequence is not fixed) would be preferable.

It is of interest to note that increasing the scale of facilities (increasing their operating speeds) only compensates partly for the disdvantages inherent in multiple products even when there is no set-up or changeover times. If facility utilization is high then scale-up is not particularly beneficial. Economies of scale in capital or operating costs may give a cost advantage to scale-up, but if service is highly weighted in comparison with other costs then there will always be niches for the specialized producer, particularly when facility utilization at the multiple product producer is high. However, when utilization is low, either because of low demand or because a new scaled-up facility has just been installed and has not yet obtained reasonable load, then the specialized producer is in a weak competitive position.

Division of Labour.

Our analysis shows that subdivision of work or series expansion is most attractive if the variability of processing times is low, e.g. $C_{Si}^2 \approx 0.01$. In these circumstances synchronous transfer can be preferable to asynchronous transfer. Note that fixed transfer is always worse than free synchronous transfer even apart from the quality problems that it introduces. However, low variability of processing times is difficult to achieve unless product variety is very low. Also it can conflict with attempts to give workers responsibility for quality because finding and repairing defects would introduce an added source of variability in processing times. If this is considered desirable then it would be better to use asynchronous transfer, such as can be achieved with power- and-free conveyors or AGV based assembly systems.

Subdivision of processing into smaller tasks may make it easier to mechanize or automate production. This in turn is likely to reduce the variability of processing times, however, it can introduce a further source of variability due to machine breakdowns and their subsequent repair. So it is necessary to break the system up into segments or stages with sufficient space for interstage buffers so that the segments operate asynchronously. Some operations, such as test and rework are inherently variable and thus have to be buffered from the rest of the system.

However, if individual processes exhibit a high degree of variability, whether it is due to product mix, machine failures, or inherent in the requirements of the operation, in particular if $C_{Si}^2 > 1$, then central dispatch systems should be considered as it smooths out the impact of variabiity from whatever source and enables available storage space to be used more effectively.

Work assignment.

Independent random assignment does not perform as well as other work assignments. As would be expected a work assignment rule that either is based on agreement or cooperation between the facilities on sharing work, such as cyclic assignment, or a rule that uses information about the current work load, such as assigning to the facility with the shortest queue, or that defers assignment until necessary to avoid idle facilities when there is available work, such as assigning to the facility with the earliest completion time, perform better. There is the implication that any collaboration between producers is desirable unless they have specialized to unique products because of the diversity of products. Thus it is not surprising that coöperatives have developed in a number of craft and small scale industry sectors. Further, such systems can achieve equity between facilities which is sometimes a desirable goal in spite of individual differences. However, note also that allocations that are based on achieving equity over the long run, such as earliest completion time assignment, are better in our model than assignments that require equity to be maintained over the short run, such as cyclic assignment. Note, however, that allocations that aim at equity over the long run can result in incentives to slow down working rate (an aspect not included in our models).

In summary, our models support the trends towards increasing parallelism in manufacturing system structure combined with increasing cooperation and coordination of the parallel facilities as an effective means of coping both with inherent variability within the system and variability due to the product mix. Nevertheless the models do not in themselves answer the question as to whether the current North American tendency to use manufacturing system structures that in themselves cope with variability is preferable to the Japanese tendency to seek to eliminate sources of variability so that series systems with subdivided work become attractive.

ACKNOWLEDGEMENTS

This research was supported by the Natural Sciences and Engineering Research Council of Canada. Many of the ideas arose out of discussions and other joint work with George Shanthikumar of the University of California, Berkeley, but the author takes sole responsibility for the use (or abuse) of the stochastic models.

REFERENCES

Boucher, T. O. (1988), "Adam Smith and the humanists: An enquiry into the productivity of labor controversy," IIE Trans., Vol. 20, 73-82.

Buzacott, J. A. (1990), "Abandoning the moving assembly line: Models of human operators and job sequencing," Int. J. Prod. Res., Vol. 28, 821-835.

Buzacott, J. A., and Chaouch, A. B. (1988), "Capacity expansion with interrupted demand growth," European J. of Operational Res. Vol. 34, 19-26.

Fine, C. H., and Freund, R. M. (1990), "Optimal investment in product-flexible manufacturing capacity," Management Science, Vol. 36, 449-466.

Freidenfels, J. (1980), "Capacity expansion when demand is a birth-death process," Operations Res., Vol. 28, 712-721.

Goldhar, J. D., and Jelinek, M. (1983), "Plan for economies of scope," Harvard Business Review, Vol. 61, November-December, 141-148.

Jaikumar, R. (1986), "Post-industrial manufacturing," Harvard Business Review, Vol. 64, November-December, 301- 308.

Marchal, W. G. (1975), "An approximate formula for waiting time in single server queues," AIIE Trans., Vol. 8, 473.

Petrov, V. A. (1968), *Flowline Group Production Planning*, London: Business Publications.

Shanthikumar, J. G., and Buzacott, J. A. (1981), "Open queueing network models of dynamic job shops," Int. J. Prod. Res., Vol. 19, 255-266.

Whitt, W. (1982), "Approximating a point process by a renewal process. 1: Two basic methods," Operations Res., Vol. 30, 125-147.

Performance Evaluation of Flexible Manufacturing Systems with Starving

HORST TEMPELMEIER, HEINRICH KUHN and ULRICH TETZLAFF

1. Introduction

Real-life flexible manufacturing systems (FMS) often suffer to a considerable extent under utilization losses which are due to limited local and/or central buffer space and the limited velocity of the material handling system (MHS). In particular two forms of performance deterioration are distinguished: ***blocking and starving***. A workstation is blocked if it cannot dispose of a finished piece because there is no buffer space - neither local nor in the central buffer. Blocking has been studied by several researchers (Akyildiz 1988a, 1988b, Altiok and Perros 1986, Suri and Diehl 1986, Gershwin and Schick 1983, Perros 1984, Yao and Buzacott 1985, 1986). In an earlier paper, the authors have made a proposal to estimate the influence of blocking on the performance of an FMS (Tempelmeier, Kuhn and Tetzlaff 1989) using standard closed queueing network algorithms. According to our experience, in industrial practice blocking often does not seem to be a severe problem during the configuration phase of an FMS, because in many cases it is possible to allow for enough central buffer space at moderate costs to prevent blocking. However even if the central buffer space is large enough, the limited velocity of the material handling system (MHS) in combination with limited local buffer space may lead to a form of reduction of machine utilization that is known as *starving*.

Moreover, in real-life FMS *two additional factors* must be considered that have a direct influence on the workload of the MHS and thus indirectly upon the extent of starving. These are the ***trips to and from the central buffer*** and the ***empty trips*** of the MHS carts. Both types of trips constitute a form of "workload" for the MHS that is not known in classical closed queueing network models widely used to evaluate the performance of an FMS. In industrial practice however they may play an important role. Therefore a system planner often spends a considerable amount of time estimating the mean transportation time of a workpiece as a function of the process plans and the layout of the FMS under consideration.

2. Starving in a flexible manufacturing system

21. General description of the starving phenomenon

In order to analyze the process of starving it is assumed that the central buffer is unlimited and therefore blocking is impossible. If the local buffer space at a particular downstream machine is limited, the situation may arise that a workpiece to be processed on this machine next must wait in the local buffer of the upstream station or is stored in the central buffer until the downstream machine is ready to take over the piece. Starving begins at the point in time when the downstream machine is ready to process the piece *and* the piece is still located elsewhere in the system, e.g. waiting or in transit to the machine. Note that starving must be distinguished from a machine's normal waiting for the transportation of a piece that in the classical closed queueing network model is allowed for by modeling the MHS as a central server with "processing time" equal to the mean transportation time.

The scheduling logic used in an FMS considered is as follows. A piece has been processed on a machine and after a waiting time *has been allocated to a MHS cart* and now wants to be transported to its next downstream station. However due to lack of local buffer space at the downstream machine, access is denied and the piece is transported to the central buffer or remains waiting in the actual station's local buffer, in which case the cart is immediately released. Now, at the point in time, when the downstream station has made room free to take over the piece considered, the piece first must be transported to the downstream station. Before processing of the piece can start, the machine in the downstream station may suffer from several delays: there may occur a ***waiting time*** before allocation of a cart, an ***idle trip time*** of the cart and the ***transportation time*** of the piece to the machine. All or parts of these times may occur while the machine is waiting for the workpiece to arrive.

To clarify the conditions of starving of a machine, two situations must be distinguished:

- ***Case 1: starving of a machine with one local buffer space***

  If a station has only one local buffer space (this is the slot reserved for piece exchange), starving begins immediately after a workpiece has been allocated to a machine in the station. The machine is starved between the instant of the piece's allocation to the machine and its arrival at the station. Figure 1 illustrates the process of starving in this case. Initially at the instant t0 when machine 2 completes processing of piece B the downstream machine 1 considered is busy with workpiece A until time t1 and denies access to piece B. Therefore the MHS transports piece B to the central buffer. In time t1 piece B is allocated to machine 1. Now, as piece B is located in the central buffer, it first must be transported from the central buffer to the machine, while the finished piece A has to be transported from machine 1 to its downstream station. If the MHS is heavily utilized, it is possible that both pieces must wait in the MHS queue while the MHS is busy transporting other pieces somewhere else in the FMS. In this case the starving time of machine 1 consists of the *waiting time* [t1,t2], the time for the disposal trip of piece A [t2,t3] and the *transportation time* [t3,t4] of piece B from the central buffer to its downstream machine 1.

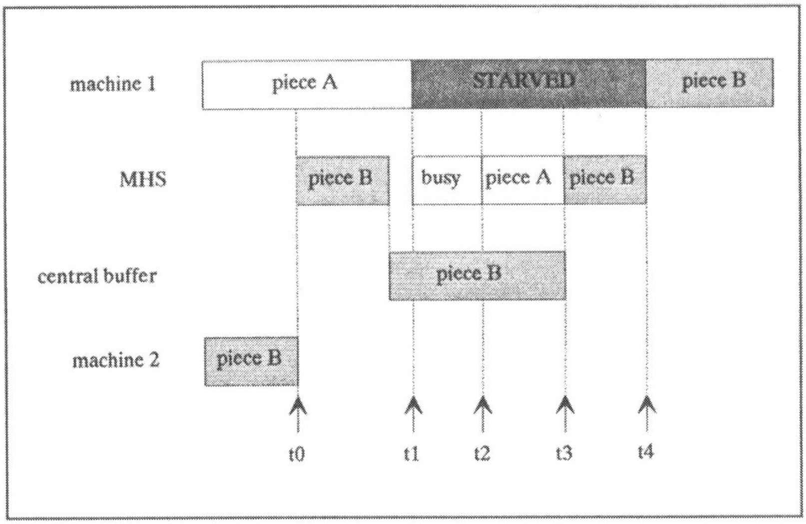

Fig. 1: Starving of a machine with one local buffer

- ***Case 2: starving of a machine with several local buffer spaces***

If there are several local buffer spaces at a station, then there may be some unprocessed pieces waiting in these buffers and the machines may have a certain amount of workload to process until a newly allocated piece arrives at the station. It is assumed that the allocation of this next piece to its downstream station immediately takes place when a local buffer space is freed and the station therefore is ready to take over the next unprocessed piece. In this case, starving of the machine does not begin with the entering of the piece in the MHS queue but it is delayed until the instant the work content of the downstream station is empty. Fig. 2 illustrates the situation considered. Suppose that piece B has been stored in the central buffer some time ago (t0), when its downstream station (machine 1) refused access. According to the logic of the FMS scheduling system, at time t1, when piece

C frees a local buffer place, piece B is allocated to the downstream station (machine 1) and enters the MHS queue. However, as the MHS is busy transporting the just finished piece A to its next machine, piece B must wait in the MHS queue until time t2. In the meantime machine 1 processes piece C which was already waiting in the local buffer. In the situation depicted, however, the work contents of station 1 resulting from piece C [t1,t3] is less than the waiting and transportation time of piece B [t1,t4]. Consequently, machine 1 is starving from the time it is released by piece C (t3) until the arrival of piece B (t4).

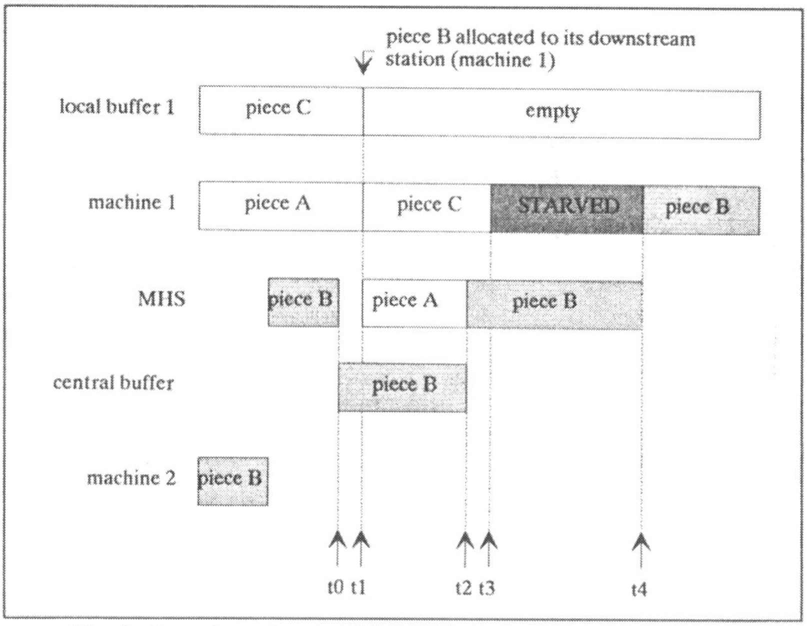

Fig. 2: Starving of a machine with several local buffers

Obviously in both cases considered the duration of starvation is influenced by the *velocity* of the MHS, its *utilization* and the *number of local buffer spaces* at the machines in the FMS. If the mean transportation time is low compared to the mean processing time of the machines, then starving will rarely be observed. However, if the MHS is highly utilized, then the pieces often queue up for transportation and the machines are frequently starved.

22. Idle trips

A MHS cart has to perform an idle trip, when its position at the moment of allocation to a piece to be transported is not identical with the pieces position. An idle trip is not necessary, if the cart can take over a new piece at the same station where it just delivered a piece. The ratio of idle trips to the total MHS utilization mainly depends on the *number of pallets* circulating in the FMS, the *number of stations* and the *number of MHS carts*.

23. Trips between the machines and the central buffer

If there is a central buffer in the FMS, then an additional workload of the MHS arises due to trips to and from the central buffer. In the sequel we will call trips from a productive station to the central buffer ***disposal trips*** and trips to a productive station ***supply trips.*** The latter may start at a productive station or at the central buffer. This type of MHS workload is completely ignored in the classical CQN approach which only knows supply trips between productive stations. The number of trips between the central buffer locations and the productive stations of the FMS depends on the necessity to use the central buffer as an overflow region which in turn is mainly influenced by the number of local buffer places at the machines.

If the central buffer is large enough (unlimited) and there is no space available at the downstream station, then according to the scheduling logic a piece always is transferred to the central buffer. Therefore a station M_i is fed with raw material via the central buffer, if its local buffer space is filled. If the number of central buffer places is limited, then it is possible that with a certain number of pallets circulating in the system a piece is refused access to its downstream station ***and*** to the central buffer with the consequence of no need to transport the piece at all - at least at that point in time. This situation leads to a reduction of the MHS workload described by the mean transportation time. As a consequence, in an FMS with *limited central buffer* space the production rate may first decrease and then increase with the number of pallets circulating in the FMS.

3. Starving in the classical CQN model

So far we have characterized the factors influencing the performance of an FMS in situations where starving may occur. In what follows a two-stage procedure to quantify these factors is developed. Because of its heavy influence on the total "real" workload of the MHS the ***idle trip times*** and the ***disposal trip times to the central buffer*** first are approximated and used to modify the mean transportation time as employed in the classical CQN model. Then the station-specific mean starving times are estimated and added to the mean processing times at the stations. With the transportation and processing times modified in this way a classical CQN algorithm is employed to compute the performance measures of an FMS under consideration.

The MHS activities may be ordered in three groups as follows

- *supply trips* (loaded trips between two productive stations; supply trips from the central buffer to a productive station)
- *disposal trips* (trips from a productive station to the central buffer)
- *idle trips*.

Each of these activities uses up a certain amount of the total available time of the MHS.

31. Idle trips

An idle trip of a MHS cart is necessary when the location of the cart is not identical with the location of the piece to be transported at the instant of allocation. In an FMS with only **one pallet** circulating idle trips cannot occur because the cart accompanies the pallet through the system. In an FMS with **two pallets** (and one cart) and an even spatial distribution of trips between the two pieces the cart is with probability 0.5 located at the station of the piece to be transported. With increasing number of pallets in the FMS the probability of undertaking an idle trip further reduces. If the number of stations is smaller than the number of pallets N the necessity of idle trips is determined by the number of stations. Based on these arguments we estimate an idle trip factor as follows:

$$F_L(N,M) = 1-\max\{1/N, 1/M_M\} \tag{1}$$

↑ idle trip factor (as a function of the number of pallets and stations)

M_M is the number of stop points in the FMS (productive stations and central buffer places) while M is the number of productive stations. If there are several MHS carts (S_1) in the FMS, then the idle trip factor further reduces as follows:

$$F_L(N,M,S_1) = \max\{1-\max[1/N, 1/M_M]\cdot S_1, 0\} \tag{2}$$

↑ number of MHS carts (S_1)

↑ idle trip factor (as a function of the number of pallets, stations and MHS carts) (F_L)

The *mean transportation time including the estimated idle times* is approximated with eqn. (3):

mean idle trip time ↓ ($[B_1\cdot F_L(N,M,S_1)]$)

$$T_L = B_1+[B_1\cdot F_L(N,M,S_1)] \tag{3}$$

↑ idle trip factor ($F_L(N,M,S_1)$)

↑ mean transportation time (load trips) (B_1)

↑ mean transportation time (incl. idle trips) (T_L)

In an analogous way the *disposal trip time* introduced in the next chapter is inflated by the idle trip factor [see eqn. (4)].

$$B_{ZL} = B_Z + [B_Z \cdot F_L(N,M,S_1)] \tag{4}$$

↑ mean disposal trip time (incl. idle trips)
↑ mean disposal trip time (load trips)
↑ idle trip factor

32. Disposal trips between productive stations and the central buffer

Disposal trips from a productive station to the central buffer are necessary, when the local buffer capacity of a downstream station is used up. The mean total *number of central buffer places needed*, U, is approximated by eqn. (5).

$$U = \sum_{j=2}^{M} \left\{ \sum_{n=1}^{N} \text{Prob}\{n_j=n\} \cdot \max\left[0, [n-(N_{max(j)})]\right] \right\} \tag{5}$$

↑ total number of central buffer places needed
↑ number of pallets at station M_j
↑ available space at station M_j

The mean *number of disposal trips* induced by to limited local buffer space at a station M_i is given by the product of the visiting frequency r_i of pieces and the probability, that the local buffer in this station is occupied [$\text{Prob}\{n_i > N_{max(i)}\}$].
In case of *unlimited central buffer space* the mean proportion of disposal trips to the central buffer depends only on the population of the productive stations, as expressed by eqn. (6).

$$F_{Z'} = \sum_{i=2}^{M} [r_i \cdot \text{Prob}\{n_i > N_{max(i)}\}] \qquad U \le N_{max(ZP)} \tag{6}$$

↑ proportion of disposal trips (unlimited central buffer)
↑ visiting frequency at station M_i
↑ probability that downstream station M_i is full

However, if the *central buffer space is limited* and fully utilized at a certain point in time, then a disposal trip is not possible because in this case a necessary condition for blocking is satisfied. To reflect this possibility, the proportion of disposal trips described in eqn. (6) is multiplied with the probability, that the central buffer is able to take over the piece considered. However this probability cannot be derived from the solution of a classical CQN model. Therefore a central buffer availability factor P_Z is introduced [see eqn. (7)].

$$P_Z = \begin{cases} 1, & N \le N_{max(ZP)} + \min\{N_{max(i)}, i=2,\ldots,M\} \\ N_{max(ZP)}/N & N > N_{max(ZP)} + \min\{N_{max(i)}, i=2,\ldots,M\} \end{cases} \tag{7}$$

↑ availability of the central buffer

This factor assumes the value 1, if the number of pallets in the FMS is low, such that an eventually existing limitation of the central buffer space does not influence the number of disposal trips. With increasing number of pallets circulating in the FMS the central buffer availability factor reduces accordingly.

Using the factors introduced, the *proportion of disposal trips* is approximated by eqn. (8):

$$F_Z = \sum_{i=2}^{M} [r_i \cdot Prob(n_i > N_{max(i)})] \cdot P_Z \qquad (8)$$

- P_Z: availability of the central buffer
- $Prob(n_i > N_{max(i)})$: probability that the downstream station M_i is full
- r_i: visiting frequency of station M_i
- F_Z: proportion of disposal trips (limited central buffer)

In order to reflect the influence of disposal trips to the central buffer the mean transportation time T_L is incremented by the mean disposal trip time [see eqn. (9)].

$$T_{ZL} = T_L + (F_Z \cdot B_{ZL}) \qquad (9)$$

- B_{ZL}: mean disposal transportation time (incl. idle trips)
- F_Z: disposal trip factor
- T_{ZL}: total transportation time (incl. idle trips and disposal trips)

The estimation procedure described was tested with numerous simulation models depicting several FMS of varying complexities and structures and found to give satisfying results.

33. Quantification of the starving conditions and computation of the mean starving time

So far we have only discussed preliminary aspects to the intended approximation of starving times. In what follows the process of starving will be discussed in detail. As has been pointed out starving occurs when a piece allocated to a machine physically is not available at the machine. In order to approximate the starving time of a machine, the **work contents at a station** at the beginning of the piece's MHS request must be compared to the **mean transportation delay**.

a) *work contents at a station*

The work contents at a station depends on the *mean processing time* and the *local buffer space*. In case of only one local buffer space the local work contents is equal to zero, because according to the scheduling strategy the buffer is to be held free to allow for part exchange between the MHS cart and the station.

As pointed out earlier, starving occurs, when the *downstream station* of a piece at the instant of the piece's end of processing at the upstream station *is not able to take over* the piece which then is forced to wait somewhere in the FMS for the downstream station to become free. If starving occurs, a station must be fully occupied, and the relative *work contents* V_i waiting for a machine at station M_i can be described by eqn. (10).

$$V_i = (P_i - 1) \cdot B_i / S_i \qquad M_i = 2, \ldots, M \qquad (10)$$

- S_i: number of machines (servers) in station M_i
- B_i: mean processing time in station M_i
- P_i: number of local buffer spaces in station M_i available for unprocessed parts
- V_i: local work contents at station M_i

Note that it is assumed, that the local buffer is fully occupied by *unprocessed* pieces. However, if the central buffer has limited capacity, then it may happen, that in the local buffer there are *processed* workpieces too. For simplicity this possibility is neglected.

b) *Time between piece exchange and arrival of the next unprocessed piece (transportation delay)*

Immediately after the processing of a piece at a station the FMS control computer creates an order for the MHS to transport the processed piece (A) to its next destination. In addition there is already a second MHS order waiting in the MHS queue to transport an unprocessed part (B) to the station under consideration. We assume that the order (A) is placed in the MHS queue immediately in front of the order (B). The total time between the piece exchange at the station and the arrival of the next unprocessed piece (B) depends on the *mean transportation delay* and the *waiting time* of the MHS orders (A and B) in the MHS queue. If first MHS order A is processed and then MHS order B, the following elements of the ***transportation delay*** may be observed:

- ***waiting time of the processed piece*** (A) for an MHS cart;
- ***transportation time of the processed piece*** (A) to the downstream station (normal load trip) or to the central buffer (disposal trip);
- ***transportation time of the unprocessed piece*** (B) to the station under consideration (supply trip).

α) ***waiting time***
The *waiting time* for the MHS cart is the product of the mean transportation time and the mean number of MHS orders located *in front of* the two orders A and B in the MHS queue. If the MHS is heavily utilized, then the MHS queue will become very long and it is possible that two unprocessed pieces (B) waiting for transportation *to the same station* are residing in the MHS queue. As starving ends when the *first* unprocessed piece arrives at the station, we only have to consider the head of the MHS queue. At a certain point in time the MHS might look as depicted in figure (3).
Let NS_1 be the mean total number of MHS orders waiting in the MHS queue, including the two orders A and B located somewhere in the queue. In

order to compute the mean waiting time of order A, we have to compute the mean length of the *head of the queue* - not considering the orders A and B. If we assume, that the MHS workload is uniformly distributed over all stations of the FMS, then we can approximate the maximal length of the relevant queue (including A and B), W_T, by eqn. (11):

$$W_T = \min[NS_1, M_M]/2 \qquad (11)$$

↑ W_T: length of the relevant MHS queue (from the top of the MHS queue to order B)
↑ NS_1, M_M: number of productive stations and central buffer positions in the FMS

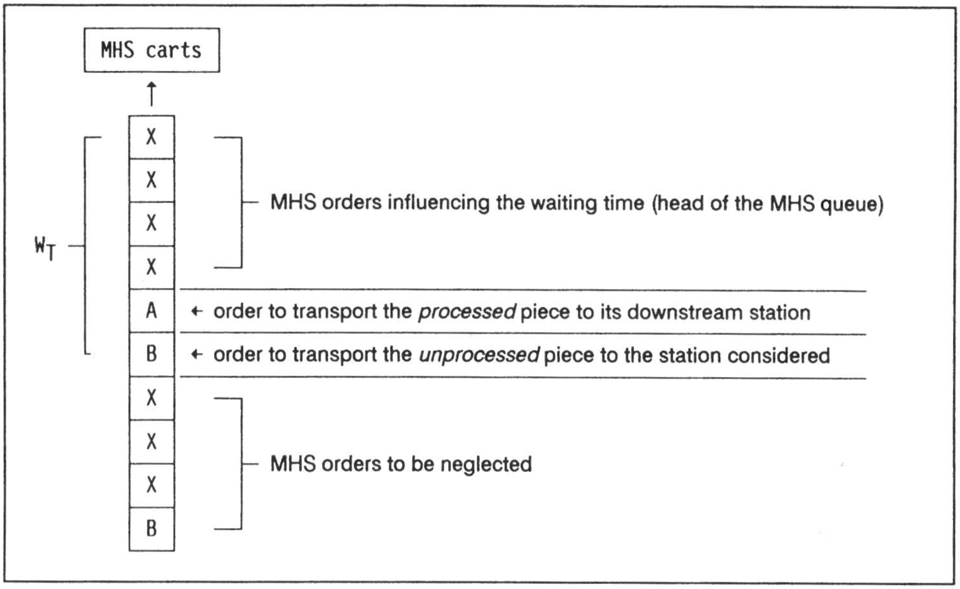

Fig. 3: MHS queue

The length of the MHS queue from the top to the order immediately located in front of piece B (which is located at station i) is approximated by eqn. (12), which excludes piece B from W_T:

$$W_{TAi} = W_T \cdot (1 - r_i) \qquad (i=2,\ldots,M) \qquad (12)$$

r_i: visiting frequency at station M_i
$(1-r_i)$: proportion of pieces *not* to be transported to station M_i (including piece A)
W_T: length of the relevant MHS queue
W_{TAi}: mean number of pieces in the relevant MHS queue not to be transported to station M_i

W_{TAi} contains the piece A to be transported to its downstream station. As we shall account for this piece's transportation time directly, we correct W_{TAi} with respect to this piece by means of eqn. (13).

$$W_{Ti} = \frac{(W_{TAi})^2}{W_{TAi}+1} \qquad (i=2,\ldots,M) \qquad (13)$$

W_{Ti} approximates the mean number of orders waiting for an MHS cart in front of the orders A and B. Now the mean waiting time of piece (A) for transportation can be approximated as follows [eqn. (14)]:

$$T_{Wi} = T_{ZL} \cdot W_{Ti} / S_1 \qquad (i=2,\ldots,M) \qquad (14)$$

- T_{ZL}: mean total transportation time

β) **Transportation time of the processed piece (A)**

Upon completion of processing at a machine, a part is transferred to its downstream station or to the central buffer. The *mean transportation time* is given by eqn. (15):

$$T_A = T_L \cdot (1-F_Z) + B_{ZL} \cdot F_Z \qquad (15)$$

- proportion of pieces shipped to the central buffer
- mean disposal trip time (incl. idle trip)
- proportion of pieces directly shipped to the downstream station
- mean transportation time to the downstream station (incl. idle trip)
- mean transportation time of the processed piece

γ) **Transportation time of the unprocessed part (B)**

The mean transportation time of the unprocessed piece to the station M_i is given by eqn. (3).

The **mean transportation delay** of the next piece enroute to station M_i is given by the sum of all these components [see eqn. (16)].

- transportation time of the processed piece A, T_A
- waiting time of piece A

$$T_i = T_L + T_L \cdot (1-F_Z) + B_{ZL} \cdot F_Z + T_{ZL} \cdot W_{Ti} / S_1 \qquad (i=2,\ldots,M) \qquad (16)$$

- number of MHS carts
- length of the relevant MHS queue
- mean total transportation time (incl. idle and disposal trips)
- disposal trip factor
- disposal trip time (incl. idle trip time)
- proportion of pieces directly shipped to the downstream station
- mean transportation time of the unprocessed piece B (supply trip)
- *mean total transportation delay* of a piece enroute to station M_i

c) machine idle time

The *starving time observed at a station* is given with eqn. (17) as the positive difference between the two stochastic variables t_i and v_i.

$$l_i = \begin{cases} t_i - v_i & \text{if } t_i > v_i \\ 0 & \text{if } t_i \le v_i \end{cases} \qquad i=2,\ldots,M \qquad (17)$$

Under the assumption that t_i and v_i are exponentially distributed the expected value of the expression described by eqn. (17) is given by eqn. (18).

$$L_i = \frac{T_i^2}{T_i+V_i} \tag{18}$$

34. Implementation in the CQN model

So far we estimated the total transportation time (incl. idle trips and disposal trips) and the starving time of a station M_i. These times are used to modify the mean processing times at the stations, which in turn are employed to compute performance measures of an FMS. The numerical quantification of the modifications presented is based on the stationary probability distribution of the number of pallets at the stations in the FMS as computed by the well-known convolution algorithm. In particular we proceed as follows:

- **idle trip times**
  The idle trip times depend on M (number of stations), N (number of pallets) and S_1 (number of MHS carts). They are computed without use of the CQN model.

- **disposal trip times to the central buffer**
  The mean disposal trip time to the central buffer is influenced by the distribution of pallets among stations, that can be described by the probability distribution of pallets at the station. This probability distribution is computed by use of the classical convolution algorithms with mean transportation times modified to account for idle trips. Based on this probability distribution the disposal trip factor is computed as defined above.

- **modification of the mean processing times**
  After the mean starving time of a station has been computed, the "normal" processing time of a machine is inflated by this starving time [eqn. (19)].

  $$B_{i*} = B_i + L_i \qquad M_i=2,\dots,M \tag{19}$$

  As starving only is possible if a piece was denied access to a downstream station the processing times are only inflated for system states in which the station M_i is fully occupied ($n_i > N_{max(i)}$). The stationary probability distribution is then given by eqn. (20) and can be numerically evaluated with the convolution algorithm.

$$\text{Prob}\{\underline{n}\} = \frac{1}{G(N,M)} \cdot \prod_{i=1}^{M} \frac{\prod_{j=1}^{n_i} X_i}{A_i(n_i)} \tag{20}$$

with

$$X_i = \begin{cases} r_i \cdot B_i, & \text{if } n_i \le N_{max}(i) \\ r_i \cdot B_{i*}, & \text{if } n_i > N_{max}(i) \end{cases} \tag{21}$$

$$A_i(0) = 1 \qquad i=1,2,\ldots,M \tag{22}$$

$$A_i(k) = \prod_{j=1}^{k} a_i(j) \qquad k>0;\ i=1,2,\ldots,M \tag{23}$$

$$a_i(j) = \begin{cases} j, & \text{if } j \le S_i \\ S_i, & \text{if } j > S_i \end{cases} \qquad i=1,2,\ldots,M \tag{24}$$

The modification scheme of the mean transportation time influences the stationary probability distribution Prob{$\underline{n}$}. The computation is therefore performed in an iterative way until convergence.

35. Numerical results

The approximation scheme presented has been tested by means of numerous simulation runs with respect to several FMS configurations. In what follows, we present the results found considering some of the FMS structures depicted in [Tempelmeier, Kuhn, Tetzlaff (1989)]. We refer to one machine in the FMS. For each number of pallets in the FMS the results computed with the algorithm presented (CQNBLK) and the results computed with the classical CQN model (CQN) without any modifications are compared to the simulation results (SIM) observed.

- **System 4:** FMS with 4 machines; unlimited central buffer; local buffer sizes $P_i = (25,1,1,2)$

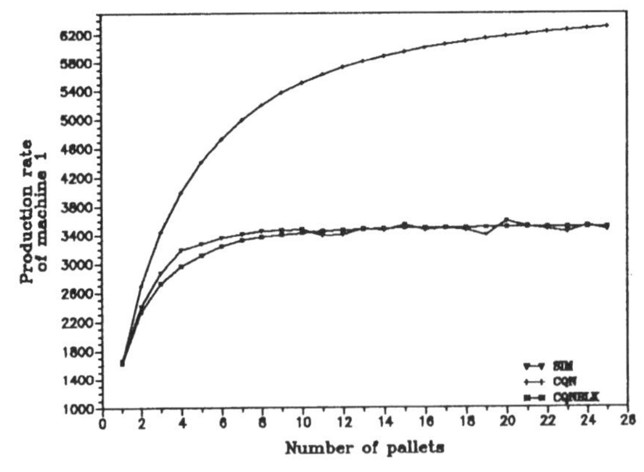

Fig. 4: Results for FMS 4

- **System 12:** FMS with 12 machines; unlimited central buffer; local buffer sizes $P_1 = 1$, $P_i = 2$ $(2 \leq i \leq 11)$, $P_{12} = 40$

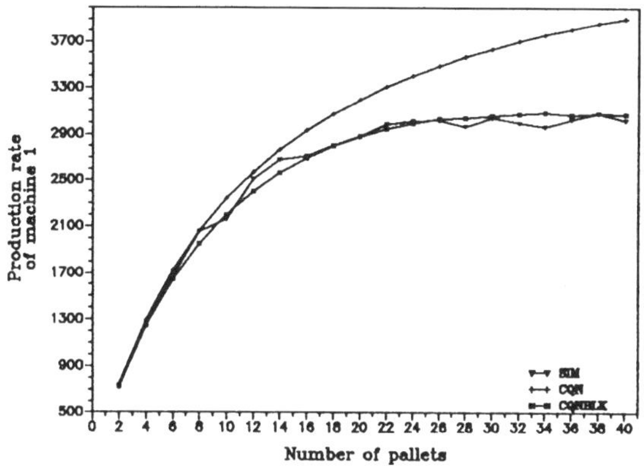

Fig. 5: Results for FMS 12

4. Extension to the special pallet case

So far we have presented results for FMS with universal pallets. The approximation scheme may be used to compute performance measures for FMS with special pallets as follows. First the starving times are estimated as described above (assuming an FMS with universal pallets). Next the mean pallet-type specific processing times are inflated by the mean starving times at the stations and fed into a heuristic MVA-algorithm to compute pallet-type specific results.

Fig. (6) shows the layout of an FMS configuration with a L/UL station with two servers and 4 machining stations, one of which with two servers (4a and 4b). In the FMS considered 3 types of pallets circulate. The processing times are *deterministic* according to prespecified routing sheets. The central buffer is assumed to be large enough such that only starving is possible.

Fig. (7) shows the results obtained (column labeled CQNBLK/MVA). In the column labeled MVA-O the results of the classical CQN model for special pallets without any modification are reported. In the column labeled MVA-Z the idle and disposal trip times are computed and the modified transportation time is fed into the MVA-algorithm.

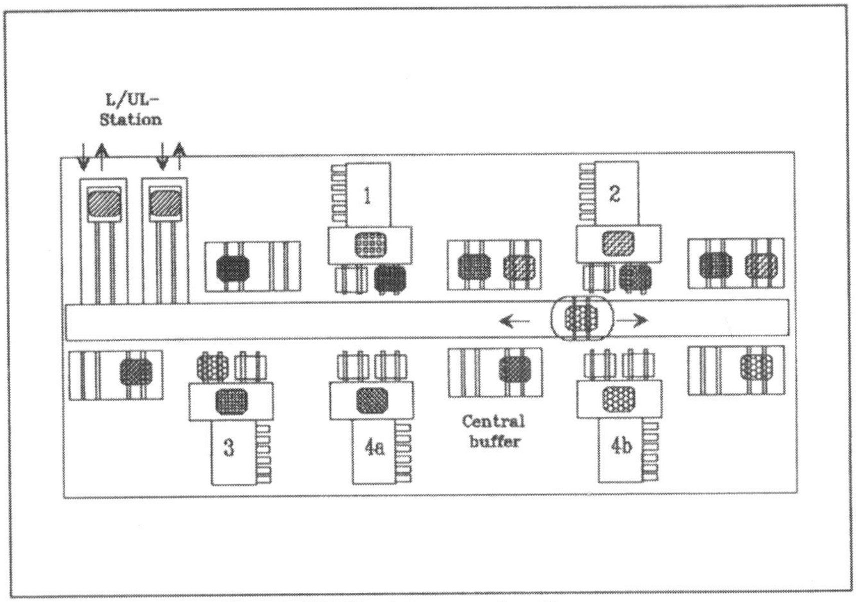

Fig. 6: FMS with several pallet types

8/2/4 pallets (1 MHS cart)				
	Simulation	CQNBLK/MVA	MVA-0	MVA-Z
STIM1	289	283	213	240
STIM2	255	266	186	220
STIM3	414	386	275	317
util-M1	0.48	0.49	0.67	0.59
util-M2	0.23	0.23	0.32	0.27
util-M3	0.18	0.19	0.27	0.23
util-M4	0.44	0.45	0.60	0.53
util-L/UL	0.69	0.69	0.94	0.82
util-MHS	0.76	0.78	0.70	0.93
10/4/6 pallets (2 MHS carts)				
	Simulation	CQNBLK/MVA	MVA-0	MVA-Z
STIM1	343	340	284	285
STIM2	335	330	269	270
STIM3	487	473	385	386
util-M1	0.53	0.54	0.66	0.65
util-M2	0.36	0.36	0.45	0.44
util-M3	0.24	0.25	0.31	0.31
util-M4	0.47	0.47	0.57	0.57
util-L/UL	0.81	0.81	0.98	0.98
util-MHS	0.51	0.46	0.37	0.56
STIM = system time; util = utilization				

Fig. 7: FMS with several pallet types

The results show, that it is not sufficient to improve the estimation of the mean transportation time alone (MVA-Z), but that starving times may have a major influence on the performance of the FMS.

5. Conclusion

In this paper a procedure to modify the input data of the classical CQN algorithm to account for starving has been presented. The results obtained during a large simulation experiment considering several configurations of FMS show, that using the procedure leads to good approximations of the performance of an FMS.

The approximation is based on a detailed analysis of the processes and states of an FMS. Special consideration is given to the conditions leading to starving. During the development of the approximation scheme many heuristic correction factors have been introduced. These factors have been tested with numerous simulation results and found to work well for the different configuration types of FMS considered.

Symbols

B_1	mean transportation time (load trips)
B_i	mean processing time at station M_i
B_{i*}	modified mean processing time at station M_i
B_Z	mean disposal trip time (load trips)
B_{ZL}	mean disposal trip time (incl. idle trips)
F_L	idle trip factor
F_Z	disposal trip factor
G(N,M)	normalization constant
L_i	mean starving time of station M_i
M	number of productive stations in the FMS
M_i	station
M_M	number of stop points in the FMS (productive stations and central buffer places)
N	number of workpieces (pallets) in the FMS
n_i	number of workpieces in station M_i
$N_{max(i)}$	maximal number of workpieces in station M_i
$N_{max(ZP)}$	capacity of the central buffer
P_i	number of local buffer places at station M_i
P_Z	availibitity of the central buffer
r_i	visiting frequency of station M_i
S_1	number of MHS carts
S_i	number of machines (servers) in station M_i
T_A	mean transportation time of the processed piece A
T_i	mean total transportation delay of a piece enroute to station M_i
T_L	mean transportation time (incl. idle trips)
T_{Wi}	mean waiting time of processed piece A at station M_i for transportation
T_{ZL}	mean transportation time (incl. idle trips and disposal trips)
U	total number of central buffer places needed
V_i	local work content at station M_i (times units)
W_T	length of the relevant MHS queue
W_{TAi}	mean number of pieces in the relevant MHS queue not to be transported to station M_i

w_{Ti} corrected mean number of pieces in the relevant MHS queue not to be transported to station M_i (not considering piece A)

Literature

Akyildiz, I.F., Mean Value Analysis for Blocking Queueing Networks, in: IEEE Transactions on Software Engineering 14(1988a)4, pp. 418-428

Akyildiz, I.F., On the Exact and Approximate Throughput Analysis of Closed Queueing Networks with Blocking, in: IEEE Transactions on Software Engineering 14(1988b)1, pp. 62-70

Altiok, T. and Perros, H.G., Open Networks of Queues with Blocking: Split and Merge Configurations, in: IIE Transactions (1986), pp. 251-261

Bruell, S.C. and Balbo, G., Computational Algorithms for Closed Queueing Networks, New York (North-Holland) 1980

Co, H.C. and Wysk, R.A., The robustness of CAN-Q in modelling automated manufacturing systems, in: International Journal of Production Research 24(1986)6, pp. 1485-1503

Gershwin, S.B. and Schick, I.C., Modeling and Analysis of Three-Stage Transfer Lines with Unreliable Machines and Finite Buffers, in: Operations Research 31(1983), pp. 345-380

Perros, H.G., Queueing Networks with Blocking: A Bibliography, in: Performance Evaluation Review 12(1984)2, pp. 8-12

Solberg, J.J., A mathematical model of computerized manufacturing systems, in: Proceedings of the Fourth International Conference on Production Research, Tokyo, 1977

Suri, R. and Diehl, G.W., A Variable Buffer-Size Model and its Use in Analyzing Closed Queueing Networks with Blocking, in: Management Science 32(1986)2, pp. 206-224

Tempelmeier, H., Kapazitätsplanung für Flexible Fertigungssysteme, in: Zeitschrift für Betriebswirtschaft 58(1988)9, pp. 963-980

Tempelmeier, H., Verfahren zur Kapazitätsplanung für flexible Fertigungssysteme, Technische Hochschule Darmstadt, Fachgebiet Fertigungs- und Materialwirtschaft, Arbeitspapier, Darmstadt, September 1987

Tempelmeier, H., Kuhn, H. and Tetzlaff, U., Performance evaluation of flexible manufacturing systems with blocking, in: International Journal of Production Research 27(1989)11, pp. 1963-1979

Yao, D.D. and Buzacott, J.A., Modeling a class of state-dependent routing in flexible manufacturing systems, in: Annals of Operations Research 3(1985), pp. 153-167

Yao, D.D. and Buzacott, J.A., Models of flexible manufacturing systems with limited local buffers, in: International Journal of Production Research 24 (1986)1, pp. 107-118

Non Delay - Revisited

JANEZ DEKLEVA

INTRODUCTION

A Flexible Manufacturing System (FMS) is capable of simultaneous production of a medium sized volume of a variety of products. This production system is an integrated Computer Controlled system composed of CNC machine tools /centers and automated devices for workpiece manipulation and control.

A FMS represents a significant investment and therefore requires meticulous planning in marketing, technology and operations scheduling.

The introduction of a FMS could be promising when it is carried out as the final step in the process of modernisation of a production system based on group technology. We would recommend production flow analysis as the best tool and method for the proposed modernisation. First, it simplifies the overall flow of material in the production system, and second, it converts the production shop into a cellular production system with groups of machins (cells) dedicated to the manufacturing of groups of similar parts. The emerging cells are in permanent evolution due to marketing requirements, improvements in design and technology, and management decisions. The cell system tends to adapt gradually its capacities to time, space and quality requirements. This cell system of production represents then the best basis for the development of FMS structure.

This article will explain a two-phase procedure for designing a FMS with the goal to obtain its best possible performance if not optimal. Job shop simulation is proposed for FMS modeling or more precisely a modified non-delay schedule is offered.

In the original agorithm of non delay schedule generation conflicts arise e.g. in the machine m^* for every operation with $\sigma^* = \sigma_j$; where σ^* determines the earliest time in which an operation could be realized (index j refers to the earliest time for part j). The original algorithm resolves this conflict by enumeration /1/,/2/.

On the other hand, a heuristic procedure that is designed to generate only one full schedule need resolve a conflict in only one way in the course of constructing a schedule. This means that the procedure must specify a priority rule for selecting one operation from among the conflicting operations.

Using this modification, we replaced the original non delay schedule algorithm with the new algorithm named the heuristic schedule generation (HSG).

The evaluation of a two-phase design procedure of a FMS will be obtained through its comparison with Mean Value Analyses of Queues (MVAQ) and discrete simulation /3/.

The usefulness of the HSG algorithm for the fine planning (scheduling) of a FMS and for activity control will be presented. Finally, its value for the rescheduling will be shown assuming that the systems states are known due to the presence of an adequate diagnostic system.

DESIGN AND PERFORMANCE EVALUATION OF FMS

The performance of a manufacturing system is to a great extent determined by decisions made in the design phase of this system. Of course, the production program and its modifications, the reliability of manufacturing system, and the quality of scheduling the system are among additional factors which can influence its value and determine its outcome.

In this respect we would like to mention the recent article written by Leung and Suri /4/ where a systematic account of tools for performance evaluation is given, together with the activities of the systems design phase. In the same place the reader is reminded that the applied tools are helpful for performace evaluation and not performance optimization. However it would be very convenient to dispose of tools which would solve both tasks at the same time.

We believe that the choice and the application of non delay schedule generation for job shop simulation generates solutions with very good performances if not optimal.

In this section a procedure for designing a FMS is proposed (Fig.1). The procedure tries to advance a best possible performance at a minimum investment cost. However a known production program for the FMS under consideration is assumed.

In the proposed procedure, the FMS is divided into two subsystems:
- the manufacturing subsystem and
- the auxiliary subsystem.

The manufacturing subsystem consists of CNC machine tools /centers and working tools. This subsystem represents the available capacities based on the previously calculated (estimated) required capacities. The latter are defined by the planned production program, route sheets (cards) for all parts within this program and the reliability data of CNC machine tools/centers incorporated into this subsystem. To select adequate CNC machine tools/centers in addition to required capacity, the data on geometry and quality requirements are needed.

The auxiliary subsystem consists of load/unload stations (L/U), automated manipulation units (robots), control units, transportation units, pallets, fixtures etc. It is this subsystem which in union with the manufacturing subsystem exemplifies the physical part of a FMS.

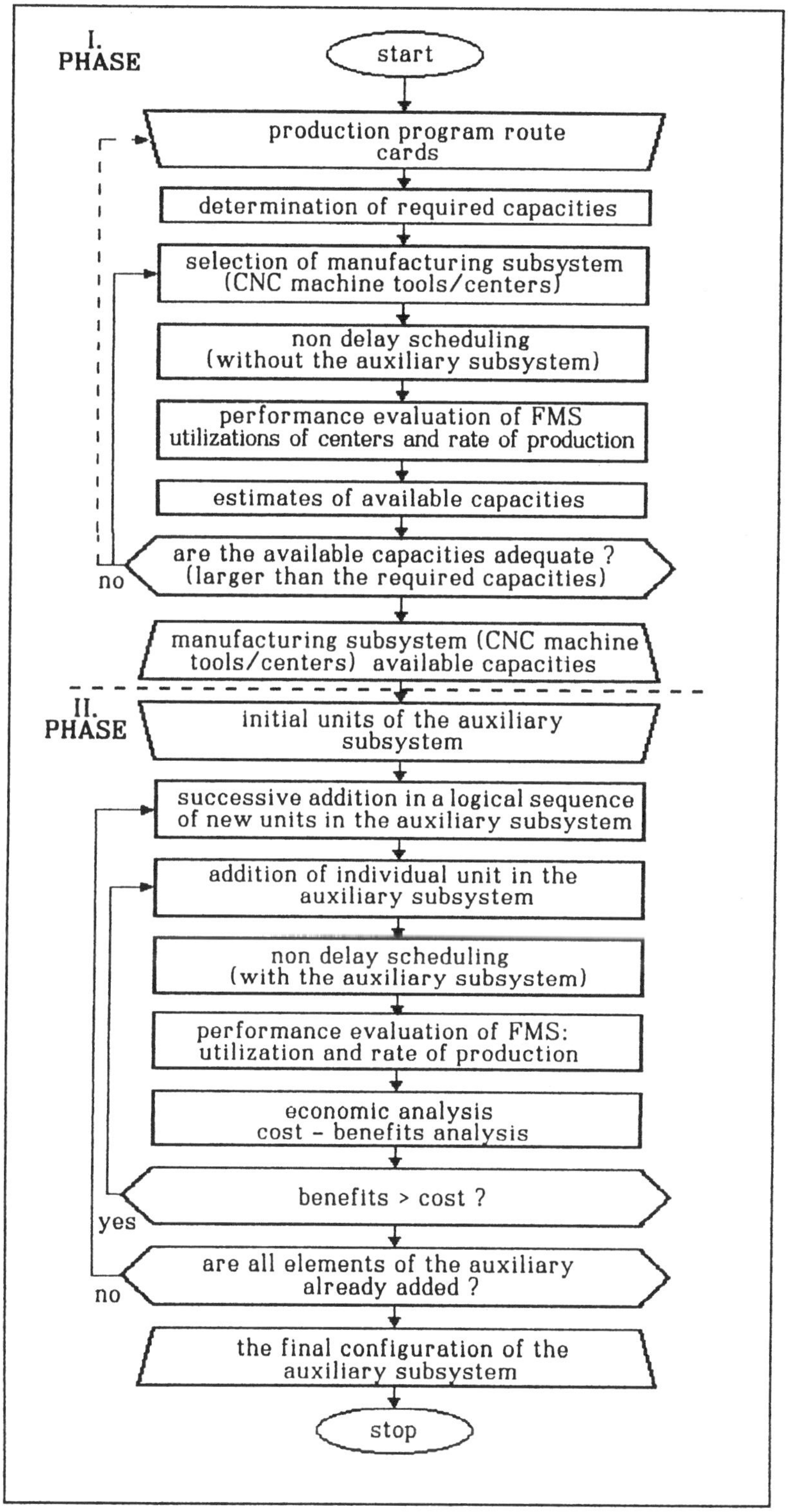

Fig.1: Flowchart of the two-phase procedure for designing FMS

Let us discuss the motives for the proposed division of a FMS into two subsystems.

First, the manufacturing subsystem which represents the required manufacturing capacities has to accomplish the best possible performance levels, measured (expressed) with a high utilization rate of all units, with a minimum average time spent in the subsystem and with the high production rate. The use of the HSG algorithm for the simulation of a manufacturing subsystem is aiming at realizing the above goal (Fig.2).

Second, the auxiliary subsystem has to execute different manipulation functions from the entrance of the part into the FMS up to its exit from the system. An unproperly assembled auxiliary subsystem represents a bottleneck for the manufacturing subsystem which may cause a substantial reduction of its original performance. The elimination of the bottleneck requires logical and sequential additions of properly selected units into the auxiliary subsystem which results in progressive improvements of performance toward its original values. This assymptotical approach is brought to the end when the cost of new units outweighs the effect of improvements.

With the above explanations the division of a FMS into two subsystems becomes logical (Fig. 1).

For the first subsystem the decision on the number and the type of CNC machine tools /centers is of highest importance. This represents the first part of the first phase in our procedure.

In the second part of the first phase the selected manufacturing subsystem is modelled by use of job shop simulation (i.e. HSG) and the performance of the system is calculated. We search for the best possible utilization of CNC machine tools/centers under the assumption that the manipulation and transportation times equal zero.

If the available capacities of a manufacturing subsystem do not meet the requirements, we have to look for a new solution of the manufacturing subsystem. We have to iterate. In the first part of our procedure all three activities discussed by Leung and Suri /4/ are included: System Planning, Agregate Analysis and Detailed Analysis. Two models were applied for performance evaluation: static allocation model in the first part of the first phase and discrete event simulation in the second part. With the use of HSG for discrete event simulation one may hope to obtain good performance if not optimum.

Unfortunately, the auxiliary subsystem is absent within the first phase and its design starts within the second phase. The auxiliary subsystem has to be built with logically selected decisions. Let us start with the L/U station, with the pallet, workpiece and fixture. After the workpiece is loaded, it has to be transferred. The automatic vehicle is used for the transportation of the workpiece from the L/U station to the CNC machining center. The workpiece is transported to the CNC center and unloaded into the rotary pallet shuttle (outher position). If this position is occupied the workpiece is unloaded into the pallet stand where it has to wait for the transportation to the CNC centre etc. This scenario is one out of the many which you may imagine just looking at the Gantt chart.

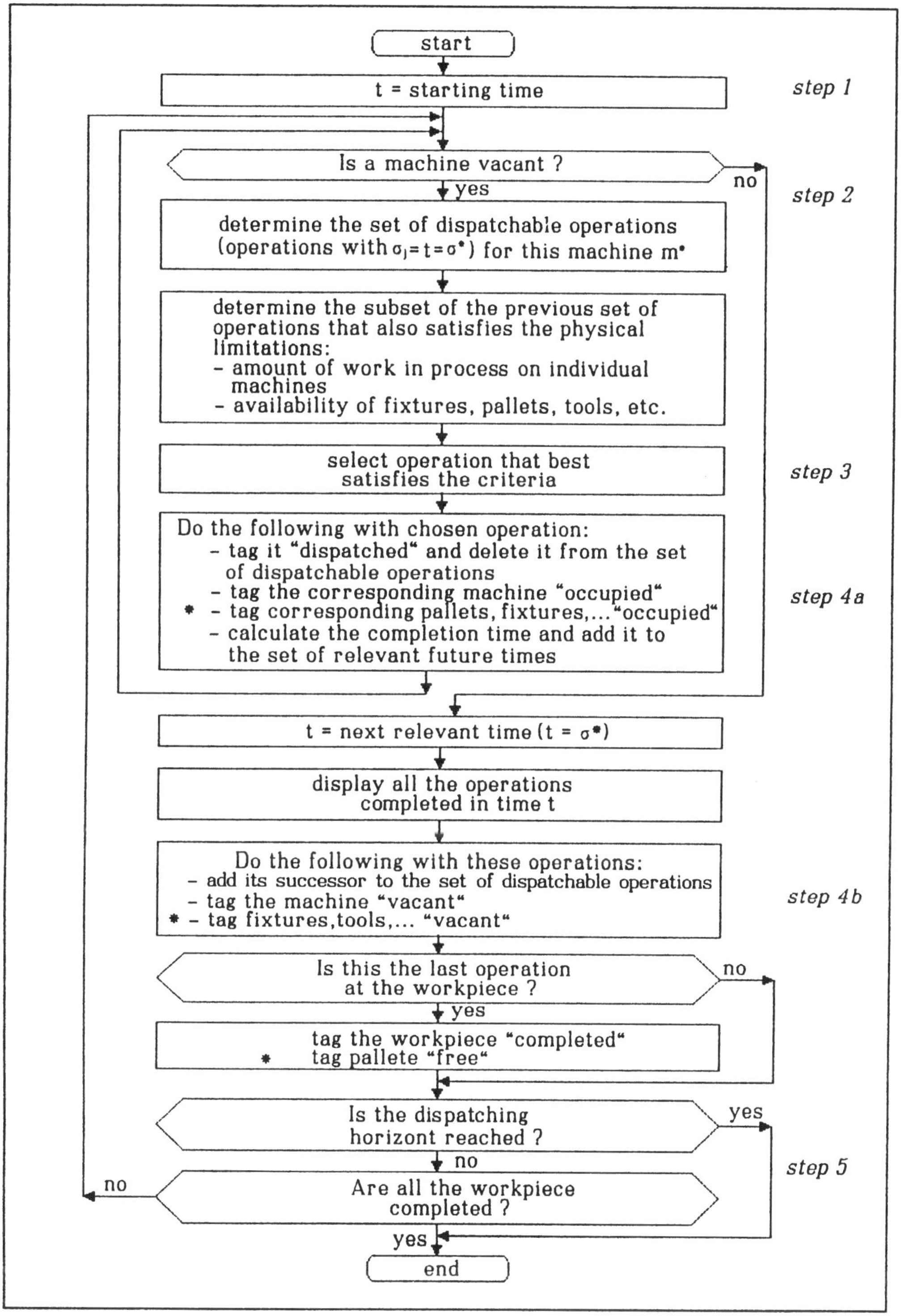

Fig. 2 Flowchart of the modified non delay algorithm (HSG)
(• — activities of second phase of simulation)

The addition of every new element (L/U station, vehicle, waiting place-pallet stand, pallet, fixture etc.) represents a new entity i.e. the time axis, which ought to be added (included) into the model. The implementation of elements must be logical e.g. the first L/U station and the first vehicle could join the system at the same time, but additional L/U stations may be added without the simultaneous increase of the number of vehicles. For each individual model the necessary strategy for the extension of the auxiliary system by sequientially increasing the number of particular units must be elaborated. Otherwise the performance evaluation may be time consuming. And now, what do we want to gain with the auxiliary subsystem. First, the periphery of the manufacturing subsystem is bound to answer all transportation and manipulation requirements. It is supposed to offer the entrance and the exit potentialities of the whole system with the purpose to *preserve or at least to approach assymptotically the performance of the first subsystem as defined in the first phase* (i.e. when the auxiliary subsystem was still absent). The goal of the second phase is to decide how much one should invest into the auxiliary subsystem if the efficiency of the whole system (i.e. FMS) measured with benefits of performance improvement should outweigh the investment cost of the added unit.

COMPARISON OF JOB-SHOP SIMULATION (NON DELAY) WITH MVAQ AND DISCRETE SIMULATION /3/

We have tested the usefulness of the HSC for FMS simulation by comparing its results with the results of Mean Value Analyses of Queues (MVAQ) and discrete simulation as found in reference /3/,/5/,/6/,/7/. The simulated system is schematically presented in Fig. 3. The input data used were the same, Table 1.

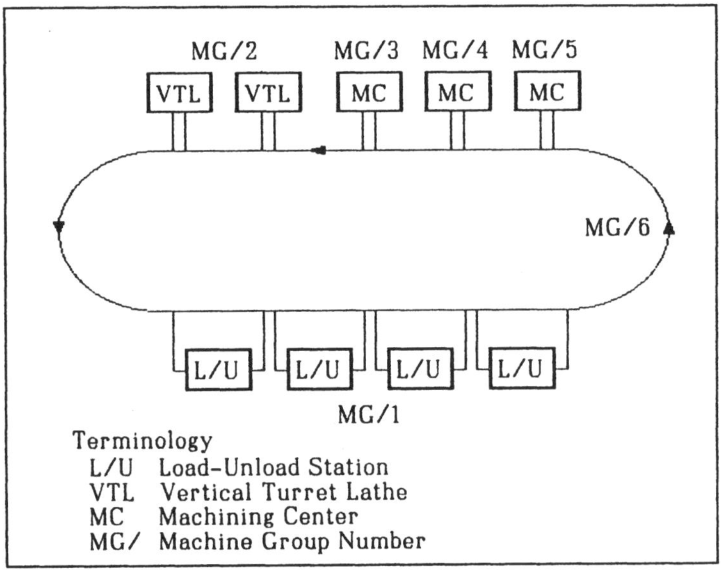

Fig. 3: FMS according to reference /3/.

SYSTEM SIZE	
NUM MACH. GROUPS	NUM PARTS
6	3

MACHINE GROUP INFORMATION	
MACHINE GROUP NO.	NO. OF MACHINES
1	2
2	2
3	1
4	1
5	1
6	8

TOTAL NUMBER OF PALLETS/FIXTURES FOR EACH PART	
PART	NO.
12340001	4
12340002	1
12340003	2

PART ROUTING INFORMATION			
PART NO.	OPER. NO.	MACHINE GROUP	MACHINE TIME
12340001	1	1	10.0
	2	6	2.5
	3	2	28.0
	4	6	2.5
	5	1	5.0
	6	1	10.0
	7	6	2.5
	8	3	12.0
	9	6	2.5
	10	1	5.0
12340002	1	1	10.0
	2	6	2.5
	3	4	30.0
	4	6	2.5
	5	1	5.0
12340003	1	1	10.0
	2	6	2.5
	3	5	15.0
	4	6	2.5
	5	1	5.0

Table 1: MVAQ input data /3/.

As can be seen in Table 2 and Table 3, our results differ at most by 4.2% from the results of discrete simulation (rem. discrete simulation refers to the unspecified discrete simulation in referenc /3/), which means that the difference is much smaller than that between MVAQ and discrete simulation. Besides, our results give better production performances than discrete simulation where the difference is biggest. On the basis of the data from the mentioned article it can be concluded that the program using non delay is simpler and faster than discrete simulation. The results are obtained in a few seconds (in the mentioned example the duration of the program was 3.5 sec.) after the input of the data . This is approximately the same duration as that of MVAQ. (The non-delay algorithm is programmed in C on PC-AT).

PART NUMBER	MVAQ	SIM	% DIFF
PRODUCTION RATE			
12340001	15.46	17.60	13.5
12340002	7.10	8.00	12.7
12340003	16.71	19.00	13.7
AVERAGE TIME IN SYSTEM			
PART NUMBER			
12340001	124.19	109.09	12.2
12340002	67.61	60.00	11.3
12340003	57.45	50.53	12.0
UTILIZATION			
MACHINE GROUP			
1	0.86	0.98	13.9
2	0.45	0.52	16.7
3	0.39	0.44	12.8
4	0.44	0.50	13.6
5	0.52	0.60	15.4

Table 2: MVAQ and Simulation Comparison /3/.

PART NUMBER	NON DELAY	SIM	% DIFF.
PRODUCTION RATE			
12340001	17.6	17.60	1.1
12340002	8.0	8.00	0.0
12340003	19.8	19.00	4.2
AVERAGE TIME IN SYSTEM			
PART NUMBER			
12340001	110.34	109.09	0.4
12340002	60.00	60.00	0.0
12340003	48.48	50.53	4.2
UTILIZATION			
MACHINE GROUP			
1	0.98	0.98	0.0
2	0.51	0.52	2.0
3	0.44	0.44	0.0
4	0.50	0.50	0.0
5	0.62	0.60	3.2

Table 3 : NON DELAY and Simulation Comparison.

Finally we want to illustrate the design and the performance evaluation of FMS using the input data from reference /3/ and following the proposed two-phase procedure described in the last section of this article. This means that the basic configuration of the FMS remains the same - we have the same number and type of machine tools, the same configuration of the auxiliary subsystem (but not the same number!) and the same workpieces. We increase the number of the individual elements of the auxiliary subsystem until we approach the utilization of the machines which we have determined within the first phase of simulation. It is understood that, in the second part of the first phase of simulation, we consider only those operations that pertain to the machine tools (Table 4).

MACHINE GROUP INFORMATION

MACHINE GROUP		NUMBER OF MACHINES
1	VERTICAL TURRET LAHES	2
2	MACHINING CENTER 1	1
3	MACHINING CENTER 2	1
4	MACHINING CENTER 3	1

PART ROUTING INFORMATION

PART NO.	OPER. NO.	MACHINE GROUP	MACHINE TIME
12340001	1	1	28.0
	2	2	12.0
12340002	1	3	30.0
12340003	1	4	15.0

Table 4: Input data of the NON DELAY program for FMS simulation which are used in the first phase of simulation.

AVERAGE UTILIZATON OF THE GROUPS OF MACHINES

MACHINE GROUP		NUMBER OF MACHINES	UTILIZATION
1	VERTICAL TURRET LAHES	2	1.00
2	MACHINING CENTER 1	1	0.81
3	MACHINING CENTER 2	1	1.00
4	MACHINING CENTER 3	1	1.00

Table 5: Average utilization of the groups of machines calculated in the first phase of simulation.

With this data we obtain the following utilization of the machine tools (Table 5): groups of machines 1, 3 and 4 are utilized 100%, while the utilization of group 2 is lower. In the first shift it is 81% and later approaches 86% (rem.: 100% utilization is possible because operation time includes setup on the machine tool center).

The next phase is the selection of the auxiliary subsystem. Let us start with L/U stations. Initially we have included two of them into the model, Table 6.

AVERAGE UTILIZATON OF THE GROUPS OF MACHINES			
MACHINE GROUP		NUMBER OF MACHINES	UTILIZATION
1	LOAD/UNLOAD STATIONS	2	1.00
2	VERTICAL TURRET LAHES	2	0.41
3	MACHINING CENTER 1	1	0.30
4	MACHINING CENTER 2	1	0.98
5	MACHINING CENTER 3	1	0.66

Table 6: Average utilization of the groups of machines after the addition of load/unload stations.

Table 6 indicates that two L/U stations are not sufficient, so their number must be increased until machine utilization approaches the values obtained within the first phase of simulation (see Table 5). The same procedure is followed when determining the optimal number of automated guided vehicles (AGVs).

The final configuration and the resulting machine utilizations are shown in Table 7.

AVERAGE UTILIZATON OF THE GROUPS OF MACHINES			
MACHINE GROUP		NUMBER OF MACHINES	UTILIZATION
1	LOAD/UNLOAD STATIONS	4	0.91
2	VERTICAL TURRET LAHES	2	0.97
3	MACHINING CENTER 1	1	0.75
4	MACHINING CENTER 2	1	0.97
5	MACHINING CENTER 3	1	0.97
6	TRANSPORTATION SYSTEM	2	0.59

Table 7: Average utilization of the groups of machines after the addition of transportation means.

SCHEDULING AND RESCHEDULING

The load of a FMS is decided on a higher level through the master schedule with the decision what and how much a FMS should produce and when. Two different levels are identified within a FMS; the first (higher) level is responsible for the conversion of data obtained through the master schedule into the schedule (fine schedule) which ought to be implemented in a FMS. The next task of this level is to eliminate the inconsistencies in this conversion, when the implementation of master schedule cannot be fullfilled.

The second (lower) level within a FMS is responsible for the implementation of the fine schedule within a FMS i.e. it is liable for activity control in the FMS in accordance with the fine schedule.

The topic of our consideration is connected with this lower level. But before we proceed with the discussion of our main task i.e. the implementation of fine schedule, a secondary task should be brought to our attention and joined to the first one. The control mechanism ought to cope with all kinds of deficiencies which we may meet within the FMS.

Deficiences and disturbances met within a FMS can be divided into: organisational disturbances and technical disturbances.

Within the first group one can find:
- changes of orders volume
- changes of priority
- changes and supplements within the technological database.

It is our belief that the organisational disturbances fall under the control levels which reside above the FMS levels.Therefore we shall ignore them for the time being.

Technical deficiences are caused by faults and errors within the FMS. A great variety of technical disturbances exists and they can be subdivided into:
- disturbances within the manufacturing subsystem
- disturbances within the auxiliary subsystem
- disturbances on parts deliveries
- disturbances on tool deliveries and maintenance
- errors within L/U stations
- errors caused by system control, software bugs
- communication errors etc.

Although all types of deficiencies are of importance when controlling the FMS, they have to be treated in a systematic way. The diagnostic system, wich is an integral part of any FMS, is responsible for permanent control of system status and for systematic and permanent deliveries of: alarms, warnings and messages. This system is required to produce a carefully maintained and preserved database with all possible faults which can occur in the different parts of a FMS. The statistics of faults is of greatest importance (e.g. MTBF - Mean Time Between Failures, MTTR - Mean Time To Repair).

Let us concentrate on our main task of controlling the FMS. This undertaking i.e. the completion of fine schedule could be accomplished in two different ways. First, we could try to realize (implement) the same fine schedule which has been generated as a proof of master schedule correctness. In this case we have the realization of the schedule which is generated as a single decision, but where all events which could possibly happen after the decision was made, were neglected.

Secondly, we can control the FMS by generating schedules with the so called dispatching procedure. With this procedure the scheduling decisions need not be made all at once but only as they are needed. In this case the latest data on a FMS can be taken into account and no real difference between scheduling and rescheduling exists.

There are some additional objections with the first type of control. Namely, the schedule, which was generated as a single decision, is based on the ideal data or states: i.e. ideal parameters of materials, tools, etc. It is well known that the parameters of (input) materials differ, that means the processing times differ around the mean value.

In order to learn the effects of discrepancies from ideal values we generated schedules with the ideal processing times and then compared them with the schedules generated with the processing times perturbed up to ± 10% around the ideal values. The manipulation and transportation times were in both cases kept ideal.

It can be seen by comparing the first and second Gantt chart (Fig.4a and Fig.4b) that parts 10 and 11 are interchanged on the axes 202 and 204 in the second Gantt chart (Fig.4b).

This example indicates that the first control procedure ought to be modified. We can still retain the same sequence of parts on the CNC machine tools/centers as it was obtained by the original schedule although the processing times may differ. But because of these time variations, the dispatching procedure should be applied for all other activities. Technical disturbances will be solved in rescheduling mode, which we shall discuss later on. In conclusion, it must be stated that with the described mode of control weaker performance values may be expected.

This finding indicates that the control mechanism of a FMS should be based on the dispatching procedure. This procedure generates schedules which are easily adapted to real processing times, technical disturbances within the system and all other factors that affect the FMS status over time.

Although we are quite familiar with the results of schedule generation through the use of the Gantt chart, we have to define the basic problems of control more in detail. Gantt chart is the set of time axes, each one for every individual CNC machine tool/center, L/U station, vehicle, pallet stand etc. The activity blocks (operations, manipulations, transportations blocks) are planned on the corresponding axes of the Gantt chart according to the route cards and priority rules.

The emphasis of the Gantt chart is on activities.Each activity is marked by two events: the start and the end. Both events are of decisive importance for the process of system control. The only exception is the occurrance of a deficiency, an event which can happen at any time on any time axis.

The control activities of the FMS can take place only if all its elements are connected to the LAN (hardware) and if the proper communication protocols are available (software). The set of communication protocols provides the ability to communicate between different station in a FMS , to interpret the contents of the frames, to transfer different files etc. (Many systems of this type are available e.g. Ethernet, MAP etc.)

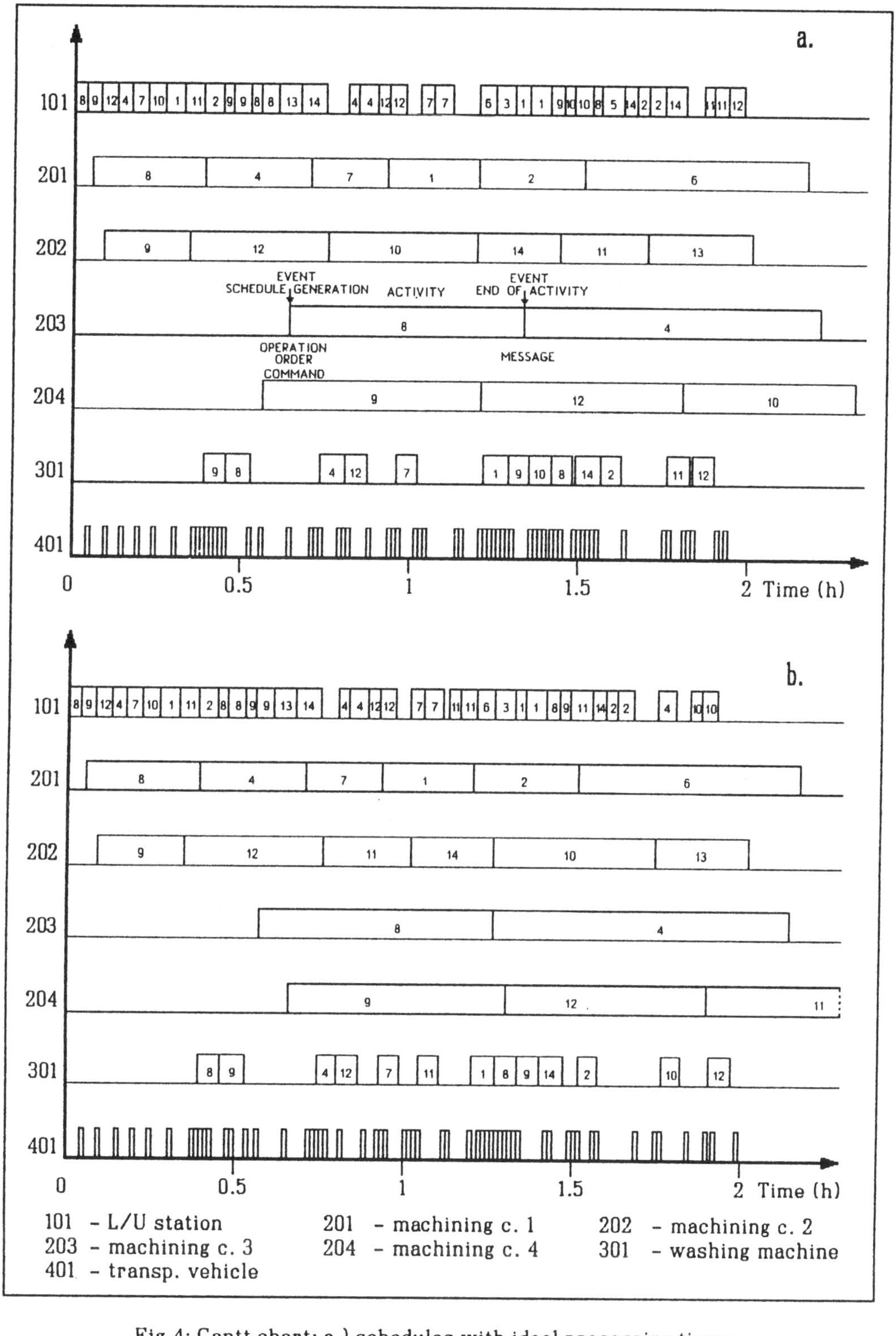

Fig.4: Gantt chart: a.) schedules with ideal processing times.
b.) schedules with perturbed processing times.

Two typical information items are present in the communication system of the FMS:
- the command
- the message on the termination of activity.

A command is generated by the event, which represents the starting point of an activity. A message is the information on the termination of an activity.

The starting events are the beginning of:
- loading of the part / unloading of the part
- movement of a transportation vehicle
- operation on a CNC tool / center
- manipulation of a manipulator (robot).

As we know HSG specifies the selection of operation (activity) which should be added on the time axis m^* i.e. to the partial schedule PS_t.

But the added activity could be implemented only by a simultaneous generation of the corresponding command and by a prompt delivery of this information to the station which must execute it.

Now look at the output of the above algorithm. It represents the sequence of partial schedules i.e. activities. In this connection every starting event in generation of PS_t ought to be followed by the generation of corresponding command which produces the execution of the activity. The main commands are the following:

PC –> L/U; loading/unloading command for the workpiece (part)
(id. no. of the workpiece, pallet no., fixture no.)

PC –> V; transportation command
(id. no. of the workpiece, starting station - destination station)
Note: starting/destination station could be L/U station, CNC machine tool/center or pallet stand

PC –> MC; operation command for machining center
(id. no. of the CNC machine tool/centre, id. no. of the workpiece, id. no. of the control program)
There are two posibilities:
- the workpiece is on the rotary pallet shuttle (inner position) and the operation can start immediately
- the wokpiece is on the rotary pallet shuttle (outer position); the operation can start only after the termination of the preceding operation

Equally important are the messages on the terminations of activities. Without them no execution of step 2 of Algorithm (Fig.2) could be initiated. The messages are generated by individual stations responsible for completion of particular activities. Let us have a look at some of the messages:

L/U —> PC; message on the termination of loading/unloading

confirm the termination of activity; this message precedes two possible commands:

- transportation command of the workpiece to some other station
- transportation command of the workpiece into the warehouse

* - I/O axis is free, immediately after the vehicle drives off

V —> PC; message on the termination of transportation

1) - I/O station is free
   or rotary pallet shuttle (outher position) is free
   or pallet stand is free

2) - vehicle is free

** (V —> PC; message on the beginning of transportation
   - I/O station is free
   or rotary ...)

MC —> PC; message on the termination of operation

- CNC machine tool/center can receive a new command for the operation of a new part
- V can receive recommand for transportation of the terminated part

Some additional information - not related to the operations control - is also present in the communication network: e.g. information connected with the communication protocols.

Quite often we have to face some extraordinary events caused by deficiencies in the CNC machine tools/centers. These are random events (in general). Let us suppose that a deficiency is caused by a fault of a CNC machine tool/center which at the same time stops working. We can expoct a question: how should we treat this extraordinary (random) event? Let us assume the presence of a diagnostic system with the responsability to follow up the state of the system (FMS) and to maintain the database of faults, statistics of faults, commands for repair activities , priority rules for commands and rules for stopping individual units of the FMS.

After we obtain the data on the cause of a deficiency and the statistics of MTTR, we can immediately place a high priority activity for the repair of the damaged subsystem of a CNC machine tool/center. This activity is immediately added to the PS_t via step 2 of our Algorithm. All other activities on the same axis (CNC machine tool) have to wait until the message of the termination of the repair appears. If the repair time is long i.e. longer than the critical time (this is being defined by fine schedule) the rescheduling takes place. Within this process a new master schedule must be generated and then converted into the fine schedule according to a well known procedure.

The control of lateness is an automatic process on all axes that represent CNC machine tools/centers.

And finally, let us look at tool handling and management. Although we try to minimize the requirements for automatic tooling through grouping (GT) of parts, generation of adequate schedules etc. it is impossible to avoid the problem and its complexity. We will try to solve this problem with the same algorithm i.e. with the heuristic non delay schedule generation. Hopefully this will be the subject of one of our next papers.

REFERENCES

/1/ CONWAY R.W., MAXWELL W.L. and MILLER L.W.: Theory of Scheduling, Addison-Wesley, Reading, Mass., 1967

/2/ BAKER K. R.: Introduction to Sequencing and Scheduling, John Wiley & Sons, New York, 1974

/3/ SURI R., HILDEBRANT R.: Modelling FMS Using Mean-Value Analysis, SME Journal of Manufacturing Systems, Vol. 3, No.1, 1984

/4/ LEUNG Y. and SURI R.: Performance Evaluation of Discrete Manufacturing Systems, Control Systems, Vol. 10, No. 4, 1990

/5/ DEKLEVA J., GABERC M. and KUŠAR J.: Simulacija fleksibilnih obdelovalnih sistemov (Simulation of Flexible Manufacturing Systems), Industrial system - IS 90, Proceedings of the Conference, Novi Sad, 1990

/6/ DEKLEVA J. and GABERC M.: Modeliranje fleksibilnih obradnih sistema pomoću job-shop simulacije (Modeling of Flexible Manufacturing Systems by job-shop simulation), in print, Strojarstvo, 1990

/7/ GABERC M. (private communication)

A Heuristic Approach to FMS Process Planning

EDWIN CHENG

Introduction

A flexible manufacturing system (FMS) is an integrated system of numerically-controlled machine tools and an automated material handling system under the supervision and control of an on-line host computer (Draper, 1984). An FMS can simultaneously process a number of different part-types of medium batch sizes in a random manner. FMS is fast becoming an appropriate production method for mid-volume batch manufacturing. This is due to the fact that FMS has dual advantages over conventional production methods, especially the job shop production system. The advantages are, namely, higher production efficiency and greater flexibility with regard to part-type variety and production.

FMS has unique structures, characteristics and constraints which are vastly different from those of the conventional production systems. These necessitate the development of new and appropriate planning and control methods to deal with the operational problems arising from the use of this new production technology. Some typical FMS operational problems include grouping of machine tools, selection of part-types, part-type mix and scheduling of part-types and material handling devices (Stecke, 1981).

This paper focuses on the process planning problem in FMS. It is a key problem in FMS design and planning because process planning is the interface between the computer and the machine tool sub-system in an FMS in regard to part-type manufacturing. More important, efficient process planning is a prerequisite for an FMS to attain a high production rate while maintaining good machine load balance and utilization (Chang and Wysk, 1985).

Problem Formulation

The process planning and job scheduling problem in a general production system can be formulated as a mathematical program (Lawrence and Zanakis, 1984; Jacobs and Mabert,

1986; Johnson and Montgomery, 1974). Over the years, formulation and solution of production planning problems for FMS have been undertaken by various researchers (see, for example, Falk, 1980; Hayes et al, 1981; Kusiak, 1984; Stecke, 1983; Suri and Whitney, 1984; Toyoda, 1975; Whitney and Gaul, 1984). It appears that none of these studies explicitly considers the cost of transporting part-types in the system. In this study, we use an objective function that considers the *total cost* of part-type production and part-type transportation among the machines. The constraint sets include part-type assignment restriction, machine capacity as well as machine loads balancing. Specifically, our formulation of the FMS processing planning problem is as follows:

$$\text{Minimize } K = \sum_{k=1}^{m}\sum_{i=1}^{n}\sum_{j=1}^{g(i)} C_k p_{ijk} x_{ijk} + C_t \sum_{i=1}^{n}\sum_{j=2}^{g(i)}\sum_{k=1}^{m} x_{ijk}\left(\sum_{s=1}^{m} t_{sk} x_{i,j-1,s}\right) \quad (1)$$

subject to

$$\sum_{k=1}^{m} x_{ijk} = 1 \qquad \text{for } i = 1, 2, ..., n \text{ and } j = 1, 2, ..., g(i) \quad (2)$$

$$x_{i11} = 1 \qquad \text{for } i = 1, 2, ..., n \quad (3)$$

$$x_{i,g(i),1} = 1 \qquad \text{for } i = 1, 2, ..., n \quad (4)$$

$$\sum_{i=1}^{n}\sum_{j=1}^{g(i)} p_{ijk} x_{ijk} \leq N_k \qquad \text{for } k = 1, 2, ..., m \quad (5)$$

$$\left|\sum_{i=1}^{n}\sum_{j=1}^{g(i)} p_{ijk} x_{ijk} - \sum_{i=1}^{n}\sum_{j=1}^{g(i)} p_{ijs} x_{ijs}\right| \leq \delta \qquad \text{for } k = 1, 2, ..., m-1,\ s = k+1, ..., m \quad (6)$$

where

i = 1, 2, ..., n is an index for the part-types and n is total number of part-types in the system,

j = 1, 2, ..., g(i) is an index for the operations of part-type i and g(i) is the total number of operations of part-type i,

k = 1, 2, ..., m is an index for the machines with $k = 1$ being the loading/unloading station on which operations 1 and g(i) of any part-type i must be performed and m is the total number of machines in the system,

p_{ijk} = processing time of operation j of part-type i on machine k,

$$x_{ijk} = \begin{cases} 1 \text{ if operation j of part - type i is assigned to machine k} \\ \\ 0 \text{ otherwise} \end{cases},$$

t_{ks} = transportation time of a part-type from machine k to machine s,

C_k = cost per unit processing time on machine k,

C_t = cost per unit transportation time of the material handing system,

N_k = total number of machine hours available on machine k,

δ = maximum machine load imbalance allowed.

Generally there are two approaches to solving a non-linear mathematical program. First the problem can be solved directly using a standard non-linear programming computer software such as EUREKA, GINO and STORM. However solving a complex non-linear mathematical program such as our process planning problem is a very difficult, if not impossible, task because of the prohibitive computational time and memory storage requirements. Alternatively, simple heuristics can be developed to obtain good feasible solutions.

In this study a simulation experiment is designed to generate various FMS configurations and some simple heuristics are developed to solve the process planning problems in each of the randomly generated FMS environments. The performance of each heuristic is evaluated by comparing its solution with the optimal solution obtained by some direct solution method. This enables identification of the FMS characteristics that determine the appropriate heuristics to use.

Heuristics

For each part-type i, we compute a quantity E_i as follows:

$$E_i = p_{i11} + \sum_{j=1}^{g(i)-1} \left\{ \min_{1 \le s \le m} \left(p_{i,j+1,s} + t_{k(i,j),s} \right) \right\} \tag{7}$$

where k(i,j) denotes the machine selected to perform operation j of part-type i; $p_{i,g(i),s} = p_{i,g(i),1}$; $t_{k(i,1),s} = t_{1s}$ and $t_{k(i,g(i)-1),s} = t_{k(i,g(i)-1),1}$, for s = 1, 2, ..., m. Intuitively, E_i is a *crude* estimate of the minimum amount of time (both processing and transportation times) necessary to process a unit of part-type i under the condition of no capacity limits. In fact, the problem of finding the minimum total processing and transportation times is a travelling salesman problem which is notoriously hard to solve (Garey and Johnson, 1979; Rinnooy Kan, 1976).

Three heuristics used to find an approximation solution to the non-linear programming problem (1) are considered in this study. The part-types are first arranged in ascending order of E_i, for i = 1, 2, ..., n and are re-indexed accordingly. Then a processing sequence (i.e. manufacturing routing) for each part-type is determined by the heuristics discussed below:

<u>Heuristic 1 (Processing and Transportation Times, PTT)</u>:

Suppose the current operation is $1 \le j < g(i)$ of part-type i being performed on machine k(i,j). Heuristic 1 selects the machine, k(i,j+1) = u, for the next operation j+1, such that

$$\min_{1 \le s \le m} \left(p_{i,j+1,s} + t_{k(i,j),s} \right) = p_{i,j+1,u} + t_{k(i,j),u}$$

subject to the capacity constraint on the selected machine u. If the capacity constraint on u is violated, it then selects another machine v such that

$$\min_{1 \le s \ne u \le m} \left(p_{i,j+1,s} + t_{k(i,j),s} \right) = p_{i,j+1,v} + t_{k(i,j),v}$$

This process can be repeated until a machine for operation j+1 is found. If no machine can be found to process operation j+1 without violating the capacity constraint, then the heuristic stops and declares that the FMS process planning problem has no feasible solution.

Heuristic 2 (Processing Time Only, PTO):

Suppose the current operation is $1 \leq j < g(i)$ of part-type i being performed on machine k(i,j). Heuristic 2 selects the machine, k(i,j+1) = u, for the next operation j+1, such that

$$\min_{1 \leq s \leq m} \left(p_{i,j+1,s}\right) = p_{i,j+1,u}$$

subject to the capacity limit on the selected machine u. If the capacity available on u is exceeded, it then selects another machine v such that

$$\min_{1 \leq s \neq u \leq m} \left(p_{i,j+1,s}\right) = p_{i,j+1,v}$$

This process can be repeated until a machine for operation j+1 is found. If no machine can be found to process operation j+1, then the heuristic stops and declares that the FMS process planning problem has no feasible solution.

Heuristic 3 (Transportation Time Only, TTO):

Suppose the current operation is $1 \leq j < g(i)$ of part-type i being performed on machine k(i,j). Heuristic 3 selects the machine, k(i,j+1) = u, for the next operation j+1, such that

$$\min_{1 \leq s \leq m} \left(t_{k(i,j),s}\right) = t_{k(i,j),u}$$

subject to the capacity constraint on the selected machine u. If the capacity constraint on u is violated, it then selects another machine v such that

$$\min_{1 \leq s \neq u \leq m} \left(t_{k(i,j),s}\right) = t_{k(i,j),v}$$

This process can be repeated until a machine for operation j+1 is found. If no machine can be found to process operation j+1, then the heuristic stops and declares that the FMS process planning problem has no feasible solution.

Experimental Design

A simulation model of an FMS is constructed and used to evaluate the performances of the heuristics under different simulated operating conditions. For the sake of simplicity, we assume both C_k and C_t are equal to unity and relax the machine balancing constraint in the problem formulation, i.e. we let $\delta \rightarrow \infty$ in constraint (6). However, we use the ratio of average processing time to average transportation time as a control variable in the simulation experiments to study the effect of processing and transportation times on the performances of the heuristics.

Let H_T be the total processing and transportation times required to produce all part-types under heuristic T, for T = 1, 2, 3. Let $E = \sum_{i=1}^{n} E_i$ be the *estimated* minimum total time required to produce all part-types. The performance of heuristic T is P_T, defined as

$$P_T = \left(\frac{H_T - E}{E}\right) x100\% \tag{8}$$

Note that since E is only an estimate of the minimum time, P_T can be any real number. In general, the smaller the value of P_T implies the better the performance of a heuristic.

Let M_T be the total processing time required to produce all part-types under heuristic T and $N = \sum_{k=1}^{m} N_k$ the total number of machine hours available in the system. Another criterion used to measure the effectiveness of a heuristic T is the relative makespan (i.e. machine utilization rate) U_T, defined as

$$U_T = \frac{M_T}{N} x100\% \tag{9}$$

Since a smaller value of U_T means that it takes fewer machine hours to produce all part types under heuristic T, the smaller the value of U_T implies the more effective of a heuristic for more part-types can be produced with the same level of machine capacity.

Different FMS configurations are generated according to different combinations of the following program parameters installed in the simulation model.

(a) The number of machines, m, is set at three levels, 5, 10 and 15.

(b) The number of part-types, n, is also set at three levels, 5m, 10m and 15m.

(c) The ratio of average processing time to average transportation time, r, is set at three levels, 1/5, 1 and 10.

For each possible combination of the above parameters, the following processing characteristics of the part-types are generated:

(a) The transportation time between any pair of machines k and s, t_{ks}, for $k \neq s$, is randomly generated from a discrete uniform distribution U(1, 9) with $t_{ks} = 0$ if $k = s$.

(b) The number of operations for each part-type, g(i), is randomly generated from a discrete uniform distribution U(3, 2m-1).

(c) The processing time of each operation, p_{ijk}, is randomly generated from a discrete uniform distribution U(1, 10r-1).

Finally, the total number of machine hours available on machine k = 1, 2, ..., m is determined by

$$N_k = \frac{qn\overline{pg}}{m} \tag{10}$$

where q, expressed as a percentage, is a simulation control factor for machine capacity; $\overline{p}$ is the average processing time per operation; $\overline{g}$ is the average number of operations per job. It should be noted that q, $\overline{p}$ and $\overline{g}$ are all functions of a particular FMS configuration generated.

For each FMS configuration, we generate three replicas using a different stream of (pseudo)random numbers from the computer, so a total of $3 \times 3 \times 3 \times 3 = 81$ simulation runs are performed for each of the three heuristics. Since there are three cases in each of the FMS configurations, we can find an average value of P_T and U_T for the heuristic T = 1, 2, 3. At this point, we can compare the performances of the heuristics under $3 \times 3 \times 3 = 27$ identical FMS configurations. We are also able to see how their performances change as these configurations vary.

Results and Discussions

In order to compare the performances of the three heuristics, we have compiled Tables 1, 2 and 3 showing the values of P_T and U_T resulting from various FMS configurations, for respective levels of the ratio of average processing time to average transportation time, r = 1/5, 1 and 10. We can readily make several observations from the data shown in these Tables.

(i) The average values of U_T are lowest at r = 10 and increase with increase in r, attaining the highest level at r = 1/5. This is so because of the way in which the machine capacity is determined in the simulation experiments. As r increases, the average processing time increases and so does the machine capacity, which is determined by eq. (10), thus leading to a lower value of U_T.

(ii) There are only very slight variations in U_T for the different heuristics at r = 1 while the U_T values are identical for r = 1/5 and 10. This suggests that the machine capacity used in the simulation experiments may not be limiting enough to have a significant impact on the performance of the heuristics. Specifically, we use q = 80% for r = 1 and r = 10 and q = 300% for r = 1/5. It is conjectured that if we had used smaller values of q to determine N_k in eq. (10), the limiting effect of machine capacity on the effectiveness of the heuristics would have been more explicitly revealed.

(iii) At r = 1/5, the performance of the heuristics in ascending order of P_T is TTO, PTT, and PTO. This should be an expected result because when the transportation time is dominant the heuristic that selects machines on the basis of transportation times between machines

# of Machines	# of Parts	Trial #	Heuristic 1 (PTT) UT	Heuristic 1 (PTT) PT	Heuristic 2 (PTO) UT	Heuristic 2 (PTO) PT	Heuristic 3 (TTO) UT	Heuristic 3 (TTO) PT
5	25	1	73.30%	-55.40%	73.30%	5.60%	73.30%	-55.40%
		2	81.70%	-46.10%	81.70%	-8.50%	81.70%	-41.20%
		3	85.00%	63.20%	85.00%	56.50%	85.00%	56.50%
		Average	**80.00%**	**-12.77%**	**80.00%**	**17.87%**	**80.00%**	**-13.37%**
	50	1	81.10%	-40.70%	81.10%	-24.90%	81.10%	-48.60%
		2	77.50%	-55.40%	77.50%	-15.00%	77.50%	-51.50%
		3	85.80%	-56.90%	85.80%	0.00%	85.80%	41.80%
		Average	**81.47%**	**-51.00%**	**81.47%**	**-13.30%**	**81.47%**	**-19.43%**
	75	1	83.10%	-46.90%	83.10%	-8.20%	83.10%	-42.90%
		2	83.30%	-65.70%	83.30%	-1.60%	83.30%	-65.30%
		3	80.20%	-45.49%	80.20%	-15.50%	80.20%	-41.60%
		Average	**82.20%**	**-52.70%**	**82.20%**	**-8.43%**	**82.20%**	**-49.93%**
10	50	1	79.20%	-71.00%	79.20%	49.00%	79.20%	-73.30%
		2	85.80%	-65.50%	85.80%	-14.80%	85.80%	-64.00%
		3	79.40%	-25.70%	79.40%	-11.60%	79.40%	-72.00%
		Average	**81.47%**	**-54.07%**	**81.47%**	**7.53%**	**81.47%**	**-69.77%**
	100	1	86.20%	67.40%	86.20%	17.20%	86.20%	-66.90%
		2	89.40%	-37.80%	89.40%	12.30%	89.40%	-40.10%
		3	82.20%	-66.90%	82.20%	-11.80%	82.20%	-64.70%
		Average	**85.93%**	**-12.43%**	**85.93%**	**5.90%**	**85.93%**	**-57.23%**
	150	1	84.60%	-75.60%	84.60%	-12.20%	84.60%	-74.50%
		2	85.90%	-79.60%	85.90%	4.80%	85.90%	-91.00%
		3	88.50%	-63.30%	88.50%	-11.80%	88.50%	-60.40%
		Average	**86.33%**	**-72.83%**	**86.33%**	**-6.40%**	**86.33%**	**-75.30%**
15	75	1	86.40%	-73.20%	86.40%	-1.10%	86.40%	-72.90%
		2	88.20%	-78.60%	88.20%	-3.60%	88.20%	-78.30%
		3	78.70%	-18.60%	78.70%	12.80%	78.70%	-19.10%
		Average	**84.43%**	**-56.80%**	**84.43%**	**2.70%**	**84.43%**	**-56.77%**
	150	1	83.20%	-68.90%	83.20%	-11.46%	83.20%	-66.90%
		2	81.20%	-54.70%	81.20%	-5.90%	81.20%	-55.40%
		3	88.40%	-72.30%	88.40%	28.60%	88.40%	-74.00%
		Average	**84.27%**	**-65.30%**	**84.27%**	**3.75%**	**84.27%**	**-65.43%**
	225	1	84.60%	-76.40%	84.60%	16.70%	84.60%	-78.10%
		2	84.20%	-80.80%	84.20%	22.20%	84.20%	-82.10%
		3	80.10%	-75.80%	80.10%	10.50%	80.10%	-76.90%
		Average	**82.97%**	**-77.67%**	**82.97%**	**16.47%**	**82.97%**	**-79.03%**
Grand Average			**83.23%**	**-50.62%**	**83.23%**	**2.90%**	**83.23%**	**-54.03%**

Note: / denotes infeasible schedule
Machine Capacity = 300%*(average processing time*average # of operations per job* # of jobs/ # of machines)

Table 1 Simulation Results
Ratio of Average Processing Time to Average Transportation Time is r = 1/5

			Heuristic 1 (PTT)		*Heuristic 2 (PTO)*		*Heuristic 3 (TTO)*	
# of Machines	**# of Parts**	**Trial #**	**UT**	**PT**	**UT**	**PT**	**UT**	**PT**
5	25	1	/	/	/	/	/	/
		2	/	/	/	/	/	/
		3	/	/	/	/	/	/
		Average	/	/	/	/	/	/
	50	1	/	/	/	/	/	/
		2	/	/	/	/	/	/
		3	/	/	/	/	/	/
		Average	/	/	/	/	/	/
	75	1	/	/	/	/	/	/
		2	/	/	/	/	/	/
		3	/	/	/	/	/	/
		Average	/	/	/	/	/	/
10	50	1	85.10%	-0.30%	85.20%	2.50%	85.50%	-6.00%
		2	79.50%	0.70%	79.50%	0.00%	79.70%	13.90%
		3	89.00%	-0.20%	88.90%	1.77%	91.70%	21.20%
		Average	**84.53%**	**0.07%**	**84.53%**	**1.42%**	**85.63%**	**9.70%**
	100	1	82.60%	1.10%	82.60%	2.30%	82.30%	0.70%
		2	79.80%	0.00%	79.80%	1.40%	79.80%	13.50%
		3	86.90%	0.70%	86.70%	3.70%	87.80%	15.50%
		Average	**83.10%**	**0.60%**	**83.03%**	**2.47%**	**83.30%**	**9.90%**
	150	1	82.60%	0.90%	82.40%	0.80%	82.60%	18.90%
		2	88.30%	-0.80%	88.00%	0.20%	88.20%	11.40%
		3	83.00%	0.00%	83.10%	2.00%	83.10%	16.40%
		Average	**84.63%**	**0.03%**	**84.50%**	**1.00%**	**84.63%**	**15.57%**
15	75	1	72.90%	0.50%	72.90%	0.40%	72.80%	-2.30%
		2	75.90%	-1.21%	75.80%	1.73%	75.80%	-3.90%
		3	77.10%	0.20%	77.10%	3.99%	77.60%	26.80%
		Average	**75.30%**	**-0.17%**	**75.27%**	**2.04%**	**75.40%**	**6.87%**
	150	1	72.30%	0.50%	72.30%	0.80%	72.40%	1.08%
		2	70.70%	-1.90%	70.80%	2.39%	70.70%	36.00%
		3	72.80%	1.60%	72.90%	3.92%	72.90%	5.60%
		Average	**71.93%**	**0.07%**	**72.00%**	**2.37%**	**72.00%**	**14.23%**
	225	1	72.90%	-0.40%	72.90%	0.70%	72.90%	0.00%
		2	72.90%	0.10%	72.90%	0.57%	73.00%	40.80%
		3	74.80%	-1.20%	74.70%	0.31%	75.00%	-8.30%
		Average	**73.53%**	**-0.50%**	**73.50%**	**0.53%**	**73.63%**	**10.83%**
	Grand Average		**78.84%**	**0.02%**	**78.81%**	**1.64%**	**79.10%**	**11.18%**

Note: / denotes infeasible schedule
Machine Capacity = 80%*(average processing time*average # of operations per job* # of job/ # of machines)

Table 2 Simulation Results
Ratio of Average Processing Time to Average Transportation Time is r = 1

			Heuristic 1 (PTT)		*Heuristic 2 (PTO)*		*Heuristic 3 (TTO)*	
# of Machines	# of Parts	Trial #	UT	PT	UT	PT	UT	PT
5	25	1	/	/	/	/	/	/
		2	/	/	/	/	/	/
		3	/	/	/	/	/	/
		Average	/	/	/	/	/	/
	50	1	/	/	/	/	/	/
		2	/	/	/	/	/	/
		3	/	/	/	/	/	/
		Average	/	/	/	/	/	/
	75	1	/	/	/	/	/	/
		2	/	/	/	/	/	/
		3	/	/	/	/	/	/
		Average	/	/	/	/	/	/
10	50	1	65.40%	-0.97%	65.40%	0.00%	65.40%	-3.40%
		2	60.80%	0.00%	60.80%	0.00%	60.80%	1.90%
		3	67.60%	1.77%	67.60%	0.00%	67.60%	4.60%
		Average	**64.60%**	**0.27%**	**64.60%**	**0.00%**	**64.60%**	**1.03%**
	100	1	64.10%	2.30%	64.10%	0.00%	64.10%	1.30%
		2	61.60%	1.40%	61.60%	0.00%	61.60%	4.30%
		3	66.70%	3.70%	66.70%	0.00%	66.70%	4.70%
		Average	**64.13%**	**2.47%**	**64.13%**	**0.00%**	**64.13%**	**3.43%**
	150	1	64.30%	0.80%	64.30%	0.00%	64.30%	5.60%
		2	68.40%	0.20%	68.40%	0.00%	68.40%	3.30%
		3	64.90%	2.00%	64.90%	0.00%	64.90%	5.50%
		Average	**65.87%**	**1.00%**	**65.87%**	**0.00%**	**65.87%**	**4.80%**
15	75	1	50.20%	0.40%	50.20%	0.00%	50.20%	-2.70%
		2	52.70%	1.73%	52.70%	0.00%	52.70%	0.00%
		3	53.10%	3.99%	53.10%	0.00%	53.10%	8.00%
		Average	**52.00%**	**2.04%**	**52.00%**	**0.00%**	**52.00%**	**1.77%**
	150	1	50.20%	0.80%	50.20%	0.00%	50.20%	1.40%
		2	49.40%	2.39%	49.40%	0.00%	49.40%	8.90%
		3	50.50%	3.92%	50.50%	0.00%	50.50%	-1.30%
		Average	**50.03%**	**2.37%**	**50.03%**	**0.00%**	**50.03%**	**3.00%**
	225	1	50.60%	0.70%	50.60%	0.00%	50.60%	1.10%
		2	50.60%	0.57%	50.60%	0.00%	50.60%	11.10%
		3	51.30%	0.31%	51.30%	0.00%	51.30%	-1.90%
		Average	**50.83%**	**0.53%**	**50.83%**	**0.00%**	**50.83%**	**3.43%**
	Grand Average		**46.70%**	**0.83%**	**46.70%**	**0.00%**	**46.70%**	**4.15%**

Note: / denotes infeasible schedule

Machine Capacity = 80%*(average processing time*average # of operations per job* # of jobs/ # of machines)

Table 3 Simulation Results
Ratio of Average Processing Time to Average Transportation Time is r = 10

will work best. However, it is interesting to note that PTT, which considers both transportation and processing times, is a close second to TTO, the best performer. This suggests that, even when transportation time is dominant, processing time should also be taken into consideration in selecting machines.

(iv) At r = 1, the performance of the heuristics in ascending order of P_T is PTT, PTO, and TTO. While we expect PTT will work best under this situation, the results indicate that PTO achieves a marginally inferior performance than PTT. However, as anticipated, the performances of both PTO and PTT are significantly better than that of TTO.

(v) At r = 10, the performance of the heuristics in ascending order of P_T is PTO, PTT and TTO. This result is expected because when the processing time is dominant the heuristic that selects machines on the basis of operation processing times will work best. Once again, we see that a heuristic that considers both processing and transportation times is very effective although the situation is dominated by processing times.

Conclusion

We have formulated the FMS process planning problem as a non-linear 0-1 integer program. In view of the difficulty of solving the resulting highly complex non-linear program analytically, we have developed three simple heuristics in hopes of obtaining good approximate solutions to the problem. A simulation model of a hypothetical FMS has been constructed and simulation experiments have been performed to evaluate the effectiveness of the heuristics under a variety of simulated FMS environments. The simulation results obtained seem to indicate that the most effective heuristic, even under vastly different FMS configurations, will be the one that considers both processing times and transportation times in determining the processing sequences for part-types. Thus, the findings of this study suggest that the class of 'composite' heuristics should be the focus of future research on heuristic approach to FMS processing planning.

Acknowledgement

This research was supported in part by a grant from the Associates Fund, Faculty of Management, University of Manitoba.

References

Chang, T.C. and Wysk, R.A. 1985. *An Introduction to Automated Process Planning Systems*. Prentice-Hall, New Jersey.

Falk, P.G. 1980. Experiments in mixed integer programming in a manufacturing system. *OMEGA*, **8**, 149-160.

Garey, M.R. and Johnson, D.S. 1979. *Computers and Intractability: A Guide to the Theory of NP-Completeness*. W. H. Freeman, New York.

Hayes, G.M., Davis, R.P. and Wysk, R.A. 1981. A dynamic programming approach to machine requirements planning. *AIIE Transactions*, **13**, 89-95.

Lawrence, K.D. and Zanakis, S.H. 1984. *Production Planning and Scheduling: Mathematical Programming Applications*. Industrial Engineering and Management Press, Georgia.

Jacobs, F.R. and Mabert, V.A. (eds.). 1986. *Production Planning, Scheduling, and Inventory Control: Concepts, Techniques and Systems*. Industrial Engineering and Management Press, Georgia.

Johnson, L.A. and Montgomery, D.C. 1974. *Operations Research in Production Planning, Scheduling and Inventory Control*. John Wiley & Sons, New York.

Kusiak, A. 1984. The part families problem in flexible manufacturing systems. *Proceedings of 1st ORSA/TIMS Special Interest Conference on FMSs*, 237-243.

Rinnooy Kan, A.H.G. 1976. *Machine Scheduling Problems: Classification, Complexity and Computations*. Martinus Nijhoff, the Hague.

Stecke, K.E. 1981. *Production Planning Problems for Flexible Manufacturing Systems*. Unpublished Ph.D. Dissertation, Purdue University.

Stecke, K.E. 1983. Formulation and solution of nonlinear integer production planning problems for flexible manufacturing systems. *Management Science*, **29**, 273-290.

Suri, R. and Whitney, C.K. 1984. Decision support requirements in flexible manufacturing.

Journal of Manufacturing Systems, **3**, 61-68.

The Charles Stark Draper Laboratory. 1984. *Flexible Manufacturing Systems Handbook.* Noyes Publications, New Jersey.

Toyoda, Y. 1975. A simplified algorithm for obtaining approximate solutions to zero-one programming problems. *Management Science*, **21**, 1417-1427.

Whitney, C.K. and Gaul, T.S. 1984. Sequential decision procedures for batching and balancing in FMSs. *Proceedings of 1st ORSA/TIMS Special Interest Conference on FMSs*, 243-256.

A Review of Research on AGVS Vehicle Management

C.G. CO and JOSE TANCHOCO

0. Preliminary Remark

There are various interacting elements that affect the performance of an AGVS. Because the effectiveness of an AGVS is very sensitive to operational design parameters, several researchers have addressed issues related to network layout [Gaskins and Tanchoco 1987, Gaskins et al. 1989, Kaspi and Tanchoco 1989, Bozer and Srinivasan 1989], location of pick-up/delivery stations [Usher et al. 1988, Rabeneck et al. 1989], fleet sizing [Egbelu 1987, Leung et al. 1987, Tanchoco et al. 1987], unit load sizing [Egbelu 1987, Steudel and Moon 1987], and vehicle management. This paper outlines the vehicle management control strategies discussed in previous studies on AGVS.

1. AGVS Vehicle Management Strategies

AGVS control terminology used by various researchers have not been consistent. For purposes of this paper, the primary AGVS vehicle management functions are defined below.

- Dispatching is the process of selecting and assigning tasks to vehicles.
- Routing is the selection of the specific paths taken by vehicles to reach their destinations.
- Scheduling is the determination of the arrival and departure times of vehicles at each segment along their prescribed paths to ensure collision-free journeys.

1.1. Dispatching Strategies

Dispatching policies determine which vehicle should be assigned to perform a task in order to satisfy objectives set by the system planner. The following discussion covers only dispatching strategies applicable to "pick and drop" type systems where the AGV tasks are limited to transporting loads from their current locations to the next workcenters in their process routes. "Stop and go" AGV systems such as those found in assembly applications will not be considered.

Dispatching conditions have been characterized in the AGV literature by the different combinations of the factors outlined below.

A. Level of Control Automation
 1. manual
 2. automatic

B. Pickup and Delivery Points of Loads Served by a Vehicle
 1. one-to-one
 2. one-to-many
 3. many-to-one
 4. many-to-many

C. Status of Vehicles
 1. unassigned and busy vehicles considered for task assignment
 2. only currently unassigned vehicles are dispatched
  a. one task for several vehicles
  b. several tasks for one vehicle

D. Nature of Transport Demand
 1. static or steady-state
 2. stochastic

E. Carrying Capacity of Vehicles
 1. single load
 2. multiple load

Under the manual mode of dispatching, an operator programs the AGV destination for each trip. Empty vehicles either wait at workcenters until a new load is ready or circle around the network until they are flagged down for a load pickup. The manual mode of operation can be effective if the number of pickup and delivery stations is limited, the distances between transfer stations are long, or the transport requests occur infrequently. Manual dispatching is simple and can be performed using on-board control equipment. However, a dispatcher always has to be present at the workcenters to program the vehicles' next destinations. Furthermore, task allocation is not optimized because each vehicle is assigned independently of the other vehicles [Wirth 1983].

In more complex systems where there is a large number of vehicles transporting loads over the whole network, a higher level of automation is necessary for efficient dispatching. A central controller receives and prioritizes all transport orders. Information on the location and status of each vehicle facilitates the selection of the best vehicle to satisfy a given task [Wirth 1983]. This mode of operation also allows more options for empty vehicle distribution. Empty vehicle management is discussed in more detail later.

Manual control and centralized control present two extremes in control automation. A more detailed discussion of these two and other methods can be found in Koff [1987].

The following discussion on the classification of pickup/delivery configurations is based on the paper by Wirth [1983]. The one-to-one pickup/delivery load service design is suitable for layouts where the rate that transport requests are made is stable and well-distributed across the network. Under this system, each vehicle is assigned to pick up loads from a predefined pick-up station and deliver loads to a predefined drop-off station. In such environments, it is possible to create specialized routes for the AGVs without penalizing vehicle utilization.

The one-to-many plan is implemented in facilities for distribution of goods where loads are picked up from a central location and delivered to many destinations. The many-to-one design is found in facilities where loads are picked up from many locations and delivered to a single centralized destination. In job shops or flexible manufacturing facilities where transport patterns are unpredictable, many-to-many assignment for AGVs is more common.

Dispatching for one-to-one and one-to-many AGVS can be trivial because all load carriers can be identically programmed to return to the singular source of materials after every dropoff activity. Wirth [1983] suggested some simple strategies for many-to-one dispatching such as arranging the stations in a circular layout or permanently assigning AGVs to specific pickup stations. Centralization and automation provide better solutions to the many-to-many dispatching problems.

The set of vehicles which can feasibly satisfy a transport order should be determined before a dispatching decision can be made. The type, size or status of the AGV are possible restrictions. Most papers describe systems where only one vehicle type is used. Thus, an assumption is made that all carriers can transport any unit load regardless of type and size. Even in the model presented by Leung et al. [1987] where several vehicle types are accommodated, the model was formulated such that only one type of vehicle is used to serve any given pair of stations. In effect, the network can be decomposed into smaller areas within which identical AGVs operate.

In most models, vehicle status is the only factor that dictates whether an AGV is a possible candidate for task assignment or not. There are several methods of identifying vehicle status. For purposes of dispatching, identifying that AGVs are busy or unassigned is more informative than labeling them as loaded or empty. In this way, an empty AGV traveling to pick up a load is differentiated from an empty AGV waiting for an assignment. A vehicle in a busy state can either be traveling with a load or be traveling empty to pick up an assigned load.

There are two possible approaches to task allocation with respect to vehicle status. One approach is to allocate a vehicle to a transport order as soon as the request is received. The second approach is to defer the task assignment until a vehicle has completed its current task. In the more general case where multiple tasks are assigned to vehicles the task assignment is deferred until the vehicle completes its last assigned task.

Gunsser [1983] referred to the first approach as pre-planning where every vehicle can be allocated a sequentially ordered list of jobs to accomplish. With this strategy, the vehicle with the most favorable connecting run for a certain order is selected even while it is still performing a previous order [Beisteiner and Moldaschl 1983]. The prototype controller described in Taghaboni and Tanchoco [1988] can conceptually dispatch a vehicle regardless of its status. Considering that currently busy vehicles can begin a new task as soon as they are released from their last assignment and that currently idle vehicles can start a new task immediately, the vehicle that can accomplish the new task soonest is selected.

Pre-planning can be implemented in a different form. Instead of allocating tasks sequentially, the set of all outstanding transport orders can be assigned simultaneously. Each AGV is designated to a "tour" which is a sequence of move transactions that start and end at a given node. The tours are formed such that empty travel distances between the delivery of one load to the pickup station of the next load is minimized [Maxwell and Muckstadt 1982, Blair et al. 1987]. Again, more than one transport order can be associated with an AGV.

The advantage of pre-planning is that all vehicles compete to satisfy a given task [Beisteiner and Moldaschl 1983]. The concept of tour formation contributes further to minimizing empty vehicle travel in that different permutations of the transport orders are evaluated. However, there is an implicit assumption that the control system is able to store and monitor the task lists for all vehicles.

The second approach where only currently unassigned vehicles can be allocated to a task is more common. Under this mode of operation, two situations are possible when a move request is received by the central dispatcher: (a) one or more vehicles are immediately available, or (b) all vehicles are busy and are not available. When one or more vehicles are available, then a vehicle can be selected based on the workcenter-initiated task assignment rule in effect. On the other hand, if all vehicles are busy, the move request will be put in a queue of pending transport orders. When a vehicle completes the delivery of a load, the dispatcher selects a transport order from the queue and assigns this to the vehicle. When two or more workcenters have outstanding requests, the transport order is selected using vehicle-initiated task assignment rules [Egbelu 1982, Egbelu and Tanchoco 1984].

Heuristic rules for both workcenter-initiated and vehicle-initiated task have been proposed by various researchers. However, empirical results show that with reasonably high AGV utilization levels, only the vehicle-initiated task assignment rule is significant [Egbelu 1982, Egbelu and Tanchoco 1984].

The basic idea behind vehicle-initiated task assignment rules is prioritization of workcenters with outstanding move requests based on some parameter such as distance from vehicle, length of the outgoing load queue, outgoing buffer capacity or elapsed time since a move request was transmitted. In some cases, weighted combinations of these parameters can be computed as in the heuristic rule developed by Hodgson et al. [1987] using Markov decision processes. Using this rule, a workcenter becomes desirable if it is close to the current location of the vehicle, if the volume of transport orders from this workcenter is high or if the load to be picked up from this workcenter is to be delivered to another workcenter where a load is waiting. Workcenters are assigned a priority value based on the selected parameter. When a vehicle is released from its last assignment, it is dispatched to the workcenter with the highest score.

Bakkalbasi and McGinnis [1988] suggested the use of worklists so that an unassigned vehicle only looks at a prespecified list of workcenters for possible load pickups. The reduced number of alternatives can facilitate dispatching.

Variations of the nearest-workcenter-first (NWF) rule are the most commonly cited rule. This rule is referred to as best-follow-up-order by Beisteiner and Moldaschl [1983], shortest-travel-time-first (STTF) by Egbelu and Tanchoco [1984], or vehicle-looks-for-work (VLFW) by Newton [1985]. Under this rule, a vehicle is sent to the closest workcenter with a pending transport order. "Closeness" can be defined in terms of travel time or distance. The popularity of this rule probably stems from the fact that it lucidly attempts to minimize empty travel time. This rule could prove detrimental in layouts where a pickup station is not close to any delivery station [Egbelu and Tanchoco 1984].

Interest in the first-encountered-first-served (FEFS) rule is fueled by its simplicity. It is one of the few strategies that do not require centralized control. This rule is denoted by Egbelu and Tanchoco [1984] as the *sequential dispatching strategy* where the AGVs move continuously along a predefined sequence of nodes forming a closed loop and check every station for a "ready" load. An unassigned vehicle picks up the first "ready" load it encounters in the loop.

In a model where all transport orders are known beforehand and stations are arranged in a single loop layout, the FEFS rule is equivalent to the NWF rule. On the other hand, in an environment where transport calls occur intermittently, different dispatching selections under the two rules will be observed. This is because under the NWF rule only loads already available at the time an assignment decision is required are considered. Once a decision is made, the vehicle ignores any loads that may subsequently become ready along its path towards the first assignment [Egbelu and Tanchoco 1984].

The FEFS rule is most suitable for layouts where a single circular loop that covers all pickup and delivery stations can easily be found [Bartholdi and Platzman 1989]. An illustration of the effectiveness of this rule was shown using an FMS facility evaluated by Ozden [1988] where four workcenters were located along the perimeter of a rectangular area. Bozer and Srinivasan [1988] also used the FEFS rule in developing a performance model for a single loop, single vehicle system. For more complex layouts, the critical issue is to find a guide path layout that will allow vehicles to complete a round of visits to all pickup and delivery stations without traveling along some aisles more than once. Without such a network, the performance of the FEFS rule deteriorates [Egbelu 1982].

A more complete discussion of the effects of vehicle-initiated task assignment rules on overall system performance is found in Egbelu and Tanchoco [1984]. The rules discussed above and some additional ones are summarized in Bozer [1988].

Most dispatching rules assume that transport orders are triggered by loads in the outgoing queue of the pickup station. In a later work, Egbelu [1987] investigated the benefits of demand-driven dispatching in just-in-time operations. With demand-driven dispatching, a transport request is sent by a workcenter waiting for parts. The vehicles are dispatched to pick up the parts from the outgoing queue of another workcenter. The difficulty in implementing this is that a sophisticated part tracking mechanism is necessary.

Another question in dispatching is where to send vehicles waiting for transport orders. The easiest alternative in a system with random transport patterns is to park the AGV close to the workcenter where it last unloads. Doing so may cause congestion in the area around the workcenter. The key to effective idle vehicle disposition is to distribute vehicles so that they are strategically located for future demands. One strategy is to assign a cruising loop for idle vehicles. An unassigned AGV travels around this loop until a transport order is received. Another option is to send the unassigned AGV to a particular parking station [Egbelu and Tanchoco 1984, Duffau and Bardin 1985]. The parking station is usually prescribed as such because the rate of movement from this point is high or because high demand centers are located in its vicinity. Again, congestion near the parking station is a potential problem.

If the amount of parts to be transported from one station to another is predetermined, then empty vehicle travel time can be minimized. The number of empty vehicles that will be needed at a particular station is equal to the number of loads that have to be moved from that station. The number of empty vehicles leaving a delivery station is equal to the loads dropped off at that station. Therefore using the classical assignment model, the number of vehicle movements between each pair of stations can be found. In effect, transport orders are sequenced to form tours. The AGVs run the same tours continuously until all transport demands are satisfied. It should be emphasized that in such static models, factors such as times at which loads become available and blocking or congestion in the system are not weighed [Maxwell and Muckstadt 1982, Blair et al. 1987, Leung et al. 1987].

Dynamic models of production shops are more realistic. The typical approach to solving a dynamic problem is to treat the problem as a series of static models solved on a rolling horizon basis [Rachamadugu et al. 1986]. Thus, one possibility is to use exact algorithms for static models to determine optimal dispatching decisions with the current status of the system. Then, the solution can be updated at regular intervals or whenever a change in the status of the system occurs. Because this methodology can be computationally impractical in real-time operations, heuristic dispatching rules are preferred by practitioners.

The usual mode of heuristic dispatching is that once a vehicle assignment is made, it cannot be changed. This poses a heavy disadvantage in very dynamic environments. For example, a vehicle is dispatched following the nearest-workcenter-first (NWF) rule to a particular workcenter. As mentioned earlier, any additional loads that become available along a vehicle's path are ignored even though this policy contradicts the very definition of the NWF rule. Hodgson et al. [1987] avoid this weakness by recommending in their algorithm that the weighting formula for the workcenters be updated at every station the vehicle encounters. This ensures that the selection of a workcenter is made based on up-to-date information and all changes such as new transport orders are accounted. Again, the main implementation issue is computational practicality.

So far, only dispatching rules for single-load carriers have been reviewed. Despite the potential benefits and market availability of vehicles with multiple-load capacity, very few papers on multiple-load AGVs have been published. FEFS can be implemented for multiple-load carriers where the AGVs pick up any load they encounter as long as there are empty positions on board. Loads are dropped off when their destinations are encountered. Like the single-load version of the FEFS, this rule using multiple-load AGVs is effective for single loop networks [Bartholdi and Platzman 1989] or in simple layouts [Ozden 1988]. Hodgson et al. [1987] provided an extension of their heuristic for two load vehicles. The modified weighting factor for workcenters is biased for dispatching AGVs so that both loads on board will have identical destinations. This control rule is evaluated against a multiple load version of the VLFW rule which the authors referred to as the expanded VLFW heuristic (eVLFW). Similar to VLFW, eVLFW designates empty vehicles to the closest workcenter with an outstanding move request. Loaded vehicles deliver the part whose destination is closest to their current locations. While traveling, carriers stop to pick up all loads encountered that can be accommodated on board. In Tanchoco and Co [1989], alternative control strategies for dispatching AGVs with multiple load carrying capacity are examined. In their work, the multiple load vehicles are allowed to deviate from the original paths in order to pick up additional loads.

Acknowledging the dynamism of the environment, the idea of incorporating look-ahead functions into dispatching has been suggested by numerous researchers [Egbelu 1982, Bozer 1988, Taghaboni and Tanchoco 1988]. Two questions will have to be answered before this direction can be pursued. First, how far into the future should the planning horizon be in order to improve significantly the performance of dispatching policies? Second, how will the future conditions of the system be forecast reliably without unreasonably taxing the computer system?

An alternative to "looking ahead" which has not been explored by AGV planners is "waiting it out". This means that assignments are deferred until more information is available.

Finally, a comment should be made that no specific dispatching strategy is consistently the best. The performance of dispatching rules is highly dependent on the guide path layout, the fleet size and the transport patterns in the network. In addition, each strategy addresses the task assignment problem from different angles. Therefore, the ratings would vary depending on the criteria used for evaluation.

1.2. Routing and Scheduling

An AGVS network can be modeled as a graph consisting of nodes connected by a set of arcs. The nodes represent points on the network such as intersections and loading or unloading stations. At these nodes, the next immediate segment in the path is selected by an AGV traversing a route. Traffic conditions are also checked to evaluate the feasibility of allowing the AGV to proceed. The arcs define the system of physical or virtual guide paths showing directed segments that vehicles can pass in moving from node to node. Associated with each arc is a cost which denotes either the distance between the two end points of the segment or the time required by an AGV to cross the arc.

This graph is the primary input to the routing function of an AGV control system. Given the location of an AGV and its prescribed destination, the route manager can find the sequence of nodes that specify the path of the vehicle. Alternative routes can be evaluated based on the aggregate cost of traversing the sequence of arcs.

Routing can be static or dynamic [Hodgson et al. 1987]. When routing is static, the path taken by an AGV between any two given nodes is fixed. The simplest solution is to follow the shortest path all the time.

When routing is dynamic, different paths can be taken by an AGV in moving from node to node. Taking into consideration the current traffic configuration, the route manager selects a path for the AGV at the time when the vehicle is dispatched.

There are essentially two aspects in preventing collisions. First, rear end collisions should be avoided. Second, the flow of vehicles into an intersection or junction should be controlled [Egbelu and Tanchoco 1982, Wirth 1983, Taghaboni and Tanchoco 1988].

Sensing elements on the vehicles can keep the AGVs from running too close to each other [Wirth 1983]. However, zone control is a more popular and reliable technique. In implementing zone control, vehicles travel from one zone to another and are prohibited from entering a zone already occupied by another vehicle [Koff 1987].

Priority right-of-ways for vehicles arriving at an intersection should be established. Usually, vehicles are allowed to cross the intersection in the same order in which they arrived at the junction, i.e. on a first-come-first-served basis. Egbelu and Tanchoco [1982] suggested alternative rules for intersection control. In Taghaboni and Tanchoco [1988], a subroutine is incorporated in the routing procedure to check if more than one vehicle can pass an intersection simultaneuosly without crossing each other's paths.

Developing such heuristic rules is one approach to reducing vehicle waiting time at intersections. Another approach is minimizing the frequency of vehicles meeting at an intersection. This may be achieved by balancing traffic flows across the network or modifying the guide path layout. The ideal case with zero intersections is of course the circular or single-loop layout.

In simple networks with minimal intersection points, vehicles can trace their routes autonomously using on-board transmitters and receivers. In more complex layouts, traffic management is more efficient if each vehicle can compare its position with the position of every other vehicle [Bose 1986]. Thus, route managers should be able to derive where each traveling vehicle will be at any point in the future.

This forward scheduling capability of the route manager is not necessary but beneficial. Without the scheduling facility, intersection conflicts can be resolved instantaneously when the situation arises and collisions are still prevented. However, with intelligent control systems that can reroute vehicles in order to avoid problem areas, scheduling vehicle movements provides a basis for projecting traffic flow and congestion. Scheduling also predicts when a busy vehicle will be released from its last task. This information is essential if the dispatching strategy adopted allows the selection of busy vehicles.

The scheduling procedures described by Walker et al. [1985] and Taghaboni and Tanchoco [1988] have the same basic framework. The main idea is that no two vehicles should attempt to occupy the same node simultaneously. The route manager initially assumes that the shortest path will be taken. A timetable of arrival and departure times at each node in the route is maintained for all current and future trips. In the event that a conflict arises, i.e. the newly routed vehicle will enter a node at the same time that a previously assigned vehicle is scheduled to arrive at that node, then an alternative route is examined. All alternative routes that do not include the conflict segment are evaluated. If no alternative route is feasible, then the new vehicle is scheduled to slow down. In both these systems, routing is dynamic. However, once a vehicle commences the trip, neither the route nor the schedule can be revised. To resolve conflicts, it is always the latest journey itinerary being inserted to the timetable that is adjusted.

The main difference in the scheduling system for the two works cited is that the controller of Taghaboni and Tanchoco allowed the modeling of multiple lanes between nodes. This flexibility creates situations where vehicles entering a node from different directions can be permitted to do so simultaneously. They proposed a new algorithm based on geometric methods for resolving collision among many vehicles crossing an intersection simultaneously.

2. Summary

The review given in this paper has been limited to the AGVS vehicle management problem. The AGVS vehicle management problem is only one of the many important aspects of the overall AGVS design and control problem. But as we have seen, there has been a great deal of interest in this aspect alone during the last few years. We hope that this review will further spur the growth of this very challenging area of research.

Acknowledgement

The work reported in this paper was supported in part by NSF/Purdue University Engineering Research Center for Intelligent Manufacturing Systems under Grant No. CDR8803017.

References

Bakkalbasi, O. and L. F. McGinnis, "ABC's of Preliminary In-House Planning and Analysis of AGVS Applications," *AGVS '88 Forum and Exhibit,* pp. 13-47, Material Handling Institute, Cincinnati, Ohio, September, 1988.

Bartholdi, J. J. and L. K. Platzman, "Decentralized Control of Automated Guided Vehicles on a Simple Loop," *IIE Transactions,* Vol. 21, pp. 76-81, March, 1989.

Beisteiner, F. and J. Moldaschl, "Strategies for the Employment of Vehicles in an Automated Guided Transportation System," *Proceedings for the 2nd International Conference on AGVS,* pp. 119-128, IFS Conferences Ltd., Stuttgart, W. Germany, June, 1983.

Blair, E. L., P. Charnsethikul, and A. Vasquez, "Optimal Routing of Driverless Vehicles in a Flexible Material Handling System," *Material Flow,* Vol. 4(1-2), pp. 73-83, 1987.

Bose, P. P., "AGV Systems," *American Machinist and Automated Manufacturing,* pp. 106-122, McGraw-Hill Inc., March, 1986.

Bozer, Y. A., "Guided Vehicle Systems: Information/Control System Implications of Alternative Design and Operation Strategies," *Proceedings,* NATO Advanced Research Workshop on Advanced Information Technologies for Industrial Material Flow Systems, Grenoble, France, June 13-17, 1988.

Bozer, Y. A. and M. M. Srinivasan, "Tandem Configurations for AGV Systems Offer Simplicity and Flexibility," *Industrial Engineering,* pp. 23-27, February, 1989.

Duffau, B. and C. Bardin, "Evaluating AGV Circuits By Simulation," *Proceedings of the 3rd International Conference on AGVS,* pp. 229-246, IFS Conferences Ltd., Stockholm, Sweden, October, 1985.

Egbelu, P. J. and J. M. A. Tanchoco, "Operational Considerations for the Design of Automatic Guided Vehicle Based Material Handling Systems," Technical Report No. 8201, Virginia Polytechnic Institute & State University, Blacksburg, Virginia, 1982.

Egbelu, P. J., "A Design Methodology For Operational Control Elements for Automatic Guided Vehicle Based Material Handling Systems," PhD Dissertation, Virginia Polytechnic Institute and State University, Blacksburg, Virginia, July, 1982.

Egbelu, P. J. and J. M. A. Tanchoco, "Characterization of Automatic Guided Vehicle Dispatching Rules," *International Journal on Production Research,* Vol. 22(3), pp. 359-374, 1984.

Egbelu, P. J., "Pull Versus Push Strategy for Automated Guided Vehicle Load Movement in a Batch Manufacturing System," *Journal of Manufacturing Systems*, Vol. 6(3), pp. 209-221, 1987.

Egbelu, P. J., "The Use of Non-simulation Approaches to Estimating Vehicle Requirements in an Automated Guided Vehicle Based Transport System," *Material Flow*, Vol. 24(1-2), pp. 33-51, 1987.

Egbelu, P. J., "Integration of Unit Load Selection in the Design of AGV Based Automated Manufacturing Shop," *Proceedings of the IXth International Conference on Production Research*, pp. 2159-2166, Cincinnati, Ohio, August, 1987.

Gaskins, R. J. and J. M. A. Tanchoco, "Flow Path Design for Automated Guided Vehicle Systems," *International Journal of Production Research*, Vol. 25(5), pp. 667-676, 1987.

Gaskins, R. J., J. M. A. Tanchoco, and F. Taghaboni, "Virtual Flow Path for Free-Ranging Automated Guided Vehicle Systems," *International Journal of Production Research*, Vol. 27(1), pp. 91-100, 1989.

Gunsser, P., "Control Techniques in Guided Vehicle Systems: Examples of Applications and Various Control Concepts," *Proceedings of the 2nd International Conference on AGVS*, pp. 85-113, IFS Conferences Ltd., Stuttgart, W. Germany, June, 1983.

Hodgson, T. J., R. E. King, and S. K. Monteith, "Developing Control Rules for an AGVS Using Markov Decision Processes," *Material Flow*, Vol. 4(1-2), pp. 85-96, 1987.

Kaspi, M. and J.M.A. Tanchoco, "Optimal Flow Path Design of Unidirectional AGV Systems," Technical Report, School of Industrial Engineering, Purdue Univ., West Lafayette, IN, February, 1989, to be published in International Journal of Production Research.

Koff, G.A., "Automatic Guided Vehicle Systems: Applications, Controls and Planning," *Material Flow*, Vol. 4(1-2), pp. 3-16, 1987.

Leung, L. C., S. K. Khator, and D. L. Kimbler, "Assignment of AGVS with Different Vehicle Types," *Material Flow*, Vol. 4(1-2), pp. 65-72, 1987.

Maxwell, W. L. and J. A. Muckstadt, "Design of Automatic Guided Vehicle Systems," *IIE Transactions*, Vol. 14(2), pp. 114-124, 1982.

Newton, D., "Simulation Model Helps Determine How Many Automatic Guided Vehicles are Needed," *Industrial Engineering*, Vol. 17, pp. 68-74, February 2,1985.

Ozden, M., "A Simulation Study of Multiple-Load-Carrying Automated Guided Vehicles in a Flexible Manufacturing System," *International Journal of Production*

Research, Vol. 26(8), pp. 1353-1366, August, 1988.

Rabeneck, C.W., J.S. Usher, and G.W. Evans, "An Analytical Model for AGVS Design," *Proceedings,* 1989 International Industrial Engineering Conference, Toronto, Canada, pp. 191-195, May, 1989.

Rachamadugu, R. V., N. Raman, and F. B. Talbot, "Real Time Scheduling of an Automated Manufacturing Center," *Real Time Optimization in Automated Manufacturing Facilities,* pp. 293-315, NBS, Gaithesburg, Maryland, January, 1986.

Steudel, H. J. and H. K. Moon, "Selection of Unit Load Size for Automated Guided Vehicles in Large Cellular Manufacturing Environments," *Proceedings of the IXth International Conference on Production Research,* pp. 2189-2209, Cincinnati, Ohio, August, 1987.

Taghaboni, F., and J. M. A. Tanchoco, "A LISP-Based Controller for Free-Ranging Automated Guided Vehicle Systems," *International Journal of Production Research,* Vol. 26 (1988), No. 2, pp. 173-188.

Tanchoco, J. M. A., P. J. Egbelu, and F. Taghaboni, "Determination of the Total Number of Vehicles in an AGV- Based System," *Material Flow,* Vol. 4(1-2), pp. 33-51, 1987.

Tanchoco, J.M.A. and C.G. Co, "A Controller for Automated Guided Vehicles with Multiple Load Carrying Capacities," *1988-89 CIDMAC Annual Report,* Schools of Engineering, Purdue University, West Lafayette, IN, pp. 237-238, 1989.

Usher, J., G. Evans, and M. Wilhelm, "AGV Flow Path Design and Load Transfer Point Location," *Proceedings of the 1988 IIE Conference,* pp. 174-179, 1988.

Walker, S. P., S. K. Premi, C. B. Besant , and A. J. Broadbent, "Free-ranging AGV and Scheduling System," *Proceedings of the 3rd International Conference on AGVS,* IFS Conferences Ltd., Stockholm, Sweden, October, 1985.

Wirth, P., "Control System for Robovehicle Installations: Application-Oriented In Spite of Standardization," *Proceedings of the 2nd International Conference on AGVS,* pp. 61-72, IFS Conferences Ltd., Stuttgart, W. Germany, June, 1983.

Flexible Assembly and Shortest Queue Problems

IVO ADAN, JAAP WESSELS and HENK ZIJM

1. Introduction

For the design of flexible assembly systems an iterative approach seems advisable. Such an approach would require the evaluation of several alternative designs. Therefore it is necessary that individual designs can be evaluated very efficiently. In this paper it is argued that extended shortest queue models can play an important role in designing flexible assembly systems. Consequently, it would be essential to have efficient methods to evaluate shortest queue models. The paper presents a new approach for shortest queue models and shows how this approach can be used for several variants. The approach uses mixtures of product forms as approximative solutions for the equilibrium distribution of the Markov chain. Although no rigorous proofs are presented in this paper, it is indicated that the exact solution can be approximated deliberately good in this way with the particular property that the approach also generates upper and lower bounds.

The modelling and design of flexible assembly systems has many aspects. In this paper we will treat one of these aspects, namely the use of queueing models of a shortest queue type. The motivation for our efforts on shortest queue problems consists of two different parts. The first part is that the design of flexible assembly systems is usually a highly iterative procedure which works best if prospective designs can be evaluated easily and efficiently. Only then, the design process can be structured as a search procedure. The second part of the motivation is the observation that in several cases an essential part of the design can be modelled as a queueing problem which is strongly related to the classical shortest queue problem as introduced by Haight in [14]. Usually, however, the models are related to the shortest queue model, but in fact also much more complicated.

The relation between the shortest queue problem and the modelling and design of flexible assembly systems will be worked out a little bit in the next section. In the present section a little bit more will be said about the shortest queue problem and our way of solving it.

In its simplest and classical formulation the shortest queue model consists of two identical and exponential servers with a Poisson instream of one type of customers. Incoming customers join the server with the shortest queue. In case both queue lengths are equal, an incoming customer selects between both servers with equal probabilities.

For this model the main analystical results have been published by Kingman [16], Flatto and McKean [11], Cohen and Boxma [9] and Fayolle and Iasnogorodski [10]. The basis of the analysis is always the two-dimensional random walk performed by the couple of queue lengths. Kingman gives a generating function for the equilibrium probabilities of this random walk and derives some of its properties. This analysis is refined in [11]. In [9] and [10] it is shown that the evaluation of the random walk can be reduced to the evaluation of a Riemann-Hilbert boundary value problem. None of these approaches leads to an explicit solution or to efficient numerical procedures. Nevertheless, particularly the generating function approach gives a lot of insight in the structure and the asymptotics of the equilibrium distribution. Our approach is different by nature, but the results are related to the results of the generating function approach and, actually, extend these results considerably. For the simplest form of the shortest queue problem our approach has been described extensively in [2], but in section 3 of the present paper we will give a short outline of the idea and the results. As said before, we need an efficient numerical procedure and, indeed, that is our main result. In section 4 it will be demonstrated that our approach can be extended to several related problems and section 5 gives conclusions and some comments.

2. Why shortest queue models?

In this paper we understand by "designing a flexible assembly system" all activities which lead to the specification of a working system. That means it comprises on the highest level the selection of the machine configuration of workstations and also the internal transport system. On a lower level it may comprise the allocation of tools and components to machines. On a still lower level it may comprise the exact location of tools and components. Then it may comprise the clustering and routing of products and finally the detailed execution of the assembly operations.

A typical structure which is encountered in several types of situations is the one depicted in figure 1.

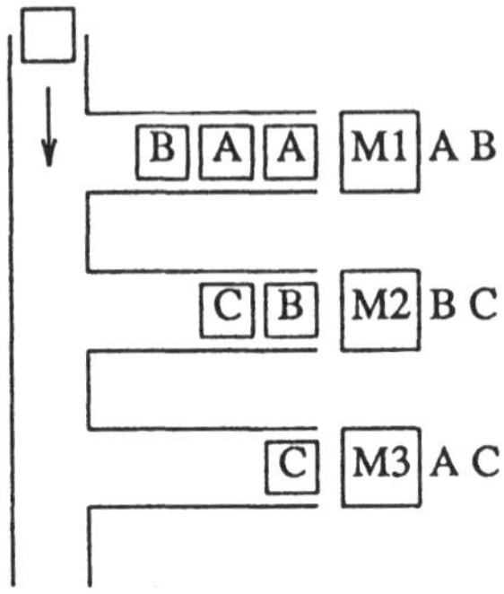

fig. 1: A typical production structure. $M1$, $M2$, $M3$ denote machines which treat the job type A, B, C. Each machine treats a restricted set of jobs only.

As an example one might think of a machine for the mounting of electronic components on printed circuit boards. For a detailed technical description of such machines the reader is referred to the paper of Zijm in this volume [20]. In this structure there is an incoming jobstream with several (usually many) types of jobs for a set of machines. Each machine can handle a restricted set of jobs and each incoming job joins one of the feasible machine queues according to some routing procedure.

This structure may occur if one is interested in the selection of machines (types and numbers), but also if one is interested in the allocation of tools and parts to be assembled. Quite often the allocation of tools and parts determines which jobs can be handled by which machines. Further decision aspects may be the releasing of jobs and the routing.

One of the most relevant routings seems to be:

Join the feasible machine which promises the shortest delay.

Or, simpler, if operations on different job types are of the same order of magnitude:

Join the feasible machine with the shortest queue.

This observation leads to the conclusion that we should be able to evaluate very efficiently models which contain in some way the features mentioned above.

A simple model containing these features is the following (compare fig. 2):

N types of jobs enter a system according to independent Poisson streams. The system consists of M identical exponential servers, each being able to treat a subset of the jobtypes. Jobs join the queue of the feasible machine with the shortest queue. In case of equally short queue lengths, ties are broken with equal probabilities.

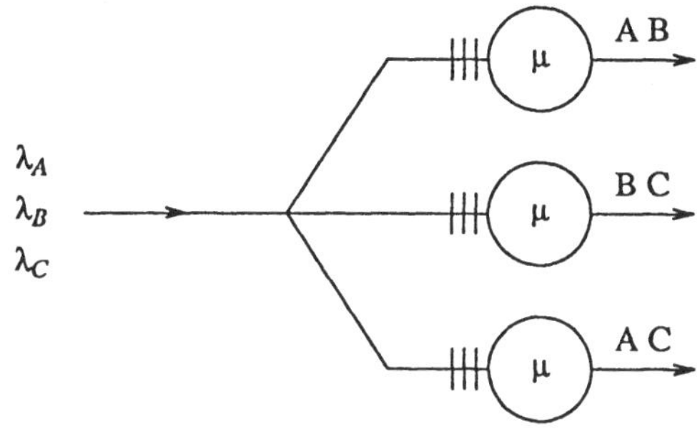

fig. 2: A queueing model for the structure of fig. 1.

Of course this model is a strong simplification of reality. In reality, we may encounter:

- A job release mechanism not leading to Poisson arrivals,
- non-exponential service time distributions,
- service time distributions which are job and/or machine dependent,
- more complicated routing procedures,
- priority rules at the servers.

Moreover, one should be aware of the fact that the number of product types might be substantial.

If one really wants to be able to evaluate realistic models efficiently with a reasonable accurateness, then one should definitely be able to evaluate the simple model of fig. 2 very efficiently and very accurately.

Impired by the argument of this section we have tried in the years 1987-1988 to develop heuristic evaluation methods for the model of fig. 2. For an example of such a method we refer to [1]. However, the results were not fully successful. In fact we did not succeed in developing a heuristic which was both very efficient and very accurate.

On the other hand our numerical experiments showed some nice features which made us believe that a satisfactory evaluation method should be feasible.

Therefore, we came to the conclusion that we should understand the process better and we decided to try to obtain that better understanding by considering a further simplified process. By restricting ourselves to one job type only and to two severs we arrived at the classical shortest queue model which has been mentioned in the introduction.

The next section is devoted to our efforts with respect to the classical shortest queue model.

3. The classical shortest queue model

Let us start the evaluation of the shortest queue model by introducing some notation (see fig. 3). For convenience, we suppose that one type of jobs arrives in a Poisson stream with intensity 2ρ $(0<\rho<1)$ at a system of two identical exponential servers with service rate 1. An incoming job joins the server with the shortest queue. Ties are solved by chosing each server with probability $\frac{1}{2}$.

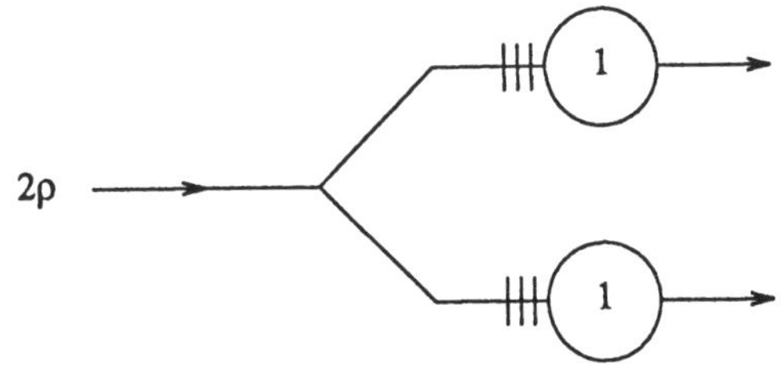

fig. 3: The classical shortest queue model

The natural state space for describing the behaviour of this system is the set of 2-tuples (m, n) with $m, n = 0, 1, 2, \cdots$, where m and n denote the queuelengths of the first and second server respectively. A job in service is also counted as being in queue. In the states (m, n) the system performs a continuous-time Markov chain. In most states it behaves in fact as a random walk.

We are interested primarily in the equilibrium probabilities $p_{m,n}$. By symmetry we easily see that

$$p_{m,n} = p_{n,m} \ .$$

It will be clear that for the process particularly the difference of the queue lengths and therefore we introduce the new state definition (m, r)

$$\text{with} \quad r \equiv n - m.$$

The equilibrium probabilities for this new state (m, r) are denoted by $q_{m,r}$.

Trivially, we have

$$q_{m,r} = p_{m,m+r}$$

$$q_{m,r} = q_{m,-r} \ .$$

In the previous section it was stated that the model of fig. 2 shows empirically some nice properties. In table 1, one can find the similar findings for the shortest queue problem. It appears that the quotients

$$\frac{q_{m+1,r}}{q_{m,r}} \text{ and } \frac{q_{m,r+1}}{q_{m,r}}$$

are numerically constant for most values of m and r. The first quotient is practically constant except for $m < 2$ (table 1a). This appears to be a general feature.

(a) The ratios $q_{m+1,r} / q_{m,r}$

r ↑	$m=0$	1	2	3	4	5
6	0.19	0.24	0.25	0.25	0.25	0.25
5	0.19	0.24	0.25	0.25	0.25	0.25
4	0.19	0.24	0.25	0.25	0.25	0.25
3	0.19	0.24	0.25	0.25	0.25	0.25
2	0.20	0.24	0.25	0.25	0.25	0.25
1	0.28	0.25	0.25	0.25	0.25	0.25
0						

(b) The ratios $q_{m,r+1} / q_{m,r}$

r ↑	$m=0$	1	2	3	4	5
6	0.10	0.10	0.10	0.10	0.10	0.10
5	0.10	0.10	0.10	0.10	0.10	0.10
4	0.10	0.10	0.10	0.10	0.10	0.10
3	0.10	0.10	0.10	0.10	0.10	0.10
2	0.11	0.10	0.10	0.10	0.10	0.10
1	0.15	0.11	0.10	0.10	0.10	0.10
0						

Table 1: The ratios $q_{m+1,r}/q_{m,r}$ and $q_{m,r+1}/q_{m,r}$ for $\rho = .5$

So, empirically, we find an asymptotic solution for the equilibrium probabilities of the form

$$q_{m,r} \sim K\alpha^m\beta^r \quad (m,\ r \to \infty)$$

with, for the case $\rho = .5$, $\alpha = .25$, $\beta = .10$.

The first question is whether we can give general expressions for α and β.

We see that

$$q_{m+1,r} \sim \alpha \ ,$$

but also

$$\frac{q_{m+1,r}}{q_{m,r}} = \frac{p_{m+1,m+1+r}}{p_{m,m+r}} \ .$$

Hence the state $(m+1,\ r)$ differs from $(m,\ r)$ by 2 jobs. This suggests as a guess for α:

$$\alpha = \rho^2 \ .$$

Indeed, for $\rho = .5$ we found empirically $\alpha = .25$. In order to make a guess for β, we have to consider the equilibrium equations.

We formulate the equilibrium equations for $r \geq 0$ by considering for each state the rate of entrances and leavings. In the formulation below, we have eliminated $q_{m,0}$ from equations (2)-(4).

(1) $$2(\rho+1)\, q_{m,r} = 2\rho q_{m-1,r+1} + q_{m,r+1} + q_{m+1,r-1} \qquad \text{for } m \geq 1,\ r \geq 2.$$

(2) $$2(\rho+1) q_{m,1} = 2\rho q_{m-1,2} + q_{m,2} + \frac{2\rho}{1+\rho} q_{m,1} + q_{m,2} + \frac{2\rho^2}{1+\rho} q_{m-1,1} + \frac{\rho}{1+\rho} q_{m,1} \qquad \text{for } m \geq 1$$

(3) $$(2\rho+1)\, q_{0,r} = q_{0,r+1} + q_{1,r-1} \qquad \text{for } r \geq 2$$

(4) $$(2\rho+1)\, q_{0,1} = q_{0,2} + \frac{2\rho}{1+\rho} q_{0,1} + \frac{1}{1+\rho} q_{1,1} + q_{0,1} \ .$$

Moreover we have the defining equations:

(5) $$(\rho+1)\, q_{m,0} = 2\rho q_{m-1,1} + q_{m,1} \qquad \text{for } m \geq 1$$

(6) $$\rho q_{0,0} = q_{0,1}$$

In order to be an asymptotic solution, $K\alpha^m\beta^2$ has to satisfy (1). Inserting $K\alpha^m\beta^r$ into (1) and dividing by $K\alpha^{m-1}\beta^{r-1}$ gives the following condition:

Lemma: For $K\alpha^m\beta^r$ to satisfy the equation (1) for $m \geq 1$, $r \geq 2$, it is necessary that α and β satisfy

(7) $$2(\rho+1)\,\alpha\beta = 2\rho\beta^2 + \alpha\beta^2 + \alpha^2 \ .$$

If we insert $\alpha = \rho^2$ in (7), then we get for β:

(8) $$2\rho(\rho+1)\beta = 2\beta^2 + \rho\beta^2 + \rho^3$$

leading to $\beta = \rho$ or $\dfrac{\rho^2}{2+\rho}$.

The solution $\alpha = \rho^2$, $\beta = \rho$ is exactly the solution for two independent $M\,|M\,|\,1$-queues, each with utilization ρ, and therefore, not very likely.

Hence we choose as solution

$$\alpha_0 = \rho^2,\ \beta_0 = \frac{\rho^2}{2+\rho}$$

which fits empirically (note that $\beta_0 = .10$ for $\rho = .5$).

This asymptotic solution is a productform solution and it is also clear it does not fit completely. If we consider the set of equations (1)-(4), then we see that the asymptotic solution satisfies (1), but also (2).

Therefore form a new solution by compensating the old one with another product form in such a way that the new solution satisfies (1) and (3).

Form the linear combination

$$\alpha_0^m \beta_0^r + c_0 \alpha^m \beta^r$$

and try to choose c_0, α and β such that this linear combination satisfies (3) and (1).

Inserting in (3) gives for **all r > 1**

$$(\beta_0^r + c_0 \beta^r)(2\rho + 1) = (\beta_0^{r+1} + c_0 \beta^{r+1}) + (\alpha_0 \beta_0^{r-1} + c_0 \alpha \beta^{r-1}) \ .$$

Hence

$$\beta = \beta_0 \ .$$

For $c_0 \alpha^m \beta_0^r$ to satisfy (1) requires

$$\alpha = \alpha_0 \ \text{ or } \ 2\frac{\rho^3}{(2+\rho)^2} \qquad \text{(Lemma)}$$

so take for α the second solution, which we call α_1. In order to satisfy (3) choose: $c_0 = -\frac{\alpha_1 - \beta_0}{\alpha_0 - \beta_0}$.

In table 2, it is depicted how this solution repairs the first asymptotic solution.

This asymptotic solution now satisfies (1) and (3) but violates (2) because the extra term does not satisfy (2). Therefore form the linear combination

$$\alpha_0^m \beta_0^r + c_0 \alpha_1^m \beta_0^r + d_1 \alpha^m \beta^r$$

and try to choose d_1, α and β such that this linear combination satisfies (2) and (1).

Or:

Try to find d_1, α and β such that

$$c_0 \alpha_1^m \beta_0^r + d_1 \alpha^m \beta^r$$

satisfies (1) and (2).

Result:

Choose $\alpha = \alpha_1$

$\uparrow$ r						
6	0.19	0.24	0.25	0.25	0.25	0.25
5	0.19	0.24	0.25	0.25	0.25	0.25
4	0.19	0.24	0.25	0.25	0.25	0.25
3	0.19	0.24	0.25	0.25	0.25	0.25
2	0.19	0.24	0.25	0.25	0.25	0.25
1	0.19	0.24	0.25	0.25	0.25	0.25
0						
	0	1	2	3	4	5

$m \rightarrow$

The ratios $\dfrac{(\alpha_0^{m+1}+c_0\alpha_1^{m+1})\beta_0^r}{(\alpha_0^m+c_0\alpha_1^m)\beta_0^r}$

$\uparrow$ r						
6	0.10	0.10	0.10	0.10	0.10	0.10
5	0.10	0.10	0.10	0.10	0.10	0.10
4	0.10	0.10	0.10	0.10	0.10	0.10
3	0.10	0.10	0.10	0.10	0.10	0.10
2	0.10	0.10	0.10	0.10	0.10	0.10
1	0.10	0.10	0.10	0.10	0.10	0.10
0						
	0	1	2	3	4	5

$m \rightarrow$

The ratios $\dfrac{(\alpha_0^m+c_0\alpha_1^m)\beta_0^{r+1}}{(\alpha_0^m+c_0\alpha_1^m)\beta_0^r}$

Table 2: The analogous quotients are given as in table 1, but now for the second asymptotic solution ($\rho = .5$).

$$\beta = \frac{\rho^3}{(2+\rho)(2+2\rho+\rho^2)} \quad \text{(call this solution } \beta_1\text{)}$$

$$d_1 = -\frac{\dfrac{\alpha_1+\rho}{\beta_1}-(\rho+1)}{\dfrac{\alpha_1+\beta}{\beta_0}-(\rho+1)}\, c_0 \; .$$

In table 3 the quality of this third asymptotic solution is depicted and, indeed, it nearly fits the exact solution.

Indeed, this approach of alternately compensating the current asymptotic solution in order to satisfy (2) and (3) respectively can be generalized easily and leads to a series of product forms.

In fact we can rigorously prove

(9) $$q_{m,r} = c^{-1} \sum_{i=0}^{\infty} d_i(\alpha_i^m + c_i\alpha_{i+1}^m)\,\beta_i^r \quad \text{for } m \geq 0,\ r \geq 1$$

where the series is alternating and absolutely convergent. Moreover, the d_i, c_i, α_i and β_i have several nice (asymptotic) properties.

With regard to their values we can compute them recursively.

α_i and α_{i+1} are the roots of

r ↑						
6	0.19	0.24	0.25	0.25	0.25	0.25
5	0.19	0.24	0.25	0.25	0.25	0.25
4	0.19	0.24	0.25	0.25	0.25	0.25
3	0.19	0.24	0.25	0.25	0.25	0.25
2	0.20	0.24	0.25	0.25	0.25	0.25
1	0.29	0.25	0.25	0.25	0.25	0.25
0						
$m \rightarrow$	0	1	2	3	4	5

The ratios $\dfrac{\beta_0^r\alpha_0^{m+1}+(c_0\beta_0^r+d_1\beta_1^r)\alpha_1^{m+1}}{\beta_0^r\alpha_0^m+(c_0\beta_0^r+d_1\beta_1^r)\alpha_1^m}$

r ↑						
6	0.10	0.10	0.10	0.10	0.10	0.10
5	0.10	0.10	0.10	0.10	0.10	0.10
4	0.10	0.10	0.10	0.10	0.10	0.10
3	0.10	0.10	0.10	0.10	0.10	0.10
2	0.11	0.10	0.10	0.10	0.10	0.10
1	0.15	0.11	0.10	0.10	0.10	0.10
0						
$m \rightarrow$	0	1	2	3	4	5

The ratios $\dfrac{\beta_0^{r+1}\alpha_0^m+(c_0\beta_0^{r+1}+d_1\beta_1^{r+1})\alpha_1^m}{\beta_0^r\alpha_0^m+(c_0\beta_0^r+d_1\beta_1^r)\alpha_1^m}$

Table 3: The analogous quotients are given as in table 1, but now for the third asymptotic solution.

$$2(\rho+1)\alpha\beta_i = 2\rho\beta_i^2 + \alpha\beta_i^2 + \alpha^2 \quad \text{(compare the lemma)} \ .$$

Similarly, β_i and β_{i+1} are the roots of

$$2(\rho+1)\ \alpha_{i+1}\beta = 2\rho\beta^2 + \alpha_{i+1}\beta^2 + \alpha_{i+1}^2 \ .$$

This leads to the following recursions

$$\alpha_{i+1} = \frac{2\rho\beta_i^2}{\alpha_i}$$

$$\beta_{i+1} = \frac{\alpha_{i+1}^2}{(2\rho+\alpha_{i+1})\beta_i} \ .$$

Also the c_i and d_i may be computed recursively:

$$c_i = -\frac{\alpha_{i+1}-\beta_i}{\alpha_i-\beta_i}$$

$$d_{i+1} = -\frac{(\alpha_{i+1}+\rho)/\beta_{i+1}-(\rho+1)}{(\alpha_{i+1}+\rho)/\beta_i-(\rho+1)}\ c_i d_i \ .$$

Moreover, for the normalization constant c we find:

$$c = \frac{\rho(2+\rho)}{2(1+\rho)(1-\rho)(2-\rho)} \ .$$

In this way we have constructed an approximation method which is very efficient and also very accurate. Moreover, the alternation and monotonicity properties also provide upper- and lowerbounds in each step.

Discussion: The approach as outlined in this section has been worked out in detail in [2]. However, not all the results are new. In fact, Kingman [16] and Flatto and McKean [11] have analysed the generating function for the equilibrium probabilities. They show that the generating function is memomorphic. Then, it can be show by decomposition into partial fractions that the equilibrium probabilities can be written as infinite sum of productforms. However, the actual decomposition is cumbersome and in fact only the first two terms had been found.

There are other publications on numerical analysis of the classical shortest queue problem. Most of those studies deal with the evaluation of approximating models. In [2] we give a short overview of these efforts. Probably the best other approach is based on a power series method to calculate the stationary queue length distribution for fairly general multidimensional exponential queueing systems. This method has been developed by Hooghiemstra et.al in [15] and applied on the shortest queue problem by Blanc in [7] and [8]. The theoretical basis of this method, however, is still not understood and it does not provide upper- and lowerbounds or any other quality characteristic other than practical experience.

4. Extensions

In section 2 it was argued that shortest queue type models have some relevance for the design of flexible assembly systems. It has also been argued that it should be helpful if a wide variety of models could be evaluated very efficiently. In section 3 an outline has been given of a method that provides an elegant, accurate, and efficient evaluation of the shortest queue problem. The remaining questions regard the possibility of extension to practically more relevant models. Or, if that is not possible, the possibilities to base heuristics on the approach.

It is not clear at the moment how far the approach can be extended, however, there is some experience already in this direction. This section will be devoted to the experience so far with efforts to extend the approach to related models.

A. Unequal routing probabilities in case of a draw

We first consider a minor change in the model of section 3, namely, the equal probabilities for joining both queues in case of equal queue length are replaced by q and $1-q$ for the m- and the n-queue respectively.

This problem is not symmetric any more. However, its non-symmetry is rather weak, therefore, a symmetrization operation might help.

Define:

$$\tilde{q}_{m,r} \equiv \frac{1}{2}\left(q_{m,r} + q_{m,-r}\right) \ .$$

These $\tilde{q}_{m,r}$ satisfy the equilibrium equations (1)-(6) of the symmetric problem

Hence

$$\tilde{q}_{m,r} = C^{-1} \sum_{i=0}^{\infty} d_i(\alpha_i^m + c_i\alpha_{i+1}^m)\, \beta_i^r \qquad \text{for } m \geq 0,\ r \geq 0$$

$$\tilde{q}_{m,\,0} = C^{-1} \sum_{i=0}^{\infty} d_i\left(\frac{\alpha_i^{m+1}}{\alpha_i+\rho} + c_i\, \frac{\alpha_{i+1}^{m+1}}{\alpha_{i+1}+\rho}\right) \quad \text{for } m \geq 0\ .$$

In [6] it has been worked out how this solution can be exploited to find the $q_{m,r}$. Indeed, it has been proved that the $q_{m,r}$ can again be written in a form very closely to the form for $\tilde{q}_{m,r}$:

$$q_{m,r} = c^{-1} \sum_{i=0}^{\infty} d_i^+(\alpha_i^m + c_i\alpha_{i+1}^m)\, \beta_i^2 \qquad \text{for } m \geq 0,\ r \geq 1$$

$$\text{and } q_{m,-r} = c^{-1} \sum_{i=0}^{\infty} d_i^-(\alpha_i^m + c_i\alpha_{i+1}^m)\, \beta_i^r \qquad \text{for } m \geq 0,\ r \geq 1\ .$$

For $r = 0$ similar results are obtained. For the d_i^+ and d_i^- again simple recursive expressions have been obtained.

All nice numerical properties are maintained in this case.

B. Unequal service rates

An essentially more complicated case is the case with unequal service rates for both servers. For convenience we suppose that the service rates are γ_1 and γ_2 respectively with $\gamma_1, \gamma_2 > 0$ and $\gamma_1 + \gamma_2 = 2$. Again q is the probability of sending a job to the first queue in case of equal queue length.

Now the equilibrium probabilities loose the symmetry property completely. Essentially, we could do the analysis of section 3 (or better: of extension A) for the regions $r > 0$ and $r < 0$ separately. However, the remaining problem is how to fit them on the line $r = 0$. Therefore, the iterative compensation for both regions should be executed simultaneously, fitting them on $r = 0$.

The sketched procedure has been worked out in detail in [3]. We just mention here that the generation procedure of section 3, which has the structure

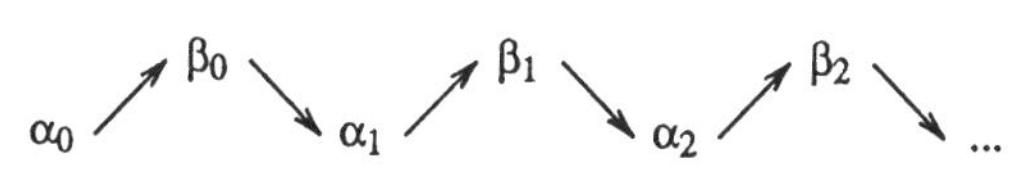

is replaced by a generation procedure with the form of a binary tree as depicted in fig. 4.

In both regions of the state space the solution is an infinite sum of product forms each based on its own α and β from the tree. For a detailed description of the results and for rigorous proofs the reader is referred to [3].

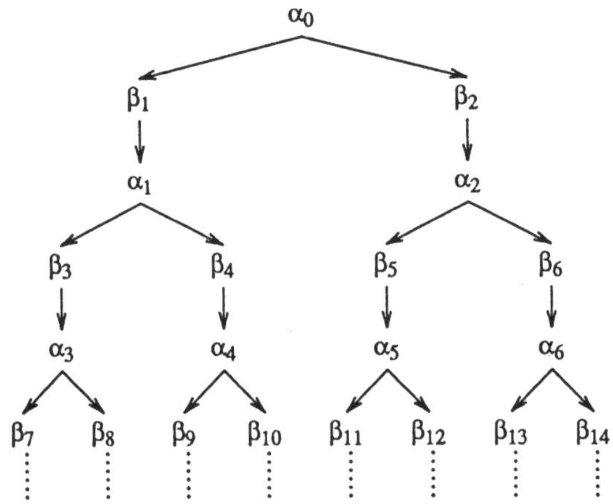

fig. 4: The parameter tree for the asymmetric shortest queue problem.

The resulting numerical procedure has the same nice properties as in the symmetric case.

For this model no other analytical results are available than those of [3]. Also the number of numerical approaches is rather restricted. We might mention here a paper of Grassmann ([13]) and a paper by Rao and Posner ([18]), both based on approximating models.

C. Shortest queue routing with jockeying

A simple extension of the previous models is obtained if one allows the job to jump to the other queue if that one is essentially shorter. In the model this is worked out by saying that as soon as

$$|n - m| > N$$

for some given number N, then a job jumps from the longer to the shorter queue.

This model has been proposed by Gertsbakh in [12] as an approximation for the classical shortest queue model of section 3.

The jockeying model has a much simpler state space than the original model, since r only varies between $-N$ and N.

It appears that the first productform already satisfies the equilibrium equations for all states with max $\{m, n\} > N$. So for those states no compensation is needed at the boundaries. If one chooses this product form for those "larger" states, the remaining finite set of equilibrium equations can be solved exactly. This works since the larger states can only be reached from the smaller states via one state viz. (N, N). Therefore $p_{N,N} = q_{N,0}$ may be used as a normalization parameter.

For an extensive treatment of this model the reader is referred to [4], where it is also shown that the resulting procedure may be viewed as a matrix-geometric type of procedure as advocated by Neuts (see e.g. [17]). Actually, Gertsbakh's approach is also of the matrix- geometric type, however, his results are less explicit and his procedure less efficient. The reason for this difference is a difference in the partitioning of the state space: Gertsbakh's partitioning is based on the lines $\min\{n, m\} = \text{constant}$ and our's on $\max\{n, m\} = \text{constant}$.

Our procedure solves the jockeying model exactly and very efficiently.

D. Erlangian servers

One of the main weak points in the models so far from a practical point of view is the use of exponential service time distributions.

If one supposes an l-Erlang distribution for the service time in both servers, than an arriving job may be considered as being equivalent to l subjobs each requiring a negative-exponentially distributed service time.

Let us suppose that an incoming job joins the queue with the lowest number of subjobs. Now the system can be characterized by the state (m, n) where m and n denote the numbers of subjobs for the first and second server respectively.

For the resulting Markov chain the equilibrium equations can be derived as in section 3.

In [5] it is shown that, indeed, the same approach may be used for this model. Although the analysis is rather delicate, the resulting numerical procedure is again efficient and accurate.

It may be noted here that the model is quite close to the model of group arrivals. The analysis of the latter model is simpler however, at least if arriving jobs are supposed to join the shortest queue subsequently.

5. Conclusions and comments

The preceding two sections show that the compensation method, as introduced in [2] for the classical shortest queue problem is very promising as a method for a much wider class of shortest queue type problems. Typical for the method in its present state is that on one hand the analysis requires a lot of skill but on the other hand the resulting procedures are simple, fast and accurate.

It is very likely that the method can be extended further. A first step could be to replace the Erlang distributions for the service times by finite mixtures of Erlang distributions with the same scale parameter. The procedure can definitely be extended to that case and that would mean that the approach can be applied for general service time distributions (compare Schassberger [19]).

The essential step would be the introduction of the job-dependent parallel structure as depicted in section 2. Empirically, these models exhibit similar features as those depicted in table 1 for the classical shortest queue problem. Therefore, it is quite likely that the approach can be extended in the direction mentioned.

In the sections 3 and 4, the emphasis lies on the computation of the equilibrium probabilities of the continuous time Markov chain. These probabilities are usually not the characteristics of primary interest if one evaluates a queueing model. However, because of the nice form of the solution, it is easily shown that the relevant characteristics can be computed equally efficient and accurate. As an example consider the expected system time W for the classical shortest queue problem:

$$W = 2 \sum_{m=1}^{\infty} \sum_{r=1}^{\infty} m q_{m,r} + \sum_{m=1}^{\infty} m q_{m,0}$$

$$\underset{=}{(5)} \sum_{m=1}^{\infty} m\left(\frac{2\rho}{1+\rho} q_{m-1,1} + \frac{1}{1+\rho} q_{m,1} + \sum_{r=1}^{\infty} q_{m,r}\right)$$

$$\underset{=}{(9)} C^{-1}\{2 \sum_{i=0}^{\infty} d_i\left(\frac{\alpha_i}{(1-\alpha_i)^2} + \frac{c_i \alpha_{i+1}}{(1-\alpha_{i+1})^2}\right) \frac{\beta}{1-\beta_i} +$$

$$+ \frac{1}{1+\rho} \sum_{i=0}^{\infty} d_i\left(\frac{2\rho+\alpha}{(1-\alpha_i)^2} + \frac{c_i(2\rho+\alpha_{i+1})}{(1-\alpha_{i+1})^2}\right) \beta\} .$$

Again, in these series the terms are alternating and monotonously decreasing in absolute value. Hence, the partial sums can be computed with known accuracy in the same way as the equilibrium probabilities themselves.

For higher moments and other characteristics the results are comparable. Even for the higher moments, the number of terms of the sum to be taken into account to reach a certain accuracy is surprisingly low.

References

1. ADAN, I.J.B.F. , WESSELS, J. , AND ZIJM, W.H.M. , "Queuing analysis in a flexible assembly system with a job-dependent parallel structure ," in *Operations Research Proceedings 1988* , pp. 551-558 , Springer-Verlag , Berlin , 1989 .

2. ADAN, I.J.B.F. , WESSELS, J. , AND ZIJM, W.H.M. , "Analysis of the symmetric shortest queue problem ," *Stochastic Models* , vol. 6 , 1990, (to appear) .

3. ADAN, I.J.B.F. , WESSELS, J. , AND ZIJM, W.H.M. , "Analysis of the asymmetric shortest queue problem ," , Eindhoven University of Technology, Dep. of Math. and Comp. Sci. , 1990 . Submitted for publication in Queueing Systems.

4. ADAN, I.J.B.F. , WESSELS, J. , AND ZIJM, W.H.M. , "Analysis of the asymmetric shortest queue problem with threshold jockeying ," Memorandum COSOR 90-20 , Eindhoven University of

Technology, Dep. of Math. and Comp. Sci., 1990. Submitted for publication.

5. ADAN, I.J.B.F., WESSELS, J., AND ZIJM, W.H.M., "Analysis of the symmetric shortest delay problem for Erlangian servers," Memorandum COSOR, Eindhoven University of Technology, Dep. of Math. and Comp. Sci., 1990. In preparation.

6. ADAN, I.J.B.F., WESSELS, J., AND ZIJM, W.H.M., "An asymmetric shortest queue problem," in *Operations Research Proceedings 1989*, pp. 434-441, Springer-Verlag, Berlin, 1989.

7. BLANC, J.P.C., "On a numerical method for calculating state probabilities for queueing systems with more than one waiting line," *J. Comput. Appl. Math.*, vol. 20, pp. 119-125, 1987.

8. BLANC, J.P.C., "The power-series algoritm applied to the shortest-queue model," Memorandum 379, Tilburg University, Department of Economics, 1989.

9. COHEN, J.W. AND BOXMA, O.J., *Boundary value problems in queueing system analysis*, North-Holland, Amsterdam, 1983.

10. FAYOLLE, G. AND IASNOGORODSKI, R., "Two coupled processors: the reduction to a Riemann-Hilbert problem," *Z. Wahrsch. Verw. Gebiete*, vol. 47, pp. 325-351, 1979.

11. FLATTO, L. AND MCKEAN, H.P., "Two queues in parallel," *Comm. Pure Appl. Math.*, vol. 30, pp. 255-263, 1977.

12. GERTSBAKH, I., "The shorter queue problem: A numerical study using the matrix-geometric solution," *Eur. J. Oper. Res.*, vol. 15, pp. 374-381, 1984.

13. GRASSMANN, W.K., "Transient and steady state results for two parallel queues," *OMEGA Int. J. of Mgmt Sci.*, vol. 8, pp. 105-112, 1980.

14. HAIGHT, F.A., "Two queues in parallel," *Biometrica*, vol. 45, pp. 401-410, 1958.

15. HOOGHIEMSTRA, G., KEANE, M., AND REE, S. VAN DE, "Power series for stationary distributions of coupled processor models," *SIAM J. Appl. Math.*, vol. 48, pp. 1159-1166, 1988.

16. KINGMAN, J.F.C., "Two similar queues in parallel," *Ann. Math. Statist.*, vol. 32, pp. 1314-1323, 1961.

17. NEUTS, M.F., *Matrix-Geometric Solutions in Stochastic Models*, Johns Hopkins University Press, Baltimore, 1981.

18. RAO, B.M. AND POSNER, M.J.M., "Algorithmic and approximation analysis of the shorter queue model," *Naval Res. Log. Q.*, vol. 34, pp. 381-398, 1987.

19. SCHASSBERGER, R.S., *Warteschlangen*, Springer-Verlag, Berlin, 1973.

20. ZIJM, W.H.M., "Operational Control of Automated PCB Assembly Lines,". (this volume).

A Quantitative Model for the Analysis of Distribution Network Scenario's

JAN V. DOREMALEN and HEIN FLEUREN

1 INTRODUCTION

1.1 Multi-echelon distribution networks

Multi-echelon distribution networks exist to support customer service and to provide transportation economics. For an overview of the managerial problems and possible solutions see for example Tersine [1] and Vollmann, Berry and Whybark [2]. For an introduction to the mathematical analysis of multi-echelon network models and solution techniques we refer to Schwarz [3] and Axsäter, Schneeweiss and Silver [4] and Silver and Peterson [5]. In this paper we discuss managerial and inventory control problems starting from the description of a real-life problem.

1.2 Description of the situation

In the sequel of this paper we will refer to an abstracted problem situation to illustrate the line of argument. A manufacturer produces a family of electronic devices at a production centre in the USA. Western Europe is an important market and Holland (NL), Germany (DL) and the United Kingdom (UK) are the key countries. We concentrate on the market in these key countries.

The electronic devices are composed of four basic components: body, control panel, monitor and technical manual. Different combinations of components make different endproducts. Currently body and control panel are manufactured at a central production centre in the USA. There also the endproduct is bundled from these subassemblies and the monitors and technical manuals which are purchased at third party manufacturers. The products are transported from the production centre to a central supply centre in Holland by ship and truck. From the supply centre the products are distributed to the sales organisations by truck.

Product structure and distribution network thus allow for (1) the postponement of the bindling of the four components into an endproduct and (2) the purchasing of monitors and manuals at other suppliers.

In each of the key countries a sales organisation is responsible for sales orientation, sales and service of the product range. Every month the sales organisations give forecasts at endproduct level for the next twelve months on a rolling horizon basis. The forecasts are unreliable and the realized demand thus strongly deviates from the forecasts.

At a central planning department the long term forecasts are translated into raw material purchasing plans and production schedules for the production centre. The long term forecast errors (three months up to a year) lead to problems at the production centre as some basic and expensive raw material is supplied with lead times varying from three up to six months.

Customer orders are accepted at the sales organisations. The order is met from the stock at the sales organisation or the central stock at the supply centre. Backorders are delivered from the production centre via the supply centre. The short term forecast errors (one up to three months) in combination with the required customer service levels lead to high safety stock levels in the distribution network.

1.3 Description of the problems

The actual situation is characterized by high stock levels and long pipeline lead times. Nevertheless the customer service level is unsatisfactory.

The objective is to have a customer order cycle time (COCT) of two weeks where the COCT is defined as the time elapsing between acceptation of an order and delivery (including installation) of the equipment at the customer. In reality the COCT varies from one week up to eight weeks.

Discussions with people in the organization of the client revealed that the following factors were greatly influencing the performance:

1 The operational process is characterized by a large range of high value endproducts, long and unreliable lead times at the production centre and long and fluctuating transportation lead times;
2 The planning and control is characterized by a large discrepancy between forecast and actual demand and insufficient information on availability of products in the pipeline;
3 The organisation is characterized by long and tedious procedures and, consequently, organisational slack in the logistic chain.

High stock levels and long pipeline lead times are a serious problem as high capital costs are associated with keeping endproducts in the pipeline. The main capital cost components are an interest rate of about 5% on year basis, a price erosion of approximately 15% and an obsolesence risk of about 5%.

1.4 Restructuring the distribution network

From the above it will be evident that there is a strong urge to improve the production process and the distribution network. We were asked to design and implement (in close cooperartion with the client) a quantitative model to support the analysis of a number of alternative designs for the distribution network.

Key issues are:

1 Where should we produce our products or components;
2 Should we stock products at production centre, supply centre or sales organisations;
3 How are we to replenish stocks at the stock points;
4 What mode of transport should we choose and;
5 Should we postpone the assembly of components into endproducts to supply centre or sales organisations.

Important design dependent and independent parameters are:

1 Market characteristics;
2 Costs of manufacturing, assembly and transportation;
3 Value of products and components;
4 Endproduct range and product structure and component diversity;
5 Manufacturing and transportation lead times;
6 Capital costs as price erosion, obsolescence risk and interest and;
7 Targets for customer service levels.

Given the design of the network and the values of parameters the model should evaluate:

1 Goods flow costs;
2 Stock levels;
3 Stock costs;
4 Distribution lead times and;
(5) Customer service levels.

With respect to the performance of the network we have defined a two step customer service requirement. The COCT is build up from information lead times (how long does it take to get the order to the distribution point and to prepare the shipment, including waiting for a backorder) and the physical distribution and installation (how long does it take to get the product to the customer and to install it). In the following we will split the service requirements into two parts. Firstly, it will be demanded that the physical lead time of decoupling point to customer is smaller than a prerequisite number of days. Secondly, at the decoupling point a service degree will be required expressed as percentage of the orders which can be met from stock immediately.

1.5 Overview of the paper

In section 2 we describe the alternative distribution network designs, in section 3 a cost calculation model, in section 4 the stock level calculations and in section 5 an example. In section 6 we conclude with some final remarks and attention points for further research.

2 BASIC STRUCTURES FOR THE DISTRIBUTION NETWORK

In this section a number of designs for the distribution network are discussed. We will start with a brief description of two important logistics principles underlying the design choices, namely the principle of the customer order decoupling point and the principle of postponement.

2.1 The principle of the customer order decoupling point

The customer order decoupling point indicates how far a customer order penetrates the logistics chain, i.e. it defines a point in the logistics chain separating the organization of production and distribution process in a forecast driven and a customer order driven trajectory (cf. Hoekstra and Romme [5] for more details).

In a distribution network the location of the decoupling point will be determined by (1) a trade-off between costs of transportation (high costs are associated with fast and frequent transport) and holding stocks (high costs are associated with long pipeline lead times and high safety stock levels) and (2) restrictions determined by the required customer order cycle time and delivery performance.

Our client had three options for the decoupling point in mind, namely (1) at the supply centre (the actual situation), (2) at the production centre and (3) at the sales organisations.

2.2 The principle of postponement

The principle of postponement (cf. Zimm and Bowersox [6]) may be formulated as: delaying the movement or final formulation of a product until after the customer order has been received. Five types of postponement may be distinghuished: (1) labelling, (2) packaging, (3) final assembly, (4) manufacturing and (5) time of shipment.

Our client wanted to consider the postponement of the final assembly and of the time of shipment.

The postponement of the bundling of components into an endproduct (final assembly) until after a customer order has been received is possible, as this activity can be executed at production centre, supply centre or sales organisations. The economic trade-off is between an increase of assembly costs (extra activities in supply centre or sales organisations) and a decrease in transportation costs (better density ratio) and storage costs (lower safety stocks due to commonality, lower value and lower obsolescence risk).

It is also possible to postpone the time of shipment. As capital cost factors on the high value products form an important cost factor, it is worthwhile to consider a faster albeit more expensive mode of transport. The alternative that the client wants to consider is to transport the endproducts from production centre to sales organisations by air after a customer order has been received. This mode of transport ensures that the required customer order cycle time of two weeks can be realized.

2.3 Basic designs of the distribution network

As a result of the discussions four alternative designs remained open.

I Decoupling point at supply centre and bundling at production centre

This describes an idealized version of the actual situation. The components are bundled into an endproduct at a production centre in the USA and the endproducts are shipped to a central supply centre in Europe. The customer orders are accepted at the sales organisations and are supplied from the stock at the supply centre. From the supply centre the ordered products are transported by truck to the sales organisation which is responsible for the installation of the products at the customer (see figure 1: scenario I).

II Decoupling point and bundling at supply centre

The bundling of components into an endproduct is a relatively simple activity and the first alternative is therefore to create a bundling activity at a central supply centre in Europe. The components are shipped from the USA to Europe by ship. The products of a customer orders are bundled at the supply centre and then transported to the sales organisation. As some of the components are manufactured at third suppliers one might consider the purchasing of these components at other suppliers (see figure 1: scenario II).

III Decoupling point and bundling at the production centre

One of the reasons to have a supply centre in Europe is the long transportation time of about three weeks from the USA to Europe. Management wants to evaluate the consequences of transportation by air. The decoupling point in that case would be at the production centre in the USA from where customer orders are shipped after bundling to the sales organisation (see figure 1: sencario III).

IV Decoupling point and bundling at the sales organisations

As an alternative to the previous situation it was suggested to move the decoupling point downstream to the sales organisations and to leave out a central supply centre. Components are produced in the USA and transported by ship and truck to the sales organisations on the basis of forecast information. Customer orders are bundled at and distributed from the sales organisations (see figure 1: scenario IV).

3 A COST CALCULATION MODEL

In this section we describe a cost calculation model. It forms the basis for our quantitative model. In the next section we describe the algorithms to determine required safety stocks levels at the relevant stock points in the respective scenario's. The cost calculation model gives insight in quantitative performance measures associated with pipeline and safety stocks.

3.1 Framework

The cost model concentrates on the evaluation of the pipeline costs, where the pipeline encompasses the entire goods flow trajectory between the end of the production lines at the production centres and the departure of endproducts from the sales organisation.
The pipeline does not include the manufacturing process at the production centres and the delivery of goods from sales organisation to customer. It includes the transportation lead times, the safety stock holding times and handling and bundling times (even if this activity takes place at the production centres).

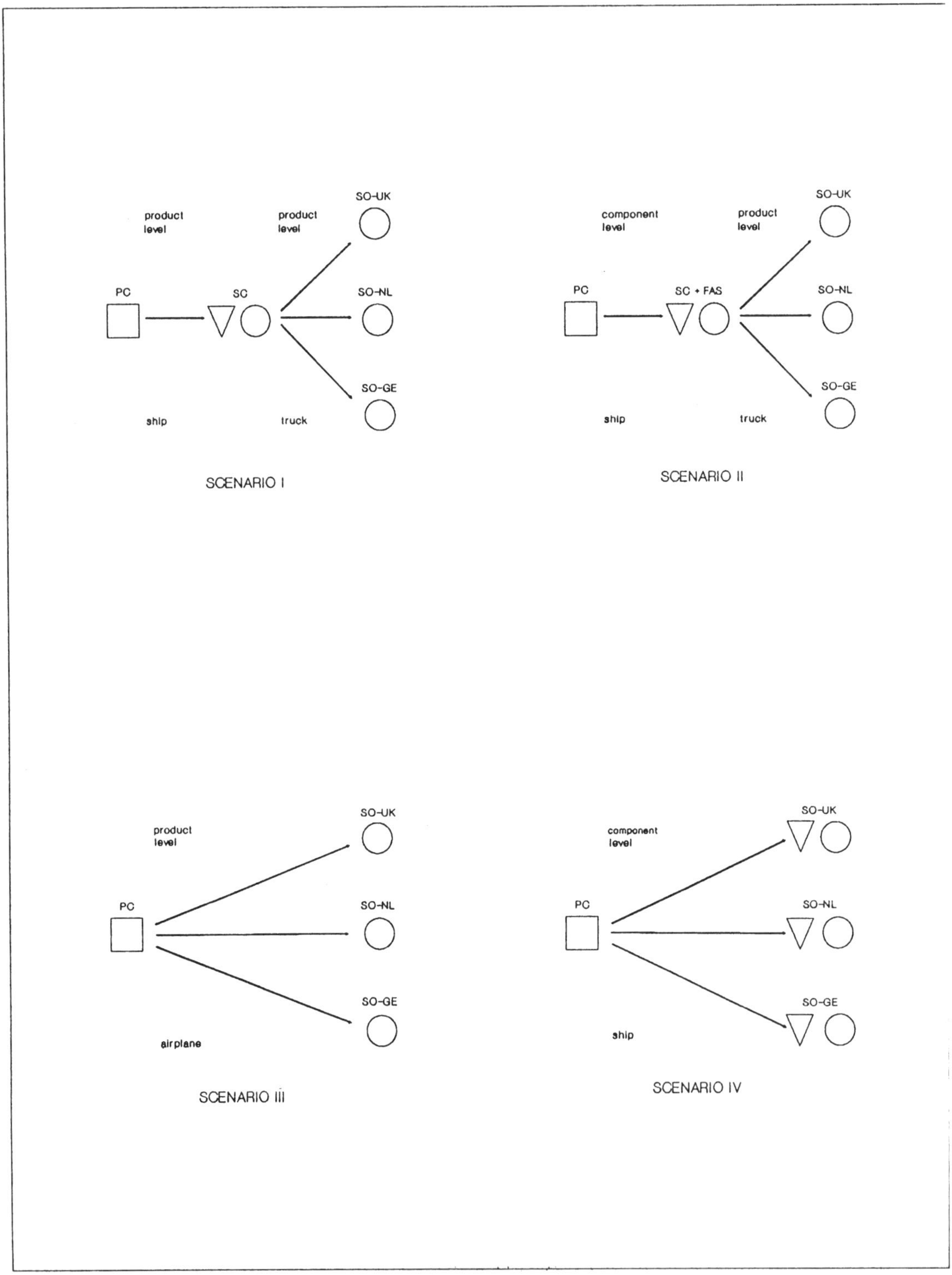

Figure 1 Four distribution network designs

The cost factors are divided in activity based costs and time based costs.

The activity based costs are the costs associated with transportation from production centre to supply centre or sales organisation, distribution from supply centre to sales organisations and bundling of components into endproducts. These costs depend on the structure of the network, the mode of transport and the number of endproducts sold and are independent of lead times and capital loss rates.

The time based costs are costs due to capital loss based on interest rate, price erosion and obsolescence risk. High costs of holding stock emerge as a consequence of having high value and high risk products in a pipeline with long transportation lead times and high safety stock levels.

It will be clear that for each of the designs discussed in section 2 we need a dedicated cost breakdown with a associated set of input data. In figure 2 the breakdown for scenario II is pictured.

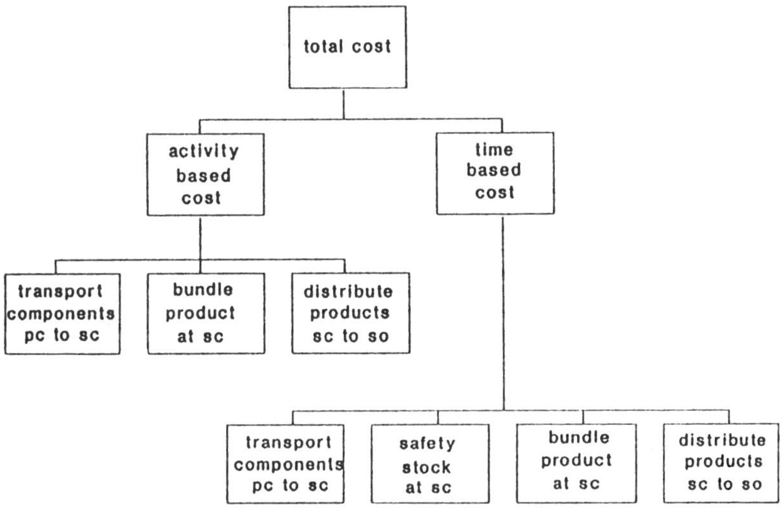

Figure 2 Cost Breakdown for scenario II

3.2 Information needed for the cost calculation model

To run the cost calculation model we need information on (1) product and product structure, (2) market characteristics (3) lead times (4) average safety stock levels and (5) cost factors.

1 The product information concentrates on a description of endproducts, components and product structure;
2 The market characteristics concentrate on required customer service level, length of review period, forecast and realized demand per product per NSO;
3 Depending on the scenario the lead time information concentrates on manufacturing lead times, bundling lead times at production centres, supply centre or sales organisations, transportation lead times from production centres to central supply centre or sales organisations and transportation lead times from supply centre to sales organisations;
4 Depending on the scenario the safety stock levels at the decoupling points are expressed in average numbers of endproducts or components. How these are determined we discuss in section 4;
5 The cost factor information concentrates on percentages for interest, obsolescence risk and price erosion of pipeline stock, costs of transportation of products or components, costs of bundling at production centres, supply centre or sales organisations, the costs of handling at supply centre and sales organisations and the value of products and components in the pipeline.

3.3 Output on financial and performance characteristics

The financial information concentrates on the value of products and components in the pipeline, goods flow costs (i.e. transportation, bundling and handling) and the costs of pipeline stock (i.e. in transport, handling, bundling and holding stock).

The performance information concentrates on lead time information for the whole pipeline and for the part of the chain from decoupling point to sales organisation, on average safety stock levels and on achieved customer service levels.

The financial information makes trade-offs possible as it gives the costs for a specific distribution network structure with given product structure and demand, lead time and cost figures. The performance information indicates whether the requirements w.r.t. customer service levels and customer order cycle time can be met for the specific distribution network.

4 STOCK LEVEL CALCULATIONS

In this section we discuss how we have determined average safety stock levels at the corresponding decoupling points (at endproduct or component level) for the different distribution network scenario's.

First we summarize the important assumptions underlying our analysis. Then, we describe the basic inventory control model used as a basis for our calculations. Eventually, we indicate how we have transformed the network information into input parameters for the inventory control model.

4.1 Assumptions

Underlying the mathematical evaluation of the different scenario's are a set of assumptions. We mention the most important ones:

1 There are only safety stocks at the decoupling point and safety stocks are replenished on a periodic review basis to a order-up-to-level;
2 The customer service levels used in determining the safety stock levels are formulated at endproduct level at the sales organisations and may be expressed as (1) the fraction of demand delivered from stock at the decoupling point and (2) the probability that no stock-out occurs in a review period. We shall concentrate on criterium (1);

As we have pointed out in subsection 1.4 the customer service level is a combination of delivery performance from stock and the customer order cycle time. For the evaluation of the safety stocks we use as performance criterium the fraction of demand delivered from stock at the decoupling point.

4.2 A periodic-review order-up-to-level inventory-control model

The basis for the approximation of average safety stock levels is a single-item periodic-review order-up-to-level inventory-control model, a so-called (R,S)-strategy (for details see Silver and Peterson [5]).

Every R units of time (the review period) a replenishment order is placed of sufficient magnitude to raise the inventory position to an order-up-to-level S. It is assumed that R is fixed by environmental, organisational and economical considerations.

Consider a point t_0 in time at which a replenishment order is placed. With L_0 we denote the lead-time of this replenishment order. With L_1 we denote the lead-time of the replenishment order placed at time $t_0 + R$, i.e. the next replenishment order. The replenishment order placed at time t_0 is meant to meet the requirements of the time period $(t_0 + L_0, t_0 + R + L_1)$, so this time period plays an essential role in the determination of the size of the replenishment order at time t_0.

Based upon information on demand pattern and lead-time behaviour we may evaluate an order-up-to-level S in such a way that the criterium is met (i.e. the fraction of demand delivered from stock). In Van Doremalen and Fleuren [8] we discuss this analysis in more detail. Here we suffice with describing the input and output of this analysis.

The input for the analysis is:

μ_0	:	mean demand over $(t_0, t_0 + L_0)$;
σ_0	:	standard deviation of demand over $(t_0, t_0 + L_0)$;
μ_1	:	mean demand over $(t_0, t_0 + R + L_1)$;
σ_1	:	standard deviation of demand over $(t_0 + R + L_1)$;
α	:	fraction of demand delivered from stock.

The output of the analysis is:

S_α	:	order-up-to-level for criterium α;
V_α	:	average safety stock at inventory-point for criterium α.

4.3 Influence of forecasting information

In our analysis we have used a simple stochastic model for forecasts and actual demands. Per endproduct and per sales organisation we have given per fixed period (in our example a week) mean and standard deviation of the forecast and a measure for the deviation of the realized demand from the forecast.

The model that we use is as follows. The forecast over period n is an independent stochastic variable, F(n) say, with mean μ_F and standard deviation σ_F. The realized demand, D(n) say, is defined as the product of the forecasted demand F(n) and a independent stochastic variable U(n) with mean 1 and standard deviation σ_D denoting the deviation from the forecast, i.e. D(n) = F(n)*U(n).

In our module we have used a transformation of forecast, demand and lead-time information into information for the model of subsection 4.2 in order to obtain estimates for the average order-up-to-levels S_α.

The input for the transformation is:

R	:	length of review period in days;
μ_F	:	mean forecasted demand per review period;
σ_F	:	standard deviation;
σ_D	:	deviation of demand from forecast per review period;
μ_L	:	average leadtime in days;
σ_L	:	standard deviation.

The output of the transformation is:

$$\mu_0 = \mu_F \, \mu_L/R;$$
$$\sigma_0^2 = (\sigma_F^2 + \mu_F^2)\, \sigma_D^2\, (\mu_L/R) + \mu_F^2(\sigma_L/R)^2;$$
$$\mu_1 = \mu_F((\mu_L/R) + 1);$$
$$\sigma_1^2 = (\sigma_F^2 + \mu_F^2)\, \sigma_D^2\, ((\mu_L/R) + 1) + \mu_F^2(\sigma_L/R)^2.$$

4.4 The distribution network information

In this subsection we give a more formal description of the input required for the analysis of the distribution network.

Starting point is the product structure described by:

J	:	number of endproduct types (j=1,2,...,J);
I	:	number of component types (i=1,2,...,I);
P_j	:	{i | component i is part of product j}, j=1,2,...,J;
S_i	:	{j | product j contains component i}, i=1,2,...,I.

So P_j is the set of components forming product j and S_i is the set of products which contain component i.

Review period, forecasts, demands and performance criteria are formulated as:

R	:	length of review period in days;
$\mu_F(j,k)$	:	mean forecasted demand per review period for product j at SO k;
$\sigma_F(j,k)$	:	standard deviation;
$\sigma_D(j,k)$	:	deviation of realized demand from forecasted demand for product j at SO k.
$\alpha(j,k)$	:	required service level for product j at SO k.

Eventually information on lead times of manufacturing, bundling and transportation (in days) is given by:

$\mu_M(m)$	:	mean manufacturing lead time at production centre m;
$\sigma_M(m)$	:	standard deviation;
$\mu_B(i)$	:	mean bundling time at place i in the network;
$\sigma_B(i)$	:	standard deviation;
$\mu_T(i,j)$	:	mean transportation lead time from place i to place j in the network;
$\sigma_T(i,j)$	:	standard deviation;

4.5 Transformation per scenario

One of the difficulties in determining safety stock levels in multi-echelon distribution networks is that one has to translate the information at the endproduct and sales organisation level into relevant information at the decoupling point. Relevant means that it must be possible to evaluate safety stock levels at endproduct or component level in such a way that at the sales organisation level the required customer service level is met.

We follow a straightforward approximation based on a simple aggregation of the forecast and demand information over sales organisations and endproducts. We intend to improve the models and approximations by using the ideas and heuristics formulated in Hax and Meal [9] and De Kok [10].

In the following we describe the aggregation for scenario II. For more details on the other scenario's we refer to Van Doremalen and Fleuren [8].

In scenario II we need safety stock levels per component type at the supply centre. Per component type i we first evaluate aggregates of the mean and variance of the demand of the components:

$$\mu_F(i) = \sum_{j \in S_i} \sum_{SO} \mu_F(j,SO);$$

$$\sigma_F^2(i) = \sum_{j \in S_i} \sum_{SO} \sigma_F^2(j,SO);$$

$$\sigma_D^2(i) = \Big(\sum_{j \in S_i} \sum_{SO} (\mu_F^2(j,SO) + \sigma_F^2(j,SO)) \; \sigma_D^2(j,SO)/\mu_F^2(i);$$

Then we evaluate an internal customer service level for the delivery performance of components at the supply centre:

$$\alpha(i) = \max_{j \in S_i} \max_{SO} \{ \alpha(j,SO) \};$$

Note that for the customer service level we use a maximum as the aggregation of the given service level targets. This simple heuristic is based on the idea that to obtain a desired service level for a specific component the maximum over the service levels per endproduct and per sales organisation seems reasonable. To estimate the dependency between the different components is difficult and we have chosen for an optimistic rule, set the target as if the dependency is one-to-one. Note that for example a complete independency would have led to a product of the component targets to guarantee the endproduct target. These heuristic is not very satisfactory and further research is needed here.

Eventually we evaluate per component the mean and variance of the summed lead times from production centres to supply centre:

$$\mu_L(i) \quad = \quad \mu_M(PC) + \mu_B(SC) + \mu_T(PC,SC);$$

$$\sigma_L^2(i) \quad = \quad \sigma_M(PC) + \sigma_B^2(SC) + \sigma_T^2(PC,SC).$$

where SO, PC and SC are sales organisation, production centre and supply centre respectively.

5 AN EXAMPLE

In this section we discuss an example. The following designs have been evaluated:

- I Shipment of endproducts by ship to a supply centre and distribution by truck from supply centre to customers via sales organisation;
- II Shipment of components by ship to supply centre and distribution by truck from supply centre to customers via sales organisations after bundling at the supply centre;
- III Shipment of endproducts by air to customers via sales organisations;
- IV Shipment of components by ship to sales organisations and bundling at and distribution from sales organisations to customers;

For all designs we need information on product structure (see figure 3), endproduct composition (see table 1) and demand figures (see table 2). We work with a review period of a week and the fraction of demand that has to be deliverded from stock is 98% for all endproducts at all sales organisations.

The capital loss rates are 5% interest rate, 15% price erosion and 5% obsolescence risk on a year basis.

In tables 3 and 4 we have pictured for all designs the information with respect to the lead times is given in the form "mean (standard deviation)". In tables 5 and 6 information is given for the costs. The tables have to be read with the description of the designs (see figure 1) in mind.

The endresults are pictured in table 7. Note that scenario III has high physical transportation costs ($3.92 million) but that capital loss in transport ($1.04 million) is relatively low. The economic trade-off shows that scenario III has the best overall cost performance. If we look at the customer order cycle times from decoupling point to customer (COCT (DP to CST), then we see that for all scenario's the distribution of the products will give no problem. It must be remarked, however, that the 8.9 days for scenario III are rather tricky as we also have to reckon with an organisational delay of several days. A smooth and fast organisation is a prerequisite for the success of scenario III.

In [8] we discuss the example in more detail and there we look at a number of other alternatives as well.

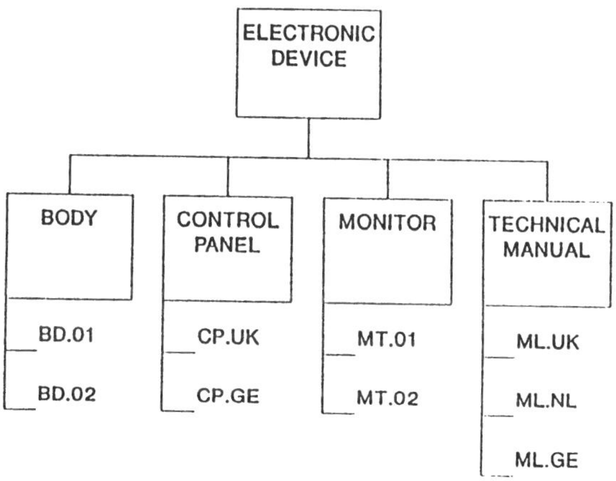

Figure 3 Product structure

End Product	Body	Control Pannel	Monitor	Technical Manual
ED0101NL	BD.01	CP.UK	MT.01	ML.NL
ED0101UK	BD.01	CP.UK	MT.01	ML.UK
ED0101GE	BD.01	CP.GE	MT.01	ML.GE
ED0102NL	BD.01	CP.UK	MT.02	ML.NL
ED0102UK	BD.01	CP.UK	MT.02	ML.UK
ED0102GE	BD.01	CP.GE	MT.02	ML.GE
ED0202NL	BD.02	CP.UK	MT.02	ML.NL
ED0202UK	BD.02	CP.UK	MT.02	ML.UK
ED0202GE	BD.02	CP.GE	MT.02	ML.GE

Table 1 Endproducts and components

Sales Organisation	End Product	μ_F	σ_F	σ_D
NL	ED0101NL	300	100	0.20
	ED0102NL	400	200	0.20
	ED0202NL	100	80	0.20
UK	ED0101UK	200	80	0.30
	ED0102UK	250	100	0.30
	ED0202UK	100	50	0.30
GE	ED0101GE	500	200	0.25
	ED0102GE	800	200	0.25
	ED0202GE	300	100	0.25

Table 2 Forecast and demand information

	PRODUCTION CENTRE	SUPPLY CENTRE	SO UK	SO NL	SO GE
ED0101	0 (0)	2.5 (0.5)	4 (2)	4 (2)	4 (2)
ED0102	0 (0)	2.5 (0.5)	4 (2)	4 (2)	4 (2)
ED0202	0 (0)	2.5 (0.5)	4 (2)	4 (2)	4 (2)

Table 3 Bundling times in the format mean (standard deviation)

	PRODUCTION LEAD TIME	TRANSPORT FROM PC TO SC	FROM SC TO CST			FROM PC TO CST		
			UK	NL	GE	UK	NL	GE
ED0101	4 (2)	24 (1)	4 (1)	2 (1)	2 (1)	10 (2)	8 (2)	9 (2)
ED0102	4 (2)	24 (1)	4 (1)	2 (1)	2 (1)	10 (2)	8 (2)	9 (2)
ED0202	4 (2)	24 (1)	4 (1)	2 (1)	2 (1)	10 (2)	8 (2)	9 (2)
BD.01	4 (2)	24 (1)				26 (2)	24 (2)	25 (2)
BD.02	4 (2)	24 (1)				26 (2)	24 (2)	25 (2)
CP.UK	3 (1)	24 (1)				26 (2)	24 (2)	25 (2)
CP.GE	3 (1)	24 (1)				26 (2)	24 (2)	25 (2)
MT.01	2 (1)	24 (1)				26 (2)	24 (2)	25 (2)
MT.02	2 (1)	24 (1)				26 (2)	24 (2)	25 (2)
ML.UK	1 (0.5)	24 (1)				26 (2)	24 (2)	25 (2)
ML.NL	1 (0.5)	24 (1)				26 (2)	24 (2)	25 (2)
ML.GE	1 (0.5)	24 (1)				26 (2)	24 (2)	25 (2)

Table 4 Lead times of manufacturing and transportation in the format mean (standard deviation)

	FACTORY STANDARD PRICE	TRANSPORT FROM PC TO SC	FROM SC TO CST			FROM PC TO CST		
			UK	NL	GE	UK	NL	GE
ED0101	1000	11	3.00	0.80	1.70	28	24	26
ED0102	1200	12	4.00	1.00	2.00	30	26	27
ED0202	1400	13	4.50	1.10	2.50	31	28	29
BD.01	400	4				7.5	6.5	7.0
BD.02	600	5				8.5	7.5	8.0
CP.UK	200	2				4.5	3.5	4.0
CP.GE	200	2				4.5	3.5	4.0
MT.01	200	4				7.5	6.5	7.0
MT.02	400	4				7.5	6.5	7.0
ML.UK	100	1				4.5	3.5	4.0
ML.NL	100	1				4.5	3.5	4.0
ML.GE	100	1				4.5	3.5	4.0

Table 5 Factory Standard Prices and transportation costs

	PRODUC-TION CENTRE	SUPPLY CENTRE	SO UK	SO NL	SO GE
ED0101	1.0	7.5	8.5	8.5	8.5
ED0102	1.0	7.5	8.5	8.5	8.5
ED0202	1.0	7.5	8.5	8.5	8.5

Table 6 Bundling costs

	SCENARIO I	SCENARIO II	SCENARIO III	SCENARIO IV
ACTIVITY BASED COSTS				
TRANSPORTATION	1.71	1.61	3.92	2.31
BUNDLING	0.14	1.08	0.14	1.22
DISTRIBUTION	0.30	0.30		
TIME BASED COSTS				
TRANSPORTATION	2.79	2.44	1.04	2.37
BUNDLING		0.29		0.47
SAFETY STOCK	0.89	0.35		0.62
DISTRIBUTION	0.31	0.31		
TOTAL COSTS	6.20	6.40	5.09	7.00
PIPELINE VALUE	16.00	13.60	7.00	13.80
PIPELINE LEAD TIME	26.7	29.2	8.9	28.0
COCT (DP TO CUSTOMER)	2.7	5.2	8.9	4.0

Table 7 Endresults for the four scenario's

6 FINAL REMARKS AND CRITICAL ISSUES

We have described a real-life case of the situation where a new distribution network structure had to be considered. Important aspects of the model are the cost calculations and the stock level model.

As we see in the example the costs of transport (both activity and time based) do have a very large effect on the final result. However, we think that the following points need further attention as they seem to be critical for the analysis of this type of models:

1 In general the behaviour of the demand will be highly non-stationary as products in this area have a product life-cycle of less than two years and the demand is highly fluctuating due to seasonal effects, uncertain market trends and a severe competition. This implies that we probably should use non-stationary models for forecast and demand information and dynamic models for the inventory control mechanisms (for example a dynamic order-up-to-level).

2 The translation of customer service levels at the sales organisation and endproduct level into service levels at endproduct or component level at the decoupling point needs further analysis. In fact even the definition of customer service levels needs attention.

3 A managerial question is how to manage the stocks at the decoupling point (especially in the case of a central supply centre) in order to meet the customer service level of all sales organisations. The formulation of appropiate models is a first step.

REFERENCES

[1] R. Tersine, "Principles of inventory and materials management", North Holland, New York (1988).

[2] T.E. Vollmann, W.L. Berry and D.C. Whybark, "Manufacturing planning and control systems", Richard D. Irwin Inc., Homewood (1984).

[3] "Multi-level Production/Inventory Control Systems: Theory and Practice", Ed. L.B. Schwarz, North Holland, New York (1981).

[4] "Multi-Stage Production PLanning and Inventory Control", Eds. S. Axsäter, Ch. Schneeweiss and E. Silver, Lecture Notes in Economics and Mathematical Systems, Springer Verlag, Berlin (1986).

[5] E. Silver and R. Peterson, "Decision Systems for Inventory Management and Production Planning (Second Edition)", John Wiley and Sons, New York (1979).

[6] Sj. Hoekstra and J. Romme, "Op weg naar integrale logistieke structuren" (Towards integral logistics structures), Kluwer, Deventer (Nederland) (1985) (in Dutch).

[7] W. Zimm and D.J. Bowersox, "Planning Physical Distribution with the Principle of Postponement", Journal of Business Logistics, Vol 9, No 2.

[8] J. van Doremalen and H. Fleuren, "A quantitative model for the analysis of distribution network scenario's", To appear as CQM-Report, Centre for Quantitative Methods, Philips, Eindhoven (1990).

[9] Hax and Meal, "Hierarchical integration of production planning and scheduling" In: Logistics (Studies in Management Sciences, Vol.1), Ed. M.A. Geisler, North Holland, New York (1975).

[10] A.G. de Kok, "Hierarchical production planning for consumer goods", CQM-Note nr. 71, Centre for Quantitative Methods, Philips, Eindhoven (1988).

Systematic Development and Implementation of CIM Systems

GÜNTER GRODITZKI

Preface

Until a few years ago only specialists used to know the term and the meaning of 'Computer Integrated Manufacturing' (CIM), now it is known widely. This is because further development in this area has wide-ranging effects not only on the manufacturing industry, but also on the future sociological and political conditions of the human society in general. Therefore the industrial enterprises all over the world realize that it is of primary importance for them to implement CIM. However, as this involves extraordinarily extensive and complex efforts, implementing CIM is a big challenge for each enterprise.

In the following we will show how large application systems like CIM can be systematically developed, implemented and continuously adapted to the fast changing environment of the marketplace.

The concept decribed deals with three key information processing technologies:

1. Computer Aided Software Engineering (CASE), i.e. engineering methods and tools that can be used to build and maintain complex application systems
2. Artificial Intelligence (AI), especially expert systems, in order to adapt generic systems intelligently to particular user requirements, and
3. Database technology used to bridge the gap between a generic CIM system and that operating in a vendor-specific, e.g. IBM hard- and software environment

The following aspects will be considered:

- Why do we need CASE to develop CIM systems?

  This section will show why the development of complex application systems like CIM requires a systematic software development approach utilizing CASE methods and tools.

- What does a CIM model contain?

  This section gives an overview of the complex requirements a CIM model must fulfil, and of the components which form such a model.

- What are CASE methods in the context with CIM?

  Computer Aided Software Engineering methods used to build complex CIM systems will be introduced.

- What are CASE tools in the context with CIM?

  This section gives an overview of Computer Aided Software Engineering tools and their application on the development and implementation of CIM systems.

- How could the CASE future in context with CIM look like?

  This section gives an overview of the efforts made to provide generic CIM systems which then can be adapted to particular enterprise requirements.

Each section will contain of a summary for a short information and of a more detailed description.

Why do we need CASE to develop CIM systems?

Although every enterprise, at least in the industrialized countries of the West, which wants to be successful or, at least, to stay in business, must take up the CIM challenge, the implementation of CIM provides most companies with great problems, because

- the functional scope of CIM is so large that
  - the personal capacity and
  - the necessary knowledge

  are not available in a single enterprise,
- the progress in the area of the manufacturing and information technology is so fast that the satisfactory solutions of today will be outdated tomorrow. Due to this fact there is a strong necessity to adapt the application software systems continuously to the fast changing market environment.

But, even if the above problems can be solved, there is still the problem of coordinated development. In contrast to the traditional development of application systems, the implementation of CIM demands an incomparably larger consideration of the integration aspects, because the desired rationalisation effect can only be achieved through integration of application functions. In order to be able to implement CIM with multiple partners involved in this effort, the enterprise therefore requires an overall concept with the integration aspects in mind from the very beginning, so that the coordinated development of concurrently implemented CIM subprojects is possible.

Of course it is possible to define this overall concept verbally, but the large functional scope of CIM and the fast technological progress in this area demand a much more systematical approach. The basis of the implementation concept of CIM must be formed around a flexible CIM enterprise model, developed with CASE methods and tools and supporting all development stages, from the definition of requirements to system implementation. A model, e.g. a scale-down model of a car, is a construct which contains all the essential aspects of an object from a certain viewpoint, in the example of the car for instance all the basic exterior aspects as seen by the bodywork designer. By analogy a CIM model in a computer is a representation of the actual CIM system containing all the essential aspects relating to the development and installation of such a system from the viewpoint of the system implementers.

Such a model must be developed in a stepwise fashion with the goals and objectives of the enterprise in mind, starting from a global viewpoint and progressing to the specific by the work of knowledgable professionals in the different CIM areas, until it represents a complete enterprise-wide CIM system.

In the following paragraphs the contents of such a CIM model and the CASE methods and tools used in it's construction will be discussed.

What does a CIM model contain?

Summary

A complete enterprise-wide CIM system has a much larger functional scope and thus is much more complex than all previous application systems. Therefore the first part of this section gives an overview of the increased requirements on CIM compared to present systems. Aspects to be considered include:

- *Integration of application functions*, along with the larger range of applications,
- *Enterprise-wide information processing*, along with the distribution of application functions in a computer network,
- *Individual data processing*,
- the role and importance of *Artificial Intelligence* in the future, and
- the role and importance of the *CIM Architecture*

The second part of this section will describe which contents a CIM model must have in order to allow a systematic development and implementation of an operating CIM system. From the business requirements the necessary application functions and activities will be derived. Data flow models as well as application building blocks and application models are needed to complement the *logical* CIM model. The *physical* CIM model then consists of application program building blocks and modular application programs.

Which requirements must a CIM system fulfil?

It is often asked 'What is so new about CIM, that today's application systems, correspondingly modified, can't fulfil the CIM requirements?', because many of these systems are functionally very powerful and cover a large part of CIM. Though in some cases it might be feasible to modify current systems in order to meet CIM requirements, a complete CIM system represented by a complete CIM model goes so far beyond all previous systems in so many aspects, that it should be designed from scratch. The following paragraphs give an overview of these extended requirements.

Integration of application functions, larger application scope

Nearly all the application software packages currently available have not been implemented with the CIM integration aspects in mind. Nevertheless, in a complete CIM model all these aspects (i.e. the integration of CAD/CAE, B&PS and and CAM functions) must have been considered.

New applications within the functional scope of CIM, e.g. Business Planning or Cost-Driven Design, are scarcely implemented in current systems, the interfaces with adjacent application areas like Finance & Accounting or Office Administration are weak in most of the cases.

In the area of Production and Business Planning (P&BS) present solutions cover only certain production environments, mainly

- small and medium scale production in
- job shop organisation.

Large scale and make-to-order, continuous flow and repetitive production are rarely supported. The same is true relating to the support of management and plant, customer and supplier personell, because current systems usually support primarily the business professional in his daily routine activities like buying, customer order processing etc.

In contrast, a CIM system must cover all CIM application areas in an integrated manner, all production environments and all kinds of users.

Enterprise-wide distributed application architecture

Enterprise-wide integrated application systems like CIM should enable every user to get access to the information necessary to do his job regardless of who 'owns' the data within the enterprise and where and how it is stored. I a manufacturing enterprise this information normally is distributed over a hierarchically organized computer and communication network.

Another necessity is to think in terms of enterprise-wide business processes, because in an integrated business function environment it is absolutely necessary that business processes can be performed and controlled across personal and organizational boundaries.

Individual data processing, integrated user services

The information processing environment has changed dramatically over the last few years due to effective personal computing. The combination of excellent price/performance on the hardware side together with ease-of-use software from countless vendors has opened application opportunities as never before. This development has provided the end user with the opportunity to build his applications on his own, i.e. without having to wait on IS staff support. 4th generation languages, spreadsheets, text and graphics editors, list generators etc. enable the end user to get personally adjusted computer support immediately and directly. This way of working is so superior to traditional computing, that it should be preferred to 'hard coding' implemented by the company's IS organization wherever possible.

Artificial Intelligence (AI)

In future systems AI applications, mainly expert systems, will play a leading role in several aspects:

1. In every manufacturing enterprise there are countless applications where expert systems and other AI methods (e.g. robotics) can be used. The field is also wide open for totally new solutions, from which in the long term no enterprise can escape.
2. The man/machine relationship between the IS user and the computer terminal will be much more user-friendly in the future with the aid of expert systems together with natural language support, image processing etc.
3. Expert systems can be used in the future with the user determining the requirements and deriving the particular hardware, software and production equipment requirements. This AI aspect will be discussed later in some more detail, because this technology can be effectively used to adapt generic CIM systems to enterprise-specific CIM requirements.

CIM architecture, CIM standards

When many parties are working on different parts of a CIM system and are, in addition, using CIM-based hardware and software systems from different vendors, it is obvious that a common scheme is necessary. Therefore a CIM architecture, i.e. the intellectual concept on which a system is built, together with CIM standards is absolutely necessary.

Consequently, there are various efforts to progress in this area. The EEC for instance sponsores an ESPRIT project called CIM-OSA (Computer Integrated Manufacturing - Open System Architecture), with 21 leading edge companies in Europe (including IBM) involved with the aim to define a CIM reference architecture together with CIM standards. Even CIM models developed by individual enterprises without the help of third parties should utilize this architecture in the interest of longevity of their development. For software firms attention to the CIM Reference Architecture and Standards is an absolute necessity, guaranteeing that enterprise-specific developed software can be integrated with vendor-supplied software into a common system.

In addition, there are application independent standards like OSI, MAP etc. to be considered. There are also vendor-specific standards which should be followed. In an IBM environment for instance it is strongly recommended to follow the System Application Architecture (SAA) standard, which provides common presentation, communication and data management support across all strategic IBM computer platforms (in IBM /370, AS400 and PS/2 hardware architecture).

This all may look very complicated, but in reality the development, implementation and adaptation of CIM systems is eased enormously, because complex systems can be developed and implemented systematically and therefore time- and cost-effectively, providing a problem-free growth together with a perfect utilization of common software built to the same specifications.

What are the components of a CIM model?

The exact contents of a CIM model have not been described so far. Therefore this aspect will be considered more closely now, because the CIM implementation concept depends to a large extent on the contents of the model.

When we talk about a CIM model, we usually understand a CIM information system (IS) model. This is fine, because this indeed is the most important component of a CIM system, at least from an IS point of view. Nevertheless, a complete CIM model could contain some more components, because

- the *technology available* at the time of implementation and the corresponding hardware and software components and other factors like production equipment, enterprise organisation, personell qualifications etc., fundamentally influence the concrete solution.
- From the *enterprise management* point of view the implementation of CIM is not a goal in itself, but a strategy to achieve the business goals and objectives, together with other strategies (e.g. the product, the competitive and the financial policy). A CIM model with built-in knowledge-based methods to achieve the enterprise goals and objectives, taking in mind the critical success factors of the enterprise, would ensure the strategic direction of the CIM system implementation.

The kernel of a CIM system, however, is the CIM information system model. Therefore the components of such a model will be described in some more detail.

CIM requirements

The development of a CIM application system starts with the definition of CIM requirements, ideally derived from the strategic enterprise goals and objectives. As in all other development stages, this work is done by cooperation of system analysts and end users, starting from global aspects and progressing in a stepwise fashion to the details.

CIM business functions

The next logical step is to define the CIM business functions necessary to fulfil the defined requirements. This work also starts with the global definition of functional areas (like business planning, production logistics etc.) and ends with the detailed functions (like release of manufacturing orders).

CIM business processes

Business functions are executed by business processes, i.e. by activity chains needed to support a particular process, e.g. the release of a new product to manufacturing. Business processes performed across personal and organizational boundaries are essential in a CIM system, because the basic characteristic of CIM is *integration* across those boundaries.

CIM activities

The next logical step is to define the activities needed to execute a business process, for instance those the users have to perform in order to process a customer order or to release a new product to manufacturing.

CIM data flow model

The next step is the definition of the information flow needed to support the activities, the data flow analysis. In this step the data flowing between the activities is defined together with the transformation processes performed by the activities. The final result is an enterprise-wide data flow model encompassing all data requirements of all users in the enterprise.

CIM application building blocks

Although each activity chain must be modelled separately, the resultant CIM data flow model will contain many parts which are similar, if not identical. It therefore makes sense to define generic application building blocks, for instance to support planning activities, to be utilized in multiple parts of the system.

CIM application models

Generic application models support a particular application, for instance Production Planning, and could themselves utilize application building blocks. These generic models are built in that way that they can be adapted to the needs of a particular enterprise.

Logical and physical CIM models

The CIM components described so far concern primarily the 'what?', not the 'how?' of CIM implementation. They do not consist of program code, but of logical constructs forming a generic logical CIM model, which considers the concept of distribution of application functions in an hierarchically organized computer network.

In contrast to the logical CIM model the physical CIM model considers the information and manufacturing technologies available at the time of implementation.

The separation into the logical and the physical aspects of a system is useful, because the logic of a system changes much slower than the physical implementation due to the fast progress in the information and manufacturing technologies.

The following components of a CIM information system model belong already to the physical CIM model, because their implementation depends on the physical components available.

CIM application program building blocks

To the generic application building blocks decribed above correspond generic CIM application program building blocks, i.e. the programmed version of the defined logical constructs. The focus in this case is on 'how' the system works in a concrete computer environment rather than on 'what' has to be done, on program control rather than on data flow.

CIM application programs

The generic application models described above correspond to generic CIM application programs, which are the programmed version of the logical CIM application models. These application programs are modular application software packages, which can be adapted to the needs of a particular enterprise.

Of course the creation of such a voluminous generic CIM system is not an effort to be performed overnight. Nevertheless, there are large worldwide efforts underway which work into this direction, as will be described later.

What are CASE methods?

Summary

The importance of Computer Aided Software Engineering (CASE) methods in the development of complex systems has been widely recognized. From the beginning some 20 years ago a series of methodologies has been developed, of course with various strengths and weaknesses relating to the different phases in the life cycle of a system. Well known examples are Structured Analysis (SA), Warnier-Orr or the Structured Analysis and Design Technique (SADT).

In the following paragraphs there will be discussed how such methods could be used in the development of CIM systems. As the focus in this case lies on the understanding of the principle, not on the evaluation of the different methods available, the methodolgy developed by Yourdon/DeMarco is portrayed. This methodology is very well accepted, and the requirements of the manufacturing industry (e.g. real-time processing) are especially well covered by it.

As many readers may not be familiar with the concept of *structured techniques*, these will be described first, although a certain redundancy with the following text is thereby unavoidable. In the second part of this section, the structured life cycle of application systems, in which these structured techniques are used, is discussed. The focus lies thereby on the early phases of development, because these are the key phases in the development of complex systems like CIM.

What are Structured Techniques?

Originally structured techniques arose in order to make programming more effective, but then broadened to include other activities within the application life cycle and consist in the meantime of

- Structured Analysis
- Structured Design
- Structured Programming
- Top-down Development
- Programming Teams
- Structured Walkthoughs

Structured Analysis

Structured Analysis basically serves to build a logical model of the new system. To this aim first the current system is analyzed in cooperation between system analysts and end users and then migrated into the new system by means of a so-called 'Structured Specification', consisting of the following components:

With the help of Data Flow Diagrams (DFDs) the data flowing through an application system is analyzed and captured (regardless of whether an activity is handled manually or computer-based) and systematically documented. Figure 1 shows the four basic elements of such a DFD: data sources and sinks from outside the system (symbolized by rectangles), data flows into and out of the processes (symbolized by directed arrows), processes processing the data (symbolized by circles) and data stores utilized by the processes (symbolized by double lines).

The complete description of a data flow demands several hierarchically (from global to detail) performed analysis steps, with each process decomposed into more detailed processes, until the greatest possible degree of detail, the so-called functional primitive level, is achieved.

A Data Dictionary (DD) is utilized to store all DFD data in a systematic form. Thus a DD is a meta data system containing 'data about data' like data element names, formats, relationships etc.

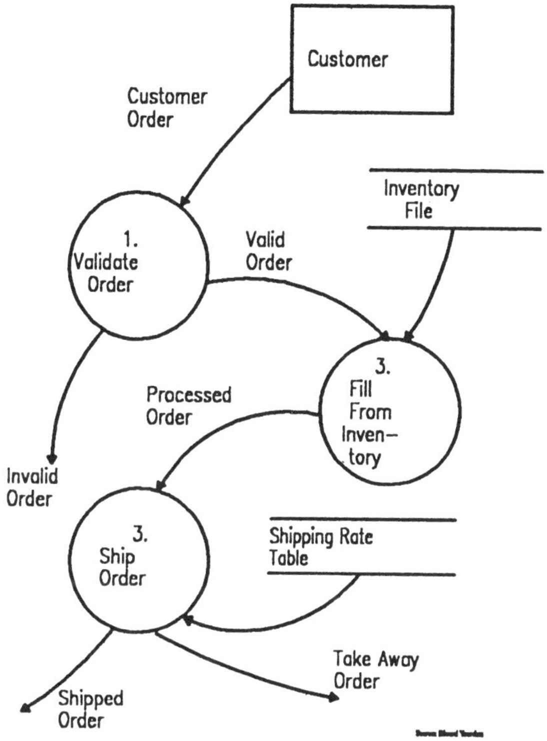

Figure 1. Example of a Data Flow Diagram

In Data Structure Diagrams (DSDs) the logical data structure is described in graphical format. Figure 2 shows as an example the relationships between department, project and employee data. DSD's are mainly needed to develop the real data interrelationships in cooperation with the end users by utilizing a graphical format, which is a common base between the more technically oriented system analysts and the application oriented end users.

Mini Specifications describe the processes in the most detailled form using a limited set of verbs and nouns (usually the Structured Programming techniques with IF-THEN-ELSE and DO WHILE constructs forming pseudocode are used) in order to instruct the programmers how to implement a system in detail.

Structured Design

The logical application model, defined by Structured Analysis techniques, describes the new system exactly and in detail. Intentionally it does not contain any program control logic, which is a matter of a particular type of computer implementation, namely 'hard coded' programs. Therefore the next logical step is the Structured Design of the application system. The essential thing in this methodology consists of designing programs in a hierarchical model structure with as few connections between program modules as possible in order to ease program implementation and maintenance. Program Structure Diagrams can be derived from Data Flow Diagrams, as shown in Figure 3.

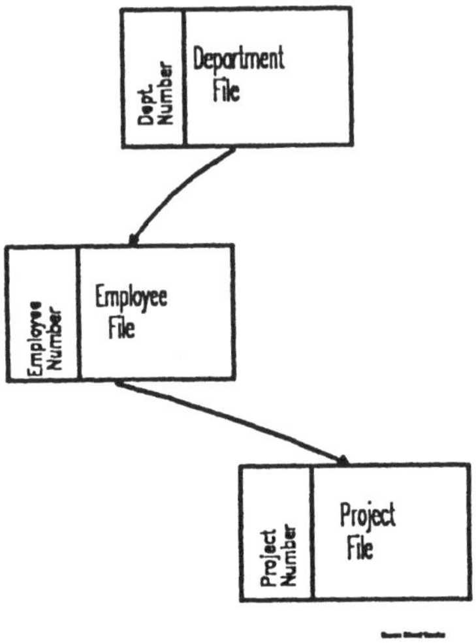

Figure 2. Example of a Data Structure Diagram

By utilizing Structured Programming techniques the Mini Specifications developed in the analysis phase and the Structured Design Specifications developed during system design are transformed into real programs. This technique uses for reason of simplicity only three basic constructs - sequence of commands, if-then-else and do-while - to define the program logic.

Top-Down-Development, Programming Teams, Structured Walkthroughs

The other Structured Techniques concern the further implementation and can be neglected here. Top-down Development, in this narrow sense, is a method to test programs hierarchically from top to bottom. The organisation of a Programming Team is build around a chief programmer, including besides programmers a data base administrator, a toolsmith and a librarian in order to provide effective programming. A Structured Walkthrough is a way to ensure the correctness and quality of a system by a team of all kind of people involved in the project - end users, systems analysts and programmers.

The Structured Life Cycle of application systems

The Structured Techniques described should cover the whole life cycle of a system (from the first blueprint through to installation and maintenance). This structured lifecycle subdivides into phases, as represented in so-called waterfall models. Figure 4 shows such a model in a simplified form.

The waterfall model describes a sequence of logical phases basically following on each other, in reality of course being performed incrementally and iteratively, i.e. the development usually is going from global to specific within each phase (with possible phase overlap), and findings of the tasks performed in later phases can change the model descriptions defined in earlier phases.

As in performing complex tasks like implementing a CIM system human errors cannot be avoided, this human imperfection must be considered from the beginning in the concept. Therefore the CASE methods and tools utilized to implement CIM must allow for permanent correction without problems.

The most important phases in the development of CIM systems are the early phases, especially the analysis phase, because here the logical CIM model is developed as the basis for all further realistic implementation. In these phases the Structured Analysis method is applied. In several steps the new system is developed in

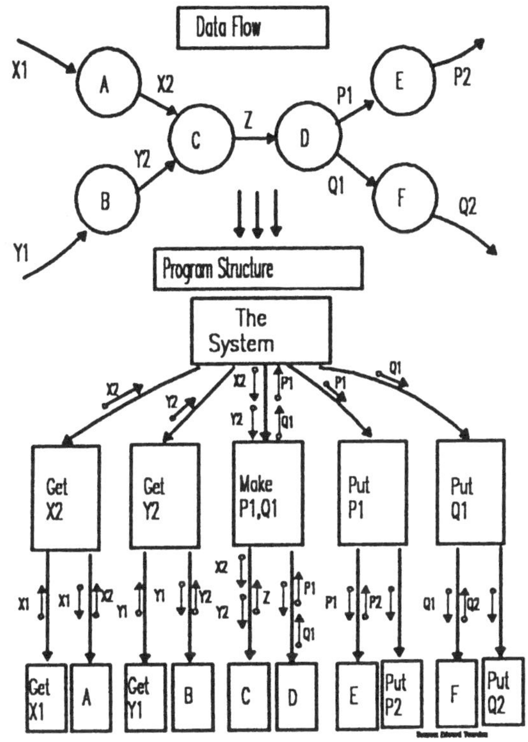

Figure 3. Derivation of a Data Structure from a Data Flow Diagram

cooperation between system analysts and end users. Proceeding from CIM requirements the important CIM business functions are defined, from these CIM business processes and CIM activities are derived, and eventually the data flow and process models representing the system are created. The result is a concise and redundant free model representing an enterprise-wide CIM system.

Already Tom De Marco has differentiated between between the logical and the physical aspects of a system. This concept has been extended by Stephan McMenamin and John F. Palmer. In order to achieve the essentials of a system, they assume a perfect technology without any restriction. This way a logical model, called the Essential System Model, is achieved. In a further step implementation limitations are considered and modelled in an implementation model called the Incarnation Model.

This methodology is especially important in the context with CIM, because the fast technological change requests an easy change to new market conditions, and logical models are of course much more stable than implementation models.

Another also very important extension of structured techniques in the context of the manufacturing industry requirements has been made by Paul T. Ward and Stephan J. Mellor by introducing a methodology to define real-time systems. This methodology allows the construction of dynamic models, which meet the requirements of real-time manufacturing processing. By utilizing this methodology, real-time processes can be modelled and integrated into the enterprise-wide information system, forming a real complete system encompassing all aspects of a manufacturing enterprise.

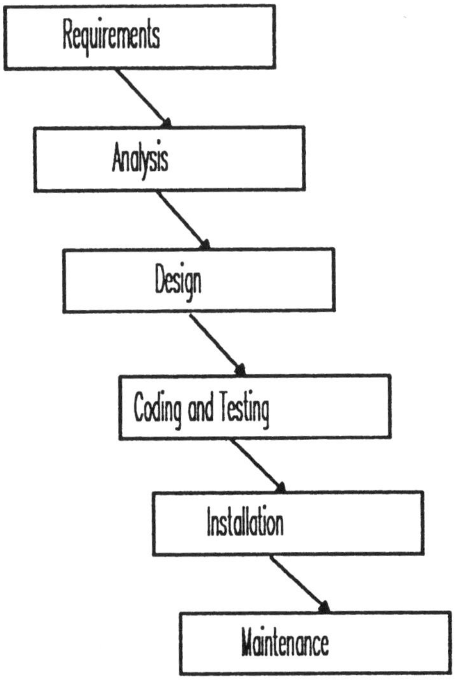

Figure 4. Waterfall model of a system life cycle

What are CASE Tools?

Summary

In the last few years a whole range of CASE tools has emerged supporting methodologies like those developed by Yourdon/DeMarco and others. This is still in progress, with large new opportunities opened up by micro- and minicomputers. But as, again, the focus of this paper lies on the use of this technology in the context with CIM and not on the evaluation of available CASE tools, a specific set of tools around PSL/PSA (Problem Statement Language/Problem Statement Analyzer) is decribed in this context.

PSL/PSA is a CASE package for the systematic development of information systems, which was developed with the support of leading-edge US companies (including IBM) under the leadership of Daniel Teichroew at the University of Michigan, Ann Arbor/USA, and which is installed several hundred times worldwide. PSL/PSA is host-oriented and runs on computers of several vendors including IBM, DEC and Hewlett Packard. The originally developed kernel program has been complemented in the meantime by several further program offerings, part of these running on microcomputers.

PSL/PSA and complementary program products

PSL/PSA

The kernel program package PSL/PSA consists of two parts:

- *PSL (Problem Statement Language)* is a language for the unambiguous description of information systems.
- With the aid of PSL statements a central system encyclopaedia is built-up, which can then be analysed through PSA (Problem Statement Analyzer).

PSL is a so-called non-procedural language allowing the description of a system by the use of keywords, without being bound to a specific methodology (like Structured Analysis, Warnier-Orr or SADT) relating to analysis, design or implementation.

These descriptions are based on entity-relationship (E-R) models, by which entities (e.g. personal, departments, processes) are connected by relations (e.g. departments 'employ' personal). The entities and relations are characterized by attributes (in the case of the entity 'person' for instance by the attributes sex, birthdate, skill etc.). PSL recognizes 19 different entity types (e.g. process, input, output) and 102 different relations (e.g. receives, generates).

With the aid of PSL, complex systems can be described in a concise manner. Thus CIM models can be built incrementally, starting from a global view and progressing to the details. In the context with CIM it is specifically important that the language also supports entities and relations that allow the modelling of real-time systems.

PSA is the program which first checks the PSL statements entered and then stores them as logical constructs in the central encyclopaedia. From there the contents then can be analyzed and displayed or printed. PSA is able to generate 37 different report types which provide the user with the means to check the completeness, integrity and consistency of the system.

Structured Architect, Structured Architect Integrator

PSL/PSA is host-oriented and has no components to model a system using graphics. This gap is filled by Structured Architect (SA), a microcomputer program package running under Microsoft Windows on IBM-standard PCs or PS/2s, respectively. The graphics and the other features (e.g. windowing technique, local data store) allow DFDs to be outlined in graphical form in cooperation between system analysts and end users. As the program employs its own data enclyclopaedia on the PC, the contents can be analyzed, displayed and printed locally. Eventually, PSL statements can be generated and uploaded to the host system in order to be processed by PSA.

A package of this type is especially important in the context with CIM because a graphical representation provides a common basis of understanding between the system analysts and the end users which would otherwise be difficult to establish. (The phrase 'A picture says more than 1000 words' is very appropriate in this context.)

This product allows too the modelling of real time systems.

Structured Architect Integrator (SA-I) is a host-oriented program package which serves to consolidate SA inputs from multiple PC's. As this product too employs its own data base, the contents can again be analyzed, displayed and printed , before eventually consolidated PSL statements are generated and loaded into the PSA encyclopaedia.

Such a product is important for the modelling of a complex system, as it allows the subdivision of large projects by employing multiple developers without losing the overall view.

QuickSpec

QuickSpec is also a program package running under Microsoft Windows on an IBM-standard PC or PS/2, respectively. It allows the user to define systems in PSL notation without having to learn the language in much detail, because the input and output can be defined in the 'language' of the user and then transformed into PSL notation.

In the context with CIM a product like this one is very useful, because it allows the involvement of people not so familar with CASE techniques. It also opens up the ability to extend a system with new entities and relationships in an easy way.

View Integration System

View Integration System (VIS) is a host-oriented program package used to design relational data bases in normalized form. With the aid of SA, user views (i.e. information requirements of a specific user or department) can be defined and uploaded to the host system in the form of PSL statements. VIS processes this input, extracts the corresponding statements previously entered from the PSA encyclopaedia and then forms from the new input together with that already defined by earlier user views a relational data base design in Third Normal Form (3NF).

A product like VIS is especially important in the context with the development of complex CIM models, because it supports the integration of all user requirements in a company into an enterprise-wide integrated information system.

Report Specification Interface

RSI is a 4th generation language used to extract information from the PSA encyclopaedia, which then can be formatted in various ways. By means of this product, a user can generate his own reports (in addition to those offered by PSA), but also prepare output in a format which can be further processed by other systems.

A program of this type is undoubtedly very useful during CIM implementation, as reports not provided by PSA can be generated in an easy way. Also the easy transformation from the PSA-world to the data base environment of a specific vendor (like IBM) could be of great importance.

Reverse Engineering

Reverse Engineering is a host-oriented program package, which analyzes the source code of existing COBOL, FORTRAN or other programs in order to determine the logical structure, and generate from that PSL statements. PSA can then be used to check these programs on completeness and consistency in order to provide the basis to add missing program documentation or to integrate old and new systems.

A product of this type is especially important in the context with CIM implementation, because every potential CIM user has existing information systems which must be integrated with the new system. Also, a part of the enormous amount of money invested into the current systems could probably be saved, if 'knowledge' imbedded in current programs could be transferred into the new system in an easy way.

How could the future look like?

Summary

It is clear that the development and implementation of a complete and correct CIM system is no easy task, even if the CASE methods and tools described here are used. The question is therefore, how an enterprise can be helped to overcome its personal, qualification and capital limitations. It is often claimed 'CIM is a concept, not a product; it therefore cannot be bought'. Today that may be true, but what about in the future? Why could not, similar to the modular application software packages of today, generic CIM systems be created, which then can be adapted to the requirements of a particular enterprise? The ESPRIT project CIM-OSA mentioned earlier for instance is based on this idea.

The following paragraphs deals with this concept, utilizing the previously described CASE methods and tools for the development of complex systems.

Basic Concept

Every enterprise has requirements which are really specific. Nevertheless, studies have shown that almost every requirement is valid for many enterprises, although they may appear differently, based for instance on different terminology. At least the logical CIM model is very similar for many enterprises. Therefore it should be able to create generic CIM models valid for many enterprises which thus can be utilized by the total CIM user community.

Figure 5 shows the schema of such an implementation concept, based on the following five components:

1. **Generic CIM model**

   This model consists of a generic definition of a CIM system, from which enterprise-specific models can be derived. This model has been built utilizing the previously described (or comparable) CASE methods and tools on the modelling of various enterprises with different business characteristics.

2. **Derived CIM model**

   This model consists of those parts of the generic model which apply to a particular enterprise. Figure 6 shows this relationship.

   The outer circle in the lower part shows the application scope needed in a particular enterprise, the inner line those parts of the generic model which apply to this particular enterprise. The area between both lines indicates the part which must be added by the particular enterprise in order to provide a complete enterprise-specific CIM model.

3. **Enterprise-specific CIM model**

   This model represents the actual target model of an enterprise. It encompasses the derived CIM model plus the enterprise-specific parts, built with the same CASE methods and tools.

4. **Procedures to complete the derived CIM model**

   These procedure serve to complete the derived models with those parts that are not derivable from the generic model. Using the same CASE methods and tools as described before, the specific requirements of the enterprise are incorporated into the model.

5. **Procedures to Extend the Generic CIM Model**

   These procedures deal with the analysis of the extensions a particular enterprise has added to the derived model in order to evaluate, if these extensions in reality are not of a generic nature. If this is the case, those extensions are in turn added to the generic model which in this way becomes more and more complete. This analysis is of course a task to be performed by highly qualified application specialists in a kind of project office, maybe supported by advanced analysis methods and tools, for instance expert systems.

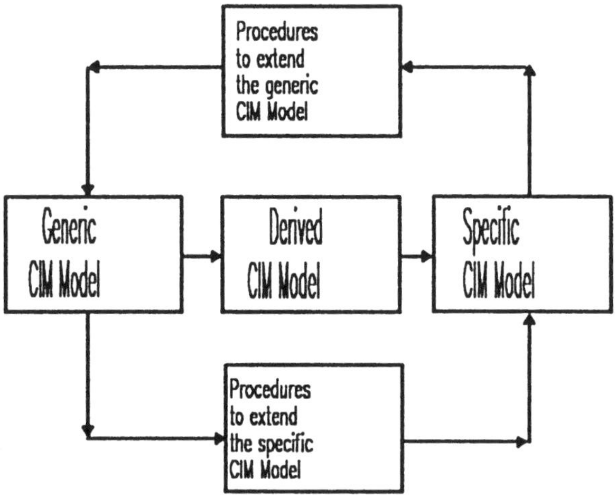

Figure 5. Generic CIM implementation concept

The ESPRIT CIM-OSA Approach

The ESPRIT project CIM-OSA mentioned earlier has the objective to develop an open CIM architecture together with CIM standards valid for all kinds of manufacturing enterprises. The project involves 21 leading-edge European companies including IBM Germany. The following paragraphs give an overview of the modelling concept of CIM-OSA, because this approach is rather similar to the concept described earlier in this section.

The basic idea is to provide generic CIM models which then can be adapted to the requirements of particular enterprises. Figure 7 shows this concept of definition and derivation of CIM-OSA models.

Compared to Figure 5, the CIM-OSA modelling approach is more differentiated, in that there are three modelling levels:

- The horizontal level shows the degree of commonality, from generic to enterprise-specific. The first column shows generic CIM-OSA building blocks which can be used to build CIM models. The partial models in the second column correspond basically to the generic enterprise model, and the particular models in the third column correspond to the enterprise-specific CIM model in figure 5. The process to proceed from generic to particular is called 'instantiation'.
- The vertical level shows the degree of realisation from conceptual to implementation (i.e. from logical to physical). The first row depicts building blocks and models from a conceptual requirements definition view. The second row (with design specification building blocks and models) shows an intermediate step towards implementation, with more system-oriented constructs, but without assuming specific hard- and software. The third row shows building blocks and models regarding all hardware, software and other technology facts valid at implementation time. The particular implementation model is basically the

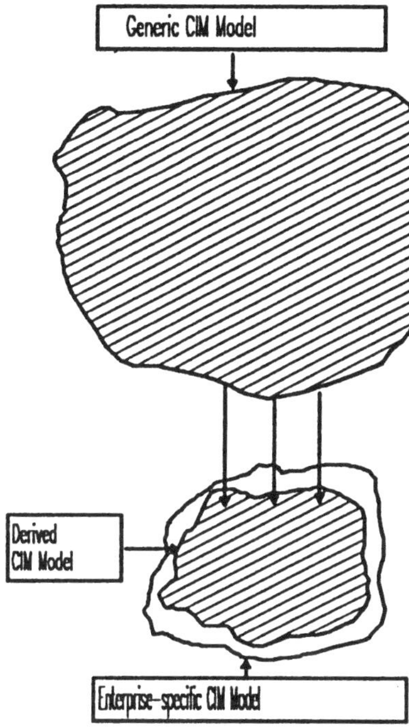

Figure 6. Relationship of the CIM model types

same as the specific model in figure 5. The process to proceed from conceptual to implementation is called 'derivation'.

- The third level shows the four different views which can be applied on CIM building blocks and models. The Function View describes the CIM business processes, enterprise functions and activities, the Information View all the enterprise-wide data, the Resource View all resources of the enterprise, and the Organisation View the personal responsibilities in the enterprise. The process to proceed from the functional to the organizational view is called 'generation'.

It will surely take still some time, until complete CIM-OSA deliverables will be available to the marketplace. Nevertheless, it is strongly recommended to follow the concepts and standards developed so far already today, because these are well designed and provide a safe strategy of CIM development and implementation.

Example of a development concept

In order to check the feasibility of the concepts described, we developed a computer based prototype which is described in the following paragraphs.

The build-time environment of the implementation concept

Figure 8 shows the build-time environment scheme of the prototype.

As stated throughout this paper, complex systems like CIM can be developed reliably only with the deployment of functionally powerfull CASE methods and tools. In this case the model has been created using PSL/PSA and its complementary products. The result is a generic CIM model, the details of which are stored in the PSA system encyclopaedia in a complete and consistent form.

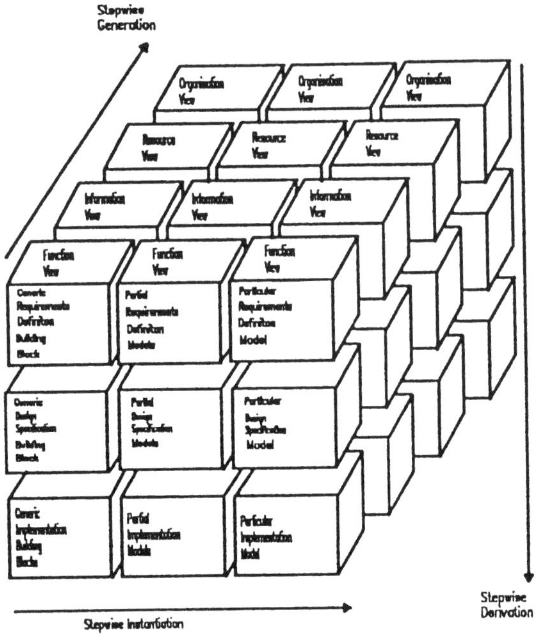

Figure 7. CIM-OSA concept to define and derive CIM models

The derivation component has the actual task of recognizing the requirements of the users and configuring on this basis the CIM system of a particular enterprise. In this concept this task is technically performed by an expert system, the software used is a so-called expert system shell, IBM Expert System Environment (ESE). This expert system ascertains, in the language of the user, the requirements and infers (by means of a so-called inference engine) from the answers the particular requirements and from these the other parts of the model.

In the PSA system encyclopaedia all constructs of the model are stored, but not in a format to be used at run-time in a vendor-specific (e.g. IBM) environment. We therefore need a component that provides the transformation from the vendor-neutral to the vendor-specific format. As mentioned before, the Report Specification Interface program product allows the derivation of data base schemata from the PSA system encyclopaedia. Therefore, for this prototype, we plan to use RSI to transfer the model contents from the PSA encyclopaedia into the repository system of the IBM application development toolset AD/Cycle.

In the IBM CIM Architecture concept, the data repository (as a meta data system holding not only the data descriptions themselves, but also all other model information about business processes and functions, user views etc.) controls the data access of all users. That means that all access to shared data on real data stores is routed through the repository. In this prototype, we therefore plan to have AD/Cycle perform this control function (at build-time) with the data itself stored in distributed DB2 data bases. The users then will access this data through the repository at program run-time.

As in this concept the physical CIM system utilizes a building block approach, we will need also a library of generic application program building blocks and application programs selectable for the particular enterprise.

All components of this concept have a common, decidedly important aspect: They are open systems in the sense that model changes and extensions can be incorporated without problems. Therefore the CIM model

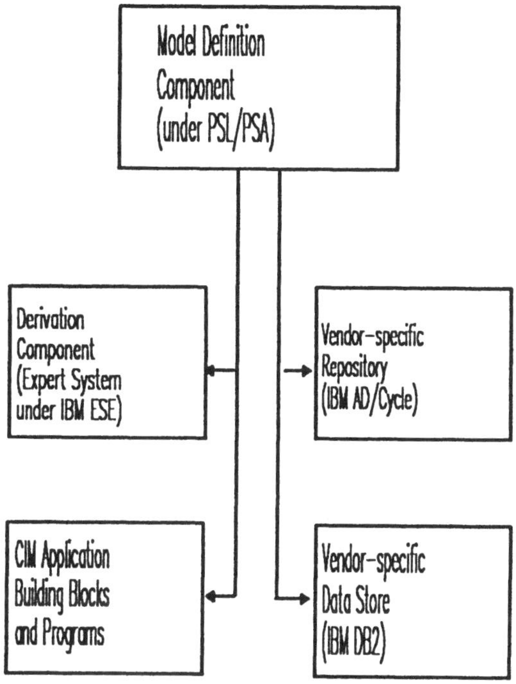

Figure 8. Build-time environment of the prototype

can also be developed starting with this prototype and then continuously extended and adapted to changing market conditions, without the effort becomming too large and too expensive.

The concept is also open in the sense that the model could serve many implementation partners as a common, consistent development base.

The concept to derive enterprise-specific CIM models

The implementation of CIM in an enterprise will always be difficult, not only because of the complexity of the subject, but also because of the various implementors involved. Enterprise-specific implementation must always be based on the actual requirements of the users, not on technical information system aspects. For these reasons, in our concept the endusers define their system with systems analysts using an expert system as an 'intelligent consultant' (see figure 9).

The user dialogue in the prototype has been built from a CIM questionnaire, which asks questions regarding

1. type of business, e.g. size, number of plant sites etc,
2. type of products marketed, e.g. standard, assemble-to-order or make-to-order products,
3. type of production, e.g. discrete or bulk production, job shop, repetitive or continoous flow production,
4. type of material sourcing, e.g. characteristics of the supplier market, importance of electronic data interchange with suppliers,
5. type of sales, e.g. market characteristics, importance of electronic data interchange with customers, and
6. type of distribution, e.g. use of distribution centers, kind of transportation etc.

Dependent from the answers of the users in the expert system dialog, the system selects from the complete generic list of requirements those that apply to the particular user. Technically, the requirements are cur-

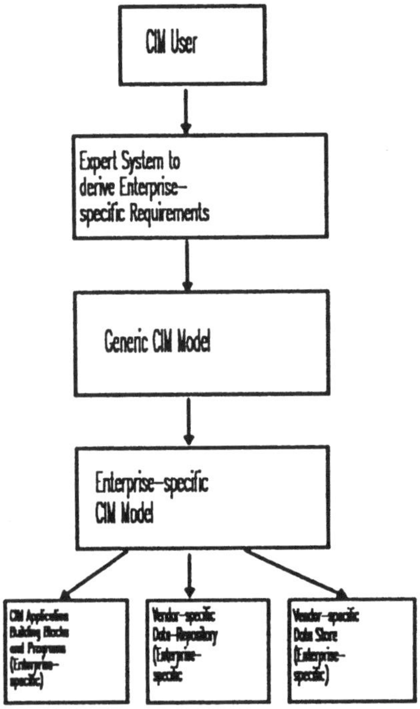

Figure 9. Concept to derive enterprise-specific CIM models

rently put into a CMS file which is handed over to PSL/PSA. The particular requirements are then extracted from the PSA data encyclopaedia and printed on a page printer.

In a future complete system (i.e. a system with a complete CIM model), the particular enterprise using this concept would still have the task to add its requirements not covered by the generic model. But, if the derived model is used as a basis and the same CASE methods and tools are used as those used to build the generic CIM model, this task should not be too difficult.

Bibliography

1. Benson, Robert J. and Parker, Marilyn M.: Enterprise-wide Information Management, An Introduction to the Concepts, IBM Los Angeles Scientific Center Report G320-2768, 1985
2. Chen, Peter P.: Entity-Relationship Approach to Systems Analysis and Design, University of California, Los Angeles 1979
3. DeMarco, Tom: Structured Analysis and System Specification, Prentice-Hall, Englewood Cliffs 1979
4. ESPRIT Project nr 688: Reference Architecture Specification, Veroeffentlichung des CIM OSA-Konsortiums, Brussels 1987
5. Ives, Blake and Learmonth, Gerard P.: The Information System as a Competitive Weapon, in: Communications of the ACM, Vol. 27 No. 12, Seiten 1193 - 1201, 1984
6. McMenamin, Stephen M. and Palmer, John F.: Essential Systems Analysis, Yourdon Press, New York 1984
7. Porter, Michael E.: Competitive Strategy, The Free Press, New York 1980
8. Porter, Michael E.: Competitive Advantage, The Free Press, New York 1985
9. Problem Statement Language/Problem Statement Analyzer (PSL/PSA): Systemdokumentation, Volume I - IV, Meta Systems, Ann Arbor/Michigan 1987
10. Rockhart, John F.: Current Uses of the Critical Success Factors Process, in: Strategic Planning and Information Management Conference Proceedings of the Society for Information Management, Seiten 17 - 22, 1982
11. Scheer, A.-W.: CIM - Der computergesteuerte Industriebetrieb, Springer-Verlag, Berlin Heidelberg New York 1987
12. Scheer, A.-W.: Informationssysteme in der Industrie, Springer-Verlag, Berlin Heidelberg New York 1988
13. Ward, Paul T. and Mellor, Stephen J.: Structured Development for Real-Time Systems, Yourdon Press, New York 1986
14. Yourdon, Edward: Managing the System Lyfe Cycle, Yourdon Press, New York 1982

New Approaches to CIM Specification

ALLAN CARRIE

INTRODUCTION

In the past there have been many examples of both computer systems and manufacturing systems which have delivered less than they promised. Frequently, the fault is due to inadequate specification of the system at the outset (Carrie, 1990). Too often aspects of the system, which it is known will need to be addressed at some time have been left over until later. At the commissioning stage it may be found that they are critical to the specification and efficient operation of the system. The result is that either the efficiency of the system is compromised throughout its life, or can only be achieved through additional expenditure which was not budgeted for at the start of the project.

In recent years several techniques have emerged in the field of software engineering which can be exploited in the design of manufacturing systems. In particular this paper refers to the application of computer aided software engineering (CASE) tools in Manufacturing Systems Engineering. It is suggested that using these tools it is possible to specify both computer software systems and also physical manufacturing systems, ie both information flow systems and material flow systems, and to develop simulation models of both types.

The paper gives an overview of several projects in this field, concerned with manufacturing integration, integration of information and material flow, using CASE tools to develop simulation models, and profiling small enterprises for CIM.

CASE TOOLS

Computer Aided Software Engineering techniques have matured considerably in recent years. They provide a discipline which can ensure that a thorough software system specification will be developed. Martin and McClure (1988) give a comprehensive review of the various methodologies and CASE tools available.

According to Bowen (1989), any system comprises three main characteristics - data, dynamics and process - and provides tools for modelling each of these "views" of the system:

The "Data viewpoint" models the complexity of the data in the system. Thus it will distinguish between for example the engineering data bases of a company manufacturing a narrow range of products assembled from few parts from that of a manufacturer of complex products involving wide and deep bills of material. Entity Relationship Diagrams can be used to model the structure of the data.

The "Process viewpoint" models the complexity of the processes to be carried out on the data. Thus a production control system would be recognised as much more complex than a wages package in which relatively simple calculations are performed. Data Flow Diagrams may be used to model the processes involved.

The "Dynamics viewpoint" models the dynamic behaviour of the system from the time-dependent aspect. It measures how quickly the data must be processed and how frequently that processing is required. Thus a real-time control system necessitates a much more rapid and frequent data processing than would a master scheduling system in a company manufacturing long lead time products. State Transition Diagrams are used to model the dynamic nature of the system.

The complexity of the system measured on these three viewpoints can be combined to give a so-called system "footprint", figure 1.

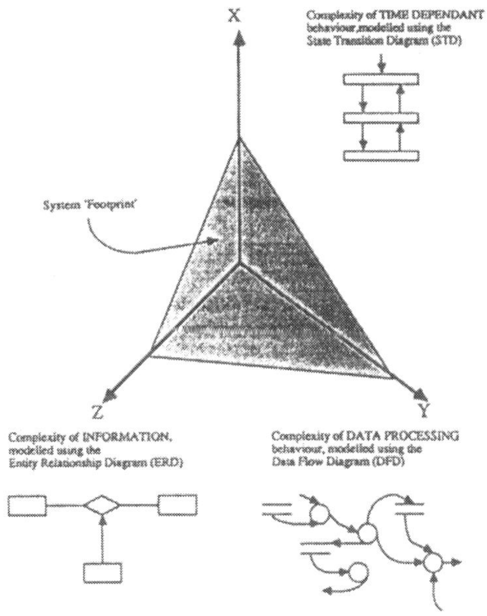

Figure 1 System Footprint concept, from Bowen, 1989

A substantial number of software engineering methodologies have been developed, although most provide broadly equivalent tools. All recognise that several tools are necessary to capture all aspects of a system, and incorporate several other diagrammatic and textual tools in addition to those mentioned.

One of the key principles of CASE tools is that of hierarchical development of a system specification. Thus a top level definition, giving very little detail, may be exploded into several diagrams, in successively greater detail, at a number of lower levels. This corresponds to the realities of manufacturing systems, which are commonly considered to comprise several layers, such as plant, shop, cell and workcentre.

The models at all layers can be constantly checked for consistency, so that any change at one point will be automatically taken into account at all other points. The leading CASE packages now have facilities to generate executable code automatically.

MANUFACTURING SYSTEMS INTEGRATION

Many manufacturing companies have invested in advanced systems, but are increasingly identifying the lack of integration between sub-systems as a limitation to improving business performance. The issue of manufacturing integration was addressed by Moniot and Waterlow (1986) in a review of the problems of implementing computer aided production management (CAPM).

The problem is due partly to the lack of an effective, systematic methodology which enables companies to measure their existing level of integration, identify the degree of integration which is necessary to support business objectives and provide a staged programme for measurable improvement.

The importance of this subject area has promoted the establishment, by the Application of Computers in Manufacturing Directorate (ACME) of the UK's Science and Engineering Research Council (SERC), of a collaborative research programme involving the Universities of Liverpool, Loughborough and Strathclyde, and a number of industrial companies. The primary objectives of the programme are to develop a generalised approach to manufacturing systems integration and to propose an effective methodology and to develop the methods and tools through which improved integration can be established.

The research hypothesis is that improved manufacturing systems integration is brought about through the effective coordination of Information (data), Material (parts) and Organisation (people and culture).

Although the common theme of the research programme is the improvement in manufacturing integration, each of the collaborating institutions will focus upon a different element of the problem:

- Strathclyde, in collaboration with Caledonian Airmotive, Prestwick: Material flow systems and the links with information systems.

- Liverpool, in collaboration with JCB Transmissions, Wrexham: Company-wide information systems.

- Loughborough, in collaboration with Davy Morris, Loughborough: Organisational structures in relation to information and material flow systems.

Each of the collaborating research institutions will provide methods for defining the characteristic system "footprints", of their particular views, from which a composite modelling methodology will be formulated.

Whilst each of the academic institutions will work most closely with its specific collaborator, as the programme develops there will be substantial interchange between research teams and industrial sites, so that the resulting methodology will be fully tested in all three environments. In addition to the collaborating companies, the programme is supported by a number of manufacturing systems consultancy organisations and software houses.

The main outputs from the research programme will be:

- A formalised methodology for developing the information systems and organisational procedures necessary to improve the level of manufacturing integration.

- A set of integrated computer-based analysis tools incorporating material flow analysis, information system analysis and organisation analysis.

- Detailed industrial case studies of the application of the manufacturing systems integration methodology, in the three collaborating companies.

These projects commenced in June 1990, therefore it is too early to report results.

METHODOLOGY FOR IMPROVED INTEGRATION OF MATERIAL AND INFORMATION FLOW

Bullinger (1989) asserted that there is a real danger that companies will find that they have information systems and material flow systems which have been designed in isolation and cannot be integrated efficiently.

A contributory factor is that the culture of manufacturing engineers and software engineers have been quite separate in the past. Over the past fifty years or so, manufacturing engineers have developed a multitude of diagramming tools to represent the engineering requirements of the products to be made and to map the flow of materials on to the factory floor. They have also developed more sophisticated heuristic procedures, for effective manufacturing systems engineering. However, in general these tools lack any formal way of being integrated with information system design techniques, and do not adequately address organisational issues. They often assume that the information system can be built around the material flow system.

Similarly, computing techniques for designing information systems have been developed in isolation of the physical realities of manufacturing. Recently there has been much work into designing CIM architectures, such as CIM-OSA (Vernadat, 1989) and GRAI (Doumeingts, 1989) projects, figures 3 and 4. These address the organisational and conceptual issues in information system design, but, to date largelly ignore material flow system design issues.

The Strathclyde research project, in the above collaborative programme, will establish a methodology for jointly defining the architectures of the physical manufacturing system and the supporting information and control system, and thereby improving their integration.

The hypothesis underlying the research is that CASE tools provide a basis for modelling and specifying both the information system and the material flow system. It postulates that a manufacturing information system comprises several levels, in each of which a number of information flows link together a number of decision nodes, figure 2. These levels could be the plant, shop, cell and workcentre levels, and can be mapped onto the hierarchical approach of CASE tools. At these decision nodes, data processes use this information to control various sub-functions. At the lowest level, shop floor activities within the material flow system are being controlled. By describing the decision mechanisms of these nodes in an appropriate way, it will be possible to establish more definitive relationships between the information system and the material flow system.

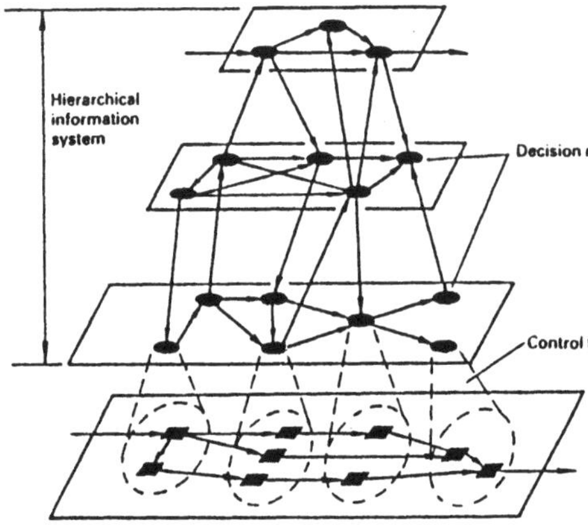

Figure 2 Hierarchical model of information system interfacing with material flow system

Figures 3, 4 and 5, from Bititci and Carrie (1990), illustrate the equivalence of CASE diagramming techniques and the needs of manufacturing engineering techniques. Assembly relationships, which may be extracted from the company's engineering data-base, can be modelled by Entity Relationship Diagrams, as in figure 3, which uses the diagram to define an assembly structure and the tools require for the

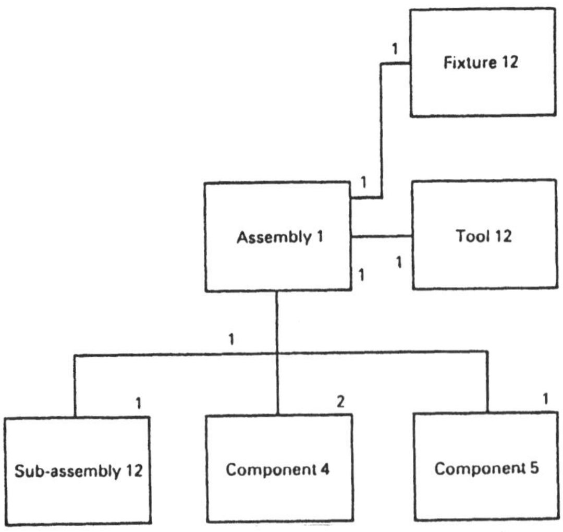

Figure 3 The Use of entity relationship diagrams to illustrate material relationships

assembly operation. Material flow information can be depicted by Data Flow Diagrams, as in figure 4. DFDs can show not only the physical flow of material, but also the information needed for carrying out the task. State Transition Diagrams, which indicate the various states, either active or passive, in which any entity may be found, and the conditions necessary for the transitions to take place can be used to

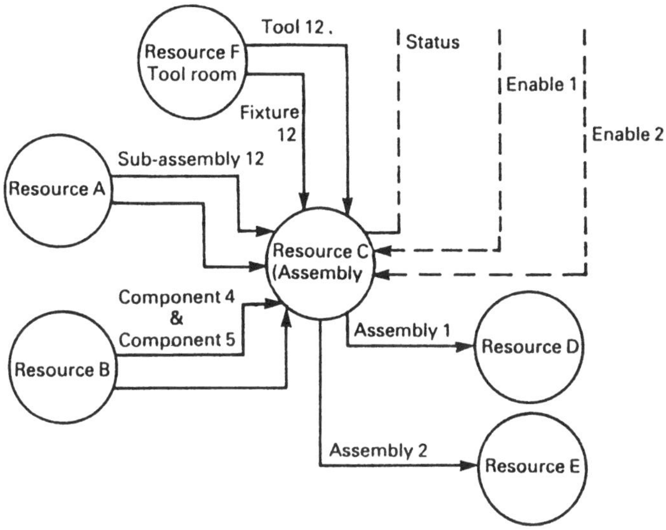

Figure 4 Use of Data Flow Diagram to illustrate material processing requirements

define the dynamics of a system in figure 5. Carrie and Bititci (1990) show the correspondence between STDs and various simulation approaches, especially Activity Cycle Diagrams, which underlie several

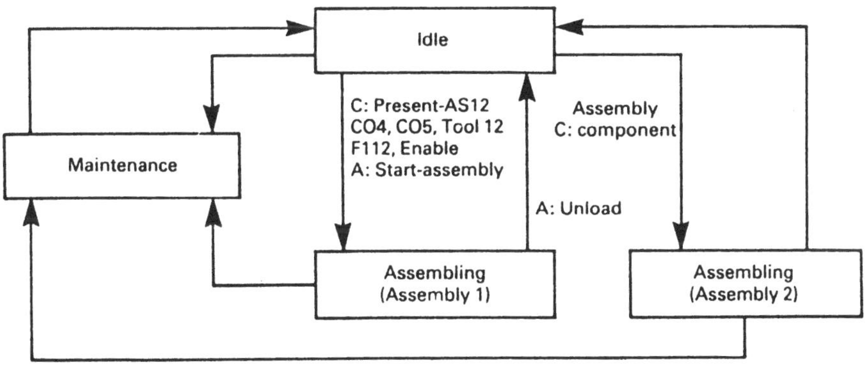

Figure 5 Use of state transition diagrams to illustrate the control view.

simulation approaches, especially Activity Cycle Diagrams, which underlie several simulation packages (Carrie, 1988). Table 1 summarises the equivalence of CASE and manufacturing engineering techniques:

Function	Manufacturing approach	Case approach
Product Data	Bills of Materials	Entity-Relationship Diagrams
Process data	Production Flow Analysis	Data flow Diagrams
Dynamics and controls	Simulation Modelling	State Transition Diagrams

Table 1: Examples of the equivalence of CASE and manufacturing engineering techniques.

The project is beginning with developing a simple top-level model of the complete manufacturing operation of the collaborating company, which overhauls aero-engines. From this a rough-cut capacity planning model will be developed. Due to the nature of the industry, the number of engines which may be received for overhaul can vary from less than five to over twenty per week. The work content in an engine can vary from 100 manhours to 7000 manhours, depending on the duty to which it has been subjected. Whatever the workload, the competitive situation in the business requires that the engine is overhauled and returned to the customer in less than two months. Capacity management is therefore a difficult but key function in Caledonian Airmotive.

Later, the model will be expanded to incorporate greater detail, and will be integrated with the activities at Liverpool and Loughborough.

LINKING CASE TOOLS TO SIMULATION MODELLING

As observed by Carrie (1990), inadequate specification of systems is a cause of the achievements of investment in advanced manufacturing systems being less than anticipated.

Another project specifically addresses the relationship between CASE tools and simulation modelling. The objective of the project is to develop a CASE tool-based decision support system which brings within one methodology three previously separate functions:

- developing the specification of a manufacturing system,
- automatically creating simulation models of the manufacturing system, at any stage during the specification process.
- developing the control software to manage the system once commisioned.

The project exploits the equivalence between the techniques available within CASE tools and the activities, events, queues and decision rules commonly used in simulation packages. There is no point in developing the specification of the physical system using one tool, the control software using another and the simulation model of the system using a third approach, if one system can provide for the needs of all three purposes.

The project is establishing a generalised software link between a CASE tool and a simulation package, taking a CIM Cell in the University of Strathclyde as a test bed. The cell contains a conveyor, a lathe and a milling machine and a robot to load and to unload the machines. The lathe and the milling machine are CNC controlled and can get their machine code from the FANUC postprocessor of the CAM package Pathtrace, which converts design data from AutoCAD into machine code.

The CASE tool Excelerator/RTS (Index Technology, 1989) is being used for this project. The RTS (real time systems) version provides enhanced facilities for defining the dynamic nature of a system, and is based on the Ward & Mellor technique (Ward & Mellor, 1985). Excelerator provides an integrated set of analysis and design tools that automate the development of specifications of software systems, and can also generate code in C.

SIMAN (Pegden, 1986) is the simulation package used on the project. It is a combined discrete-continuous simulation analysis language, which contains a number of features that make it particulary useful for modelling manufacturing systems.

The intention is to generate code using Excelerator which is understood by the simulation package SIMAN. This code should be the result of a specification of the manufacturing system. The output should contain simulation-specific data.

Using the different diagram and other structured techniques every data and control aspect can be described in detail, and a data dictionary developed. The lowest level of the whole specification contains only signals and elements which can be described in a primitive process specification, figure 9, from which computer code may be generated. They also can be listed in a Record or a Control Table linked with logical operators. These contain all the resources of the model and all activities and conditions using logical rules.

To run a simulation, SIMAN merges two separate files, an experimental file and a model file, which will be linked together with a "Linker". The output file of Excelerator contains most of the necessary information of the model file, but only the parameters, resources and the transport facilities the experimental file can be defined, figure 6. Simulation-dependent data, such as all discrete data information for the simulation run, can be provided by a postprocessor in a dialogue menu.

PROFILING SMEs FOR CIM

In its workprogramme for ESPRIT II, the European Community included "methodologies for evaluating and customising of CIM tools applicable to a wide range of small and medium enterprise manufacturing

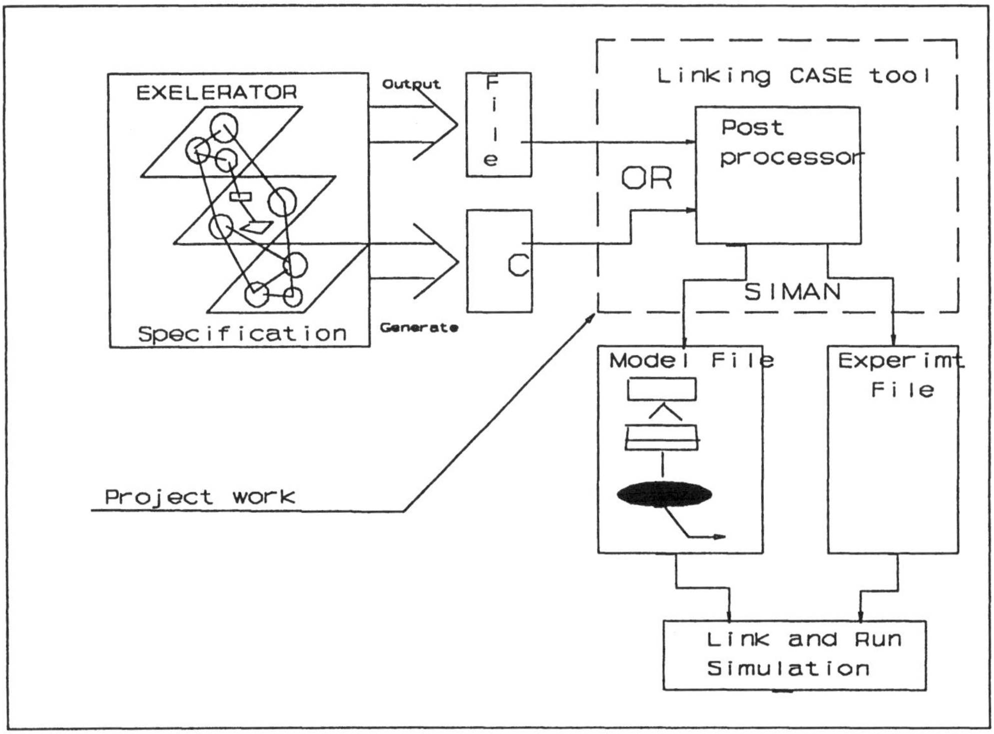

Figure 6 CASE - Simulation Link

sectors". We believe that it would be wrong to approach this customisation from the viewpoint of existing tools. Existing tools are too closely linked to large enterprise environments and take inadequate cognisance of the human manager in small and medium sized enterprises (SMEs), where procedures, organisational structures and skills are less formal than in larger bodies. Instead, it is necessary to look at the needs of SMEs to ascertain their true requirements for integrated manufacturing, whether by computer or manually.

A further project within this field concerns the study of SMEs to ascertain their true needs for CIM. This study will identify certain principal classes of SME, which will have similar requirements for CIM. It should further be possible to develop the CIM requirements specification for the principal types of SME, and hence develop the specification of those CIM modules appropriate to the SME. Another outcome of the study woulld be the identification of those functions which should remain within the human domain, given the current level of technology.

Before such specifications can be developed it will be necessary to develop a method of classifying SMEs. This will necessitate studying a large sample of SMEs. The data from the survey will be analysed using both the CASE tool "system footprint" as described above, and also an extensions known as "positioning tools" and "Stereo graphs".

McBeth, et al (1989), in their work on supply chain management, developed a positioning tool, figure 7. This extends into the multi-dimensional domain the three dimensional approach of Yourdon.

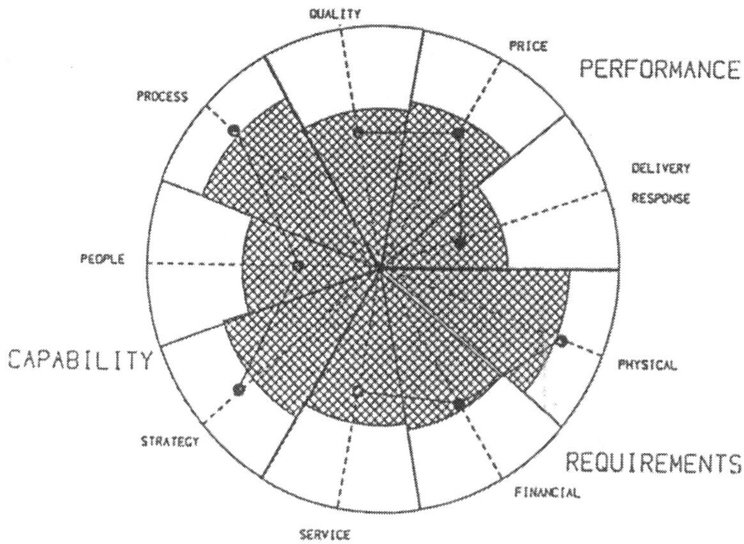

Figure 7 Positioning Tool, from McBeth, et al 1987

McBeth postulated that there were three main dimensions of the supplier-customer relationship, namely requirements, capability and performance. Each dimension could be rated on a number of points in relation to some ideal. For example, performance could be rated in terms of quality, price and delivery response.

A very similar diagrammatic method is in use in Japan to evaluate CIM, which Furukawa (1989) refers to as a "stereo graph", Figure 8. This approach rates a potential CIM investment in terms of its divisional and characteristics evaluation on a numerical scale. As with McBeth's approach each of these facets has a number of features on which the system under consideration is rated. The ratings may be combined to give an overall numerical score, although for the work described here the shape of the profiling diagram will be of more interest.

The survey of SMEs will be done using these tools, looking at both the company as a whole and specific applications within the information system heirarchy, to establish company "profiles". The SMEs will be classified according to their profiles. Within each category of SME the requirements for CIM will be identified, and the specifications of CIM modules developed. The functions which should remain within the human domain will also be identified for each category of SME. A software tool, probably knowledge based, will be developed implement the methodology.

Once the general methodology has been established it will be applied in a small number of collaborating SMEs so that their CIM requirements may be identified, the modules specified and installed as prototype demonstrators of the application of the methodology.

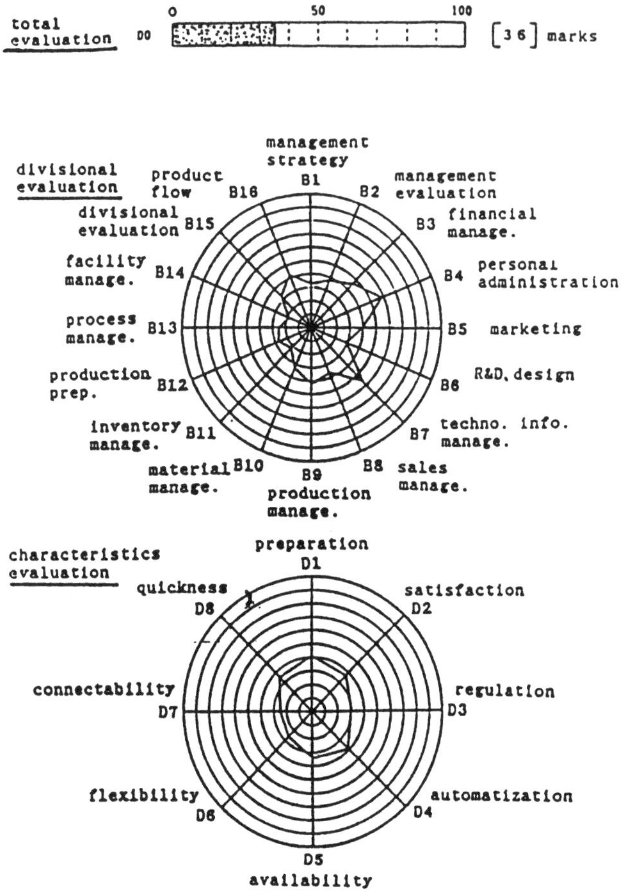

Figure 8 Stereo Graph, from Furakawa, 1989

The outcomes of the research, which will be of considerable benefit to the IT community and software vendors, will be:

- A methodology for categorising SMEs for CIM, and for profiling any SME to evaluate its CIM requirements; a software package, probably an IKBS, for carrying out the methodology; a substantial database of survey results.

- A specification of the CIM module requirements for each class of SME, and a library of CIM module specifications for each class of SME.

- Case studies of the use of the methodology with one or more prototype installations.

European partners are sought since the value of the work may be enhanced by extending the survey of SMEs to several countries within Europe.

CONCLUSION

This paper has described several projects which exploit computer aided software engineering tools for planning and modelling manufacturing systems and for assessing the CIM needs of small enterprises. These will illustrate and refine the use of CASE tools to develop improved methods for manufacturing systems engineering and CIM.

ACKNOWLEDGEMENTS

The sponsorship is gratefully acknowledged of the ACME Directorate of the SERC and of Caledonian Airmotive, in particular of the late Mr Syd Greaves, formerly Operations Director at the company. Acknowledgement is also due to Jürgen Huber, who is working on the CASE-simulation link project.

REFERENCES

Bowen J, 1989, "Yourdon in Manufacturing", Conference on Methods in Manufacturing Systems Engineering, Southport, November 1989.

Bititci US and Carrie AS, 1990, "Information Material Flow Mapping", Logistics Information Management, vol 3, number 1, March 1990, p31-36.

Bullinger H-J, 1989, "Information Technology: Strategic Weapon for Factory of the Future", 10th International Conference on Production Research, Nottingham, August 1989.

Carrie A S, 1988, "Simulation of Manufacturing Systems", John Wiley and Sons.

Carrie A S, 1990, "Manufacturing Systems - Simulation or Specification" invited seminar presentation, University of Dundee, January 1990.

Carrie, A S and Bititci U S, 1990, "Tools for Integrated Manufacturing", Proceedings 28th International MATADOR Conference, Manchester, April 1990, pp217-223.

Doumeingts G, 1989, "GRAI approach to designing and controlling advanced manufacturing systems in a CIM environment", in Advanced Information Technology for Industrial Material Flow Systems, S Y Nof and C L Moodie, editors, NATO ASI Series volume 53, Springer-Verlag, pp461 - 529.

Furukawa, Y, 1989, "The Factory of the Future", International Conference on Strategic Manufacturing, Gleneagles, December 1989.

Index Technology Corporation, 1989, Excelerator documentation set, Cambridge, Massachusetts.

McBeth, D K, Baxter, L F, Ferguson, N, Neil, G C, 1989, "Buyer/Supplier Relationships: A Tool for Analysis and Improvement", Fifth Conference on Exploitable Research for Manufacturing industry, IEE, London.

Martin J and McClure C, 1988, "Structured Techniques: The basis for CASE", Prentice Hall.

Moniot, J and Waterlow, G, 1986, "A Study in the State-of-the-Art in Computer Aided Management", SERC.

Pedgen, CD, 1986, "Introduction to SIMAN ", Systems Modeling Corp.

Vernadat, F, DiLeva, A and Giolito, P, 1989, "Organisation and Information System Design of Manufacturing environments: the new M* approach", Computer-Integrated Manufacturing Systems, Vol 2, No 2, May 1989, pp69-81.

Ward, P and Mellor, S J "Structured Development Techniques for Real-Time Systems", Volumes 1 to 3, Prentice Hall, 1985.

New Looks on CIMS Modelling

LUCAS PUN

1. INTRODUCTION

Computerised Production Management Systems (PMS) bear the form of Decision - Support Systems (DSS). For the DSS-design, modelling is essential. It has two major roles : (a) It must help the understanding (and the structuring) of the PMS - problems, and (b) It must facilitate the DSS-design tasks. The INTEGRATION aspect in the framework of Computer Integrated Manufacturing Systems (CIMS) involves a much higher degree of complexity (Hordeski 1988, Browne et al 1988). It calls for new approaches to modelling problems. In this paper, we present the basis of an INtegrated COnceptual REference MOdel (INCOREMO). It is :

(a) integrated, because it must cover all types of system structures and activities;

(b) conceptual, because it must be generic to cover large classes of problems;

(c) reference, because it must show typical ways to establish models of specific applications;

(d) model, in the sense of organising and structuring the reality.

This model evolves from some experience gained by the utilisation of the GRAI -method (Doumeingts 1987, Roboan et al 1989, PUN 1987) ; (a) GRAI - grid, a reference function-horizon model of decision centers, which allows the depiction of inconsistencies in existing management structures, and (b) GRAI - nets, an activity - modelling framework which aids the specification and design of the DSS. To establish this model, we take into account the following modelling - problems formulated according to recent developments in PMS - scope , and DSS-design methodology.

Problem 1 : The model must cover many types of physical objects (parts, products, equipment), many types of procedures (Technical Process Plannings (TPP), quality control, maintenance), and many types of transformation law proper to each type of activity.

Problem 2 : The model must cover three interrelated structures belonging to three different universes : technical, economical, and societal.

Problem 3 : The model must cover PMS strategies, tactics, situations and conditions for decision-makings.

Problem 4 : Establish a set of basic concepts capable of describing uniformly the functions of all the sub-systems involved in a PMS: production physical system PS, management system MS, decision-support system DSS. Use the experience gained in GRAI, IDEF-SADT, MERISE. (Figure 1).

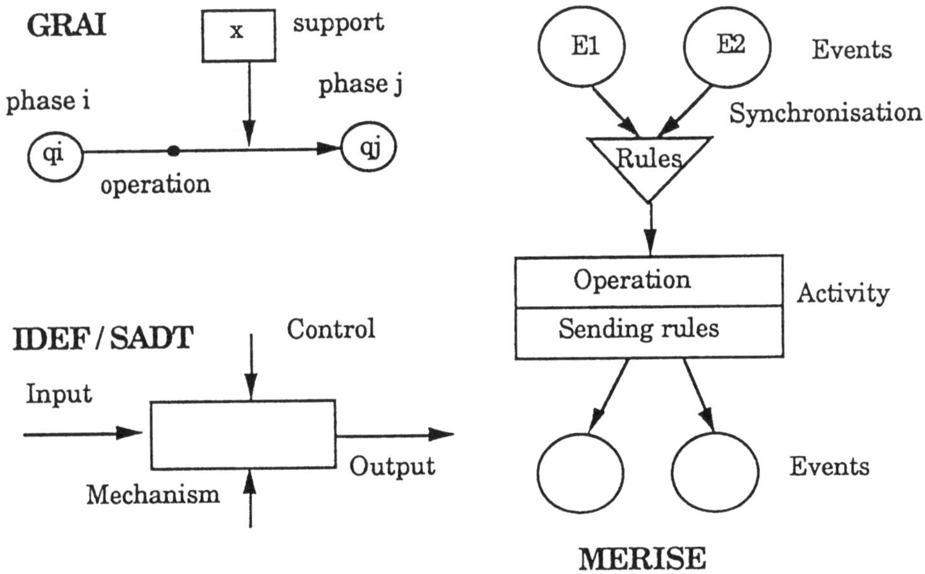

Figure 1 - Basic elements in several methods for analysis

Problem 5 : Associate, with this set of concepts, capacities for formulating problems, and problem-solving procedures.

Problem 6 : Connect easily the elements of this language to those of suitably chosen mathematical and programming languages for the DSS design. (Predicate logics, advanced logics, object-oriented programming languages) (Thayse 1988, Sombé 1988).

Problem 7 : Favour the use of advanced logic, not only to represent strategies and tactics, but especially to create more efficient PMS strategies and tactics. (For example, for treating dynamic perturbations).

Problem 8 : Incorporate AI techniques in the representation of knowledge and reasoning. (Pun 1984, Laurière 1986, 1988).

Problem 9 : Incorporate interesting features of the concept "object", and of object-oriented languages.(Meyer 1989, Roche & Laurent 1989).

Problem 10 : The DSS- design comprises three main tasks : problem analysis and formulation, knowledge formalisation for calculation, and information structuring for computerisation. Each task has its specific aim, and requires a specific modelling language. The general framework must therefore contain methodologies to connect easily the individual models.

Problem 11 : For each DSS - design task, there must be a reference model (reference structure, reference framework, reference organising rules) and a methodology to derive models for specific applications.

From Problems 1 to 11, we derive four principles which guide the establisement of our INCOREMO.

Principle 1 : Separation of organic and behavioural characterisation. - The organic characterisation aims to understand the functional roles of the organs of the systems. The behavioural characterisation aims to describe the capabilities of the system organs, so that we can understand how they behave when they accomplish the system goal.

Principle 2: Aggregation of sub-structure modelling. - Use appropriate modelling language to suit each specific organ. Link (aggregate) suitably the sub-models for global understanding.

Principle 3 : Macro/micro inference. - In practice, we need different levels of detail to state our problems. We use therefore macro and micro models. The micro models must be derived methodologically from the macro models.

Principle 4 : Transitivity of the models for the DSS-design. - The purpose of each design-activity requires :

- formulation model M1 to understand
- formalisation model M2 to calculate
- information model M3 to computerise.

We use a conceptual global model : knowledge - tree (KN - tree), and translation tables to enable transitions between the KN-trees of M1, M2, and M3.

The adopted approach to INCOREMO is finally a group of models structured in the following way : (Figure 2) The macro-INCOREMO comprises two dual macro models :

INCOREMO- L for organic characterisation

- Warehouse sub-models for structural analysis
- activity-net sub-models for functional analysis.

INCOREMO-R for behavioural characterisation

- conceptual kn-tree for problem formulation
- formal kn-tree for calculations
- information kn-tree for computerisation.

2. MACRO - INCOREMO

The general INCOREMO consists of two macro models :

INCOREMO=(INCOREMO-L,INCOREMO-R)

L : Left, mother, organ to receive, to form, YING in YI-KING.

R : Right, father, impulse, inspiration, YANG in YI-KING.

The macro INCOREMO-L represents the organic characterisation, the macro INCOREMO-R represents the behavioural characterisations. (Figure 2)

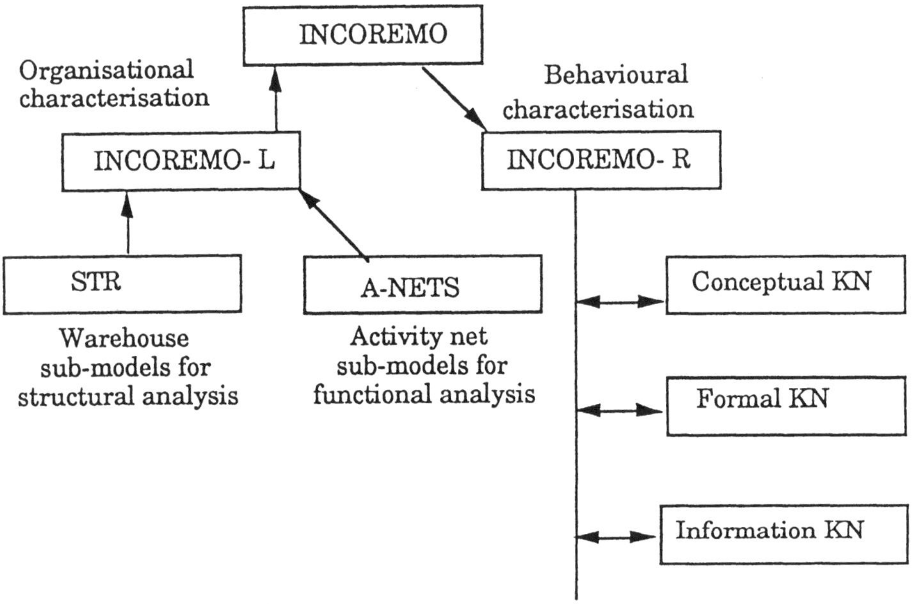

Figure 2 - Group of Integrated Conceptual Reference Models

2.1 - INCOREMO - L. - (Figure 3)

The INCOREMO-L considers three interrelated sub-systems: physical production system PS, management system MS, and decision-support system DSS. The organs of each sub-system are defined functionally.

(a) finality : why we have this sub-system

(b) functional roles and activities : which permit the attainment of this finality

(c) supports : required by the funcional activities to accomplish their roles. Two basic types : intellectual supports and material supports.

(d) flows; on which the activities create events (phases, steps, results,etc)

(e) structures of groups of organs : (for supports and for flowing elements) from which individual organs are extracted to effect specific activities.

(f) unexpected internal and external events.

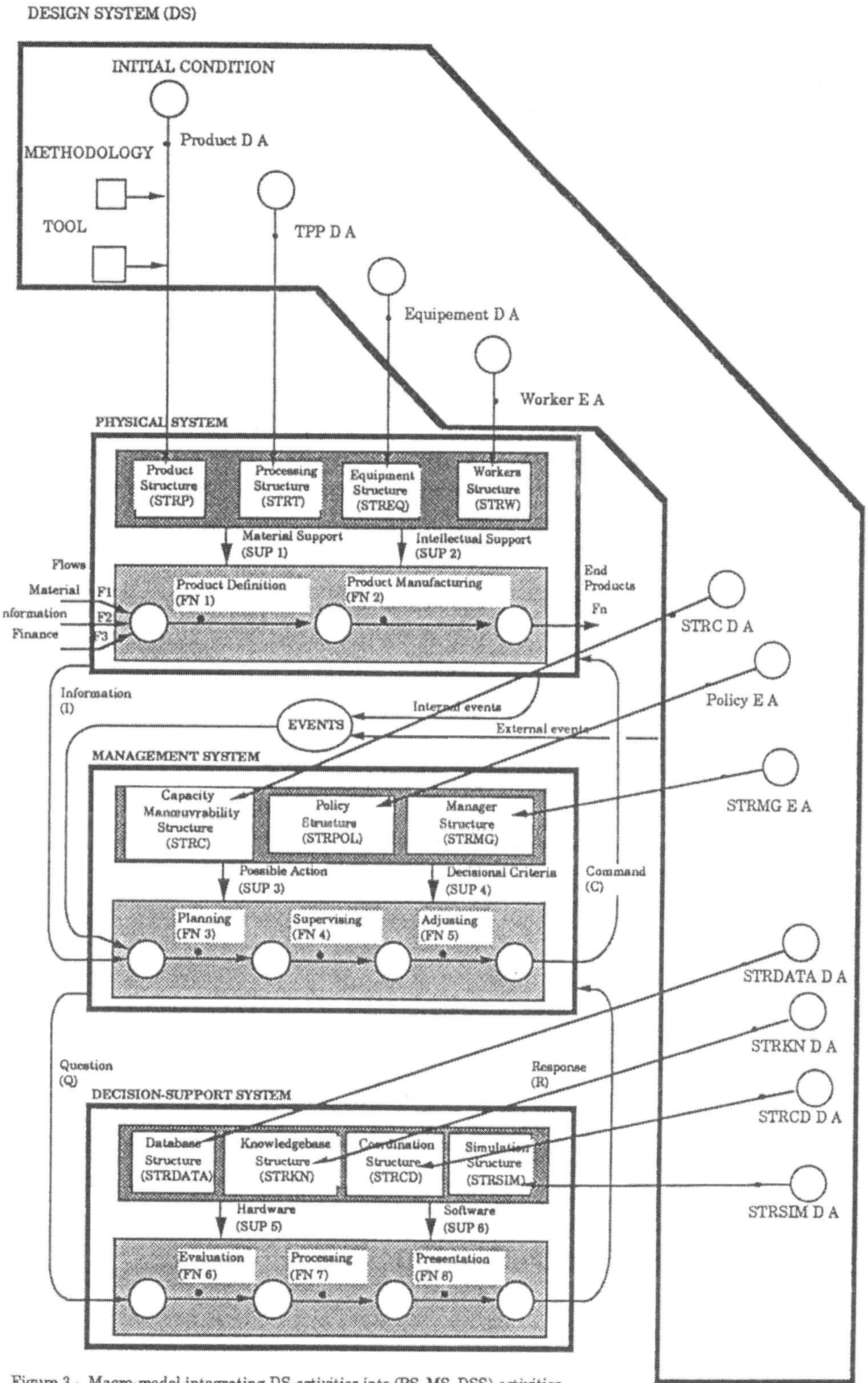

Figure 3 - Macro-model integrating DS-activities into (PS_MS_DSS)-activities
(D=design, E=elaboration, A=activity)

The interrelations between the sub-systems are :

(a) PS to MS. Information I on activity-execution states and situations (state = how is one support, one flow-event; situation = how the activity is effected)

(b) MS to PS : Control C : how to adjust the flow and the support situations.

(c) MS to DSS : questions Q

(d) DSS to MS : answers R.

In INCOREMO - L, we define two levels of generality :

General system generality : the same set of expressions (finality, flows, functional roles and activities, supports, sets of structures) is applicable to all sub-systems.

Specific system generality : for each sub-system, PS, MS and DSS, we define corresponding expressions. They are summarised in TABLE 1.

Types of systems	PS : Physical Production system	Sub-systems MS : Management system	DSS : Decision support system
1. Finality	Product with highest tech. abd econo. efficiency	Organise PS-activities with best coordination & synchronisation	Supply best decisional conditions to MS-activities
2. Functional roles & activities	FN1 : Product definition FN2 : product manufacturing	FN3 : planning FN4 : supervision FN5 : adjusting	FN6 : evaluation FN7 : processing FN8 : rextraction
3. Flows	f1: materials f2 : informat. f3 : finances (f1, f2,f3) to fn	I : states, situations C: commands I to C	Q : questions R : answers Q to R
4. Material type support	SUP1 : articles machines tools, equipments	SUP3 : possible actions in controlling PS flows & supports	SUP5 : hardware
5. Intellectual type	SUP2 : TPP, drawings	SUP4 : opportunity, criteria for decision	SUP6 : software
6. Sub-structures	P : Product T : TPP EQ : equipment W : workers	C : capacity maneuvrability POL : policy MG : managers	DATA : data KN : knowledge CD : coordination SIM : simulation
7. Unexpected & events failure	Machine fai. Bottlenecks programme change	Changes in strategies and tactics	dynimical data knowledge in harware & software

TABLE 1 :
Generic expressions of the ortgans in the specific sub-systems : PS, MS and DSS

2.2 - INCOREMO- R. - (Figure 4)

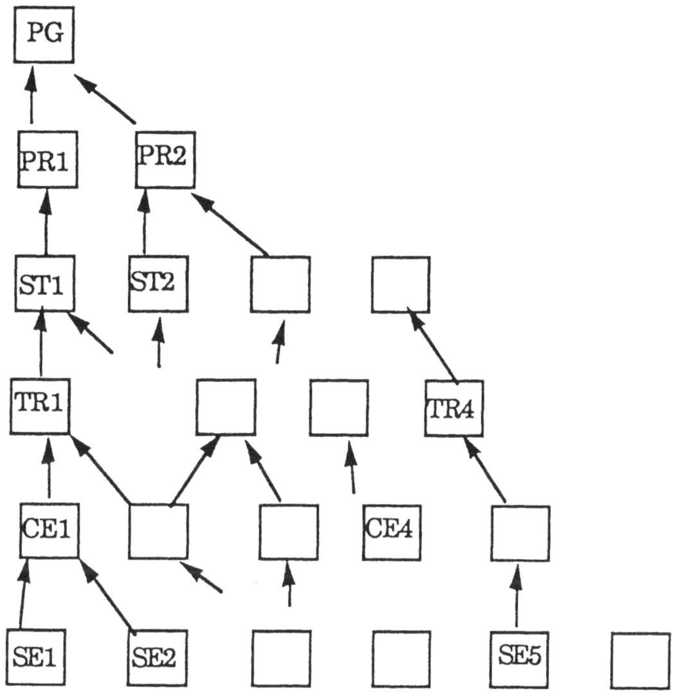

Figure 4 - A Standardised knowledge-tree

The macro INCOREMO - R is an integrated reference model intended to represent behavioural characteristics of system-organs. It is a knowledge - tree (in the semantic tree sense). It is established according to the classification of knowledge levels. Each knowledge-level needs a sub-modelling. The sub-modelling language to be used depends on the purpose of the design - activity : formulation, formalisation, or computerisation. To clarify the understanding of our vocabulary, we adopt the following definitions.

KNOWLEDGE : something the existence of which we are aware. Know_HOW is therefore knowledge. (frequent confusion)

INFORMATION : form taken by knowledge for communicating, between Men and/or Machine.

DATA : form taken by knowledge for computing, or for processing, and often based in some equipment (frequent confusion between information and data)

(a) Knowledge classification. - (Pun 1984, Laurière 1986,1988)

Level KN(SE) : SE is a simple entity , physical or abstract

Level KN(CE): CE is a complex entity, formed from several SE.The formation operators can be arithmetical, algebraic, logical, vectorial or linguistic (or relational in entity-association database techniques)

Level KN(TR) : TR is a transformation over several SE or CE. The transformation operator can be a function, an equation, an application, a morphism (category theory), a predicate.

Level KN(ST) : ST is a statement which announces a tautology by connecting several TR. The connecting operator can be a modus ponens, a modus tolens, etc. In mathematics, axioms and theorems are statements.

Level KN(PR) : PR is a procedure which solves a problem. A procedure is formed by several ST. The connecting operators are various types of induction, deduction, condition definitions, value measuring, etc. In mathematics, PR is an algorithm.

Level KN(PG) : PG is a programme obtained by connecting several PR. A programme is aimed at solving several correlated problems. The connecting operators are for instance : put in series, put in parallel, jump to.In Object-oriented programming languages, the so-called object = (instance, message) is a connector between procedures.

In the sense of common languages, KN(SE), KN(CE), KN(TR) are descriptions, KN(ST), KN(PR), KN(PG) are know-hows.

(b) Knowledge tree : KN-tree.

A KN-tree is a standardised form (Figure 4) to represent all the knowledge associated with the behavioural characteristics of the organs of a system. The nodes represent various levels of knowledge. The links represent various types of operators grouping a lower-lever entity to a higher - level entity.

(Note : There is an important difference between a KN-tree and a solution - tree used in AI-techniques. In a KN-tree, the incoming arrows to an element are AND-connected. In a solution - tree, the outgoing arrows due to production rules are OR-connected.

(c) Object-oriented definition of classes of knowledge.

In our PMS problems, we have to understand and analyse the behaviour of the system organs. Studying a behaviour involves defining organ characteristics (qualitative study) , and measuring some value (quantitative study). Some of the system organs lead to knowledge levels SE, CE, TR. Some others like programmes of activities, strategies and tactics for management, decisional processes, lead to higher levels. Therefore, we adopt an object-oriented definition for all levels of knowledge :

KN(x) = (x, attributes, values, links)

The semantics of the components of the various levels of knowledge are summarised in TABLE 2. The KN-tree, associated with the classification and the object-oriented definition, will play two roles : (a) It will provide a methodology for extracting behavioural characteristics from the organic representation INCOREMO - L. and (b) It will act as a translator support between conceptual, formal and informational models.

Class of knowledge	Knowlegde componenl			
	Element	Attributes	Values	Links
KN (SE) Simple entity	physical or abstract element	state of a support, phase, event	Real, binary, fuzzy, linguistical	Arithmetic, logical, vectorial
KN (CE) complex entity	complex element	State-vector	same	predicate
KN (TR) Transformatior	Transfor mation laws	truth	binary, ternary, quertary	modus ponens, modus tolens
KN (ST) Statement	Statement of a truth, of a situation	consistency, necessity, complete-ness	binary, conditional valuation	Algorithmic operator, manipulator functor
KN (PR) Procedure	Procedure for solving one problem	Appropriate-ness	same	coordination operator
KN (PG) Programme	Programme for solving several corre-lated problems	same	complete, partial, conditional valuation	contextual operators

TABLE 2 :
Object-oriented definition of knowledge - semantics of the knowledge components

3. REFERENCE ORGANIC SUB-MODELS FOR INCOREMO-L

3.1. - Needs for Reference Sub-models.

The general models INCOREMO-L and INCOREMO-R are general integrated representations. If we wish to show deeper insights in the organic roles, and the behavioural characteristics, we need finer models. For instance, for the management systems (human-structure), we have used a reference functional-horizon grid to study the coherence of existing management structures. Similar needs may occur for the workshops, product structure, TPP, or for some given strategy-tactical structures. As regarding knowledge representation, we need formal or mathematical representations for calculating, and information representations for computers. Therefore, we suggest :

- reference sub-models for INCOREMO-L
- reference operational models for INCOREMO-R.

We discuss the knowledge reference operational models in the next section. For the INCOREMO-L, we suggest two types of sub-models.

(a) Warehouse reference sub-models, which represent the organic structures of the supports and flowing entities. This concept is similar to :

- set X of variables x (x X) in predicate logics
- world of instances in AI languages
- database and knowledgebase in Expert Systems.

(b) Activity-net reference sub-models, which represent the organic structure of entities around the functional roles and activities (flow-events, phases, supports, operations). This concept corresponds to "algorithms" in formal mathematics, and to "processing modules" in computer software.

32. - Warehouse reference sub-models in INCOREMO-L.

The sub-systems to be represented are those named STRP, STRT,.....STRSIM in Figure 3. What we have to represent are :

- what they are
- what is their utility with respect to the functional roles and activities of the system
- how are they utilised ; one single element alone, or several elements grouped together
- how to find them.

In the case of our PMS, the functional roles of the sub-systems are.

STRP .-Product structure :the set of (nomenclature and part lists) for material requirement calculation and planning, part decomposition for the manufacturing processes.

STRT. - Technical Processing Planning: operating guide for machining, part assembling, sub-system assembling, testing, etc.

STREQ. - Equipment structure. Machine-tools, robots, pallets with loaders, automatic guided vehicles (AGVs) in FMS, stock yards, various equipment for heat treatment, cleaning, etc., for the manufacturing operations.

STRW. - Workers who execute manufacturing operations.

STRC. - Structure of possible actions, for the management. They cover the manoeuvrability for the flow control, and for the utilisation of the supports.

STRPOL. - Structures of policies established according to defined management strategies and tactics. The 3-horizon type of planning (long, medium, and short) is a kind of policy.

STRMG. - Structure of the managing staff; general director, department managers, section managers..

STRDATA. - Structure of all the data required for treating the PMS problems. They must be organised to facilitate the data introduction, and the data processing.

STRKN . - Structure of domain knowledge which describes the operating laws of the controlled objects, and decisional knowledge which helps the managers to do their work. This must be organised to facilitate dynamic introduction, modification, and utilisation.

STRCD. - Structure of a set of procedures , having if possible, some strategic views, and aiming to coordinate the operations on STRDATA, STRKN.

SYTRSIM . - Structure of the elements of the simulator, in the case when the DSS is supposed to play the role of "simulation analysis".

It appears that the HBDS (Hypergraph Based Data Structuring) (Bouille 1977, Pun 1984) modelling language is appropriate for establishing a Reference Model for all these sub-systems. Each sub-system will be modelled by :

- definitions of classes, each class is :

  class = (objects, attributes, values, links)

- groups of classes linked together hierarchically with the same attributes
- group of classes with different attributes, and linked by special purpose.

Example : (Figure 5) Two classes of objects, C1 = machines; C2 = tools.

- Hierarchical linking : for C1 for example, we have successively : different types of machining operations, different sizes, different powers, etc.
- Purpose linking : link between one milling machine and one milling tool which must be used together for certain types of parts.

Note : all the sub-structures have a common configuration : a number of hierarchical structures linked together by purpose.

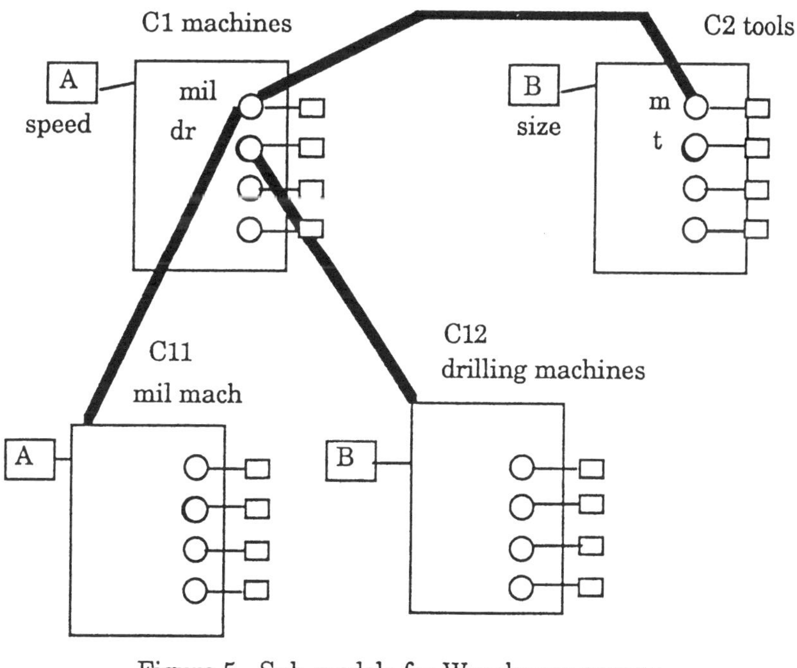

Figure 5 - Sub-models for Warehouse organs

33. - Activity - net sub-models.. - (Doumeingts 1983, Pun 1988)

An activity - net comprises the following basic elements :

- the model representing one activity
- bifurcating and cumulating configurations for phases and for supports
- serial configurations of activities defining one branch
- parallel configurations between branches, with a systematised connection, ie : the support of an activity comes always from the result of another activity.

In the INCOREMO - framework, we suggest a new representation of the basic activity, which evolves from the old representation to take into account the management aspect. (Figure 6)

The symbolic definition of this representation is the following :

a = (qi, xS, dH, xE, (†S, †H , †E) ta, qj, (CD), (SS))

The old GRAI - elements are :

qi : event at the initial phase

qj : event at the final phase

xS : support SKY, intellectual - type support, such as YANG, sky, inspiration, father, impulse, opportunity, method, operating manual...

xE : support EARTH; material-type support, such as YING,earth, energy, machines, equipments, tools...

dH : support HARMONISER , operator - type support, such as worker, numerical controls, robots, controlling computer). Note : The symbolic essentiality of these elements to characterise an activity comes from YIKING, which states that an activity is successful, if the SKY is opportune, if the EARTH is sharp (as for a knife), and if the OPERATOR harmonises SKY and EARTH.

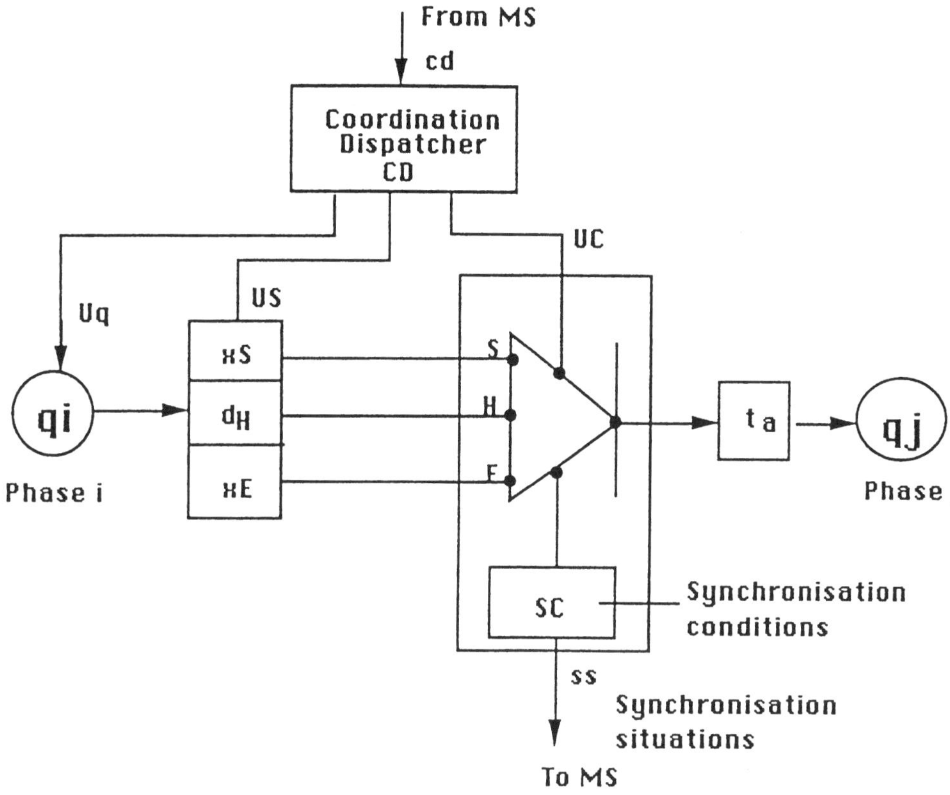

Figure 6 - A new GRAI model of an activity

The new GRAI elements are :

(†S, †H, †E) : dates of availability of the supports xS, dH, xE

t a : activity execution time (inspired from PERT)

SC : one box named "Synchronisation condition", with a "semi-conductor" graphical symbol (inspiration from PETRI)

ss : synchronisation situation, information to be sent to the management

CD : one box named "Cordinating Dispatcher" (inspiration from Robot activity planning)

cd : control decision received from the management

(uqq, uc, ua) : commands to modify phase, supports, and to start the activity.

The vocabulary used in this model is chosen to facilitate knowledge elicitation from the industrial reality. Therefore, when this model is used in practice, the involvement of mathematical and programming languages must be carefully avoided. Mathematicians (talking variables, parameters, theorems, axioms,..) and programmers (talking with database, inference engine, oriented objects...) are bad elicitors of industrial knowledge. The elements of this model are determined to reflect essentially and teleologically the functional roles of the organs. Two functional roles are considered :

- Internal functional role related to the goal-accomplishment of the system itself. This is the case of (qi, xS, dH, xE, qj)

- Connecting functional role, indicating the relations between two systems : System 1 acts, System 2 organises the activities of System 1. Note that , abstractly, this generic parity may represent many problem situations :

- (System 1/System2) = (PS/MS) : production management problems
- (SYSTEM 1/SYSTEM 2) = (MS/DSS) :management decision-support problems
- (System 1/System 2) = (DSS/ DSS-designer) : DSS design problems.

4. REFERENCE KNOWLEDGE OPERATIONAL MODELS FOR INCOREMO-R

In INCOREMO-L, reference models permit the analysis of the coherence of organic structures (management staff, plant lay-out, equipment in a FMS). In INCOREMO-R, reference models are also needed for representing the knowledge. At the formalisation-step of the DSS-design, we need formal or mathematical reference models. At the computerisation-step , we need programming reference models. It is however more difficult to talk about models, in this compact form. We are rather guided by some framework, defined by a specific vocabulary and grammatical laws

Formulation (KN defined as objects)	Formalisation (predicate-logical language)	Computerisation (object - oriented language)	Formulation (KN defined as objects)	Formalisation (predicate-logical language)	Computerisation (object - oriented language)
KN (SE) Simple element	Variable, constant, (time dependent or not)	data	KN (ST) Statement	clauses, axioms, theorems (conventional math)	rules (domain, decisional)
KN (CE) complex element	vectors, functions (one or several arguments) logical relations	relational data (entity-association)	KN (PR) Procedure	unification procedure	inference engine
KN (TR) Transfor-mation	predicate,transfer function (control) morphism (category theory)	premisces	KN (PG) Programme	a programme for theorem proving	declarative, procedural programmes

TABLE 3: Operational knowledge modelling at various steps of the DSS- design

41. - Knowledge representation in mathematical and programming languages. (TABLE 3) At the present time, we are uncertain , for PMS - problems, about the choice of appropriate mathematical and programming languages. For illustration, we summarise in TABLE 3, the operational modelling of various levels of knowledge. For mathematical represention, we try the first order predicate logics. For the programming language, we try a combination of object-oriented language and expert-system shells.

42. - Connexions between organic and behavioural characteristics.

For the DSS-design, we need KN-trees. To obtain the elements of the KN-trees, we need translating methods and tools, converting organic characteristics into behavioural characteristics. For our PMS problems, we have done this work on the basis of the elements contained in Figure 3. The results are summarised in TABLE 4. Note : The present approach in object - oriented languages (Meyer 1988, chapters 4 and 11) is the following : (a) determine data-object structure, (b) determine functional-module object structure, and (c) determine appropriate links within and between these structures. The INCOREMO - R supplies direct connexions to this approach.

Original concepts in PMS In activity-net sub-models	Knowledge behavioural characterisation	Original concepts in PMS In warehouse sub-models	Knowledge behavioural characterisation
state q	KN(SE)	STRP (products)	KN(SE)
situation (q,xS, H, xE)	KN(ST)	STRT(TPP)	KN(TR)
state of xS	KN(SE)	STREQ(equipment)	KN(SE)
state of xE	KN(SE)	STRW (workers)	KN(SE)
state of operator H	KN(TR)	STRC(capacity manoeuvrabiity)	KN(TR)
		STRPOL (policy)	KN(ST)
		STRMG (managers)	KN(SE)
		STRDATA (data)	KN(SE)
		STRKN (knowledge)	KN(ST)
		STRCD (coordination)	KN(PR)
		strsim (simulation)	KN(PG)
		I information on situations	KN(SE) or KN(ST)
		C commands	KN(TR)
		Q question	KN(ST)
		R answer	KN(ST)

TABLE 4 : Connexions between organic and behavioural characteristion

5. CONCLUSIONS. -

In this paper, we have presented one method for defining some Integrated Conceptual Reference models. (INCOREMO). We suggest forming this model with two parts. One organic part, INCOREMO-L, representing the functional structures of the organs which enter in a PMS. The aim is to permit coherence analysis of the organic structures. One behavioural part; INCOREMO-R, representing the operational knowledge of the organs. The aim is to facilitate the DSS-design tasks of problem formulation, problem formalisation and solving, and computerisation of the DSS.

Perspectives : From INCOREMO-L, establish reference functional - models for FMS, TPP, management strategies, database, etc., to study coherence or appropriateness, and (b) Establish methodologies to help formalisation and computerisation.

REFERENCES

Barathy B. H., (1988), Matching Design Methods to System Types. Systems research, v5, n1, 27 - 34.

Bayart M., (1988), Conception d'un système de pilotage temps réel pour une cellule d'assemblage automatisée. Thèse Docteur d'université, Université des Sciences et Technologies de Lille - Flandres-Artois.

Browne J, J. Hahren & J. Shivinan, (1988), Production Management Systems - A CIM -perspective, Addison-Wesley, 248p.

Doumeingts G, D. Breuil & L. Pun, (1983), Gestion de production assistée par ordinateur, Hermès, Paris.

Doumeingts G., D. Darricau, B. Valespir & M. Roboam, (1987), Design Methodology for Advanced Manufacturing Systems, Computer in Industry 9(1987), 271-296.

Ducanau R. & M. Habb,(1989), La multiplicité de l'héritage dans les langages objets. TSI:AFCET, v8, n1, 41 - 62

Filip F.G., G. Neagu & D.A. Donciulescu, (1983), Jobshop Scheduling Optimisation in Real-time Production, Computer in Industry, v4, 395-403.

Hordeski M, (1988), Computer Integrated Manufacturing - Techniques and Applications. TAB Books Inc, Blue-bridge Summit, Pa, USA.

Laurière J.L., (1986, 1988) , Intelligence artificielle. Tome 1 : Résolution de problèmes par l'homme et la machine. Tome 2 : Représentation des connaissances. Eyrolles, Paris.

Meyer B., (1988), Object-oriented Solftware Construction. Prentice-Hall, New York.

Pun L., (1984), Systèmes industriels industriels d'intelligence artificielle. Editests, Paris.

Pun L., (1986), Static and Dynamical Production Plannings. Survey paper, 4th IFAC Symposium onf Large Scale Systems, August 1986, Zürich.

Pun L., (1987), Computer-aided Management in Developing Countries, Survey paper, IFAC World Congress X,Munich, Aujust 1987

Pun L., (1988), Intelligent Distributed Structure for Production Management Systems, Survey paper, IFAC Symposium on Distributed Intelligent Systems, Varna.

Pun L., (1988), From GRAI to Prolog, in "Intelligent Manufacturing Systems", Milacic Editor, Nr 1, North Holland.

Roboam M., M. Zanettin & L. Pun, (1989), Utility of advanced design methodologies for Advanced Manufacturing Systems. CIMS, FEBRUARY 1989.

Roboam M., M. Zanettin & L. Pun, (1989), GRAI-IDEFO-MERISE,; Integrated Methodology to Analyse and Design Manufacturing Systems. CIMS, may 1989.

Roche C. & J.P. Laurent, (1989), Les approches objets et le langage LRO2 (keops), TSI/AFCET, V8, N1, 21-39.

Sombé L., (1988), Inférences non classiques en intelligence artificielle. Ebauche de comparaison sur un exemple. Journées PRC:AI, TOULOUSE.

Thayse A. & al., (1988), Approche logique classique à la programmation logique. Dunod, Paris.

Tsichritzis D. Editor, (1988), Objects and Things, =centre Universitaire Informatique, Genève.

Design of Interfactory Computer Network Interconnection

NIOVI-FOTINI PAVLIDOU and BYRON PAPATHANASSIU

1. Introduction

During the last decade, the expansion of local area computer networks has been impressive, regarding industrial applications as well as office implementation. This developping trades do not seem to stop in the near future [1]. On the other hand, the public data and telephone networks are continuously changing from amalog to digiatal techniques, reinforcing their reliability, ensuring the transimission of any kind of information (voice, data, image) over large distances with very low transmission error probabilities. New kinds of telecommunications networks have been appeared, for example, Metropolitan Area Networks (MANs), Cable Television Networks, Videotex Networks et.c. As a result, the necessity to interconnect all these telecommunications facilities into one integrated network has become apparent. This enormous ability of information transfer angments the productivity and the automation of any job giving to people the possibility to be supervisors rather and not hardworkers.

Computer networks into organisations or factories, offen called Interorganisation Networks (ION's) [2], support person-to-person communications via electronic mail, exchange of CAD/CAM data software products, file transfers, input to an accounting system via a database query, data acquisition and processing in industrial applications e.t.c. Even more, through the interconnection of LAN to WAN (wide Area Networks) the organisation attain the chance to participate in video conferencing meetings, to access public videotex systems, to communicate by fax worldwide.

In order to interconnect computer networks the International Standards Organisation has proposed some evaluated techniques. These techniques are compatible to the seven layer open systems interconnection model (OSI) which has been specified some years ago aiming to put an end to the nummerous proprietary commercial products of computer networks. The target of the OSI model is to define some standards and to clasify all the services offered by an operating network into hierarhical layers, and thus to treat changes in operation concerning one layer separetely, without affecting the whole network structure. In parallel, standards for the succesful cooperation of various manufacturers systems have been stated. The OSI model is given in fig. 1, and explanations about the exact services of each layer can be found in the literature [3,4].

The layer we are interested about is the third, the network layer, where the services of routing, congestion, flow control, and error control are placed. ISO propose that the interconnection of networks is made at that layer, through specially developped devices called routers or gateways. These devices can implement extended routing algorithms, flow control techniques, fragmentation and sequencing of packets. They are designed in such a manner that they translate any differences of the interconnected networks in a form understood by every part.

Application Layer	
Presentation Layer	
Session Layer	
Transport Layer	
Network Layer	interconection layer
Data Link Layer	
Physical Layer	

Fig 1: OSI reference model

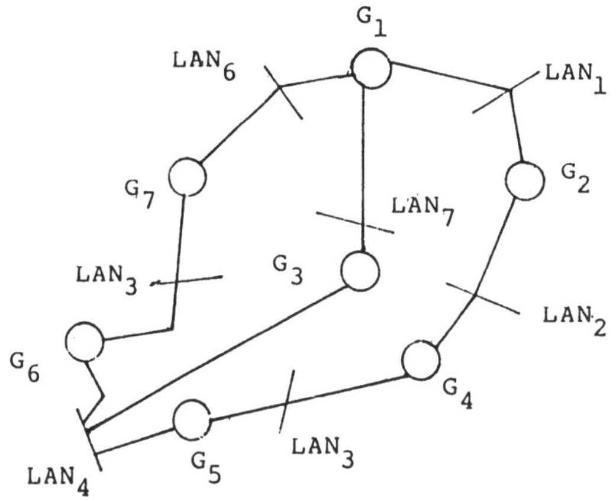

Fig. 2: LANS interconnected through gateways (G)- Internet.

The resulting network of the interconnection of various LANs or WANs is a very complex network and classic techniques used to existent public networks can not sufficiently be applied, thus need extensions [5]. In fig. 2 an internet (interconnected network) of LANs is given. Its analysis and design consists a complex subject and must treated with great attention. In the following study some features of this design process and especially routing items will be analyzed and solutions will be proposed.

2. Internet design and routing algorithms

In the design of a computer network a funtamental problem is to select the best paths for message routing. This selection is always made under a specified criterion and in case of communications networks this criterion is the av erage message delay from transmitter to receiver (source to destination). The path between source and destination may be of various types, and thus, the first design problem is what kind of transmission media will be used. Telephone lines as transmission facilities are very common, but they suffer of high error rates and the new teleservices require great bandwidths they cannot support. Data cables are not subject to outside interferences and are able to get a higher transmission rate, but they are expensive. Radio transmission is characterized by a high data-transmission rate, and it is unique by its savings of network-start up costs, the installation flexibility and movability of the stations, but there will be collisions, whenever two stations want to broadcast simultaneously. In purpose to apply very fast routing algorithms, we shall make the assumption that the media used have high bandwidth and low error probabilities. Generally speaking, the routing techniques that can be applied on an internet can be classified in two broad categories. First, these techniques under which the nodes of the network must possess information about the topology of the whole network and the state of the paths, like shortest-paths algorithms, Dijkstra algorithms e.t.c., and second, these techniques under which the nodes are responsible for a restricted knowledge of their nearby area (neighbor nodes), this knowledge must be updated in frequent time intervals. In this category there are a lot of interesting propositions and the more fertil e area routing subjects [6,7].

Especially, when dealing with interconnected networks, a key factor to their performance is the maintainance of the very high speeds the LANs provide (of the order of several mbps). Thus, the first decision made for an internet design is the selection of a routing algorithm of the second category, which leads to medium size routing tables at each nodegateway, and medium buffer requirements. It must be mentioned that a gateway is a node which forwards packets in two directions simultaneously, arranging all the necessary protocol translations, and so the routing tables it has to scan must be of managable size. (fig. 3).

A second severe decision we shall make for the network design is to select an hierarchical structure for the network which also leads to routing tables restricted in size and time requirements. A classic approach to hierarchical network structures is given in fig. 4. Nodes are grouped over a geographic area and they communicate straightly to each other. But nodes which belong to different geographic groups must be addressed through a node of higher level in the hierarchy. A lot of approach have been stated for the determination of these areas [8,9]. But all these approaches have the pitfall that two nearby nodes cannot address to each other if they belong to different groups, and this matter couses serious problems in high speed networks. Therefore, we come to choose another approach of hierarchic routing which has been stated at [10] and guarantees the direct communication of adjacent nodes.

An hierarchical network under this approach is given in fig 5, where we consider to exist two levels of hierarchy. In the first level any simple node belongs and it can communicate with its neighbors. In the second level belong only the two nodes encircled with large circles, they are called landmarcks and they, carry on the traffic from one group of simple nodes to farmost groups of other nodes. Thus, if node 1 and node 10 want to echange data they can do it only through node 3 or 7. The levels or hierarchy can be numerous and they depend on the size and traffic loading of the internet.

In fig. 5 we note that an interconnected network can be considered a bipartite graph, the two kinds of nodes being the LANs and the gateways. For the application of routing algorithms we can assume the two kinds of nodes to be of the some type or we can consider LANs as simple links between the gateways. The simulation results will be the same.
In order to clarify the structure of the algorithm we give in table 1, the routing table of node 2. The routing table of each node will contain entries for all the adjacent nodes and entries for the lamdamark nodes belonging to the higher level, which connect the mode with the rest of the network. For node 2, the landmarck node is node 3, and it is accessed through node 1 which is the shortest-path node to 3. Several cost criteria can be studied as link cost-distances, in order to chose the shortest-path to the landmark. Here, for simplicity, we suppose that the costs are equal to unit, thus the minimum number of hops to the Landmark is attained.

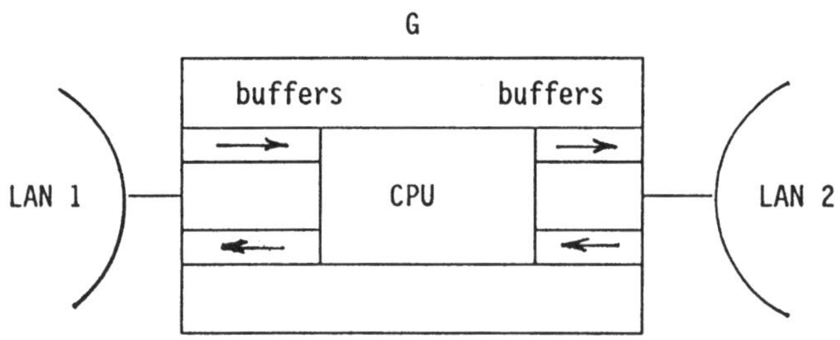

Fig. 3. A simple gateway model

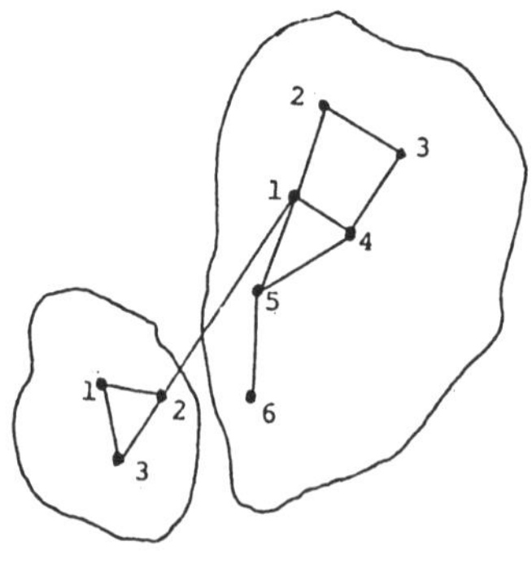

Fig. 4. Geographic hierarchy network

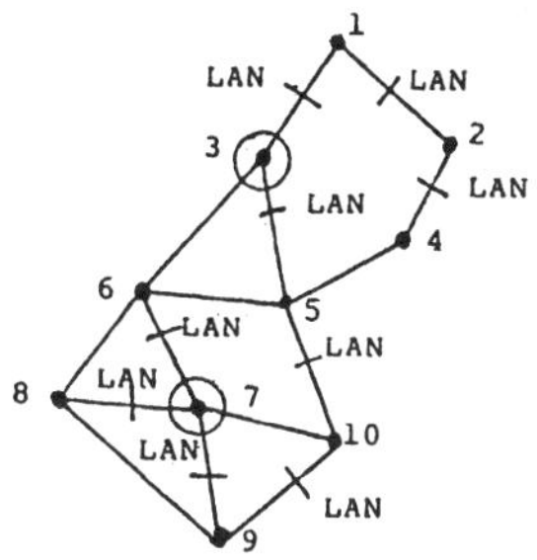

Table 1: Routing table of node 2

Level	Node	Next node cost
1	1	1/1
1	4	4/1
2	3	1/2

Fig. 5. Landmark hierarchy internet

3. Design criteria and algorithm performance.

In previous studies [10] has been stated that the landmark selection does not affect the network performance. But in case of LANs internet it is proper to indicate some criteria for the selection of nodes to be landmarks, because of speed requirements. Some criteria we believe to be adequate are the following: a) Landmarks are suppose to be nodes with many outgoing links for reasons of reliability, as well as reachability by the other nodes, b) The Landmark's own traffic to the neighbor nodes it is better to be kept in low levels because of the network traffic concentration on its links. Really in fig. 6 it is observed the increase in average delay D versus average message arrivals (messages per second) if nodes 2 and 10 are chosen as lamdmarks instead of 3 and 5. Traffic requirements on links were taken to be the same all over the network.

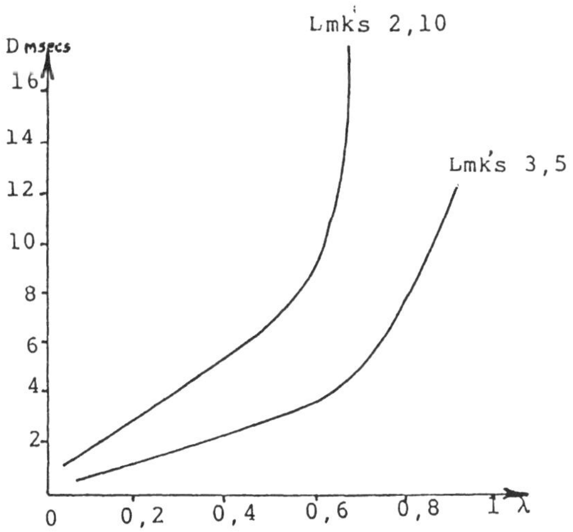

fig. 6. Time delay versus packet arrivals λ for various lamdmarck selection

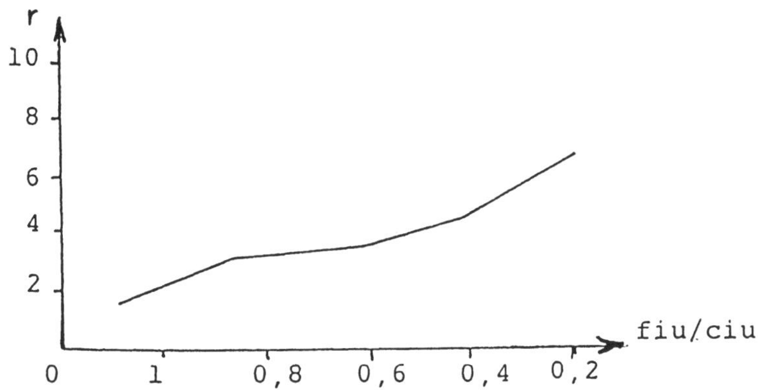

fig. 7. Increase in radius r when fiu/cin (links adjacent to lamdmarcks) becomes small.

It is evident that the lower the traffic on lamdmarks links the larger the radius it can cover. As radius r of a landmark we define the greatest-hops node which has entry for that landmark in its routing table. In fig. 7 is given the increase in radius r when D is kept constant and the traffic from landmark to adjacent nodes becomes lower.

Another factor interesting to internet performance is the number of nodes grouped around each landmark and consequently the number of landmarks at each level. Since the dominate factor for the internet is the law delay, we came to use a few nodes in each group, of about a dozen, to keep the routing tables in a manageable size because the relationship of this size and the size of the groups is almost linear. All the simulation results we have obtained are based on this choice. Furthermore, a trial has been made to divide in each group the same average. LAN speeds in order to avoid great time delays for a specific network area.

When dealing with routing problems a subject that we try to avoid is the situations of congestion, that means, heavy traffic conditions on the network which can cause bottlenecks and failures. Especially in case of high speed networks congestion avoidance becomes a very sensitive subject. Thus, we have enforced our algorithm with the ability to prevent unpleasent situations by evaluating the ratio fin/cin and send messages to lans to halt their traffic whenever this ratio becomes dangerous.

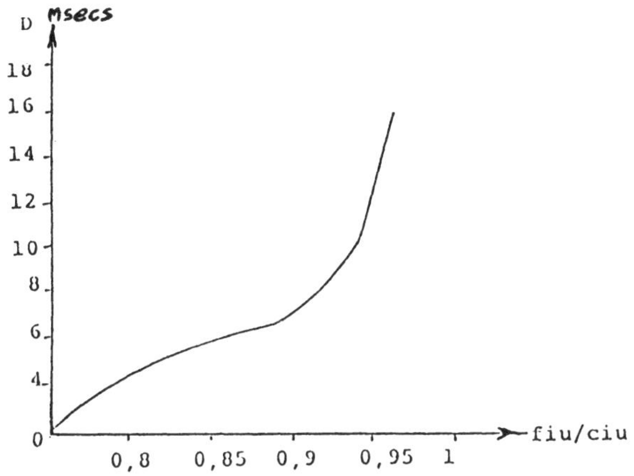

fig. 8. Relationship between critical values of fin/cin and average time delay D.

The impact of different critical values of this ratio on delay measurement is given in fig. 8. We note, that fin is the load on each link measured in messages per second and cin is the capacity of the link measured, too, in messages per second. Traffic regulation can be applied by each gateway-node on the internet, but the critical ration fin/cin may have different values on landmark links and on other simple links.

The hierarchical structure proposed for the internet can offer some advantages at a very high problem of interconnected networks, data security [11]. This problem becomes very important when dealing with LANs belonging to different organisations which frequently make their internal computing facilities accessible to off-site employees. These external connections sometimes require added security control in order to support access to certain internal machines, services, processes, while, at the same time, preventing access to other facilities. Generally, the physical isolation of these two groups of facilities are impossible, therefore software control must be exercised to assign access rights to each external LAN. The complex this software structure is the largest the time delays on the internet. An hierarchical structure of this software protocols can be tried giving lower control capabilities to every gateway on the internet and enforced to the landmarks. The study of the relationship between different control nechanisms and the performance of the network is one of the possible extension of the proposed algorithm.

4. Conclusions

Several design factors for interconnected LANs have been studied and especially hierarchical routing techiques aiming in performance evalu ation of the internet. Some factors affecting the routing techniques have been analyzed. Special emphasis has been given in keeping the routing tables at a reasonable size, something very essential for high speed network operation. Further extension of the technique to encompass security controls can be achieved and it is under study.

References

1. C. Sunshine "Network Interconnection and gateways", IEEE Selected Areas in Communications, vol. 8, no. 1, 1990, pp. 4-12.

2. D. Estrin "Interconnection Protocols for Interorganisation Networks" IEEE Selected Areas in Communications, vol. 5, no 9, 1987 p.p. 1480-1492.

3. M. Gien, H. Zimmerman "Design principles for network interconnection" Proc. VI Data Comm. Symp. (IEEE/ACM), 1979, p.p. 109-119.

4. P. Green "Protocol Conversion" IEEE Trans Commun. vol.34, 1986. p.p. 257-268.

5. V. Warrier, C. Sunshine "A Platform for Heterogeneous interconnection Network Management" IEEE Selected Areas in Communications, vol. 8, no 1., 1990, pp. 119-127.

6. J. Jaffe, F. Moss "A Responsive Distributed Routing Algorithm for Computer Network" IEEE Trans on Comm. vol. 30, no. 7, 1982, pp. 1758-1762.

7. J. Garcia-Luna-Aceves "A New Minimum Hop Routing Algotithm" Proc. IEEE Infocom '87 pp. 170-180.

8. R. Perlman "Hierarchical Networks and the Subnetwork Partition Problem" Computer Networks and ISDN Systems, 1985, vol. 9, pp. 297-303.

9. P. Bell et al "Review of Point-to-Point Routing Algorithms" IEEE Commun Magazine, vol. 24, 1986, pp. 34-38

10. P. Tsuchiya "The landmark hierarchy: A new hierachy for routing in very large networks", ACM SIGCOMM, 1988, pp. 8-14.

11. V. Voydock "Security Mechanisms in High-Level Network Protocols" ACM Computing Surveys., vol. 15, no. 2, 1983 pp 135-171.

List of Contributors

Adan, Ivo	University of Technology Department of Mathematics and Computing Science Den Dolech 2 P.O. Box 513 5600 MB Eindhoven The Netherlands
Ahn, Jaekyoung	The University of Iowa Industrial and Management Engineering College of Engineering Iowa City, Iowa 52242 USA
Axsäter, Sven	Luleå University of Technology Division of Transportation and Material Flow Systems 951 87 Luleå Sweden
Blackburn, Joseph	Vanderbilt University Owen Graduate School of Management Nashville, Tennessee 37203 USA
Buzacott, John	University of Waterloo Faculty of Engineering Department of Management Sciences Waterloo, Ontario N2L 3G1 Canada
Carlson, John	University of Southern California School of Business Administration Departement of Decision Systems Los Angeles, California 90089-1421 USA
Carrie, Allan	University of Strathclyde Department of Design, Manufacture and Engineering Management Manufacture and Engineering Management Division James Weir Building 75 Montrose Street Glasgow G1 1XJ Scotland

Cheema, Preetinder	University of Wales Cardiff School of Electrical, Electronic and Systems Engineering P.O. Box 904 Cardiff, CF1 3YH Wales
Cheng, Edwin	The University of Manitoba Faculty of Management Department of Actuarial & Management Sciences Winnipeg, Manitoba R3T 2N2 Canada
Co, C.G.	Purdue University School of Industrial Engineering Michael Golden Labs, Room 1307A West Lafayette, Indiana 47907 USA
Dekleva, Janez	University of Ljubljana p.p 394 Faculty of Mechanical Engineering Murnikova 2 61000 Ljubljana Yugoslavia
Donselaar, Karel v.	University of Technology Den Dolech 2 P.O. Box 513 5600 MB Eindhoven The Netherlands
Doremalen, Jan v.	Nederlandse Philips Bedrijven B.V. Centre for Quantitative Methods Building HCM-720 P.O. Box 218 5600 MD Eindhoven The Netherlands
Dudek, Richard	Texas Tech University Department of Industrial Engineering Lubbock, Texas 79409-3061 USA
Ebrahimpour, Maling	The University of Rhode Island Department of Management Science and Information Systems College of Business Administration Kingston, RI 02881-0802 USA

Fabrycky, Wolter
Virginia Polytechnic Institute and State University
Blacksburg
College of Engineering
Department of Industrial Engineering
and Operations Research
Virginia 24061-0118
USA

Fandel, Günter
Fernuniversität Hagen
Lehrstuhl für Produktion und Investition
Feithstr.140/AVZ II
5800 Hagen
FRG

Fleuren, Hein
Nederlandse Philips Bedrijven B.V.
Centre for Quantitative Methods
Building HCM-720
P.O. Box 218
5600 MD Eindhoven
The Netherlands

Groditzki, Günter
IBM Deutschland
Industrie und Technik
Poccistr.11
8000 München 2
FRG

Hall, Randolph
University of California, Berkeley
Engineering Systems Research Center
Department of Engineering and Operations Research
3115 Etcheverry Hall
Berkeley, California 94720
USA

Harhalakis, George
University of Maryland
Department of Mechanical Engineering and
System Research Center
USA

Hars, Alexander
Universität des Saarlandes
Institut für Wirtschaftsinformatik
Im Stadtwald, Gebäude 14
6600 Saarbrücken 11
FRG

Hiraki, Shusaku
Hiroshima University
Faculty of Economics
1-1-89 Higashi-senda, Naka-ku
Hiroshima 730
Japan

Huang, Philip	Virginia Polytechnic Institute and State University Department of Management Science The R.B. Pamplin College of Business Blacksburg, Virginia 24061 USA
Inderfurth, Karl	Universtät Bielefeld Fakultät für Wirtschaftswissenschaften Universitätsstraße Postfach 8640 4800 Bielefeld 1 FRG
Ishii, Kazuyoshi	Kanazawa Institute of Technology Department of Managerial Engineering 7-1 Ohgigaoka, Nonoichi P.O. Kanazawa-South Ishikawa 921 Japan
Jalali, Mohammad	Texas Tech University Department of Industrial Engineering Lubbock, Texas 79409 USA
Janiak, Adam	Technical University of Wroclaw Institute of Engineering Cybernetics ul. Janiszewskiego 11/17 50 372 Wroclaw Poland
Jönsson, Henrik	Gothenburg School of Economics Department of Business Administration Vasagatan 3 41124 Gothenburg Sweden
Joshi, Sanjay	The Pennsylvania State University Department of Industrial and Management Systems Engineering College of Engineering 207 Hammond Building University Park, PA 16802 USA
King, Barry	Marquette University College of Business Administration Milwaukee, Wisconsin 53233 USA

Kistner, Klaus-Peter | Universität Bielefeld
Fakultät für Wirtschaftswissenschaften
Universitätsstraße
Postfach 8640
4800 Bielefeld 1
FRG

Knolmayer, Gerhard | Universität Bern
Institut für Wirtschaftsinformatik
Hallerstr. 6
3012 Bern
Switzerland

Kongrit, Anukul | Asian Institute of Technology
Division of Industrial Engineering and Management
P.O. Box 2754
Bangkok 10501
Thailand

Konovalov, Nick | Manager of Operations
Baxter/Bentley Laboratory
USA

Krajewski, Lee | The Ohio State University
Academic Faculty of Management Sciences
301 Hagerty Hall
1775 College Road
Columbus, OH 43210-1399
USA

Kühn, Martin | Kreuzstr. 4
7521 Kronau
FRG

Küpper, Hans-Ulrich | Ludwig-Maximilians-Universität München
Fakultät für Betriebswirtschaft
Institut für Produktionswirtschaft and Controlling
Rosenheimer Str. 139
8000 München
FRG

Kuhn, Heinrich | Technische Universität Braunschweig
Fachgebiet Produktionswirtschaft
Abteilung Betriebswirtschaftslehre
Pockelsstraße 14
3300 Braunschweig
FRG

Kusiak, Andrew | The University of Iowa
Industrial and Management Engineering
College of Engineering
Iowa City, Iowa 52242
USA

Laftit, Said	Inria-Cescom Technopôle Metz 2000 57070 Metz Cedex France
Miltenburg, John	McMaster University Faculty of Business 1280 Main Street Hamilton, Ontario Canada, L8S 4M4
Moras, Rafael	Texas Tech University Department of Industrial Engineering Lubbock, Texas 79409 USA
Osgood, William	VP Operations Baxter/Edwards Laboratory USA
Papathanassiu, Byron	Aristotle's University Thessaloniki Faculty of Technology Civil Engineering Department Transport and Organization Section P.O. Box 452 540,06 Thessaloniki Greece
Pavlidou, Niovi-Fotini	Aristotle's University Thessaloniki Faculty of Technology Civil Engineering Department Transport and Organization Section P.O. Box 1508 540,06 Thessaloniki Greece
Proth, Jean-Marie	Inria-Lorraine Cescom Technopôle Metz 2000 4, Rue Marconi 57070 Metz France
Pun, Lucas	Université Bordeaux 1 Laboratoire Grai 351 Cours Liberation 33405 Talence France
Ramani, Sankaran	National Institute for Training in Industrial Engineering Vihar Lake Bombay 400087 India

Reese, Joachim
Rheinische Friedrich-Wilhelms Universität Bonn
Institut für Gesellschafts- und Wirtschaftswissenschaften
Betriebswirtschaftliche Abteilung 1
Adenauerallee 24-42
5300 Bonn 1
FRG

Ritzman, Larry
The Ohio State University
Academic Faculty of Management Sciences
Operations Management
301 Hagerty Hall
Columbus, Ohio 43210-1399
USA

Scheer, August-Wilhelm
Institut für Wirtschaftsinformatik
Universität des Saarlandes
Im Stadtwald, Gebäude 14
6600 Saarbrücken 11
FRG

Schneeweiss, Christoph
Universität Mannheim
Lehrstuhl für Betriebswirtschaftslehre und
Unternehmensforschung
Schloß
6800 Mannheim
FRG

Sheu, Chwen
The Ohio State University
Academic Faculty of Management Sciences
301 Hagerty Hall
1775 College Road
Columbus, OH 43210-1399
USA

Silver, Edward
The University of Calgary
Faculty of Management
2500 University Dr. NW
Calgary T2N 1N4
Canada

Soshiroda, Michio
Waseda University
School of Science and Engineering
3-4-1 Ohkubo, Shinjuku-ku
Tokyo 169
Japan

Steven, Marion
Universität Bielefeld
Fakultät für Wirtschaftswissenschaften
Postfach 8640
Universitätsstraße
4800 Bielefeld 1
FRG

Suminokura, Toshihiko
Kanazawa Institute of Technology
Department of Managerial Engineering
7-1 Ohgigaoka, Nonoichi
P.O. Kanazawa-South
Ishikawa 921
Japan

Tabucanon, Mario
Asian Institute of Technology
Division of Industrial Engineering and Management
P.O. Box 2754
Bangkok 10501
Thailand

Takahashi, Katsuhiko
Hiroshima Prefectural University
Department of Management Information Systems
Faculty of Management
562 Nanatsuka, Shobara
Hiroshima 727
Japan

Tanchoco, Jose
Purdue University
School of Industrial Engineering
Michael Golden Labs, Room 1307A
West Lafayette, Indiana 47907
USA

Tempelmeier, Horst
Technische Universität Braunschweig
Fachgebiet Produktionswirtschaft
Abteilung Betriebswirtschaftslehre
Pockelsstraße 14
3300 Braunschweig
FRG

Tetzlaff, Ulrich
University of California at Berkeley
Walter A. Haas School of Business
350 Barrows Hall
Berkeley, CA 94720
USA

Towill, Denis
University of Wales
Cardiff School of Electrical,
Electronic and Systems Engineering
P.O. Box 904
Cardiff, CF1 3YH
Wales

Vitt, Lawrence
Manufacturing Planning Horizons
Software and Planning Programs
1808 Upper Chelsea Road
Columbus, Ohio 43212
USA

Wessels, Jaap	University of Technology Departement of Mathematics and Computing Science Den Dolech 2 P.O. Box 513 5600 MB Eindhoven The Netherlands
Wijngaard, Jacob	University of Technology Department of Industrial Engineering Den Dolech 2 P.O. Box 513 5600 MB Eindhoven The Netherlands
Woerlee, Auke	Erasmus University Rotterdam Econometric Institute P.O. Box 1738 3000 DR Rotterdam The Netherlands
Wysk, Richard	Texas A&M University Royce Weisenbaker Chair in Innovation Department of Industrial Engineering Zachry Engineering Center College Station, TX 77843 USA
Zäpfel, Günther	Universität Linz Institut für Industrie und Fertigungswirtschaft Altenbergerstr. 69 4020 Linz Austria
Zijm, Henk	University of Twente University for Technical and Social Sciences Faculty of Mechanical Engineering Laboratory of Production Management and Logistics P.O. Box 217 7500 AE Enschede The Netherlands

List of Sponsors

Fernuniversität Hagen

Deutsche Forschungsgemeinschaft, Bonn-Bad Godesberg

Ministerium für Wissenschaft und Forschung des Landes
Nordrhein-Westfalen, Düsseldorf

Sparkasse Hagen

Der Beauftragte für die Pflege und Förderung der Beziehungen
zwischen den Hochschulen des Landes Nordrhein-Westfalen und
des Königreiches der Niederlande und des Königsreiches Belgien, Aachen

IBM Deutschland GmbH, Stuttgart

Krupp Brünninghaus GmbH, Werdohl

Varta Batterie AG, Hannover

Elektromark AG, Hagen

Andreas Brauerei, Hagen

Fischer Metroplast GmbH, Lüdenscheid

Autohaus Jürgens GmbH, Hagen

SPRINGER NATURE

GPSR Compliance

The European Union's (EU) General Product Safety Regulation (GPSR) is a set of rules that requires consumer products to be safe and our obligations to ensure this.

If you have any concerns about our products, you can contact us on ProductSafety@springernature.com

In case Publisher is established outside the EU, the EU authorized representative is:

Springer Nature Customer Service Center GmbH
Europaplatz 3
69115 Heidelberg, Germany

Kolibri 360 GmbH
Ferdinand-Jühlke-Straße 7
99095 Erfurt, Deutschland
produktsicherheit@kolibri360.de